Game-Theoretical Control Problems

N.N. Krasovskiĭ A.I. Subbotin

Game-Theoretical Control Problems

Translated by Samuel Kotz

With 23 Illustrations

Springer-Verlag
New York Berlin Heidelberg
London Paris Tokyo

N.N. Krasovskiĭ and A.I. Subbotin
Institute of Mathematics and Mechanics
Ural Scientific Center
Sverdlovsk 620219
U.S.S.R.

Translator:
Samuel Kotz
College of Business and Management
University of Maryland
College Park, MD 20742
U.S.A.

This volume is part of the Springer Series in Soviet Mathematics.
Advisors: L.D. Fadeev (Leningrad); R.V. Gamkrelidze (Moscow).

AMS Classification: 90D25

Library of Congress Cataloging in Publication Data
Krasovskiĭ, N.N. (Nikolaĭ Nikolaevich)
Game-theoretical control problems.
Translation from Russian.
Bibliography: p.
Includes index.
1. Differential games. 2. Control theory.
I. Subbotin, A.I. (Andreĭ Izmaĭlovich) II. Title.
QA272.K73 1987 519.3 86-26299

Translation of the original Russian edition:
Pozitjionnii differentjial'nie igri, Nauka 1974

Softcover reprint of the hardcover 1st edition 1988

9 8 7 6 5 4 3 2 1

ISBN-13: 978-1-4612-8318-8 e-ISBN-13: 978-1-4612-3716-7
DOI: 10.1007/978-1-4612-3716-7

PREFACE

This book is a substantially revised version of our monograph *Positional Differential Games* (Nauka, Moscow 1974), which was based on results obtained by the authors and their associates in the late 1960s and early 1970s. Since then, intensive investigations were conducted, which extended the theory of differential games and substantially altered its makeup. However, the concepts presented in our monograph turned out to be quite viable, and they continue to serve as a sound basis for new developments. Therefore, when preparing the English translation, it was decided to retain the structure of the above-mentioned monograph but to incorporate the changes which have taken place in the theory of differential games.

In this book, problems of control are investigated; they can be stated as follows. Let a control system be given, the dynamics of which are described by ordinary differential equations. It is necessary to construct a strategy, i.e. a control based on the feedback principle which will assure a definite quality for the controlled process under the conditions when the process is subject to the action of uncertain factors. These factors may be either disturbance or a controlling action of the opposite side. In the framework of the theory of differential games, this type of problem is considered in conjunction with the inverse problem. Joint consideration of the two inverse problems turns out to be useful not only in situations when the two opposing sides (two players) control the system, but also in cases where the controlled system (by a single player) is subject to an uncertain disturbance.

In this book, a formalization of differential games is presented, existence theorems for equilibrium are proved, and certain properties of value functions are derived. A substantial amount of attention is devoted to the construction of optimal strategies. Special classes of differential games are singled out in which the construction of solving strategies requires a relatively small number of computations. Problems of stability of the proposed solutions relative to the informational errors are considered. The presentation is carried out within the framework of a unified concept and is based on a small number of basic notions. Formal definitions and assertions are accompanied by suitable motivation and discussion, indicating the transition to those control procedures and computational algorithms which can be implemented in practice.

The material is organized as follows. In Chapter 1, positional strategies and quality criterions of strategies are defined, the notions of a saddle point and the value of a differential game are introduced. A system of these definitions constitutes a formalization of differential games. In Chapter 2 the differential game of approach-evasion is considered. When investigating this game, stable bridges and strategies extremal to these bridges are defined. Constructions introduced herein are utilized in the course of the whole book. The basic result of Chapter 2 is the theorem on an alternative for the differential game of approach-evasion, which asserts that this game possesses an equilibrium situation. As is shown in Chapter 3, an investigation of differential games with various types of payoff functionals reduces to consideration of the corresponding game of approach-evasion. Therefore, the existence of the value for different types of differential games follows from the theorem on an alternative. Thus the first three chapters contain a formalization of differential games and an existence theorem for equilibrium situations.

In Chapter 4, properties of the value function are investigated. Here inequalities are obtained for the directional derivatives of the value function. The inequalities supplemented by corresponding boundary conditions form a group of necessary and sufficient conditions to be satisfied by the value function. At the points of differentiability of the value function, these obtained inequalities become the well-known Isaacs-Bellman equation. Results presented in Chapter 4 can be viewed as a formal basis for the method of dynamic programming in the theory of differential games.

Evaluation of value functions and construction of optimal strategies usually requires an extremely large number of computations. Therefore, it is of interest to single out special types of differential games for which relatively simple computational algorithms can be devised. These problems are considered in the next three chapters. In Chapter 5, linear differential games with a convex payoff functional are investigated. For these games the so-called regularity conditions are stated, under which the value of a positional differential game coincides with the programmed maximin. In regular cases, evaluation of the value and construction of an optimal strategy reduces to the solution of relatively simple problems of mathematical programming. The approach suggested herein is called the method of extremal aiming.

In Chapter 6, the method of extremal aiming is carried over to the case of a nonlinear control system. As in the preceding chapter, conditions are specified which assure the equality of the value of the positional differential game and the programmed maximin. Under these conditions, positional strategies are constructed by synthesizing solutions of auxiliary problems of programmed control. In order to reveal to its full extent the possibilities of the method of extremal aiming in Chapter 6, in place of the usual programmed controls, generalized (or equivalently, relaxed) programmed controls are utilized. In Chapter 7, controlled systems are considered in which the possibilities of one of the players grossly prevail over the possibilities of the opponent. In this case the solution of a problem for this player actually reduces to the solution of a problem of programmed control which allows us to determine the trajectory -- the so-called stable path -- on which the movement of a conflict-controlled system can be contained. In addition to stable paths, stable integral manifolds are considered in Chapter 7, which are formed by the sets of solutions of differential inclusions. Conditions are indicated under which the method of stable paths and the method of stable integral manifolds complement each other and determine a pair of optimal strategies.

In Chapter 8, qualitative problems of the theory of differential games are considered. It is shown here that a solution obtained in the class of positional strategies may be incorrect, i.e., unstable with respect to informational errors. To regularize this solution, the so-called control procedure with a guidance is proposed. Several modifications of control procedures with a guidance are considered; in particular, stabilized control procedures are described. In the conclusion of this chapter, universal optimal strategies are defined -- these are strategies optimal for any initial position in a given region.

In Chapters 1-8, control systems which satisfy a special condition are considered: this special condition is called the condition of a saddle point in a small game. The condition is fulfilled, for example, for a system with an additively partitioned control of players. In Chapters 9 and 10, differential games in which this condition may be violated are studied; thus in the class of pure positional strategies an equilibrium situation may not exist in general. For such systems one can extend the notion of a strategy and define a differential game either in the class of mixed positional strategies for both players or in the class of positional strategies of one player and the class of counterstrategies of the second.

In Chapter 9, mixed positional strategies are defined and the existence of an equilibrium situation in the class of these strategies is proved. It is also shown in this chapter how the results of preceding chapters are carried over to the case of a differential game considered in the class of mixed strategies. Considerable space in this chapter is devoted to a description of stochastic control procedures which approximate mixed strategies.

In Chapter 10, counterstrategies are introduced. Counterstrategy is a method of control based on information about the position of a game and control of an opponent chosen at the current time. It is established that the equilibrium situation exists in the class of positional strategies of one player and the counterstrategies of the other. To compute the value of this differential game and to construct optimal strategies and counterstrategies, one can utilize methods proposed in preceding chapters. In the conclusion of Chapter 10, a construction is described which allows us to unify the investigation of various types of differential games.

In the last chapter, differential games are studied for which knowledge of the current position of a game is not sufficient, and it is necessary to store the realized motion of the system. For these games the existence of an equilibrium is proved in an appropriate class of strategies, and some properties of the value functional are considered.

In this book, a number of examples illustrating basic results are presented. We note that for purposes of illustration, relatively simple model examples were chosen. To accommodate references to the most often used facts, a short appendix is included which contains statements of certain results related to convex analysis, multivalued mappings, and differential inclusions. For facts of a general mathematical nature utilized in the book, we cite papers which contain detailed exposition of these topics. An appropriate bibliography is presented at the end of the book. The bibliography also contains a small number of works dealing with the theory of differential games. We decided against the compilation of a comprehensive bibliography with comments. In the course of exposition, we refer only to those works which deal directly with the problems under consideration.

The following is the order of enumeration adopted. Sections are numbered by two indices, the first indicating the chapter number and the second the number of the section in the chapter. Theorems, lemmas, etc., are denoted by three indices: the first two are the number of the section and the third is the number of the theorem in that section. Within a section, formulas are enumerated by a single digit. When referring to formulas in other sections, the three-index enumeration is used. Finally, note that sections are subdivided into subsections which are also enumerated by three indices.

The authors take pleasure in thanking their professional colleagues for their great assistance in compiling this book. We wish to acknowledge with gratitude Mrs. Yakovenko and Miss Elena Sidorova who prepared the typescript.

N. N. Krasovskii
A. I. Subbotin

CONTENTS

Chapter I
GAME-THEORETICAL CONTROL

1.1. Minimax Control Problems

The purpose of this section is to present a general idea of the problems studied in this book. The subject of this book, differential game theory, is a branch of mathematical optimal control theory. Examples of controlled processes may be found in various fields, from beer brewing to space flight. The control of such systems can be modeled as a mathematical problem. The various types of problems that arise are studied in control theory. In this section, we will informally describe special features of the problems that are studied in this book. Precise definitions of the concepts discussed below will be presented in following sections.

1.1.1. In the problems we consider, the behavior of a controlled system will be described by *ordinary differential equations*. The phase space is chosen to be finite dimensional. A large number of control problems satisfy this condition. For example, according to the laws of classical mechanics the motion of a mechanical system, subject to external forces, is described by ordinary differential equations. Components of the vector of geometric coordinates and the velocity vector serve as the phase vector of the system. External forces play the role of controlled actions. Various controlled processes in economics, technology, and biology can also be described by means of ordinary differential equations. In particular, one can consider the problem of controlling movements of an aircraft. The actual motion of an aircraft can be described with sufficient precision by a system of differential equations. The phase variables here are the coordinates of the center of mass, the components of the velocity vector, and a few other quantities which characterize the angles between the axis of an aircraft and the velocity vector. The controlling parameters characterize the aileron angles, the mode of operation of the engines, and so on.

In order that the above mentioned examples should not incorrectly suggest that every controlled process can be described by ordinary differential equations, we shall mention as a counterexample the problem of the controlled heating of a rod. The phase state for this problem is the distribution of temperatures along the rod. This distribution is infinite dimensional and its evolution in time is described by a partial differential equation.

1.1.2. An important feature of the problems studied below is the positional nature of the control. Under positional control the choice of controlling actions depends on the information about the realized state of the system. *Positional control* is often called *feedback control.* In order to emphasize the specific features of positional control, it is useful to contrast it with *programmed control.* Programmed control is identified with a time dependent function which is defined a priori over the whole control interval and specifies the controlling action dependent only on the current time. The difference between positional and programmed controls can be illustrated by the following two examples. To drive a car (especially in heavy city traffic) it is necessary to constantly take into account the changing traffic conditions. This is an example of feedback control. However, to boil an egg lightly it is sufficient to measure the time during which the egg is in the boiling water. It can be said that here we have programmed control.

1.1.3. We shall now explain why the control problems discussed in this book are called *minimax* or *game-theoretic.* To do this we shall discuss, in general terms, the automatic landing of an aircraft. In this problem it is required to construct a positional control which will guarantee the safe descent and touchdown of the plane at the beginning of the landing strip. It is assumed that in the process of motion, several variable quantities are measured, the distance to the ground, the distance to the landing strip, the velocity, and so on. These variables, taken together, constitute the information that forms the basis for automatic changes in the aileron angles and the mode of operation of the engines. In actual flight the aircraft is subject to disturbance, mainly in the form of wind gusts. This fact can be taken into account in the mathematical model by adding parameters (to the equation of motion) that reflect the action of the disturbance. For these parameters, bounds within which the parameters can vary are set. The required control must guarantee a safe landing under any possible disturbance.

We shall now specify how one can formally define the notion of a safe landing. Assume that some nominal trajectory is defined, as well as a metric which provides a measure of the deviation of an arbitrary motion from the nominal trajectory. Thus a functional is defined on the set of motions which can be called a miss. Motions can be compared by means of this functional. If, for one of these motions, the value of the functional is smaller than the value of the miss for another motion, the first motion is then "better" than the second. We shall assume that some critical value of the miss is defined. If the miss is smaller than this critical value, then technical requirements for a safe landing are fulfilled, i.e. the touchdown occurs in the admissible part of the landing strip, the vertical velocity is not large, and so on.

We shall now define the index which will qualify the positional control. This is carried out as follows: Let some positional control be chosen. Consider the set of motions which can be realized under this positional control and the possible admissible disturbances. We shall define the maximal value of a miss on this set. The value obtained will characterize the quality of the control. If it is less than the critical value, then the control under consideration assures a safe landing. Otherwise, if the value exceeds the critical one, it follows that for some admissible disturbance the miss will exceed the critical value. In this case the control does not guarantee a safe landing.

Furthermore, it is natural to consider a control optimal if the index corresponding to it is minimal, i.e. the miss is minimal under the maximally unfavorable disturbance. We thus obtain that the optimal control is determined here by the minimax condition in which the maximum is calculated over the set of admissible disturbances and the minimum is calculated over the set of admissible positional controls. Note that the situation described here can be interpreted as a game in which two conflicting sides participate: one chooses the control while the other selects the disturbance, and the payoff is the miss functional. This interpretation partially explains the use of the term "game-type control". Another reason for the use of this term is given below.

Concluding the discussion of a minimax control problem, we note that for a system with disturbance, other formulations of control problems are known. For example, *stochastic control problems* are considered in which the statistical distribution of the disturbance is assumed known. The quality of control in these problems is usually estimated by either the mathematical expectation of the functional defined on random motions or by the probability of reaching the given target. Stochastic control problems have a large number of applications. We are, however, interested mainly in *deterministic* problems of minimax control. The need to consider these problems arises in those cases when either the stochastic description of the disturbance is not available or when it is required to assure the preassigned quality of the controlled process for any possible disturbance.

1.1.4. It was stated previously that this book deals with the theory of differential games. However, up until now, it has not been shown how the problems under consideration are connected with *differential games.* We shall try to correct this omission.

The game of pursuit-evasion is a typical example of a differential game. In this case two opposing players participate – the pursuer and the evader. Each one of them controls the motion of its system. The game is terminated when, for the first time, the two controlled objects close to within a given distance. The pursuer is interested that this closing in should occur as soon as possible, while the evader tries, on the contrary, to prevent the approach; if it is impossible then he (she) maximizes the duration of the game. We thus deal in this game with two problems – a problem of pursuit and a problem of evasion. Each one of these problems can and should be formulated independently of the other. This is achieved by means of the concept of minimax (maximin) control. For example, in a pursuit problem we do not possess the information on the future behavior of the evader. In this situation the optimal strategy of the pursuer is to minimize the duration of the game under the least favorable motion of the evader system. It is important to emphasize that as far as the pursuer is concerned, it is irrelevant whether the movement of the evader object is the result of rational behavior, or whether the evader moves completely randomly.

In this connection the following problem arises. Suppose it is required to study the pursuit problem. Since the statement of this problem is not connected with the evasion problem, is there any useful information that can be extracted as a result of studying a differential game which is obtained by combining the pursuit and evasion problems? As will be shown in succeeding sections, the answer is affirmative. We shall note, in particular, that when studying the minimax control problem, the question of choosing a class of controls which contains a solution of this problem that cannot be improved under the given nature of information is indeed a difficult one. An answer to this question can be obtained by considering the corresponding differential game. It is important to emphasize that an investigation of differential games turns out to be useful not only in those cases when there exists an explicit confrontation between two sides, but also in control problems with disturbance of the type described previously.

To conclude this section we note that minimax control problems arise not only when systems subject to perturbations are considered or when problems comprising differential games are studied. Other situations are possible which result in problems of minimax control. We note for example, *control problems under the condition of incomplete information.* When studying these problems, the notion of an informational region is often used. An informational region is a set in the phase space containing the true state of the system. In certain cases, evolution of informational regions can be described be described by means of a system of ordinary differential equations by choosing for phase variables a collection of parameters which characterize the informational region. Indeterminancy caused by imprecision of current measurements is expressed in terms of disturbances which appear in the right-hand side (r.h.s.) of the evolution equation. Thus, the problem of informational regions controlling can be incorporated within the framework of the minimax control scheme described above. In certain problems for which the controlled system is very complex or contains parameters which are not known exactly, approximate equations of motion are utilized. Errors involved in such a model can be compensated by introducing into the r.h.s. of equations indeterminate residuals which can be interpreted as disturbances. Such a device allows us to utilize the concept of minimax control once again.

1.2. Equations of Motion

For a mathematical statement of a minimax control problem it is required to define the following three basic components: the differential equations which describe the motion of the controlled system, the class of admissible controls, and the target functional. In this section problems connected with the motion equation are considered.

1.2.1. Let $x(t)$ be an n-dimensional phase vector which characterizes the state of the controlled system at time t. It is assumed that motion of the phase vector is described by the differential equation of the following form:

$$\frac{dx(t)}{dt} = f(t,x(t),u(t),v(t)). \tag{1}$$

Here $u(t)$ is a p-dimensional vector which characterizes the controlling action realized at time t. The q-dimensional vector $v(t)$ corresponds to the disturbance acting on the system at time t. It is assumed that the controlling actions $u(t)$ and the disturbance $v(t)$ satisfy the restrictions

$$u(t) \in \mathcal{P}, \ v(t) \in \mathcal{Q}, \tag{2}$$

where $\mathcal{P}$ and $\mathcal{Q}$ are compact sets in Euclidean spaces R^p and R^q respectively. We note that when dealing with differential games the vectors $u(t)$ and $v(t)$ will be called the controls of the first and second players, respectively.

Properties of the function f in the r.h.s. of equation (1) will be discussed below. Firstly, however, we shall specify the formation of the control $u(t)$.

We shall assume that at each and every current time t, the pair $(t,x(t))$ is known (here $x(t)$ is the realized phase vector of the system).

This pair is called the *position* of the controlled system at time t. Note that this term is seldom used in control theory but is widely applied in the theory of differential games. For uniformity we shall use this term in problems of conflicting control as well as in problems of controlling systems with an *indefinite disturbance*.

Thus, let the control process start at time t_0 and terminate at some time $\theta > t_0$. The time interval $[t_0,\theta]$ is subdivided into half-intervals $[\tau_i,\tau_{i+1})$, where

$$t_0 = \tau_0 < \tau_1 < \dots < \tau_{k+1} = \theta .$$

Let a function $(t,x) \to U(t,x)$ be selected which takes on values on a compact set $\mathcal{P}$ and is defined for all numbers t in the interval $[t_0,\theta]$ and for all vectors x in the space R^n. We shall assume that the function U forms the control $u(t)$ according to the rule

(3) $$u(t) = U(\tau_i,x(\tau_i)),\ \tau_i \leqslant t < \tau_i,\ i = 0, 1, \dots, k .$$

Thus the piecewise-constant control $u(t)$ is formed on the basis of the *feedback principle* depending on the realized position of the system. Note that feedback acts here with a delay. There would have been no delay if the control were defined by the equality

(4) $$u(t) = U(t,x(t)),\ t_0 \leqslant t \leqslant \theta .$$

However, as will be shown, one cannot exclude from consideration discontinuous functions U, and for these functions the system (1), (4) may turn out to be incompatible. We shall discuss this point in more detail in the next section; meanwhile we shall assume for definiteness that the control is formed according to the rule (3).

Under the conditions stated below there exists a solution of the system (1), (3). This solution is called *step-by-step motion of the controlled system*, and control of the form (3) is called *step-by-step positional control.* Control procedures of this type have been known for a long time and are so natural it is hardly possible to pinpoint the author who was the first to utilize them. We also note that the constructions considered herein are in their essence closely related to Euler's broken lines. This fact serves as justification to utilize the term "Euler's broken lines" in our book [42] to denote step-by-step motions.

We shall now dwell a bit on the function $t \to v(t)$. Unlike the case of control, we are not at all interested in how disturbances are formed. The only thing of importance is that a measurable function $t \to v(t)$ is realized, with values on the compact set $\mathcal{Q}$. The realization of a particular disturbance may be correlated in some manner with the realized state of the system. In a differential game the control of an opponent $v(t)$ may be formed based on the feedback principle in the same manner as the control $u(t)$ is formed. However, we emphasize once more that we do not specify the nature of the rules which form the disturbance or the opponent's control.

In place of differential equation (1), one can utilize a differential inclusion which does not involve a specific disturbance but only indicates the bounds of indeterminancy generated by a possible disturbance. We shall show briefly how this can be done. Let the phase vector $x(t)$ be realized at some time t and the control $u(t)$ be chosen. Since the disturbance can be realized by any vector belonging to the compact set $\mathcal{Q}$, any vector belonging to the set $\{f(t,x(t),u(t),v)\colon\ v \in \mathcal{Q}\}$ can be implemented as the vector of phase velocity. We thus obtain that the motion of the controlled system must satisfy the differential inclusion

$$(5) \qquad \frac{dx(t)}{dt} \in \{f(t,x(t),u(t),v)\colon\ v \in \mathcal{Q}\}$$

where, as above, the control $u(t)$ is formed by the rule (3). Under the conditions stipulated below, the differential inclusion (5) will have a solution and moreover, the set of solutions will, as a rule, be infinite. The definitions of motion in terms of equation (1) or inclusion (5) differ only in form rather than in essence. When differential inclusion is utilized, the problem of controlling consists of choosing a method of control which will guarantee the required property for all solutions of the differential inclusion.

1.2.2. Consider now the function f in the r.h.s. of equation (1). First we shall explain the symbols which will often be utilized in the sequel. The symbols $\langle y,z\rangle$ denote the scalar product of two m-dimensional vectors y and z, i.e.

$$\langle y,z\rangle \stackrel{\text{def}}{=} y_1 z_1 + \dots + y_m z_m = y^T z.$$

Vectors in the Euclidean space are interpreted as column vectors. The upper index T denotes transposition. Thus the column vector y can be written as $(y_1, \dots, y_m)^T$. The symbol $\|y\|$ denotes the Euclidean norm of the vector y, i.e. $\|y\| \stackrel{\text{def}}{=} \langle y,y\rangle^{1/2}$. Finally we note that the derivatives are denoted both as $dx(t)/dt$ and as $\dot{x}(t)$.

We shall thus assume that the function f satisfies the following conditions:

(i) The function $f\colon\ T \times R^n \times \mathcal{P} \times \mathcal{Q} \to R^n$ is continuous. Here T is an interval on which the time t varies in the control problems under consideration. On the whole, we shall consider cases where $T = [t_{\#},\vartheta]$ or $T = (-\infty,\vartheta]$.

(ii) For any bounded region $\mathcal{D} \subset T \times R^n$ the following Lipschitz condition is fulfilled:

$$(6) \qquad \|f(t,x^{(1)},u,v) - f(t,x^{(2)},u,v)\| \leqslant \lambda(\mathcal{D})\|x^{(1)} - x^{(2)}\|$$

for any $(t,x^{(i)}) \in \mathcal{D}$ $(i = 1,2)$, $(u,v) \in \mathcal{P} \times \mathcal{Q}$.

(iii) For any $(t,x,u,v) \in T \times R^n \times \mathcal{P} \times \mathcal{Q}$ the inequality

$$(7)\qquad \langle x, f(t,x,u,v)\rangle \leqslant \xi(1 + \|x\|^2)$$

is valid where ξ is a constant.

(iv) For any $(t,x) \in T \times R^n$ and $s \in R^n$ the equality

$$(8)\qquad \min_{u \in P} \max_{v \in Q} \langle s, f(t,x,u,v)\rangle = \max_{v \in Q} \min_{u \in P} \langle s, f(t,x,u,v)\rangle$$

is valid.

The last condition differs in its essence from the first three conditions. Condition (iv) will be called the *saddle point condition in a small game.* The meaning of this name and the role played by this condition will become evident in subsequent chapters. In Chapters 9 and 10 control systems are considered for which condition (iv) may be violated. Methods of solving minimax control problems and formulation of differential games substantially depend on whether or not the controlled systems satisfy the condition of saddle point in a small game. Observe that condition (iv) is fulfilled in particular in the case when function f is of the form

$$(9)\qquad f(t,x,u,v) = f^{(1)}(t,x,u) + f^{(2)}(t,x,v).$$

Unlike condition (iv), conditions (i) – (iii) will be assumed to be satisfied throughout the entire book. Conditions of this type are often considered in control theory. We shall now discuss the meaning of these conditions. In the theory of ordinary differential equations the so-called Cauchy problem is studied. In the case which is relevant to us, this problem is as follows. It is required to obtain an absolutely continuous function $t \to x(t)$ defined on the interval $[t_0,\theta]$ which satisfies the initial condition $x(t_0) = x_0$, and for almost all $t \in [t_0,\theta]$ satisfies equations (1) where $u(t)$ is formed in accordance with equality (3). By virtue of well-known theorems (cf., e.g., [13, 34, 74]) the solution of this problem exists and is unique. Thus when conditions (i) – (iii) are fulfilled it is meaningful to talk about the step-by-step motion of the controlled system (1), (3), identifying it with the solution of the corresponding Cauchy problem. This motion is uniquely determined for any choice of the initial point (t_0,x_0) in the region $T \times R^n$ and for any measurable realization $t \to v(t)$. Note that condition (i) assures the existence of solutions $x(t)$ on some interval $[t_0,t\ )$; when condition (ii) is fulfilled this solution is unique while when (iii) is satisfied the solution can be extended up to the time $t = \theta$, up to which the r.h.s. of equation (1) is defined. Concerning the extendability condition, we note that various forms of this condition are available; for example, it may be required that

$$(10)\qquad \|f(t,x,u,v)\| \leqslant \xi(1 + \|x\|)$$

for all $(t,x,u,v) \in T \times R^n \times P \times Q$. For our purposes the mere fact of extendability of motions is of importance while the conditions which assure extendability are of no relevance. Therefore one may require the fulfillment of inequality (10) instead of bound (7) or may replace it by some other condition. Concluding the discussion of conditions (i) – (iii) we note that the step-by-step motion of the controlled system (1), (3) is also a step-by-step motion of the system (5), (3); thus the set of step-by-step motions of the last system is not empty.

1.2.3. Consider a particular case of the controlled system – the linear controlled system. The motion of this system is described by either an equation of the form

$$\dot{x}(t) = A(t)x(t) + B(t)u(t) + C(t)v(t) \tag{11}$$

or by equation

$$\dot{x}(t) = A(t)x(t) + g(t,u(t),v(t)). \tag{12}$$

Here $A(t)$,$B(t)$, and $C(t)$ are $n \times n$, $n \times p$, and $n \times q$ matrices, respectively. It is assumed that components of these matrices depend continuously on the parameter t. In equation (12) the function $g: T \times P \times Q \to R^n$ is continuous. It is easy to verify that for equations (11) and (12) conditions (i) – (iii) are satisfied. For equation (11), condition (iv) is also satisfied while for equation (12) equality (8) becomes

$$\min_{u \in P} \max_{v \in Q} \langle s,g(t,u,v)\rangle = \max_{v \in Q} \min_{u \in P} \langle s,g(t,u,v)\rangle, \tag{13}$$

which should be satisfied for all $s \in R^n$ and $t \in T$.

When considering systems of the form (11) or (12) we shall sometimes be using the formula of variation of constants [13, 17, 57]. According to this formula the motion of system (12) with the initial condition $x(t_0) = x_0$ satisfies

$$x(t) = \Phi(t,t_0)x_0 + \int_{t_0}^{t} \Phi(t,\tau)g(\tau,u(\tau),v(\tau))d\tau. \tag{14}$$

Here $\Phi(t,\tau)$ is an $n{\times}n$-matrix, $\tau \to \Phi(t,\tau)$ satisfies the matrix differential equation

$$\frac{d\Phi(t,\tau)}{d\tau} = -\Phi(t,\tau)A(\tau), \tag{15}$$

and condition $\Phi(t,t) = E$ where E is the unit matrix. In the case when $A(t) = A$ (for all $t \in T$) the equality $\Phi(t,\tau) = \Phi_*(t - \tau)$ is valid, where the function $r \to \Phi_*(r)$ satisfies the equation

$$\frac{d\Phi_*(r)}{dr} = A\Phi_*(r) \tag{16}$$

and condition $\Phi_*(0) = E$. Note that in this case the symbol e^{Ar} is often used to denote the matrix $\Phi_*(r)$.

1.2.4. Example. Consider the equation of motion of controlled systems in a differential game. A solution of game-theoretical problems comprising this game will be obtained in §2.5.

Let two mass points with masses m_1 and m_2 move in a horizontal plane. It is assumed that some coordinate system is chosen on this plane. Let (y_1,y_2) be the coordinates of the first point and (z_1,z_2) be the coordinates of the second. On each point a frictional force is applied proportional to the velocity of this point. Let α_1 and α_2 be friction coefficients for the first and second points, respectively. In addition to friction forces on the first and second points, controlling forces u and v, respectively, are applied. These forces are identified with two-dimensional vectors whose norms are bounded, i.e. $\|u\| \leqslant \beta_1$, $\|v\| \leqslant \beta_2$ where β_1 and β_2 are given numbers.

Consider the equation of motion of the first point. In accordance with Newton's law, the equality

$$(17) \qquad m_1\ddot{y} = -\alpha_1\dot{y} + u$$

is valid where $\dot{y}$ is the velocity vector and $\ddot{y}$ is the acceleration vector. We introduce the following notation: $\dot{y}_1 = y_3$, $\dot{y}_2 = y_4$. Then the equality (17) will be rewritten as follows:

$$(18) \qquad \dot{y}_1 = y_3,\ \dot{y}_2 = y_4,\ \dot{y}_3 = (-\alpha_1 y_3 + u_1)m_1^{-1},\ \dot{y}_4 = (-\alpha_1 y_4 + u_2)m_1^{-1}.$$

Analogously, for the second point we obtain

$$(19) \qquad \dot{z}_1 = z_3,\ \dot{z}_2 = z_4,\ \dot{z}_3 = (-\alpha_2 z_3 + v_1)m_2^{-1},\ \dot{z}_4 = (-\alpha_2 z_4 + v_2)m_2^{-1}.$$

In a differential game, controlled systems (18), (19) are considered jointly, therefore components of their phase vectors should be combined and an eight-dimensional phase vector $x = (x_1, \ldots, x_8)^{\mathrm{T}} = (y_1, \ldots, y_4, z_1, \ldots, z_4)^{\mathrm{T}}$ formed. The motion of the phase point $x(t)$ is described by the equations

$$(20) \qquad \begin{aligned} &\dot{x}_1 = x_3, \quad \dot{x}_2 = x_4, \\ &\dot{x}_3 = (-\alpha_1 x_3 + u_1)m_1^{-1}, \quad \dot{x}_4 = (-\alpha_1 x_4 + u_2)m_1^{-1}, \\ &\dot{x}_5 = x_7, \quad \dot{x}_6 = x_8, \\ &\dot{x}_7 = (-\alpha_2 x_7 + v_1)m_2^{-1}, \quad \dot{x}_8 = (-\alpha_2 x_8 + v_2)m_2^{-1}. \end{aligned}$$

We thus obtain equations of the form (11) where matrices $A(t)$, $B(t)$, and $C(t)$ do not depend on time t. These matrices are not written down explicitly here since it would be too cumbersome.

In what follows, the formula of variation of constants for system (20) will be required. We shall thus define the corresponding matrix $\Phi_*(r)$ here. This matrix is of the form

$$\Phi_*(r) = \begin{bmatrix} \Phi_1(r) & 0 \\ 0 & \Phi_2(r) \end{bmatrix} \tag{21}$$

where Φ_1, Φ_2, and 0 are (4×4)-matrices and all the entries of the matrix 0 are zeros. Matrices $\Phi_1(r)$ and $\Phi_2(r)$ are determined by the equality

$$\Phi_i(r) = \begin{bmatrix} 1 & 0 & \varphi_i(r) & 0 \\ 0 & 1 & 0 & \varphi_i(r) \\ 0 & 0 & \psi_i(r) & 0 \\ 0 & 0 & 0 & \psi_i(r) \end{bmatrix} \quad (i = 1,2). \tag{22}$$

Here

$$\begin{aligned} \varphi_i(r) &= [1 - \exp(-\alpha_i m_i^{-1} r)]\alpha_i^{-1} m_i \ , \\ \psi_i(r) &= \exp(-\alpha_i m_i^{-1} r) \ . \end{aligned} \tag{23}$$

It can be directly verified that the function $r \to \Phi_*(r)$ defined by the relations (21) – (23) indeed satisfies the differential equation (16) and the condition $\Phi_*(0) = E$.

1.3. Positional Strategies

In this section positional strategies and motions generated by these strategies will be defined. The motions introduced are important components of the proposed formalization of differential games and minimax control problems. They will be suitable for the proofs of existence theorems for equilibrium situations in differential games and for the study of the structure of solutions of minimax control problems. The step-by-step positional control of the form (1.2.3) serves as a constructive basis for the proposed definitions. We shall return to a step-by-step positional control for a physically realizable interpretation of theoretical results.

1.3.1. We shall begin with formal constructions. Let the function U: $T \times R^n \to P$ and the initial position $(t_0, x_0) \in T \times R^n$ be given.

Choose a partition

$$\Delta \overset{\text{def}}{=} \{t_0 = \tau_0 < \tau_1 < \ldots < \tau_{k+1} = \theta\}. \tag{1}$$

Denote by $X_1(t_0, x_0 U, \Delta)$ the set of step-by-step motions of the system (1.2.5), (1.2.3). The elements of this set are absolutely-continuous functions $x(\cdot)$: $[t_0, \theta] \to R^n$ which satisfy the initial condition $x(t_0) =$

x_0 and differential inclusion (1.2.5.) for almost all $t \in [t_0,\theta]$ where the control $u(t)$ is formed by the rule (1.2.3.). Define the quantity

(2) $$\operatorname{diam} \Delta \overset{\text{def}}{=} \max_i (\tau_{i+1} - \tau_i) \quad \text{for } i = 0, 1, ..., k.$$

We are interested in the limits of the sequences of step-by-step motions

(3) $$x_\ell(\cdot) \in X_1(t_0,x_0U,\Delta_\ell), \quad \ell = 1, 2, ...,$$

where $\operatorname{diam}\Delta_\ell \to 0$ as $\ell \to \infty$. The totality of these limits is denoted by the symbol $X_1(t_0,x_0,U)$. Thus the function $x(\cdot)$: $[t_0,\theta] \to R^n$ belongs to the set $X_1(t_0,x_0,U)$ if and only if there exists a sequence of step-by-step motions of the form (3) converging to it. This can be expressed by the formula

(4) $$X_1(t_0,x_0,U) \overset{\text{def}}{=} \{x(\cdot) | \exists\, x_\ell(\cdot) \in X_1(t_0,x_0,U,\Delta_\ell),$$
$$\ell = 1,2,3, ..., \ x_\ell(\cdot) \to x(\cdot), \operatorname{diam}\Delta_\ell \to 0 \text{ for } \ell \to \infty\}.$$

The symbol $x_\ell(\cdot) \to x(\cdot)$ denotes uniform convergence, i.e.

$$\max_{t_0 \leqslant t \leqslant \theta} \|x_\ell(t) - x(t)\| \to 0 \quad \text{as } \ell \to \infty.$$

The meaning of the index $_1$ which is included in the notation of the sets $X_1(t_0,x_0,U,\Delta)$ and $X_1(t_0,x_0,U)$ will be explained below (cf. §1.5.1.).

Definition. *A function* U: $T \times R^n \to P$ *is called a positional strategy and the elements of the set* $X_1(t_0,x_0,U)$ *determined by the relation* (4) *are called constructive motions, generated by a positional strategy* U *and emerging from the initial position* (t_0,x_0).

Observe that the set $X_1(t_0,x_0,U)$ can be defined as follows: the function $x(\cdot)$ belongs to the set $X_1(t_0,x_0,U)$ if and only if there exists a sequence of step-by-step motions converging to it:

$$x_\ell(\cdot) = x(\cdot,t_0,x_0,U,v_\ell(\cdot),\Delta_\ell), \quad \ell = 1, 2, 3, ...,$$

where $\operatorname{diam}\Delta_\ell \to 0$ as $\ell \to \infty$, $v_\ell(\cdot)$: $[t_0,\theta] \to Q$ is an arbitrary sequence of measurable disturbances. Here the symbol $x(\cdot,t_0,x_0,U,v(\cdot),\Delta)$ denotes the step-by-step motion of the system (1.2.1), (1.2.3) corresponding to some measurable disturbance $v(\cdot)$.

The equivalence of the two definitions presented above follows from the equality

(5) $$X_1(t_0,x_0,U,\Delta) = \bigcup_{v(\cdot)} \{x(\cdot,t_0,x_0,U,v(\cdot),\Delta)\}$$

where the union is taken over all measurable disturbances.

Evidently this union is contained in the set $X_1(t_0,x_0,U,\Delta)$. The converse inclusion can be proved utilizing Filippov's theorem [22, 74]. However, we shall omit the detailed proof of equality (5) since this equality is not needed in the formal proofs below.

1.3.2. We shall now discuss the definitions presented above. Recall that any function $U\colon \mathrm{T} \times R^n \to P$ can be chosen to be a positional strategy. It is not required that such a function be continuous or satisfy the Lipschitz condition, etc. As will be shown below, for any positional strategy U the set of corresponding constructive motions is not empty and these motions are absolutely continuous functions. The idea behind the definition of constructive motions is quite simple: the set $X_1(t_0,x_0,U)$ consists of functions $x(\cdot)$ which can be approximated by the corresponding step-by-step motions for sufficiently refined partitions Δ.

Note that one does not require that constructive motions $x(\cdot) \in X_1(t_0,x_0,U)$ satisfy, for almost all $t \in [t_0,\theta]$, the differential inclusion

(6) $$\dot{x} \in \{f(t,x,U(t,x),v)\colon\ v \in Q\}$$

or the differential equation

(7) $$\dot{x} = f(t,x,U(t,x),v(t))$$

for some measurable disturbance $v(\cdot)$. This assertion can be demonstrated by the following example.

Let x,u, and v be scalars. Set

(8) $$f(t,x,u,v) = u + v,\ P = \{u \in R^1\colon\ |u| \leqslant 1\},\ Q = \{0\},$$

i.e. the motion equation is of the form $\dot{x} = u$ and there is no disturbance. Consider the function

(9) $$U(t,x) = \begin{cases} 1 \text{ for } t \text{ irrational}, x \in R^1, \\ -1 \text{ for } t \text{ rational}, x \in R^1. \end{cases}$$

Assume for definiteness that $t_0 = 0$, $x_0 = 0$, $\theta = 1$. In this case the differential inclusion (6) becomes the equation

(10) $$\dot{x}(t) = U(t,x(t)),$$

which possesses a unique classical solution $x(t) = t$, $0 \leqslant t \leqslant 1$. On the other hand, the set of constructive motions $X_1(t_0,x_0,U)$ contains all absolutely-continuous functions $x(\cdot)$, satisfying the equality $x(0) = 0$ and the inequality $|\dot{x}(t)| \leqslant 1$ for almost all $t \in [0,1]$. Note that although there is no disturbance in this example, the set of constructive motions is infinite. All constructive motions except for the motion $x(t) = t$, $0 \leqslant t \leqslant 1$, are not solutions of equation (10), i.e.

they do not satisfy this equality for almost all $t \in [0,1]$.

It is easy to give an example for which differential inclusion (6) and differential equation (7) do not have classical solutions while the set of constructive motions is not empty. This holds for the example presented above (8), (10) when the function $U(t,x)$ is chosen to be

$$U(t,x) = \begin{cases} 1 & \text{for } 0 \leqslant t \leqslant 1, \ x \leqslant 0, \\ -1 & \text{for } 0 \leqslant t \leqslant 1, \ x > 0. \end{cases} \tag{11}$$

Finally, we note that to define the set $X_1(t_0,x_0,U)$ it is sufficient to determine function U on the region $[t_0,\theta] \times R^n$. However, it is more convenient to define a class of positional strategies which do not depend on the parameter t_0 whose range of variation is the interval T. We shall therefore identify positional strategies with functions U defined on the whole space of positions $T \times R^n$.

1.3.3. Two important properties of the set $X_1(t_0,x_0,U)$ will be stated as lemmas.

Denote by the symbol $C_n([t_0,\theta])$ the space of continuous functions $x(\cdot)$: $[t_0,\theta] \to R^n$. In this space the norm

$$|x(\cdot)| \overset{\text{def}}{=} \max_{t_0 \leqslant t \leqslant \theta} \|x(t)\| \tag{12}$$

is defined.

Lemma 1.3.1. *For any initial position $(t_0,x_0) \in T \times R^n$ and any positional strategy U, the set of constructive motions $X_1(t_0,x_0,U)$ is a non-empty compact set in the space $C_n([t_0,\theta])$. Moreover, any constructive motion $x(\cdot) \in X_1(t_0,x_0,U)$ is an absolutely continuous function.*

Proof. Consider an arbitrary step-by-step motion $x(\cdot) \in X_1(t_0,x_0,U,\Delta)$. It follows from condition (1.2.7) that

$$\frac{d\|x(t)\|^2}{dt} = 2 \langle x(t),\dot{x}(t)\rangle \leqslant 2 \xi(1 + \|x(t)\|^2).$$

Integrating this inequality we obtain

$$\|x(t)\|^2 \leqslant r^2(t) \overset{\text{def}}{=} (1 + \|x_0\|^2)\exp 2 \xi(t - t_0) - 1, \ t \in [t_0,\theta].$$

This bound can be rewritten in the form

$$(t,x(t)) \in \mathfrak{D}, \quad t_0 \leqslant t \leqslant \theta, \tag{13}$$

where $\mathfrak{D}$ is a bounded set in $T \times R^n$ defined by the equality

$$\mathfrak{D} \overset{\text{def}}{=} \{(t,x): t_0 \leqslant t \leqslant \theta, \ \|x\| \leqslant r(t)\}.$$

Thus the inclusion (13) is valid for any step-by-step motion $x(\cdot) \in X_1(t_0,x_0,U,\Delta)$ and any choice of the partition Δ.

Set

$$(14) \qquad \Lambda \overset{\text{def}}{=} \max \|f(t,x,u,v)\| \quad \text{for } (t,x) \in \mathcal{D}, \ (u,v) \in P \times Q.$$

Consider the sequence of step-by-step motions of the form (3). It follows from the discussion above that any function $x_\ell(\cdot)$ in this sequence satisfies the Lipschitz condition

$$\|x_\ell(t_1) - x_\ell(t_2)\| \leqslant \Lambda|t_1 - t_2|, \quad t_i \in [t_0,\theta], \quad i = 1, 2.$$

Therefore, in view of the Ascoli-Arzela lemma [18, 36, 74] one can select from this sequence a convergent subsequence whose limit will be an element of the set $X_1(t_0,x_0,U)$ (here the convergence is in the norm (12) all over). It is thus proved that $X_1(t_0,x_0,U) \neq \phi$.

Let $y(\cdot) \in X_1(t_0,x_0,U)$ and $x_\ell(\cdot)$ ($\ell = 1, 2, \ldots$) be a sequence of the form (3), convergent to the function $y(\cdot)$. Since each one of the functions $x_\ell(\cdot)$ satisfies the above stated Lipschitz condition, the limiting function $y(\cdot)$ satisfies the Lipschitz condition

$$\|y(t_1) - y(t_2)\| \leqslant \Lambda|t_1 - t_2|, \quad t_i \in [t_0,\theta], \quad i = 1, 2$$

as well. Consequently the function $y(\cdot)$ is absolutely continuous.

It remains to verify that from any sequence $y_\ell(\cdot) \in X_1(t_0,x_0,U)$, $\ell = 1, 2, \ldots$ a convergent subsequence can be selected whose limit is contained in $X_1(t_0,x_0,U)$. By definition of constructive motions there exist step-by-step motions $x_\ell(\cdot) \in X_1(t_0,x_0,U,\Delta_\ell)$ such that

$$(15) \qquad \operatorname{diam}\Delta_\ell \leqslant \epsilon_\ell = 1/\ell, \quad \|x_\ell(\cdot) - y_\ell(\cdot)\| \leqslant \epsilon_\ell.$$

From the sequence $x_\ell(\cdot)$, $\ell = 1, 2, \ldots$ a convergent subsequence $x_{\ell_j}(\cdot)$, $j = 1, 2, \ldots$ can be selected. By definition (4) we obtain from (15)

$$\lim_{j\to\infty} x_{\ell_j}(\cdot) = \lim_{j\to\infty} y_{\ell_j}(\cdot) \in X_1(t_0,x_0,U).$$

Lemma 1.3.1 is thus proved.

The second property follows directly from the definition of the set $X_1(t_0,x_0,U)$.

Lemma 1.3.2. *For any number $\epsilon > 0$, the initial position $(t_0,x_0) \in T \times R^n$ and the positional strategy U, one can find a $\delta > 0$ such that the inclusion*

$$X_1(t_0,x_0,U,\Delta) \subset X_1^\epsilon(t_0,x_0,U)$$

is fulfilled for any choice of the partition Δ, satisfying $\operatorname{diam}\Delta \leqslant \delta$.

Here $\mathcal{X}_1^\epsilon(t_0,x_0,U)$ *is an* ϵ*-neighborhood of the set* $\mathcal{X}_1(t_0,x_0,U)$*, i.e.*

$$\mathcal{X}_1^\epsilon(t_0,x_0,U) \overset{\text{def}}{=} \{y(\cdot) + z(\cdot): y(\cdot) \in \mathcal{X}_1(t_0,x_0,U),\ z(\cdot) \in C_n([t_0,\theta]), |z(\cdot)| < \epsilon\}.$$

In what follows, the basic facts will be stated for constructive motions. Lemma 1.3.2 permits us to obtain a physically realizable interpretation of these facts for corresponding step-by-step motions.

1.3.4. The function $(t,x) \to U(t,x)$ defines the feedback in the system indicated above. In Section 1.3.1 we have defined, for an arbitrary function U, motions which we call constructive. However, other approaches to the definition of motions of systems with a discontinuous feedback are available. We shall describe one of them which has been widely used [3, 14, 23, 76]. Here we consider a somewhat weakened version which is sufficient for our purposes.

Let a function $U: T \times R^n \to P$ be defined. Select a point $(t_*,x_*) \in T \times R^n$. Set

$$B(t_*,x_*;\epsilon) \overset{\text{def}}{=} \{(t,x) \in T \times R^n : |t - t_*| \leqslant \epsilon,\ \|x - x_*\| \leqslant \epsilon\},$$

$$U(B(t_*,x_*;\epsilon)) \overset{\text{def}}{=} \{U(t,x): (t,x) \in B(t_*,x_*;\epsilon)\},$$

i.e. the second set contains all the values of function U taken by this function in a closed ϵ-neighborhood of the point (t_*,x_*). We adjoin to the set $U(B(t_*,x_*;\epsilon))$ all its limit points, i.e. we consider the closure of this set denoted by $cl\ U(B(t_*,x_*;\epsilon))$. Now define the set

$$(16) \qquad \mathcal{U}(t_*,x_*) \overset{\text{def}}{=} \bigcap_{\epsilon>0} cl\ U(B(t_*,x_*;\epsilon)).$$

This set is not empty since $U(t_*,x_*) \in \mathcal{U}(t_*,x_*)$. Moreover, $\mathcal{U}(t_*,x_*)$ is closed and imbedded in P since this set is an intersection of closed subsets of the compact set P.

Thus a multivalued mapping $\mathcal{U}$ is constructed which places the points $(t,x) \in T \times R^n$ into correspondence with non-empty closed subsets of the compact set P. The mapping obtained is upper semicontinuous (cf. §A3).

We now present a proof. Let $(t_\ell,x_\ell) \in T \times R^n$, $u_\ell \in \mathcal{U}(t_\ell,x_\ell)$, $\ell = 1, 2, \ldots$ and let $(t_\ell,x_\ell) \to (t_*,x_*)$, $u_\ell \to u_*$ as $\ell \to \infty$. It is required to verify that $u_* \in \mathcal{U}(t_*,x_*)$. Set

$$B(u_*,\alpha) \overset{\text{def}}{=} \{u \in P: \|u - u_*\| \leqslant \alpha\}.$$

We show that

$$(17) \qquad B(u_*,\alpha) \cap cl\ U(B(t_*,x_*;\epsilon)) \neq \phi \quad \text{for any } \alpha > 0 \text{ and } \epsilon > 0.$$

Choose ℓ so that

$$u_\ell \in B(u_*;\alpha), \quad |t_\ell - t_*| \leqslant \epsilon/2, \quad \|x_\ell - x_*\| \leqslant \epsilon/2.$$

Then $\mathcal{B}(t_\ell,x_\ell;\epsilon/2) \subset \mathcal{B}(t_*,x_*;\epsilon)$ and thus

$$cl\ U(\mathcal{B}(t_\ell,x_\ell;\epsilon/2)) \subset cl\ U(\mathcal{B}(t_*,x_*;\epsilon)).$$

Since $u_\ell \in \mathcal{U}(t_\ell,x_\ell)$, it follows that

$$u_\ell \in cl\ U(\mathcal{B}(t_\ell,x_\ell;\epsilon/2)) \subset cl\ U(\mathcal{B}(t_*,x_*;\epsilon)).$$

Hence, $u_\ell \in \mathcal{B}(u_*;\alpha) \cap cl\ U(\mathcal{B}(t_*,x_*;\epsilon))$, which proves relation (17). Since (17) is valid for any $\alpha > 0$ and the sets $\mathcal{B}(u_*;\alpha)$ and $clU(\mathcal{B}(t_*,x_*;\epsilon))$ are compact sets, we obtain that $u_* \in cl\ U(\mathcal{B}(t_*,x_*;\epsilon))$. This inclusion is valid for any $\epsilon > 0$. Hence $u_* \in \mathcal{U}(t_*,x_*)$ and the proof is complete.

Now consider the set

(18) $$\mathcal{F}_U(t,x) \overset{\text{def}}{=} co\{f(t,x,u,v): u \in \mathcal{U}(t,x),\ v \in \mathcal{Q}\}.$$

The symbol $co\ G$ denotes the convex hull of the set G. The lower index in the designation of the set (18) indicates that it is constructed for a given function U. Observe that the set $\mathcal{F}_U(t,x)$ can also be defined as follows:

$$\mathcal{F}_U(t,x) \overset{\text{def}}{=} \bigcap_{\epsilon>0} \overline{co}\{f(\tau,y,U(\tau,y),v): (\tau,y) \in \mathcal{B}(t,x;\epsilon),\ v \in \mathcal{Q}\}$$

where the symbol $\overline{co}\ G$ denotes the convex closed hull of the set G.

It follows from the continuity of the function f, the inequality (1.2.7), and the upper semi-continuity of the mapping $(t,x) \to \mathcal{U}(t,x)$, that the multivalued mapping $(t,x) \to \mathcal{F}_U(t,x)$ satisfies all the conditions stipulated in Theorem A.4.1. Denote by $X^0(t_0,x_0,U)$ the set of absolutely-continuous functions $x(\cdot)$: $[t_0,\theta] \to R^n$ which satisfy the initial condition $x(t_0) = x_0$ and the differential inclusion

(19) $$\dot{x}(t) \in \mathcal{F}_U(t,x(t))$$

for almost all $t \in [t_0,\theta]$. The elements of the set $X^0(t_0,x_0,U)$ will be called generalized motions generated by the strategy U. According to Theorem A.4.1 the set $X^0(t_0,x_0,U)$ is a non-empty compact set in the space $C_n([t_0,\theta])$.

We now compare the definitions of generalized and constructive motions. It turns out that for any initial position $(t_0,x_0) \in T \times R^n$ and any function U: $T \times R^n \to P$, the set of constructive motions is embedded into the set of generalized motions, i.e.

(20) $$X_1(t_0,x_0,U) \subset X^0(t_0,x_0,U).$$

This embedding means that the existence of generalized motions can be proved by a limiting transition applied to step-by-step motions. Such a proof is in many aspects similar to the existence proof of ordinary differential equations using the limiting transition of Euler's broken lines.

The question arises: is the inverse inclusion valid? Consider the following simple example. Assume that in the system (8), (10) the function U is defined by the equality

$$U(t,x) = \begin{cases} +1 & \text{for } 0 \leqslant t \leqslant 1, \; x \geqslant 0 \\ -1 & \text{for } 0 \leqslant t \leqslant 1, \; x < 0. \end{cases}$$

For $t_0 = 0$, $x_0 = 0$, the set $X_1(t_0,x_0,U)$ contains the unique constructive motion $x(t) = t$, $0 \leqslant t \leqslant 1$. Consider the set of generalized motions. In the example

$$F_U(t,x) = \mathcal{U}(t,x) = \begin{cases} \{1\} & \text{for } x > 0, \; t \in [0,1], \\ [-1,1] & \text{for } x = 0, \; t \in [0,1], \\ \{-1\} & \text{for } x < 0, \; t \in [0,1] \end{cases}$$

and the set $X^0(t,x,U)$ consists of functions of the form

$$x(t) = 0 \;\text{ for }\; 0 \leqslant t \leqslant \tau, \quad x(t) = \pm(t - \tau) \;\text{ for }\; \tau \leqslant t \leqslant 1,$$

where the parameter τ can take on any value in the interval [0,1].

This example shows that not every generalized motion is a constructive motion. It turns out that the availability of "extra" generalized motions can substantially deteriorate the result guaranteed for constructive motions. However, for some type of problems, optimal results obtained for constructive and generalized motions do coincide.

1.3.5. The positional step-by-step procedure considered above specified a control $u(t)$ which was constant on the half-intervals $[\tau_i,\tau_{i+1})$. Moreover, the partition of the segment $[t_0,\theta]$ into these half-intervals was selected at the initial time $t = t_0$. Other step-by-step control procedures are also known. For example, one can choose the function $(t,x) \to \delta(t,x) > 0$ and by means of this function form the intervals of constant control. This is accomplished in the following manner. At the initial time a number $\delta_0 = \delta(t_0,x_0)$ and a constant control $u(t) = U(t_0,x_0)$, $t_0 \leqslant t < \tau_1 = t_0 + \delta_0$ is determined. At time $t = \tau_1$, depending on the realized position a number $\delta_1 = \delta(\tau_1,x(\tau_1))$ and control $u(t) = U(\tau_1,x(\tau_1))$, $\tau_1 \leqslant t < \tau_1 + \delta_1$, are selected and so on. It may, however, occur that the sequence $\tau_{i+1} = \tau_i + \delta(\tau_i,x(\tau_i))$, $i = 0, 1, \ldots$ will not reach time $t = \theta$ in a finite number of steps. In order to obtain a motion defined on the whole interval $[t_0,\theta]$ it is necessary to explicitly specify how to continue the motion in such situations (we shall omit discussion of this point). Step-by-step procedures may form not a piecewise constant but rather a piecewise-programmed control. In this case, depending on the realized position $(\tau_i,x(\tau_i))$ a measurable function $u(\cdot)$: $[\tau_i,\tau_{i+1}) \to P$ is assigned, while the times τ_i may be defined a priori and may be formed depending on the positions. Finally, one can consider various control procedures in which, in addition to the information concerning the realized position, other data are also used. For many modifications of step-by-step control procedures, one can prove existence theorems of situations of ϵ-equilibrium for definite types of differential games. However, we prefer to work with step-by-step positional control (1.2.3.) and the corresponding step-by-step and constructive motions since these motions are most simple and convenient.

1.4. Quality Criterion

In order to define the notion of an optimal strategy it is necessary to introduce the quality criterion for an arbitrary positional strategy. This is done in accordance with the scheme indicated in §1.1.3. Along with the general formulation of minimax control problems in this section, several particular types of such problems are considered. A comparison of results – which are guaranteed for constructive motions and approximating step-by-step motions – is presented. It is explained why one should *not* confine himself to continuous strategies when formulating problems of positional control.

1.4.1. Let an initial position $(t_0,x_0) \in T \times R^n$ be given. Assume that a certain functional γ is defined on the space $C_n([t_0,\theta])$ of continuous functions $x(\cdot)$: $[t_0,\theta] \to R^n$. In particular, for any motion $x(\cdot)$ originating from the initial position (t_0,x_0), the value $\gamma(x(\cdot))$ of this functional is defined. We assume that the functional γ allows us to compare the quality of motions: if for two motions $x_1(\cdot)$ and $x_2(\cdot)$ the inequality $\gamma(x_1(\cdot)) < \gamma(x_2(\cdot))$ is fulfilled then the motion $x_1(\cdot)$ is "better than" $x_2(\cdot)$; if $\gamma(x_1(\cdot)) \leqslant \gamma(x_2(\cdot))$ then the motion is "not worse" than the motion $x_2(\cdot)$. The functional γ is called a *target functional.*

Recall the statement of an optimal control problem for a system without disturbance. The motion of such a system is described by the equation $\dot{x}(t) = f(t,x(t),u(t))$. Any measurable function $u(\cdot)$: $[t_0,\theta] \to P$ can be taken for a control, and moreover this function (programmed control) can be prescribed at the initial time t_0. The quality of a programmed control $u(\cdot)$ can be estimated by the quantity $\Gamma(u(\cdot)) = \gamma(x(\cdot))$, where $x(\cdot)$ is the motion of the system under consideration generated by control $u(\cdot)$. For a given initial position (t_0,x_0) such a motion is unique. In this case it is natural to define the optimal control $u^0(\cdot)$ by the condition $\Gamma(u^0(\cdot)) = \min\Gamma(u(\cdot))$, where $u(\cdot)$ are admissible programmed controls.

We are, however, concerned with a controlled system affected by disturbance. The only thing that is known about the disturbance is the set Q in which it takes its values; a specific disturbance realized in the course of a control process is unknown. It is assumed that at each time t the realized position $(t,x(t))$ is known. This information should not be neglected, and therefore positional strategies are utilized for controlling the system. Let a positional strategy U be chosen. This strategy, together with all kinds of disturbance, generates the set of motions. For a chosen strategy U, any motion from the set $X_1(t_0,x_0,U)$ can be realized, in particular the worst possible motion $x_*(\cdot)$ such that $\gamma(x_*(\cdot)) = \Gamma_1(U) \overset{\mathrm{def}}{=} \max\gamma(x(\cdot))$ for $x(\cdot) \in X_1(t_0,x_0,U)$. Evidently $\Gamma_1(U)$ is an exact estimate of the result guaranteed for constructive motions, i.e. $\gamma(x(\cdot)) \leqslant \Gamma_1(U)$ for $x(\cdot) \in X_1(t_0,x_0,U)$; on the other hand, for any number $\Gamma_* < \Gamma_1(U)$ there exists a motion $x(\cdot) \in X_1(t_0,x_0,U)$ such that $\gamma(x(\cdot)) > \Gamma_*$. The quantity $\Gamma_1(U)$ is called *the quality criterion of a positional strategy.* Optimal strategy is - by definition - such a strategy for which the quantity $\Gamma_1(U)$ takes on the smallest value.

Before we finally state the problem of optimal strategy we note that the definition of the quantity $\Gamma_1(U)$ presented above is well-defined in the case of a continuous functional. In this case the maximum of the functional γ on the set $X_1(t_0,x_0,U)$ is actually attained since $X_1(t_0,x_0,U)$ is a compact set in $C_n([t_0,\Theta])$. In the general case, when the continuity of the functional γ is not assumed, the quantity $\Gamma_1(U)$ is defined by the equality

$$(1) \qquad \Gamma_1(U) \overset{\mathrm{def}}{=} \sup_{x(\cdot)} \gamma(x(\cdot)) \text{ for } x(\cdot) \in X_1(t_0,x_0,U).$$

The above arguments lead us to the following problem.

Problem 1.4.1. *It is required to determine a positional strategy such that*

$$(2) \qquad \Gamma_1(U^0) = \min \Gamma_1(U)$$

where the minimum is taken over all possible positional strategies.

Relations (1) and (2) can be combined as follows:

$$\sup\{\gamma(x(\cdot)): x(\cdot) \in X_1(t_0,x_0,U^0)\} = \min_U \sup\{\gamma(x(\cdot)): x(\cdot) \in X_1(t_0,x_0,U)\}.$$

This notation justifies naming Problem 1.4.1 as a problem on *minimax strategy.* Strategy U^0 will also be called *optimal strategy.* It may turn out, however, that the minimum in (2) is not attained. In this case we shall consider the problem of ϵ-*optimal strategy* U^ϵ.

Problem 1.4.1$^\epsilon$. *It is required to determine positional strategy* U^ϵ *such that*

$$(3) \qquad \Gamma_1(U^\epsilon) \leqslant \inf \Gamma_1(U) + \epsilon,$$

where the inf *is taken over all positional strategies, and* ϵ *is a positive parameter.*

1.4.2. Consider a few types of target functionals. We start with the functional of the form

$$(4) \qquad \gamma(x(\cdot)) \overset{\text{def}}{=} \sigma(x(\theta)).$$

Here $\sigma: \quad R^n \to R^1$ is a continuous function and $x(\theta)$ is a phase vector which is realized at the last instance of time $t = \theta$. Often $\sigma(x) = \operatorname{dist}(x,M)$ is the distance from the point x to the set M. In this case, Problem 1.4.1 has the following meaning: it is required to choose a strategy U^0 which will guarantee the smallest deviation of the phase point $x(\theta)$ from the set M. If it turns out that $\Gamma_1(U^0) = 0$, then any motion $x(\cdot) \in X_1(t_0,x_0,U^0)$ falls at the last time instance $t = \theta$ into the set M.

Another type of functional is of the form

$$(5) \qquad \gamma(x(\cdot)) \overset{\text{def}}{=} \min_{t_0 \leqslant t \leqslant \theta} \rho(t,x(t)).$$

Here $\rho: \ T \times R^n \to R^1$ is a continuous function. Consider again the particular case when $\rho(t,x) = \operatorname{dist}(x,M)$, $t \in T$, $x \in R^n$. For such a choice of target functional the quantity $\Gamma_1(U)$ is the guaranteed distance between any constructive motion $x(\cdot) \in X_1(t_0,x_0,U)$ and the set M. Unlike the case of functionals γ (4), the times of "approach" to M may be different here for different motions. If for some strategy U the quantity $\Gamma_1(U) = 0$, then for any motion $x(\cdot) \in X_1(t_0,x_0,U)$ a time $t \in [t_0,\theta]$ can be found when the phase point $x(t)$ will get into the set M.

Observe that functionals of the type (4) and (5) are continuous. We shall consider now a functional which, in general, does not possess this property. Let M be a closed set in $T \times R^n$ and $x(\cdot)$: $[t_0,\theta] \to R^n$ be a continuous function. We set

$$(6) \qquad T_0(x(\cdot)) \overset{\text{def}}{=} \{t \in [t_0,\theta]\colon \ (t,x(t)) \in M\}.$$

Consider the functional

$$(7) \qquad \gamma(x(\cdot)) = \tau^\circ(x(\cdot)) \overset{\text{def}}{=} \begin{cases} \min\{t\colon t \in T_0(x(\cdot))\} & \text{for } T_0(x(\cdot)) \neq \phi \\ \theta + 1 & \text{for } T_0(x(\cdot)) = \phi. \end{cases}$$

Thus if $T_0(x(\cdot)) \neq \phi$, i.e. the graph of the function $x(\cdot)$ intersects with the set M, then $\tau^0(x(\cdot))$ is the instant when the point $(t,x(t))$ falls into M for the first time. If, however, $T_0(x(\cdot)) = \phi$, we set $\tau^0(x(\cdot)) = \theta + 1$. Observe that in the second case any number larger than θ can be used as the value of $\tau^\circ(x(\cdot))$, in particular the improper number $+\infty$ can serve this purpose. We show that the functional τ_0 is lower semicontinuous. Let $x_k(\cdot) \to x_*(\cdot)$, $\tau^0(x_k(\cdot)) \to \tau_*$ as $k \to \infty$. It is required to show that $\tau^0(x_*(\cdot)) \leqslant \tau_*$.

If $\tau_* = \theta + 1$, then this inequality is fulfilled. Consider the case $\tau_* \leqslant \theta$. Set $\tau_K = \tau^0(x_k(\cdot))$. One can also assume that $\tau_k \leqslant \theta$ ($k =$

1,2, ...). Then we have, in view of (7), that $(\tau_k, x_k(\tau_k)) \in M$. The sequence $(\tau_k, x_k(\tau_k))$ $k = 1, 2, \ldots$ converges to the limit $(\tau_*, x_*(\tau_*))$ which belongs to the closed set M. Consequently, $\tau_* \in T_0(x_*(\cdot))$ and $\tau_* \geqslant \tau^0(x_*(\cdot))$ q.e.d.

So far we have discussed those target functionals which are functionals defined on the space $C_n([t_0,\theta])$. Certain functionals which formally are not of this nature can, by means of simple transformations, be reducd to this form. We note, for example, the functional χ defined by the equality

$$(8) \qquad \chi(x(\cdot),u(\cdot),v(\cdot)) \overset{\text{def}}{=} \int_{t_0}^{\theta} f_{n+1}(t,x(t),u(t),v(t))dt.$$

Here f_{n+1}: $T \times R^n \times P \times Q \to R^1$ is a continuous function. We shall be interested in the triples $(x(\cdot),u(\cdot),v(\cdot))$ which can be realized under positional step-by-step control by the system (1.2.1), (1.2.3). In this case control $u(\cdot)$ will be piecewise constant, the disturbance $v(\cdot)$ will be measurable, and the motion $x(\cdot)$ will be an absolutely continuous function satisfying the equation (1.2.1). The functional χ depends explicitly on control and the disturbance.

We introduce a new phase variable x_{n+1}, add it to an n-dimensional phase vector x, and obtain a new $(n + 1)$-dimensional vector $x^{\#} \overset{\text{def}}{=} (x_1, \ldots, x_n, x_{n+1})^T$ whose motion will be described by the equation

$$(9) \qquad \dot{x}_i(t) = f_i(t,x(t),u(t),v(t)), \quad i = 1, 2, \ldots, n + 1.$$

We assume that $x_{n+1}(t_0) = 0$. We shall also assume that the function $f^{\#}$: $T \times R^n \times P \times Q \to R^{n+1}$ in the r.h.s. of equation (9) satisfies all the conditions (i) - (iv) indicated in Section 1.2.2. We obtain that for the functional χ the equality

$$(10) \qquad \chi(x(\cdot), u(\cdot),v(\cdot)) = \chi^{\#}(x^{\#}(\cdot)) \overset{\text{def}}{=} x_{n+1}(\theta)$$

is valid. Here $\chi^{\#}$ is a functional of the form (4) which is defined for functions $x^{\#}(\cdot)$: $[t_0,\theta] \to R^{n+1}$ with values in the extended phase space.

All the functionals discussed above belong to a special type of functional which we call *positional functionals.* A quite general description of the structure of positional functionals is known [37]. However, we are not going to present constructive definitions here, but only state several properties of positional functionals. These are as follows.

Let the target functional be a functional of the positional type. Then the optimal result in the minimax control problem which is considered in the class of positional strategies cannot be improved under any control method which is based on additional information on the course of the controlled process. We shall assume, for definiteness, that in a step-by-step procedure this method forms a piecewise-constant control $u(t) = u(\tau_i)$, $\tau_i \leqslant t < \tau_{i+1}$, $i = 0, 1, \ldots, k$

dependent on the realized motion $x(\cdot)$: $[t_0,\tau_i] \to R^n$, control $u(\cdot)$: $[t_0,\tau_i) \to P$, and the disturbance $v(\cdot)$: $[t_0,\tau_i] \to Q$. Furthermore, as in the case for positional strategies, one can introduce constructive motions defined by a limiting transition applied to the corresponding sequences of step-by-step motions. Next the result is determined, guaranteed by such a control method on the set of generated constructive motions.

If in an antagonistic differential game the payoff functional is a functional of the positional type, then an equilibrium situation exists in the class of positional strategies of the players. Exact definitions of these concepts will be presented in Section 1.5.

To state yet another property of positional functionals, it is necessary to describe the statement of an (M,N)-approach problem. Let two closed sets M and N be given in the space of positions. In the problem of (M,N)-approach it is required to determine a strategy U^0 such that for any motion $x(\cdot) \in X_1(t_0,x_0,U^0)$, the graph of the function $x(\cdot)$ intersects with the set M and belongs to the set N. The set M is the target set and the set N is the phase constraint. It turns out that for any positional functional γ one can choose sets M and N such that the strategy U^0 which solves the (M,N)-approach problem will be optimal for Problem 1.4.1. This fact allows us to substantially unify the analysis of problems of minimax control with various target functionals of the positional type and reduce the investigation of these problems to the study of an (M,N)-approach problem.

We note that the functionals mentioned above hardly exhaust the class of positional functionals.

1.4.3. In the statement of Problems 1.4.1 and $1.4.1^{\epsilon}$ constructive motions were considered. We now discuss an approximate form of these problems where step-by-step motion will be present.

Consider the first case when the functional γ is continuous. Denote by X_Λ the totality of functions $x(\cdot)$: $[t_0,\theta] \to R^n$ satisfying the initial condition $x(t_0) = x_0$ and the Lipschitz condition $\|x(t_1) - x(t_2)\| \leqslant \Lambda|t_1 - t_2|$, $t_i \in [t_0,\theta]$, $i = 1, 2$, where the quantity Λ is determined by the equality (1.3.14). It was proved in §1.3.3 that all the constructive and step-by-step motions are contained in X_Λ. Observe that X_Λ is a compact set in $C_n([t_0,\theta])$. Therefore the functional γ is uniformly continuous in X_Λ, i.e. for any $\epsilon > 0$ one can find $\alpha(\epsilon) > 0$ such that the inequality

$$(11) \qquad |\gamma(x(\cdot)) - \gamma(y(\cdot))| \leqslant \epsilon$$

will be fulfilled for any $x(\cdot)$ and $y(\cdot)$ belonging to the compact set X_Λ such that $|x(\cdot) - y(\cdot)| \leqslant \alpha(\epsilon)$. Let U^0 be an optimal strategy in the minimax Problem 1.4.1. According to Lemma 1.3.2 one can choose a positive number $\delta = \delta(\epsilon)$ in such a manner that the inclusion

$$X_1(t_0,x_0,U^0,\Delta) \subset X_1^{\alpha}(t_0,x_0,U^0) \text{ for } \alpha = \alpha(\epsilon), \text{ diam}\Delta \leqslant \delta(\epsilon)$$

is fulfilled. Consequently, for any step-by-step motion $x(\cdot) \in X_1(t_0,x_0,U^0,\Delta)$, there corresponds a constructive motion $y(\cdot) \in X_1(t_0,x_0,U^0)$ such that $|x(\cdot) - y(\cdot)| \leqslant \alpha(\epsilon)$. Therefore we obtain from (1) and (2) that $\gamma(x(\cdot)) \leqslant \gamma(y(\cdot)) + \epsilon \leqslant \Gamma_1(U^0) + \epsilon$.

Thus we arrive at the following assertion. Let γ be a continuous target functional. Then for any $\epsilon > 0$ one can find $\delta(\epsilon) > 0$ such that for any step-by-step motion $x(\cdot) \in X_1(t_0,x_0,U^0,\Delta)$ where diam$\Delta \leqslant \delta(\epsilon)$ the inequality

$$\gamma(x(\cdot)) \leqslant \Gamma_1(V^\circ) + \epsilon \tag{12}$$

is assured.

We shall now discuss the case when the target functional is not continuous. In this case one cannot express in terms of the inequality (12) the approximating property of step-by-step motions. Consider the form of this property in Problem 1.4.1 with the functional τ^0 (7). Let U^0 be the optimal strategy for this problem (it will be shown below that such a strategy does exist). Assume that $\theta_* \overset{\text{def}}{=} \Gamma_1(U^0) < \theta + 1$, i.e. for any constructive motion $x(\cdot) \in X_1(t_0,x_0,U^0)$ the position $(t,x(t))$ falls into the set M not later than at time $t = \theta_*$. Utilizing Lemma 1.3.2 we obtain immediately that for any $\epsilon > 0$ a $\delta(\epsilon) > 0$ can be found such that for any step-by-step motion $x(\cdot) \in X_1(t_0,x_0,U^0,\Delta)$ where diam $\Delta \leqslant \delta(\epsilon)$, the position $(t,x(t))$ falls into an ϵ-neighborhood of the set M not later than at time θ_*. This constitutes the approximating property of step-by-step motions for the given problem.

In connection with the discussion of approximating properties of step-by-step motions we note that the quality criterion of the positional strategy can be introduced utilizing only the step-by-step motions without defining the constructive motions. We shall show by means of the example of the previous problem how this can be accomplished. Denote by τ_ϵ the functional defined similarly to the functional τ^0 but which differs from τ^0 by replacing the set M with a closed ϵ-neighborhood of this set. Let U be the positional strategy. We set

$$\Gamma_*(U) \overset{\text{def}}{=} \sup_{\epsilon>0} \overline{\lim_{\text{diam}\Delta\to 0}} \sup_{x(\cdot)} \tau_\epsilon(x(\cdot)) \,, \tag{13}$$

(Here $x(\cdot) \in X_1(t_0,x_0,U,\Delta)$).

Let $\Gamma_1(U)$ be a quantity defined by the equality (1) where it is set that $\gamma = \tau^0$. It follows from Lemma 1.3.2 that $\Gamma_*(U) = \Gamma_1(U)$. It should, however, be emphasized once again that for the sake of simplicity and convenience we would prefer definition (1) over the definition of the form (13).

1.4.4. We now compare the possibility of solving minimax control problems in the classes of discontinuous and continuous strategies. Consider an example in which the result guaranteed for a discontinuous strategy cannot be approximated by continuous strategies.

Let the motion of the controlled system be described by the equation

$$\text{(14)} \qquad \dot{x}(t) = u(t) + (2 - t)v(t).$$

Here x,u, and v are two-dimensional vectors. We set $t_0 = 0$, $\theta = 2$. The control and the disturbance satisfy the constraints

$$\text{(15)} \qquad u(t) \in \mathcal{P} = \mathcal{B},\ v(t) \in \mathcal{Q} = \mathcal{B}, \qquad 0 \leqslant t \leqslant 2$$

where $\mathcal{B} \overset{\text{def}}{=} \{y \in R^2\colon\ \|y\| \leqslant 1\}$. We assume that the initial position $t_0 = 0$, $x_0 = (0,0)^T$, and the target functional

$$\text{(16)} \qquad \gamma(x(\cdot)) = -\|x(2)\|$$

are given.

Let $U\colon\ [0,2] \times R^2 \to \mathcal{B}$ be a continuous function. Moreover we shall first assume that it satisfies the Lipschitz condition with respect to the variable x, i.e.

$$\|U(t,x^{(1)}) - U(t,x^{(2)})\| \leqslant \lambda\|x^{(1)} - x^{(2)}\|, \quad (t,x^{(i)}) \in [0,2] \times R^2,\ i = 1,2.$$

We choose a point $v \in \mathcal{Q} = \mathcal{B}$ and consider the differential equation

$$\text{(17)} \qquad \dot{x}(t) = U(t,x(t)) + (2 - t)v.$$

This equation possesses a unique solution which we denote by $x(\cdot,v)$. This solution is determined here in the usual manner: function $t \to x(t,v)$ is continuously differentiable, and satisfies the initial condition $x(0,v) = (0,0)^T$ and equation (17) for all $t \in [0,2]$.

All the conditions of the theorem on continuous dependence of solutions of differential equations on a parameter [see, e.g., 13, 17, 57, 74] are satisfied for equation (17) (in this case the vector v is the parameter). Therefore the mapping $v \to x(2,v)$ which places the points $v \in \mathcal{Q}$ into correspondence with the values of the function $x(\cdot,v)$, at the right end, will be continuous. We consider the function

$$y(v) \overset{\text{def}}{=} v - x(2,v)/2,\ v \in \mathcal{B}.$$

This function is continuous. We shall estimate the region of its values. For any $v \in \mathcal{B}$ we have

$$2y(v) = 2v - \int_0^2 U(t,x(t))dt - \int_0^2 (2 - t)v\,dt = -\int_0^2 U(t,x(t))dt.$$

Consequently

$$\|y(v)\| = \frac{1}{2}\|\int_0^2 U(t,x(t))dt\| \leqslant 1.$$

We thus obtain that the continuous function $v \to y(v)$ maps the circle B onto itself. By the fixed point theorem [18, 34, 74] there exists a point $v_* \in B$ such that $y(v_*) = v_*$. It follows from the construction of function $y(\cdot)$ that the solution $x(\cdot,v_*)$ (where v is chosen to be equal to v_*) satisfies the equality $x(2,v_*) = (0,0)^T$.

Denote by $X(U)$ the set of motions generated by the strategy U and emerging from the initial position $t_0 = 0$, $x_0 = (0,0)^T$. This set can be determined in two ways: either as a union of absolutely continuous solutions of differential equations

$$\dot{x}(t) = U(t,x(t)) + (2 - t)v(t), \quad 0 \leqslant t \leqslant 2 \tag{18}$$

where $v(\cdot)$: $[0,2] \to B$ are all possible measurable functions; or as a set of constructive motions generated by the positional strategy U. In the given case when the function U is continuous and satisfies the Lipschitz condition with respect to the variable x, these two definitions are equivalent. The constructed solution $x(\cdot,v_*)$ belongs to the set $X(U)$, and in view of (16), $x(2,v_*) = (0,0)^T$ implies $\gamma(x(\cdot,v_*)) = 0$. We thus obtain

$$\Gamma_1(U) \overset{\text{def}}{=} \sup\gamma(x(\cdot)) = 0 \text{ for } x(\cdot) \in X(U). \tag{19}$$

Furthermore, it can be shown that equality (19) remains valid for any continuous function U: $[0,2] \times R^2 \to B$. We shall sketch the proof. Let U_k $(k = 1, 2, \ldots)$ be a sequence of Lipschitz functions convergent to the function U. For each function U_k there exists a motion $x_k(\cdot)$ such that $\gamma(x_k(\cdot)) = 0$. One can assume that the sequence of functions $x_k(\cdot)$ $(k = 1, 2, \ldots)$ converges to the function $x_*(\cdot)$ which, in view of the continuity of the functional γ satisfies the equality $\gamma(x_*(\cdot)) = 0$. Observe that the set of absolutely continuous solutions of equations (18) and the set of constructive motions generated by the strategy U may differ now (provided the function U does satisfy the Lipschitz condition). However, the function $x_*(\cdot)$ is simultaneously a solution of equation (18) as well as a constructive motion generated by the strategy U. Therefore the equality (19) is valid for any one of the two ways of defining the set $X(U)$.

Consider a discontinuous positional strategy: U_*: $[0,2] \times R^2 \to B$ such that

$$\Gamma_1(U_*) \overset{\text{def}}{=} \sup\gamma(x(\cdot)) = -\frac{1}{2} \text{ for } x(\cdot) \in X_1(t_0,x_0,U_*). \tag{20}$$

Here $X_1(t_0,x_0,U_*)$ is a family of constructive motions of the system (14) generated by this strategy and emanating from the initial position $t_0 = 0$, $x_0 = (0,0)^T$. This strategy is defined by the equality

$$U_*(t,x) = \begin{cases} x/\|x\| & \text{for } \|x\| \neq 0,\ t \in [0,2], \\ (1,0)^T & \text{for } \|x\| = 0. \end{cases}$$

Note that any unit vector can be chosen as the value of $U_*(t,x)$ for $\|x\| = 0$.

Let a partition $\Delta = \{0 = \tau_0 < \tau_1 < \ldots < \tau_{k+1} = 2\}$ be chosen. Consider an arbitrary step-by-step motion $x(\cdot) \in X_1(t_0,x_0,U_*,\Delta)$. This motion satisfies equation (14) where $u(t) = U_*(\tau_i,x(\tau_i))$, $\tau_i \leqslant t < \tau_{i+1}$, $i = 0, 1, \ldots, k$ and $t \to v(t)$ is a measurable function with values in B. On any interval $[\tau_i,\tau_{i+1}]$ the equality

$$x(\tau_{i+1}) = x(\tau_i) + (\tau_{i+1} - \tau_i)U_*(\tau_i,x(\tau_i)) + \int_{\tau_i}^{\tau_{i+1}} (2 - \tau)v(t)dt$$

is valid. Therefore

$$\|x(\tau_{i+1})\| \geqslant \|x(\tau_i) + (\tau_{i+1} - \tau_i)U_*(\tau_i,x(\tau_i))\| - \left\|\int_{\tau_i}^{\tau_{i+1}} (2 - t)v(t)dt\right\|.$$

We have by the definition of U_* that

$$\|x(\tau_i) + (\tau_{i+1} - \tau_i)U_*(\tau_i,x(\tau_i))\| = \|x(\tau_i)\| + \tau_{i+1} - \tau_i.$$

Also the inequality

$$\left\|\int_{\tau_i}^{\tau_{i+1}} (2 - t)v(t)dt\right\| \leqslant 2(\tau_{i+1} - \tau_i) - (\tau_{i+1}^2 - \tau_i^2)/2$$

is valid. We thus arrive at the following inequality:

$$\|x(\tau_{i+1})\| \geqslant \|x(\tau_i)\| + (\tau_{i+1}^2 - \tau_i^2)/2 - (\tau_{i+1} - \tau_i).$$

Successively utilizing this inequality, starting from some time τ_j up to the final time $\tau_{k+1} = 2$, we obtain

$$\|x(2)\| = \|x(\tau_{k+1})\| \geqslant \|x(\tau_j)\| + (\tau_{k+1}^2 - \tau_j^2)/2 - (\tau_{k+1} - \tau_j) \geqslant \tau_j - \tau_j^2/2.$$

We choose a number τ_j from the condition $|1 - \tau_j| = \min|1 - \tau_i|$ for $i = 0, 1, \ldots, k + 1$. Set $\alpha \overset{\text{def}}{=} (1 - \tau_j)$, then the preceding inequality can be rewritten as

$$\|x(2)\| \geqslant 1/2 - \alpha^2/2.$$

This inequality is valid for any step-by-step motion. Note that $|\alpha| \leqslant \operatorname{diam}\Delta$. Therefore, for an arbitrary constructive motion $x(\cdot) \in X_1(t_0,x_0,U_*)$ which is obtained by a limiting transition applied to step-by-step motions as $\operatorname{diam}\Delta \to 0$, the inequality $\|x(2)\| \geqslant 1/2$ will be valid, or equivalently the inequality $\gamma(x(\cdot)) \leqslant -1/2$. It is thus proved that $\Gamma_1(U_*) \leqslant -1/2$. By tracing the course of the preceding inequality one can verify that the inequality obtained is exact, i.e. the equality (20) is indeed valid.

Comparing (20) and (19) we conclude that the result guaranteed by a discontinuous strategy U_* cannot be approximated by continuous strategies. Continuous strategies do not utilize to the fullest extent the possibilities of the controlled system for the given nature of the information. In what follows in Section 5.5 and in Chapter 8, we shall continue the discussion of qualitative problems associated with informational possibilities and the nature of the feedback.

1.4.5. In the statement of the problem of an optimal strategy (cf. §1.4.1), the initial position was assumed to be fixed. The strategy U^0 and the optimal result $\Gamma_1(U^0)$ guaranteed by this strategy were determined for a given initial position (t_0,x_0). A complete designation of the value of the optimal result should be $\Gamma_1(t_0,x_0,U^0)$. In what follows, we shall be concerned with the function which associates with the initial position $(t_0,x_0) \in T \times R^n$ the optimal result in the minimax control system, i.e. the function

$$(t_0,x_0) \to \Gamma_1(t_0,x_0, U^0) \overset{\text{def}}{=} \min_U \Gamma_1(t_0,x_0,U)$$
$$= \min_U \sup\{\gamma(x(\cdot)): \ x(\cdot) \in X_1(t_0,x_0,U)\}.$$

In order to consider such a function, it is necessary to complete the definition of a target functional γ. We shall assume that the functional γ is defined on the set

$$C \overset{\text{def}}{=} \bigcup_{t_0 \in T} C_n([t_0,\theta]),$$

which consists of all possible continuous functions $x(\cdot)$: $[t_0,\theta] \to R^n$ where the parameter t_0 may vary within the bounds of the interval T. We can also proceed in a different manner, assuming that a family of functionals $\gamma_{t_0}(t_0 \in T)$ is defined, where for each t_0 the functional γ_{t_0} is defined on the space $C_n([t_0,\theta])$. In all the examples discussed above ((4), (5), (7), (8)), it is easy to complete the definition of a target functional in any one of the above indicated ways.

Another question which arises in connection with the variation of the initial position is as follows: can the optimal strategy be defined in such a manner that it will be universal? A universal optimal strategy is, by definition, a function U^0: $T \times R^n \to P$ which, for any choice of initial position $(t_0,x_0) \in T \times R^n$, is optimal in the sense of definitions (1), (2). Below, optimal universal strategies will be characterized in more detail. Problems will be shown for which an optimal universal strategy can be constructed. On the other hand, it will be determined that optimal universal strategies may not be available even for simple examples. It will be shown how to modify the definition of a positional strategy in order that optimal universal strategies will exist in a sufficiently wide class of problems.

1.5. Differential Games

Two players participate in a differential game, which we shall define in this section. We assume that the first player may select any positional strategy U, while the second player has the choice of choosing positional strategy V. The player's goals are conflicting: the first player strives to minimize the value of the target functional γ, while the second, on the contrary, seeks to maximize its value. Neither of the players possess reliable information concerning the control to be realized by the opponent. The players are cautious and estimate the quality of their strategies by a guaranteed result. We therefore assume that the first player solves a minimax problem and the second player solves a maximin problem which is stated in this section. Next, the concepts of the value of the game and of the saddle point are introduced – these are basic notions in the theory of differential games by means of which the connection between the problems of the first and second players manifests itself. Along with the formal definitions in this section, motivations will also be presented.

1.5.1. Consider the statement of a maximin problem for the second player. We now present the necessary definitions.

A function $V\colon\ T \times R^n \to Q$ is called a *positional strategy for the second player*. As was the case with the strategy U, it is not required that the function V be continuous, satisfy the Lipschitz condition, etc. Constructive motions generated by a strategy V can be defined in two ways.

First Approach. Let an initial position (t_0,x_0) be given. Choose a partition $\Delta = \{t_0 = \tau_0 < \tau_1 < \dots < \tau_{k+1} = \theta\}$. Let $u(\cdot)\colon [t_0,\theta] \to P$ be a measurable function. Consider an absolutely continuous solution $x(\cdot)\colon\ [t_0,\theta] \to R^n$ which satisfies the initial condition $x(t_0) = x_0$ and the differential equation (1.2.1) for almost all $t \in [t_0,\theta]$. It is assumed here that the control $v(t)$ appearing in equation (1.2.1) is formed by the rule

$$(1)\qquad v(t) = V(\tau_i, x(\tau_i)),\ \tau_i \leqslant t < \tau_{i+1},\quad i = 0, 1, \dots, k.$$

Observe that we do not specify here in what manner the function $t \to u(t)$ is formed; it is only important that the function be measurable. The solution under consideration, $x(\cdot)$, will be denoted by $x(\cdot,t_0,x_0,u(\cdot),V,\Delta)$ and is called a step-by-step motion. Now consider a sequence of step-by-step motions

$$(2)\qquad x_\ell(\cdot) = x(\cdot,t_0,x_0,u_\ell(\cdot),V,\Delta_\ell),\quad \ell = 1, 2, \dots,$$

where $\operatorname{diam}\Delta_\ell \to 0$ as $\ell \to \infty$, $u_\ell(\cdot)\colon\ [t_0,\theta] \to P$ are measurable functions. The totality of limits of all possible sequences of the form (2) is denoted by $X_2(t_0,x_0,V)$ and will be called the set of constructive motions generated by the positional strategy V of the second player.

Second Approach. Consider the differential inclusion

$$\dot{x}(t) \in \{f(t,x(t),u,v(t)):\ u \in P\} \tag{3}$$

where the control $v(t)$ is formed by rule (1). Denote by $X_2(t_0,x_0,V,\Delta)$ the totality of absolutely continuous functions $x(\cdot)$: $[t_0,\theta] \to R^n$ which satisfy the initial condition $x(t_0) = x_0$ and the differential inclusion (3) for almost all $[t_0,\theta]$. The elements of the set $X_2(t_0,x_0,V,\Delta)$ are called step-by-step motions generated by the strategy of the second player. The set of constructive motions corresponding to the positional strategy V is defined by the equality

$$\begin{aligned} X_2(t_0,x_0,V) \overset{\text{def}}{=} \{x(\cdot) \mid \exists\, x_\ell(\cdot) \in X_2(t_0,x_0,V,\Delta_\ell), \\ \ell = 1, 2, \ldots,\ \operatorname{diam}\Delta_\ell \to 0,\ x_\ell(\cdot) \to x(\cdot) \text{ as } \ell \to \infty\}. \end{aligned} \tag{4}$$

It is easy to see that the definitions of the sets $X_1(t_0,x_0,U)$ and $X_2(t_0,x_0,V)$ are essentially the same. Therefore the sets $X_2(t_0,x_0,V)$ possess all the properties stated in §1.3 for the sets $X_1(t_0,x_0,U)$. In particular, by a direct replacement of sets $X_1(t_0,x_0,U)$ and $X_1(t_0,x_0,U,\Delta)$ by the sets $X_2(t_0,x_0,V)$ and $X_2(t_0,x_0,V,\Delta)$ we obtain assertions analogous to Lemmas 1.3.1 and 1.3.2.

It is assumed – as in §1.4 – that a target functional γ is defined in the space $C_n([t_0,\theta])$. However, unlike the situation in the preceding section, it is assumed here that the second player aims to maximize the value of the functional γ, i.e. for this player the motion $x_1(\cdot)$ will be "better" than $x_2(\cdot)$ provided $\gamma(x_1(\cdot)) > \gamma(x_2(\cdot))$. Following the outline presented in §1.4.1 for the statement of the minimax problem, we obtain that the quality of the positional strategy V must be estimated by the quantity

$$\Gamma_2(V) \overset{\text{def}}{=} \inf_{x(\cdot)} \gamma(x(\cdot)) \text{ for } x(\cdot) \in X_2(t_0,x_0,V). \tag{5}$$

We thus arrive at the following problem:

Problem 1.5.1. *It is required to determine a positional strategy* V^0 *such that*

$$\Gamma_2(V^0) = \max_V \Gamma_2(V) \tag{6}$$

where the maximum is taken over all positional strategies of the second player.

The strategy V^0 chosen from condition (6) is called an *optimal strategy for the second player.* If the maximum in (6) is not attained, a problem about the ϵ-optimal strategy is considered.

Problem 1.5.1$^{\epsilon}$. *It is required to determine the positional strategy V^{ϵ} such that*

(7) $$\Gamma_2(V^{\epsilon}) \geqslant \sup_V \Gamma_2(V) - \epsilon,$$

where ϵ is a positive number.

Combining relations (5) and (6) we obtain the following condition which defines an optimal strategy V^0:

$$\inf_{x(\cdot)\in X_2(t_0,x_0,V^0)} \gamma(x(\cdot)) = \max_V \inf_{x(\cdot)\in X_2(t_0,x_0,V)} \gamma(x(\cdot)).$$

From here it is obvious why Problems 1.5.1 and 1.5.1$^{\epsilon}$ are called maximin.

Below we shall adhere to the following agreement. When considering differential games we shall assume that the first player solves the minimax problem while the second player solves the maximin problem. The strategies of the first player will be denoted by the letter U and that of the second by V. Note also the following. If the functions U and V corresponding to the strategies of the first and second players coincide, i.e. $U(t,x) = V(t,x)$, $(t,x) \in T \times R^n$, this equality does not necessarily imply the equalities $X_1(t_0,x_0,U,\Delta) = X_2(t_0,x_0,V,\Delta)$ and $X_1(t_0,x_0,U) = X_2(t_0,x_0,V)$. This explains the reason why indices 1 and 2 were introduced in the designation of the sets of step-by-step and constructive motions.

1.5.2. Minimax and maximin problems turn out to be dual problems in a certain sense. The duality is expressed by specific notions (saddle point, the value of the game) related to the theory of differential games. We shall define these concepts shortly, but first we shall consider the problem of compatibility of positional strategies of the first and second players.

Two players participating in a differential game can choose any pair of positional strategies U and V. It is therefore required to define the motion generated by an arbitrary pair of strategies (U,V). As usual, we shall start with the definition of step-by-step motions. Let an initial position (t_0,x_0) be given. We assume that the first and second players have chosen partitions $\Delta^{(1)}$ and $\Delta^{(2)}$ respectively:

$$\Delta^{(1)} \overset{\text{def}}{=} \{t_0 = \tau_0^{(1)} < \tau_1^{(1)} < \dots < \tau_{k_1+1}^{(1)} = \theta\},$$

$$\Delta^{(2)} \overset{\text{def}}{=} \{t_0 = \tau_0^{(2)} < \tau_1^{(2)} < \dots < \tau_{k_2+1}^{(2)} = \theta\}.$$

Consider an absolutely continuous solution $x(\cdot)$: $[t_0,\theta] \to R^n$ of the equation (1.2.1) in which the controls $u(t)$ and $v(t)$ are formed by the rules

$$u(t) = U(\tau_i^{(1)},x(\tau_i^{(1)})),\ \tau_i^{(1)} \leqslant t < \tau_{i+1}^{(1)},\quad i = 0, 1, ..., k_1,$$

$$v(t) = V(\tau_j^{(2)},x(\tau_j^{(2)})),\ \tau_j^{(2)} \leqslant t < \tau_{j+1}^{(2)},\quad j = 0, 1, ..., k_2.$$

It is assumed that the solution under consideration satisfies the initial condition $x(t_0) = x_0$. This solution will be denoted by the symbol $x(\cdot,t_0,x_0,U,V,\Delta^{(1)},\Delta^{(2)})$ and will be called a step-by-step motion generated by the pair of positional strategies (U,V) and corresponding partitions $\Delta^{(1)}$ and $\Delta^{(2)}$ which were chosen by the first and the second players. We note that for a given initial position (t_0,x_0) and the chosen $U,V,\Delta^{(1)}$ and $\Delta^{(2)}$ the step-by-step motion $x(\cdot,t_0,x_0,U,V,\Delta^{(1)},\Delta^{(2)})$ exists and is unique.

We now proceed to constructive motions. Consider a sequence of step-by-step motions

(8)
$$x_\ell(\cdot) = x(\cdot,t_0,x_0,U,V,\Delta_\ell^{(1)},\Delta_\ell^{(2)})\ (\ell = 1, 2, ...),$$
$$\operatorname{diam}\Delta_\ell^{(1)} \to 0,\ \operatorname{diam}\Delta_\ell^{(2)} \to 0 \text{ as } \ell \to \infty$$

Denote by $X(t_0,x_0,U,V)$ the set of all absolutely continuous functions $x(\cdot)$: $[t_0,\theta] \to R^n$ for each one of which there exists a sequence convergent to it of step-by-step motions of the form (8). The elements of the set $X(t_0,x_0,U,V)$ are called *constructive motions* generated by a pair of positional strategies U and V. It is easy to verify that the set $X(t_0,x_0,U,V)$ is a non-empty compact set in $C_n([t_0,\theta])$. This assertion is proved in the same manner as Lemma 1.3.1.

Next it follows directly from the definitions of the sets $X_1(t_0,x_0,U)$, $X_2(t_0,x_0,V)$, and $X(t_0,x_0,U,V)$ that

(9) $$X(t_0,x_0,U,V) \subset X_1(t_0,x_0,U) \cap X_2(t_0,x_0,V),\quad X(t_0,x_0,U,V) \neq \phi.$$

We say that a pair consisting of the strategies of the first and second players is *compatible*, if for this pair of strategies one can define motions of the system generated by them. It was shown above that *any pair of positional strategies* (U,V) *is compatible.*

Note that in addition to positional strategies, many other types of strategies are considered in the theory of differential games. We present a simple example of an incompatible pair of strategies. Let x,u, and v be scalars. The motion of the system is described by equation $\dot{x} = uv$. The controls of the players satisfy the constraints $|u| \leqslant 1$, $|v| \leqslant 1$. Consider the strategy

$$U_*(v) = \begin{cases} 1 & \text{for } v \geqslant 0, \\ -1 & \text{for } v < 0. \end{cases}$$

Note that the strategies which are identified with Borel measurable functions $(t,v) \to U(t,v)$ are called countercontrols. In some publications they are called stroboscopic strategies. The countercontrol U_* is compatible with any measurable control $t \to v(t)$. The corresponding motion is determined by the equation $\dot{x}(t) = v(t)U_*(v(t)) = |v(t)|$. It is also possible to define motions generated by the pair (U_*,V), where V is an arbitrary positional strategy of the second player. However, if we consider the countercontrol of the second player,

$$V_*(u) = \begin{cases} 1 & \text{for } u \leqslant 0, \\ -1 & \text{for } u > 0, \end{cases}$$

then the pair of countercontrols (U_*,V_*) turns out to be incompatible. Indeed, the system of equations $u = U_*(v)$ and $v = V_*(u)$ does possess a solution; therefore it is not possible to define a motion generated by the pair (U_*,V_*).

We return to the consideration of positional strategies. Utilizing the compatibility of positional strategies U and V, one can compare quantities $\Gamma_1(U)$ and $\Gamma_2(V)$. Below, symbols $\sup\gamma(\mathcal{X})$ and $\inf\gamma(\mathcal{X})$ are used to denote the quantities

$$\sup_{x(\cdot)\in\mathcal{X}} \gamma(x(\cdot)) \quad \text{and} \quad \inf_{x(\cdot)\in\mathcal{X}} \gamma(x(\cdot))$$

respectively. It follows from (1.4.1), (5), and (9) that

$$\Gamma_1(U) = \sup\gamma(\mathcal{X}_1(t_0,x_0,U)) \geqslant \sup\gamma(\mathcal{X}(t_0,x_0,U,V))$$
$$\geqslant \inf\gamma(\mathcal{X}(t_0,x_0,U,V)) \geqslant \inf\gamma(\mathcal{X}_2(t_0,x_0,V)) = \Gamma_2(V).$$

Thus for any pair of positional strategies U and V the inequality $\Gamma_1(U) \geqslant \Gamma_2(V)$ is valid. Therefore

(10) $$\inf_U \Gamma_1(U) \geqslant \sup_V \Gamma_2(V).$$

Taking (1.4.1) and (5) into account, relation (10) can be rewritten as

$$\inf_U \sup\gamma(\mathcal{X}_1(t_0,x_0,U)) \geqslant \sup_V \inf\gamma(\mathcal{X}_2(t_0,x_0,V)).$$

We obtain an analog of a well known inequality in the theory of games

(11) $$\inf_{g\in G} \sup_{h\in\mathcal{H}} \varphi(g,h) \geqslant \sup_{h\in\mathcal{H}} \inf_{g\in G} \varphi(g,h).$$

The case of equality in (10) is of special interest. In this case, the quantity

(12) $$\omega^0 \overset{\text{def}}{=} \inf_U \Gamma_1(U) = \sup_V \Gamma_2(V)$$

is called *the value of a positional differential game.* When the equality in (10) is fulfilled, we say that there exists *a situation of* ϵ*-equilibrium* in a differential game. A pair of positional strategies (U^ϵ, V^ϵ) satisfying the conditions

$$\Gamma_1(U^\epsilon) - \epsilon \leqslant \inf \Gamma_1(U) = \sup \Gamma_2(V) \leqslant \Gamma_2(V^\epsilon) + \epsilon$$

is called an ϵ*-saddle point.* If there exists a pair of positional strategies (U^0, V^0) satisfying the equalities

$$(13) \qquad \Gamma_1(U^0) = \min_U \Gamma_1(U) = \omega^0 = \max_V \Gamma_2(V) = \Gamma_2(V^0),$$

then this pair is called a *saddle point of a positional differential game.* We note that in the theory of differential games the target functional is usually called *the payoff functional.*

In the next two chapters, theorems on the existence of values will be proved for various types of differential games. First we shall prove the existence of an equilibrium situation for the differential game of approach and evasion. Next, as a corollary to this theorem, the existence of the value for those differential games in which the payoff functional is determined by one of the equalities (1.4.4), (1.4.5), and (1.4.7) will be obtained.

1.5.3. We shall discuss the definitions of the value, the ϵ-saddle point, and the saddle point presented above and compare these definitions with the notions used in game theory [51, 55].

In game theory an antagonistic game is determined by the triple $(G, \mathcal{H}, \varphi)$. Here G is the set of strategies for the first player, $\mathcal{H}$ is the corresponding set for the second player, and $\varphi\colon G \times \mathcal{H} \to R$ is a payoff function. It is assumed that the first player aims to minimize the payoff while the second aims to maximize it. We adhere now to the tradition developed in the theory of differential games where, as a rule, the first player minimizes the target functional and the second maximizes it. In game theory the roles of the players are usually interchanged: the first player maximizes φ while the second minimizes it.

It is known that the inequality (11) is always valid. If in (11) the equality holds, then the quantity

$$(14) \qquad \omega^0 = \inf_{g \in G} \sup_{h \in \mathcal{H}} \varphi(g,h) = \sup_{h \in \mathcal{H}} \inf_{g \in G} \varphi(g,h)$$

is called the value of an antagonistic game. The pair of strategies $(g^0, h^0) \in G \times \mathcal{H}$ is called a saddle point if

$$\varphi(g^0, h) \leqslant \varphi(g^0, h^0) \leqslant \varphi(g, h^0)$$

for any $g \in G$ and $h \in \mathcal{H}$.

The pair of strategies $(g^\epsilon, h^\epsilon) \in G \times \mathcal{H}$ is called an ϵ-saddle point provided

(15) $\varphi(g^{\epsilon},h) - \epsilon \leqslant \omega^0 \leqslant \varphi(g,h^{\epsilon}) + \epsilon.$

We now show how one proceeds from definitions presented in §1.5.2 to the standard notions of game theory. This can be accomplished as follows. Let the pairs $(U,\Delta^{(1)})$ be the elements of the set G, each one of which consists of a positional strategy U and a partition $\Delta^{(1)}$ selected by the first player. Analogously, let the set $\mathcal{H}$ be the totality of pairs $(V,\Delta^{(2)})$, where V is a positional strategy of the second player and $\Delta^{(2)}$ is the partition selected by this player. The function φ is defined by the equality (the initial position (t_0,x_0) is fixed)

(16) $$\varphi(g,h) \overset{\text{def}}{=} \gamma(x(\cdot,t_0,x_0,U,V,\Delta^{(1)},\Delta^{(2)}).$$

For any $g = (U,\Delta^{(1)})$ and $h = (V,\Delta^{(2)})$ the step-by-step motion $x(\cdot,t_0,x_0,U,V,\Delta^{(1)},\Delta^{(2)})$ exists and is unique. Consequently, for any $g \in G$ and $h \in \mathcal{H}$ the quantity $\varphi(g,h)$ is defined and is the value of the payoff functional of this motion. Thus the triple $(G,\mathcal{H},\varphi)$ is defined.

Assume now that there exists a saddle point (U^0,V^0) in the sense of definition (13). Assume also that the payoff functional γ is continuous. We shall utilize the approximating property of step-by-step motions stated in §1.4.3. According to this property, for any $\epsilon > 0$ one can find a partition $\Delta_{\epsilon}^{(1)}$ such that for any step-by-step motion $x(\cdot) \in X_1(t_0,x_0,U^0,\Delta_{\epsilon}^{(1)})$ the inequality

(cf.(13)) $\gamma(x(\cdot)) \leqslant \Gamma_1(U^0) + \epsilon = \omega^0 + \epsilon$

holds. It follows from the definition of the set $X_1(t_0,x_0,U,\Delta^{(1)})$ and the step-by-step motion $x(\cdot,t_0,x_0,U^0,V,\Delta^{(1)},\Delta^{(2)})$ (cf. §1.3.1 and §1.5.2) that $x(\cdot,t_0,x_0,U^0,V,\Delta^{(1)},\Delta^{(2)}) \in X_1(t_0,x_0,U^0,\Delta^{(1)})$ for any V and $\Delta^{(2)}$. Therefore for any pair $(V,\Delta^{(2)})$ the inequality

$$\gamma(x(\cdot,t_0,x_0,U^0,V,\Delta_{\epsilon}^{(1)},\Delta^{(2)})) \leqslant \omega^0 + \epsilon$$

is valid.

Taking (16) into account we obtain

$$\varphi(g^{\epsilon},h) - \epsilon \leqslant \omega^0$$

where $g^{\epsilon} = (U^0,\Delta_{\epsilon}^{(1)})$, $h = (V,\Delta^{(2)})$ is an arbitrary pair. Arguing analogously one can arrive at $\varphi(g,h^{\epsilon}) + \epsilon \geqslant \omega^0$ where $h^{\epsilon} = (V^0,\Delta_{\epsilon}^{(2)})$, $g = (U,\Delta^{(1)})$ is an arbitrary pair, $\Delta_{\epsilon}^{(2)}$ is a sufficiently refined partition. It follows from the inequalities obtained that $(g^{\epsilon},h^{\epsilon})$ is an ϵ-saddle point in the sense of the definition (15).

Thus the connection between the definitions introduced in §1.5.2 and the usual notions of game theory is as follows: Let the sets G, $\mathcal{H}$ and the function φ be defined in an antagonistic game $(G,\mathcal{H},\varphi)$ in the manner indicated above. Let ω^0 be the value of a positional differential game in the sense of the definition (12). Then this quantity ω^0 will also be the value of an antagonistic game $(G, \mathcal{H}, \varphi)$. Let (U^0, V^0) be a saddle point of a positional differential game in the sense of definition (13). Then $g^\epsilon = (U^0, \Delta_\epsilon^{(1)})$ and $h^\epsilon = (V^0, \Delta_\epsilon^{(2)})$ form an ϵ-saddle point of an antagonistic game $(G, \mathcal{H}, \varphi)$, where $\Delta_\epsilon^{(1)}$ and $\Delta_\epsilon^{(2)}$ are sufficiently refined partitions of the segment $[t_0, \theta]$. It is important to note that the functions U^0 and V^0 do not depend on the parameter ϵ.

The formalization of a differential game presented in 1.5.2 is not the only one possible. In the theory of differential games, different definitions of the value of the game and various proofs of its existence are known; for example, the widely used approach based on approximation of differential games by multistage games (cf. e.g. [20, 24, 27, 56, 73]). Usually two types of multistage games are considered – the upper and the lower multistage games. In each one of these games, the time interval $[t_0, \theta]$ is subdivided into half-intervals $[\tau_i, \tau_{i+1}), i = 0, 1, ..., k$. In a lower multistage game, the first player at time $t = \tau_i$ knows the position of the game which was realized as well as the control which will be realized by the second player on the interval $[\tau_i, \tau_{i+1})$. Thus, in this game the first player possesses an informational advantage. In an upper multistage game it is the second player who holds the informational advantage. Let ω_Δ and ω^Δ be the values of the lower and upper multistage games, respectively. The index Δ denotes the partition corresponding to the multistage games under consideration. The quantities ω_Δ and ω^Δ satisfy the inequality $\omega_\Delta \leqslant \omega^\Delta$. Next, the limits $\omega_- = \lim \omega_\Delta$ and $\omega^+ = \lim \omega^\Delta$ for $\operatorname{diam}\Delta \to 0$ are defined. These limits are called, respectively, the lower and upper values of a differential game. Under certain conditions, one can prove that the values ω_- and ω^+ coincide. In such a case, the differential game possesses a value $\omega_0 = \omega_- = \omega^+$. The definition of the value in the sense indicated here is equivalent to the definition of the value as given by equality (12), i.e. the quantities ω_0 and ω^0 are identical. We note that in the investigations cited above, the basic emphasis is on the proof of the existence theorems for the value of the game. In this book, however, we try to devote more attention to the investigation of optimal strategies and the problems of their structure.

1.5.4. We return to a discussion of the minimax control problem. We shall show that the results of the theory of differential games permit us to give an answer to basic questions arising in the formulation of this problem.

Consider again a controlled system (1.2.1) where $u(t)$ is a control and $v(t)$ is an undetermined disturbance. It is required to minimize the guaranteed value of a target functional γ. However, unlike the case in Problem 1.4.1 where it was required to determine a solution among positional strategies, here the class of admissible strategies is not fixed, and various methods for forming the control $u(t)$ can be tried out.

One can start, for example, with programmed controls $u(\cdot)$: $[t_0,\theta] \to P$ (it is assumed that these functions are measurable). In this case the minimax control problem will consist of choosing a control $u^0(\cdot)$ satisfying the condition

$$(17) \qquad \max_{v(\cdot)} \gamma(x(\cdot,t_0,x_0,u^0(\cdot),v(\cdot))) = \min_{u(\cdot)} \max_{v(\cdot)} \gamma(x(\cdot,t_0,x_0,u(\cdot),v(\cdot))).$$

Here $x(\cdot,t_0,x_0,u(\cdot),v(\cdot))$ is the motion of the system (1.2.1) generated by the control $u(\cdot)$ and the disturbance $v(\cdot)$.

The result obtained can be compared with the solution of the minimax problem in the class of positional strategies. It turns out that as a rule the quantity $\Gamma_1(U^0)$ (1.4.1) is less than the quantity (17). This indeed makes sense: additional information allows us to improve on the guaranteed result. Next, one can proceed to even more complex strategies which, for example, utilize at time t information concerning realized motions $x(\cdot)$: $[t_0,t] \to R^n$, controls $u(\cdot)$: $[t_0,t) \to P$, and the disturbance $v(\cdot)$: $[t_0,t] \to Q$. We shall, however, not attempt to enumerate all possible types of strategies. We mention only the single requirement imposed on a strategy U. It is as follows: for any positional strategy V and partition Δ there exists a motion $x(\cdot)$ generated by a strategy U and step-by-step positional control $v(t)$ (1); this motion is contained in the set $X_2(t_0,x_0,V,\Delta)$. In other words, the strategy U should be compatible with the positional step-by-step control (1). Such strategies U will be called admissible. Apparently only a strategy which assigns at time t a control $u(t)$ dependent on the disturbance $v(\tau)$ to be realized at $\tau > t$ will be an inadmissible one. Such strategies should definitely be excluded from consideration by the nature of the problem under consideration.

We commenced the discussion of a minimax control problem by considering its solution in the class of programmed controls. Next we proceed to positional strategies which improve on the guaranteed result. Now we have introduced a sufficiently wide class of admissible strategies and would like to find out whether it is possible among admissible strategies to choose a strategy for which the guaranteed result will be better than the result guaranteed by an optimal positional strategy.

For this purpose, consider a differential game composed of Problems 1.4.1 and 1.5.1. Assume that the value exists for this game. For the convenience of further discussions, we shall also assume that the functional γ is continuous and possesses a saddle point (U^0,V^0). In view of (13) we have $\Gamma_2(V^0) = \omega^0$. Utilizing the approximating property of step-by-step motions and the continuity of the functional γ for every $\epsilon > 0$, one can find a partition Δ_ϵ such that the inequality $\gamma(x(\cdot)) > \Gamma_2(V^0) - \epsilon = \omega^0 - \epsilon$ will be valid for any motion $x(\cdot) \in X_2(t_0,x_0,V^0,\Delta_\epsilon)$. Let U be an arbitrary admissible strategy. Among the motions which this strategy generates there is a motion belonging to the set $X_2(t_0,x_0,V^0,\Delta_\epsilon)$. Consequently, the strategy U cannot guarantee a result smaller than $\omega^0 - \epsilon$ (here ϵ is an arbitrarily small number). On the other hand, for the positional strategy U^0 the equality $\Gamma_1(U^0) = \omega^0$ is valid (cf. (13)). We thus conclude: the result guaranteed by a positional strategy U^0 cannot be improved by transition to an arbitrary admissible strategy.

Thus, by introducing differential games into consideration, one can substantiate the statement of a minimax control problem. As will be seen below, the theory of differential games also suggests methods for solving minimax problems.

1.5.5. Consider the following example of a differential game. Let (y_1,y_2) be coordinates of an inertia-free point Y whose motion is controlled by the first player. The motion of this point is described by the equation

$$\dot{y}_1 = u_1, \quad \dot{y}_2 = u_2 \,. \tag{18}$$

Here $u = (u_1,u_2)^{\mathrm{T}}$ is the velocity vector chosen by the first player. The velocity of the point may change in a discontinuous manner. However, it is required that the bound

$$u_1^2 + u_2^2 \leqslant 1$$

be valid. The second player controls the motion of a mass point Z whose coordinates are (z_1,z_2). The controlling action of the second player is materialized by an external force v applied at this point. The motion of the mass point is described by the equations

$$\dot{z}_1 = z_3, \ \dot{z}_2 = z_4, \ \dot{z}_3 = v_1, \ \dot{z}_4 = v_2. \tag{19}$$

It is assumed that the mass of the point Z equals one, and the controlled forces are bounded by

$$v_1^2 + v_2^2 \leqslant 1.$$

The game is considered on the interval [0,2]. The payoff function is given by the equality

$$\gamma(y(\cdot),z(\cdot)) = -[(y_1(2) - z_1(2))^2 + (y_2(2) - z_2(2))^2]^{1/2}. \tag{20}$$

Thus in the game under consideration, the first player desires that at the last instance of time $t = 2$ the distance between the points Y and Z be as large as possible. Conversely, the second player aims to bring points Z and Y nearer. Choose the following initial data:

$$y_1(0) = -2, \ y_2(0) = 1, \ z_1(0) = 0, \ z_2(0) = -1, \ z_3(0) = -1, \ z_4(0) = 1. \tag{21}$$

Consider the strategies

$$(22) \qquad U^0(t,y,z) = \begin{cases} x(t,y,z)/\|x(t,y,z)\| & \text{for } \|x(t,y,z)\| > 0, \\ (1,0)^T & \text{for } \|x(t,y,z)\| = 0, \end{cases}$$

$$(23) \qquad V^0(t,y,z) = \begin{cases} x(t,y,z)/\|x(t,y,z)\| & \text{for } \|x(t,y,z)\| > 0, \\ (0,0)^T & \text{for } \|x(t,y,z)\| = 0, \end{cases}$$

where $x(t,y,z)$ is a two-dimensional vector with components

$$(24) \qquad \begin{aligned} x_1(t,y,z) &= y_1 - z_1 - (2-t)z_3, \\ x_2(t,y,z) &= y_2 - z_2 - (2-t)z_4. \end{aligned}$$

We shall show that the pair of positional strategies (U^0,V^0) forms a saddle point for the game under consideration.

Let the first player choose a partition $\Delta = \{0 = \tau_0 < \tau_1 < \dots < \tau_{k+1} = 2\}$ and the control $u(t)$ be formed by the rule

$$(25) \qquad u(t) = U^0(\tau_i,y(\tau_i),z(\tau_i)), \ \tau_i \leqslant t < \tau_{i+1}, \quad i = 0, 1, \dots, k.$$

In addition, a measurable control $t \to v(t)$ is realized in the system (19). Let $y(t)$ and $z(t)$ be the corresponding motions of systems (18) and (19). We calculate the derivative of the function $t \to x(t) = x(t,y,(t),z(t))$. In view of (18), (19), and (24) we have

$$(26) \qquad \dot{x}_1(t) = u_1(t) - (2-t)v_1(t), \quad \dot{x}_2(t) = u_2(t) - (2-t)v_2(t).$$

It follows from (24) and (21) that $x_1(0) = x_2(0) = 0$. Note that the controlled system (26) was considered in §1.4.4. It was shown there that under the choice of the control (22), (25), the motion of the system (26) will satisfy the inequality

$$\|x(2)\| \geqslant 1/2 - (\mathrm{diam}\Delta)^2/2.$$

In view of (20), (24) the equality

$$-\|x(2)\| = \gamma(y(\cdot),z(\cdot))$$

is valid. Hence the control of the form (22), (25) assures that the inequality $\gamma(y(\cdot),z(\cdot)) \leqslant -1/2 + (\mathrm{diam}\Delta)^2/2$ is fulfilled. Approaching the limit as $\mathrm{diam}\Delta \to 0$ we conclude that in the differential game (18) – (20) the strategy U^0 satisfies the inequality

$$(27) \qquad \Gamma_1(U^0) \leqslant -1/2.$$

We shall estimate now the result guaranteed by the strategy V^0. Let the control $v(t)$ be formed by the rule

(28) $$v(t) = V^0(\tau_i, y(\tau_i), z(\tau_i)), \quad \tau_i \leqslant t < \tau_{i+1}, \quad i = 0, 1, ..., k$$

and let some measurable function be realized as the control $u(t)$. As in the preceding case we shall consider the system (26). It follows from the equality

$$x(\tau_{i+1}) = x(\tau_i) - v(\tau_i) \int_{\tau_i}^{\tau_{i+1}} (2 - t)dt + \int_{\tau_i}^{\tau_{i+1}} u(t)dt$$

and relations (23), (28) that

(29) $$\|x(\tau_{i+1})\| \leqslant \big| \|x(\tau_i)\| - 2(\tau_{i+1} - \tau_i) + \tau_{i+1}^2/2 - \tau_i^2/2 \big| + (\tau_{i+1} - \tau_i).$$

Assume that

(30) $$\|x(\tau_i)\| \geqslant 2(\tau_{i+1} - \tau_i) - \tau_{i+1}^2/2 + \tau_i^2/2 \text{ for } i = j, j+1, ..., k$$

(31) $$\|x(\tau_{j-1})\| < 2(\tau_j - \tau_{j-1}) - \tau_j^2/2 + \tau_{j-1}^2/2.$$

Then

$$\|x(\tau_{i+1})\| \leqslant \|x(\tau_i)\| - (\tau_{i+1} - \tau_i) + \tau_{i+1}^2/2 - \tau_i^2/2$$
$$\text{for } i = j, j+1, ..., k.$$

Hence

$$\|x(2)\| = \|x(\tau_{k+1})\| \leqslant \|x(\tau_j)\| - (\tau_{k+1} - \tau_j) + \tau_{k+1}^2/2 - \tau_j^2/2$$
$$= \|x(\tau_j)\| + \tau_j - \tau_j^2/2 \leqslant \|x(\tau_j)\| + 1/2.$$

From (29), (31) we obtain that

$$\|x(\tau_j)\| \leqslant 2(\tau_j - \tau_{j-1}) - \tau_j^2/2 + \tau_{j-1}^2/2 - \|x(\tau_{j-1})\|$$
$$+ \tau_j - \tau_{j-1} < 5(\tau_j - \tau_{j-1}) \leqslant 5\operatorname{diam}\Delta.$$

Thus, if conditions (30), (31) are fulfilled then

(32) $$\|x(2)\| \leqslant 1/2 + 5\operatorname{diam}\Delta.$$

Otherwise we have

$$\|x(\tau_k)\| \leqslant 2(\tau_{k+1} - \tau_k) - \tau_{k+1}^2/2 + \tau_k^2/2.$$

From this inequality and inequality (29) we obtain

$$\|x(2)\| = \|x(\tau_{k+1})\| \leqslant 5\operatorname{diam}\Delta.$$

Recall that $\gamma(y(\cdot),z(\cdot)) = - \|x(2)\|$. Therefore, it follows from the inequalities obtained that the step-by-step positional control (23), (28) assures the fulfillment of the inequality $\gamma(y(\cdot),z(\cdot)) \geqslant - 1/2 - 5\mathrm{diam}\Delta$. In the limit as $\mathrm{diam}\Delta \to 0$, we obtain the inequality

$$\Gamma_2(V^0) \geqslant - 1/2. \tag{33}$$

It follows from (10), (27), and (33) that

$$\Gamma_1(U^0) = \Gamma_2(V^0) = \omega^0 = - 1/2,$$

i.e. the pair of positional strategies U^0 (22) and V^0 (23) form a saddle point of the differential game (18) - (20). The value of this game is $-1/2$.

Figure 1.5.1 shows the trajectories of the controlled system (18), (19) which were obtained by the choice of strategy U^0 (22) – paired with some admissible controls of the second player. These trajectories are expressed in the coordinate system $\xi_1 = y_1 - z_1$, $\xi_2 = y_2 - z_2$. In view of (21) we have $\xi_1(0) = -2$, $\xi_2(0) = 2$. In all the trajectories presented in Fig. 1.5.1 the endpoints are located outside

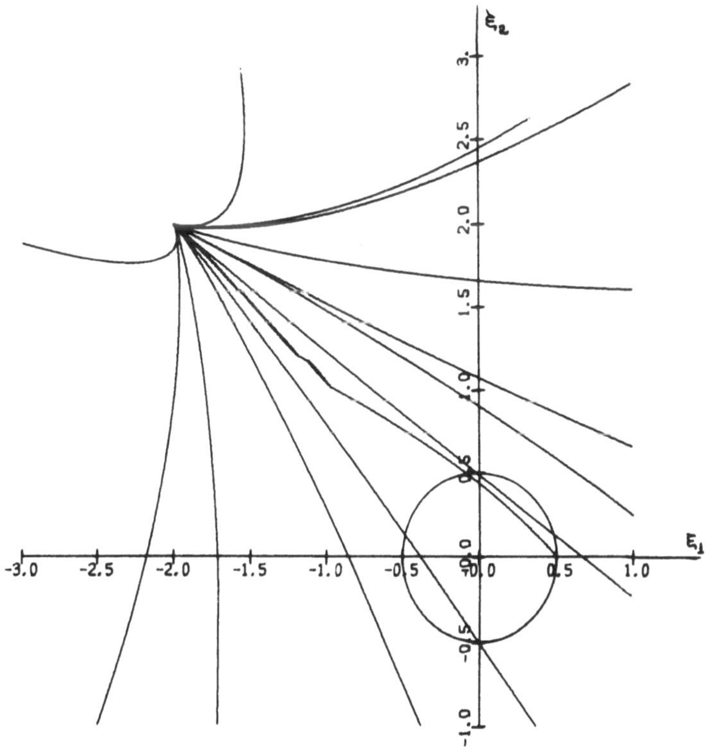

Fig. 1.5.1

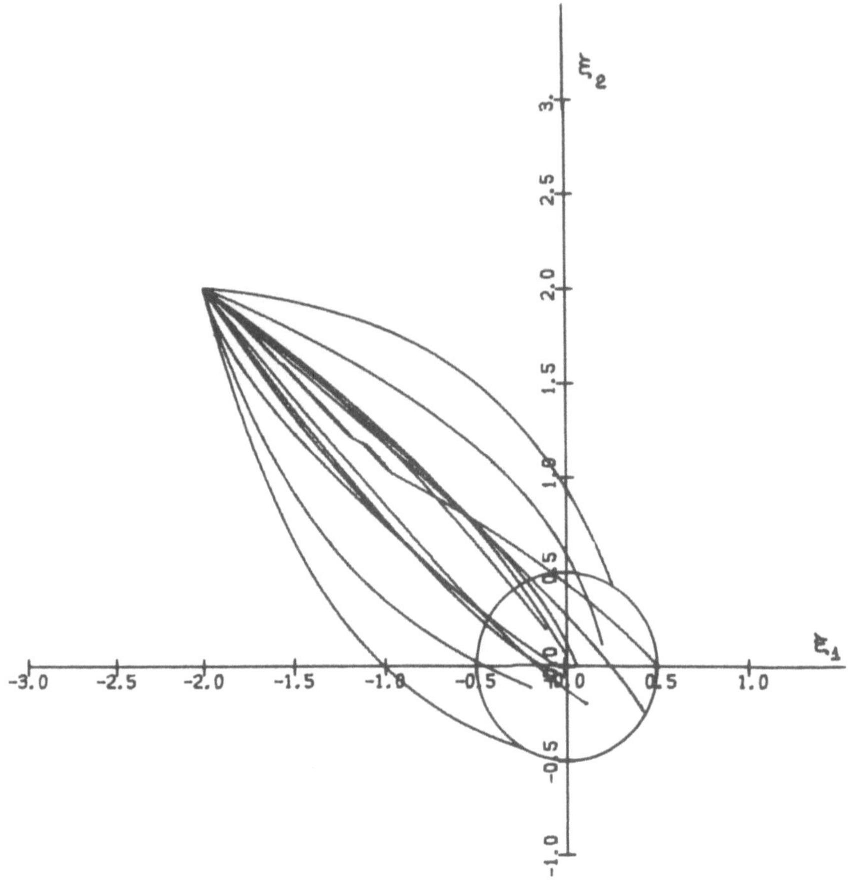

Fig. 1.5.2

the circle $K^0 = \{(\xi_1,\xi_2): (\xi_1^2 + \xi_2^2)^{1/2} < 1/2\}$. In Fig. 1.5.2 the trajectories of the controlled system (18), (19) are presented, obtained when the strategy V^0 (23) paired with some admissible control of the second player is realized. Here all the endpoints are located in the circle $K = \{(\xi_1,\xi_2): (\xi_1^2+\xi_2^2)^{1/2} \leqslant 1/2\}$.

In conclusion we compare solutions of the minimax problem for the first player in the class of positional strategies and programmed controls. Let $\mathcal{B} \overset{\text{def}}{=} \{x \in R^2: \|x\| \leqslant 1\}$. Denote by $\mathfrak{L}([0,2],\ \ )$ the set of measurable functions defined on the interval [0,2] and taking on values in the circle $\mathcal{B}$. Consider the sets

$$G_1 \overset{\text{def}}{=} \left\{\int_0^2 u(t)dt: \ u(\cdot) \in \mathfrak{L}([0,2], \mathcal{B})\right\},$$
$$G_2 \overset{\text{def}}{=} \left\{\int_0^2 (2-t)v(t)dt: \ v(\cdot) \in \mathfrak{L}([0,2], \mathcal{B})\right\}.$$

It can be shown that

$$G_1 = G_2 = \{x \in R^2: \ \|x\| \leqslant 2\}.$$

Observe that the boundary points of the sets G_1 and G_2 are obtained under the constant controls $u(t) = u_*$, $v(t) = v_*$, $0 \leqslant t \leqslant 2$, $\|u_*\| = 1$, $\|v_*\| = 1$.

Let a programmed control $u(\cdot) \in \mathfrak{L}([0,2], \mathcal{B})$ be selected. Since $G_1 = G_2$ one can find $v(\cdot) \in \mathfrak{L}([0,2], \mathcal{B})$ such that

$$\int_0^2 u(t)dt = \int_0^2 (2-t)v(t)dt. \tag{34}$$

Consider once more the function $t \to x(t,y(t),z(t))$, where $y(t)$ and $z(t)$ are motions of the system (18) and (19) generated by the controls $u(\cdot)$ and $v(\cdot)$. In view of (26), (34), and the initial condition $x(0) = 0$ we obtain

$$x(2) = \int_0^2 u(t)dt - \int_0^2 (2-t)v(t)dt = 0.$$

Since $\gamma(y(\cdot),z(\cdot)) = -\|x(2)\|$ we conclude that for any programmed control $u(\cdot)$ the guaranteed result is equal to zero. Comparing this result with the result $\Gamma_1(U^0) = -1/2$ obtained for the positional strategy U^0 we verify that the positional strategy is indeed superior over programmed control.

Observe that the function U^0 is discontinuous. For any continuous function $(t,y,z) \to U(t,y,z)$ the guaranteed result will be zero. To verify this, one could transform the controlled system (18), (19) to a system of the form (26) and then utilize the corresponding result from §1.4.4. It thus turns out that the result guaranteed in the class of continuous positional strategies coincides with the result which can be achieved in the class of programmed controls. This is not a chance coincidence but a reflection of a regularity to be discussed in §5.5.

Chapter 2
DIFFERENTIAL GAME OF APPROACH-EVASION

2.1. Game-Theoretical Problems of Approach and Evasion

The basic result of this chapter is the theorem on an alternative for the differential game of approach-evasion. From this theorem we obtain the existence of the value for various types of differential games. In the course of the proof of the theorem, the notions of stable bridges and extremal strategies are introduced. In this section we formulate the game-type problem of approach and evasion and the statement and proof of the theorem on an alternative given in the next section is outlined.

2.1.1. Let two sets $\mathcal{M}$ and $\mathcal{N}$ be defined in the space of positions $T \times R^n$. The approach problem posed for the first player consists of choosing a positional strategy U^0 such that for any motion $x(\cdot) \in X_1(t_0,x_0,U^0)$ the point $(t,x(t))$ should contact the set $\mathcal{M}$ at some time $\tau \in [t_0,\theta]$ and in the process of its transition from the initial position (t_0,x_0) into set $\mathcal{M}$, it should remain in the set $\mathcal{N}$. In this problem $\mathcal{M}$ is the target set, $\mathcal{N}$ is the phase restriction or the life space of the first player.

We now specify the statement of the approach problem. We shall assume that the sets $\mathcal{M}$ and $\mathcal{N}$ are closed and the inclusion $\mathcal{M} \subset \mathcal{N}$ is valid. Let $x(\cdot)$: $[t_0,\theta] \to R^n$ be a continuous function. In §1.4.2 the functional

$$\tau^0(x(\cdot)) \overset{\text{def}}{=} \begin{cases} \min\{t:\ t \in T_0(x(\cdot))\} & \text{for } T_0(x(\cdot)) \neq \phi, \\ \theta + 1 & \text{for } T_0(x(\cdot)) = \phi \end{cases} \tag{1}$$

was defined where

$$T_0(x(\cdot)) \overset{\text{def}}{=} \{t \in [t_0,\theta]:\ (t,x(t)) \in \mathcal{M}\}. \tag{2}$$

The set $T_0(x(\cdot))$ is closed, thus the minimum in (1) is actually attained.

We say that the function $x(\cdot)$ satisfies the condition of (M, N)-approach if

$$(3) \qquad \tau^0(x(\cdot)) \leqslant \theta, \quad (t,x(t)) \in N \text{ for } t_0 \leqslant t \leqslant \tau^0(x(\cdot)).$$

In Figure 2.1.1 the solid lines represent the sets M and N, the broken lines, the graph of the function $x(\cdot)$ satisfying the condition of (M, N)-approach.

Thus the problem of (M, N)-approach is stated in the following manner.

Problem 2.1.1. *For a given initial position (t_0,x_0) it is required to obtain a positional strategy U^0 such that for any motion $x(\cdot) \in X_1(t_0,x_0,U^0)$ the condition of (M, N)-approach holds.*

We now turn to the statement of the evasion problem. Consider the first approach. Again let $x(\ ): [t_0,\theta] \to R^n$ be a continuous function. We say that this function satisfies the condition of (M, N)-evasion if either

$$(4) \qquad \tau^0(x(\cdot)) > \theta, \quad \text{or } \{(t,x(t)): t_0 \leqslant t \leqslant \tau_0(x(\cdot))\} \not\subset N.$$

Note that the condition of (M, N)-evasion is a negation of the

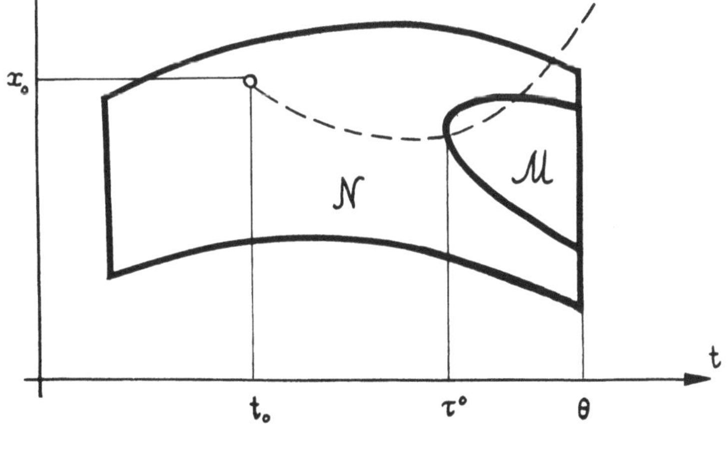

Fig. 2.1.1

condition of (M, N)-approach. The (M, N)-evasion problem can be formulated as follows: it is required to determine a positional strategy V^0 such that for any motion $x(\cdot) \in X_2(t_0,x_0,V^0)$ the condition (4) is valid.

Consider now the other statement of the evasion problem. We introduce some additional notation. Let $\epsilon > 0$ and the open ϵ-neighborhoods for the sets M and N be denoted by M^ϵ and N^ϵ, i.e.

(5) $$M^\epsilon \overset{\text{def}}{=} \{(t,x) \in T \times R^n: \ |t - t_*| + \|x - x_*\| < \epsilon, \ (t_*,x_*) \in M\},$$

(6) $$N^\epsilon \overset{\text{def}}{=} \{(t,x) \in T \times R^n: \ |t - t_*| + \|x - x_*\| < \epsilon, \ (t_*,x_*) \in N\}.$$

For a continuous function $x(\cdot)$: $[t_0,\theta] \to R^n$ we define the set

(7) $$T^\epsilon(x(\cdot)) \overset{\text{def}}{=} \{t \in [t_0,\theta]: \ (t,x(t)) \notin N^\epsilon\}$$

and the number

(8) $$t^\epsilon(x(\cdot)) = \begin{cases} \min\{t: \ t \in T^\epsilon(x(\cdot))\} & \text{for } T^\epsilon(x(\cdot)) \neq \phi, \\ \infty & \text{for } T^\epsilon(x(\cdot)) = \phi \end{cases}$$

Note that the set $T^\epsilon(x(\cdot))$ is closed, thus the minimum in (7) is attained.

We say that a continuous function $x(\cdot)$: $[t_0,\theta] \to R^n$ satisfies *the condition of* $(M, N)^\epsilon$*-evasion* if

(9) $$(t,x(t)) \notin M^\epsilon \text{ for } t_0 \leqslant t \leqslant \min\{t^\epsilon(x(\cdot)),\theta\}.$$

We thus arrive at the formulation of the following problem, which we shall call the $(M, N)^\epsilon$-evasion problem.

Problem 2.1.2. *For a given initial position* (t_0,x_0) *and a number* $\epsilon > 0$ *it is required to obtain a positional strategy* V^ϵ *such that for any motion* $x(\cdot) \in X_2(t_0,x_0,V^\epsilon)$ *the condition of* $(M, N)^\epsilon$*-evasion holds.*

Let us clarify the sense of the condition of $(M, N)^\epsilon$-evasion. Suppose first that the graph of a function $x(\cdot)$ is not located entirely in the ϵ-neighborhood of the set N, i.e. $T^\epsilon(x(\cdot)) \neq \phi$. In this case $t^\epsilon(x(\cdot))$ is the moment when the point $(t,x(t))$ leaves the set N^ϵ for the first time. The condition (9) signifies that the point $(t,x(t))$ does not encounter the set M^ϵ on the interval $[t_0,t^\epsilon(x(\cdot))]$. If the graph of a function $x(\cdot)$ entirely lies in the set N^ϵ then the point $(t,x(t))$ does not encounter the set M^ϵ on the whole interval $[t_0,\theta]$. In Fig. 2.1.2 the heavy and light lines represent the sets M, N and M^ϵ, N^ϵ respectively. The broken lines represent graphs of the two functions satisfying the condition of $(M, N)^\epsilon$-evasion. For one of these functions $T^\epsilon(x(\cdot)) \neq \phi$, and for the other $T^\epsilon x(\cdot)) = \phi$.

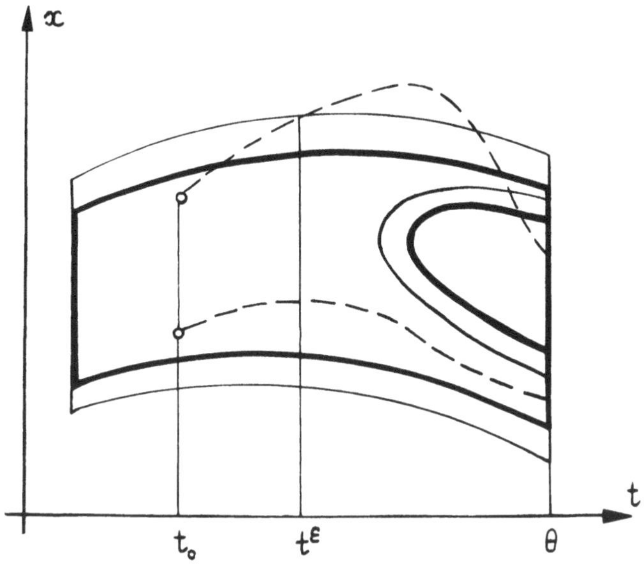

Fig. 2.1.2

It is easy to establish the connection between the solutions of the problems of (M, N)-evasion and that of $(M, N)^{\epsilon}$-evasion.

Let the strategy V^{ϵ} yield a solution for the $(M, N)^{\epsilon}$-evasion problem; then this strategy will also yield a solution for the (M, N)-evasion problem. Let the strategy V^0 yield a solution for the (M, N)-evasion problem. Then there exists an $\epsilon > 0$ such that the strategy V^0 will yield a solution of the $(M, N)^{\epsilon}$-evasion problem. The last assertion follows from the compactness of the set $X_2(t_0,x_0,V^0)$ and the cloasure of the sets M and N.

Define the sets

$$K_{\epsilon}^{(1)} \stackrel{\text{def}}{=} (T\times R^n)\setminus N^{\epsilon},\ K_{\epsilon}^{(2)} \stackrel{\text{def}}{=} \{(\theta,x):\ (\theta,x) \notin M^{\epsilon}\},$$

(10)

$$K_{\epsilon} \stackrel{\text{def}}{=} K_{\epsilon}^{(1)} \cup K_{\epsilon}^{(2)},\ \mathfrak{L}_{\epsilon} \stackrel{\text{def}}{=} (T\times R^n)\setminus M^{\epsilon}$$

Observe that the set K_{ϵ} and $\mathfrak{L}_{\epsilon}$ are closed and that $K_{\epsilon} \subset \mathfrak{L}_{\epsilon}$. Consider the condition of $(K_{\epsilon},\mathfrak{L}_{\epsilon})$-approach which is obtained if in (1) - (3) the sets M and N are replaced by the sets K_{ϵ} and $\mathfrak{L}_{\epsilon}$, respectively. One can verify directly that the conditions of $(K_{\epsilon},\mathfrak{L}_{\epsilon})$-approach and $(M, N)^{\epsilon}$-evasion are equivalent. Therefore, the problem of $(M, N)^{\epsilon}$-evasion can be viewed as a problem of the $(K_{\epsilon}\mathfrak{L}_{\epsilon})$-approach posed to the second player. Below we shall utilize this fact quite often.

Moreover, in what follows, we shall consider auxiliary problems of $(K,\mathfrak{L})$-approach not only for the sets $K = K_{\epsilon}$ and $\mathfrak{L} = \mathfrak{L}_{\epsilon}$ but also for other closed sets $K \subset \mathfrak{L} \subset T \times R^n$. The problem of the

$(\mathcal{K}, \mathcal{L})$-approach is stated as follows:

Problem 2.1.3. *Let $\mathcal{K} \subset \mathcal{L}$ be closed sets in $T \times R^n$. For a given initial position $(t_0,x_0) \in T \times R^n$ it is required to obtain a positional strategy V_* such that for any motion $x(\cdot) \in X_2(t_0,x_0,V_*)$ the condition of $(\mathcal{K}, \mathcal{L})$-approach will be valid, i.e.*

$$\tau_*(x(\cdot)) \leqslant \theta \text{ and } (t,x(t)) \in \mathcal{L} \text{ for } t_0 \leqslant t \leqslant \tau_*(x(\cdot)), \tag{11}$$

where

$$\tau_*(x(\cdot)) \overset{\text{def}}{=} \begin{cases} \min\{t\colon\ t \in T_*(x(\cdot))\} \text{ for } T_*(x(\cdot)) \neq \phi, \\ \theta + 1 \text{ for } T_*(x(\cdot)) = \phi, \end{cases} \tag{12}$$

$$T_*(x(\cdot)) \overset{\text{def}}{=} \{t \in [t_0,\theta]\colon\ (t,x(t)) \in \mathcal{K}\}. \tag{13}$$

Note that each time an $(\mathcal{M}, \mathcal{N})$-approach problem ($(\mathcal{K}, \mathcal{L})$-approach problem) is mentioned it is posed in the class of positional strategies for the first (second) player.

2.1.2. The differential game of approach-evasion can be considered in the framework of the general scheme described in §1.5.2. To do this a payoff functional should be introduced. We assume that the functional is defined as follows: $\gamma(x(\cdot)) = 0$ provided $x(\cdot)$ satisfies the condition of $(\mathcal{M}, \mathcal{N})$-approach and $\gamma(x(\cdot)) = 1$ otherwise. Next, the corresponding minimax and maximin problems should be stated. It turns out that a differential game defined in this manner possesses the value and a saddle point. This fact follows immediately from the theorem on an alternative stated below.

Theorem 2.1.1. *For any closed sets $\mathcal{M}$ and $\mathcal{N}$ ($\mathcal{M} \subset \mathcal{N} \subset T \times R^n$) and for any initial position $(t_0,x_0) \in T \times R^n$, one and only one of the following assertions is valid:* (I) *there exists a positional strategy U^0 which yields a solution of an $(\mathcal{M}, \mathcal{N})$-approach problem;* (II) *there exists an $\epsilon > 0$ and a positional strategy V^ϵ which yields a solution of an $(\mathcal{M}, \mathcal{N})^\epsilon$-evasion problem.*

The proof of this theorem is presented in §§2.2 - 2.4. Recall that in this and in the immediately following chapters, it is assumed that all the conditions stipulated in §1.2.2, including condition (iv), are fulfilled.

We shall now determine the value and the saddle point of a differential game of approach-evasion. Let the first assertion of the alternative be fulfilled, then the value is zero and the saddle point will be the pair of strategies (U^0,V), where V is an arbitrary positional strategy of the second player. If the second assertion of the alternative is fulfilled then the value of an approach-evasion game is equal to one and the pair (U,V^{ϵ}) will be a saddle point where U is an arbitrary strategy of the first player.

A statement of the alternative for step-by-step motions is presented below. In this statement the following notation is used for closed ϵ-neighborhoods of the sets M and N:

$$
(14)\qquad \begin{aligned}
\bar{M}^{\epsilon} &\overset{\text{def}}{=} \{(t,x) \in T \times R^n : |t - t_*| + \|x - x_*\| \leqslant \epsilon, (t_*,x_*) \in M\},\\
\bar{N}^{\epsilon} &\overset{\text{def}}{=} \{(t,x) \in T \times R^n : |t - t_*| + \|x - x_*\| \leqslant \epsilon, (t_*,x_*) \in N\}.
\end{aligned}
$$

Thus for step-by-step motions the alternative consists of fulfilling one and only one of the following assertions:

(a) For an arbitrary small $\alpha > 0$ there exists $\delta, 0$ such that for any step-by-step motion $x(\cdot) \in X_1(t_0,x_0,U^0,\Delta)$ where $\operatorname{diam}\Delta \leqslant \delta$, the condition of $(\bar{M}^{\alpha}, \bar{N}^{\alpha})$-approach is fulfilled. (This condition is obtained from relations (1) - (3) by replacing the sets M and N with their closed α-neighborhoods.)

(b) For any number $\beta \in (0,\epsilon)$ as close to ϵ as desired there exists a $\delta > 0$ such that for any step-by-step motion $x(\cdot) \in X_2(t_0,x_0,V^{\epsilon},\Delta)$ where $\operatorname{diam}\Delta \leqslant \delta$, the condition of $(M, N)^{\beta}$-evasion will be fulfilled.

Here strategies U^0,V^{ϵ} and the number ϵ are the same as in Theorem 2.1.1. It is natural to assume that the smaller the parameter α is in the condition of (M^{α}, N^{α})-approach which is satisfied by the step-by-step motions, the better it is for the first player. Conversely, the larger the parameter β is, in the condition of $(M, N)^{\beta}$-evasion, the better it is for the second player. This fact is reflected in the above statement where it is specifically stated that parameter α can be chosen as small as possible while the parameter β can be chosen as close to the number ϵ as desired.

2.1.3. In the next sections, a complete proof of Theorem 2.1.1 will be presented. We shall now sketch the course of the proof.

Let a subset W be selected from the set N. For clarity this set can be represented as a tube situated along the t axis. We shall assume that the right-hand end of this tube is contained in the set M. We thus say that the set W is cut short on the set M (cf. Fig. 2.1.3). To simplify the exposition we shall use expressions of the type "motion $x(\cdot)$ remains on the set W," which would mean that the corresponding positions $(t,x(t))$ belong to the set W or, equivalently, that the graph of the function $x(\cdot)$ is contained in the set W. We shall assume that W possesses the following important property.

Let the position (t_*,x_*) be contained in the set W. Let the second player communicate his (her) control $v(t)$ $(t_* \leqslant t < t^*)$ to the first player at time t_*, which will be realized on the interval $[t_*,t^*)$. We assume that the first player can choose the control $u(t)$ $(t_* \leqslant t < t^*)$ such that the motion will remain on the set W either for all $t \in (t_*,t^*]$ or up to its encounter with the set M. In this condition

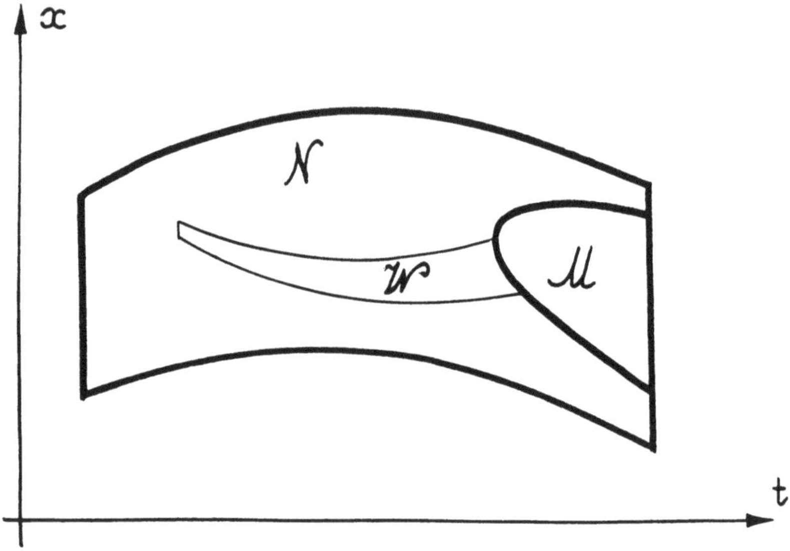

Fig. 2.1.3

(t_*,x_*) is an arbitrary point in the set W, t^* is an arbitrary number in the interval $(t_*,\theta]$, and $v(\cdot)$: $[t_*,t^*) \to Q$ is an arbitrary measurable function.

Such a condition is called the condition of u-stability of the set W. A set W possessing all the properties indicated above is called a u-stable bridge in an (M,N)-approach problem. A formal definition of a u-stable bridge is presented in §2.2.1.

Consider a game which is implemented according to the following rules. The second player chooses a partition $\Delta = \{t_0 = \tau_0 < \tau_1 < \ldots < \tau_{k+1} = \theta\}$. As each instant of time $t = \tau_i (i = 0, 1, \ldots, k)$, the second player chooses the control $v(t)$ $(\tau_i \leqslant t < \tau_{i+1})$ which will be realized on the interval $[\tau_i,\tau_{i+1})$. At the same instant $t = \tau_i$ the control selected by the second player becomes known to the first. Possessing this information as well as the information about the realized position $(\tau_i,x(\tau_i))$ the first player chooses his (her) control $u(t)$ $(\tau_i \leqslant t < \tau_{i+1})$. The selected controls $u(t)$ and $v(t)$ generated the motion of the system on the interval $[\tau_i,\tau_{i+1})$. Next, at time $t = \tau_{i+1}$ the procedure described above is repeated. Thus, in the game under consideration, the first player possesses an *informational advantage* or in other words there exists an *informational discrimination* against the second player.

Assume that the initial position (t_0,x_0) is located on a u-stable bridge W. It follows directly from the condition of u-stability that in the game under consideration the first player may successively choose his (her) control $u(t)(\tau_i \leqslant t < \tau_{i+1}, i = 0, 1, \ldots, k)$ such that the motion will remain on the bridge W up to its encounter with the set M. This encounter is bound to occur since the set W is cut short on

$\mathcal{M}$. Since $\mathcal{W} \subset \mathcal{N}$ the motion will satisfy the condition of an $(\mathcal{M},\mathcal{N})$-approach. Thus, if the initial position (t_0,x_0) belongs to a u-stable bridge then the first player can assure an $(\mathcal{M},\mathcal{N})$-approach.

However, this conclusion is obtained under the condition of information discrimination of the second player. The question arises: Is it not possible to drop this unnatural condition and construct a positional strategy for the first player which will retain, as before, the motions on the u-stable bridge? An affirmative answer to this question is given in §2.4 where a positional strategy U^e is defined which is called a strategy extremal to the bridge $\mathcal{W}$. The following assertion is proved in §2.4: If the initial position (t_0,x_0) is contained in a u-stable bridge $\mathcal{W}$, then any constructive motion $x(\cdot) \in X_1(t_0,x_0,U^e)$ remains on the bridge $\mathcal{W}$ up until this motion encounters the target set $\mathcal{M}$. It is thus established that the positional strategy U^e yields a solution for the $(\mathcal{M},\mathcal{N})$-approach problem. In the proof of this assertion, step-by-step motions $x(\cdot) \in X_1(t_0,x_0,U^e,\Delta)$ are considered. It is shown that they remain in the ϵ-neighborhood of the set $\mathcal{W}$ while $\epsilon \to 0$ as $\operatorname{diam}\Delta \to 0$. In turn, the proof of the latter fact is based on a local estimate of the variation in the distance up to the bridge on a small time interval $[\tau_i,\tau_{i+1}]$. This estimate is derived in §2.3.

We turn now to the problem of $(\mathcal{M},\mathcal{N})^\epsilon$-evasion. Recall that this problem is equivalent to the problem of $(K_\epsilon,\mathfrak{L}_\epsilon)$-approach. In §2.2 the definition of a v-stable bridge $\mathcal{W}$ contained in $\mathfrak{L}_\epsilon$ and cut short on K_ϵ is presented. The following assertion is valid (cf. §2.4): let $\mathcal{W}$ be a v-stable bridge in the problem of $(\mathcal{M},\mathcal{N})^\epsilon$-evasion and let the initial position (t_0,x_0) be contained in the set $\mathcal{W}$. Then the problem of $(\mathcal{M},\mathcal{N})^\epsilon$-evasion is solvable and its solution is given by the positional strategy V^e extremal to the set $\mathcal{W}$.

To complete the proof of Theorem 2.1.1, it is sufficient to verify that for any initial position (t_0,x_0) will either always fall into a u-stable bridge in the $(\mathcal{M},\mathcal{N})$-approach problem or there exists a parameter $\epsilon > 0$ and a v-stable bridge in the $(\mathcal{M},\mathcal{N})^\epsilon$-evasion problem containing the point (t_0,x_0). This fact is established in §2.2 where it is shown that the space of positions $T \times R^n$ can be represented as a union of two sets $\mathcal{W}_0^1$ and $\mathcal{W}_0^2$. The set $\mathcal{W}_0^1$ is a u-stable bridge in the problem of $(\mathcal{M},\mathcal{N})$-approach. The set $\mathcal{W}$ is defined as follows. For different values of the parameter $\epsilon > 0$ the $(\mathcal{M},\mathcal{N})^\epsilon$-evasion problems and the corresponding v-stable bridges are considered. The union of all such bridges forms the set $\mathcal{W}_0^2$.

2.2. Stable Bridges

In this section, formal definitions of u-stable and v-stable bridges for the problems of approach and evasion are presented. Some properties of these bridges are considered. The basic result obtained in this section is Theorem 2.2.1, which establishes that the space of the positions can be subdivided into two parts: one of these parts being a u-stable bridge while the other is a union of v-stable bridges.

The proof of Theorem 2.2.1 is based on Lemmas 2.2.1 - 2.2.6. We present preliminary explanations. In §2.2.1 u-stable bridges for the $(\mathcal{M}, \mathcal{N})$-approach problem and for the auxiliary $(\mathcal{M}^\epsilon, \mathcal{N}^\epsilon)$-approach problem are presented. In §2.2.2, Lemmas 2.2.1 and 2.2.2, which describe the simplest properties of u-stable bridges, are stated and proved. In particular, it is shown that the union of u-stable sets will be a u-stable bridge. In §2.2.3, v-stable bridges for the $(\mathcal{K}, \mathfrak{L})$-approach problem are defined. Next, Lemma 2.2.3 states that any position should be contained on either u-stable bridge in the $(\mathcal{M}^\epsilon, \mathcal{N}^\epsilon)$-approach problem or on a v-stable bridge in a $(\mathcal{K}_\epsilon, \mathfrak{L}_\epsilon)$-approach problem. In §2.2.4 it is utilized for the first time tht the sets $\mathcal{M}$, $\mathcal{N}$, $\mathcal{K}_\epsilon$, and $\mathfrak{L}_\epsilon$ are closed. Here Lemmas 2.2.4 - 2.2.6 are proved which state topological properties of u-stable and v-stable bridges.

2.2.1. We introduce notation to be used in the proof of Theorem 2.2.1.

Let $(t,x) \in T \times R^n)$, $u \in \mathcal{P}$, $v \in \mathcal{Q}$. Set

$$\mathcal{F}_1(t,x,u) \overset{\text{def}}{=} co\{f(t,x,u,v):\ v \in \mathcal{Q}\},$$

(1)

$$\mathcal{F}_2(t,x,v) \overset{\text{def}}{=} co\{f(t,x,u,v):\ u \in \mathcal{P}\}.$$

Recall that the symbol $co\ G$ denotes the convex hull of the set G.

Let $(t_*,x_*) \in T \times R^n$. The symbol $X_1(t_*,x_*,u)$ denotes the set of all absolutely continuous solutions $x(\cdot)$: $[t_*,\theta] \to R^n$ of the differential inclusion

$$\dot{x}(t) \in \mathcal{F}_1(t,x(t),u) \tag{2}$$

satisfying the initial condition $x(t_*) = x_*$.

The set of all the absolutely continuous functions $x(\cdot)$: $[t_*,\theta] \to R^n$ which satisfy the initial condition $x(t_*) = x_*$ and for almost all $t \in [t_*,\theta]$ satisfy the differential inclusion

$$\dot{x}(t) \in \mathcal{F}_2(t,x(t),v) \tag{3}$$

will be denoted as $X_2(t_*,x_*,v)$.

The following assertion is well-known (cf., e.g., [22, 74]): if the vectogram

$$f(t,x,\mathcal{P},v) \overset{\text{def}}{=} \{f(t,x,u,v):\ u \in \mathcal{P}\} \tag{4}$$

is convex for all $(t,x) \in T \times R^n$, then the set $X_2(t_*,x_*,v)$ coincides with the set

$$\{x(\cdot,t_*,x_*,u(\cdot),v):\ u(\cdot) \in \mathfrak{L}([t_*,\theta], \mathcal{P})\} \tag{5}$$

where $x(\cdot,t_*,x_*,u(\cdot),v)$ is a solution of the differential equation

(6) $$\dot{x}(t) = f(t,x(t),u(t),v), \quad x(t_*) = x_*$$

and $\mathfrak{L}([t_*,\theta], \mathcal{P})$ is the set of all measurable functions $u(\cdot)$: $[t_*,\theta] \to \mathcal{P}$. If, however, the vectogram $f(t,x, \mathcal{P},v)$ is not convex then the set $X_2(t_*,x_*,v)$ will coincide with the closure of the set (5). An analogous assertion is also valid for the set $X_1(t_*,x_*,u)$.

Let $(t_*,x_*) \in T \times R^n$, $t^* \in [t_*,\theta]$, $u \in \mathcal{P}$ and $v \in \mathcal{Q}$. Set

(7) $$\begin{aligned} Z_1(t_*,x_*,t^*,u) &\overset{\text{def}}{=} \{(t,x(t)): t_* \leqslant t \leqslant t^*, \ x(\cdot) \in X_1(t_*,x_*,u)\}, \\ Z_2(t_*,x_*,t^*,v) &\overset{\text{def}}{=} \{(t,x(t)): t_* \leqslant t \leqslant t^*, \ x(\cdot) \in X_2(t_*,x_*,v)\}. \end{aligned}$$

The $Z_1(t_*,x_*,t^*,u)$ is a set in the space $T \times R^n$ which is defined as an intersection of the strip $[t_*,t^*] \times R^n$ with the union of the graphs of all possible functions $x(\cdot) \in X_1(t_*,x_*,u)$. The set $Z_2(t_*,x_*,t^*,v)$ is interpreted analogously.

We shall also utilize the following notation. Let $\mathcal{D} \subset T \times R^n$, $t^* \in T$. Set

(8) $$\mathcal{D}(t^*) \overset{\text{def}}{=} \{(t,x) \in \mathcal{D} : \ t = t^*\}.$$

We now proceed to the definition of the concept of the u-stable bridge.

Definition. *The set $W \subset T \times R^n$ is called a u-stable bridge in an $(\mathcal{M}, \mathcal{N})$-approach problem if it satisfies the following conditions:*

(1_u) $W \subset \mathcal{N}$, $W(\theta) \subset \mathcal{M}$

(2_u) *for any $(t_*,x_*) \in W$, $t^* \in (t_*,\theta]$ and $v \in \mathcal{Q}$ there exists a solution $x(\cdot) \in X_2(t_*,x_*,v)$ such that either*

$$(t^*,x(t^*)) \in W \quad \textit{or} \quad \{(t,x(t)): t_* \leqslant t \leqslant t^*\} \cap \mathcal{M} \neq \phi .$$

Condition (2_u) is called the *condition of u-stability of the set W*. Below, we shall comment on this definition, but first we present another definition of the condition of u-stability.

Let $\mathcal{D}$ be a set in the space $T \times R^n$, let $t_* \in T$ and $t^* \in [t_*,\theta]$. Define the set

(9) $$\Pi_2(t_*,t^*,\mathcal{D}) \overset{\text{def}}{=} \bigcap_{v \in \mathcal{Q}} \{(t_*,x_*): \ Z_2(t_*,x_*,t^*,v) \cap \mathcal{D} \neq \phi\}.$$

The set $\Pi_2(t_*,t^*,\mathcal{D})$ is called the set of programmed absorption of the target set $\mathcal{D}$. In the case when the vectogram $f(t,x,\mathcal{P},v)$ is convex, equality (9) defines the set of all initial positions (t_*,x_*) possessing the following property: for any constant control $v(t) = v$ $(t_* \leqslant t < t^*)$ there exists a programmed control $u(\cdot) \in \mathfrak{L}([t_*,t^*], \mathcal{P})$ such that for the motion $x(\cdot)$ of the system (6) the encounter with the set $\mathcal{D}$ will occur not later than at time $t = t^*$, i.e. the condition $(\tau,x(\tau)) \in \mathcal{D}$ will be fulfilled for some $\tau \in [t_*,t^*]$. Utilizing the set Π_2, the condition of u-stability can be expressed as follows:

(2_u) for any numbers $t_* \in T$ and $t^* \in (t_*,\theta]$ the inclusion

$$(10) \qquad \mathcal{W}(t_*) \subset \Pi_2(t_*,t^*, \mathcal{M} \cup \mathcal{W}(t^*))$$

is valid.

We shall discuss the definition of a u-stable bridge. Consider first the condition of u-stability. Assume once again that the vectogram $f(t,x,\mathcal{P},v)$ (4) is convex for all $(t,x,v) \in T \times R^n \times \mathcal{Q}$. In this case the set $X_2(t_*,x_*,v)$ coincides with the set (5). Thus the condition of u-stabiity can be expressed as follows: Let position (t_*,x_*) be located on the set $\mathcal{W}$, let an instant of time $t^* \in (t_*,\theta]$ be given, and let the second player choose a constant control $v(t) = v$ $(t_* \leqslant t < t^*)$. At the starting time $t = t_*$ this control is known to the first player. It is assumed that the first player can select his (her) control $u(t)$ $(t_* \leqslant t < t^*)$ in such a manner that for the corresponding motion of the system (6), position $(t,x(t))$ either falls into the set $\mathcal{W}$ at time $t = t^*$ or the encounter $(\tau,x(\tau)) \in \mathcal{M}$ took place at some $\tau \in [t_*,t^*]$.

Note that the definition of u-stability presented here differs from the definition presented in §2.1.3. There, all possible measurable controls $v(t)$ $(t_* \leqslant t < t^*)$ were considered, while here controls $v(t)$ are constant on $[t_*,t^*)$. Here it is required that position $(t,x(t))$ falls into $\mathcal{W}$ at $t = t^*$ while in §2.1.3 it was stated that $(t,x(t)) \in \mathcal{W}$ for $t_* \leqslant t \leqslant \tau$, where τ is the time of encounter with $\mathcal{M}$. We shall see that this difference is not a basic one. For a preliminary clarifying illustration, the definition as given in §2.1.3 is more appropriate while for formal investigations, the definitions presented herein are more suitable.

Consider now condition (1_u). The inclusion $\mathcal{W} \subset \mathcal{N}$ probably does not require clarification. The second inclusion $\mathcal{W}(\theta) \subset \mathcal{M}$ is equivalent to the inclusion $\mathcal{W}(\theta) \subset \mathcal{M}(\theta)$ and the case $\mathcal{M}(\theta) = \phi$ is not excluded. In this case condition $\mathcal{W}(\theta) \subset \mathcal{M}(\theta)$ implies that $\mathcal{W}(\theta) = \phi$. One can suppose that inclusion $\mathcal{W}(\theta) \subset \mathcal{M}$ means that the bridge $\mathcal{W}$ must be cut short on the set $\mathcal{M}$. We are not going to specify in the geneal case how this formal condition is connected with the intuitive representation given in §2.1.3. We consider only the particular case when the set $\mathcal{W}$ is closed and bounded. Let $\mathcal{W}(t_0) \neq \phi$ and $[t_0,\tau_{\#}]$ be the maximal interval on which $\mathcal{W}(t) \neq \phi$, i.e. for any $\tau > \tau_{\#}$ there exists $t \in (\tau_{\#},\tau)$ such that $\mathcal{W}(t) = \phi$. One can show that such a number $\tau_{\#}$ actually exists. Moreover, it turns out that $\mathcal{W}(\tau_{\#}) \subset \mathcal{M}(\tau_{\#})$. This inclusion corresponds to the intuitive representation given in §2.1.3 as to how a set $\mathcal{W}$ is cut short on $\mathcal{M}$. In the proof of the inclusion $\mathcal{W}(\tau_{\#}) \subset \mathcal{M}(\tau_{\#})$, condition $\mathcal{W}(\theta) \subset \mathcal{M}$ and the u-stability condition are utilized. However, the fact mentioned above is not required for subsequent arguments and we shall not present its proof.

In §2.5, examples of u-stability bridges are presented and constructive definitions are considered.

The concept of a u-stable bridge will appear in all basic assertions of this chapter. Along with the concept, the following auxiliary notion is utilized in §2.2.2 – §2.2.4.

The set $W \subset T \times R^n$ is called a u-stable bridge in an (M^ϵ, N^ϵ)-approach problem if it satisfies the following conditions:

$(1_{u,\epsilon})$ $\quad W \subset N^\epsilon, \quad W(\theta) \subset M^\epsilon,$

$(2_{u,\epsilon})$ for any $(t_*,x_*) \in W$, $t^* \in (t_*,\theta]$ and $v \in Q$ there exists a solution $x(\cdot) \in X_2(t_*,x_*,v)$ such that either

$$(t^*,x(t^*)) \in W , \quad \text{or } \{(t,x(t)): \ t_* \leqslant t \leqslant t^*\} \cap M^\epsilon \neq \phi .$$

Here M^ϵ and N^ϵ are open ϵ-neighborhoods of the set M and N (cf. (2.1.5)).

Condition $(2_{u,\epsilon})$ can be rewritten in the following equivalent form:

$$(11) \qquad W(t_*) \subset \Pi_2(t_*,t^*, M^\epsilon \cup W(t^*)).$$

It is easy to note that conditions $(1_{u,\epsilon})$ and $(2_{u,\epsilon})$ differ from conditions (1_u) and (2_u) by replacing M and N with their ϵ-neighborhoods.

Note also that the (M^ϵ, N^ϵ)-approach problem was not formulated; moreover, this problem will not be considered in what follows. It is mentioned here by analogy with the basic definition to show that a u-stable bridge is connected in a certain manner with the sets M^ϵ and N^ϵ.

2.2.2. Two lemmas stated below follow directly from the definition of a u-stable bridge. For completeness, we shall present their proofs. The lemmas will be utilized in §2.2.3 where the duality of u-stable and v-stable bridges is considered.

Let $\epsilon > 0$. Denote by $\mathbf{W}^1_\epsilon$ the family of u-stable bridges in an (M^ϵ, N^ϵ)-approach problem. Denote by $\mathbf{W}^1_0$ the set of u-stable bridges in an (M, N)-approach problem. Note that the classes of sets $\mathbf{W}^1_\epsilon$ and $\mathbf{W}^1_0$ are non-empty, since $M^\epsilon \in \mathbf{W}^1_\epsilon$ and $M \in \mathbf{W}^1_0$. The following lemma is valid.

Lemma 2.2.1. *Let the sets* W^1_0 *and* W^1_ϵ *be defined by the equalities*

$$(12) \qquad W^1_0 \overset{\text{def}}{=} \bigcup_{W \in \mathbf{W}^1_0} W, \quad W^1_\epsilon \overset{\text{def}}{=} \bigcup_{W \in \mathbf{W}^1_\epsilon} W .$$

Then $W^1_0 \in \mathbf{W}^1_0$ *and* $W^1_\epsilon \in \mathbf{W}^1_\epsilon$.

Proof. We show that $W \in \mathbf{W}_0^1$. Let $D_a \subset T \times R^n$, $a \in A$ where A is a set of indices. It follows from (9) that

$$(13) \qquad \Pi_2(t_*,t^*, \bigcup_{a\in A} D_a) \supset \bigcup_{a\in A} \Pi_2(t_*,t^*,D_a).$$

Let $W_a \in \mathbf{W}_0^1$, $a \in A$. From (10) and (13) we have

$$\bigcup_{a\in A} W_a(t_*) \subset \bigcup_{a\in A} \Pi_2\left[t_*,t^*,M \cup W_a(t^*)\right) \subset \Pi_2\left[t_2,t^*, M \cup \left[\bigcup_{a\in A} W_a(t^*)\right)\right]$$

Thus the set $\bigcup_{a\in A} W_a$ satisfies the condition (2_u). Clearly this set also satisfies condition (1_u). In particular, it now follows that $W_0^1 \in \mathbf{W}_0^1$. The assertion that $W_\epsilon^1 \in \mathbf{W}_\epsilon^1$ is proved analogously.

Lemma 2.2.2. *Let* $(t_1,y) \in T \times R^n$, $t_2 \in (t_1,\theta]$, $u_* \in P$, $\epsilon > 0$, $W_* \in \mathbf{W}_\epsilon^1$. *Set* $Z \overset{\text{def}}{=} Z_1(t_1,y,t_2,u_*)$ *(cf. (7)). Assume that*

$$(14) \qquad Z \subset N^\epsilon \text{ and } Z(t_2) \subset W_* \,.$$

Define the set $W^* \subset [t_1,\theta] \times R^n$ *by the relation*

$$(15) \qquad W^*(t) \overset{\text{def}}{=} \begin{cases} Z(t) & \text{for } t_1 \leqslant t < t_2, \\ W_*(t) & \text{for } t_2 \leqslant t \leqslant \theta \,. \end{cases}$$

The set W^* *belongs to the class* $\mathbf{W}_\epsilon^1$.

The set W^* is presented in Fig. 2.2.1. It consists of two subsets. The left subset locted in the strip $[t_1,t_2] \times R^n$ is formed by the trajectories of the differential inclusion (2) emanating from the position (t_1,y). The right subset is a part of a u-stable bridge W_* located in the strip $[t_2,\theta] \times R^n$.

Proof. It follows directly from the definition of the set W^* that it satisfies condition $(1_{u,\epsilon})$. It remains to verify that the set W^* also satisfies condition $(2_{u,\epsilon})$.

Let $(t_*,x_*) \in W^*, t^* \in (t_*,\theta]$ and $v \in Q$. It is required to show that there exists motion $x(\cdot) \in X_2(t_*,x_*,v)$ such that either

$$(16) \qquad (t^*,x(t^*)) \in W^* \text{ or } \{(t,x(t)) \colon t_* \leqslant t \leqslant t^*\} \cap M^\epsilon \neq \phi \,.$$

The following three cases are possible:

(1) $t_2 \leqslant t_* < t^*$, (2) $t_1 \leqslant t_* < t^* \leqslant t_2$, (3) $t_1 \leqslant t_* < t_2 < t^*$.

Case (1). In this case $(t_*,x_*) \in W_*$. Therefore the existence of a motion $x(\cdot)$ satisfying (16) follows from condition $(2_{u,\epsilon})$ which is fulfilled for the set W_*.

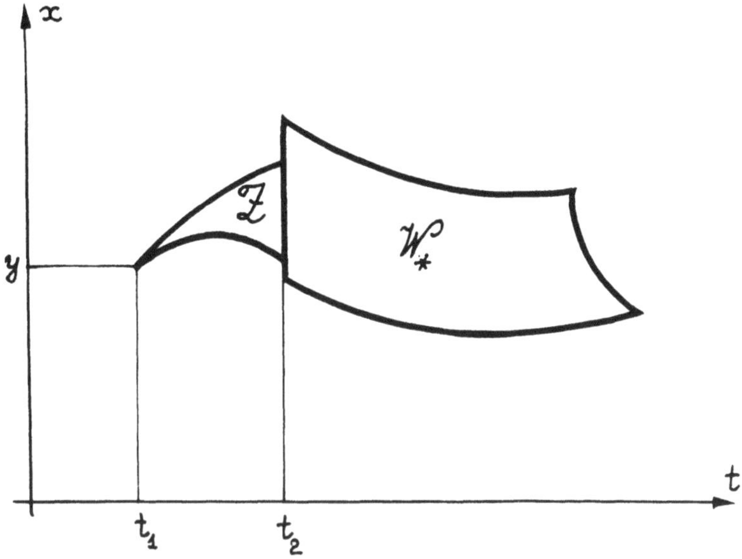

Fig. 2.2.1

Case (2). In this case $(t_*,x_*) \in \mathcal{Z}$. Consider the solution $x(\cdot)$: $[t_*,\theta] \to R^n$ of the differential equation

$$\dot{x}(t) = f(t,x(t),\ u_*,v),\ x(t_*) = x_* \ .$$

Clearly $x(\cdot) \in X_2(t_*,x_*,v) \cap X_1(t_*,x_*,u_*)$. Since $(t_*,x_*) \in \mathcal{Z}$ and $x(\cdot) \in X_1(t_*,x_*,u_*)$, it follows from the definitions of the set $\mathcal{Z}$ (cf. (7)) that $(t^*,x(t^*)) \in \mathcal{Z} \subset \mathcal{W}^*$. Thus in this case there exists a motion $x(\cdot) \in X_2(t_*,x_*,v)$ satisfying (16).

Case (3). In this case we have again $(t_*,x_*) \in \mathcal{Z}$. As it follows from the considerations in case (2), there exists a motion $x^{(1)}(\cdot) \in X_2(t_*,x_*,v)$ such that $(t_2,x^{(1)}(t_2)) \in \mathcal{Z}$. It follows from (14) that $(t_2,x^{(1)}(t_2)) \in X_2$. Since the set $\mathcal{W}_*$ is u-stable there exists a motion $x^{(2)}(\cdot) \in X_2(t_2,x^{(1)}(t_2),v)$ such that either $(t^*,x^{(2)}(t^*)) \subset \mathcal{W}_*(t^*) \subset \mathcal{W}^*$ or $\{(t,x^{(2)}(t)):\ t_2 \leq t \leq t^*\} \cap M^{\epsilon} \neq \phi$. Consider the function $x(t) = (x^{(1)}(t)$ for $t_* \leq t \leq t_2$, $x^{(2)}(t)$ for $t_2 \leq t \leq \theta)$. This function belongs to the set $X_2(t_*,x_*,v)$ and satisfies condition (16). Lemma 2.2.2 is thus proved.

2.2.3. We shall define a v-stable bridge and show that v-stable and u-stable bridges can be defined as mutually complementary sets.

Let K and $\mathfrak{L}$ be closed sets in the space $T \times R^n$ and $K \subset \mathfrak{L}$. We introduce the following notion.

Definition. *The set* $\mathcal{W} \subset T \times R^n$ *is called a* v*-stable bridge in a* $(\mathcal{K}, \mathfrak{L})$*-approach problem provided it satisfies the following two conditions:*

(1_v) $\qquad \mathcal{W} \subset \mathfrak{L}, \ \mathcal{W}(\theta) \subset \mathcal{K}$,

(2_v) $\qquad$ *for any* $(t_*, x_*) \in \mathcal{W}$, $t^* \in (t_*, \theta]$, $u \in \mathcal{P}$,

there exists a motion $x(\cdot) \in \mathcal{X}_1(t_*, x_*, u)$ *such that either* $(t^*, x(t^*)) \in \mathcal{W}$ *or* $\{(t, x(t)):\ t_* \leqslant t \leqslant t^*\} \cap \mathcal{K} \neq \phi$.

Condition (2_v) is called the *condition of* v*-stability* of the set $\mathcal{W}$. This condition can be rephrased as follows:
(2_v) for any $t_* \in T$ and $t^* \in [t_*, \theta]$ the inclusion

$$\mathcal{W}(t_*) \subset \Pi_1(t_*, t^*, \mathcal{K} \cup \mathcal{W}(t^*))$$

is satisfied where the operator Π_1 is defined by the relation

$$\Pi_1(t_*, t^*, \mathcal{D}) \overset{\text{def}}{=} \bigcap_{u \in \mathcal{P}} \{(t^*, x_*):\ \mathcal{Z}_1(t_*, x_*, t^*, u) \cap \mathcal{D} \neq \phi\}.$$

Here $\mathcal{D}$ is a set in $T \times R^n$.

Evidently the definition of a *v*-stable bridge is analogous to the definition of a *u*-stable bridge.

Denote by $\mathbf{W}^2_\epsilon$ the family of *v*-stable bridges in a $(\mathcal{K}_\epsilon, \mathfrak{L}_\epsilon)$-approach problem. Here $\epsilon > 0$ and $\mathcal{K}_\epsilon$ and $\mathfrak{L}_\epsilon$ are closed sets defined in accordance with (2.1.9). Note that inclusion $\mathcal{M} \subset \mathcal{N}$ implies the inclusion $\mathcal{K}_\epsilon \subset \mathfrak{L}_\epsilon$. The class $\mathbf{W}^2_\epsilon$ is not empty since $\mathcal{K}_\epsilon \in \mathbf{W}^2_\epsilon$. Define the set

$$(17) \qquad \mathcal{W}^2_\epsilon \overset{\text{def}}{=} \bigcup_{\mathcal{W} \in \mathbf{W}^2_\epsilon} \mathcal{W}\ .$$

We now prove the following assertion:

Lemma 2.2.3. *The set* $\mathcal{W}^2_\epsilon$ *is a* v*-stable bridge in a* $(\mathcal{K}_\epsilon, \mathfrak{L}_\epsilon)$*-approach problem. The equality*

$$(18) \qquad \mathcal{W}^1_\epsilon \cup \mathcal{W}^2_\epsilon = T \times R^n$$

is valid where the set $\mathcal{W}^1_\epsilon$ *is defined in* (12).

Proof. The first assertion is proved in the same manner as Lemma 2.2.1.

We now prove equality (18). Consider the set

$$(19) \qquad \mathcal{W}^\# \overset{\text{def}}{=} (T \times R^n) \setminus \mathcal{W}^1_\epsilon\ .$$

We shall show that

$$(20) \qquad \mathcal{W}^{\#} \in \mathbf{W}_\epsilon^2.$$

We verify the fulfillment of condition (1_v). Since $\mathcal{M}^\epsilon \in \mathbf{W}_\epsilon^1$ it follows that $\mathcal{W}_\epsilon^1 \supset \mathcal{M}^\epsilon$. Therefore

$$\mathcal{W}^{\#} \overset{\text{def}}{=} (T \times R^n) \setminus \mathcal{W}_\epsilon^1 \subset (T \times R^n) \setminus \mathcal{M}^\epsilon \overset{\text{def}}{=} \mathfrak{L}_\epsilon.$$

Now (2.1.9) yields that $\mathfrak{L}_\epsilon(\theta) = K_\epsilon(\theta)$. Therefore we obtain from the preceding inclusion that

$$\mathcal{W}^{\#}(\theta) \subset \mathfrak{L}_\epsilon(\theta) = K_\epsilon(\theta).$$

Thus it is shown that the set $\mathcal{W}^{\#}$ satisfies condition (1_v) (for $K = K_\epsilon$ and $\mathfrak{L} = \mathfrak{L}_\epsilon$). We shall show that this set also satisfies the condition (2_v) (for $K = K_\epsilon$). Assume the contrary. Let there exist $(t_1,y) \in \mathcal{W}^{\#}$, $t_2 \in (t_1,\theta]$, $u_* \in P$ such that for the set (cf. (7))

$$Z \overset{\text{def}}{=} Z_1(t_1,y,t_2,u_*)$$

the relations

$$(21) \qquad Z \cap K_\epsilon = \phi,$$

$$(22) \qquad Z \cap \mathcal{W}^{\#}(t_2) = \phi$$

are fulfilled. It follows from (2.1.10) and (21) that $Z \subset N^\epsilon$. From (19) and (22) we obtain that $Z(t_2) \subset \mathcal{W}_\epsilon^1(t_2)$. Set $\mathcal{W}_* = \mathcal{W}_\epsilon^1$. In view of Lemma 2.2.1 this set belongs to the class $\mathbf{W}_\epsilon^1$. Consequently, in the situation under consideration all the conditions of Lemma 2.2.2 are fulfilled, and according to this lemma the set $\mathcal{W}^*$ defined by relation (15) also belongs to the class $\mathbf{W}_\epsilon^1$. It follows from (15) and (12) that $(t_1,y) \in \mathcal{W}^* \subset \mathcal{W}_\epsilon^1$. Thus (19) implies that $(t_1,y) \in \mathcal{W}^{\#}$. The contradiction obtained proves relation (20).

From (17) and (20) we have $\mathcal{W}^{\#} \subset \mathcal{W}_\epsilon^2$. It follows from (19) that

$$\mathcal{W}^{\#} \cup \mathcal{W}_\epsilon^1 = T \times R^n.$$

Therefore we arrive at (18). Lemma 2.2.3 is proved.

We have thus shown that the space of positions can be subdivided into two sets $\mathcal{W}_\epsilon^1$ and $\mathcal{W}_\epsilon^2$. The first set is a u-stable bridge in an $(\mathcal{M}^\epsilon, N^\epsilon)$-approach problem. The second set is a v-stable bridge in a $(K_\epsilon,\mathfrak{L}_\epsilon)$-approach problem, where the sets K_ϵ and $\mathfrak{L}_\epsilon$ are defined according to (2.1.10).

2.2.4. To improve on the result obtained in Lemma 2.2.3, we consider topological properties of stable bridges.

Define the set $\mathcal{W}^1_*$ which consists of all limit points of sequences of the form $(t_k,x_k) \in \mathcal{W}^1_{\epsilon_k}$ $(k = 1, 2, ...)$, where $\epsilon_k \to 0$ as $k \to \infty$. We thus set

(23) $$\mathcal{W}^1_* \overset{\text{def}}{=} \Big\{(t_*,x_*) \in T\times R^n\colon\ \exists\ \epsilon_k > 0,\ \exists\ (t_k,x_k) \in \mathcal{W}^1_{\epsilon_k}$$
$$(k = 1, 2, ...),\ \epsilon_k \to 0,\ (t_k,x_k) \to (t_*,x_*) \text{ for } k \to \infty\Big\}.$$

In the next lemma we use for the first time that sets $\mathcal{M}$ and $\mathcal{N}$ are closed.

Lemma 2.2.4. *The set* $\mathcal{W}^1_*$ is a u-*stable bridge in an* $(\mathcal{M}, \mathcal{N})$*-approach problem.*

Proof. Let $(t_*,x_*) \in \mathcal{W}^1_*$. Then there exist sequences $\epsilon_k > 0$, (t_k,x_k) $(k = 1, 2, ...)$ such that

(24) $$(t_k,x_k) \in \mathcal{W}^1_{\epsilon_k},\quad \lim_{k\to\infty} \epsilon_k = 0,\quad \lim_{k\to\infty} (t_k,x_k) = (t_*,x_*).$$

Since $\mathcal{W}^1_\epsilon \subset \mathcal{N}^\epsilon$, $(t_k,x_k) \in \mathcal{N}^{\epsilon_k}$ $(k = 1, 2, ...)$. Approaching the limit as $k \to \infty$ and taking into account that the set $\mathcal{N}$ is closed, we obtain $(t_*,x_*) \in \mathcal{N}$. We have thus proved that $\mathcal{W}^1_* \subset \mathcal{N}$. The inclusion $\mathcal{W}^1_*(\theta) \subset \mathcal{M}$ is proved analogously.

We now show that the set $\mathcal{W}^1_*$ satisfies condition (2_u). Let $(t_*,x_*) \in \mathcal{W}^1_*$, $t^* \in (t_*,\theta]$ and $v \in \mathcal{Q}$. By the definition of the set $\mathcal{W}^1_*$ there exists a sequence (t_k,x_k) $(k = 1,2, ...)$ satisfying conditions (24). One can assume that $t_k < t^*$ $(k = 1,2, ...)$. Since $(t_k,x_k) \in \mathcal{W}^1_{\epsilon_k} \in \mathbf{W}^1_{\epsilon_k}$ there exists a motion $x_k(\cdot) \in X_2(t_k,x_k,v)$ such that one of the following two conditions is satisfied:

(25) $$(t^*,x_k(t^*)) \in \mathcal{W}^1_{\epsilon_k}$$

or

(26) $$\{(t,x_k(t))\colon\ t_k \leqslant t \leqslant t^*\} \cap \mathcal{M}^{\epsilon_k} \neq \phi$$

In accordance with Theorem A.4.1 one can select a subsequence $x_{k_i}(\cdot) \in X_2(t_{k_i},x_{k_i},v)$ $(i = 1, 2, ...)$ convergent to the function $x_*(\cdot) \in X_2(t_*,x_*,v)$. Moreover, this subsequence can be selected in a manner so that either for any $k = k_i$ $(i = 1, 2, ...)$ condition (25) will be fulfilled, or for each $k = k_i$ $(i = 1, 2, ...)$ condition (26) will be fulfilled. In the first case it follows from (25) and the definition of the set $\mathcal{W}^1_*$ that

(27) $$(t^*,x_*(t^*)) \in \mathcal{W}^1_*.$$

In the second case it follows from (26) and the closure of the set $\mathcal{M}$

that

(28) $\{(t,x_*(t)):\ t_* \leqslant t \leqslant t^*\} \cap \mathcal{M} \neq \phi.$

Thus the existence of a motion $x_*(\cdot) \in X_2(t_*,x_*,v)$ satisfying one of the two conditions (27) or (28) is verified, i.e. it is shown that the set W^1_* satisfies condition (2_u). Thus Lemma 2.2.4 is proved.

Let $W \subset T \times R^n$. Denote by $cl\, W$ the closure of the set W, i.e.

$$cl\ W \overset{\text{def}}{=} \Big\{(t_*,x_*) \in T \times R^n\!:\ \exists\ (t_k,x_k) \in W,\ \ (k = 1,2,\ \ldots), \tag{29}$$
$$(t_k,x_k) \to (t_*,x_*) \text{ for } k \to \infty\Big\}.$$

The following lemma is valid.

Lemma 2.2.5. *Let* W_* *be a* u-*stable bridge in an* $(\mathcal{M},\mathcal{N})$-*approach problem; then the set* $cl\ W_*$ *will also be a* u-*stable bridge in an* $(\mathcal{M},\mathcal{N})$-*approach problem. Let* K *and* $\mathfrak{L}$ *be closed sets in* $T \times R^n$ *and* W^* *be a* v-*stable bridge in a* $(K,\mathfrak{L})$-*approach problem; then the set* $cl\ W^*$ *will also be a* v-*stable bridge in a* $(K,\mathfrak{L})$-*approach problem.*

Proof: Consider the first assertion of the lemma. Since $\mathcal{M}$ and are closed, $cl\ W_*$ satisfies condition (1_u). We verify that the set $cl\ W_*$ satisfies condition (2_u). We utilize the arguments applied in Lemma 2.2.4 to establish the u-stability of the set W^1_*. Replacing $W^1_{\epsilon_k}$ in these arguments by W_*, $\mathcal{M}^{\epsilon_k}$ by $\mathcal{M}$, and W^1_* by $cl\ W_*$, we obtain that the set $cl\ W_*$ satisfies condition (2_u). The second assertion of the lemma is proved analogously.

In view of Lemma 2.2.1 the set W^1_0 defined in accordance with (12) is the *maximal u-stable bridge* in an $(\mathcal{M},\mathcal{N})$-approach problem, i.e. any set $W \in \mathbf{W}^1_0$ will be embedded in the set W^1_0. It follows from Lemma 2.2.5 that the set W^1_0 is closed.

Analogously it follows from Lemma 2.2.3 that the set W^1_ϵ defined by the equality (17) is the *maximal v-stable bridge* in a $(K_\epsilon,\mathfrak{L}_\epsilon)$-approach problem. Since the sets K_ϵ and $\mathfrak{L}_\epsilon$ are closed, it follows from Lemma 2.2.5 that so is the set W^2_ϵ.

We thus arrive at the following assertion:

Lemma 2.2.6. *The set* W^1_0 *defined by equality* (12) *is closed and is the maximal* u-*stable bridge in an* $(\mathcal{M},\mathcal{N})$-*approach problem. The set* W^2_ϵ *defined by the equality* (17) *is closed and is the maximal* v-*stable bridge in a* $(K_\epsilon,\mathfrak{L}_\epsilon)$-*approach problem.*

We are now ready to state and to prove the basic assertion of this section.

Theorem 2.2.1. *The equality*

$$(30) \qquad T \times R^n = \mathcal{W}_0^1 \cup (\bigcup_{\epsilon>0} \mathcal{W}_\epsilon^2)$$

is valid.

Proof. Let (t_*,x_*) be an arbitrary point in $T \times R^n$. It is required to show that either $(t_*,x_*) \in \mathcal{W}_0^1$ or there exists $\epsilon > 0$ such that $(t_*,x_*) \in \mathcal{W}_\epsilon^2$. Define the quantity

$$\epsilon_0 \overset{\text{def}}{=} \inf\{\epsilon > 0: \ (t_*,x_*) \in \mathcal{W}_\epsilon^1\}$$

where the set $\mathcal{W}_\epsilon^1$ is defined in accordance with (12). Two cases are possible, $\epsilon_0 = 0$ and $\epsilon_0 > 0$.

In the case $\epsilon_0 = 0$ it follows from the definition of the set $\mathcal{W}_*^1$ (23) that $(t_*,x_*) \in \mathcal{W}_*^1$. As shown in Lemma 2.2.4, the set $\mathcal{W}_*^1$ is a u-stable bridge in an $(\mathcal{M}, \mathcal{N})$-approach problem. Therefore $\mathcal{W}_*^1 \subset \mathcal{W}_0^1$ (actually $\mathcal{W}_*^1 = \mathcal{W}_0^1$). Thus, in the case when $\epsilon_0 = 0$ we have $(t_*,x_*) \in \mathcal{W}_0^1$.

Assume now that $\epsilon_0 > 0$, i.e. for any $\epsilon \in (0,\epsilon_0)$ we have $(t_*,x_*) \notin \mathcal{W}_\epsilon^1$. In accordance with (18) we obtain $(t_*,x_*) \in \mathcal{W}_\epsilon^2$ for all $\epsilon \in (0,\epsilon_0)$. The theorem is thus proved.

Note that in Section 2.4 equality (30) is complemented by the relation

$$\mathcal{W}_0^1 \cap (\bigcup_{\epsilon>0} \mathcal{W}_\epsilon^2) = \phi.$$

It is thus verified that the space of positions is subdivided into two disjoint sets, one of which is the maximal u-stable bridge in an $(\mathcal{M}, \mathcal{N})$-approach problem and the other is the union over $\epsilon > 0$ of the maximal v-stable bridges in a $(\mathcal{K}_\epsilon, \mathfrak{L}_\epsilon)$-approach problem.

2.3 Local Estimates

In the next section positional strategies are defined extremal to a given set, and Theorem 2.4.1 is proved. This theorem asserts that a strategy extremal to a stable bridge retains its constructive motions up to their encounter with the target set. In the proof of the theorem the estimate obtained in this section is used.

2.3.1. Consider first the estimate in the following particular case. Assume that the function f on the r.h.s. of the system (1.2.1) is of the form

$$f(t,x,u,v) = f(u,v),$$

i.e. the function f does not depend on the variables $t \in T$ and $x \in R^n$.

Let two initial positions be given: (t_*,x_*) and (t_*,w_*). Consider the motions $x(t)$ and $w(t)$ $(t_* \leqslant t \leqslant \theta)$ satisfying the equations

(1) $$\dot{x}(t) = f(u_0,v), \quad \dot{w}(t) = f(u,v_0)$$

and initial conditions $x(t_*) = x_*$ and $w(t_*) = w_*$. In equation (1) u and v are arbitrary points belonging to the sets $\mathcal{P}$ and $\mathcal{Q}$ respectively, and points u_0 and v_0 are chosen from the conditions

(2) $$\max_{v\in\mathcal{Q}} \langle x_* - w_*, f(u_0,v)\rangle = \min_{u\in\mathcal{P}} \max_{v\in\mathcal{Q}} \langle x_* - w_*, f(u,v)\rangle,$$
$$\min_{u\in\mathcal{P}} \langle x_* - w_*, f(u,v_0)\rangle = \max_{v\in\mathcal{Q}} \min_{u\in\mathcal{P}} \langle x_* - w_*, f(u,v)\rangle.$$

Recall that the notation $\langle g,h\rangle$ denotes the scalar product of vectors g and h.

We shall be concerned with the variation of the distance between the points $x(t)$ and $w(t)$. Introduce the following notation:

(3) $$s(t) = x(t) - w(t), \quad s_* = s(t_*) = x_* - w_* .$$

In accordance with the saddle point condition in a small game, the minimax and maximin coincide in relations (2), and we shall denote their common value by φ, i.e. we set

(4) $$\varphi \overset{\text{def}}{=} \min_{u\in\mathcal{P}} \max_{v\in\mathcal{Q}} \langle s_*, f(u,v)\rangle = \max_{v\in\mathcal{Q}} \min_{u\in\mathcal{P}} \langle s_*, f(u,v)\rangle.$$

We now calculate the derivative of the function $t \to \|s(t)\|^2 = \langle s(t),s(t)\rangle$. It follows from (1) and (3) that

(5) $$\frac{d\|s(t)\|^2}{dt} = 2\langle s(t),\ f(u_0,v) - f(u,v_0)\rangle.$$

We estimate the r.h.s. of this equality first for $t = t_*$. Since $s(t_*) = x_* - w_*$, we obtain from (2) and (4) that

$$\langle s_*, f(u_0,v)\rangle \leqslant \varphi, \quad \langle s_*, f(u,v_0)\rangle \geqslant \varphi,$$

where u and v are arbitrary points belonging to $\mathcal{P}$ and $\mathcal{Q}$, respectively. Consequently,

(6) $$\langle s_*, f(u_0,v) - f(u,v_0)\rangle \leqslant 0.$$

To arrive at an estimate for an arbitrary $t \in [t_*,\theta]$, we shall utilize the fact that for any $(u,v) \in \mathcal{P}\times\mathcal{Q}$ the inequality $\|f(u,v)\| \leqslant \Lambda$ where Λ is constant is valid. Thus

$$\|f(u_0,v) - f(u,v_0)\| \leqslant 2\Lambda, \quad \|s(t) - s_*\| \leqslant 2\Lambda(t - t_*).$$

Utilizing (6) we obtain

$$\langle s(t), f(u_0,v) - f(u,v_0)\rangle = \langle s_*, f(u_0,v) - f(u,v_0)\rangle$$
$$+ \langle s(t) - s_*, f(u_0,v) - f(u,v_0)\rangle \leqslant 4\Lambda^2(t - t_*).$$

Consequently,

$$\frac{d\|s(t)\|^2}{dt} \leqslant 8\Lambda^2(t - t_*). \tag{7}$$

Integrating this inequality on the interval $[t_*, t_* + \delta]$ we arrive at the bound

$$\|s(t_* + \delta)\|^2 \leqslant \|s_*\|^2 + 4\Lambda^2\delta^2 . \tag{8}$$

Thus, on the interval $[t_*, t_* + \delta]$ the square of the distance between points $x(t)$ and $w(t)$ cannot increase by more than a quantity of the second order of smallness relative to δ. Observe that in deriving these estimates the condition of a saddle point in a small game was used substantially.

2.3.2. Consider now the general case. The estimate to be obtained below will be utilized when considering step-by-step motions on small time invervals $[\tau_i, \tau_{i+1}]$. An estimate is required which does not depend on the location of the position $(\tau_i, x(\tau_i))$. To obtain such a uniform estimate we shall utilize the fact that step-by-step motions emanate from a fixed initial position (t_0, x_0). Therefore, graphs of such step-by-step motions remain in a bounded set $\mathfrak{D}_0$. The definition of the set $\mathfrak{D}_0$ can be made precise in the following manner.

Consider the set $X(t_0, x_0)$ of all possible absolutely continuous functions: $x(\cdot)$: $[t_0, \theta] \to R^n$ satisfying the differential inclusion

$$\dot{x}(t) \in co\{f(t, x(t), u, v):\ u \in \mathcal{P},\ v \in \mathcal{Q}\}$$

and the initial condition $x(t_0) = x_0$. Next we define

$$\mathfrak{D}_0 \overset{\text{def}}{=} \{(t, x(t)) \in T \times R^n:\ t_0 \leqslant t \leqslant \theta,\ x(\cdot) \in X(t_0, x_0)\}. \tag{9}$$

In view of Corollary A.4.1 the set $\mathfrak{D}_0$ is compact in $T \times R^n$.

Along with step-by-step motions, we shall consider concomitant motions in §2.4 which are rather closely related to them. Graphs of concomitant motions can be imbedded in the set

$$\mathfrak{D}_1 \overset{\text{def}}{=} \{(t, x + y):\ (t, x) \in \mathfrak{D}_0,\ \|y\| \leqslant 1\}. \tag{10}$$

Now after these preliminary explanations concerning the choice of the sets $\mathfrak{D}_0$ and $\mathfrak{D}_1$, we shall introduce the notation required for the statement and proof of Lemma 2.3.1.

From conditions (i), (ii) stated in §1.2.2 we obtain the following estimate:

$$(11) \qquad \|f(t_1,x_1,u,v) - f(t_2,x_2,u,v)\| \leqslant \lambda\|x_1 - x_2\| + \zeta(\delta).$$

Here λ is a constant, the function $\delta \to \zeta(\delta)$ satisfies the conditions $\zeta(\delta) > 0$ for $\delta > 0$ and $\lim\zeta(\delta) = 0$ as $\delta \to +0$. The estimate (11) is valid for any $(u,v) \in \mathcal{P} \times \mathcal{Q}$ and $(t_i,x_i) \in \mathcal{D}_1$ $(i = 1,2)$ such that $|t_1 - t_2| \leqslant \delta$ (i.e. the estimate (11) is uniform in the region $\mathcal{D}_1 \times \mathcal{P} \times \mathcal{Q}$). In order to obtain this estimate from conditions (i), (ii) (cf. 1.2.2) it suffices to note that

$$\|f(t_1,x_1,u,v) - f(t_2,x_2,u,v)\| \leqslant \|f(t_1,x_1,u,v) - f(t_1,x_2,u,v)\|$$
$$+ \|f(t_1,x_2,u,v) - f(t_2,x_2,u,v)\|.$$

Set

$$(12) \qquad \Lambda \overset{\text{def}}{=} \max\|f(t,x,u,v)\| \quad \text{for} \quad (t,x) \in \mathcal{D}_1, \quad (u,v) \in \mathcal{P} \times \mathcal{Q}.$$

Introduce the quantities

$$(13) \qquad \begin{aligned} \psi_1(t,x,s,u) &\overset{\text{def}}{=} \max_{v\in\mathcal{Q}} \langle s,f(t,x,u,v)\rangle, \\ \psi_2(t,x,s,v) &\overset{\text{def}}{=} \min_{u\in\mathcal{P}} \langle s,f(t,x,u,v)\rangle. \end{aligned}$$

It follows from (11) and (12) that for any $(t_i,x_i) \in \mathcal{D}_1$, $s_i \in R^n$ $(i = 1, 2)$, $(u,v) \in \mathcal{P} \times \mathcal{Q}$, $|t_1 - t_2| \leqslant \delta$ the following inequalities are valid:

$$(14) \qquad \begin{aligned} \psi_1(t_1,x_1,s_1,u) - \psi_1(t_2,x_2,s_2,u) &\leqslant \|s_1\|[\lambda\|x_1-x_2\| + \zeta(\delta)] + \Lambda\|s_1-s_2\|, \\ \psi_2(t_1,x_1,s_1,v) - \psi_2(t_2,x_2,s_2,v) &\geqslant - \|s_1\|[\lambda\|x_1-x_2\| + \zeta(\delta)] - \Lambda\|s_1-s_2\|. \end{aligned}$$

We shall verify the first estimate. Let a point $v_1 \in \mathcal{Q}$ be selected from the condition

$$\langle s_1,f(t_1,x_1,u,v_1)\rangle = \psi_1(t_1,x_1,s_1,u).$$

Then in accordance with (13) we have

$$\psi_1(t_1,x_1,s_1,u) - \psi_1(t_2,x_2,s_2,u) \leqslant \langle s_1,f(t_1,x_1,u,v_1)\rangle - \langle s_2,f(t_2,x_2,u,v_1)\rangle$$
$$= \langle s_1,f(t_1,x_1,u,v_1) - f(t_2,x_2,u,v_1)\rangle + \langle s_1 - s_2,f(t_2,x_2,u,v_1)\rangle.$$

Utilizing the estimates (11) and (12) we obtain the first inequality in (14). The second inequality in (14) is verified analogously.

Define

$$(15) \qquad \alpha(\delta) \overset{\text{def}}{=} 2[2\Lambda\delta + 1][\zeta(\delta) + \lambda\Lambda\delta\,] + 2\Lambda^2\delta,$$

$$(16) \qquad \beta(t,\delta) \overset{\text{def}}{=} [\exp 2\lambda(t - t_0) - 1]\,\alpha(\delta)/\lambda\,.$$

Note that $\alpha(\delta) \to 0$ as $\delta \to 0$. Therefore there exists $\delta_0 > 0$ such that the estimate

$$(17) \qquad \beta(t,\delta) < 1 \quad \text{for all } \delta \in (0,\delta_0] \text{ and for all } t \in [t_0,\theta]$$

is valid. It is useful to observe that the function $\beta(\cdot,\delta)$ satisfies the differential equation

$$(18) \qquad \dot{\beta}(t,\delta) = 2\lambda\beta(t,\delta) + 2\alpha(\delta)$$

and the initial condition $\beta(t_0,\delta) = 0$.

Let $t_* \in T$, $x_* \in R^n$, and $w_* \in R^n$. Let the points $u_0 \in P$ and $v_0 \in Q$ be chosen from the conditions

$$(19) \qquad \max_{v \in Q} \langle x_* - w_*, f(t_*,x_*,u_0,v)] = \min_{u \in P} \max_{v \in Q} \langle x_* - w_*, f(t_*,x_*,u,v)\rangle,$$

$$(20) \qquad \min_{u \in P} \langle x_* - w_*, f(t_*,x_*,u,v_0)\rangle = \max_{v \in Q} \min_{u \in P} \langle x_* - w_*, f(t_*,x_*,u,v)\rangle.$$

The following lemma is valid (*a local estimate*).

Lemma 2.3.1. *Let the set D_0 and the quantities $\beta(t,\delta),\delta_0$ be defined as indicated above. Assume that*

$$(t_*,x_*) \in D_0, \quad \|x_* - w_*\|^2 \leqslant \beta(t_*,\delta), \quad \delta \in (0,\delta_0).$$

Let $x(\cdot)$: $[t_*,\theta] \to R^n$ and $w(\cdot)$: $[t_*,\theta] \to R^n$ be solutions of differential inclusions (cf. (2.2.1))

$$(21) \qquad \dot{x}(t) \in F_1(t,x(t),u_0) \overset{\text{def}}{=} co\{f(t,x(t),u_0,v): \quad v \in Q\},$$

$$(22) \qquad \dot{w}(t) \in F_2(t,w(t),v_0) \overset{\text{def}}{=} co\{f(t,w(t),u,v_0): \quad u \in P\},$$

satisfying the initial conditions $x(t_*) = x_*$ and $w(t_*) = w_*$. Then the estimate

$$(23) \qquad \|x(t) - w(t)\|^2 \leqslant \beta(t,\delta) \quad \textit{for} \quad t_* \leqslant t \leqslant t_* + \delta$$

is valid.

Proof. Assume the contrary. Let time $\tau \in [t_*,t_*+\delta]$ exist such that

$$(24) \qquad \|x(\tau) - w(\tau)\|^2 > \beta(\tau,\delta).$$

In accordance with (17) for $\delta \leqslant \delta_0$ the inequality $\beta(t,\delta) < 1$ is fulfilled. Therefore it can be assumed that along with (24) the inequality

$$(25) \qquad \|x(t) - w(t)\| < 1 \quad \text{for all} \quad t \in [t_*, \tau]$$

is also valid. It follows from the definition of the set $\mathcal{D}_0$ and the inclusion $(t_*, x_*) \in \mathcal{D}_0$ that $(t, x(t)) \in \mathcal{D}_0$ for $t \in [t_*, \theta]$. From (25) and the definition of the set $\mathcal{D}_1$ we obtain that $(t, w(t)) \in \mathcal{D}_1$ for $t \in [t_*, \tau]$. Hence one can utilize estimates (14) below.

Set $s(t) = x(t) - w(t)$, $\dot{x}(t) = h_*(t)$, $\dot{w}(t) = h^*(t)$. Using this notation we have

$$(26) \qquad \frac{d\|s(t)\|^2}{dt} = 2\langle s(t),\ h_*(t) - h^*(t)\rangle.$$

For almost all $t \in [t_*, \theta]$ the inclusions

$$h_*(t) \in \mathcal{F}_1(t, x(t), u_0) \overset{\text{def}}{=} co\{f(t, x(t), u_0, v):\ v \in \mathcal{Q}\},$$

$$h^*(t) \in \mathcal{F}_2(t, w(t), v_0) \overset{\text{def}}{=} co\{f(t, w(t), u, v_0):\ u \in \mathcal{P}\}$$

are valid. Note that for any vector s and any compact set $\mathcal{H}$ the equalities

$$\max_{h \in \mathcal{H}} \langle s, h\rangle = \max_{h \in co\mathcal{H}} \langle s, h\rangle, \qquad \min_{h \in \mathcal{H}} \langle s, h\rangle = \min_{h \in co\mathcal{H}} \langle s, h\rangle$$

hold. Therefore in accordance with notation (13) we obtain from the inclusions above, the inequalities

$$(27) \qquad \begin{aligned} &\langle s(t), h_*(t)\rangle \leqslant \psi_1(t, x(t), s(t), u_0), \\ &\langle s(t), h^*(t)\rangle \geqslant \psi_2(t, w(t), s(t), v_0). \end{aligned}$$

Since $(t, x(t))$ and $(t, w(t)) \in \mathcal{D}_1$ for $t \in [t_*, \tau]$, in accordance with (12) and (25) we have

$$\|s(t)\| \leqslant \|s(t_*)\| + 2\Lambda(t - t_*) \leqslant 2\Lambda\delta + 1,$$

$$\|x(t) - x_*\| \leqslant \Lambda\delta, \quad \|s(t) - s(t_*)\| \leqslant 2\Lambda\delta \ (t_* \leqslant t \leqslant \tau).$$

Therefore, utilizing the first inequality (14) and notation (15) we obtain

$$(28) \qquad \begin{aligned} \psi_1(t, x(t), s(t), u_0) &\leqslant \psi_1(t_*, x_*, s(t_*), u_0) + \|s(t)\|[\lambda\|x(t) - x_*\| \\ &+ \zeta(\delta)] + \Lambda\|s(t) - s(t_*)\| \leqslant \psi_1(t_*, x_*, s(t_*), u_0) + \alpha(\delta)/2. \end{aligned}$$

Analogously it follows from (14) and (15) that

$$(29) \qquad \begin{aligned} \psi_2(t, w(t), s(t), v_0) &\geqslant \psi_2(t, x(t), s(t), v_0) - \lambda\|s(t)\|^2 \geqslant \psi_2(t_*, x_*, s(t_*), v_0) \\ &- \lambda\|s(t)\|^2 - \alpha(\delta)/2. \end{aligned}$$

Since $s(t_*) = x_* - w_*$, in view of (14), (19), and (20) we obtain

$$\text{(30)}\qquad \begin{aligned} \psi_1(t_*,x_*,s(t_*),u_0) &= \min_{u\in P}\max_{v\in Q} \langle x_* - w_*, f(t_*,x_*,u,v)\rangle \overset{\text{def}}{=} \varphi_1 \\ \psi_2(t_*,x_*,s(t_*),v_0) &= \max_{v\in Q}\min_{u\in P} \langle x_* - w_*, f(t_*,x_*,u,v)\rangle \overset{\text{def}}{=} \varphi_2 . \end{aligned}$$

The condition of a saddle point in a small game (cf. (1.2.8)) implies that $\varphi_1 = \varphi_2 = \varphi$. From relations (27) – (30) we have

$$\langle s(t),h_*(t)\rangle \leqslant \varphi + \alpha(\delta)/2, \quad \langle s(t),h^*(t)\rangle \geqslant \varphi - \alpha(\delta)/2 - \lambda \| s(t)\|^2$$

for almost all $t \in [t_*,\tau]$. Therefore, returning to the equality (26) we arrive at the differential inequality

$$\text{(31)}\qquad \frac{d\| s(t)\|^2}{dt} \leqslant 2\lambda \| s(t)\|^2 + 2\alpha(\delta).$$

Integrating this inequality we have

$$\| s(\tau)\|^2 \leqslant \| s(t_*)\|^2 \exp 2\lambda(\tau - t_*) + [\exp 2\lambda(\tau - t_*) - 1]\alpha(\delta)/\lambda.$$

From the lemma's condition, $\| s(t_*)\|^2 \leqslant \beta(t_*,\delta)$. Therefore, using notation (16) the preceding inequality can be written as

$$\begin{aligned} \| s(t)\|^2 &\leqslant \beta(t_*,\delta)\exp 2\lambda(\tau - t_*) + [\exp 2\lambda(\tau - t_*) - 1]\alpha(\delta)/\lambda \\ &= [\exp 2\lambda(t_* - t_0) - 1](\alpha(\delta)/\lambda)\exp 2\lambda(\tau - t_*) \\ &\quad + [\exp 2\lambda(\tau - t_*) - 1]\alpha(\delta)/\lambda = \beta(\tau,\delta). \end{aligned}$$

We thus arrive at the estimate $\| s(\tau)\|^2 = \| x(\tau) - w(\tau)\|^2 \leqslant \beta(\tau,\delta)$, which contradicts the assumption (24). The contradiction obtained proves the lemma.

We note once again that in the proof of Lemma 2.3.1, condition (1.2.8) was used. This is the only place in the proof of Theorem 2.1.1 where the condition of a saddle point in a small game is used.

2.4. Extremal Strategies

In this section the proof of the theorem on an alternative for the differential game of approach-evasion will be completed. Below, positional strategies extremal to stable bridges are defined. The basic results are contained in Theorems 2.4.1 and 2.4.2, which establish that extremal strategies retain motions on stable bridges.

2.4.1. Define a positional strategy U^e extremal to the set $\mathcal{W}$. Let $\mathcal{W}$ be a closed set in the space of positions $T \times R^n$ and let $t \in T$. Consider the set

(1) $$\mathcal{W}[t] \overset{\text{def}}{=} \{w \in R^n : (t,w) \in \mathcal{W}\}.$$

Recall that in §2.2 sets $\mathcal{W}(t)$ were considered which are defined as the intersection of the set $\mathcal{W}$ with the hyperplane $\{t\} \times R^n$ in the space R^{n+1}. Clearly the sets $\mathcal{W}(t)$ and $\mathcal{W}[t]$ are connected by the relation $\mathcal{W}(t) = \{t\} \times \mathcal{W}[t]$. Since the set $\mathcal{W}$ is closed in $T \times R^n$ the set $\mathcal{W}[t]$ is closed in R^n.

Let $(t,x) \in T \times R^n$. Define

(2) $$\mathcal{W}[t,x] \overset{\text{def}}{=} \begin{cases} \{w^0 \in \mathcal{W}[t]: \|x - w^0\| = \text{dist}(x, \mathcal{W}[t])\}, & \text{if } \mathcal{W}[t] \neq \phi, \\ \{x\}, & \text{if } \mathcal{W}[t] = \phi. \end{cases}$$

Here

(3) $$\text{dist}(x, \mathcal{W}[t]) \overset{\text{def}}{=} \min \|x - w\| \quad \text{for } w \in \mathcal{W}[t].$$

Thus if $\mathcal{W}[t]$ is non-empty then $\mathcal{W}[t,x]$ is the totality of points of the set $\mathcal{W}[t]$ closest to the point x. Since the set $\mathcal{W}[t]$ is closed, it follows that the minimum in (3) is attained, therefore the set $\mathcal{W}[t,x]$ is non-empty. If, however, $\mathcal{W}[t]$ is empty we formally set $\mathcal{W}[t,x] = \{x\}$.

In certain particular cases, for any $x \in R^n$, the point $w^0 \in \mathcal{W}[t]$ closest to the point x will be unique. This is, for example, the case when the set $\mathcal{W}[t]$ is convex. In the general case the mapping $(t,x) \to \mathcal{W}[t,x]$ will be multivalued. Select a single valued function e: $T \times R^n \to R^n$ satisfying the condition

(4) $$e(t,x) \in \mathcal{W}[t,x] \quad \text{for all} \quad (t,x) \in T \times R^n.$$

There are infinitely many such functions. For further constructions any functions e satisfying condition (4) will be suitable. Note that function e may be discontinuous.

Finally, define the function U^e: $T \times R^n \to \mathcal{P}$. We shall assume that for any position $(t,x) \in T \times R^n$ the vector $U^e(t,x) \in \mathcal{P}$ is chosen from the condition

(5) $$\max_{v \in \mathcal{Q}} \langle x - e(t,x), f(t,x,U^e(t,x)v) \rangle = \min_{u \in \mathcal{P}} \max_{v \in \mathcal{Q}} \langle x - e(t,x), f(t,x,u,v) \rangle.$$

Here again an element $U^e(t,x)$ at which the minimum is attained, is not unique. In what follows we shall consider an arbitrary function U^e satisfying condition (5). Note also that the function U^e may turn out to be discontinuous.

Definition. *A function U^e: $T \times R^n \to \mathcal{P}$ defined by the relations* (1) – (5) *is called a strategy extremal to the set $\mathcal{W}$.*

In this definition we do not specify that U^e is the strategy of the first player since the use of the letter U clearly indicates that it belongs to the first player.

We now discuss the definition of an extremal strategy. Condition (5) may be interpreted as follows. When choosing the point $U^e(t,x)$ the first player solves a minimax problem in the so-called small game, where the strategies of the players are the choice of points $u \in \mathcal{P}$ and $v \in \mathcal{Q}$, and the payoff is the scalar product of the vectors $s = x - e(t,x)$ and $f(t,x,u,v)$. The step-by-step procedure in which the control of the first player is formed by the rule $u(t) = U^e(\tau_i,x(\tau_i))$, $\tau_i \leqslant t < \tau_{i+1}$, $i = 0, 1, ..., k$, will consist of a sequential solution of the small minimax games indicated above. An analogous remark is valid also for extremal strategies for the second player. It can therefore be said that the differential game is composed of a sequence of small games.

In the case $\mathcal{W}[t] = \phi$ as well as in the case $x \in \mathcal{W}[t]$ the equality $x = e(t,x)$ is valid, since condition (5) degenerates, and within the bounds of the set $\mathcal{P}$ the point $U^e(t,x)$ can be chosen arbitrarily. If, however, $\mathcal{W}[t] \neq \phi$ and $x \notin \mathcal{W}[t]$ then the vector $x - e(t,x)$ will be a non-zero vector and the definition of the point $U^e(t,x)$ can be illustrated in the following manner. Let the controlled system be at the point x at time t. Assume that the first player chooses the constant control u on a small interval of time $[t, t+\delta]$ and the second player may choose any measurable control. Then for the phase point $x(t + \delta)$ the equality

$$x(t + \delta) = x + \delta g + h$$

is valid where $g \in co\{f(t,x,u,v): \ v \in \mathcal{Q}\}$, $\|h\| \leqslant o(\delta)$ and $o(\delta)$ is a quantity of a higher order of smallness relative to δ. The third summand in this equality may be neglected and one can consider the set $G_\delta(u) \overset{def}{=} co\{x + \delta f(t,x,u,v): \ v \in \mathcal{Q}\}$ which, in a linear approximation with respect to δ, approximates the totality of all points $x(t + \delta)$ of the type indicated above. In Fig. 2.4.1 a point x and the point closest to it, $w^0 = e(t,x) \in \mathcal{W}[t]$, is presented. Here also the set $G_\delta(u)$ is represented as well as the supporting hyperplane $\mathcal{H}_\delta(u) \overset{def}{=} \{y \in R^n: \ \langle x - w^0,y\rangle = \max\langle x - w^0,y\rangle\}$ for $y \in G_\delta(u)$. Varying the control $u \in \mathcal{P}$, one can change the position of this hyperplane. Condition (5) means that when the control $u^e = U^e(t,x)$ is chosen, the hyperplane will be maximally shifted in the direction towards the point w^0. One can thus assert that the extremal control u^e directs the motion of the system towards the closest point of the set $\mathcal{W}[t]$.

2.4.2. We now prove the following assertion:

Theorem 2.4.1. *Let $\mathcal{W}$ be a closed u-stable bridge in an $(\mathcal{M}, \mathcal{N})$-approach problem. Assume that $(t_0,x_0) \in \mathcal{W}$. Then the strategy U^e extremal to the set $\mathcal{W}$ is a solution for the*

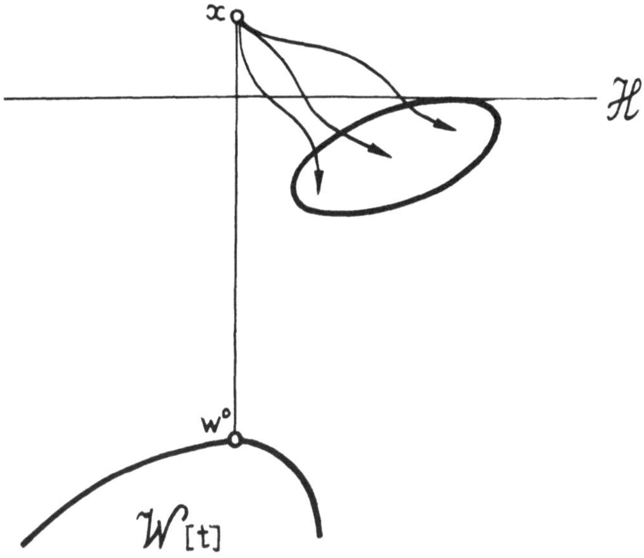

Fig. 2.4.1

(M, N)-approach problem.

Proof. Let a partition Δ be chosen. Assume that

$$(6) \qquad \delta = \mathrm{diam}\Delta \leqslant \delta_0$$

where the number $\delta_0 > 0$ is chosen in such a manner that the inequality (2.3.17) is fulfilled. Below, it will be shown that for any step-by-step motion $x(\cdot) \in X_1(t_0, x_0, U^e, \Delta)$ the condition of

$(M^{\bar{\epsilon}}, N^{\bar{\epsilon}})$-approach is satisfied. Here

$$M^{\bar{\epsilon}} \overset{\mathrm{def}}{=} \{(t,x): \ (t - t_*) + \|x - x_*\| \leqslant \epsilon, \quad (t_*, x_*) \in M\},$$

$$(7) \qquad N^{\bar{\epsilon}} \overset{\mathrm{def}}{=} \{(t,x): |t - t_*| + \|x - x_*\| \leqslant \epsilon, \ (t_*, x_*) \in N\},$$

$$\epsilon = \epsilon(\delta) \overset{\mathrm{def}}{=} \sqrt{\beta(\theta,\delta)} + (\Lambda + 1)\delta,$$

where the quantities β and Λ are defined in accordance with (2.3.16) and (2.3.12). Relation (2.3.15) and definition (7) imply that $\epsilon(\delta) \to 0$ as $\delta \to 0$.

Utilizing this assertion, it is easy to verify that any constructive motion $x(\cdot) \in X_1(t_0, x_0, U^e)$ satisfies the condition of (M, N)-approach. Indeed, let $x_\ell(\cdot) \in X_1(t_0, x_0, U^e, \Delta_\ell)$ ($\ell = 1, 2, \ldots$) be a sequence of step-by-step motions converging to some constructive motion $x_*(\cdot) \in X_1(t_0,x_0,U^e)$. It is assumed that $\delta_\ell = \operatorname{diam}\Delta_\ell \to 0$ as $\ell \to \infty$. Let $\epsilon_\ell = \epsilon(\delta_\ell)$. The motion $x_\ell(\cdot)$ satisfies the condition of $(\overline{M^{\epsilon_\ell}}, \overline{N^{\epsilon_\ell}})$-approach, i.e. there exists a number $\tau_\ell \in [t_0,\theta]$ such that

(8) $$(\tau_\ell,x_\ell(\tau_\ell)) \in \overline{M^{\epsilon_\ell}}, \quad (t,x_\ell(t)) \in \overline{N^{\epsilon_\ell}} \text{ for } t_0 \leqslant t \leqslant \tau_\ell .$$

It can be assumed that $(\tau_\ell,x_\ell(\tau_\ell)) \to (\tau_*,x_*(\tau_*))$ as $\ell \to \infty$. In relations (8) one can approach the limit as $\ell \to \infty$ and utilize the fact that $\epsilon_\ell \to 0$ as $\ell \to \infty$. Then the closures of the sets M and N yield that

$$(\tau_*,x_*(\tau_*)) \in M, \quad (t,x_*(t)) \in N \text{ for } t_0 \leqslant t \leqslant \tau_*.$$

We have thus shown that the constructive motion $x_*(\cdot)$ satisfies the (M, N)-approach condition.

Thus the proof of the theorem reduces to verification of the $(\overline{M^\epsilon}, \overline{N^\epsilon})$-approach condition (see (7)) for step-by-step motions $x(\cdot) \in X_1(t_0,x_0,U^e,\Delta)$.

Let

$$\Delta \overset{\text{def}}{=} \{t_0 = \tau_0 < \tau_1 < \dots < \tau_{k+1} = \theta\} = \{\tau_i\colon \ i \in I\}.$$

Here $I = \{0, 1, \dots, k + 1\}$ is a set of indices used to enumerate the times τ_i. Let W be a u-stable bridge which appears in the conditions of the theorem. For each $i \in I$ we define the set $W[\tau_i] \overset{\text{def}}{=} \{w \in R^n\colon (\tau_i,w) \in W\}$. If $W[\tau_i] = \phi$ for some $i \in I$, one can select from the set I a subset $I_* = \{0, 1, \dots, j\}$ such that

(9) $$W[\tau_i] \neq \phi \text{ for } 0 \leqslant i \leqslant j, \quad W[\tau_{j+1}] = \phi.$$

If, however, $W[\tau_i] \neq \phi$ for all $i \in I$, we then set $I_* = I$. Observe that in view of the conditions of the theorem, $x_0 \in W[\tau_0]$. Therefore the set I_* is non-empty.

Let $(t,x) \to e(t,x)\colon\ T \times R^n \to R^n$ be the function utilized in the construction of the strategy U^e under consideration (cf. (4), (5)). We select a function $V_0\colon\ T \times R^n \to Q$ satisfying the condition

$$\min_{u\in P} \langle x - e(t,x), f(t,x,u,V_0(t,x))\rangle = \max_{v\in Q} \min_{u\in P} \langle x - e(t,x), f(t,x,u,v)\rangle.$$

Consider an arbitrary step-by-step motion $x(\cdot) \in X_1(t_0,x_0,U^e,\Delta)$ Define points $w_{(i)} = e(\tau_i,x(\tau_i))$, $v_{(i)} = V_0(\tau_i,x(\tau_i))$ $(i \in I)$. In accordance with the preceding constructions we have

$$\min_{u\in P} \langle x(\tau_i) - w_{(i)}, f(\tau_i,x(\tau_i),u,v_{(i)})\rangle \tag{10}$$
$$= \max_{v\in Q} \min_{u\in P} \langle x(\tau_i) - w_{(i)}, f(\tau_i,x(\tau_i),u,v)\rangle.$$

Assume that $i \in I_*$. Then $W[\tau_i] \neq \phi$. Therefore it follows from the definition of function e (cf. (3), (4)) that $e(\tau_i,x(\tau_i)) = w_{(i)} \in W[\tau_i]$ or, equivalently, $(\tau_i,w_{(i)}) \in W$. The set W satisfies the condition of u-stability (cf. §2.2.2). We set in this condition $t_* = \tau_i$, $x_* = w_{(i)}$, $t^* = \tau_{i+1}$, $v = v_{(i)}$. In accordance with the condition of u-stability there exists a solution $w^{(i)}(\cdot)$: $[\tau_i,\tau_{i+1}] \to R^n$ of the differential inclusion

$$\dot{w}^{(i)}(t) \in F_2(t,w^{(i)}(t),v_{(i)}), \quad w^{(i)}(\tau_i) = w_{(i)}, \tag{11}$$

which satisfies one of the relations

$$w^{(i)}(\tau_{i+1}) \in W[\tau_{i+1}], \tag{12}$$

$$(\tau,w^{(i)}(\tau)) \in M \quad \text{for some} \quad \tau \in [\tau_i,\tau_{i+1}]. \tag{13}$$

Denote by I_{**} the set of indices $i \in I_*$ for which the relation (13) is valid. We show that $I_{**} \neq \phi$. Two cases are possible: $j < k + 1$ and $j = k + 1$ (cf. (9)). In the first case we have $W[\tau_{j+1}] = \phi$. Therefore for $i = j$ the condition (12) is not fulfilled. Consequently, for $i = j$ condition (13) is valid, i.e. in this case $j \in I_{**}$. Now consider the second case. We shall show that $k \in I_{**}$. Recall that $\tau_{k+1} = \theta$ and $W(\theta) = \{(\theta,w)\colon (\theta,w) \in W\} \subset M$ (cf. §2.2.2). Therefore for $i = k$, condition (12) implies condition (13). Thus $k \in I_{**}$.

Denote by m the minimal element of the set I_{**}. It follows from the definition of index m that

$$w^{(i)}(\tau_{i+1}) \in W[\tau_{i+1}], \quad i = 0, 1, \dots m - 1, \tag{14}$$

$$(\tau,w^{(m)}(\tau) \in M \quad \text{for some} \quad \tau \in [\tau_m,\tau_{m+1}]. \tag{15}$$

Note that for $m = 0$ (this case is not excluded) these relations are reduced to a single condition $(\tau,w^{(0)}(\tau)) \in M$ for some $\tau \in [\tau_0,\tau_1]$.

We now proceed to the principal step of the proof. Consider the case $m > 0$. We shall show that for the step-by-step motion under consideration $x(\cdot) \in X_1(t_0,x_0,U^e,\Delta)$, the estimate

$$\text{dist}^2(x(\tau_i), W[\tau_i]) \leqslant \beta(\tau_i,\delta) \quad (i = 0, 1, \dots, m) \tag{16}$$

is valid. The proof is carried out by induction. For $i = 0$ the condition $(t_0,x_0) \in W$ implies that $x_0 = x(\tau_0) \in W[\tau_0]$. Therefore, taking (2.3.16) into account we obtain $\text{dist}^2(x[t_0], W[\tau_0]) = 0 = \beta(\tau_0,\delta)$. Thus for $i = 0$ the estimate (16) is valid. Assume now that the estimate (16) is satisfied for some $i \leqslant m - 1$. Consider the motions

$x(\cdot)$ and $w^{(i)}(\cdot)$. We shall show that on the interval $[\tau_i, \tau_{i+1}]$ for these motions, all the conditions of Lemma 2.3.1 are satisfied. By the definition of the set $\mathcal{D}_0$ (9) we have $(\tau_i, x(\tau_i)) \in \mathcal{D}_0$. From the definition of the function e (4) and the point $w_{(i)} = e(\tau_i, x(\tau_i))$ we obtain from the inequality (16)

$$\operatorname{dist}^2(x(\tau_i), \ \mathcal{W}[\tau_i]) = \|x(\tau_i) - w_{(i)}\|^2 \leqslant \beta(\tau_i, \delta).$$

Assumption (6) implies that $\tau_{i+1} - \tau_i \leqslant \delta \leqslant \delta_0$. Finally we note that the motion $w^{(i)}(\cdot)$ satisfies the differential inclusion (11) where the control $v_{(i)}$ is chosen from the condition (10) and the motion $x(\cdot)$ satisfies the differential inclusion

$$\dot{x}(t) \in \{f(t,x(t),u^e,v): \ v \in \mathcal{Q}\} \subset \mathcal{F}_1(t,x(t),u^e)$$

on the interval $[\tau_i, \tau_{i+1}]$ where the fixed control $u^e = U^e(\tau_i, x(\tau_i))$ is chosen in accordance with (5); therefore

$$\max_{v \in \mathcal{Q}} \langle x(\tau_i) - w_{(i)}, \ f(\tau_i, x(\tau_i), u^e, v) \rangle$$
$$= \min_{u \in \mathcal{P}} \max_{v \in \mathcal{Q}} \langle x(\tau_i) - w_{(i)}, \ f(\tau_i, x(\tau_i), u, v) \rangle.$$

Thus all of the conditions of Lemma 2.3.1 are fulfilled and we obtain that $\|x(\tau_{i+1}) - w^{(i)}(\tau_{i+1})\|^2 \leqslant \beta(\tau_{i+1}, \delta)$. Next it follows from (14):

$$\operatorname{dist}^2(x(\tau_{i+1}), \ \mathcal{W}[\tau_{i+1}]) \leqslant \|x(\tau_{i+1}) - w^{(i)}(\tau_{i+1})\|^2 \leqslant \beta(\tau_{i+1}, \delta),$$

q.e.d.

Utilizing (16) we shall show that

$$(17) \qquad (\tau_m, x(\tau_m)) \in \mathcal{M}^{\bar{\epsilon}}, \ (t, x(t)) \in \mathcal{N}^{\bar{\epsilon}} \ \text{ for } t_0 < t \leqslant \tau_m,$$

where the quantity ϵ is defined in (7).

In view of the choice of index m there exists a number $\tau \in [\tau_m, \tau_{m+1}]$ such that $(\tau, w^{(m)}(\tau)) \in \mathcal{M}$. It follows from the preceding discussion and the definition of the set $\mathcal{D}_1$ (2.3.10) that $(t, w^{(m)}(t)) \in \mathcal{D}_1$ for $\tau_m \leqslant t \leqslant \tau_{m+1}$. Therefore in accordance with (2.3.12), the estimate

$$\|w^{(m)}(\tau) - w^{(m)}(\tau_m)\| \leqslant \Lambda(\tau - \tau_m) \leqslant \Lambda\delta$$

is valid. Consequently,

$$|\tau_m - \tau| + \|w^{(m)}(\tau_m) - w^{(m)}(\tau)\| \leqslant \delta + \Lambda\delta.$$

Since $(\tau, w^{(m)}(\tau)) \in \mathcal{M}$ we obtain, in view of (2.1.13) that

$$(18) \qquad (\tau_m, w_m(\tau_m)) \in \mathcal{M}^{\bar{\epsilon}_1}, \ \bar{\epsilon}_1 = \Lambda\delta + \delta.$$

From (16) we have for $i = m$

$$\text{(19)} \qquad \operatorname{dist}(x(\tau_m), \mathcal{W}[\tau_m]) = \|x(\tau_m) - w^{(m)}(\tau_m)\| \leqslant \sqrt{\beta(\theta,\delta)} = \epsilon_2.$$

Here the inequality $\beta(\tau_m,\delta) \leqslant \beta(\theta,\delta)$ is used. The first inclusion in (17) thus follows from (18) and (19).

We now verify the second inclusion in (17). From (16) and the inclusion $\mathcal{W} \subset N$ (cf. condition (1_u) in §2.2.1), we obtain

$$\text{(20)} \qquad (\tau_i, x(\tau_i)) \in N^{\bar{\epsilon}_2}, \text{ where } \epsilon_2 = \sqrt{\beta(\theta,\delta)}, \quad i = 0, 1, \ldots, m.$$

Moreover for any $t \in [\tau_i, \tau_{i+1}]$ the inequalities

$$\text{(21)} \qquad (t - \tau_i) \leqslant \delta, \ \|x(t) - x(\tau_i)\| \leqslant \Lambda\delta$$

are valid. Relations (20) and (21) yield the second inclusion in (17). Thus relations (17) are proved. The meaning of (17) is that the step-by-step motion $x(\cdot)$ satisfies the $(M^{\bar{\epsilon}}, N^{\bar{\epsilon}})$-approach condition where ϵ is as defined in (7).

It remains to note that this conclusion is valid not only in the case $m > 0$ but also for $m = 0$. Indeed, it follows from the conditions of the theorem that $(t_0, x_0) \in \mathcal{W} \subset N$, i.e. for m = 0 relations (19) and (20) are fulfilled. Therefore, inclusions (17) follow in view of the arguments above.

Theorem 2.4.1 is thus proved.

2.4.3. We define a strategy for the second player extremal to the closed set $\mathcal{W} \subset T \times R^n$. This definition, in essence, does not differ from the definition of an extremal strategy for the first player.

Thus, let $\mathcal{W}$ be a closed set on the space of positions $T \times R^n$. Assume that the sets $\mathcal{W}[t]$, $\mathcal{W}[t,x]$, and function e: $T \times R^n \to R^n$ are defined in accordance with (1), (2), and (4). Consider the function V^e: $T \times R^n \to \mathcal{Q}$ such that for all $(t,x) \in T \times R^n$ a point $V^e(t,x) \in \mathcal{Q}$ satisfies the condition

$$\text{(22)} \qquad \begin{aligned} &\min_{u \in \mathcal{P}} \langle e(t,x) - x, f(t,x,u,V^e(t,x))\rangle \\ &\qquad = \max_{v \in \mathcal{Q}} \min_{u \in \mathcal{P}} \langle e(t,x) - x, f(t,x,u,v)\rangle. \end{aligned}$$

Definition. *A function V^e: $T \times R^n \to \mathcal{Q}$ defined by relations* (4) *and* (22) *is called a strategy extremal to the set* $\mathcal{W}$.

Thus by choosing the point $V^e(t,x)$ the second player solves the maximin problem in a small game in which the payoff is the scalar product $\langle e(t,x) - x, f(t,x,u,v)\rangle$. Observe that in condition (5) the vector $x - e(t,x)$ appears, while here we have the opposite vector

$e(t,x) - x$. This inessential change is made to assure that in a small as well as in a "large" differential game, the second player solves the maximin problem. There is no need to dwell upon the meaning of this definition since it is the same as that of an extremal strategy for the first player.

The following theorem is valid.

Theorem 2.4.2. *Let* $\epsilon > 0$, $\mathcal{K}_\epsilon$ *and* $\mathcal{L}_\epsilon$ *be closed sets defined by the relations* (2.1.9). *Let* $\mathcal{W}$ *be a* v*-stable bridge in a* $(\mathcal{K}_\epsilon,\mathcal{L}_\epsilon)$*-approach problem. Assume that* $(t_0,x_0) \in \mathcal{W}$. *Then the strategy* V^ϵ *extremal to the set* $\mathcal{W}$ *solves the* $(\mathcal{M},\mathcal{N})^\epsilon$*-evasion problem.*

Recall that the condition of $(\mathcal{M},\mathcal{N})^\epsilon$-evasion is equivalent to the condition of $(\mathcal{K}_\epsilon,\mathcal{L}_\epsilon)$-approach. Therefore the assertion of Theorem 2.4.2 is that for any constructive motion $x(\cdot) \in X_2(t_0,x_0,V^e)$ the condition of $(\mathcal{K}_\epsilon,\mathcal{L}_\epsilon)$-approach is fulfilled. The proof of this fact coincides with the proof of Theorem 2.4.1 (except for difference in notation).

2.4.3. Theorem 2.1.1 (*the theorem on an alternative*) follows from Theorems 2.2.1, 2.4.1, and 2.4.2. Indeed, let (t_0,x_0) be an arbitrary point in the space of positions. In accordance with Theorem 2.2.1 two cases are possible: either $(t_0,x_0) \in \mathcal{W}_0^1$ or there exists $\epsilon > 0$ such that $(t_0,x_0) \in \mathcal{W}_\epsilon^2$. Recall that $\mathcal{W}_0^1$ is a closed u-stable bridge in an $(\mathcal{M},\mathcal{N})$-approach problem. Therefore, in the first case it follows from Theorem 2.4.1 that the strategy U^e extremal to the set $\mathcal{W}_0^1$ will solve the problem of an $(\mathcal{M},\mathcal{N})$-approach. In the second case $\mathcal{W}_\epsilon^2$ is a closed v-stable bridge in a $(\mathcal{K}_\epsilon,\mathcal{L}_\epsilon)$-approach problem. Therefore in this case, in view of Theorem 2.4.2, the strategy V^e extremal to the set $\mathcal{W}_\epsilon^2$ solves the $(\mathcal{M},\mathcal{N})^\epsilon$-evasion problem.

It remains to note that for a given initial problem (t_0,x_0) there can exist no pair of strategies U^0 and V^ϵ which will solve the $(\mathcal{M},\mathcal{N})$-evasion and $(\mathcal{M},\mathcal{N})^\epsilon$-evasion problems, respectively. Should such a pair of strategies exist, the motions $x(\cdot) \in X_1(t_0,x_0,U^0) \cap X_2(t_0,x_0,V^\epsilon)$ (this intersection is not empty) would simultaneously satisfy the two mutually exclusive conditions of the $(\mathcal{M},\mathcal{N})$-approach and $(\mathcal{M},\mathcal{N})^\epsilon$-evasion. Thus Theorem 2.1.1 is proved.

Since in the alternative just proved only one of the two propositions can be fulfilled, the point (t_0,x_0) cannot belong simultaneously to the sets $\mathcal{W}_0^1$ and $\mathcal{W}_\epsilon^2$. Consequently, one can adjoin the following

$$\mathcal{W}_1^0 \cap \Big(\bigcup_{\epsilon>0} \mathcal{W}_\epsilon^2 \Big) = \phi \tag{23}$$

to the relation (2.2.30) obtained in Theorem 2.2.1. Thus the maximal u-stable bridge in an $(\mathcal{M},\mathcal{N})$-approach problem and the union of the maximal v-stable bridges in a $(\mathcal{K}_\epsilon,\mathcal{L}_\epsilon)$-approach problem form a subdivision of the space of positions into two disjoint subsets. Note that this conclusion is valid under the condition of the saddle point in a small game (cf. (1.2.8)).

In conclusion, we present several theorems which are simple corollaries of the results established above. We state them in the form of theorems for the sake of convenience of subsequent references.

Theorem 2.4.3. *Let* $(t_0,x_0) \in T \times R^n$ *be an initial position. The following three assertions are equivalent:*

(a) *for a given initial position an* $(\mathcal{M}, \mathcal{N})$*-approach problem (Problem 2.1.1) possesses a solution.*

(b) $(t_0,x_0) \in \mathcal{W}_0^1$, *where* $\mathcal{W}_0^1$ *is the maximal* u*-stable bridge in an* $(\mathcal{M}, \mathcal{N})$*-approach problem.*

(c) *For a positional strategy* V: $T \times R^n \to \mathcal{Q}$ *arbitrarily chosen by the second player, there exists a motion* $x(\cdot) \in \mathcal{X}_2(t_0,x_0,V)$ *which satisfies an* $(\mathcal{M}, \mathcal{N})$*-approach condition.*

Proof. Theorem 2.4.1 implies that (b) $\Rightarrow$ (a). To verify the implication (a) $\Rightarrow$ (b) assume the contrary. Let an $(\mathcal{M}, \mathcal{N})$-approach problem possess a solution and let $(t_0,x_0) \notin \mathcal{W}_0^1$. It folows from (2.2.30) that $(t_0,x_0) \in \mathcal{W}_\epsilon^2$ where ϵ is a positive number and $\mathcal{W}_\epsilon^2$ is the maximal v-stable bridge in a $(\mathcal{K}_\epsilon, \mathcal{L}_\epsilon)$-approach problem. In accordance with Theorem 2.4.2 an $(\mathcal{M}, \mathcal{N})^\epsilon$-evasion problem is solvable. We thus obtain that for a given initial position (t_0,x_0) both an $(\mathcal{M}, \mathcal{N})$-approach and an $(\mathcal{M}, \mathcal{N})^\epsilon$-evasion problem are solvable. This contradiction verifies the implication.

Now let U^0 be a strategy which yields a solution of the $(\mathcal{M}, \mathcal{N})$-approach problem. Then for any strategy V the motion $x(\cdot) \in \mathcal{X}_2(t_0,x_0,V) \cap \mathcal{X}_1(t_0,x_0,U^0)$ (this intersection is non-empty) satisfies the condition of an $(\mathcal{M}, \mathcal{N})$-approach. Thus the implication (a) $\Rightarrow$ (c) has been proved.

It remains to verify implication (c) $\Rightarrow$ (a). Assume the contrary. Let assertion (c) be fulfilled but let the $(\mathcal{M}, \mathcal{N})$-approach problem possess no solution. Then in accordance with a theorem on alternatives (cf. Theorem 2.1.1) there exists an $\epsilon > 0$ and a strategy V^ϵ which yields a solution of the $(\mathcal{M}, \mathcal{N})^\epsilon$-evasion problem. We thus arrive at a contradiction to assertion (c). Theorem 2.4.3 is thus proved.

We introduce the notion of a set of positional absorption in an $(\mathcal{M}, \mathcal{N})$-approach problem. This set is a collection of initial positions (t_0,x_0) for which assertion (c) stated in Theorem 2.4.3 is valid. As was shown in Theorem 2.4.3, the set of positional absorption coincides with the maximal u-stable bridge.

The proof of the theorem on an alternative can be carried out as follows. First, the set of positional absorption is defined. Then it is verified that this set possesses the u-stability property. Finally, an extremal strategy is introduced which retains the motion on the set of positional absorptions. The proof of the theorem on an alternative was carried out in this manner in the book [42].

The following theorem is an analog of Theorem 2.4.3, but

differs from it by the mere interchange of the roles of the first and second players.

Theorem 2.4.4. *Let* K *and* $\mathfrak{L}$ *be closed sets in* $T \times R^n$, *and assume that* $K \subset \mathfrak{L}$. Let $(t_0,x_0) \in T \times R^n$ *be a given initial position. The following three assertions are equivalent:*

(a) *for a given initial position* (t_0,x_0) *the* $(K,\mathfrak{L})$*-approach problem (Problem* (2.1.3)) *possesses a solution in the class of positional strategies for the second player.*

(b) $(t_0,x_0) \in W^2$, *where* W^2 *is the maximal* v*-stable bridge in a* $(K,\mathfrak{L})$*-approach problem.*

(c) *For any positional strategy* U *chosen by the first player, there exists a motion* $x(\cdot) \in X_1(t_0,x_0,U)$ *satisfying the condition of a* $(K,\mathfrak{L})$*-approach.*

In the proof of Theorem 2.4.1 the following assertion was actually established.

Theorem 2.4.5. *Let* W *be a closed* u*-stable bridge in an* (M, N)*-approach problem. Assume that* $(t_0,x_0) \in W$. *Let* U^e *be a strategy extremal to the set* W. *Then for any motion* $x(\cdot) \in X_1(t_0,x_0,U^e)$,

$$T_0(x(\cdot)) \overset{\text{def}}{=} \{t \in [t_0,\theta]: \ (t,x(t)) \in M\} \neq \phi \text{ and}$$

$$(t,x(t)) \in W \quad \text{for} \quad t_0 \leqslant t \leqslant \tau^0(x(\cdot)), \text{ where}$$

$$\tau^0(x(\cdot)) \overset{\text{def}}{=} \min\{t: t \in T_0(x(\cdot))\}.$$

An analogous assertion is valid for a $(K,\mathfrak{L})$-approach problem.

Theorem 2.4.6. *Let* $K \subset \mathfrak{L}$ *be closed sets in* $T \times R^n$. *Let* W *be a closed* v*-stable bridge in a* $(K,\mathfrak{L})$*-approach problem. Assume that* $(t_0,x_0) \in W$. *Let* V^e *be a strategy extremal to the set* W. *Then for any motion* $x(\cdot) \in X_2(t_0,x_0,V^e)$,

$$T_*(x(\cdot)) \overset{\text{def}}{=} \{t \in [t_0,\theta]: \ (t,x(t)) \in K\} \neq \phi,$$

$$(t,x\ t)) \in W \ \text{ for } t_0 \leqslant t \leqslant \tau_*(x(\cdot)), \textit{ where}$$

$$\tau_*(x(\cdot)) \overset{\text{def}}{=} \min\{t: \ t \in T_*(x(\cdot))\}.$$

2.5. Construction of Stable Bridges

In §2.2 descriptive definitions of u-stable and v-stable bridges were given. Constructive definitions of stable bridges are also known. Some of these definitions are discussed in this section. Also, construction of stable bridges in simple examples of a differential game of approach-evasion is considered.

2.5.1. Consider the construction of stable bridges in the case when the phase space is one-dimensional, i.e. $n = 1$, x is a scalar. We shall assume that

$$\mathcal{M} = \{(t,x) \in T \times R^1:\ t = \theta,\ m^{(1)} \leqslant x \leqslant m^{(2)}\} \tag{1}$$
$$T = (-\infty,\theta],\quad N = T \times R^1.$$

Thus in the approach problem considered here, it is required that a point $x(\theta)$ will fall into the given interval $[m^{(1)},m^{(2)}]$. The set N coincides with the whole space of positions, i.e. the phase restriction is absent. As usual we shall assume that the motion of the controlled system is described by equation

$$\dot{x} = f(t,x,u,v).$$

Here f is a scalar function. It is assumed that it satisfies conditions (i) – (iv) (cf. §1.2.2). To construct stable bridges in the game under consideration we shall introduce the following notation. Set

$$f^{(1)}(t,x) \stackrel{\text{def}}{=} \max_{u\in P} \min_{v\in Q} f(t,x,u,v), \tag{2}$$
$$f^{(2)}(t,x) \stackrel{\text{def}}{=} \min_{u\in P} \max_{v\in Q} f(t,x,u,v)\,.$$

Define solutions $t \to w^{(1)}(t)$ and $t \to w^{(2)}(t)$ of differential equations

$$\dot{w}^{(1)}(t) = f^{(1)}(t,w^{(1)}(t)),\quad \dot{w}^{(2)}(t) = f^{(2)}(t,w^{(2)}(t)), \tag{3}$$

satisfying at the right end-point the conditions

$$w^{(1)}(\theta) = m^{(1)},\quad w^{(2)}(\theta) = m^{(2)}. \tag{4}$$

Conditions (4) can be viewed as initial conditions and one can integrate equations (3) in the opposite direction. Note that functions $f^{(1)}$ and $f^{(2)}$ satisfy all the requirements which assure the existence and uniqueness of solutions $w^{(1)}(\cdot)$ and $w^{(2)}(\cdot)$ on the interval $T = (-\infty,\theta]$.

Define now the interval $T_* \subset T$ as follows: If $w^{(2)}(t) \geqslant w^{(1)}(t)$ for all $t \in T$, we set $T_* = T$. Otherwise we define a number $\tau_* = \sup\{t \in T:\ w^{(2)}(t) < w^{(1)}(t)\}$ and set $T_* = [\tau_*,\theta]$. It follows from this definition that T_* is a maximal interval possessing the properties that $\theta \in T$ and $w^{(2)}(t) \geqslant w^{(1)}(t)$ for all $t \in T_*$.

Consider the set (cf. Fig. 2.5.1)

$$\mathcal{W}_0^1 \stackrel{\text{def}}{=} \{(t,x) \in T \times R^1:\ t \in T_*,\ x \in [w^{(1)}(t),w^{(2)}(t)]\}. \tag{5}$$

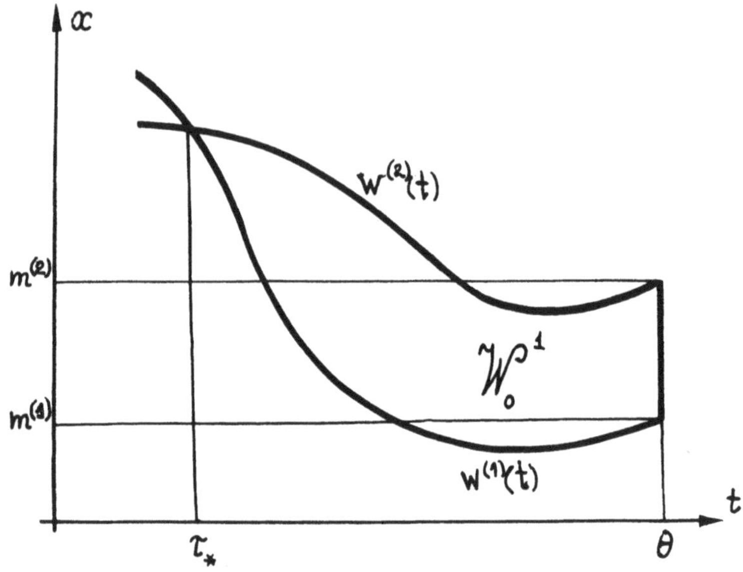

Fig. 2.5.1

We show that the set W_0^1 will be a u-stable bridge in the approach problem under consideration. Evidently $W_0^1 \subset N$ and $W_0^1(\theta) \subset M$. Therefore it is required only to verify the condition of u-stability.

Set

$$f^{(1)}(t,x,v) \overset{\text{def}}{=} \max_{u \in P} f(t,x,u,v),$$
$$f^{(2)}(t,x,v) \overset{\text{def}}{=} \min_{u \in P} f(t,x,u,v).$$

Observe that for any $(t,x) \in T \times R^n$ and $v \in Q$ the inequalities

(6) $$f^{(1)}(t,x,v) \geqslant f^{(1)}(t,x), \quad f^{(2)}(t,x,v) \leqslant f^{(2)}(t,x)$$

are valid.

Consider the multivalued mapping $(t,x,v) \to \mathcal{F}(t,x,v)$ where

$$\mathcal{F}(t,x,v) \overset{\text{def}}{=} \begin{cases} \{f^{(1)}(t,x,v)\} & \text{for } x < w^{(1)}(t), \\ co\{f(t,x,u,v):\ u \in P\} & \text{for } x \in [w^{(1)}(t), w^{(2)}(t)], \\ \{f^{(2)}(t,x,v)\} & \text{for } x > w^{(2)}(t). \end{cases}$$

Fix a point $v \in Q$ and consider the differential inclusion

(7) $$\dot{y}(t) = \mathcal{F}(t,y(t),v).$$

Denote by $Y(t_*,x_*,v)$ the set of solutions $y(\cdot) : [t_*,\theta] \to R^1$ of this differential inclusion satisfying the initial condition $y(t_*) = x_*$. Let

$\mathcal{X}_2(t_*,x_*,v)$ be the set of solutions: $x(\cdot)$: $[t_*,\theta] \to R^1$ of the differential inclusion

$$\dot{x}(t) \in \mathcal{F}_2(t,x(t),v) = co\{f(t,x(t),u,v): \ u \in \mathcal{P}\}$$

satisfying the initial condition $x(t_*) = x_*$ (cf. (2.2.1), (2.2.3)).

The following assertions are valid: (j) $Y(t_*,x_*,v) \neq \phi$; (jj) $Y(t_*,x_*,v) \subset \mathcal{X}_2(t_*,x_*,v)$; (jjj) let $(t_*,x_*) \in \mathcal{W}_0^1$, $t^* \in [t_*,\theta]$, then for any $y(\cdot) \in Y(t_*,x_*v)$ the inclusion $(t^*,y(t^*)) \in \mathcal{W}_0^1$ is valid. Note that from these assertions follows the property of u-stability of the set $\mathcal{W}_0^1$.

We shall now verify the validity of assertions (j) – (jjj). It follows from the definition of the multivalued mapping $(t,x) \to \mathcal{F}(t,x,v)$ that it satisfies conditions (a) – (c) stated in §A.4. Therefore, in accordance with Theorem A.4.1, $Y(t_*,x_*,v) \neq \phi$, i.e. assertion (j) holds. Assertion (jj) follows directly from the inclusion $\mathcal{F}(t,x,v) \subset co\{f(t,x,u,v): \ u \in P\}$. It remains to verify Assertion (jjj). Assume the contrary. Let there exist $y(\cdot) \in Y(t_*,x_*,v), y(t^*) \notin [w^{(1)}(t^*),w^{(2)}(t^*)]$. Assume, for example, that $y(t^*) < w^{(1)}(t^*)$. Since $y(t_*) \geqslant w^{(1)}(t_*)$ there exists a time $\tau_\# \in [t_*,t^*)$ such that $y(\tau_\#) = w^{(1)}(\tau_\#)$ and $y(t) < w^{(1)}(t^*)$ for $t \in (t_\#,t^*]$. On the interval $[\tau_\#,t^*]$ the motions $w^{(1)}(\cdot)$ and $y(\cdot)$ satisfy the equations $\dot{w}^{(1)} = f^{(1)}(t,w^{(1)}(t))$ and $\dot{y}(t) = f^{(1)}(t,y(t),v)$. From (6) and the Lipschitz condition for the function $w \to f^{(1)}(t,w)$ we obtain

$$|w^{(1)}(t) - y(t)| = w^{(1)}(t) - y(t) = \int_{\tau_\#}^{t} [f^{(1)}(\tau,w^{(1)}(\tau))$$

$$- f^{(1)}(\tau,y(\tau),v)]d\tau \leqslant \int_{\tau_\#}^{t} [f^{(1)}(\tau,y(\tau)) - f^{(1)}(\tau,y(\tau),v)]d\tau$$

$$+ \lambda\int_{\tau_\#}^{t} |w^{(1)}(\tau) - y(\tau)|d\tau \leqslant \lambda\int_{\tau_\#}^{t} |w^{(1)}(\tau) - y(\tau)|d\tau$$

$$(\tau_\# \leqslant t \leqslant t^*).$$

Hence the equality $w^{(1)}(t^*) = y(t^*)$ follows. We thus arrive at a contradiction to the inequality $w^{(1)}(t^*) > y(t^*)$ and assertion (jjj) is proved. Therefore it has been verified that the set $\mathcal{W}_0^1$ (5) is a u-stable bridge in the approach problem under consideration.

Next it can be shown that the constructed set $\mathcal{W}_0^1$ will be the maximal u-stable bridge. Indeed, consider the set $\mathcal{W}_\epsilon^1$ which coincides with the set $\mathcal{W}_0^1$ except that conditions (4) are replaced by conditions $w^{(1)}(\theta) = m^{(1)} - \epsilon$, $w^{(2)}(\theta) = m^{(2)} + \epsilon$. Set

$$\mathcal{W}_\epsilon^2 = cl[(T \times R^n)\setminus \mathcal{W}_\epsilon^1].$$

The set $\mathcal{W}_\epsilon^2$ will be a v-stable bridge in an $(\mathcal{M}, \mathcal{N})^\epsilon$-evasion problem in which the second player aims to prevent the hit of point $x(\theta)$ into the interval $(m^{(2)} - \epsilon,m^{(2)} + \epsilon)$. This fact is verified in the same manner as above. From the construction of the sets W_0^1 and W_ϵ^1 we obtain that

$$W_0^1 = (T \times R^n) \setminus \left(\bigcup_{\epsilon > 0} W_\epsilon^2 \right).$$

It follows from here that W_0^1 is the maximal u-stable bridge, i.e. the approach problem under consideration is solvable if and only if the initial position (t_0, x_0) belongs to the set W_0^1 (5).

The result obtained can be clarified in the following particular case. Let the motion equation be of the form

$$(8) \qquad \dot{x}(t) = r_1(t)u(t) + r_2(t)v(t).$$

Here $r_i(t)$ $(i = 1, 2)$ $u(t)$, and $v(t)$ are scalars, and it is assumed that the functions $t \to r_i(t)$ are continuous and the players' controls satisfy the constraints

$$(9) \qquad |u(t)| \leqslant \beta_1, \quad |v(t)| \leqslant \beta_2.$$

We shall also assume that

$$(10) \qquad M = \{(t,x): \ t = \theta, \ \ |x| \leqslant r_*\}.$$

In this case relations (2) – (4) determine the solutions

$$(11) \qquad w^{(2)}(t) = - \ w^{(1)}(t) = \rho_0(t) \overset{\text{def}}{=} r_* + \int_t^\theta (\beta_1 |r_1(\tau)| - \beta_2 |r_2(\tau)|) d\tau.$$

Therefore the set W_0^1 is defined by the relation

$$(12) \qquad W_0^1 = \{(t,x): \ t \in T_*, \ \ |x| \leqslant \rho_0(t)\}.$$

Here $T_* = T$, provided $\rho_0(t) \geqslant 0$ for all $t \in T$; otherwise $T_* = [\tau_*, \theta]$ where $\tau_* = \sup\{t \in T: \ \rho_0(t) < 0\}$.

Finally consider a controlled system of the form (8) where, unlike in the preceding case, $x(t)$, $u(t)$, and $v(t)$ are n-dimensional vectors. We shall assume that the controls satisfy the constraints

$$(13) \qquad \|u(t)\| \leqslant \beta_1, \quad \|v(t)\| \leqslant \beta_2 \ .$$

The target set is defined by the relation

$$(14) \qquad M = \{(t,x) \in T \times R^n: \ t = \theta, \ \ \|x\| \leqslant r_*\}.$$

It can be assumed that in the approach problem (8), (13), (14) the set

$$(15) \qquad W_0^1 \overset{\text{def}}{=} \{(t,x) \in T \times R^n : t \in T_*, \ \ \|x\| \leqslant \rho_0(t)\}$$

will be a u-stable bridge. Here the quantity $\rho_0(t)$ is defined by the equality (11) and the interval T_* is defined in the same manner as was done when the set W_0^1 (12) was defined. To verify this assumption, it is sufficient to show that the set W_0^1 (15) possesses the property of u-stability, i.e. for any t_*, t^*, x_*, v ($t_* \in T_*$, $t^* \in (t_*, \theta]$, $\|x_*\| \leqslant \rho_0(t_*)$, $v \in Q$) there exists a measurable control $t \to u(t)$ ($\|u(t)\| \leqslant \beta_1$) such that the solution of the differential equation $\dot{x}(t) = r_1(t)u(t) + r_2(t)v(t)$ ($x(t_*) = x_*$) will satisfy the condition $\|x(t^*)\| \leqslant \rho_0(t^*)$.

We introduce the following notation:

$$z = x_* + \int_{t_*}^{t^*} r_2(t) v dt, \quad \rho_1 = \int_{t_*}^{t^*} \beta_1 |r_1(t)| dt, \tag{16}$$
$$Y = \{y \in R^n: \ \|y\| \leqslant \rho_1\}.$$

Note that the set of all vectors

$$y = \int_{t_*}^{t^*} r_1(t) u(t) dt$$

– which is obtained when all the admissible controls $u(\cdot)$ are enumerated – coincides with the set Y. Therefore to verify the property of u-stability it is sufficient to show that the vector $y_* \in Y$ can be chosen so that the inequality

$$\|z + y_*\| \leqslant \rho_0(t^*) \tag{17}$$

is fulfilled. Set

$$y_* = - z \quad \text{for} \quad \|z\| \leqslant \rho_1, \tag{18}$$

$$y_* = - z\|z\|^{-1}\rho_1 \ \text{for} \ \|z\| > \rho_1. \tag{19}$$

In case (18) we have $\|z + y_*\| = 0 \leqslant \rho_0(t^*)$, i.e. the inequality (17) is fulfilled. In case (19) it follows from (11), (16), and the inequality $\|x\| \leqslant \rho_0(t_*)$ that

$$\|z + y_*\| = \|z\| - \rho_1 \leqslant \|x_*\| + \int_{t_*}^{t^*} \beta_2 |r_2(t)| dt - \int_{t_*}^{t^*} \beta_1 |r_1(t)| dt$$
$$\leqslant \rho_0(t_*) + \int_{t_*}^{t^*} (\beta_2 |r_2(t)| - \beta_1 |r_1(t)|) dt = \rho_0(t^*).$$

In this case inequality (17) is also satisfied. We have thus proved that the set W_0^1 is a u-stable bridge in the approach problem (8), (13), (14).

It is easy to verify that the set W_0^1 constructed above will be the maximal u-stable bridge in the approach problem under consideration. To do this, following the procedure described for the scalar case, it is sufficient to construct v-stable bridges W_ϵ^2 and verify that every point (t,x) which is not in W_0^1 falls into a set W_ϵ^2 for some $\epsilon > 0$. We thus obtain that in order that the problem of approach (8), (13), (14) to solvable, it is necessary and sufficient that the initial position (t_0,x_0) belong to the set W_0^1 (15) or, equivalently, the conditions

$$\|x_0\| \leqslant \rho_0(t_0), \quad \rho_0(t) \geqslant 0 \ \text{for} \ t_0 \leqslant t \leqslant \theta$$

be fulfilled.

We conclude the discussion of this problem with the construction of the strategy U^e extremal to the set W_0^1 (15). According to the definition of an extremal strategy (see §2.4.1), we determine the sets $W_0^1[t]$, $W_0^1[t,x]$ and the functions e and U^e. Let $t \in T_*$, then $W_0^1[t] = \{x \in R^n: \ \|x\| \leqslant \rho_0(t)\}$. Assume that $x \notin W_1^0[t]$ i.e. $\|x\| > \rho_0(t)$. Then the set $W_0^1[t,x]$ consists of the unique point $w^0 = e(t,x) = \dot{x}\rho_0(t)\|x\|^{-1}$. Hence in the case under consideration the vector $x - e(t,x) = x - x\rho_0(t)\|x\|^{-1}$ is collinear to the vector x. Condition (5) can be written as

$$\langle x, r_1(t)U^e(t,x)\rangle = \min_{\|u\|\leqslant\beta_1} \langle x, r_1(t)u\rangle.$$

We thus have that $U^e(t,x) = -\beta_1 x\|x\|^{-1}\text{sign}\,r_1(t)$. Consequently, the function U^e is defined for $t \in T_*$ and $x \notin W_0^1[t]$. In the remaining cases we have $x - e(t,x) = 0$ and for $U^e(t,x)$ any vector u satisfying the constraint $\|u\| \leqslant \beta_1$ can be chosen. Therefore the function U^e: $T \times R^n \to P$ can be defined as follows:

$$U^e(t,x) = \begin{cases} -\beta_1 x\|x\|^{-1}\text{sign}\,r_1(t) & \text{for } \|x\| \neq 0, \\ 0 & \text{for } \|x\| = 0. \end{cases}$$

In this simple problem it was easy to guess the solution and obtain the formulas stated above which define the approach strategy. This example is presented as an illustration of the concepts of a stable bridge and an extremal strategy. Moreover, results obtained herein will be used below for another problem whose solution is not so obvious.

2.5.2. In the examples discussed above, the structure of stable bridges was relatively simple. However, this construction was based on specific features of these examples and cannot be used in the general case. We shall discuss two methods of constructive definition of stable bridges in the general case of a differential game of approach-evasion.

To simplify the notation we shall assume that $T = [0,1]$. Subdivide this interval into 2^k parts $[\tau_{k,i+1},\tau_{k,i}]$ where

$$\tau_{k,i} = 1 - i2^{-k}, \quad i = 0,\ 1,\ \ldots,\ 2^k.$$

Note that here it is convenient to enumerate the numbers $\tau_{k,i}$ in such a manner that the numbers will decrease as the index i increases. We define recursively the sequence of sets $Z_{k,i}$ ($i =$

$0, 1, ..., 2^k$). Set

$$(20)\quad Z_{k,0} = M(\theta) = M(1),\ Z_{k,i+1} = \Pi_2(\tau_{k,i+1}, \tau_{k,i}, M \cup Z_{k,i}) \cap N(\tau_{k,i+1})$$
$$(i = 0, 1, ..., 2^k - 1).$$

Here Π_2 is an operator of programmed absorption which was introduced in §2.2.1 (cf. (2.2.9)). In accordance with the definition of this operator, the set $Z_{k,i+1}$ consists of positions $(\tau_{k,i+1}, x_*) \in N$ such that for any $v \in Q$ there exists a motion $x(\cdot) \in X_2(\tau_{k,i+1}, x_*, v)$ which satisfies either the condition $(\tau_{k,i}, x(\tau_{k,i})) \in Z_{k,i}$ or the condition $(\tau, x(\tau)) \in M$ for some $\tau \subset [\tau_{k,i+1}, \tau_{k,i}]$.

We shall discuss briefly some properties of the set $Z_{k,i}$. Let t_* be a binary-rational number in the interval [0,1], i.e. $t_* = j \cdot 2^{-k_*}$ where j is an odd number. Note that $t_* = \tau_{k_*, i_*}$ where $i_* = 2^{k_*} - j$. Clearly, for any $k > k_*$ the equality $t_* = \tau_{k,i}$ will be valid for $i = i_* 2^{(k-k_*)}$. Set

$$(21)\qquad Z_k(t_*) = Z_{k,i},\quad k \geqslant k_*, \qquad i = i_* 2^{(k-k_*)}.$$

For $t_* = 0$ the set $Z_k(t_k)$ will be defined by equality (21) for any $k = 1, 2, ...$, and for $i = 2^k$. It follows from the definition of the sets $Z_{k,i}$ that

$$Z_{k+1}(t_*) \subset Z_k(t_*) \quad \text{for any} \quad k \geqslant k_*.$$

Next, for any binary-rational number t_* one can define the set

$$(22)\qquad Z(t_*) \overset{\text{def}}{=} \bigcap_{k \geqslant k_*} Z_k(t_*).$$

It can be shown that the maximal u-stable bridge in an (M, N) - approach problem coincides with the set

$$(23)\quad Z \overset{\text{def}}{=} \{(t,x) \in [0,1] \times R^n\colon \exists\, (t_\ell, x_\ell) \in Z(t_\ell),\ t_\ell - \text{binary-rational},$$
$$\ell = 1, 2, ..., (t_\ell, x_\ell) \to (t,x) \text{ for } \ell \to \infty\}.$$

In the particular case when the phase restriction is absent and the termination time of the game is fixed, i.e. $M = M(\theta) = M(1)$, $N = [0,1] \times R^n$, the recursive procedure (20) becomes

$$(24)\qquad Z_{k,0} = M(1),\ Z_{k,i+1} = \Pi_2(\tau_{k,i+1}, \tau_{k,i}, Z_{k,i})\ (i = 0, 1, ..., 2^k - 1).$$

Note that the limit transitions (22), (23) are required for theoretical justification of the proposed procedures. For an approximate solution of game-theoretical control problems it is sufficient to construct a collection of sets $\mathcal{Z}_{k,i}$ of the form (20) or (24).

The definition of sets $\mathcal{Z}_{k,i}$ can be interpreted as follows. Consider a multistage game which begins at time $t = \tau_{k,i}$ and which consists of i stages corresponding to the time intervals $[\tau_{k,i}, \tau_{k,i-1}],[\tau_{k,i-1},\tau_{k,i-2}], ..., [1 - 2^{-k},1]$. At each time $t = \tau_{k,j}$ ($j = i, i-1, ..., 1$) the second player communicates to the first a constant control v which the second player is going to use on the interval $[\tau_{k,j},\tau_{k,j-1})$. Possessing this information and the information about the realized position $(\tau_{k,j},x(\tau_{k,j}))$, the first player chooses from the set $X_2(\tau_{k,j},x(\tau_{k,j}),v)$ a motion $x(\cdot)$: $[\tau_{k,j},\tau_{k,j-1}] \to R^n$. For simplicity consider the case $\mathcal{N} = [0,1] \times R^n$ and $\mathcal{M} = \mathcal{M}(1)$. Then the set $\mathcal{Z}_{k,i}$, defined in accordance with (24), will coincide with the set of positions $(\tau_{k,i},x_*)$ from which, in the multistage game under consideration, the first player can guarantee that the point $(\theta,x(\theta)) = (1,x(1))$ will fall into the set $\mathcal{M}$. Procedures (20) and (24) belong to a well known type of recursive procedure considered within the framework of dynamic programming. In the theory of differential games, procedures of this type are utilized in numerous papers (cf., for example, [3, 60, 62, 64, 72]).

Consider now another method of constructive definition of a stable bridge. To avoid discussion of some technical details we shall assume that for any $(t,x,v) \in [0,1] \times R^n \times \mathcal{Q}$ the vectogram $f(t,x, ,v) \overset{\text{def}}{=} \{f(t,x,u,v): u \in \mathcal{P}\}$ is convex. Denote by $\mathbf{V}_p$ and $\mathbf{U}_p$ the sets of measurable functions $v(\cdot)$: $[0,1] \to \mathcal{Q}$ and $u(\cdot)$: $[0,1] \to \mathcal{P}$ respectively. As usual, the symbol $x(\cdot,t_*,x_*,u(\cdot),v(\cdot))$ will denote the solution of the differential equation $\dot{x}(t) = f(t,x(t),u(t),v(t))$ satisfying the initial condition $x(t_*) = x_*$. We assume here that $(t_*,x_*) \in [0,1] \times R^n$, $u(\cdot) \in \mathbf{U}_p$, and $v(\cdot) \in \mathbf{V}_p$.

Consider now the operator π_* which assigns to a closed set $\mathcal{Z} \in [0,1] \times R^n$ the set of positions $(t_*,x_*) \in [0,1] \times R^n$ possessing the following property: for any $v(\cdot) \in \mathbf{V}_p$ there exists $u(\cdot) \in \mathbf{U}_p$ such that for the motion $x(\cdot,t_*,x_*,(\cdot),v(\cdot))$ the condition of $(\mathcal{M}, \mathcal{Z})$-approach is valid. The set of positions (t_*,x_*) defined herein is denoted by $\pi_*(\mathcal{Z})$. In this definition the set $\mathcal{Z}$ plays the role of phase constraint. One can verify that for any closed set $\mathcal{Z} \subset [0,1] \times R^n$ the set $\pi_*(\mathcal{Z})$ is also closed in $[0,1] \times R^n$. Note that the operator π_* as well as the operators π_1 and π_2 introduced above are operators of programmed absorption.

Utilizing operator π_*, consider the following iteration procedure:

$$(25) \qquad \mathcal{Z}_0 = \mathcal{N}, \ \mathcal{Z}_1 = \pi_*(\mathcal{Z}_0), \ \mathcal{Z}_2 = \pi_*(\mathcal{Z}_1), \ \dots .$$

From the definition of the operator π_* it follows immediately that the sets $\mathcal{Z}_k$ satisfy the inclusions $\mathcal{Z}_{k+1} \subset \mathcal{Z}_k$. Set

$$(26) \qquad \mathcal{Z}_* \overset{\text{def}}{=} \bigcap_{k=0}^{\infty} \mathcal{Z}_k .$$

One can verify that the set $\mathcal{Z}_*$ coincides with the maximal *u*-stable bridge in an $(\mathcal{M}, \mathcal{N})$-approach problem. The procedure described above is convenient in those cases when only a finite number of iterations is required, i.e. in those cases when for some k the equality $\mathcal{Z}_{k+1} = \Pi_*(\mathcal{Z}_k)$ is satisfied. The iteration procedure (25) was introduced in [10, 11], while a detailed exposition of this procedure is presented in [69].

There exist several modifications of the procedures described for the construction of stable bridges. Moreover, constructive definitions of stable bridges are available which differ significantly from the constructions considered here. It should, however, be admitted that practical implementation of universal procedures for construction of stable bridges turns out to be, as a rule, quite time consuming. Below we shall repeatedly address various problems related to the construction of stable bridges.

2.5.3. Consider a linear controlled system

$$(27) \qquad \dot{x}(t) = A(t)x(t) + B(t)u(t) + C(t)v(t).$$

Recall that here $A(t)$, $B(t)$, and $C(t)$ are $(n \times n)$, $(n \times p)$, and $n \times q)$-matrices which depend continuously on parameter t. We shall assume that the players' controls satisfy the constraints

$$u(t) \in \mathcal{P}, \quad v(t) \in \mathcal{Q}, \quad (\mathcal{P}, \mathcal{Q} \text{ are convex compacta}).$$

The assumption of convexity of the sets $\mathcal{P}$ and $\mathcal{Q}$ is not essential and it is introduced only to simplify some arguments. For the controlled system (27) consider an $(\mathcal{M}, \mathcal{N})$-approach problem. We shall assume that in this problem the phase constraint is absent and the time of completion of the game is fixed, i.e.

$$(28) \qquad \mathcal{N} = T \times R^n, \ \mathcal{M} = \{(\theta, x) : \ x \in \mathcal{M}^*\}.$$

It is also assumed that $\mathcal{M}^*$ is a closed convex set in R^n.

The following assertion is valid:

Theorem 2.5.1. *Let* $\mathcal{W}_0^1$ *be the maximal* u-*stable bridge in an* $(\mathcal{M}, \mathcal{N})$-*approach problem formulated for the system* (27) *when the sets* $\mathcal{M}$ *and* $\mathcal{N}$ *are chosen in the form* (28). *Then the sets*

$$\mathcal{W}_0^1[t] = \{x \in R^n\colon \ (t,x) \in \mathcal{W}_0^1\}$$

will be convex for all $t \in T$.

Proof. Let $t_0 \in T$, $x^{(i)} \in \mathcal{W}_0^1[t_0]$ $(i = 1, 2)$. It is required to show that $x_0^{(\alpha)} = \alpha x_0^{(1)} + (1 - \alpha)x_0^{(2)} \in \mathcal{W}_0^1[t_0]$, where $\alpha \in (0,1)$. Assume the contrary. For some $\alpha \in (0,1)$, let the relationship $x_0^{(\alpha)} \notin \mathcal{W}_0^1[t_0]$ be valid. Then, in view of Theorem 2.1.1 there exists an $\epsilon > 0$ and a positional strategy V^{ϵ} which solves an (M, N^{ϵ})-evasion problem. In our case it means that for any constructive motion $x(\cdot) \in X_2(t_0, x_0^{(\alpha)}, V^{\epsilon})$ the inequality

(29) $$\mathrm{dist}(x(\theta), M^*) \geq \epsilon$$

is valid. On the other hand, in accordance with Theorem 2.4.1 it follows from condition $x_0^{(i)} \in W_0^1[t_0]$ that for any constructive motion $x(\cdot) \in X_1(t_0, x_0^{(i)}, U^e)$ the condition $x(\theta) \in M^*$ is fulfilled (here U^e is a strategy extremal to the set W_0^1). Since constructive motions are approximated by step-by-step motions (cf. Lemma 1.3.2), one can choose a sufficiently refined partition Δ such that the inequalities

(30) $$\begin{aligned} &\mathrm{dist}(x(\theta), M^*) \geq 2\epsilon/3 \quad \text{for all} \quad x(\cdot) \in X_2(t_0, x_0^{(\alpha)}, V^{\epsilon}, \Delta), \\ &\mathrm{dist}(x(\theta), M^*) \leq \epsilon/3 \quad \text{for all} \quad x(\cdot) \in X_1(t_0, x_0^{(i)}, U^e, \Delta) \end{aligned}$$

will be fulfilled. Here the number ϵ is the same as in inequality (29).

Consider step-by-step motions satisfying the equations

$$\dot{x}^{(1)}(t) = A(t)x^{(1)}(t) + B(t)U^e(\tau_i, x^{(1)}(\tau_i)) + C(t)v(t),$$

$$\dot{x}^{(2)}(t) = A(t)x^{(2)}(t) + B(t)U^e(\tau_i, x^{(2)}(\tau_i)) + C(t)v(t),$$

$$\dot{x}^{(\alpha)}(t) = A(t)x^{(\alpha)}(t) + B(t)u(t) + C(t)V^{\epsilon}(\tau_i, x^{(\alpha)}(\tau_i))$$

for $t \in [\tau_i, \tau_{i+1})$, $i = 0, 1, \ldots, k$. It is assumed that

$$x^{(1)}(t_0) = x_0^{(1)}, \quad x^{(2)}(t_0) = x_0^{(2)}, \quad x^{(\alpha)}(t_0) = x_0^{(\alpha)}.$$

In these equation the controls $u(t)$ and $v(t)$ are chosen as follows:

$$u(t) = \alpha\, U^e(\tau_i, x^{(1)}(\tau_i)) + (1 - \alpha)U^e(\tau_i, x^{(2)}(\tau_i)),$$

$$v(t) = V^{\epsilon}(\tau_i, x^{(\alpha)}(\tau_i)), \quad t \in [\tau_i, \tau_{i+1}), \quad i = 0, 1, \ldots, k.$$

The motions under consideration, $x^{(i)}(\cdot)$ and $x^{(\alpha)}(\cdot)$, are contained in the sets $X_1(t_0, x_0^{(i)}, U^e, \Delta)$ and $X_2(t_0, x_0^{(\alpha)}, V^{\epsilon}, \Delta)$, respectively. Therefore from (30) follow the bounds

$$\mathrm{dist}(x^{(\alpha)}(\theta), M^*) \geq 2\epsilon/3, \quad \mathrm{dist}\,(x^{(i)}(\theta), M^*) \leq \epsilon/3 \quad (i = 1, 2).$$

On the other hand, it is evident from the construction of these motions that $x^{(\alpha)}(t) = \alpha x^{(1)}(t) + (1 - \alpha)x^{(2)}(t)$. Hence the convexity of M^* and the estimate $\mathrm{dist}(x^{(i)}(\theta), M^*) \leq \epsilon/3$ imply the inequality $\mathrm{dist}(x^{(\alpha)}(\theta), M^*) \leq 2\epsilon/3$. The contradiction thus obtained proves the theorem.

Note that Theorem 2.5.1 remains valid also for linear systems of the form (1.2.12) satisfying condition (1.2.13).

bridge. In particular, because of this property for linear controlled systems of small dimensionality it is feasible to develop and implement on computers computational algorithms for construction of stable bridges which are based on recursive procedures of the form (24).

2.5.4. We now describe one transformation of a linear system which will simplify the construction of stable bridges and the solution of game-theoretical control problems. Let, in the initial statement, the motion of the controlled system be described by the equation of the form (27), where x is an n-dimensional phase vector. Introduce a new m-dimensional phase vector y. Assume that the position (t,y) of the transformed system and position (t,x) of the initial system are related by the equality

$$(31) \qquad (t,y) = (t,F(t)x).$$

Here $F(t) = F\Phi(\theta,t)$, where F is a constant $(m \times n)$-matrix, $\Phi(\theta,t)$ is an $n \times n$-matrix. It is assumed that the function $t \to \Phi(\theta,t)$ satisfies the equation (cf. (1.2.15))

$$(32) \qquad \frac{d\Phi(\theta,t)}{dt} = -\Phi(\theta,t)A(t)$$

and the boundary condition $\Phi(\theta,\theta) = E$ (E is the unit matrix). We assume that the change in time of the new phase vector is described by the equation

$$(33) \qquad \dot{y}(t) = B_{\#}(t)u_{\#}(t) + C_{\#}(t)v_{\#}(t),$$

where

$$(34) \qquad B_{\#}(t) \overset{\text{def}}{=} F\Phi(\theta,t)B(t), \quad C_{\#}(t) \overset{\text{def}}{=} F\Phi(\theta,t)C(t).$$

The following assertion is valid:

Let the initial state of the system (27) and (33) be related by the equality $y(t_0) = F(t_0)x(t_0)$. Let the control in systems (27) and (33) coincide, i.e. $u(t) = u_{\#}(t), v(t) = v_{\#}(t)$ for almost all $t \in [t_0,\theta]$. Then the solution of equations (27) and (33) will be related by the equality

$$(35) \qquad y(t) = F(t)x(t).$$

To verify equality (35) it is sufficient to differentiate it and to utilize relations (32) and (34).

Assume now that in the space $T \times R^m$ sets $M_{\#}$ and $N_{\#}$ may be selected which are connected with the set M and N from the initial space of positions by the equalities

$$(36) \qquad \begin{aligned} \{(t,F(t)x)\colon\ (t,x) \in M\} &= M_{\#}, \\ \{(t,F(t)x)\colon\ (t,x) \in N\} &= N_{\#}. \end{aligned}$$

The following theorem is valid.

Theorem 2.5.2. *Let the sets* $M_\#$ *and* $N_\#$ *satisfy the equalities* (36). *Let the strategy* $U_\#$: $T \times R^n \to P$ *solve the problem of an* $(M_\#, N_\#)$*-approach from the initial position* $(t_0,y_0) = (t_0,F(t_0)x_0)$ *for the transformed system* (33). *Then for the initial system* (27) *the strategy* U^0: $T \times R^n \to P$, *defined by the equality*

$$(37) \qquad U^0(t,x) \stackrel{\text{def}}{=} U_\#(t,F(t)x),$$

solves an (M, N)*-approach problem from the initial position* (t_0,x_0).

Proof. Let $Y_1(t_0,y_0,U_\#,\Delta)$ be the set of step-by-step motions of the system (33) generated by the strategy $U_\#$. If $y_0 = F(t_0)x_0$ then the equality

$$Y_1(t_0,y_0,U_\#,\Delta) = \{t \to y(t) = F(t)x(t)\colon\ x(\cdot) \in X_1(t_0,x_0,U^0,\Delta)\}$$

is valid where $X_1(t_0,x_0,U^0,\Delta)$ is the set of step-by-step motions of the system (27) generated by the strategy U^0. This equality follows from (35) and (37). Therefore the corresponding sets of constructive motions will be connected by the analogous equality

$$Y_1(t_0,y_0,U_\#) = \{t \to y(t) = F(t)x(t)\colon\ x(\cdot) \in X_1(t_0,x_0,U^0)\}.$$

Since for any $y(\cdot) \in Y_1(t_0,x_0,U_\#)$ the condition of $(M_\#, N_\#)$-approach holds, it follows from the last equality and relation (36) that any motion $x(\cdot) \in X_1(t_0,x_0,U^0)$ will satisfy the condition of an (M, N)-approach. The theorem is proved.

Next, the following assertion can be proved.

Theorem 2.5.3. *Let conditions* (36) *be fulfilled and let* $W_\#$ *be the maximal* u*-stable bridge in an* $(M_\#, N_\#)$*-approach problem formulated for the system* (33). *Then an* (M, N)*-approach problem formulated for the system* (27) *will be solvable if and only if the initial position* (t_0,x_0) *satisfies the condition*

$$(38) \qquad (t_0,F(t_0)x_0) \in W_\#.$$

Proof. Let condition (38) be satisfied. Then, in view of Theorem 2.4.1 the strategy $U_\#$: $T \times R^m \to P$ extremal to the set $W_\#$ assures that the condition of $(M_\#, N_\#)$-approach for any motion $y(\cdot) \in Y_1(t_0,y_0,U_\#)$ where $y_0 \in F(t_0)x_0$ is fulfilled. From Theorem 2.5.2 we obtain that the strategy U^0: $T \times R^n \to P$ defined by equality (37) solves the problem of an (M, N)-approach formulated for the system (27). Sufficiency of condition (38) is thus verified.

Assume now that condition (38) is not satisfied. Set $y_0 = F(t_0)x_0$. Since the initial position (t_0,y_0) does not belong to the maximal

u-stable bridge $\mathcal{W}_{\#}$, the problem of an $(\mathcal{M}_{\#}, \mathcal{N}_{\#})^{\epsilon}$-evasion formulated for the system (33) is solvable for some $\epsilon > 0$, as shown in the preceding section. This assertion can be relaxed and one can assert that there exists a strategy $V_{\#}: T \times R^n \to \mathcal{Q}$ such that for any $y(\cdot) \in Y_2(t_0,y_0,V_{\#})$ the condition of $(\mathcal{M}_{\#}, \mathcal{N}_{\#})$-evasion is fulfilled (cf. §2.1.1). Define a strategy $V^0: T \times R^n \to \mathcal{Q}$ by the equality $V^0(t,x) \stackrel{\text{def}}{=} V_{\#}(t,F(t)x)$. Repeating the arguments presented in the course of the proof of Theorem 2.5.2, we obtain that for any motion $x(\cdot) \in X_2(t_0,x_0,V^0)$ the condition of $(\mathcal{M}, \mathcal{N})$-evasion will be fulfilled. Consequently, if the condition (38) is violated, the $(\mathcal{M}, \mathcal{N})$-approach problem formulated for the system (27) does not have a solution. The theorem is thus proved.

Usually transformation (31) is utilized in the case when in the initial $(\mathcal{M}, \mathcal{N})$-approach problem the sets $\mathcal{M}$ and $\mathcal{N}$ are of the form

$$(39) \qquad \mathcal{M} = \{(\theta,x): \ \ Fx \in \mathcal{M}_*\}, \qquad \mathcal{N} = T \times R^n,$$

where F is an $(m \times n)$-matrix, $\mathcal{M}_*$ is a closed set in R^m. Taking into account that $\Phi(\theta,\theta) = E$ we obtain $F(\theta) = F$, $\Phi(\theta,\theta) = F$. Consequently, equalities (36) will be satisfied for the sets

$$\mathcal{M}_{\#} = \{(\theta,y): \ \ y \in \mathcal{M}_*\}, \quad \mathcal{N}_{\#} = T \times R^m.$$

In accordance with Theorems 2.5.2, 2.5.3, the solution of an $(\mathcal{M}, \mathcal{N})$-approach problem formulated for the system (27) can be defined as follows. Introduce the transformation (31), (36). Construct the maximal u-stable bridge $\mathcal{W}_{\#}$ for the transformed system (33). Verify the condition of solvability of an $(\mathcal{M}, \mathcal{N})$-approach problem which is expressed by the condition (38). If this condition is satisfied we then construct a strategy $U_{\#}$ extremal to the set $\mathcal{W}_{\#}$. The required strategy $U^0: T \times R^n \to \mathcal{P}$ is defined by equality (37).

2.5.5. Example. Let two points with masses $m_1 = 1$ and $m_2 = 1$ move on a horizontal plane. The first point is subject to a frictional force proportional to the velocity of the point and to a control force u whose choice is up to the first player. On the second point, a frictional force and a control force v chosen by the second player operates. It is assumed that the game is completed at a fixed time $t = \theta = 0$. The problem formulated for the first player is to approach the first point to the second within a distance not exceeding a given number r_*. This approach should take place at a given instant of time $t = \theta = 0$. Phase restrictions are absent in this problem.

The motion equation for the problem under consideration was presented in §1.2.4 (cf. (1.2.20)). Recall that in the given differential game, the motion of the controlled system is described by equations of the form (27). In the case under consideration, x is an eight-dimensional vector, A, B, and C are (8×8), (8×2), and (8×2) constant matrices, respectively. In correspondence with the notation adopted in §1.2.4, (x_1,x_2) are the coordinates of the first point and (x_5,x_6) are the coordinates of the second point. Therefore, the quantity $[(x_1 - x_5)^2 + (x_2 - x_6)^2]^{1/2}$ determines the distance between these points.

The approach problem under consideration posed for the first player is an $(\mathcal{M}, \mathcal{N})$-approach problem where

$$\mathcal{M} = \{(\vartheta,x) = (0,x): \quad x \in R^8,\ (x_1 - x_5)^2 + (x_2 - x_6)^2 \leqslant r_*^2\},$$

$$\mathcal{N} = (-\infty,0] \times R^8, \quad T = (-\infty,0].$$

Note that the set $\mathcal{M}$ can be defined by a relation of the form (39). For this purpose set

$$F = (E,O,-E,O), \quad \mathcal{M}_* = \{y \in R^2,\ \|y\| \leqslant r_*\},$$

where E and O are, respectively, the unit and the zero matrices of dimension (2×2). We utilize transformation (31). Observe first that the sets $\mathcal{M}_\#$ and $\mathcal{N}_\#$ satisfying equalities (36) are defined as follows:

$$\mathcal{M}_\# = \{(0,y):\ y \in R^2,\ \|y\| \leqslant r_*\}, \quad \mathcal{N}_\# = (-\infty,0] \times R^2. \tag{40}$$

Now compute the matrices $B_\#(t)$ and $C_\#(t)$ (cf.(34)). Recall that in the problem under consideration $\Phi(\vartheta,t) = \Phi_*(-t)$, where the matrix $\Phi_*(-t)$ is defined by the relations (1.2.21) – (1.2.23). Substituting the corresponding matrices F, Φ, B, and C into (34) we obtain

$$B_\#(t) = r_1(t)E, \quad C_\#(t) = r_2(t)E$$

where

$$r_1(t) = [1- \exp\alpha_1 t]\alpha_1^{-1}, \quad r_2(t) = -[1 - \exp\alpha_2 t]\alpha_2^{-1}. \tag{41}$$

Consequently, in this example the auxiliary system is of the form

$$\dot{y} = r_1(t)u(t) + r_2(t)v(t), \quad \|u(t)\| \leqslant \beta_1, \quad \|v(t)\| \leqslant \beta_2. \tag{42}$$

Here y,u, and v are two-dimensional vectors. In the $(\mathcal{M}_\#, \mathcal{N}_\#)$-approach problem posed for the system (42), the sets $\mathcal{M}_\#$ and $\mathcal{N}_\#$ are given by the relations (40). Note that the auxiliary approach problem belongs to the type of problems discussed in §2.5.1 (cf. (8), (13), (14)). Therefore the maximal u-stable bridge $\mathcal{W}_\#$ for this problem is determined in accordance with (15), (11). Substituting (41) into (11) and taking into account that for $t \leqslant \vartheta = 0$, $r_1(t) \geqslant 0$, and $r_2(t) \leqslant 0$ we obtain

$$\rho_0(t) = \tau_* - \left(\frac{\beta_1}{\alpha_1} - \frac{\beta_2}{\alpha_2}\right)t - \frac{\beta_1}{\alpha_1^2}[1 - \exp\alpha_1 t] + \frac{\beta_2}{\alpha_2^2}[1 - \exp\alpha_2 t]\,. \tag{43}$$

As was shown at the end of §2.5.1, the strategy $U_{\#}$ extremal to the set

$$W_{\#} = W_0^1 = \{(t,y):\ t \in T_*,\ y \in R^2,\ \|y\| \leqslant \rho_0(t)\}$$

can be defined by the following relation:

$$U_{\#}(t,y) \overset{\text{def}}{=} \begin{cases} -\beta_1 y\|y\|^{-1} & \text{for } \|y\| \neq 0 \\ 0 & \text{for } \|y\| = 0. \end{cases}$$

We also define the vector $y(t,x) \overset{\text{def}}{=} F(t)x = F\Phi_*(-t)x$. Substituting matrices $F = (E,\ 0,\ -E,\ 0)$ and $\Phi_*(-t)$ (cf. 1.2.21) − (1.2.23)) into these equations we obtain that components of the vector $y(t,x)$ are expressed by the equalities

$$\begin{aligned} y_1(t,x) &\overset{\text{def}}{=} x_1 - x_5 + x_3[1 - \exp\alpha_1 t]\alpha_1^{-1} - x_7[1 - \exp\alpha_2 t]\,\alpha_2^{-1}, \\ y_2(t,x) &\overset{\text{def}}{=} x_2 - x_6 + x_4[1 - \exp\alpha_1 t]\alpha_1^{-1} - x_8[1 - \exp\alpha_2 t]\alpha_2^{-1}. \end{aligned} \tag{44}$$

Defining the strategy U^0 in accordance with (37) we obtain

$$U^0(t,x) = U_{\#}(t,y(t,x)) = \begin{cases} -\beta_1 y(t,x)\|y(t,x)\|^{-1} & \text{for } \|y(t,x)\| \neq 0, \\ 0 & \text{for } \|y(t,x)\| = 0. \end{cases} \tag{45}$$

In this example the result obtained in §2.4.5 can be explained as follows. Let the condition

$$\|y(t_0,x_0)\| \leqslant \rho_0(t_0),\ \rho_0(t) \geqslant 0 \text{ for } t_0 \leqslant t \leqslant \theta = 0 \tag{46}$$

be fulfilled where the quantity $\rho_0(t)$ and vector $y(t,x)$ are defined by the equalities (43) and (44). Then the positional strategy $(t,x) \to U^0(t,x)$ defined by relations (44), (45) assures the fulfillment of the inequality

$$(x_1(\theta) - x_5(\theta))^2 + (x_2(\theta) - x_6(\theta))^2 \leqslant r_*^2 \tag{47}$$

for any motion $x(\cdot) \in X_1(t_0,x_0,U^0)$, where $X_1(t_0,x_0,U^0)$ is the set of constructive motions of the system (1.2.20) generated by the strategy U^0. If condition (46) is violated, there is no positional strategy U which would assure the required "closeness" (47).

In Fig. 2.5.2 the graph of the function ρ_0 (43) is presented for the following values of the parameters:

$$r_* = 0,\ \alpha_1 = 3.5,\ \alpha_2 = 1,\ \beta_1 = 48,\ \beta_2 = 24. \tag{48}$$

We select the initial position

$$t_0 = -1,\ x_{01} = 0,\ x_{02} = 0.832,\ x_{03} = 2,\ x_{04} = 0,\ x_{05} = 0,$$
(49)
$$x_{06} = 0,\ x_{07} = 0,\ x_{08} = 0.$$

As evident from Fig. 2.5.2 (the point with coordinates $(t_0, \|y(t_0,x_0)\|)$ is denoted by a cross), condition (46) is satisfied for the parameters and initial conditions selected. Therefore the constructed strategy U^0 guarantees the fulfillment of conditions $x_1(0) = x_5(0)$, $x_2(0) = x_6(0)$ for any motion $x(\cdot) \in X_1(t_0,x_0,U^0)$. On Fig. 2.5.3 in the coordinate plane $\xi_1 = x_1 - x_5$, $\xi_2 = x_2 - x_6$, several trajectories are portrayed corresponding to the motions $x(\cdot) \in X_1(t_0,x_0,U^0)$.

The following remark could be made in connection with the choice of parameters (48). The control system considered in this example was first studied in paper [58]. In that paper, inter alia, the condition

(50) $\beta_1/\alpha_1 \geqslant \beta_2/\alpha_2,\ \beta_1 \geqslant \beta_2$

was introduced where at least one of the inequalities is strict. The meaning of this condition can be clarified as follows. If the

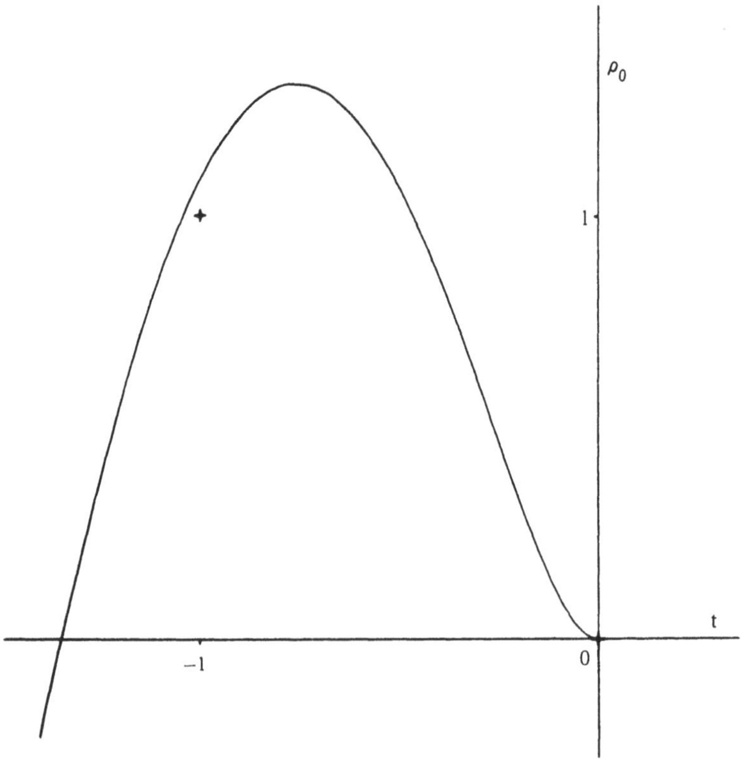

Fig. 2.5.2

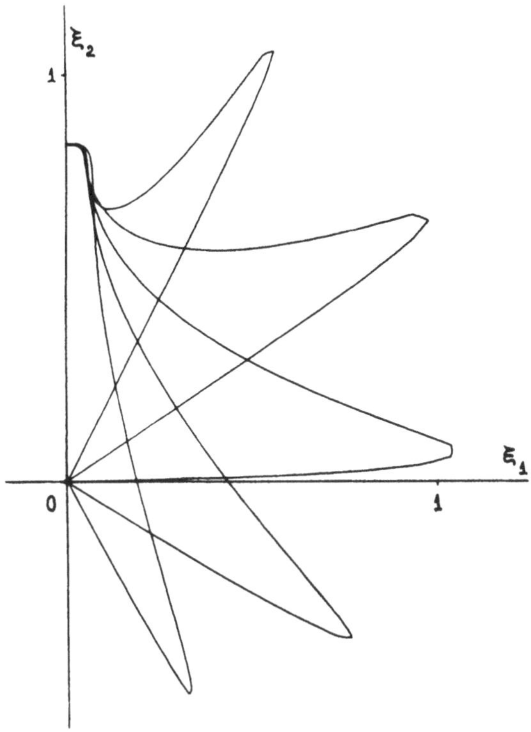

Fig. 2.5.3

inequality (50) is fulfilled then the function ρ_0 (43) satisfies the conditions

$$\rho_0(t) \to \infty \text{ as } t \to -\infty, \quad \rho(t) \geqslant 0 \text{ as } t \leqslant 0.$$

It follows from (44) that

$$\lim_{t\to-\infty} \|y(t,x)\| = [(x_1 - x_5 + x_3/\alpha_1 - x_4/\alpha_2)^2 + (x_2 - x_6 + x_4/\alpha_1 - x_8/\alpha_2)^2]^{1/2}.$$

Therefore if condition (50) is fulfilled for any point $x_0 \in R^8$, one can choose the number $t_0 \in (-\infty,0]$ in such a manner that condition (46) is fulfilled. In other words, under condition (50) and any initial condition x_0 the pursuit can be completed in a finite period of time (here it is implied that in a pursuit problem it is required that coordinates (x_1,x_2) coincide with coordinates (x_5,x_6)).

For parameters (48), condition (50) is not fulfilled. Hence the approach problem under consideration is not solvable for all points x_0. If we take, for example, the point $x_{01} = 3$, $x_{02} = x_{03} = x_{08} = 0$, then for any $t_0 \in (-\infty,0]$ the approach problem does not possess a solution.

Chapter 3
EXISTENCE OF THE VALUE FOR A POSITIONAL DIFFERENTIAL GAME

3.1. A Differential Game with a Fixed Termination Time

In this chapter, the existence of the value for a differential game with various payoff functionals will be proved. As is shown below, in the differential games under consideration minimax and maximin problems are equivalent to some problems of approach and evasion; therefore the existence of an equilibrium situation will follow from the theorem on an alternative for a differential game of approach-evasion. In this section, the existence of the value for a differential game with a fixed termination time will be proved.

3.1.1. Consider a differential game in which the motion of the controlled system is described by equation (1.2.1). As above, we shall assume that function f on the r.h.s. of equation (1.2.1) satisfies conditions (i) - (iv) (cf. §1.2.2). We suppose that in the differential game the payoff functional is of the form

(1) $$\gamma(x(\cdot)) = \sigma(x(\theta)).$$

Here $\sigma : R^n \to R^1$ is a continuous function, θ is the fixed termination time of the game, $x(\theta)$ is the phase point of the controlled system at time $t = \theta$. In accordance with the general definition of a differential game given in §1.5.2, minimax and maximin problems form the differential game under consideration. We present statements of these problems.

Problem 3.1.1. *It is required to obtain a positional strategy U^0 for the first player such that*

(2) $$\Gamma_1(U^0) = \min_U \Gamma_1(U)$$

where

(3) $$\Gamma_1(U) \overset{\text{def}}{=} \max_{x(\cdot)} \sigma(x(\theta)) \quad \textit{for} \quad x(\cdot) \in X_1(t_0,x_0,U).$$

Problem 3.1.2. *It is required to obtain a positional strategy V^0 for the second player such that*

(4) $$\Gamma_2(V^0) = \max_{V} \Gamma_2(V)$$

where

(5) $$\Gamma_2(V) \overset{\text{def}}{=} \min_{x(\cdot)} \sigma(x(\theta)) \quad \textit{for} \quad x(\cdot) \in X_2(t_0,x_0,V).$$

In these problems optimal strategies U^0 and V^0 are defined for a given fixed initial position (t_0,x_0). We also note that the maximum and minimum in (3) and (5) are actually achieved since the functional $x(\cdot) \to \gamma(x(\cdot)) = \sigma(x(\theta))$ is continuous in the space $C_n([t_0,\theta])$ and the sets $X_1(t_0,x_0,U)$ and $X_2(t_0,x_0,V)$ are compact sets in $C_n([t_0,\theta])$ (cf. Lemma 1.3.1). From the proof of Theorem 3.1.1 presented below, it will follow that the minimum and maximum in (2) and (4) are also attained.

Theorem 3.1.1. *For any choice of initial position $(t_0,x_0) \in T \times R^n$ there exists the value ω^0 and a saddle point (U^0,V^0) in a differential game with a payoff functional of the form* (1), *i.e. there exists a pair of positional strategies U^0 and V^0 such that*

(6) $$\Gamma_1(U^0) = \omega^0 = \Gamma_2(V^0).$$

Proof. Let a be a number. Set

(7) $$M_a \overset{\text{def}}{=} \{(\theta,x)\colon \ \sigma(x) \leqslant a\}, \qquad N \overset{\text{def}}{=} T \times R^n.$$

It follows directly from the definition of the condition of an (M,N)-approach (cf. §2.1.1) that for a continuous function $x(\cdot)$: $[t_0,\theta] \to R^n$ the inequality $\gamma(x(\cdot)) = \sigma(x(\theta)) \leqslant a$ will be satisfied if and only if the condition of (M_a, N)-approach is satisfied for this function.

Denote by A_1 the set of numbers a for which $M_a \neq \phi$ and such that for the given initial position (t_0,x_0) the (M_a, N)-approach problem is solvable. It is easy to verify that the set A_1 possesses the following property: if $a_* \in A_1$ then any number $a^* \geqslant a_*$ also belongs to the set A_1. Indeed, we obtain from (7) that $M_{a_*} \supset M_{a^*}$, whence the above-mentioned property follows. Therefore two cases are possible: either $A_1 = (a_1,\infty)$ or $A_1 = [a_1,\infty)$. We shall show that actually $A_1 = [a_1,\infty)$. For this purpose it is required to verify that A_1 is closed.

Let $a^{(k)} \in A_1$ ($k = 1, 2, ...,$) and let $a^{(k)} \to a_*$ for $k \to \infty$. It is required to prove that $a_* \in A_1$. We utilize Theorem 2.4.3. Let V be an arbitrary strategy for the second player. Since $a^{(k)} \in A_1$ by Theorem 2.4.3, there exists a motion $x_k(\cdot) \in X_2(t_0,x_0,V)$ which satisfies the condition of $(M_{a^{(k)}}, N)$-approach, i.e. the inequality $\sigma(x_k(\theta)) \leqslant a^{(k)}$ is satisfied. The set $X_2(t_0,x_0,V)$ is compact in $C_n([t_0,\theta])$, thus one can assume that the sequence $x_k(\cdot)$ ($k = 1, 2, ...$) converges to the limit $x_*(\cdot)$ which is contained in $X_2(t_0,x_0,V)$. The motion $x_*(\cdot)$ satisfies the condition of an (M_{a_*}, N)-approach. Indeed, taking into account that the function σ is continuous and approaching the limit as $k \to \infty$ in the inequality $\sigma(x_k(\theta)) \leqslant a^{(k)}$, we obtain $\sigma(x_*(\theta)) \leqslant a_*$. It has thus been proved that for an arbitrary strategy V there exists a motion $x_*(\cdot) \in X_2(t_0,x_0,V)$ which satisfies the condition of an (M_{a_*}, N)-approach. Therefore it follows from Theorem 2.4.3 that the problem of an (M_{a_*}, N)-approach is solvable, i.e. $a_* \in A_1$.

Let $a \notin A_1 = [a_1,\infty)$, i.e. the problem of an (M_a, N)-approach has no solution. Then, in view of Theorem 2.1.1 there exists a number $\epsilon > 0$ and a strategy V_ϵ which solves the problem of $(M_a, N)^\epsilon$-evasion. This assertion may be weakened, and one can assert that there exists a strategy V_* such that for any motion $x(\cdot) \in X_2(t_0,x_0,V_*)$ the condition of an (M_a, N)-evasion is fulfilled (in our problem this condition means that the inequality $\sigma(x(\theta)) > a$ is satisfied).

Let

$$(8) \qquad K_a \overset{\text{def}}{=} \{(\theta,x)\colon\ \sigma(x) \geqslant a\}, \qquad \mathfrak{L} \overset{\text{def}}{=} T \times R^n.$$

In view of the above, for any $a < a_1$ there exists a strategy V_* which solves the problem of a $(K_a,\mathfrak{L})$-approach. (Actually the strategy V_* assures the fulfillment of the strict inequality $\sigma(x(\theta)) > a$; however, for our purposes it is more convenient to define the set K_a by means of the non-strict inequality $\sigma(x) \geqslant a$.) Denote by A_2 the set of numbers a such that the set K_a is non-empty and for a given initial position (t_0,x_0) in the class of positional strategies for the second player the problem of $(K_a,\mathfrak{L})$-approach is solvable. It can be shown that $A_2 = (-\infty,a_2]$. Verification of this assertion is carried out in the same manner as the proof of equality $A_1 = [a_1,\infty)$. Since the inequality $a < a_1$ implies the inclusion $a \in A_2$, the inequality $a_2 \geqslant a_1$ is valid.

Let U^0 be a positional strategy for the first player which solves the problem of an (M_{a_1}, N)-approach, i.e. this strategy assures that the inequality $\sigma(x(\theta)) \leqslant a_1$ is satisfied for any motion $x(\cdot) \in X_1(t_0,x_0,U^0)$. In view of (3) we obtain that $\Gamma_1(U^0) \leqslant a_1$. Let V^0 be the positional strategy for the second player which solves the problem of $(K_{a_2}, \mathfrak{L})$-approach, i.e. the strategy V^0 assures the inequality $\sigma(x(\theta)) \geq a_2$ for any motion $x(\cdot) \in X_2(t_0,x_0,V^0)$. It follows from (5) that $\Gamma_2(V^0) \geq a_2$. Recall that for any strategies U and V the inequality $\Gamma_1(U) \geq \Gamma_2(V)$ is valid (cf. (1.5.10)). Therefore the inequalities

$$a_2 \geq a_1, \quad \Gamma_1(U^0) \leq a_1, \quad \Gamma_2(V^0) \geq a_2, \quad \Gamma_1(U^0) \geq \Gamma_2(V^0)$$

imply the equalities

$$\Gamma_1(U^0) = a_1 = \omega^0 = a_2 = \Gamma_2(V^0).$$

Theorem 3.1.1 is thus proved.

Note that in accordance with the constructions described in the preceding chapter, the value of the differential game under consideration and its saddle point can be determined as follows. Let $W_a^{(1)}$ be the maximal u-stable bridge in an (M_a, N)-approach problem and let $W_a^{(2)}$ be the maximal v-stable bridge in a $(K_a, \mathfrak{L})$-approach problem. Set

$$a_1 \overset{\text{def}}{=} \inf\{a: (t_0x_0) \in W_a^{(1)}\},$$

$$a_2 \overset{\text{def}}{=} \sup\{a: (t_0,x_0) \in W_a^{(2)}\}.$$

From the results of the preceding chapter and the proof of Theorem 3.1.1 we obtain

Corollary 3.1.1. *Quantities a_1 and a_2 coincide, the initial position (t_0,x_0) belongs to the intersection $W_{a_1}^{(1)} \cap W_{a_2}^{(2)}$. Strategies U^0 and V^0 extremal to the sets $W_{a_1}^{(1)}$ and $W_{a_2}^{(2)}$, respectively, form a saddle point of the differential game under consideration.*

Note that for any constructive motion $x(\cdot) \in X(t_0,x_0,U^0,V^0)$ generated by a pair of optimal strategies, the position $(t,x(t))$ will remain at the intersection of the bridges $W_{a_1}^{(1)} \cap W_{a_2}^{(2)}$ for all $t \in [t_0,\theta]$. At time $t = \theta$ the equality $\sigma(x(\theta)) = \omega^0 = a_1 = a_2$ holds for these motions.

In the class of step-by-step positional controls there exists in the game under consideration, an ϵ-equilibrium situation, i.e. the following assertion is valid.

For any $\epsilon > 0$ there exists a $\delta > 0$ such that for any step-by-step motion $x(\cdot) \in X_1(t_0,x_0,U^0,\Delta)$ the inequality $\sigma(x(\theta)) \leq \omega^0 + \epsilon$ is fulfilled and for any step-by-step motion $x(\cdot) \in X_2(t_0,x_0,V^0,\Delta)$ the inequality $\sigma(x(\theta)) \geq \omega^0 - \epsilon$ is valid provided $\text{diam}\Delta \leq \delta$.

3.1.2. The existence of the value of a differential game with a payoff of the form (1) can be proved by assuming that the function is either lower or upper semi-continuous. Consider, for example, the case when the function is lower semi-continuous. (The definition and properties of semi-continuous functions are presented in §A.1.)

In the case under consideration, the statements of Problems 3.1.1 and 3.1.2 are changed somewhat and are stated as follows.

It is required to find a positional strategy U^0 such that

$$\Gamma_1(U^0) = \min_U \Gamma_1(U),$$

where

$$\Gamma_1(U) \overset{\text{def}}{=} \sup_{x(\cdot)} \sigma(x(\theta)) \quad \text{for } x(\cdot) \in X_1(t_0,x_0,U).$$

For an arbitrary $\epsilon > 0$ it is required to obtain a strategy V^ϵ such that

$$\Gamma_2(V^\epsilon) \geqslant \sup_V \Gamma_2(V) - \epsilon,$$

where

$$\Gamma_2(V) \overset{\text{def}}{=} \min_{x(\cdot)} \sigma(x(\theta)) \quad \text{for } x(\cdot) \in X_2(t_0,x_0,V).$$

Note that in the equality which defines the quantity $\Gamma_2(V)$, the minimum is attained since the function $x(\cdot) \to \sigma(x(\theta))$ is lower semi-continuous on the space $C_n([t_0,\theta])$ and the set $X_2(t_0,x_0,V)$ is compact in $C_n([t_0,\theta])$.

Theorem 3.1.1 is replaced by the following assertion. For an arbitrary initial position $(t_0,x_0) \in T \times R^n$ and any number $\epsilon > 0$ there exists a pair of positional strategies (U^0,V^ϵ) such that

$$\Gamma_1(U^0) = \omega^0 \leqslant \Gamma_2(V^\epsilon) + \epsilon.$$

Here $\omega^0 = \min_U \Gamma_1(U) = \sup_V \Gamma_2(V)$ is the value of the differential game under consideration.

The proof of this assertion coincides, in essence, with the proof of Theorem 3.1.1.

3.1.3. Consider a differential game with an integral-terminal payoff. Let $(x(\cdot), u(\cdot), v(\cdot))$ be a triple consisting of the control $u(\cdot)\colon [t_0,\theta] \to P$, $v(\cdot)\colon [t_0,\theta] \to Q$ and the motion of the system (1.2.1) generated by these controls: $x(\cdot)\colon [t_0,\theta] \to R^n$. We assume that the initial position (t_0,x_0) is fixed. To each such triple we assign the value of the functional

$$\gamma(x(\cdot),u(\cdot),v(\cdot)) \overset{\text{def}}{=} \sigma(x(\theta)) + \int_{t_0}^{\theta} f_{n+1}(t,x(t),u(t),v(t))dt. \tag{9}$$

As is known, functional χ may be reduced to the form (1). For this purpose it is required to introduce an additional phase variable x_{n+1} whose motion is described by the equation $\dot{x}_{n+1}(t) = f_{n+1}(t, x(t), u(t), v(t))$. Setting $x_{n+1}(t_0) = 0$ we then obtain the equality

$$\chi(x(\cdot), u(\cdot), v(\cdot)) = \sigma(x(\theta)) + x_{n+1}(\theta) = \sigma^{\#}(y(\theta))$$

where, therefore,

$$y \overset{\text{def}}{=} (x_1, ..., x_n, x_{n+1})^{\mathrm{T}}, \quad \sigma^{\#}(y) \overset{\text{def}}{=} \sigma(x) + x_{n+1}.$$

The motion of an $(n + 1)$-dimensional phase vector y is described by the equation

$$\dot{y}(t) = f^{\#}(t, y(t), u(t), v(t)). \tag{10}$$

Here $f^{\#}(t,y,u,v)$ is an $(n + 1)$-dimensional vector with components $f_i^{\#}(t,y,u,v) = f_i(t,x,u,v)$ $(i = 1, 2, ..., n + 1)$, where $y = (x,x_{n+1})$. Note that y and x are $(n+1)$-dimensional and n-dimensional column-vectors so that the notation $y = (x^{\mathrm{T}},x_{n+1})^{\mathrm{T}}$ is appropriate. However, for simplicity, we shall omit the transposition index.

We thus obtain that the initial differential game in which the motion is described by equation (1.2.1) and the payoff is given by the equality (9), reduces to a differential game in which the motion of the $(n + 1)$-dimensional phase vector y is described by equation (10) and the payoff functional is of the form

$$\gamma^{\#}(y(\cdot)) \overset{\text{def}}{=} \sigma^{\#}(y(\theta)) = \sigma(x(\theta)) + x_{n+1}(\theta). \tag{11}$$

We shall assume that the function $\sigma: \ R^n \to R^1$ is continuous and the function $f^{\#}$: $T \times R^{n+1} \times \mathcal{P} \times \mathcal{Q} \to R^{n+1}$ satisfies all the conditions (i) – (iv) stated in §1.1.2. It immediately follows from the results of §3.1.1 that for any initial position $(t_0,y_0) = (t_0,x_0,0)$ the value ω^0 and the saddle point (U^0,V^0) exist for the differential game (10), (11), i.e.

$$\omega^0 = \min_U \Gamma_1^{\#}(U) = \Gamma_1^{\#}(U^0) = \Gamma_2^{\#}(V^0) = \max_V \Gamma_2^{\#}(V), \tag{12}$$

where

$$\Gamma_1^{\#}(U) \overset{\text{def}}{=} \max_{y(\cdot)} \sigma^{\#}(y(\theta)) \quad \text{for} \quad y(\cdot) \in \mathcal{Y}_1(t_0,y_0,U),$$
$$\Gamma_2^{\#}(V) \overset{\text{def}}{=} \min_{y(\cdot)} \sigma^{\#}(y(\theta)) \quad \text{for} \quad y(\cdot) \in \mathcal{Y}_2(t_0,y_0,V).$$

Here $\mathcal{Y}_1(t_0,y_0,U)$ and $\mathcal{Y}_2(t_0,y_0,V)$ are the sets of constructive motions of the system (10) generated by the strategies U: $T \times R^{n+1} \to \mathcal{P}$ and V: $T \times R^{n+1} \to \mathcal{Q}$, respectively. Observe that the optimal strategies U^0 and V^0 are defined on the space of positions $T \times R^{n+1}$, i.e. these functions in general depend on the new phase variable x_{n+1}.

Note that the function $f^{\#}$ does not depend on the phase variable x_{n+1} and the payoff functional is expressed in the form of the sum (11). It turns out that in this particular case, one can construct an equilibrium situation in which positional strategies of the players do not depend on the variable x_{n+1}. However, by discarding the information about the variable x_{n+1} we preclude construction of optimal strategies which form a saddle point, and we shall define ϵ-optimal strategies.

We thus state and prove the assertion concerning the existence of an ϵ-equilibrium in the class of positional strategies $(t,x) \to U(t,x)$ and $(t,x) \to V(t,x)$.

Let ω^0 be the value of the differential game (10), (11) defined in accordance with (12). It is asserted that one can construct positional strategies $(t,x) \to U^{\delta}(t,x)$ and $(t,x) \to V^{\delta}(t,x)$ such that the relations

$$\text{(13)} \qquad \Gamma_1^{\#}(U^{\delta}) \leqslant \omega^0 + \epsilon(\delta), \quad \Gamma_2^{\#}(V^{\delta}) \geqslant \omega^0 - \epsilon(\delta)$$

are valid. Here δ is a positive parameter, $\epsilon(\delta) \to 0$ as $\delta \to 0$.

We define the strategy U^{δ}. Let

$$\text{(14)} \qquad \begin{aligned} M^{\#} &\overset{\text{def}}{=} \{(\theta,y) \in T \times R^{n+1}:\ \sigma^{\#}(y) = \sigma(x) + x_{n+1} \leqslant \omega^0\}, \\ N^{\#} &\overset{\text{def}}{=} T \times R^{n+1}, \end{aligned}$$

and let W be the maximal u-stable bridge in an $(M^{\#}, N^{\#})$-approach problem formulated for the system (10). Note that W is a closed set in the space $T \times R^{n+1}$. In view of Corollary 3.1.1, $(t_0 y_0) = (t_0,x_0,0) \in W$. According to the construction of an extremal strategy (cf. §2.4.1) we define the set $W[t] = \{y \in R^{n+1} : (t,y) \in W\}$ and a function $(t,y) \to e(t,y)$; here $e(t,y)$ is the point in the set W closest to y. Note that in the problem under consideration, $W[t] \neq \phi$ for all $t_0 \leqslant t \leqslant \theta$. Indeed, it follows from (14) that for fixed (t,x) and for $x_{n+1} \to -\infty$ the point (t,x,x_{n+1}) falls into the set W.

In subsequent constructions the quantity $\beta(t,\delta)$ is utilized (cf. (2.3.16)). In the definition of this quantity which was presented in §2.3.2 the function $f^{\#}$: $T \times R^n \times P \times Q \to R^n$ should be replaced by the function $f^{\#}$: $T \times R^{n+1} \times P \times Q \to R^{n+1}$. Let $(t,x) \in T \times R^n$. Set

$$\text{(15)} \qquad G_{n+1}(t,x\delta) \overset{\text{def}}{=} \{x_{n+1} \in R^1:\ \text{dist}^2((x,x_{n+1}),W[t]) \leqslant \beta(t,\delta)\}.$$

Define the quantity

$$\text{(16)} \qquad x^*_{n+1}(t,x,\delta) \overset{\text{def}}{=} \max x_{n+1} \quad \text{for} \quad x_{n+1} \in G_{n+1}(t,x\delta).$$

The set $G_{n+1}(t,x,\delta)$ is non-empty, closed, and bounded from above; therefore the quantity $x^*_{n+1}(t,x,\delta)$ is well defined.

Denote by $s(t,x,\delta)$ the vector $y - e(t,y)$ where $y = (x,x^*_{n+1}(t,x,\delta))$, i.e. $s(t,x,\delta)$ is an $(n+1)$-dimensional vector whose components are expressed by the equalities

$$s_i(t,x,\delta) = x_i - e_i(t,x,x^*_{n+1}(t,x,\delta)),\ i = 1, 2, ..., n$$

$$s_{n+1}(t,x,\delta) = x^*_{n+1}(t,x,\delta) - e_{n+1}(t,x,x^*_{n+1}(t,x,\delta)).$$

Finally, the function U^δ: $T \times R^n \to P$ is defined from the condition

(17) $$\max_{v \in Q} \langle s(t,x,\delta), f^\#(t,y,U^\delta(t,x),v)\rangle = \min_{u \in P} \max_{v \in Q} \langle s(t,x,\delta), f^\#(t,y,u,v)\rangle$$

Recall that $y = (x,x^*_{n+1})$ and the function $f^\#$ does not depend on the variable x_{n+1}, i.e.

(18) $$f^\#_i(t,y,u,v) = f^\#_i(t,x,x_{n+1},u,v) = f_i(t,x,u,v) \quad (i = 1, 2, ..., n + 1).$$

Therefore for a fixed parameter δ condition (17) determines a function U^δ which depends only on the position (t,x).

We shall prove that for the constructed strategy U^δ the estimate $\Gamma^\#_1(U^\delta) \leqslant \omega^0 + \epsilon(\delta)$ is valid (cf. (13)). We first note the following assertion. Let $(t_*,y_*) = (t_*,x_*,x_{*n+1}) \in W$, i.e. from the initial position (t_*,y_*) the problem of the $(M^\#, N^\#)$-approach formulated for the system (10) is solvable. Then it follows from (14) and (18) that the $(M^\#, N^\#)$-approach problem will be solvable from any position (t_*,x_*,x^*_{n+1}), where $x^*_{n+1} \leqslant x_{*n+1}$. Consequently, if $(t_*,x_*,x_{*n+1}) \in W$ then the set W contains the whole half-line $\{(t_*,x_*,x_{n+1}) : x_{n+1} \leqslant x_{*n+1}\}$. Utilizing this assertion it is easy to obtain that for any $(t,x) \in T \times R^n$ and $x^{(1)}_{n+1} \leqslant x^{(2)}_{n+1}$ the inequality

(19) $$\text{dist}((x,x^{(1)}_{n+1}), W[t]) \leqslant \text{dist}((x,x^{(2)}_{n+1}), W[t])$$

will be valid.

Let a parameter $\delta \in (0,\delta_0)$ be chosen. We shall assume that for the selected number δ_0 the estimate (2.3.17) is fulfilled. Consider an arbitrary step-by-step motion $y(\cdot) \in Y_1(t_0,y_0,U^\delta,\Delta)$. Here $\Delta = \{t_0 = \tau_0 < \tau_1 < ... < \tau_{k+1} = \theta\}$ is a partition which satisfies the estimate $\text{diam}\Delta \leqslant \delta$ and $Y_1(t_0,y_0,U^\delta,\Delta)$ is a set of step-by-step motions of the system (10) generated by the strategy U^δ. We shall prove by induction that the inequalities

(20) $$\text{dist}^2(y(\tau_i), W[\tau_i]) \leqslant \beta(\tau_i,\delta),\quad i = 0, 1, ..., k + 1$$

are valid. For $\tau_0 = t_0$ we have $\beta(\tau_0,\delta) = 0$, $y(\tau_0) \in W[\tau_0]$. Therefore, for $i = 0$ the estimate (20) is fulfilled. Let the estimate (20) be fulfilled for some $i = \ell$. We shall use the notation $y(\tau_\ell) = (x(\tau_\ell), x_{n+1}(\tau_\ell))$. In accordance with (15) we obtain $x_{n+1}(\tau_\ell) \in G^*_{n+1}(\tau_\ell,x(t_\ell),\delta)$. Set $x^*_{n+1}(\tau_\ell) = x^*_{n+1}(\tau_\ell,x(\tau_\ell),\delta)$. From (16) we obtain that

(21) $$x^*_{n+1}(\tau_\ell) \geqslant x_{n+1}(\tau_\ell).$$

Let $u(\tau_\ell) = U^\delta(\tau_\ell, x(\tau_\ell))$. On the time interval $[\tau_\ell, \tau_{\ell+1}]$ the step-by-step motion $y(\cdot)$ under consideration satisfies the differential inclusion

(22) $$\dot{y}(t) \in \{f^\#(t, y(t), u(\tau_\ell), v): \; v \in Q\}.$$

Define the function $y^*(t) = (x^*(t), x^*_{n+1}(t))$ such that $x^*(t) = x(t)$, $x^*_{n+1}(t) = x_{n+1}(t) + x^*_{n+1}(\tau_\ell) - x_{n+1}(\tau_\ell)$ for $\tau_\ell \leqslant t \leqslant \tau_{\ell+1}$. It follows from (18) that the function $y^*(\cdot)$ as well as the function $y(\cdot)$ satisfies the differential inclusion (22). Set $w_* = e(\tau_\ell, y^*(\tau_\ell))$. Define a vector $v_0 \in Q$ from the condition

$$\min_{u \in P} \langle y^*(\tau_\ell) - w_*, f^\#(\tau_\ell, y^*(\tau_\ell), u, v_0)\rangle$$
$$= \max_{v \in Q} \min_{u \in P} \langle y^*(\tau_\ell) - w_*, f^\#(\tau_\ell, y^*(\tau_\ell), u, v)\rangle.$$

It follows from the condition of u-stability of the set W that there exists a solution $w(\cdot)$ of the differential inclusion

$$\dot{w}(t) \in \mathrm{co}\{f^\#(t, w(t), u, v_0): \; u \in P\}, \quad w(\tau_\ell) = w_*,$$

satisfying the condition $w(\tau_{\ell+1}) \in W[\tau_{\ell+1}]$. It also follows from the preceding constructions that the control $u(\tau_\ell)$ satisfies the condition (17) where $f^\#(t,y,u,v) = f^\#(\tau_\ell, y^*(\tau_\ell), u, v)$, $s(t,x,\delta) = y^*(\tau_\ell) - w_*$. Note also that from (15) and (16) we have the estimate $\|y^*(\tau_\ell) - w_*\|^2 \leqslant \beta(\tau_\ell, \delta)$. Thus in the situation under consideration the motions $y^*(\cdot)$ and $w(\cdot)$ satisfy all the conditions of Lemma 2.3.1. In view of this lemma we obtain

$$\|y^*(\tau_{\ell+1}) - w(\tau_{\ell+1})\|^2 \leqslant \beta(\tau_{\ell+1}, \delta).$$

Since $w(\tau_{i+1}) \in W[\tau_{\ell+1}]$ we obtain the estimate

(23) $$\mathrm{dist}^2(y^*(\tau_{\ell+1}), W[t_{\ell+1}]) \leqslant \beta(\tau_{\ell+1}, \delta).$$

Recall that the auxiliary motion $y^*(\cdot)$ and the step-by-step motion $y(\cdot)$ under consideration are related as follows:

$$y(t) = (x(t), x_{n+1}(t)), \quad y^*(t) = (x(t), x^*_{n+1}(t)),$$

$$x^*_{n+1}(t) = x_{n+1}(t) + x^*_{n+1}(\tau_\ell) - x_{n+1}(\tau_\ell) \geqslant x_{n+1}(t), \quad \tau_\ell \leqslant t \leqslant \tau_{\ell+1}.$$

(Here we utilize the inequality (21).) Thus for the case $i = \ell + 1$ inequality (20) follows from (19) and (23). Taking $\tau_{k+1} = \theta$ and $i = k+1$ in (20) we have

(24) $$\mathrm{dist}(y(\theta), [\theta]) = \|y(\theta) - e(\theta, y(\theta))\| \leqslant \epsilon_*(\delta) \overset{\mathrm{def}}{=} \beta(\theta, \delta)^{1/2}.$$

Here $e(\theta,y(\theta))$ is the point in the set $\mathcal{W}[\theta]$ closest to the point $y(\theta)$. It follows from (2.3.16) that $\epsilon_*(\delta) \to 0$ as $\delta \to 0$. From the definition of the set $M^{\#}$ (14) and a u-stable bridge it follows that $\mathcal{W}(\theta) \subset M^{\#}$, i.e. $\sigma^{\#}(w) \leqslant \omega^0$ for all $w \in \mathcal{W}[\theta]$. In particular, the inequality $\sigma^{\#}(e(\theta,y(\theta))) \leqslant \omega^0$ is fulfilled. We thus obtain that for diam$\Delta \leqslant \delta$ any step-by-step motion $y(\cdot) \in Y_1(t_0,y_0,U^{\delta},\Delta)$ satisfies relations

$$\|y(\theta) - e(\theta,y(\theta))\| \leqslant \epsilon_*(\delta), \quad \sigma^{\#}(e(\theta,y(\theta))) \leqslant \omega^0, \tag{25}$$
$$\epsilon_*(\delta) \to 0 \quad \text{as} \quad \delta \to 0.$$

Recall from the construction of the set $\mathcal{D}_1$ (2.3.10) we have $y(\theta) \in \mathcal{D}_1[\theta]$, $e(\theta,y(\theta)) \in \mathcal{D}_1[\theta]$, where

$$\mathcal{D}_1[\theta] \overset{\text{def}}{=} \{y \in R^{n+1}\colon \ (\theta,y) \in \mathcal{D}_1\}.$$

The set $\mathcal{D}_1[\theta]$ is bounded. Since the function $\sigma^{\#}$ is uniformly continuous on $\mathcal{D}_1[\theta]$, we know from (25) that there exists an $\epsilon(\delta)$ such that for diam $\Delta \leqslant \delta$ the estimate

$$\delta^{\#}(y(\theta)) \leqslant \omega^0 + \epsilon(\delta), \ \epsilon(\delta) \to 0 \quad \text{as } \delta \to 0$$

is satisfied for all step-by-step motions $y(\cdot) \in Y_1(t_0,y_0,V^{\delta},\Delta)$. Consequently the same estimate will be valid for any constructive motion $y(\cdot) \in Y_1(t_0,y_0,U^{\delta})$. Thus the first of the inequalities (13) is proved.

The construction of the strategy $(t,x) \to V^{\delta}(t,x)$ and the proof of the second inequality (13) are carried out in the same manner as was done for the strategy U^{δ}. Note that when constructing strategy V^{δ} the operation *max* which was utilized in relation (16) should be replaced by the operation *min*.

Strategies U^{δ} and V^{δ} form an ϵ-saddle point of the differential game (10), (11). This assertion is valid for those definitions of positional strategies and constructive motions presented in Chapter 1. We note that these definitions can be modified and in the game under consideration one can obtain a saddle point in the class of positional strategies which do not depend on the variable x_{n+1}. For this purpose it is required to assume that the functions U and V depend not only on the position (t,x) but also on the parameter ϵ. Constructive motions generated by these strategies are defined by a two-stage limiting transition. First the limiting motions are selected for a fixed $\epsilon > 0$ which are obtained as diam$\Delta \to 0$; next, limiting motions as $\epsilon \to 0$ are selected from the motions so obtained. Using this definition of strategies $(t,x,\epsilon) \to U(t,x,\epsilon)$ and $(t,x,\epsilon) \to V(t,x,\epsilon)$ and the corresponding constructive motions in the differential game (10), (11) there would exist a saddle point (U^0,V^0). We note that strategies of this type are considered in §8.5.

3.2. Differential Games of Pursuit-Evasion

In this section a differential game of pursuit-evasion is considered. The payoff in this game is the time elapsed before the encounter between the pursuer and evader. The existence of the value for this type of differential game is proved. This proof, as in the case of a game with a fixed termination, is based on the theorem of an alternative.

3.2.1. We present the statement of game-theoretic problems of pursuit and evasion. Let two controlled systems with the equations of motion

(1) $$\dot{y}(t) = f^{(1)}(t,y(t), u(t)),$$

(2) $$\dot{z}(t) = f^{(2)}(t,z(t), v(t))$$

be given. Here y and z are n_1-dimensional and n_2-dimensional vectors, respectively. It is assumed that the first player, the *pursuer*, controls the motion of system (1), while the motion of system (2) is controlled by the second player, the *evader*. As usual the controls of the players will satisfy the restrictions

$$u(t) \in \mathcal{P}, \quad v(t) \in \mathcal{Q},$$

where $\mathcal{P}$ and $\mathcal{Q}$ are compact sets in R^p and R^q, respectively. As far as the function $f^{(1)}$: $T \times R^{n_1} \times \mathcal{P} \to R^{n_1}$ and $f^{(2)}$: $T \times R^{n_2} \times \mathcal{Q} \to R^{n_2}$ are concerned, we shall assume that they satisfy conditions (i) – (iii) formulated in §1.2.2. However, unlike the cases considered previously we shall assume that the interval T is not bounded from the right, i.e. in this case $T = R^1 = (-\infty,\infty)$.

Instead of the pair of systems (1), (2) it is convenient to consider one system of the form (1.2.1). Thus we define $n = n_1 + n_2$, $x = (x_1, \ldots, x_n)^T = (y_1, \ldots, y_{n_1}, z_1, \ldots, z_{n_2})^T$,

(3) $$f_i(t,x,u,v) \overset{\text{def}}{=} \begin{cases} f_j^{(1)}(t,y,u) & \text{for } i = j = 1, \ldots, n_1, \\ f_j^{(2)}(t,z,v) & \text{for } i = n_1 + j, \ j = 1, \ldots, n_2. \end{cases}$$

The function f: $R \times R^n \times \mathcal{P} \times \mathcal{Q} \to R^n$, defined by equality (3), will satisfy all the conditions (i) – (iv). The motion of an n-dimensional phase vector $x(t)$ is described by the equation

(4) $$\dot{x}(t) = f(t,x(t),u(t),v(t)).$$

For an arbitrary initial position $(t_0,x_0) \in R \times R^n$ and any choice of arbitrary measurable controls $u(\cdot)$: $[t_0,\infty) \to \mathcal{Q}$, $v(\cdot)$: $[t_0,\infty) \to \mathcal{Q}$, the corresponding solution of equation (4) exists, is unique, and can be extended over the whole semi-axis $[t_0,\infty)$.

We define the payoff functional. Let a closed set M^* be selected in the phase space. Let $x(\cdot)$: $[t_0,\infty) \to R^n$ be a continuous function. Set

(5) $$T^0(x(\cdot)) \overset{\text{def}}{=} \{t \geq t_0 : \; x(t) \in M^*\}.$$

We define the functional τ^0 on the space $C_n([t_0,\infty))$ of continuous functions $x(\cdot)$: $[t_0,\infty) \to R^n$. Set

(6) $$\tau^0(x(\cdot)) \overset{\text{def}}{=} \begin{cases} \min\{t: \; t \in T^0(x(\cdot))\} \text{ for } T^0(x(\cdot)) \neq \phi, \\ \infty \qquad\qquad \text{for } T^0(x(\cdot)) = \phi . \end{cases}$$

Since the set $T^0(x(\cdot))$ is closed, the minimum is actually attained.

Consider two particular cases. In a pursuit problem let us be required to assure that the first m coordinates of the points $y(t)$ and $z(t)$ coincide. The time at which $y_i(t) = z_i(t)$ $(i = 1, \ldots, m)$ for the first time is called the capture time. The pursuer wants the capture to occur as early as possible. The evader, on the contrary, tries to either avoid the capture or at least to delay it as long as possible. In the case under consideration, after the transformations indicated above, the payoff functional should be defined in accordance with (5) and (6), where the set M^* is a linear subspace:

$$M^* = \{x \in R^n : \; x_i = x_{n_1+i}, \; i = 1, \ldots, m\}.$$

Consider now the other case. Assume that the capture occurs when the inequality

$$\sum_{i=1}^{m} (y_i(t) - z_i(t))^2 \leq r_*^2$$

is realized for the first time (here r_* is a given number − the radius of capture). In this case the set M^* is determined by the relation

$$M^* = \{x \in R^n : \; \sum_{i=1}^{m} (x_i - x_{n_1+i})^2 \leq r_*^2\}.$$

We now return to the general case. Let $x_\ell(\cdot) \in C_n([t_0,\infty))$ $(\ell = 1, 2, \ldots)$. We say that the sequence converges to the limiting function $x_*(\cdot)$: $[t_0,\infty) \to R^n$ if for any $\theta > t_0$

(7) $$\max_{t_0 \leq t \leq \theta} \|x_\ell(t) - x_*(t)\| \to 0 \text{ as } \ell \to \infty.$$

Observe that the limiting function $x_*(\cdot)$ will be continuous on $[t_0,\infty)$. The convergence (7) will be denoted by the usual symbol $x_\ell(\cdot) \to x_*(\cdot)$.

The functional τ^0 (6) is lower semi-continuous. This property means the following. Let $x_\ell(\cdot) \in C_n([t_0,\infty))$ $(\ell = 1, 2, \ldots)$, $x_\ell(\cdot) \to$

$x_*(\cdot)$, $\tau^0(x_\ell(\cdot)) \to \tau_*$ as $\ell \to \infty$ then $\tau^0(x_*(\cdot)) \leqslant \tau_*$. Note that here τ_* and $\tau^0(x_*(\cdot))$ can be the number $+\infty$. The proof of this property was presented in 1.4.2, where the lower semi-continuity of the functional τ^0 (1.4.7) was established.

Observe that in §1.4 functionals of the form $\gamma : C_n([t_0,\theta]) \to R$ were considered. Therefore we did not introduce the functional τ^0 which can take on the improper value $+\infty$, but confined ourselves to a consideration of a "truncated" functional τ^0 (1.4.7) which was defined on $C_n([t_0,\theta])$ and took on values on the interval $[t_0,\theta + 1]$.

Let U: $R \times R^n \to P$ be a *positional strategy of the pursuer*. Observe that the function $(t,x) \to U(t,x)$ is defined here for all $t \in (-\infty,\infty)$. Let $\Delta \overset{\text{def}}{=} \{t_0 = \tau_0 < \tau_1 < ...\}$ be a partition of the semi-axis $[t_0,\infty)$. As usual we introduce the set $X_1(t_0,x_0,U,\Delta)$ of *step-by-step motions*. Note here that step-by-step motions are defined on the semi-axis $[t_0,\infty)$. Set

$$\operatorname{diam}\Delta \overset{\text{def}}{=} \sup_{i=1,2,...} (\tau_{i+1} - \tau_i).$$

Let $x_\ell(\cdot) \in X_1(t_0,x_0,U,\Delta_\ell)$ (ℓ = 1, 2, ...) be a sequence of step-by-step motions. If $x_\ell(\cdot) \to x_*(\cdot)$ and $\operatorname{diam}\Delta_\ell \to 0$ as $\ell \to \infty$, then the limiting function $x_*(\cdot)$ is called a *constructive motion* generated by the strategy U. (The convergence of a sequence of functions $x_\ell(\cdot)$ is taken here in the sense (7).) The set of all the constructive motions is denoted by $X_1(t_0,x_0,U)$. The set $X_1(t_0,x_0,U)$ is non-empty and sequentially compact in $_n([t_0,\infty))$, i.e. from any sequence $x_k(\cdot) \in X_1(t_0,x_0,U)$ (k = 1, 2, ...) one can select a subsequence $x_{k_\ell}(\cdot)$ (ℓ = 1, 2, ...) convergent in the sense (7) to a limit function $x_*(\cdot)$ which is contained in the set $X_1(t_0,x_0,U)$. The proof of the property coincides in essence with the proof of Lemma 1.3.1.

The set of constructive motions $X_2(t_0,x_0,V)$ generated by the *positional strategy* V: $R \times R^n \to Q$ *of the evader* is introduced analogously. For any strategy V the set $X_2(t_0,x_0,V)$ will be non-empty and sequentially compact in $C_n([t_0,\infty))$.

Quality criteria of strategies U and V are defined in accordance with the general scheme. Namely we set

$$\Gamma_1(U) \overset{\text{def}}{=} \sup \tau^0(x(\cdot)) \quad \text{for} \quad x(\cdot) \in X_1(t_0,x_0,U), \tag{8}$$

$$\Gamma_2(V) \overset{\text{def}}{=} \min \tau^0(x(\cdot)) \quad \text{for} \quad x(\cdot) \in X_2(t_0,x_0,V). \tag{9}$$

Here the minimum is attained since the set $X_2(t_0,x_0,V)$ is sequentially compact and functional τ^0 is lower semi-continuous (cf. §A.1).

Problems of pursuit and evasion are stated as follows:

Problem 3.2.1. *For a given initial position $(t_0,x_0) \in R \times R^n$ it is required to find a positional strategy U^0 such that*

(10) $\Gamma_1(U^0) = \min_U \Gamma_1(U).$

From the proof of Theorem 3.2.1 presented below, it follows that the minimum is attained in (10).

Problem 3.2.2. *Let an initial position $(t_0,x_0) \in R \times R^n$ be given. If $\sup_V \Gamma_2(V) < \infty$, then for an arbitrary $\epsilon > 0$ it is required to obtain strategy V^ϵ such that*

$$\Gamma_2(V^\epsilon) \geqslant \sup_V \Gamma_2(V) - \epsilon.$$

If $\sup_V \Gamma_2(V) = \infty$ then for an arbitrary number θ as large as desired, it is required to obtain strategy V such that

$$\Gamma_2(V_\theta) \geqslant \theta.$$

3.2.2. The following theorem is valid.

Theorem 3.2.1. *For any initial position $(t_0,x_0) \in R \times R^n$ in a differential game of pursuit-evasion there exists the value ω^0, i.e.*

(11) $\min_U \Gamma_1(U) = \sup_V \Gamma_2(V) = \omega^0 \in [t_0,\infty].$

Proof. Let $\theta \geqslant t_0$. Set

(12)
$$M_\theta \overset{\text{def}}{=} \{(t,x):\ t \leqslant \theta,\ \ x \in M^*\},$$
$$N_\theta \overset{\text{def}}{=} \{(t,x):\ t \leqslant \theta,\ \ x \in R^n\}.$$

Denote by Θ the set of numbers $\theta \in [t_0,\infty)$ such that from a given initial position (t_0,x_0) the (M_θ, N_θ)-approach problem is solvable (cf. §2.1.1). If $\theta_1 \in \Theta$ and $\theta_2 \geqslant \theta_1$ then $\theta_2 \in \Theta$. This property follows directly from the definition of the set Θ. We shall show, using Theorem 2.4.3, that Θ is closed.

Let $\theta_\ell \in \Theta$ ($\ell = 1, 2, ...$), $\theta_\ell \to \theta_*$ for $\ell \to \infty$. We shall show that $\theta_* \in \Theta$. Let V: $R \times R^n \to \mathcal{Q}$ be an arbitrary strategy. By Theorem 2.4.3 there exists a sequence of motions $x_\ell(\cdot) \in X_2(t_0,x_0,V)$ such that the condition of (M_θ, N_θ)-approach is valid. Note that in Theorem 2.4.3, motions defined on a bounded time interval $[t_0,\theta]$ are considered. Therefore, here one should consider motions $x_\ell(\cdot)$: $[t_0,\theta_\ell] \to R^n$. However, for convenience we shall assume that these motions are extended over the whole semi-axis $[t_0,\infty)$. Clearly such an assumption does not prevent us from using Theorem 2.4.3. Thus we assume that $X_2(t_0,x_0,V) \subset C_n([t_0,\infty))$. As noted above, the set $X_2(t_0,x_0,V)$ is sequentially compact, therefore there exists a convergent subsequence whose limit is contained in $X_2(t_0,x_0,V)$. To simplify the notation, we assume that the initial sequence $x_\ell(\cdot)$ ($\ell = 1,2, ...$)

converges to the limit $x_*(\cdot)$. For any motion $x_\ell(\cdot)$ the condition of $(M_{\theta_\ell}, N_{\theta_\ell})$-approach is fulfilled, i.e.

(13) $\quad \tau^0(x_\ell(\cdot)) \leqslant \theta_\ell.$

Functional τ^0 is lower semi-continuous, therefore approaching the limit in inequality (13) we obtain $\tau^0(x_*(\cdot)) \leqslant \theta_*$. Consequently, for the limiting motion $x_*(\cdot)$ the condition of $(M_{\theta_*}, N_{\theta_*})$-approach is fulfilled. We obtain that for any strategy V there exists a motion $x_*(\cdot) \in X_2(t_0,x_0,V)$ which satisfies the condition of $(M_{\theta_*}, N_{\theta_*})$-approach. Therefore, in accordance with Theorem 2.4.3 the point θ_* is contained in the set Θ.

Assume that $\Theta \neq \phi$. It then follows from the properties proved above that $\Theta = [\theta_0,\infty)$. We shall show that in this case $\omega^0 = \theta_0$. Let U^0 be a strategy which solves the $(M_{\theta_0}, N_{\theta_0})$-approach problem, i.e. this strategy assures that the inequality $\tau^0(x(\cdot)) \leqslant \theta_0$ will be fulfilled for any motion $x(\cdot) \in X_1(t_0,x_0,U^0)$ or equivalently,

(14) $\quad \Gamma_1(U^0) \leqslant \theta_0.$

Select an arbitrary $\theta < \theta_0$. Since $\theta \notin \Theta$ the problem of an (M_θ, N_θ)-approach is not solvable. Therefore, in accordance with Theorem 2.1.1 for some $\epsilon > 0$, the problem of an $(M_\theta, N_\theta)^\epsilon$-evasion will be solvable. Consequently, there exists a strategy V_θ such that $\tau^0(x(\cdot)) > \theta$ for any $x(\cdot) \in X_2(t_0,x_0,V_\theta)$ or equivalently, $\Gamma_2(V_\theta) > \theta$. Here θ is an arbitrary number satisfying $\theta < \theta_0$. Therefore

(15) $\quad \sup_V \Gamma_2(V) \geqslant \theta_0.$

From (14), (15) and (1.5.10), we arrive at equality (11) where $\omega^0 = \theta_0$.

It remains to consider the case $\Theta = \phi$. In this case, for an arbitrary large θ there exists a strategy V_θ such that $\Gamma_2(V_\theta) > \theta$. Hence

$$\sup_V \Gamma_2(V) = \infty.$$

From (1.5.10) we have for any strategy U that

$$\Gamma_1(U) \geqslant \sup_V \Gamma_2(V) = \infty .$$

Thus, in this case we obtain the equality (11) where $\omega^0 = \infty$. Theorem 3.2.1 is thus proved.

From the proof of Theorem 3.2.1 we obtain

Corollary 3.2.1. *Let* W_θ *be the maximal* u-*stable bridge in an* (M_θ, N_θ)-*approach problem. Set*

$$\theta_0 \overset{\text{def}}{=} \min\{\theta \geqslant t_0\colon\ (t_0, x_0) \in W_\theta\}.$$

Then the strategy U^e *extremal to the set* W_{θ_0} *will be an optimal strategy in a pursuit problem.*

It also follows from the proof of Theorem 3.2.1 that for any $\theta < \omega^0$ there exists a number $\epsilon(\theta) > 0$ and a strategy V_θ which assures the evasion of motions from falling into an $\epsilon(\theta)$-neighborhood of the set M^*. This strategy can be defined as extremal to the corresponding v-stable bridge.

For step-by-step motions the following assertions are valid.

For an $\alpha > 0$ arbitrarily small there exists a $\delta > 0$ such that any step-by-step motion $x(\cdot) \in X_1(t_0,x_0,U^0,\Delta)$ falls into an ϵ-neighborhood of the set M^* not later than at a time $t = \omega^0$, provided $\operatorname{diam}\Delta \leqslant \delta$. (It is assumed here that $\omega^0 < \infty$.)

For any $\theta < \omega^0$ and $\beta \in (0,\epsilon(\theta))$ there exists a $\delta > 0$ such that any step-by-step motion $x(\cdot) \in X_2(t_0,x_0,V_0,\Delta)$ on the time interval $[t_0,\theta]$ will avoid falling into the β-neighborhood of the set M^*, provided $\operatorname{diam}\Delta \leqslant \delta$.

3.2.3. When describing the differential game of pursuit-evasion we have obtained the system (4) by combining the two controlled systems (1) and (2). One can, however, in the initial statement of the differential game of pursuit-evasion, suppose that the motion of the system is described by an equation of the form (4) where the function f satisfies all the conditions (i) – (iv) (cf. §1.2.2). Moreover, one can consider a generalized differential game of pursuit-evasion in which the target set M is defined in the space $R \times R^n$ and the phase constraint N is present. Here it should be assumed that $M \subset N$ and M and N are closed in $R \times R^n$. In this generalization of the statement of pursuit and evasion problems, only the definition of the set $T^0(x(\cdot))$ (5) is altered. Now this set is defined by the relation

$$T^0(x(\cdot)) \overset{\text{def}}{=} \{\tau \geqslant t_0\colon\ (\tau, x(\tau)) \in M,\ (t,x(t)) \in N \ \text{ for } t_0 \leqslant t \leqslant \tau\}.$$

The statement of Theorem 3.2.1 remains the same. In the proof of this theorem the sets M_θ and N_θ must be defined by the relations

$$M_\theta \overset{\text{def}}{=} \{(t,x)\colon\ t \leqslant \theta,\ (t,x) \in M\},$$

$$N_\theta \overset{\text{def}}{=} \{(t,x)\colon\ t \leqslant \theta,\ (t,x) \in N\}.$$

Otherwise, the proof remains the same.

3.3. Differential Games with a Bounded Duration

In this section, differential games are considered for which the payoff functional is defined as the minimum of a function computed on a given interval of time along the motion of the controlled system. Modifications of such a functional will also be indicated. It will be shown that for the differential games under consideration there exists an equilibrium situation.

3.3.1. We shall assume in this section that the duration of a differential game is bounded by the time $t = \theta$. It is also assumed that the function f on the r.h.s. of equation (1.2.1) is defined on $T \times R^n \times P \times Q$, where $T = (-\infty,\theta]$. Consider the differential game with the payoff functional given by equality

$$(1) \qquad \gamma(x(\cdot)) \overset{\text{def}}{=} \min_{t_0 \leqslant t \leqslant \theta} \rho(t,x(t)),$$

where $\rho: T \times R^n \to R^1$ is a continuous function. The functional γ is defined and is continuous on the space $C_n([t_0,\theta])$. This functional was considered earlier in §1.4.2 where some explanatory remarks were made. We shall therefore proceed immediately to the statement of minimax and maximin problems which form the differential game under consideration.

Problem 3.3.1. *For a given initial position* $(t_0,x_0) \in T \times R^n$ *it is required to obtain a strategy* U^0 *such that* $\Gamma_1(U^0) = \min_U \Gamma_1(U)$ *where*

$$(2) \qquad \Gamma_1(U) \overset{\text{def}}{=} \max_{x(\cdot)} \min_{t_0 \leqslant t \leqslant \theta} \rho(t,x(t)) \quad \textit{for} \quad x(\cdot) \in X_1(t_0,x_0,U).$$

Problem 3.3.2. *For a given initial position* $(t_0,x_0) \in T \times R^n$ *it is required to obtain a strategy* V^0 *such that* $\Gamma_2(V^0) = \max_V \Gamma_2(V)$ *where*

$$(3) \qquad \Gamma_2(V) \overset{\text{def}}{=} \min_{x(\cdot)} \min_{t_0 \leqslant t \leqslant \theta} \rho(t,x(t)) \quad \textit{for} \quad x(\cdot) \in X_2(t_0,x_0,V).$$

The following theorem is valid.

Theorem 3.3.1. *For any choice of initial position* $(t_0,x_0) \in T \times R^n$ *in a differential game with the payoff of the form* (1), *there exists the value* ω^0 *and the saddle point* (U^0,V^0), i.e.

$$(4) \qquad \min_U \Gamma_1(U) = \Gamma_1(U^0) = \omega^0 = \Gamma_2(V^0) = \max_V \Gamma_2(V).$$

Proof. The scheme of the proof of this theorem is the same as that of Theorems 3.1.1 and 3.2.1. Let a be a number. Set

$$(5) \qquad M_a \overset{\text{def}}{=} \{(t,x) \in T \times R^n : \rho(t,x) \leqslant a\}, \quad N \overset{\text{def}}{=} T \times R^n.$$

For any motion $x(\cdot)$: $[t_0,\theta] \to R^n$ the inequality

$$\gamma(x(\cdot)) = \min_{t_0 \leqslant t \leqslant \theta} \rho(t,x(t)) \leqslant a$$

is equivalent to the condition of an (M_a, N)-approach. Denote by A_1 the set of numbers a for which $M_a \neq \phi$ and for the chosen initial position (t_0,x_0) the (M_a, N)-approach problem is solvable. Utilizing Theorem 2.4.3 it can be shown that $A_1 = [a_1,\infty)$. Let U^0 be a strategy which solves the (M_{a_1}, N)-approach problem. For such a strategy the inequality

$$\text{(6)} \qquad \Gamma_1(U^0) \leqslant a_1$$

is valid.

Select an arbitrary $a < a_1$. Since $a \notin A_1$ the (M_a, N)-approach problem is not solvable. Therefore by Theorem 2.1.1 the (M_a, N)-evasion problem possesses a solution. Hence there exists a strategy V_a which for the chosen initial position (t_0,x_0) solves the problem of $(K_a,\mathfrak{L}_a)$-approach where

$$K_a \overset{\text{def}}{=} \{(\theta,x) : x \in R^n,\ \rho(\theta,x) \geqslant a\},\quad \mathfrak{L}_a \overset{\text{def}}{=} \{(t,x) \in T \times R^n : \rho(t,x) \geqslant a\}.$$

Observe that the condition of $(K_a,\mathfrak{L}_a)$-approach is equivalent to the inequality

$$\text{(7)} \qquad \gamma(x(\cdot)) = \min_{t_0 \leqslant t \leqslant \theta} \rho(t,x(t)) \geqslant a\,.$$

Denote by A_2 the set of numbers a such that for a chosen initial position (t_0,x_0) the $(K_a,\mathfrak{L}_a)$-approach problem is solvable. Utilizing Theorem 2.4.3 it can be shown that $A_2 = (-\infty,a_2]$. As was mentioned above, the inequality $a < a_1$ implies the inclusion $a \in A_2$. We thus obtain $a_2 \geqslant a_1$. Let V^0 be a strategy which solves the $(K_{a_2},\mathfrak{L}_{a_2})$-approach problem. For this strategy the estimate

$$\Gamma_2(V^0) \geqslant a_2$$

is valid. Taking the inequality $a_2 \geqslant a_1$ into account we obtain

$$\sup_V \Gamma_2(V) \geqslant \Gamma_2(V^0) \geqslant a_2 \geqslant a_1 \geqslant \Gamma_1(U^0) \geqslant \inf_U \Gamma_1(U).$$

On the other hand, in view of (1.5.10) we have

$$\sup_V \Gamma_2(V) \leqslant \inf_U \Gamma_1(U).$$

This implies (4), and so $\omega^0 = a_1 = a_2$. Theorem 3.3.1 is thus proved.

Corollary 3.3.1. *Let* W_a^1 *be the maximal* u-*stable bridge in an* (M_a, N)*-approach problem and let* $a_1 \overset{\text{def}}{=} \min\{a : (t_0,x_0) \in W_a^1\}$. *Then the* *strategy U^e extremal to the set $W^1_{a_1}$ will be an optimal strategy for the first player.*

Let W^2_a be the maximal v-stable bridge in a $(K_a, \mathfrak{L}_a)$-approach problem and let $a_2 \overset{\text{def}}{=} \max\{a : (t_0,x_0) \in W^2_a\}$. Then the strategy V^e extremal to the set $W^2_{a_2}$ will be optimal strategy for the second player.

This assertion follows from the results of Chapter 2 and the proof of Theorem 3.3.1 presented herein.

Consider a particular case. Let M^* be a closed set in R^n. Let

$$\rho(t,x) = \operatorname{dist}(x,M^*). \tag{8}$$

Assume that the value of a differential game with payoff of the form (1), (8) equals zero. Then the strategy U^0 assures a contact with the set M^* not later than at time $t = \theta$. The converse is also true. If a pursuer's strategy exists which assures contact not later than at time $t = \theta$, then the value of a differential game with the payoff (1), (8) will be equal to zero. Note that the payoff functional (1), (8) is continuous on $C_n([t_0,\theta])$ while in a differential game of pursuer-evader the functional (3.2.6) is only lower semi-continuous. Therefore the differential game with the payoff (1), (8) may be considered as regularization of the differential game of pursuit-evasion.

3.3.2. Let the payoff functional be defined by the equality

$$\gamma(x(\cdot)) \overset{\text{def}}{=} \max_{t_0 \leqslant t \leqslant \theta} \rho(t,x(t)), \tag{9}$$

where as above $\rho : T \times R^n \to R^1$ is a continuous function. A differential game with a payoff of the form (9) actually belongs to the type considered in §3.3.1. Indeed set $\rho^*(t,x) \overset{\text{def}}{=} -\rho(t,x)$. We thus obtain

$$\begin{aligned}
&\gamma^*(x(\cdot)) \overset{\text{def}}{=} \min_{t_0 \leqslant t \leqslant \theta} \rho^*(t,x(t)) = -\max_{t_0 \leqslant t \leqslant \theta} \rho(t,x(t)) = -\gamma(x(\cdot))\\
&\Gamma_1^*(U) \overset{\text{def}}{=} \min_{x(\cdot)} \gamma^*(x(\cdot)) = -\max_{x(\cdot)} \gamma(x(\cdot)) = -\Gamma_1(U)\\
&\qquad \text{for } x(\cdot) \in X_1(t_0,x_0,U),\\
&\Gamma_2^*(V) \overset{\text{def}}{=} \max_{x(\cdot)} \gamma^*(x(\cdot)) = -\min_{x(\cdot)} \gamma(x(\cdot)) = -\Gamma_2(V)\\
&\qquad \text{for } x(\cdot) \in X_2(t_0,x_0,V),
\end{aligned} \tag{10}$$

$$\max_U \Gamma_1^*(U) = -\min_U \Gamma_1(U), \quad \min_V \Gamma_2^*(V) = -\max_V \Gamma_2(V).$$

Consider the two problems: Find strategies U^* and V^* such that

(11) $$\Gamma_1^*(U^*) = \max_U \Gamma_1^*(U),$$

(12) $$\Gamma_2^*(V^*) = \min_V \Gamma_2^*(V).$$

In these problems the target functional γ^* is of the form (1). Problems (11) and (12) differ from Problems 3.3.1 and 3.3.2 in that the roles of the players have been interchanged. In Problem (11) the first player solves a maximin problem while in Problem (12) the second player solves a minimax problem. Theorem 3.3.1 and equalities (10) imply that the value and a saddle point exist for the differential game with payoff (9).

Consider two more payoff functionals. The first is defined as follows. Let a closed set T_* be selected on the semi-axis $T = (-\infty,\theta]$ and set

(13) $$\gamma(x(\cdot)) \overset{\text{def}}{=} \max_t \rho(t,x(t)) \quad \text{for} \quad t \in [t_0,\theta] \cap T_*.$$

Here ρ is a continuous function. Observe that in the case when $T_* = \{\theta\}$ the functional γ (13) becomes a functional of the form (3.1.1). A differential game with payoff functional of the form (13) possesses the value and a saddle point. The proof of this assertion hardly differs from the proof of Theorem 3.3.1.

We now define the second functional. Let M_* be a closed set in $T \times R^n$. Assume that the set M_* contains the hyperplane $\{(\theta,x) : x \in R^n\}$. Let $x(\cdot) : [t_0,\theta] \to R^n$ be a continuous function. Set

$$T_0(x(\cdot)) \overset{\text{def}}{=} \{t \in [t_0,\theta] : (t,x,(t)) \in M_*\},$$

$$\tau_0(x(\cdot)) = \min \{t : t \in T_0(x(\cdot))\}.$$

In this case the set $T_0(x(\cdot))$ is always non-empty, since it contains the point θ. The functional τ_0 takes on values on the interval $[t_0,\theta]$. Consider the functional

(14) $$\gamma(x(\cdot)) = \max_t \rho(t,x(t)) \quad \text{for} \quad t_0 \leqslant t \leqslant \tau_0(x(\cdot)),$$

where ρ is a continuous function as above. Observe that functionals τ_0 and γ are lower semi-continuous. The following assertion is valid.

For any choice of the initial position $(t_0,x_0) \in T \times R^n$ in a differential game with a payoff of the form (14) there exists an ϵ-equilibrium situation, i.e. for any $\epsilon > 0$ there exists a pair of strategies (U^0,V^ϵ) such that

(15) $$\Gamma_1(U^0) = \min_U \Gamma_1(U) = \omega^0 = \sup_V \Gamma_2(V) \leqslant \Gamma_2(V^\epsilon) + \epsilon$$

where

$$\Gamma_1(U) = \sup_{x(\cdot)} \gamma(x(\cdot)) \quad \text{for } x(\cdot) \in X_1(t_0,x_0,U),$$

$$\Gamma_2(V) = \min_{x(\cdot)} \gamma(x(\cdot)) \quad \text{for } x(\cdot) \in X_2(t_0,x_0,V).$$

The proof of this assertion is carried out as above; therefore we shall present only the basic points. Let

$$M \overset{\text{def}}{=} M_*, \quad N_a \overset{\text{def}}{=} \{(t,x) \in T \times R^n : \rho(t,x) \leqslant a\}.$$

The condition of (M, N_a)-approach is equivalent to the inequality $\gamma(x(\cdot)) \leqslant a$. Let A_1 be a set of numbers a such that for a chosen initial position (t_0,x_0) the (M, N_a)-approach problem is solvable. It can be shown that $A_1 = [a_1,\infty)$. Let strategy U^0 solve the problem of (M, N_{a_1})-approach. For this strategy the estimate

$$\Gamma_1(U^0) \leqslant a_1 \tag{16}$$

is valid. For any $a < a_1$ the (M, N_a)-approach problem is not solvable. Therefore in accordance with Theorem 2.1.1 there exists strategy V_a which assures the fulfillment of the condition of (M, N_a)-evasion for all motions $x(\cdot) \in X_2(t_0,x_0,V_a)$, i.e. any motion $x(\cdot) \in X_2(t_0,x_0,V_a)$ will satisfy the inequality $\gamma(x(\cdot)) > a$. Hence the estimate $\Gamma_2(V_a) > a$ is valid for the strategy V_a. Here a is an arbitrary number satisfying the inequality $a < a_1$. We thus obtain

$$\sup_V \Gamma_2(V) \geqslant a_1. \tag{17}$$

Equations (1.5.10), (16), and (17) imply (15).

Theorems of the existence of the value for various types of differential games ascertained in this chapter are proved according to a unified procedure based on the theorem on an alternative for a differential game of approach-evasion. This procedure is quite thoroughly discussed in the proofs presented above. We shall therefore confine ourselves to the existence theorems already proved, and we shall not repeat them for other types of payoff functionals.

3.4. Valuc Function

In preceding sections the existence of the value was established for various types of differential games. It was proved that the value of a differential game exists for any choice of the initial position. Therefore, one can consider a function which is defined on the space of positions and which places the initial position into correspondence with the value of a differential game.

3.4.1. In preceding sections, payoff functionals were defined for a function $x(\cdot)$ defined either on the interval $[t_0,\theta]$ or on the semi-axis

$[t_0,\infty)$. The time t_0 was assumed to be fixed. Now we shall consider values of a differential game defined for arbitrary initial positions (t_0,x_0). It will therefore be required that the payoff functional be defined for all values of parameter $t_0 \in T$ and for all continuous functions $x(\cdot) : [t_0,\theta] \to R^n$ (in a differential game of pursuit-evasion for all continuous functions $x(\cdot) : [t_0,\infty) \to R^n$). Retaining the previous notation we shall assume that functionals (3.1.1), (3.3.1), (3.3.9), (3.3.13), (3.3.14) are defined for all $t_0 \in (-\infty,\theta]$ and functional τ^0 (3.2.6) for all $t_0 \in (-\infty,\infty)$.

It was proved for differential games considered in §3.1 – 3.3 that for any choice of initial position $(t_0,x_0) \in T \times R^n$ the value of the game exists. We denote this value by $\omega^0(t_0,x_0)$. Since the value $\omega^0(t_0,x_0)$ is defined for all points $(t_0,x_0) \in T \times R^n$, we obtain the function $\omega^0 : T \times R^n \to R$ which we shall call the *value function*. Observe that in the differential game of pursuit and evasion the value function takes on values from the semi-extended line $(-\infty,\infty]$, i.e. in this game $\omega^0 : T \times R^n \to (-\infty,\infty]$.

In preceding sections continuous, lower semi-continuous, and upper semi-continuous payoff functionals were considered. We shall show that the value function inherits these properties of the payoff functional.

Consider, for example, the differential game of pursuit-evasion. As was noted in §3.2, the payoff functional in this game is lower semi-continuous. Note that this property was defined for the functional $\tau^0 : C_n([t_0,\infty)) \to (-\infty,\infty]$, where t_0 is a fixed number. We shall not specify precisely how to formulate the lower semi-continuity property for the functional

$$\tau^0 : \bigcup_{-\infty<t_0<\infty} C_n([t_0,\infty)) \to (-\infty,\infty]$$

since this specification will not be required in what follows.

Let $\omega^0 : R \times R^n \to (-\infty,\infty]$ be the value function of a differential game of pursuit-evasion. We show that this function is lower semi-continuous.

As is known, the semi-continuity property of function ω^0 is equivalent to the fact that for any $\theta \in R$ the set $\{(t,x) \in R \times R^n : \omega^0(t,x) \leqslant \theta\}$ is closed (cf. §A.1). Let θ be a number. It follows from the proof of Theorem 3.2.1 that the inequality $\omega^0(t_0,x_0) \leqslant \theta$ holds if and only if the (M_θ, N_θ)-approach problem is solvable, where the sets M_θ and N_θ are defined in accordance with (3.2.12). On the other hand, as was shown in Chapter 2, the problem of an (M_θ, N_θ)-approach is solvable if and only if $(t_0,x_0) \in W^1_\theta$, where W^1_θ is the maximal u-stable bridge for this problem. Hence

$$\{(t_0,x_0) \in R \times R^n : \omega^0(t_0,x_0) \leqslant \theta\} = W^1_\theta.$$

In accordance with Lemma 2.2.6 the set W^1_θ is closed, thus the lower semi-continuity of the function ω^0 is proved.

Consider now a differential game with a fixed termination time.

Let the payoff functional be defined by the equality (3.1.1) where $\sigma : R^n \to R$ is a continuous function. Let $\omega^0 : (-\infty,\theta] \times R^n \to R$ be the value function in this game. We shall show that the function ω^0 is continuous. For this it is sufficient to show that it is lower and upper semi-continuous. Let a be a number. From the proof of Theorem 3.1.1 the equalities

$$\{(t_0,x_0) \in T \times R^n : \omega^0(t_0,x_0) \leqslant a\} = W^1_a,$$

$$\{(t_0,x_0) \in T \times R^n : \omega^0(t_0,x_0) \geqslant a\} = W^2_a,$$

follow. Here W^1_a is the maximal u-stable bridge in an (M_a, N)-approach problem, W^2_a is the maximal v-stable bridge in a $K_a,\mathfrak{L})$-approach problem, the sets M_a, N, and $K_a, \mathfrak{L}$ are defined in accordance with (3.1.7) and (3.1.8). Since the sets W^1_a and W^2_a are closed, the function ω^0 is lower and upper semi-continuous. We thus have shown that ω^0 is continuous.

Analogously, for any one of the remaining types of payoff functionals one can prove that the value function inherits the continuity or semi-continuity property of the payoff functional.

Assume now that the function $\sigma : R^n \to R$ satisfies the Lipschitz condition in every bounded region $G \subset R^n$. It will be proved below that the value function of a differential game with payoff (3.1.1) will satisfy the Lipschitz condition in every bounded region $Z \subset T \times R^n$. For a differential game with payoff of the form (3.3.1), an analogous assertion is valid: if function $\rho : T \times R^n \to R$ satisfies the Lipschitz condition in every bounded region $Z \subset T \times R^n$, then the payoff function possesses the same property. To prove this assertion it will be necessary to first consider a property of an (M, N)-approach problem.

3.4.2. Let M and N be closed sets in the space $T \times R^n$. As usual it is assumed that $M \subset N$. Let W be the maximal u-stable bridge in an (M, N)-approach problem. Denote by U^e that strategy extremal to the set W. Assume that the initial position (t_*,x_*) is contained in the set W. Then for any motion $x(\cdot) \in X_1(t_*,x_*,U^e)$ the condition of an (M, N)-approach is fulfilled. This proposition was verified in Theorem 2.4.1. Select now for the initial position another point (t^*,x^*) which may not belong to the set W. Define for this initial position, the set of motions $X_1(t^*,x^*,U^e)$. The condition of (M, N)-approach may not be fulfilled for certain motions $x(\cdot) \in X_1(t^*,x^*,U^e)$. However, one can extend the sets M and N, replacing them by closed ϵ-neighborhoods $\overline{M^\epsilon}$ and $\overline{N^\epsilon}$ such that the condition of $(\overline{M^\epsilon}, \overline{N^\epsilon})$-approach will be valid for all motions $x(\cdot) \in X_1(t^{*0},x^*,U^e)$. The purpose of the derivations presented below is to estimate how the quantity ϵ depends on the distance between the points (t_*,x_*) and (t^*,x^*).

We now proceed to formulate the estimate just mentioned. Denote by

$\mathcal{X}(t_*,x_*)$ the set of solutions $x(\cdot)$: $[t_0,\theta] \to R^n$ of the differential inclusion

(1) $\dot{x}(t) \in \mathrm{co}\{f(t,x(t),u,v) : u \in \mathcal{P},\ v \in \mathcal{Q}\}$

satisfying the initial condition $x(t_*) = x_*$. Let $\mathcal{Z}$ be a compact set in $T \times R^n$. We set

(2)
$$\mathcal{D}(t_*,x_*) \overset{\text{def}}{=} \{(t,x(t)) \in T \times R^n : t_* \leqslant t \leqslant \theta,\ x(\cdot) \in \mathcal{X}(t_*,x_*)\},$$
$$\mathcal{D}(\mathcal{Z}) \overset{\text{def}}{=} \bigcup_{(t_*,x_*)} \mathcal{D}(t_*,x_*) \text{ for } (t_*,x_*) \in \mathcal{Z}.$$

It is known that the set $\mathcal{D}(\mathcal{Z})$ is compact in $T \times R^n$ (cf. §A.4).

From the construction of the set $\mathcal{D}(\mathcal{Z})$ follows the property: let $(t^*,x^*) \in \mathcal{D}(\mathcal{Z})$ and let $x(\cdot) : [t^*,\theta] \to R^n$ be a solution of the differential inclusion (1) satisfying the initial condition $x(t^*) = x^*$, then

(3) $(t,x(t)) \in \mathcal{D}(\mathcal{Z})$ for all $t \in [t^*,\theta]$.

If $\mathcal{W}$ is a u-stable bridge in an $(\mathcal{M},\mathcal{N})$-approach problem, then the intersection $\mathcal{W} \cap \mathcal{D}(\mathcal{Z})$ will also be a u-stable bridge in this problem. This assertion follows from the above-mentioned property of the set $\mathcal{D}(\mathcal{Z})$.

We shall utilize the contructions presented in 2.3.2. Replace the set $\mathcal{D}_1$ (2.3.10) by the set $\mathcal{D}(\mathcal{Z})$ (2). Following the replacement we shall define the quantities $\lambda,\Lambda,\zeta(\delta)$ and $\alpha(\delta)$ (cf. (2.3.11), (2.3.12), (2.3.15)). Introduce the function $t \to \beta^*(t,\delta)$ defined by the equality

(4) $$\beta^*(t,\delta) \overset{\text{def}}{=} (\beta_* + \alpha(\delta)\lambda^{-1}) \exp 2\lambda(t - \tau_*) - \alpha(\delta)\lambda^{-1},$$

where $\beta_* > 0$ and $\tau_* < \theta$ are numbers. Note that the function $\beta^*(\cdot,\delta)$ satisfies the differential equation (cf. (2.3.18))

(5) $$d\beta^*(t,\delta)/dt = 2\lambda\beta^*(t,\delta) + 2\alpha(\delta)$$

and condition $\beta^*(\tau_*,\delta) = \beta_*$. The following assertion is valid (an analog of Lemma 2.3.1).

Let $(t_*,x_*) \in \mathcal{D}(\mathcal{Z})$, $(t_*,w_*) \in \mathcal{D}(\mathcal{Z})$, and $\|x_* - w_*\|^2 \leqslant \beta^*(t_*,\delta)$. Assume that points $u_0 \in \mathcal{P}$ and $v_0 \in \mathcal{Q}$ are chosen from the conditions (2.3.19) and (2.3.20). Let $x(\cdot) :\ [t_*,\theta] \to R^n$ and $w(\cdot) :\ [t_*,\theta] \to R^n$ be solutions of differential inclusion (2.3.21) and (2.3.22) satisfying the initial conditions $x(t_*) = x_*$ and $w(t_*) = w_*$. Then the bound

(6) $$\|x(t) - w(t)\|^2 \leqslant \beta^*(t,\delta) \text{ for } t_* \leqslant t \leqslant t_* + \delta$$

is valid.

The proof of this estimate almost completely coincides with the proof of Lemma 2.3.1. The estimate (6) will be used in the proof of the following lemma.

Lemma 3.4.1. *Let* $\mathcal{W}$ *be the maximal* u-*stable bridge in an* $(\mathcal{M}, \mathcal{N})$*-approach problem. Assume that*

$$(t_1,x_1) \in \mathcal{Z}, \quad (t_2,x_2) \in \mathcal{W} \cap \mathcal{Z}, \quad \mathcal{Z} \subset [t^0,\theta] \times R^n.$$

Let U^e *be a strategy extremal to the set* $\mathcal{W} \cap \mathcal{D}(\mathcal{Z})$*. Then for any motion* $x(\cdot) \in \mathcal{X}_1(t_1,x_1,U^e)$ *the condition of* $(\overline{\mathcal{M}^r}, \overline{\mathcal{N}^r})$*-approach will be fulfilled where*

$$r = r(t_1,x_1,t_2,x_2) \overset{\text{def}}{=} (\|x_1 - x_2\| + (\Lambda+1)\,|t_1 - t_2|)\exp\lambda(\theta - t^0). \tag{7}$$

Proof. Four cases are possible.

Case 1. $t_1 \leqslant t_2$ and $(t_2,x_2) \in \mathcal{M}$.

Case 2. $t_2 < t_1$ and there exists a motion $x_*(\cdot) \in \mathcal{X}_1(t_2,x_2,U^e)$ such that $(\tau,x(\tau)) \in \mathcal{M}$ for some $\tau \in [t_2,t_1]$.

Case 3. $t_1 \leqslant t_2$ and $(t_2,x_2) \notin \mathcal{M}$.

Case 4. $t_2 < t_1$ and $(t,x(t)) \notin \mathcal{M}$ for all $t_2 \leqslant t \leqslant t_1$ and for any $x(\cdot) \in \mathcal{X}_1(t_2,x_2,U^e)$.

We shall consider each one of these cases.

Case 1. In this case we have the inequality

$$\operatorname{dist}((t_1,x_1),(t_2,x_2)) \overset{\text{def}}{=} |t_1 - t_2| + \|x_1 - x_2\| \leqslant r.$$

Since $(t_2,x_2) \in \mathcal{M} \subset \mathcal{N}$ it follows immediately from this inequality that any motion $x(\cdot) \in \mathcal{X}_1(t_1,x_1,U^e)$ satisfies the condition of an $(\overline{\mathcal{M}^r}, \overline{\mathcal{N}^r})$-approach.

Case 2. Function $x_*(\cdot)$ satisfies the Lipschitz condition with constant Λ. Therefore

$$\|x_*(\tau) - x_1\| \leqslant \|x_*(\tau) - x_*(t_2)\| + \|x_*(t_2) - x_1)\| \leqslant \|x_1 - x_2\| + \Lambda|t_1 - t_2| .$$

(Here we utilize the fact that $x_*(t_2) = x_2$.) Consequently

$$\operatorname{dist}((\tau,x_*(\tau)),(t_2,x_2)) \leqslant \|x_1 - x_2\| + (\Lambda + 1)|t_1 - t_2| \leqslant r.$$

Taking inclusion $(\tau,x_*(\tau)) \in \mathcal{M} \subset \mathcal{N}$ into account we obtain that in this case the assertion of the lemma is also valid.

Case 3. Consider an arbitrary constructive motion $y(\cdot) \in \mathcal{X}_1(t_1,x_1,U^e)$.

This motion satisfies the Lipschitz condition. Therefore

$$\|y(t) - x_2\| \leqslant \|y(t_1) - y(t)\| + \|y(t_1) - x_2\| \leqslant \Lambda|t_1 - t_2| + \|x_1 - x_2\|$$

$$\text{for } t \in [t_1, t_2].$$

(Here we utilize that $y(t_1) = x_1$.) Taking into account that $(t_2, x_2) \in \mathcal{W} \subset \mathcal{N}$ we obtain

(8) $$\operatorname{dist}((t, y(t)), \mathcal{N}) \leqslant \|x_1 - x_2\| + (\Lambda + 1)|t_1 - t_2| \leqslant r \text{ for } t_1 \leqslant t \leqslant t_2.$$

Set $\mathcal{W}_* \overset{\text{def}}{=} \mathcal{W} \cap \mathcal{D}(\mathcal{Z})$. Recall that $\mathcal{W}_*$ is a u-stable bridge in an $(\mathcal{M}, \mathcal{N})$-approach problem. In the case under consideration $(t_2, x_2) \in \mathcal{W}_* \backslash \mathcal{M}$. We select a constructive motion $x_*(\cdot) \in X_1(t_2, x_2, U^e)$. In accordance with Theorem 2.4.5 the inclusion

(9) $$(t, x_*(t)) \in \mathcal{W}_* \text{ for } t_2 \leqslant t \leqslant \tau^0$$

is valid where τ^0 is the time of the encounter of point $(t, x_*(t))$ with the set $\mathcal{M}$. Observe that $\tau^0 > t_2$ here.

Let $x_\ell(\cdot) \in X_1(t_1, x_1, U^e, \Delta_\ell)$ $(\ell = 1, 2, \ldots)$ be a sequence of step-by-step motions converging to the constructive motion $y(\cdot)$ under consideration. Here

$$\Delta_\ell = \{t_1 = \tau_{\ell,0} < \tau_{\ell,1} < \ldots < \tau_{\ell,k(\ell)+1} = \theta\},$$

$$\operatorname{diam} \Delta_\ell = \delta_\ell \to 0 \text{ as } \ell \to \infty .$$

Let $\tau_{\ell,j}$ be the smallest among the numbers $\tau_{\ell,i}$ $(i = 0, 1, \ldots, k(\ell) + 1)$ satisfying the inequality $\tau_{\ell,i} \geqslant t_2$. Observe that $\tau_{\ell,j} \leqslant t_2 + \delta_\ell$; therefore, for ℓ sufficiently large the inequality $\tau_{\ell,j} \leqslant \tau^0$ is valid. We shall estimate the squared distance: $\operatorname{dist}^2(x_\ell(\tau_{\ell,j}), \mathcal{W}_*[\tau_{\ell,j}])$. The functions $x_\ell(\cdot)$ and $x_*(\cdot)$ satisfy the Lipschitz condition with the constant Λ, thus

$$\|x_\ell(\tau_{\ell,j}) - x_*(\tau_{\ell,j})\| \leqslant \|x_\ell(\tau_{\ell,j}) - x_\ell(t_1)\| + \|x_\ell(t_1) - x_*(t_2)\|$$

$$+ \|x_*(t_2) - x_*(\tau_{\ell,j})\| \leqslant \|x_1 - x_2\| + \Lambda(|t_1 - t_2| + 2\delta_\ell) \leqslant \sqrt{\beta_*(\delta_\ell)},$$

where

(10) $$\beta_*(\delta_\ell) \overset{\text{def}}{=} [\|x_1 - x_2\| + (\Lambda + 1)(|t_1 - t_2| + 2\delta_\ell)]^2 .$$

In view of (9), for $\tau_{\ell,j} \leqslant \tau^0$ the inclusion $x_*(\tau_{\ell,j}) \in \mathcal{W}_*[\tau_{\ell,j}]$ is valid and we obtain

(11) $$\operatorname{dist}^2(x_\ell(\tau_{\ell,j}), \mathcal{W}_*[\tau_{\ell,j}]) \leqslant \beta_*(\delta_\ell).$$

Consider one of the step-by-step motions $x_\ell(\cdot)$. For simplicity of notation in the derivations below we shall omit index $_\ell$. It can be shown that

$$\text{(12)} \qquad \mathrm{dist}^2(x(\tau_i), \mathcal{W}_*[\tau_i]) \leqslant \beta^*(\tau_i,\delta) \quad (i = j, j+1, ..., m).$$

Here the quantity $\beta^*(t,\delta)$ is defined in accordance with (4), where we set $\tau_* = \tau_j$, $\beta_* = \beta_*(\delta)$. For $i = j$ it follows from (4) that $\beta^*(\tau_j,\delta) = \beta_*(\delta)$; therefore for $i = j$ the bound (12) follows from (11). Next, one can prove (12) by induction for $i = j+1, ..., m$ where the index m is determined in the same manner as in the proof of Theorem 2.4.1. The proof of the estimate (12) is almost the same as that of the estimate (2.4.16). The difference is that instead of applying Lemma 2.3.1 the estimate (6) is utilized. Thereupon it follows from the estimate (12) in the same manner as in the proof of Theorem 2.4.1 that the step-by-step motion $x(\cdot)$ on the interval $[\tau_j,\theta]$ will satisfy the condition of $(\overline{\mathcal{M}^{\epsilon(\delta)}}, \overline{\mathcal{N}^{\epsilon(\delta)}})$-approach where

$$\epsilon(\delta) \overset{\text{def}}{=} \sqrt{\beta^*(\theta,\delta)} + (\Lambda + 1)\delta.$$

We now return to the original notation. Set $\tau_j = \tau_{\ell,j}$, $\delta = \delta_\ell$. If we substitute $\delta = \delta_\ell$, $\tau_* = \tau_{\ell,j}$, $\beta_* = \beta_*(\delta_\ell)$ into (4) and approach the limit as $\ell \to \infty$, then in accordance with (7) and (10) we obtain

$$\lim \tau_{\ell,j} = t_2, \quad \lim \beta_*(\delta_\ell) = [\|x_1 - x_2\| + (\Lambda + 1)|t_1 - t_2|]^2,$$

$$\lim \beta^*(\theta,\delta_m) = r^2 = [\|x_1 - x_2\| + (\Lambda + 1)|t_1 - t_2|]^2 \exp 2\lambda(\theta - t^0).$$

Therefore $\lim \epsilon(\delta_\ell) = r$ as $\ell \to \infty$. Hence the constructive motion $y(\cdot) \in X_1(t_1,x_1,U^e)$, which is defined as the limit of the sequence $x_\ell(\cdot)$, satisfies the condition of $(\overline{\mathcal{M}^r}, \overline{\mathcal{N}^r})$-approach on the interval $[t_2,\theta]$. Taking (8) into account we conclude that for the motion $y(\cdot) \in X_1(t_1,x_1,U^e)$ the condition of $(\overline{\mathcal{M}^r}, \overline{\mathcal{N}^r})$-approach is fulfilled.

Case 4. Let $x_*(\cdot) : [t_2,\theta] \to R^n$ be an arbitrary constructive motion belonging to the set $X_1(t_2,x_2,U^e)$. Since $(t_2,x_2) \in \mathcal{W}_*$ and $(t,x_*(t)) \notin \mathcal{M}$ for $t_2 \leqslant t \leqslant t_1$, then $(t_1,x_*(t_1)) \in \mathcal{W}_*$. The function $x_*(\cdot)$ satisfies the Lipschitz condition, therefore

$$\|x_*(t_1) - x_1\| \leqslant \|x_*(t_1) - x_*(t_2)\| + \|x_*(t_2) - x_1\|$$
$$\leqslant \|x_1 - x_2\| + \Lambda|t_1 - t_2|.$$

Since $x_*(t_1) \in \mathcal{W}_*[t_1]$, we have

$$\mathrm{dist}^2(x_1, \mathcal{W}_*[t_1]) \leqslant \beta_*(\delta).$$

Here the quantity $\beta_*(\delta)$ is defined in accordance with (10) and this inequality is valid for any $\delta > 0$. Next we repeat the arguments presented in the course of Case 3 where we set $j = 0$. We obtain that any constructive motion $x(\cdot) \in X_1(t_1,x_1,U^e)$ satisfies the condition of $(\overline{M^r}, \overline{N^r})$-approach. Lemma 3.4.1 is thus proved.

Note that in Lemma 3.4.1 it was shown how the solution of an (M, N)-approach problem depends on the choice of the initial position. It was established that this dependence is uniformly continuous within the bounds of any compact set $Z \subset T \times R^n$, i.e. in this sense the solution of the (M, N)-approach problem is stable.

3.4.3. We shall utilize Lemma 3.4.1 for the proof of Theorems 3.4.1 and 3.4.2.

Theorem 3.4.1. *Let the function* $\sigma : R^n \to R^1$ *satisfy the Lipschitz condition*

$$|\sigma(x_*) - \sigma(x^*)| \leqslant L_1(G)\|x_* - x^*\| \quad (x_* \in G,\ x^* \in G), \tag{13}$$

where G is an arbitrary compact set in R^n. Then the value function ω^0: $T \times R^n \to R^1$ of a differential game with payoff of the form (3.1.1) *will satisfy the Lipschitz condition*

$$\begin{gathered} |\omega^0(t_1,x_1) - \omega^0(t_2,x_2)| \leqslant L_2(Z)[|t_1 - t_2| + \|x_1 - x_2\|], \\ (t_i,x_i) \in Z,\ i = 1, 2, \end{gathered} \tag{14}$$

where Z is a compact set in $T \times R^n$.

Proof. Let Z be a compact set in $T \times R^n$. Select arbitrarily points $(t_i,x_i) \in Z$, $i = 1, 2$. Set

$$a \overset{\text{def}}{=} \omega^0(t_2,x_2),\quad D \overset{\text{def}}{=} D(Z),\quad M_a \overset{\text{def}}{=} \{(\theta,x) : \sigma(x) \leqslant a\},$$

$$\mathcal{E}_a \overset{\text{def}}{=} M_a \cap D,\quad N \overset{\text{def}}{=} T \times R^n.$$

The problem of (M_a, N)-approach is solvable from the initial position (t_2,x_2). As it follows from the construction of the set D (cf. (2), (3)), the problem of $(\mathcal{E}_a, N)$-approach is also solvable from the initial position (t_2,x_2). We thus obtain that $(t_2,x_2) \in W_* \overset{\text{def}}{=} W \cap D$, where W is the maximal u-stable bridge in an $(\mathcal{E}_a, N)$-approach problem. Let U^e be a strategy extremal to the set W_* and let $x(\cdot) : [t_1,\theta] \to R^n$ be a motion belonging to the set $X_1(t_1,x_1,U^e)$. In accordance with Lemma 3.4.1 the inclusion

$$(\theta,x(\theta)) \in \overline{\mathcal{E}^r_a}$$

is valid. We also note that from the construction of the set $\mathcal{D}$ we have

$$(\theta, x(\theta)) \in \mathcal{D} .$$

Set $G \stackrel{\text{def}}{=} \mathcal{D}[\theta] = \{x \in R^n : (\theta, x) \in \mathcal{D}\}$. The set G is a compact set in R^n. Thus $x(\theta) \in G$ and there exist points $y \in G$ and $z \in R^n$ such that $x(\theta) = y + z$, $\|z\| \leqslant r$, $\sigma(y) \leqslant a = \omega^0(t_2, x_2)$. We thus obtain from (13) that

$$\sigma(x(\theta)) \leqslant \omega^0(t_2, x_2) + L_1(G)r.$$

This estimate is derived for an arbitrary motion $x(\cdot) \in X_1(t_1, x_1, U^e)$. In accordance with (3.1.3) we have

$$\Gamma_1(t_1, x_1, U^e) \stackrel{\text{def}}{=} \max_{x(\cdot)} \sigma(x(\theta)) \leqslant \omega^0(t_2, x_2) + L_1(G)r$$
$$\text{for} \quad x(\cdot) \in X_1(t_1, x_1, U^e).$$

From the definition of the value of a game (cf. (3.1.2) and (3.1.6)) we have

$$\omega^0(t_1, x_1) \leqslant \Gamma_1(t_1, x_1, U^e) \leqslant \omega^0(t_2, x_2) + L_1(G)r.$$

If we interchange points (t_1, x_1) and (t_2, x_2) here and substitute expression (7) we then arrive at the estimate

$$|\omega^0(t_1, x_1) - \omega^0(t_2, x_2)| \leqslant L_1(G)r$$
$$= L_1(G)[\|x_1 - x_2\| + (\Lambda + 1)|t_1 - t_2|] \exp\lambda(\theta - t^0).$$

It follows from this estimate that function ω^0 satisfies the Lipschitz condition (14). Theorem 3.4.1 is thus proved.

Theorem 3.4.2. *Let the function $\rho : T \times R^n \to R^1$ satisfy the Lipschitz condition*

$$|\rho(t_1, x_1) - \rho(t_2, x_2)| \leqslant L_1(\mathcal{D})[\|x_1 - x_2\| + |t_1 + t_2|],$$
(15)
$$(t_i, x_i) \in \mathcal{D}, \quad i = 1, 2,$$

where $\mathcal{D}$ is an arbitrary compact set in $T \times R^n$. Then the value function $\omega^0 : T \times R^n \to R^1$ of a differential game with payoff of the form (3.3.1) *satisfies the Lipschitz condition*

$$|\omega^0(t_1, x_1) - \omega^0(t_2, x_2)| \leqslant L_2(Z)[\|x_1 - x_2\| + |t_1 - t_2|],$$
(16)
$$(t_i, x_i) \in Z, \quad i = 1,2,$$

where $\mathcal{Z}$ *is an arbitrary compact set in* $T \times R^n$.

Proof. Let $(t_i,x_i) \in \mathcal{Z}$, $i = 1, 2, ...,$ where $\mathcal{Z}$ is a compact set in $T \times R^n$. Set

$$a \overset{\text{def}}{=} \omega^0(t_2,x_2), \quad \mathcal{D} \overset{\text{def}}{=} \mathcal{D}(\mathcal{Z}), \quad \mathcal{M}_a \overset{\text{def}}{=} \{(t,x) \in T \times R^n : \rho(t,x) \leqslant a\},$$

$$\mathcal{E}_a \overset{\text{def}}{=} \mathcal{M}_a \cap \mathcal{D}, \quad \mathcal{N}_a \overset{\text{def}}{=} T \times R^n.$$

We have $(t_2,x_2) \in \mathcal{W}_* \overset{\text{def}}{=} \mathcal{W} \cap \mathcal{D}$, where $\mathcal{W}$ is the maximal u-stable bridge in an $(\mathcal{E}_a, \mathcal{N})$-approach problem. Let U^e be a strategy extremal to the set $\mathcal{W}_*$ and let $x(\cdot)$ be a motion belonging to the set $\mathcal{X}_1(t_1,x_1,U^e)$. In accordance with Lemma 3.4.1 this motion satisfies the condition of $(\overline{\mathcal{E}^r_a}, \mathcal{N})$-approach, i.e. there exists $\tau \in [t_1,\theta]$ such that

$$(17) \qquad (\tau,x(\tau)) \in \overline{\mathcal{E}^r_a}\,.$$

We also have from the construction of the set $\mathcal{D}$ that

$$(18) \qquad (\tau,x(\tau)) \in \mathcal{D}\,.$$

Inclusion (17) implies that there exists a point (τ_*,y_*) such that

$$(19) \qquad \operatorname{dist}((\tau,x(\tau)),(\tau_*,y_*)) \leqslant r, \quad (\tau_*,y_*) \in \mathcal{D}\,, \quad \rho(\tau_*,y_*) \leqslant a.$$

From (15), (18), and (19) we obtain

$$\gamma(x(\cdot)) \overset{\text{def}}{=} \min_{t_1 \leqslant t \leqslant \theta} \rho(t,x(t)) \leqslant \rho(\tau,x(\tau)) \leqslant \omega^0(t_2,x_2) + L_1(\mathcal{D})r.$$

This estimate is valid for an arbitrary motion $x(\cdot) \in \mathcal{X}_1(t_1,x_1,U^e)$. Therefore

$$\Gamma_1(t_1,x_1,U^e) \overset{\text{def}}{=} \max_{x(\cdot)} \gamma(x(\cdot)) \leqslant \omega^0(t_2,x_2) + L_1(\mathcal{D})r$$
$$\text{for} \quad x(\cdot) \in \mathcal{X}_1(t_1,x_1,U^e).$$

From the definition of the value of a game we obtain

$$\omega^0(t_1,x_1) \leqslant \Gamma_1(t_1,x_1,U^e) \leqslant \omega^0(t_2,x_2) + L_1(\mathcal{D})r.$$

Interchanging the points (t_1,x_1) and (t_2,x_2) and substituting expression (7) we arrive at the estimate

$$|\omega^0(t_1,x_1) - \omega^0(t_2,x_2)| \leqslant L_1(\mathcal{D})[\|x_1 - x_2\| + |t_1 - t_2|(\Lambda + 1)]\exp\lambda(\theta - t^0),$$
$$(t_i,x_i) \in \mathcal{Z}, \quad i = 1, 2,$$

which shows that the function ω^0 satisfies the Lipschitz condition

(16). Theorem 3.4.2 is thus proved.

3.4.4. We note yet another property of the value function. Consider a differential game with a fixed termination time in which the payoff is given by the equality

(20) $\gamma_1(x(\cdot)) = \sigma_1(x(\theta))$,

where σ_1: $R^n \to R^1$ is a continuous function. Let ω_1: $T \times R^n \to R^1$ be the value function for this game.

Consider also another differential game where the payoff is defined by the equality

(21) $\gamma_2(x(\cdot)) = \sigma_2(x(\theta))$.

Assume that functions σ_1 and σ_2 are related as follows:

$$\sigma_2(x) = \psi(\sigma_1(x)), \quad x \in R^n,$$

where ψ is a continuous monotonically increasing function, i.e. $\psi(r_1) \leqslant \psi(r_2)$ for any $r_1 \leqslant r_2$. The value function for the differential game with payoff (21) will be denoted by ω_2.

It is easy to verify that

(22) $\omega_2(t,x) = \psi(\omega_1(t,x)), \quad (t,x) \in T \times R^n$.

Indeed, let (U^0,V^0) be a saddle point for the differential game with payoff (20) defined for some initial position (t_0,x_0). In accordance with (3.1.3), (3.1.5), and (3.1.5) we have

$$\sigma_1(x(\theta)) \leqslant \omega_1(t_0,x_0) \text{ for all } x(\cdot) \in X_1(t_0,x_0,U^0),$$

$$\sigma_1(x(\theta)) \geqslant \omega_1(t_0,x_0) \text{ for all } x(\cdot) \in X_2(t_0,x_0,V^0).$$

Since the function ψ is monotonically increasing we obtain from these inequalities that

$$\psi(\sigma_1(x(\theta))) = \sigma_2(x(\theta)) \leqslant \psi(\omega_1(t_0,x_0)) \text{ for } x(\cdot) \in X_1(t_0,x_0,U^0),$$

$$\psi(\sigma_1(x(\theta))) = \sigma_2(x(\theta)) \geqslant \psi(\omega_1(t_0,x_0)) \text{ for } x(\cdot) \in X_2(t_0,x_0,V^0).$$

The inequalities obtained signify that strategies U^0 and V^0 form a saddle point for the differential game with payoff (21) and the quantity $\psi(\omega_1(t_0,x_0))$ is the value of this game. Thus equality (22) is proved.

It follows from the discussion above that equality (22) will be valid for different types of payoff functionals related by the relation $\gamma_2(x(\cdot)) = \psi(\gamma_1(x(\cdot)))$, where ψ is a continuous monotonically increasing function.

As an illustration, consider the following simple example. Let

$$\sigma_1(x) \overset{\text{def}}{=} \|x\|, \quad \psi(r) \overset{\text{def}}{=} 0 \text{ for } r \leqslant 0, \quad \psi(r) \overset{\text{def}}{=} r^2 \text{ for } r > 0.$$

Function ψ is continuous and monotonically increasing. Note that $\sigma_1(x) \geqslant 0$ for all $x \in R^n$; therefore the value function ω_1 also takes on non-negative values. Hence

$$\psi(\sigma_1(x)) = \|x\|^2, \quad \psi(\omega_1(t,x)) = \omega_1^2(t,x).$$

We thus arrive at the following conclusion: let ω_1 be the function of a differential game with payoff $\gamma_1(x(\cdot)) = \|x(\theta)\|$, then the function ω_1^2 will be the value function of a differential game with the payoff $\gamma_2(x(\cdot)) = \|x(\theta)\|^2$.

3.5. Examples

In accordance with the procedure described in this chapter, it is easy to define the value of a differential game provided the solution of the corresponding problems of approach and evasion is known. Below we shall consider several simple examples in which the value of the game is constructed in this manner. We shall also consider definitions of the value function using recurrence and iterative procedures.

3.5.1. Let the motion of a controlled system be described by the differential equation

$$\dot{x}(t) = r_1(t)u(t) + r_2(t)v(t). \tag{1}$$

Here x, u, and v are n-dimensional vectors, the controls of the players satisfy the constraints

$$\|u(t)\| \leqslant \beta_1, \quad \|v(t)\| \leqslant \beta_2,$$

and functions $t \to r_1(t)$ and $t \to r_2(t)$ are continuous. The payoff functional in the differential game under consideration is defined by

$$\gamma(x(\cdot)) \overset{\text{def}}{=} \|x(0)\|, \tag{2}$$

where $\theta = 0$ is a fixed time of termination of the game. Set

$$T \overset{\text{def}}{=} (-\infty,0], \quad M_a \overset{\text{def}}{=} \{(0,x) : \|x\| \leqslant a\}, \quad N = T \times R^n.$$

Let W_a be the maximal u-stable bridge in an (M_a, N)-approach problem. As was shown in 2.5.1, the set W_a is determined by the relation (cf. (2.5.15))

$$W_a = \{(t,x): \|x\| \leqslant \rho_0(t,a), \ \rho_0(\tau,a) \geqslant 0 \ \text{ for } \ t \leqslant \tau \leqslant 0\}, \tag{3}$$

where

$$\rho_0(t,a) \stackrel{\text{def}}{=} a + r_*(t), \quad r_*(t) \stackrel{\text{def}}{=} \int_t^0 [\beta_1|r_1(\tau)| - \beta_2|r_2(\tau)|]d\tau.$$

In accordance with Corollary 3.1.1 the value of a differential game (1), (2) can be determined by the equality

(4) $$\omega^0(t,x) = \min\{a : (t,x) \in W_a\}.$$

We introduce the notation

(5) $$a_1(t,x) \stackrel{\text{def}}{=} \|x\| - r_*(t),$$
$$a_2(t) \stackrel{\text{def}}{=} -\min_{t \leqslant \tau \leqslant 0} r_*(\tau).$$

Relation (3) can be rewritten as follows

$$W_a = \{(t,x) : a_1(t,x) \leqslant a, \ a_2(t) \leqslant a\} = \{(t,x) : a \geqslant \max\{a_1(t,x),a_2(t)\}.$$

The relation (4) thus becomes the equality

(6) $$\omega^0(t,x) = \max\{a_1(t,x),a_2(t)\}.$$

Hence a formula is obtained which determines the value function of a differential game (1), (2). Observe that here $\omega^0(t,x) = \omega^*(t,\|x\|) = \max\{a^*(t,\|x\|), a_2(t)\}$ where $a^*(t,\alpha) = \alpha - r_*(t)$.

As an illustration, consider the two systems

(7) $$\dot{x}(t) = u(t) + tv(t),$$

(8) $$\dot{x}(t) = tu(t) + v(t), \quad |u(t)| \leqslant 1, \quad |v(t)| \leqslant 1.$$

Assume that x, u, and v are scalars. In Figs. 3.5.1 and 3.5.2 the level surfaces of the value function for the differential games (2), (7) and (2), (8), respectively, are presented. In Fig. 3.5.1, the heavily drawn lines represent the boundary of the region in which $\omega^0(t,x) = 0$ and in Fig. 3.5.2 the heavily drawn lines represent the boundary of the region where $\omega^0(t,x) = 1/2$.

3.5.2. Let the motion of the controlled system be described by equation (1) and the payoff in the differential game given by the equality

(9) $$\gamma(x(\cdot)) \stackrel{\text{def}}{=} \min_{t_0 \leqslant t \leqslant 0} \|x(t)\|.$$

As was shown in §3.3 the value of the differential game (1), (9) can be determined by the equality

$$\omega^0(t_0,x_0) = \min\{a : (t_0,x_0) \in W_a\}.$$

Here W_a is the maximal u-stable bridge in a problem of an (M_a, N)-approach, where

$$M_a \overset{\text{def}}{=} \{(t,x) \in (-\infty,0] \times R^n : \|x\| \leqslant a\}, \quad N \overset{\text{def}}{=} (-\infty,0] \times R^n.$$

We now describe the structure of the set W_a. Introduce the notation

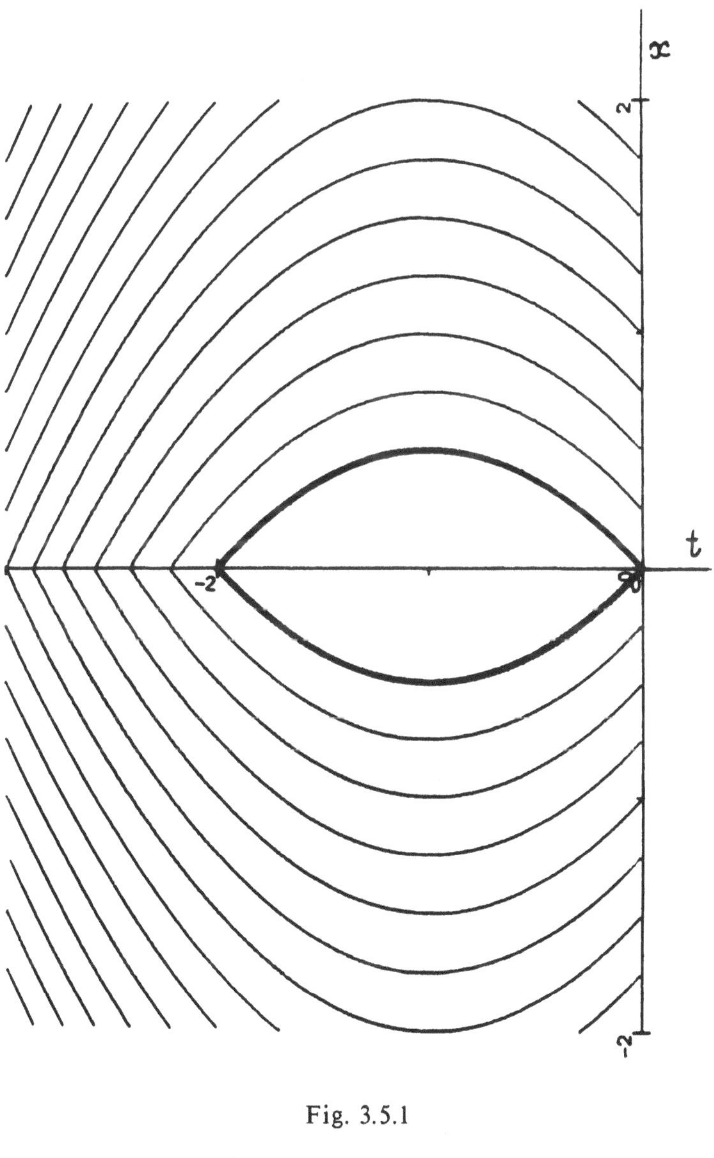

Fig. 3.5.1

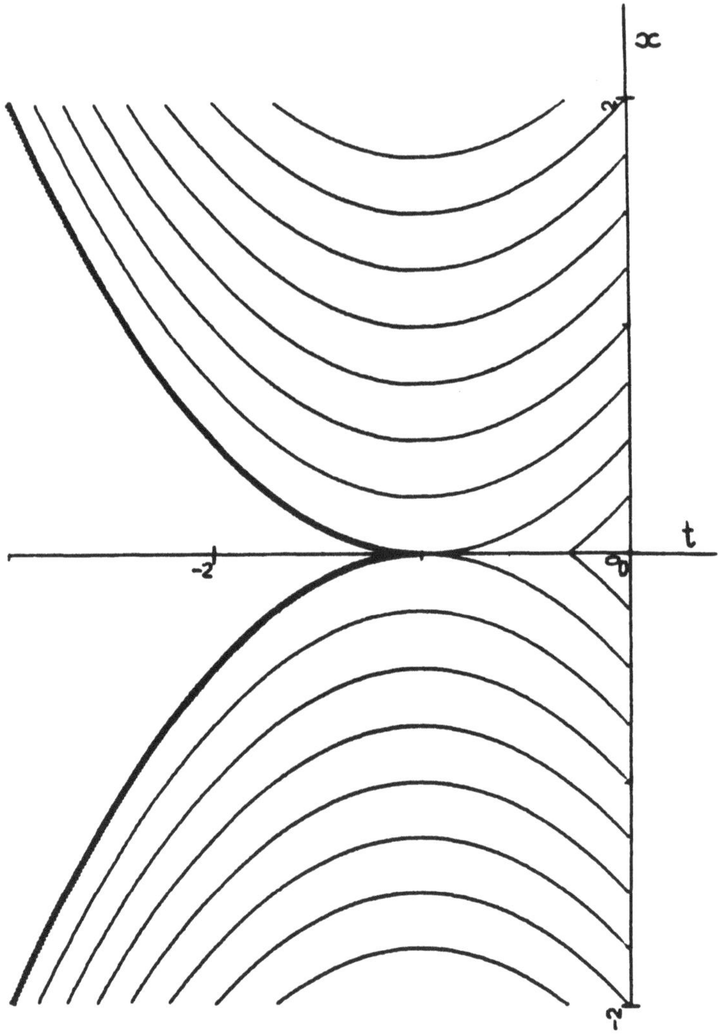

Fig. 3.5.2

$$\varphi(t) \overset{\text{def}}{=} \beta_2|r_2(t)| - \beta_1|r_1(t)|,$$

(10)
$$g_a(t,y) \overset{\text{def}}{=} \begin{cases} \varphi(t) & \text{for } y > a, \\ \varphi(t) & \text{for } y \leqslant a, \ \varphi(t) \leqslant 0, \\ 0 & \text{for } y \leqslant a, \ \varphi(t) \geqslant 0. \end{cases}$$

where $a \geqslant 0$, $y \in R^1$. Consider the differential equation

(11) $\dot{y}(t) = g_a(t,y(t))$.

Let $t \to y(t,a)$ be an absolutely continuous function satisfying the condition $y(0,a) = a$ at the right endpoint and for almost all $t \leqslant 0$ the differential equation (11). Although the function $(t,y) \to g_a(t,y)$ is discontinuous in the phase variable y it can, however, be shown that equation (11) possesses a unique solution $y(\cdot,a)$ which can be extended to the whole semi-axis $(-\infty,0]$. Note that this assertion is valid when equation (11) is integrated in reverse time. If the solution of equation (11) were considered on some interval $[t_*,0]$ with the initial condition $y(t_*) = y_*$, then in general, such a solution may not have been unique.

Consider briefly the construction of the function $y(\cdot,a)$. Assume for simplicity that on any bounded interval $[t^*,0]$ the function $\varphi(\cdot)$ either does not change its sign or the sign is changed a finite number of times. Define the interval T_1 as follows. If $\varphi(0) < 0$, then $T_1 = \{t_1\} = \{0\}$. If $\varphi(t) \geqslant 0$ for all $t \in (-\infty,0]$ then $T_1 = (-\infty,0]$. If $\varphi(t) \geqslant 0$ and $\{t : \varphi(t) < 0, t < 0\} \neq \phi$, then $T_1 = [t_1,0]$. Here t_1 is the time such that $\varphi(t) \geqslant 0$ for $t \in [t_1,0]$ and $\varphi(t) < 0$ for $t \in (t_1-\delta,t_1)$ where δ is a positive number. Set $y(t,a) = a$ for $t \in T_1$. If it turns out that $T_1 = (-\infty,0]$, then the construction of function $y(\cdot,a)$ is completed. Assume that $T_1 \neq (-\infty,0]$. Define the function

$$y_2(t) \overset{\text{def}}{=} a - \int_t^{t_1} \varphi(\tau)d\tau.$$

If $y_2(t) > a$ for all $t < t_1$, we set $T_2 = (-\infty,t_1]$. If $\{t : y_2(t) \leqslant a, t < t_1\} \neq \phi$ then $T_2 = [t_2,t_1]$. Here t_2 is the time such that $y_2(t) > a$ for $t_2 < t < t_1$ and $y_2(t_2) = a$. For $t \in T_2$ the function $y(\cdot,a)$ is defined by the equality $y(t,a) = y_2(t)$. If it turns out that $T_2 = (-\infty,t_1]$ then the construction of the function $y(\cdot,a)$ is completed. If, however, $T_2 \neq (-\infty,t_1]$ then at time $t = t_2$ a new cycle begins which coincides with the one described above. Under our assumption, any interval $[t^*,0]$ will contain a finite number of cycles. Note that the function $y(\cdot,a)$ satisfies the estimate $y(t,a) \geqslant a$ for all $t \leqslant 0$.

Utilizing the function $y(\cdot,a)$, the maximal u-stable bridge W_a for the differential game under consideration can be determined by the relation

$$W_a = \{(t,x) \in (-\infty,0] \times R^n : \|x\| \leqslant y(t,a)\}.$$

The proof of this assertion is omitted since it is basically the same as the arguments presented in §2.5.1. We thus obtain that the value of the differential game (1), (9) is determined by the equality

(12) $\omega^0(t,x) = \min\{a \geqslant 0 : y(t,a) - \|x\| \geqslant 0\}$.

On Figs. 3.5.3 and 3.5.4, level surfaces of the value function for the differential games (7), (9) and (8), (9), respectively, are presented.

Heavily drawn lines represent the boundaries of the regions where $\omega^0(t,x) = 0$.

3.5.3. Let the payoff for a differential game be defined by the functional

$$\gamma(x(\cdot)) \overset{\text{def}}{=} \max_{t_0 \leqslant t \leqslant 0} \|x(t)\|. \tag{13}$$

As in the previous cases, we assume that the motion of the controlled system is described by equation (1). Set

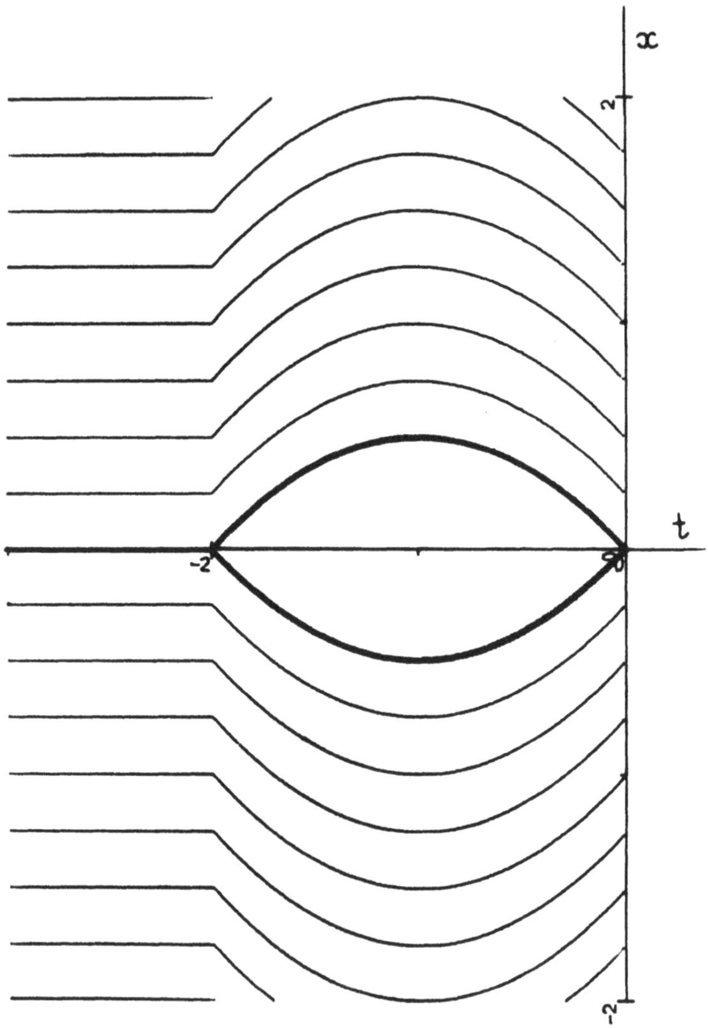

Fig. 3.5.3

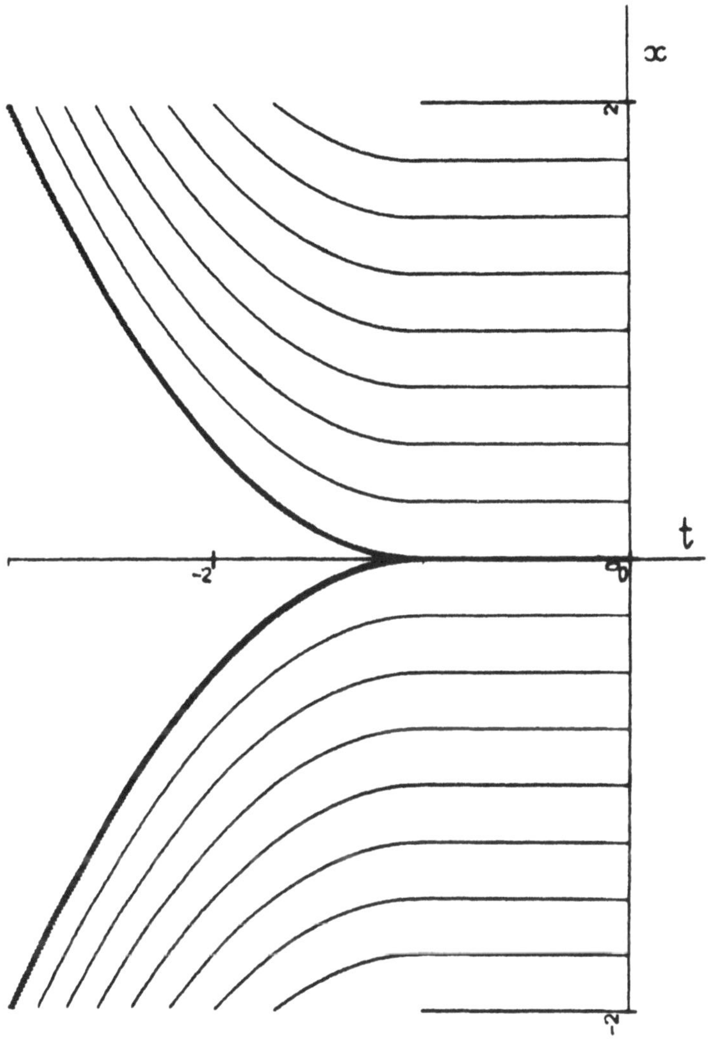

Fig. 3.5.4

$$\mathcal{M} \overset{\text{def}}{=} \{(\theta,x): x \in R^n\}, \quad \mathcal{N}_a \overset{\text{def}}{=} \{(t,x) \in (-\infty,0] \times R^n : \|x\| \leqslant a\}.$$

Let $\mathcal{W}_a$ be the maximal u-stable bridge in an $(\mathcal{M}, \mathcal{N}_a)$-approach problem. The value of the differential game (1), (13) can be determined by the relation

$$\omega^0(t_0,x_0) = \min\{a \geqslant 0 : (t_0,x_0) \in \mathcal{W}_a\}.$$

We now describe the construction of the set $\mathcal{W}_a$. Set

$$h_a(t,z) = \begin{cases} \varphi(t) & \text{for } z < a, \\ \varphi(t) & \text{for } z \geqslant a,\ \varphi(t) \geqslant 0, \\ 0 & \text{for } z \geqslant a,\ \varphi(t) \leqslant 0. \end{cases}$$

Here $a \geqslant 0$, $z \in R^1$ and the quantity $\varphi(t)$ is defined in accordance with (10). Let $t \to z(t,a)$ be an absolutely continuous solution of the differential equation

$$\dot{z}(t) = h_a(t,z(t)), \quad t \leqslant 0,$$

satisfying at the right end point the condition $z(0,a) = a$. Note that such a solution exists, is unique, and can be extended over the whole semi-axis $(-\infty,0]$. The construction of the solution $z(\cdot,a)$ is analogous to the construction of the function $y(\cdot,a)$ described in §3.5.2. The function $z(\cdot,a)$ satisfies the estimate $z(t,a) \leqslant a$ for all $t \leqslant 0$. Set

$$z_*(t,a) = \min_{t \leqslant \tau \leqslant 0} z(\tau,a).$$

Utilizing the notation introduced above, the set W_a can be defined by the relation

$$W_a = \{(t,x) \in (-\infty,0) \times R^n : \|x\| \leqslant z(t,a),\ z_*(t,a) \geqslant 0\}.$$

Therefore the value function of the differential game (1), (13) is determined by the equality

$$\omega^0(t,x) = \min\{a : \min\{z_*(t,a),\ z(t,a) - \|x\|\} \geqslant 0\}. \tag{14}$$

On Figs. 3.5.5 and 3.5.6 the level surfaces of the value function for the differential games (7), (13) and (8), (13), respectively, are presented. On Fig. 3.5.6 the heavily drawn lines denote the region where $\omega^0(t,x) = 1/2$.

3.5.4. Various definitions of the value function are known in the theory of differential games. In particular, the value function can be defined by means of a recursive procedure at every step of which the function of programmed maximin is computed. We briefly describe this procedure for a differential game with payoff of the form (3.1.1).

Let an interval $[t_0,\theta]$ be given and a partition $\Delta = \{t_0 = \tau_k < \tau_{k-1} < \dots < \tau_0 = \theta\}$ be chosen. Define the function $\omega_\Delta : [t_0,\theta] \times R^n \to R^1$. Set $\omega_\Delta(\theta,x) = \omega(\tau_0,x) = \sigma(x)$, $x \in R^n$. Assume that the function ω_Δ is defined in the region $[\tau_i,\theta] \times R^n$. For $(t_*,x_*) \in [\tau_{i+1},\tau_i] \times R^n$ this function is defined by the equality

$$(15) \qquad \omega_\Delta(t_*,x_*) \overset{\text{def}}{=} \sup_{v(\cdot)} \inf_{u(\cdot)} \omega_\Delta(\tau_i,x(\tau_i,t_*,x_*,u(\cdot),v(\cdot))).$$

Here $u(\cdot) : [t_*,\tau_i] \to P$ and $v(\cdot) : [t_*,\tau_i] \to Q$ are measurable functions (programmed controls), $x(\tau_i,t_*,x_*,u(\cdot),v(\cdot))$ is the phase point of the system (1.2.1) into which this system arrives at time $t = \tau_i$ from the initial position (t_*,x_*), for the chosen programmed controls. Using relation (15) the function ω_Δ will be defined throughout the region $[\tau_{i+1},\theta] \times R^n$. Implementing this procedure for $i = 0, 1, ..., k - 1$ we arrive at function ω_Δ defined in the region $[t_0,\theta] \times R^n$.

It can be proved that for $\text{diam}\Delta_\ell \to 0$ as $\ell \to \infty$ the sequence of

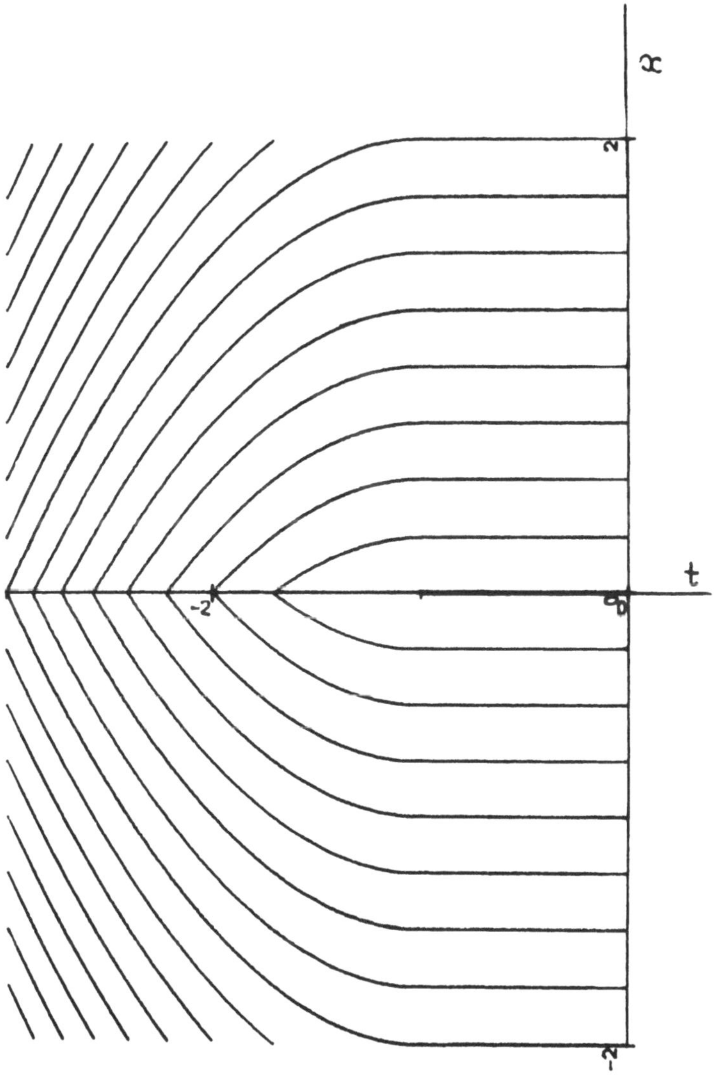

Fig. 3.5.5

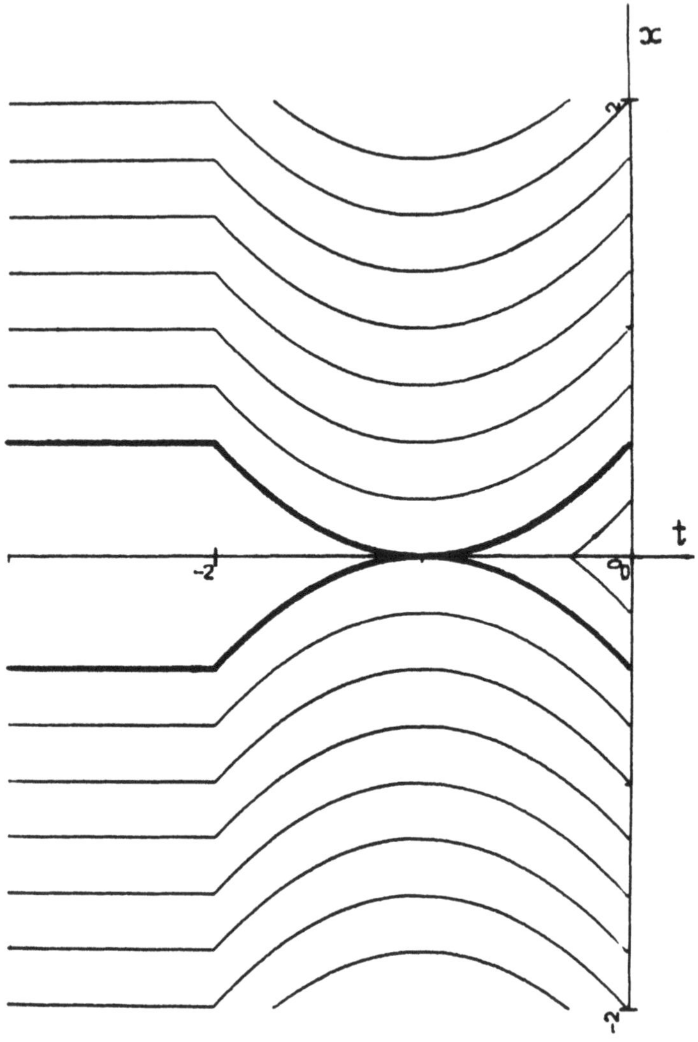

Fig. 3.5.6

functions ω_{Δ_ℓ} (ℓ = 1, 2, ...) converges to the value function ω^0 uniformly in every bounded set $D \subset [t_0, \theta] \times R^n$. For the construction just described, the estimate $\omega_\Delta \leqslant \omega^0$ is valid, i.e. the sequence ω_{Δ_ℓ} converges to the function ω^0 from below. The operation of the programmed maximin in (15) can be replaced by a maximin in which the maximum and minimum are calculated for constant controls $v \in Q$ and $u \in P$, respectively. The value function ω^0 may be approximated from above by function ω^Δ. These functions are determined by a minimax recursive procedure which differs from procedure (15) by interchanging the operations of *sup* and *inf*. Such constructions of the value function go back to paper [24] and were considered in numerous investigations (cf. for example [20, 27, 56, 73]). Observe that in linear differential games with a convex payoff function σ a maximin recursive procedure yields a better approximation than a minimax construction. In this case, when utilizing procedures of the form (15) the error at each stage is of the second order of smallness relative to the length of the interval $[\tau_{i+1},\tau_i]$. For linear systems of small dimensions this procedure can be implemented on a computer.

The construction of functions ω_Δ and ω^Δ is analogous to the recursive construction of stable bridges which was described in §2.5.2. Therein an iterative construction of stable bridges was also considered. Its analog is an iterative method of defining the value function. We shall now describe it.

Let $\omega_* : [t_0,\theta] \times R^n \to R^1$ be a continuous function. Define the operator A which transforms the function ω_* into a continuous function $\omega^* = A(\omega_*)$. The function $\omega^* : [t_0,\theta] \times R^n \to R^1$ will be defined by the equality

$$\omega^*(t_*,x_*) \overset{\text{def}}{=} \max_{t_* \leqslant \tau \leqslant \theta} \sup_{v(\cdot)} \inf_{u(\cdot)} \omega_*(x(\tau,t_*,x_*,u(\cdot),v(\cdot))). \tag{16}$$

Here, as in (15), $u(\cdot)$ and $v(\cdot)$ are programmed controls. We also define the function of programmed maximin

$$\epsilon^0(t_*,x_*) \overset{\text{def}}{=} \sup_{v(\cdot)} \inf_{u(\cdot)} \sigma(x(\theta,t_*,x_*,u(\cdot),v(\cdot))). \tag{17}$$

Consider a sequence of functions

$$\omega_1 = \epsilon^0, \quad \omega_2 = A(\omega_1), \quad \omega_3 = A(\omega_2), \ \dots .$$

One can prove that the sequence obtained converges to the value function uniformly in any bounded region $\mathfrak{D} \subset [t_0,\theta] \times R^n$.

An account of the method of programmed iterations is presented in papers [11, 12, 69].

In conclusion, we note that recursive and iterative constructions of the value function discussed herein are utilized mainly in theoretical investigations. Computational implementation of these methods is very time consuming.

Chapter 4
DYNAMIC PROGRAMMING

4.1. The Main Equation for a Smooth Value Function

In this chapter we shall continue the investigation of value functions for differential games. The results obtained below are to some extent connected with the so-called main equation of the theory of differential games. The main equation is a first order partial differential equation of the Hamilton-Jacobi type, considered by numerous authors. In the literature on the theory of differential games various terms are used to denote this equation: it is called the main equation, the Bellman–Isaacs equation, the equation of the dynamic programming method, and so on. It is known that the main equation possesses a classical (continuously-differentiable) solution satisfying the required boundary conditions if and only if the value function is continuously differentiable, and in this case the value function coincides with the solution of the main equation. However, as a rule, the value function is not everywhere differentiable. If the value function is non-smooth, then it must satisfy the main equation at all the points where it is differentiable. This necessary condition is, however, not sufficient. It therefore becomes necessary to complement the main equation by relations which describe the behavior of the value function on singular surfaces on which it is not differentiable. These problems are investigated in the present chapter.

Below, some inequalities for directional derivatives of the value function are obtained. These inequalities together with the boundary conditions form a set of necessary and sufficient conditions which should be satisfied by the value function. At those points where the value function is differentiable the inequalities obtained become the main equation. Thus they could be viewed as a generalization of the main equation.

4.1.1. Consider a differential game in which the movement of the controlled system is described by the equation (1.2.1) and the payoff is determined by the equality $\gamma(x(\cdot)) \overset{\text{def}}{=} \sigma(x(\theta))$. In other words, the game considered herein is formed by Problems 3.1.1 and 3.1.2. Let ω^0: $T \times R^n \to R$ be the value function for this game. Assume that at each point (t,x) $(t < \theta,\ x \in R^n)$ the function ω^0 is differentiable. It is known that in this region it satisfies the equation

$$(1) \qquad \frac{\partial\omega^0(t,x)}{\partial t} + \min_{u \in P} \max_{v \in Q} \left[\frac{\partial\omega^0(t,x)}{\partial x} f(t,x,u,v)\right] = 0.$$

Here $\partial\omega/\partial x \overset{\text{def}}{=} (\partial\omega/\partial x_1, \ldots, \partial\omega/\partial x_n)$ is a row-vector, thus

$$\frac{\partial\omega}{\partial x} f \overset{\text{def}}{=} \sum_{i=1}^{n} \frac{\partial\omega}{\partial x_i} f_i \,.$$

Note that in accordance with condition (1.2.8) the operations *min* and *max* in (1) can be interchanged.

For the differential game under consideration, equation (1) is called the *main equation.* It is also called the equation of the method of dynamic programming or the Bellman-Isaacs' equation. It follows directly from the definition of the value that for this game the function ω^0 satisfies the equality (boundary condition)

$$(2) \qquad \omega^0(\theta,x) = \sigma(x) \quad (x \in R^n).$$

Thus the following assertion is valid: if the value function is continuously differentiable then it is a solution of the main equation with the boundary condition (2). This necessary condition satisfied by a differentiable value function is well known in the theory of differential games and is presented in numerous sources (see, e.g., [26, 24, 32]). We note that the necessary condition stated herein follows from the results presented in §4.4. The converse assertion is also known to be valid.

To formulate this assertion we introduce some notation. Let $-\infty \leqslant a < b \leqslant \infty$. Let $\omega : T \times R^n \to R$ be a continuous function. The symbol T^0 denotes the interior of the interval T, i.e. if $T = (-\infty,\theta]$ then $T^0 = (-\infty,\theta)$, if $T = [t_*,\theta]$ then $T^0 = (t_*,\theta)$. Define the sets

$$\mathcal{E} \overset{\text{def}}{=} \{(t,x) \in T^0 \times R^n : a < \omega(t,x) < b\},$$

$$\mathcal{E}_* \overset{\text{def}}{=} \{(t,x) \in T \times R^n : a \leqslant \omega(t,x) < b\},$$

$$\mathcal{E}^* \overset{\text{def}}{=} \{(t,x) \in T \times R^n : a < \omega(t,x) \leqslant b\}.$$

Note that $\mathcal{E}$ is an open set and $cl\ \mathcal{E} \subset \mathcal{E}_* \cup \mathcal{E}^*$.

Assume that the function ω is continuously differentiable in the domain $\mathcal{E}$. Let $(t,x) \in \mathcal{E}$. Define the sets

$$\mathcal{U}^0(t,x) \overset{\text{def}}{=} \{u^0 \in \mathcal{P} : \max_{v \in \mathcal{Q}} \ (\partial\omega(t,x)/\partial x) f(t,x,u^0,v)$$

$$= \min_{u \in \mathcal{P}} \max_{v \in \mathcal{Q}} (\partial\omega(t,x)\partial x) f(t,x,u,v)\},$$

(3)

$$\mathcal{V}^0(t,x) \overset{\text{def}}{=} \{v^0 \in \mathcal{Q} : \min_{u \in \mathcal{P}} \ (\partial\omega(t,x)/\partial x) f(t,x,u,v^0)$$

$$\max_{v \in \mathcal{Q}} \ \min_{u \in \mathcal{P}} \ (\partial\omega(t,x)/\partial x) f(t,x,u,v)\}.$$

For $(t,x) \in (T \times R^n)\backslash\mathcal{E}$ we set $\mathcal{U}^0(t,x) \overset{\text{def}}{=} \mathcal{P}$, $\mathcal{V}^0(t,x) \overset{\text{def}}{=} \mathcal{Q}$. Let the functions $U^0 : T \times R^n \to \mathcal{P}$ and $V^0 : T \times R^n \to \mathcal{Q}$ satisfy the conditions

(4) $$U^0(t,x) \in \mathcal{U}^0(t,x), \quad V^0(t,x) \in \mathcal{V}^0(t,x), \quad (t,x) \in T \times R^n.$$

The following theorem is valid.

Theorem 4.1.1. *Let the continuous function $\omega : T \times R^n \to R^1$ satisfy boundary condition (2). Let this function be continuously differentiable in the region $\mathcal{E}$ and satisfy equation (1). Then in the region cl $\mathcal{E}$ the function ω coincides with the value function of the differential game under consideration. For an arbitrary initial position $(t_0,x_0) \in \mathcal{E}$ the pair of strategies (U^0,V^0) selected from the conditions (3), (4) forms a saddle point. If $(t_0,x_0) \in \mathcal{E}_*$ then*

(5) $$\sigma(x(\theta)) \leqslant \omega(t_0,x_0) \ \textit{for all} \ x(\cdot) \in \mathcal{X}_1(t_0,x_0,U^0).$$

If $(t_0,x_0) \in \mathcal{E}^*$, then

(6) $$\sigma(x(\theta)) \geqslant \omega(t_0,x_0) \ \textit{for all} \ x(\cdot) \in \mathcal{X}_2(t_0,x_0,V^0).$$

Proof. We shall show that the inequality (5) is valid. Consider a multi-valued mapping $(t,x) \to \mathcal{U}^0(t,x)$ from $T \times R^n$ into $\mathcal{P}$. It is easy to verify that it is upper semi-continuous (cf. §A.3). Let

$$\mathcal{H}(t,x) \overset{\text{def}}{=} \{f(t,x,u,v) : u \in \mathcal{U}^0(t,x), \ v \subset \mathcal{Q}\}.$$

In accordance with (1) and (3) for any $(t,x) \in \mathcal{E}$ and $h \in \mathcal{H}(t,x)$ we have

$$(\partial\omega(t,x)/\partial x)h \leqslant \min_{u \in \mathcal{P}} \ \max_{v \in \mathcal{Q}} (\partial\omega(t,x)/\partial x) f(t,x,u,v),$$

$$\partial\omega(t,x)/\partial t + (\partial\omega(t,x)/\partial x)h \leqslant 0.$$

These inequalities remain valid if the vector $h \in \mathcal{H}(t,x)$ is replaced by an arbitrary vector $g \in co\ \mathcal{H}(t,x)$. We thus obtain

$$(7) \qquad \frac{\partial \omega(t,x)}{\partial t} + \frac{\partial \omega(t,x)}{\partial x} g \leqslant 0, \qquad (t,x) \in \mathcal{E}, \;\; g \in co\, \mathcal{H}(t,x).$$

Consider the differential inclusion

$$\dot{x}(t) \in co\, \mathcal{H}(t,x(t)).$$

Denote by $X(t_0,x_0, U^0)$ the set of solutions of the differential inclusion satisfying the initial condition $x(t_0) = x_0$, where (t_0,x_0) is a point in $T \times R^n$. The multi-valued mapping $(t,x) \to co\, \mathcal{H}(t,x)$ satisfies the conditions (a) – (c) stated in §A.4. Therefore in accordance with Theorem A.4.1 the set $X(t_0,x_0, U^0)$ is a non-empty compact set in $C_n([t_0,\theta])$.

Let $(t_0,x_0) \in \mathcal{E}_*$, i.e. $\omega(t_0,x_0) \in [a,b)$. We shall show that

$$(8) \qquad \max\{\sigma(x(\theta)) : x(\cdot) \in X(t_0,x_0,U^0)\} \leqslant \omega(t_0,x_0).$$

Assume the contrary, i.e. let there exist a solution $x(\cdot) \in X(t_0,x_0, U^0)$ such that

$$\sigma(x(\theta)) = \omega(\theta,x(\theta)) > \omega(t_0,x(t_0))$$

(here the equality (2) is utilized). It follows from this inequality that numbers t_1 and t_2 ($t_0 < t_1 < t_2 < \theta$) exist such that

$$\omega(t_1,x(t_1)) < \omega(t_2,x(t_2)), \; (t,x(t)) \in \mathcal{E} \text{ for } t \in [t_1,t_2].$$

Recall that the function $\omega : \mathcal{E} \to R$ is continuously differentiable and the function $x(\cdot) : [t_0,\theta] \to R^n$ is absolutely continuous. Therefore the function $t \to \omega[t] \overset{\text{def}}{=} \omega(t,x(t)) : [t_1,t_2] \to R^1$ is also absolutely continuous. It follows from (7) that $\dot{\omega}[t] \leqslant 0$ for almost all $t \in [t_1,t_2]$. We thus have

$$\omega[t_1] = \omega(t_1,x(t_1)) \geqslant \omega[t_2] = \omega(t_2,x(t_2)).$$

The contradiction obtained proves the validity of the inequality (8).

Let $X_1(t_0,x_0,U^0)$ be the set of constructive motions generated by the positional strategy U^0 emanating from the initial position $(t_0,x_0) \in \mathcal{E}_*$. Taking into account condition (4) and utilizing Theorem A.4.2 we obtain the inclusion $X_1(t_0,x_0,U^0) \subset X(t_0,x_0, U^0)$. This inclusion and the estimate (8) imply the inequality (5).

Inequality (6) can be proved analogously. From inequalities (5) and (6) we obtain that for any initial position $(t_0,x_0) \in \mathcal{E}$ the pair of strategies U^0 and V^0 form a saddle point for the game under consideration and the quantity $\omega(t_0,x_0)$ is the value of this game.

Function ω considered in Theorem 4.1.1 and the value function ω^0 coincide in the region $\mathcal{E}$. Both functions are continuous in $T \times R^n$. Hence they coincide on the set $cl\, \mathcal{E}$. Theorem 4.1.1 is thus proved.

Thus if the value function ω^0 of a differential game is differentiable, then it can be defined as a solution of the main equation (1) with the boundary condition (2). In this and only in this case the solution of Problem (1), (2) exists and is unique. Various methods can be used to integrate the main equation. The most common is the method of characteristics.

Computing the value function, one can determine the sets $\{(t,x) : \omega^0(t,x) \leqslant a\}$ and $\{(t,x) : \omega^0(t,x) \geqslant a\}$ which will be u-stable and v-stable bridges, respectively. Next, setting $a = \omega^0(t_0,x_0)$ one can construct strategies U^e and V^e extremal to these bridges. A saddle point for the given initial position (t_0,x_0) will be defined in this manner. However, in the case under consideration it seems to be more suitable to define optimal strategies of the players in accordance with Theorem 4.1.1. Strategies U^0 and V^0 defined by the conditions (4) possess the following important properties: *for any initial position* $(t_0,x_0) \in \mathcal{E}$ the pair of strategies U^0 and V^0 form a saddle point of the game, i.e. these strategies are optimal *universal* strategies. Note that in the general case optimal strategies U^0 and V^0 defined for a fixed initial position may not possess this property of universality (this problem will be discussed in §8.5).

4.1.2. If the value function is differentiable, its calculation and the construction of optimal strategies for players are substantially simplified. As shown in §3.4 the value function ω^0 inherits the properties possessed by the payoff function σ, i.e., it is continuous and satisfies the Lipschitz condition. Assume that the payoff function σ is continuously differentiable with respect to all variables. The question arises: will the value function in this case be differentiable? To obtain an answer, consider the following example.

Let the motion of the controlled system be described by the equation

$$\dot{x}(t) = tu(t) + v(t). \tag{9}$$

Here x, u, and v are two-dimensional vectors, $t \leqslant 0$, and $\theta = 0$ is the fixed time of termination of the game. It is assumed that controls of the players satisfy the constraints

$$\|u(t)\| \leqslant 1, \quad \|v(t)\| \leqslant 1.$$

The payoff in this game is defined by the equality

$$\gamma(x(\cdot)) \overset{\text{def}}{=} \|x(0)\|^2, \tag{10}$$

i.e. here $\sigma(x) = x_1^2 + x_2^2$.

Firstly, we determine the value of the differential game in which the motion of the system is described by equation (9) and the payoff is given by equality

$$\gamma_1(x(\cdot)) \overset{\text{def}}{=} \sigma_1(x(0)) = \|x(0)\|. \tag{11}$$

We now utilize the results presented in §3.5.1. In accordance with (3.5.5) and (3.5.6) we compute the quantities

$$r_*(t) = \int_t^0 (|\tau| - 1)d\tau = -\int_t^0 (\tau + 1)d\tau = t^2/2 + t,$$

(12)
$$a_1(t,x) = \|x\| - r_*(t) = \|x\| - t^2/2 - t,$$

$$a_2(t) = -\min_{t \leqslant \tau \leqslant 0} r_*(\tau) = \begin{cases} -t^2/2 - t & \text{for } t \in [-1,0] \\ 1/2 & \text{for } t \leqslant -1 \end{cases}.$$

Denote by ω_1 the value function of the differential game (9), (11). In accordance with (3.5.6) this function is determined by the equality

(13) $$\omega_1(t,x) = \max\{a_1(t,x), a_2(t)\}.$$

Return now to the initial differential game (9), (10). Observe that the payoff functions σ (10) and σ_1 (11) are connected by the equality $\sigma = \sigma_1^2$. Therefore, as shown in §3.4.4, the corresponding value functions are connected by the relation $\omega^0 = \omega_1^2$; here ω^0 is the value of the differential game (9), (10).

We obtain that the value function of the differential game (9), (10) is determined by the equality

(14) $$\omega^0(t,x) = [\max\{a_1(t,x), a_2(t)\}]^2$$

where the quantities $a_1(t,x)$ and $a_2(t)$ are defined in accordance with (12). Observe that the level surfaces of the function ω^0 are of the same form as the level surfaces of the function ω_1 presented in Fig. 3.5.2. According to (12), (14) the equality

$$\omega^0(t,x) = [\|x\| - t^2/2 - t]^2$$

is valid for $t \in [-1,0]$. It is easy to verify directly that for $t \in (-1,0)$ the function ω^0 is not differentiable.

In the differential game (9), (10) the functions

$$(t,x,u,v) \to f(t,x,u,v) = tu + v \quad \text{and} \quad x \to \sigma(x) = \|x\|^2$$

are analytic functions. The sets P and Q are circles, i.e. the boundary of these sets is a smooth curve. However, the value function does not inherit the smoothness of functions f, σ and of the sets P and Q; this function turns out to be nondifferentiable. The situation noted in this simple example is not an exceptional one. On the contrary, the value function is rarely differentiable. Thus only in particular cases does the main equation possess a classical solution.

If we assume that the payoff function σ is locally Lipschitzian, it then follows from Theorem 3.4.2 that the value function ω^0 also possesses this property. According to the Rademacher Theorem [27, 65] the function ω^0 will be differentiable almost everywhere on $T \times R^n$. We present the definition of a generalized solution for equation (1). A locally Lipschitzian function $\omega : T \times R^n \to R$ is called a *generalized solution* of equation (1) if the function ω satisfies the equation at all points (t,x) where it is differentiable.

The following assertion is valid: the value function ω^0 is a generalized solution of the main equation satisfying the boundary condition (2).

This assertion is presented, for example, in papers [26, 27]. We shall prove it below utilizing the results of §4.4.

This assertion determines a necessary condition which the value function ω^0 satisfies. However this condition is not sufficient. In fact, the set of generalized solutions of the main equation satisfying the boundary condition (2) may be infinite, i.e. this set contains infinitely many functions which differ from the required value function. In order to select the value function from the generalized solutions of the main equation, it is necessary to obtain relations which characterize its behavior on the "singular set" i.e. where the function is not differentiable. These problems are studied in the subsequent sections of this chapter.

In conclusion, to illustrate the assertions stated in this section we return to the example of a differential game (9), (10). In this example the main equation is of the form

$$\frac{\partial \omega(t,x)}{\partial t} + (1 + t) \left[\left(\frac{\partial \omega(t,x)}{\partial x_1}\right)^2 + \left(\frac{\partial \omega(t,x)}{\partial x_2}\right)^2\right]^{1/2} = 0. \tag{15}$$

One can verify directly that the function ω^0 (14) is a generalized solution of equation (15) satisfying the boundary condition $\omega(\theta,x) = \|x\|^2$. Equation (15) possesses infinitely many generalized solutions satisfying this boundary condition. For example, the function

$$\omega(t,x) \overset{\text{def}}{=} [\|x\| - t^2/2 - t]^2, \quad t \leqslant 0, \quad x \in R^2$$

is such a solution.

4.2. Properties of Value Functions

In Chapter 2 u-stable and v-stable sets in the space of positions were considered. The properties of u- and v-stabilities can be defined also for real-valued functions defined on the space of positions. Based on these notions in this section we state necessary and sufficient conditions which should be satisfied by the value function of a differential game. The results obtained here will be used in §4.4 for the derivation of the generalized form of the main equation.

4.2.1. We define the notions of u-stable and v-stable functions. We recall and introduce some notation. Set $T \overset{\text{def}}{=} (-\infty,\theta]$ and $T^0 \overset{\text{def}}{=} (-\infty,\theta)$. Let $\mathcal{E}$ be an open set in $T^0 \times R^n$. For the continuous functions $x(\cdot) : [t_*,\theta] \to R^n$ we define the sets

$$T_*(x(\cdot)) \overset{\text{def}}{=} \{t \in [t_*,\theta] : (t,x(t)) \in (T \times R^n)\backslash \mathcal{E}\}$$

and the functional

$$\tau_*(x(\cdot)) \overset{\text{def}}{=} \min \{t : t \in T_*(x(\cdot))\}.$$

Note that $T_*(x(\cdot))$ is closed and $T_*(x(\cdot)) \neq \phi$, since $\theta \in T_*(x(\cdot))$. If $(t_*,x(t_*)) \in \mathcal{E}$, then $(t,x(t)) \in \mathcal{E}$ for $t_* \leqslant t < \tau_*(x(\cdot))$; here $\tau_*(x(\cdot))$ is the time when the point $(t,x(t))$ leaves the set $\mathcal{E}$ for the first time.

Recall (cf. §2.2) that the symbols $X_1(t_*,x_*,u)$ and $X_2(t_*,x_*,v)$ denote the sets of absolutely-continuous solutions $x(\cdot) : [t_*,\theta] \to R^n$ $(x(t_*) = x_*)$ of the differential inclusions

$$\dot{x}(t) \in \mathcal{F}_1(t,x(t),u), \quad \dot{x}(t) \in \mathcal{F}_2(t,x(t),v)$$

where

$$\mathcal{F}_1(t,x,u) \overset{\text{def}}{=} co \{f(t,x,u,v) : v \in \mathcal{Q}\},$$

$$\mathcal{F}_2(t,x,v) \overset{\text{def}}{=} co \{f(t,x,u,v) : u \in \mathcal{P}\}.$$

Below we shall consider the operations

(1) $$\min_{x(\cdot)} \max_{t} \omega(t,x(t)) \quad \text{for} \quad x(\cdot) \in X_2(t_*,x_*,v),\ t \in [t_*,\tau_*(x(\cdot))],$$

(2) $$\max_{x(\cdot)} \min_{t} \omega(t,x(t)) \quad \text{for} \quad x(\cdot) \in X_1(t_*,x_*,u),\ t \in [t_*,\tau_*(x(\cdot))].$$

Here $\omega : T \times R^n \to R$ is a continuous function. It can be shown that the minima and maxima in (1) and (2) are actually attained.

Consider, for example, operation (1). On the space $C_n([t_*,\theta])$ we define the functional

$$x(\cdot) \to \xi(x(\cdot)) \overset{\text{def}}{=} \max_{t} \omega(t,x(t)) \quad \text{for} \quad t_* \leqslant t \leqslant \tau_*(x(\cdot)).$$

Evidently the maximum is attained here. We shall show that the functional ξ is lower semi-continuous. Let $x_k(\cdot) \in C_n([t_*,\theta])$ $(k = 1, 2, \ldots)$ $x_k(\cdot) \to x^0(\cdot)$, $\xi(x_k(\cdot)) \to \xi^0$ for $k \to \infty$. Consider the sequence $\tau_k = \tau_*(x_k(\cdot))$ $(k = 1, 2, \ldots)$. Without loss of generality it may be assumed that $\tau_k \to t^0$ for $k \to \infty$. Therefore

$$\xi^0 = \lim_{k\to\infty} [\max_{t_* \leqslant t \leqslant \tau_k} \omega(t,x_k(t))] = \max_{t_* \leqslant t \leqslant t^0} \omega(t,x^0(t)).$$

Since the functional τ_* is lower semi-continuous (cf. the proof of lower semi-continuity of functional τ^0 in §1.4.2) the inequality $\tau_*(x^0(\cdot)) \leqslant t^0$ is valid. Consequently,

$$\xi(x^0(\cdot)) = \max \{\omega(t,x^0(t)) : t_* \leqslant t \leqslant \tau_*(x^0(\cdot))\}$$

$$\leqslant \max \{\omega(t,x^0(t)) : t_* \leqslant t \leqslant t^0\} = \xi^0.$$

We have thus shown that functional ξ is lower semi-continuous. The set $X_2(t_*,x_*,v)$ is compact in $C_n([t_*,\theta])$ (cf. §2.2.1 and §A.4). Therefore the lower semi-continuous functional ξ attains its minimum on the set $X_2(t,x,u)$ (cf. §A.1). We have thus shown that the minimum and maximum are attained in operation (1). Analogously, notation (2) can be justified.

Definition. *Let $\omega : T \times R^n \to R$ be a continuous function and let $\mathcal{E} \subset T^0 \times R^n$ be an open region. We say that the function ω is* u-*stable in the region $\mathcal{E}$ provided the inequality*

(3) $$\sup_{(t_*,x_*,v)} \min_{x(\cdot)} \max_{t} [\omega(t,x(t)) - \omega(t_*,x_*)] \leqslant 0$$

is fulfilled. Here $(t_*,x_*,v) \in E \times \mathcal{Q}$, $x(\cdot) \in X_2(t_*,x_*,v)$, $t \in [t_*,\tau_*,(x(\cdot)]$.

We clarify the meaning of this definition. Assume that for any $(t,x,v) \in T \times R^n \times \mathcal{Q}$ the vectogram $f(t,x,P,v)$ is convex. Then

$$X_2(t_*,x_*v) = \{x(\cdot,t_*,x_*,u(\cdot),v) : u(\cdot) \in \mathfrak{L}([t_*,\theta], P)\}$$

(cf. an analogous remark in §2.2.1). In this case the inequality (3) can be interpreted as follows: for any initial position $(t_*,x_*) \in \mathcal{E}$ and for any choice of the constant control $v \in \mathcal{Q}$ by the second player, the first player can select control $u(\cdot)$ in such a manner that for the motion $x(\cdot) = x(\cdot,t_*,x_*,u(\cdot),v)$ the estimate $\omega(t,x,(t)) \leqslant \omega(t_*,x_*)$ is valid until the point $(t,x(t))$ leaves the region $\mathcal{E}$ for the first time.

The notion of a v-stable function is introduced analogously.

Definition. *Let $\omega : T \times R^n \to R$ be a continuous function and let $\mathcal{E}$ be an open region in $T^0 \times R^n$. We say that the function ω is* v-*stable in the region $\mathcal{E}$ provided*

(4) $$\inf_{(t_*,x_*,u)} \max_{x(\cdot)} \min_{t} [\omega(t,x(t) - \omega(t_*,x_*)] \geqslant 0.$$

Here $(t_*,x_*u) \in \mathcal{E} \times P$, $x(\cdot) \in X_1(t_*,x_*,u)$, $t \in [t_*,\tau_*(x(\cdot))]$.

Below we shall consider mainly u-stable and v-stable functions on the set $T^0 \times R^n$. For these functions the maximum in (3) and minimum in (4) are computed over all $t \in [t_*,\theta]$.

The definitions introduced are meaningful in the case when $\mathcal{E} \neq \phi$.

However, for the convenience of subsequent statements we shall not exclude the case $\mathcal{E} = \phi$, assuming that in an empty region every continuous function ω is u-stable and v-stable.

Conditions of u-stability and v-stability can be stated in various equivalent forms. We shall introduce two inequalities which, as it will be shown below, will be equivalent to the inequality (3):

$$\sup_{(t_*,x_*,v)} \inf_{x(\cdot)} \overline{\lim_{\delta\to+0}} [\omega(t_* + \delta, x(t_* + \delta)) - \omega(t_*,x_*)]\delta^{-1} \leqslant 0, \tag{5}$$

$$\sup_{(t_*,x_*,v)} \inf_{x(\cdot)} \underline{\lim_{\delta\to+0}} [\omega(t_* + \delta, x(t_* + \delta)) - \omega(t_*,x_*)]\delta^{-1} \leqslant 0. \tag{6}$$

Here $(t_*,x_*,v) \in \mathcal{E} \times \mathcal{Q}$, $x(\cdot) \in X_2(t_*,x_*,v)$.

We now state two inequalities which are equivalent to the inequality (4):

$$\inf_{(t_*,x_*,u)} \sup_{x(\cdot)} \underline{\lim_{\delta\to+0}} [\omega(t_* + \delta, x(t_* + \delta)) - \omega(t_*,x_*)]\delta^{-1} \geqslant 0, \tag{7}$$

$$\inf_{(t_*,x_*,u)} \sup_{x(\cdot)} \overline{\lim_{\delta\to+0}} [\omega(t_* + \delta, x(t_* + \delta)) - \omega(t_*,x_*)]\delta^{-1} \geqslant 0, \tag{8}$$

where $(t_*,x_*,u) \in \mathcal{E} \times P$, $x(\cdot) \in X_1(t_*,x_*,u)$.

Below the following lemma will be needed.

Lemma 4.2.1. *Conditions* (3), (5), *and* (6) *are equivalent. Conditions* (4), (7), *and* (8) *are equivalent.*

Proof. We shall show the equivalence of conditions (3), (5), and (6). Implications (3) $\Rightarrow$ (5) $\Rightarrow$ (6) are obvious. It remains to verify that (6) implies (3). Assume the contrary. Let inequality (6) be fulfilled, but let there exist $(t_*,x_*) \in \mathcal{E}$ and $v \in \mathcal{Q}$ such that

$$\min_{x(\cdot)} \max_{t} [\omega(t,x(t)) - \omega(t_*,x_*)] > 2r > 0, \tag{9}$$

where $x(\cdot) \in X_2(t_*,x_*,v)$, $t \in [t_*,\tau_*(x(\cdot))]$. Set $\alpha(t) \overset{\text{def}}{=} \omega(t_*,x_*) + r\cdot(t-t_*)\cdot(\theta - t_*)^{-1}$, $\beta(t) \overset{\text{def}}{=} \alpha(t) + r$,

$$\begin{aligned} \mathcal{H}(x(\cdot)) &\overset{\text{def}}{=} \{h \in [t_*,\theta]: \omega(h,x(h)) \leqslant \alpha(h),\ \omega(t,x(t)) \leqslant \beta(t) \text{ for } t_* \leqslant t \leqslant h\} \\ h_0(x(\cdot)) &\overset{\text{def}}{=} \max\{h : h \in \mathcal{H}(x(\cdot))\}; \end{aligned} \tag{10}$$

here $x(\cdot) \in X_2(t_*,x_*,v)$. Observe that $\mathcal{H}(x(\cdot)) \neq \phi$ for any $x(\cdot) \in X_2(t_*,x_*v)$ since $t_* \in \mathcal{H}(x(\cdot))$. For an arbitrary $x(\cdot) \in X_2(t_*,x_*,v)$ the set $\mathcal{H}(x(\cdot))$ is a compact subset of the interval $[t_*,\theta]$. It can be verified that the multi-valued mapping $x(\cdot) \to \mathcal{H}(x(\cdot))$ is upper semi-continuous. Therefore the functional $x(\cdot) \to h_0(x(\cdot))$ is upper semi-continuous. Consequently, on the compact set $X_2(t_*,x_*,v)$ the functional h_0 attains its maximum. Let the motion $x_0(\cdot) \in X_1(t_*,x_*,v)$ be chosen from the condition

$$h_0(x_0(\cdot)) = \max_{x(\cdot)} h_0(x(\cdot)) \quad \text{for} \quad x(\cdot) \in X_2(t_*,x_*,v).$$

Introduce the notation $t^* \overset{\text{def}}{=} h_0(x_0(\cdot))$. According to (10) we have

$$\omega(t,x_0(t)) \leqslant \beta(t) = \omega(t_*,x_*) + r\cdot(t - t_*)\,(\theta - t_*)^{-1} + r \leqslant \omega(t_*,x_*) + 2r$$

$$\text{for } t_* \leqslant t \leqslant t^*.$$

Therefore (9) implies that $t^* < \tau_*(x_0(\cdot))$. The definition of the functional τ_* yields that $(t^*,x_0(t^*)) \in \mathcal{E}$. This inclusion allows us to utilize condition (6).

According to condition (6) there exists a motion $x^*(\cdot) \in X_2(t^*,x_0(t^*),v)$ and a number $\delta > 0$ such that

(11) $$\omega(t^* + \delta,x^*(t^* + \delta)) \leqslant \omega(t^*,x_0(t^*)) + \delta\cdot r\cdot(\theta - t_*)^{-1} \leqslant \alpha(t^* + \delta).$$

Here we use the fact

(12) $$\omega(t^*,x_0(t^*)) \leqslant \alpha(t^*).$$

Since the number δ can be taken arbitrarily small, it can be assumed that along with (11) the estimate

(13) $$\omega(t,x^*(t)) \leqslant \omega(t^*,x_0(t^*)) + r\cdot(t - t^*)\cdot(\theta - t_*)^{-1} + r \quad \text{for } t \in [t^*,t^* + \delta]$$

is also valid. Consider the motion $x^\#(t) = (x_0(t)$ for $t_* \leqslant t \leqslant t^*$, $x^*(t)$ for $t^* \leqslant t \leqslant \theta)$. From the definition of the number $t^* = h_0(x_0(\cdot)) \in \mathcal{H}(x_0(\cdot))$ we have

$$\omega(t,x_0(t)) = \omega(t,x^\#(t)) \leqslant \beta(t) \quad \text{for} \quad t_* \leqslant t \leqslant t^*.$$

From (10), (12), and (13) we obtain

$$\omega(t,x^*(t)) = \omega(t,x^\#(t)) \leqslant \alpha(t^*) + r(t - t^*)\,(\theta - t_*)^{-1} + r = \beta(t)$$

$$\text{for } t \in [t^*,t^* + \delta].$$

Taking (11) into account we conclude that $t^* + \delta \in \mathcal{H}(x^\#(\cdot))$. Observe that the motion $x^\#(\cdot)$ is contained in the set $X_2(t_*,x_*,v)$. For this motion the inequality

$$h_0(x^\#(\cdot)) \geqslant t^* + \delta > t^* = \max_{x(\cdot)} h_0(x(\cdot)) \quad \text{for} \quad x(\cdot)) \in X_2(t_*,x_*,v)$$

is valid. We have thus arrived at a contradiction.

The equivalence of the inequalities (3), (5), and (6) is proved. The equivalence of conditions (4), (7), and (8) can be proved analogously.

Stable functions and stable bridges are related concepts. The connection between these concepts is revealed in the assertions discussed below.

4.2.2. Utilizing the concepts of u-stable and v-stable functions one can formulate necessary and sufficient conditions to be satisfied by the value function of a differential game. Consider first a differential game with a fixed termination time. As usual we assume that the motion of the controlled system is described by equation (1.2.1). The payoff functional in this game is defined by the equality $\gamma(x(\cdot)) = \sigma(x(\theta))$. We shall assume that the function σ is continuous. Then the value function $\omega^0 : T \times R^n \to R$ will also be continuous (cf. §3.4.1). The following theorem is valid:

Theorem 4.2.1. *In order for a function* $\omega : T \times R^n \to R$ *to be the value function for the differential game with payoff of the form* (3.1.1) *where* σ *is a continuous function, it is necessary and sufficient that the following set of conditions be fulfilled*:

(14) (a) $\omega(\theta,x) = \sigma(x)$, $x \in R^n$,

(b) ω *is continuous on* $T \times R^n$,

(c) ω *is* u-*stable and* v-*stable in the region* $\mathcal{E} = T^0 \times R^n$.

We now state and prove the lemmas which will imply Theorem 4.2.1.

Lemma 4.2.2. *Let* $\omega^0 : T \times R^n \to R$ *be the value function of the differential game with payoff* (3.1.1). *Then the function* ω^0 *is* u-*stable in the region* $T^0 \times R^n$.

Proof. The property of u-stability is defined for continuous functions, thus we note that the function ω^0 is continuous (cf. §3.4.1)). Let $(t_*,x_*,v) \in T^0 \times R^n \times \mathcal{Q}$. It is required to show that there exists a motion $y(\cdot) \in X_2(t_*,x_*,v)$ which satisfies the inequality

(15) $$\omega^0(t^*,y(t^*)) \leqslant \omega^0(t_*,x_*)$$

for any $t^* \in [t_*,\theta]$. We set

$$a \overset{\text{def}}{=} \omega^0(t_*,x_*),\ M_a \overset{\text{def}}{=} \{(\theta,x) : \sigma(x) \leqslant a\},\quad N \overset{\text{def}}{=} T \times R^n.$$

Let W_a be the maximal u-stable bridge in an (M_a, N)-approach problem. Denote by U^e the strategy extremal to the set W_a. Consider the motion $y(\cdot) \in X_2(t_*,x_*,v) \cap X_1(t_*,x_*,U^e)$ (this intersection is not empty). Select an arbitrary $t^* \in [t_*,\theta]$. It follows from Theorem 2.4.5 that $(t^*,y(t^*)) \in W_a$. If the point $(t^*,y(t^*))$ is chosen as an initial position, then Theorem 2.4.1 implies that for any motion $x(\cdot) \in$

$\mathcal{X}_1(t^*,y(t^*),U^e)$ the condition of (M_a, N)-approach is valid, i.e.

$$\Gamma_1(t^*,y(t^*),U^e) \overset{\text{def}}{=} \max_{x(\cdot)} \sigma(x(\theta)) \leqslant a \quad \text{for} \quad x(\cdot) \in \mathcal{X}_1(t^*,y(t^*),U^e).$$

It follows from the definition of the value (cf. (3.1.2), (3.1.6)) that

$$\omega^0(t^*,y(t^*)) \leqslant \Gamma_1(t^*,y(t^*),U^e) \leqslant a.$$

Taking into account that $a = \omega^0(t_*,x_*)$ we arrive at (15). Lemma 4.2.2 is thus proved.

Lemma 4.2.3. *Let the function* $\omega : T \times R^n \to R$ *be continuous,* u-*stable in the region* $T^0 \times R^n$ *and satisfy condition* (14). *Then* $\omega^0(t_*,x_*) \leqslant \omega(t_*,x_*)$ *for any position* $(t_*,x_*) \in T \times R^n$ *where* ω^0 *is the value function of the differential game with payoff* (3.1.1).

Proof. Let $(t_*,x_*) \in T \times R^n$, $a \overset{\text{def}}{=} \omega(t_*,x_*)$. Consider the set

$$W_* \overset{\text{def}}{=} \{(t,x) \in T \times R^n : \omega(t,x) \leqslant a\}.$$

It follows from the condition of u-stability of a function ω and equality (14) that W_* is a stable bridge in an (M_a, N)-approach problem where $M_a \overset{\text{def}}{=} \{(\theta,x) : \sigma(x) = \omega(\theta,x) \leqslant a\}$, $N \overset{\text{def}}{=} T \times R^n$. Define the strategy U^e extremal to the set W_* (note that W_* is closed). The point (t_*,x_*) belongs to W_*, therefore it follows from Theorem 2.4.1 that for any motion $x(\cdot) \in \mathcal{X}_1(t_*,x_*,U^e)$ the condition of (M_a, N)-approach is valid, i.e.

$$\Gamma_1(t_*,x_*,U^e) \overset{\text{def}}{=} \max_{x(\cdot)} \sigma(x(\theta)) \leqslant a \quad \text{for} \quad x(\cdot) \in \mathcal{X}_1(t_*,x_*,U^e).$$

From the definition of the value (cf. (3.1.2), (3.1.6)) we have

$$\omega^0(t_*,x_*) \leqslant \Gamma_1(t_*,x_*,U^e) \leqslant a = \omega(t_*,x_*),$$

q.e.d.

Note that the estimate $\omega^0 \leqslant \omega$ remains valid if the equality (14) is replaced by the inequality $\omega(\theta,x) \geqslant \sigma(x)$, $x \in R^n$.

Proofs of the following two lemmas are omitted since they coincide essentially with the proofs of Lemmas 4.2.2 and 4.2.3.

Lemma 4.2.4. *The value function of a differential game with payoff* (3.1.1) *is v-stable in the region* $T^0 \times R^n$.

Lemma 4.2.5. *Let a function* $\omega : T \times R^n \to R$ *be continuous, v-stable in the region* $T^0 \times R^n$, *and satisfy equality* (14). *Then* $\omega^0 \geqslant \omega$, *where* ω^0 *is the value function of the differential game with payoff* (3.1.1).

Proof of Theorem 4.2.1. Necessity of conditions (a) – (c). The continuity of function ω^0 was proved in §3.4.1. Equality (14) follows directly from the definition of the value. The properties of u-stability and v-stability of the function ω^0 are established in Lemmas 4.2.2 and 4.2.4.

Sufficiency of conditions (a) – (c). When conditions (a) – (c) are fulfilled the estimate $\omega^0 \leqslant \omega$ follows from Lemma 4.2.3 while the estimate $\omega^0 \geqslant \omega$ follows from Lemma 4.2.5. Thus $\omega = \omega^0$. Theorem 4.2.1 is proved.

4.2.3. Consider a differential game with payoff of the form

$$\gamma(x(\cdot)) \overset{\text{def}}{=} \min_{t_0 \leqslant t \leqslant \theta} \rho(t,x(t)). \tag{16}$$

We shall assume that the function $\rho : T \times R^n \to R$ is continuous. Then, as was shown in §3.4.1, the value function of the game under consideration will be continuous on $T \times R^n$. The following theorem is valid:

Theorem 4.2.2. *In order that function $\omega : T \times R^n \to R$ be the value function of the differential game with payoff of the form* (16), *it is necessary and sufficient that the following set of conditions be fulfilled:*

(a) *function ω is continuous;*

(17) (b) $\omega(t,x) \leqslant \rho(t,x)$, $\omega(\theta,x) = \rho(\theta,x)$ for $t \in T$, $x \in R^n$;

(c) *function ω is u-stable in the region;*

$$\mathcal{E} \overset{\text{def}}{=} \{(t,x) \in T^0 \times R^n : \omega(t,x) < \rho(t,x)\}; \tag{18}$$

(d) *function ω is v-stable in the region $T^0 \times R^n$.*

The validity of this theorem follows from the lemmas stated below.

Lemma 4.2.6. *Let $\omega^0 : T \times R^n \to R$ be the value function of the differential game with payoff* (16). *The function ω^0 satisfies the condition of u-stability in the region $\mathcal{E}^0 = \{(t,x) \in T^0 \times R^n : \omega^0(t,x) < \rho(t,x)\}$.*

Proof. If $\mathcal{E}^0 = \phi$ (this is possible) then the function ω^0 will be u-stable in $\mathcal{E}$ in accordance with the convention stipulated in 4.2.1.

Let $\mathcal{E}^0 \neq \phi$. Let $(t_*,x_*) \in \mathcal{E}^0$, $v \in \mathcal{Q}$. Choose a number $t^0 \in (t_*,\theta]$ such that the inequality

$$\omega^0(t_*,x_*) < \rho(t,x(t)) \text{ for } t_* \leqslant t \leqslant t^0 \tag{19}$$

is valid for any $x(\cdot) \in X_2(t_*,x_*,v)$. We shall show that a motion $y(\cdot) \in X_2(t_*,x_*,v)$ exists such that

(20) $\quad \omega^0(t^*,y(t^*)) \leqslant \omega^0(t_*,x_*)$ for $t_* \leqslant t^* \leqslant t^0$.

This will prove that the function ω^0 satisfies inequality (5) in the region $\mathcal{E}^0$, i.e. the u-stability condition is satisfied.

Set

$$a \overset{\text{def}}{=} \omega^0(t_*,x_*),\ \mathcal{M}_a \overset{\text{def}}{=} \{(t,x) \in T \times R^n : \rho(t,x) \leqslant a\},\quad \mathcal{N} \overset{\text{def}}{=} T \times R^n.$$

Let $\mathcal{W}_a$ be the maximal u-stable bridge in an $(\mathcal{M}_a, \mathcal{N})$-approach problem. Observe that $(t_*,x_*) \in \mathcal{W}_a$. We define a strategy U^e extremal to the set $\mathcal{W}_a$. Consider the motion $y(\cdot) \in X_2(t_*,x_*,v_*) \cap X_1(t_*,x_*,U^e)$ (this intersection is not empty). Select an arbitrary $t^* \in [t_*,t^0]$. It follows from (19) that $(t,y(t)) \notin \mathcal{M}_a$ for $t \in [t_*,t^*]$. Therefore, in accordance with Theorem 2.4.5 we obtain that $(t^*,y(t^*)) \in \mathcal{W}_a$. Next it follows from Theorem 2.4.1 that for any motion $x(\cdot) \in X_1(t^*,y(t^*),U^e)$ the condition of $(\mathcal{M}_a, \mathcal{N})$-approach is valid, i.e.

$$\Gamma_1(t^*,y(t^*),U^e) \overset{\text{def}}{=} \max_{x(\cdot)} \min_{t^* \leqslant t \leqslant \theta} \rho(t,x(t)) \leqslant a$$
$$\text{for } x(\cdot) \in X_1(t,y(t^*),U^e).$$

In accordance with the definition of the value (cf. (3.3.4)) the estimates

$$\omega^0(t^*,y(t^*)) \leqslant \Gamma_1(t^*,y(t^*),U^e) \leqslant a = \omega^0(t_*,x_*)$$

are valid. We have arrived at (20). Lemma 4.2.6 is thus proved.

Lemma 4.2.7. *Let a function* $\omega : T \times R^n \to R$ *satisfy the conditions* (a), (b), *and* (c) *stated in Theorem* 4.2.2. *Then for any point* (t_0,x_0) *the inequality* $\omega(t_0,x_0) \geqslant \omega^0(t_0,x_0)$ *is valid where* ω^0 *is the value function of a differential game with payoff* (16).

Proof. Let $(t_0,x_0) \in T \times R^n$, $a \overset{\text{def}}{=} \omega(t_0,x_0)$. Consider the set

$$\mathcal{W}_* \overset{\text{def}}{=} \{(t,x) \in T \times R^n : \omega(t,x) \leqslant a\}.$$

Set

$$\mathcal{M}_* \overset{\text{def}}{=} \{(t,x) \in T \times R^n : \omega(t,x) = \rho(t,x) \leqslant a\},\quad \mathcal{N} \overset{\text{def}}{=} T \times R^n.$$

We show that $\mathcal{W}_*$ is a u-stable bridge in an $(\mathcal{M}_*, \mathcal{N})$-approach problem. Clearly $\mathcal{W}_* \subset \mathcal{N}$ and it follows from equality (17) that $\mathcal{W}_*(\theta) \subset \mathcal{M}_*$. It remains to verify that the set $\mathcal{W}_*$ is u-stable, i.e. for any $(t_*,x_*) \in \mathcal{W}_*$, $t^* \in (t_*,\theta]$ and $v \in \mathcal{Q}$ a motion $x(\cdot) \in X_2(t_*,x_*,v)$ exists such that either

(21) $\quad \omega(t^*,x(t^*)) \leqslant a$

or

(22) $\omega(\tau,x(\tau)) = \rho(\tau,x(\tau)) \leqslant a$ for some $\tau \in [t_*,t^*]$.

Two cases are possible as inequality (17) indicates: (j) $\omega(t_*,x_*) = \rho(t_*,x_*)$, (jj) $\omega(t_*,x_*) < \rho(t_*,x_*)$. The inclusion $(t_*,x_*) \in W_*$ implies that $\omega(t_*,x_*) \leqslant a$, therefore in case (j) condition (22) will be fulfilled for any motion $x(\cdot) \in X_2(t_*,x_*,v)$ for $\tau = t_*$. Consider now case (jj). The function ω is u-stable in the region $\mathcal{E}$ (18). Therefore in accordance with (3) there exists a motion $x(\cdot) \in X_2(t_*,x_*,v)$ such that

(23) $\omega(t,x(t)) \leqslant \omega(t_*,x_*) \leqslant a$ for $t_* \leqslant t \leqslant \tau_*$.

Here $\tau_* = \tau_*(x(\cdot))$. From the definitions of functional τ_*, and of region $\mathcal{E}$ it follows that

(24) $\omega(\tau_*,x(\tau_*)) = \rho(\tau_*,x(\tau_*))$.

If $\tau_* > t^*$ then (23) implies (21). If, however, $\tau_* \leqslant t^*$ then (23) and (24) yield (22) with $\tau = \tau_*$.

We have thus shown that W_* is a u-stable bridge in a problem of (M_*, N)-approach. Let U^e be a strategy extremal to the set W_*. The initial position (t_0,x_0) is contained in W_*, therefore according to Theorem 2.4.1 for any motion $x(\cdot) \in X_1(t_0,x_0,U^e)$ the condition of (M_*, N)-approach is valid. It follows from the definition of the set M_* that

$$\Gamma_1(t_0,x_0,U^e) \overset{\text{def}}{=} \max_{x(\cdot)} \min_{t_0 \leqslant t \leqslant \theta} \rho(t,x(t)) \leqslant a = \omega(t_0,x_0)$$
$$\text{for } x(\cdot) \in X_1(t_0,x_0,U^e).$$

From this estimate and equality (3.3.4) we obtain that $\omega^0(t_0,x_0) \leqslant \omega(t_0,x_0)$, q.e.d.

Lemma 4.2.8. *The value function of a differential game with payoff* (16) *satisfies the condition of v-stability in the region* $T^0 \times R^n$.

The proof of this lemma is quite similar to the proofs of Lemmas 4.2.2 and 4.2.6. We therefore present only a brief outline of it.

Let $(t_*,x_*) \in T^0 \times R^n$, $u \in P$. Set

$$a \overset{\text{def}}{=} \omega^0(t_*,x_*), \quad K \overset{\text{def}}{=} \{(\theta,x) : x \in R^n\},$$
$$\mathcal{L}_a \overset{\text{def}}{=} \{(t,x) \in T \times R^n : \rho(t,x) \geqslant a\}.$$

Let W_a be the maximal v-stable bridge of a $(K,\mathcal{L}_a)$-approach problem and let V^e be a strategy extremal to the set W_a. Consider the motion

$y(\cdot) \in X_1(t_*,x_*,u) \cap X_2(t_*,x_*,V^e)$. For any $t^* \in [t_*,\theta]$ we have $(t^*,y(t^*)) \in W_a$, therefore $\omega^0(t^*,y(t^*)) \geq \omega^0(t_*,x_*) = a$. This estimate implies that the function ω^0 is v-stable in the region $T^0 \times R^n$.

Lemma 4.2.9. *Let function* ω *satisfy conditions* (a), (b), *and* (d) *stipulated in Theorem 4.2.2. Then* $\omega \leq \omega^0$ *where* ω^0 *is the value function of a differential game with payoff* (16).

We outline the proof of this lemma. Let $(t_0,x_0) \in T \times R^n$,

$$a \overset{\text{def}}{=} \omega(t_0,x_0),\quad W_* \overset{\text{def}}{=} \{(t,x) \in T \times R^n : \omega(t,x) \geq a\},$$

$$K \overset{\text{def}}{=} \{(\theta,x) : x \in R^n\},\quad \mathfrak{L}_* \overset{\text{def}}{=} \{(t,x) \in T \times R^n : \rho(t,x) \geq a\}.$$

The set W_* is a v-stable bridge in a $(K,\mathfrak{L}_*)$-approach problem. The point (t_0,x_0) is contained in W_* and the strategy V^e extremal to the set W_* assures that the inequality

$$\min_{t_0 \leq t \leq \theta} \rho(t,x(t)) \geq a = \omega(t_0,x_0)$$

is fulfilled for any $x(\cdot) \in X_2(t_0,x_0,V^e)$. This implies that $\omega^0(t_0,x_0) \geq \omega(t_0,x_0)$.

To complete the proof of Theorem 4.2.2 it remains only to note that the value function of a differential game with payoff (16) satisfies the relations

$$\omega^0(t,x) \leq \rho(t,x),\quad \omega^0(\theta,x) = \rho(\theta,x) \quad (t \in T,\ x \in R^n).$$

These relations follow directly from equality (16) and the definition of the value. All the other assertions of Theorem 4.2.2 are contained in Lemmas 4.2.6 – 4.2.9.

From constructions presented in §4.2.2 and §4.2.3 follows the proposition:

Corollary 4.2.1. *Let* $\omega^0 : T \times R^n \to R$ *be the value function of a differential game with payoff* (3.1.1) *or* (16). *Let* U^e *be a strategy extremal to the set* $W^1 \overset{\text{def}}{=} \{(t,x) \in T \times R^n : \omega^0(t,x) \leq \omega^0(t_0,x_0)\}$, V^0 *be a strategy extremal to the set* $W^2 \overset{\text{def}}{=} \{(t,x) \in T \times R^n, \omega^0(t,x) \geq \omega^0(t_0,x_0)\}$ *where* (t_0,x_0) *is the given initial position of the game. Then the pair of strategies* U^0, V^0 *forms a saddle point.*

4.3. Directional Differentiability

In this section the definition of a directional derivative is given and several of its properties are described. Examples of functions possessing directional derivatives are presented. This material will be used in the next section where inequalities for directional derivatives of the value function are obtained.

4.3.1. We present only those facts on directional derivatives which will be required in what follows. We shall adhere to the notation to be outlined in the next section. A rather detailed exposition of results related to functions with directional derivatives is contained in papers [15, 16].

We shall consider functions $\omega : T \times R^n \to R$ satisfying the Lipschitz condition in every bounded region $\mathfrak{D} \subset T \times R^n$:

$$(1) \qquad |\omega(t_1,x_1) - \omega(t_2,x_2)| \leqslant \lambda(\mathfrak{D})[|t_1 - t_2| + \|x_1 - x_2\|], \quad (t_i,x_i) \in \mathfrak{D},$$
$$i = 1, 2.$$

We assume that $T \overset{\text{def}}{=} (-\infty,\theta]$, $T^0 \overset{\text{def}}{=} (-\infty,\theta)$. Let

$$(t,x) \in T^0 \times R^n, \quad \alpha \in R, \quad h \in R^n.$$

The function ω is said to be differentiable at the point (t,x) in the direction of (α,h) provided the limit

$$(2) \qquad D\omega(t,x)|(\alpha,h) \overset{\text{def}}{=} \lim_{\delta \to +0} \delta^{-1}[\omega(t + \delta\alpha, \; x + \delta h) - \omega(t,x)]$$

exists. In view of the Lipschitz condition this limit is finite. The number $D\omega(t,x)|(\alpha,h)$ is called *the (directional) derivative of function* ω *at the point* (t,x) *in the direction of* (α,h). If for any $(t,x) \in T^0 \times R^n$, $\alpha \in R$, $h \in R^n$ the function ω is differentiable at point (t,x) in the direction of (α,h), then this function is said to be *directionally differentiable.*

We note that directions are usually defined by means of vectors, therefore instead of (α,h) the symbol $(\alpha,h^T)^T$ should have been used (recall that the index ^{T} denotes transposition). However, to simplify the notation we omit the transposition indices.

A function ω differentiable at point (t,x) (in the ordinary sense) will be differentiable at this point in any direction. In this case the equality

$$(3) \qquad D\omega(t,x)|(1,h) = \frac{\partial\omega(t,x)}{\partial t} + \frac{\partial\omega(t,x)}{\partial x} h = \frac{\partial\omega(t,x)}{\partial t} + \sum_{i=1}^{n} \frac{\partial\omega(t,x)}{\partial x_i} h_i$$

is valid. Here $\partial\omega/\partial t$, $\partial\omega/\partial x_i$ $(i = 1, 2, ..., n)$ are partial derivatives of the function ω and $\partial\omega/\partial x \overset{\text{def}}{=} (\partial\omega/\partial x_1, ..., \partial\omega/\partial x_n)$ is a row-vector. Note that we set $\alpha = 1$ in equality (3). This is because in the results stated and derived in the next section only the direction $(1,h)$ is considered. For referencing convenience, formula (3) as well as other formulas in this section are presented for the case $\alpha = 1$.

A function ω is called non-smooth if a point $(t,x) \in T^0 \times R$ exists at which this function is not differentiable (in the usual sense). The class of directionally differentiable functions, contains non-smooth functions. This can be seen in the examples presented below.

4.3.2. *Example* 1. *Piecewise-smooth functions.* Let I and J be finite sets. For any $i \in I$ and $j \in J$, let the function

$$(t,x) \to \varphi(t,x,i,j) : T \times R^n \to R$$

be smooth, i.e. this function will be differentiable at each point $(t,x) \in T^0 \times R^n$. We define a function

$$(t,x) \to \omega(t,x) \overset{\text{def}}{=} \min_{i \in I} \max_{j \in J} \varphi(t,x,i,j). \tag{4}$$

The function $\omega : T \times R^n \to R$ of the form (4) will be called piecewise-smooth. We shall prove that a piecewise-smooth function is directionally differentiable.

Introduce the following notation:

$$\begin{aligned} I_0(t,x) &\overset{\text{def}}{=} \{i_0 \in I : \max_{j \in J} \varphi(t,x,i_0,j) = \min_{i \in I} \max_{j \in J} \varphi(t,x,i,j) = \omega(t,x)\}, \\ J_0(t,x,i) &\overset{\text{def}}{=} \{j_0 \in J : \varphi(t,x,i,j_0) = \max_{j \in J} \varphi(t,x,i,j)\}. \end{aligned} \tag{5}$$

Select a point $(t,x) \in T^0 \times R^n$. It follows from the continuity of functions $(t,x) \to \varphi(t,x,i,j)$ and the finiteness of the sets I and J that for $\epsilon > 0$ sufficiently small the inclusions

$$\begin{aligned} &I_0(\tau,y) \subset I_0(t,x), \quad J_0(\tau,y,i) \subset J_0(t,x,i) \\ &\text{for all } (\tau,y) \in B(t,x;\epsilon),\ i \in I \end{aligned} \tag{6}$$

are valid. Here

$$B(t,x;\epsilon) \overset{\text{def}}{=} \{(\tau,y) \in T^0 \times R^n : |\tau - t| + \|y - x\| \leqslant \epsilon\}.$$

From (5), (6) we have

$$\begin{aligned} &\omega(t,x) = \varphi(t,x,i_0,j_0),\ i_0 \in I_0(t,x),\ j_0 \in J_0(t,x,i_0), \\ &\omega(\tau,y) = \min_{i_0} \max_{j_0} \varphi(\tau,y,i_0,j_0) \\ &\qquad \text{for } (\tau,y) \in B(t,x;\epsilon),\ i_0 \in I_0(t,x),\ j_0 \in J_0(t,x,i_0). \end{aligned} \tag{7}$$

Let $h \in R^n$. Select a number $\delta_0 > 0$ such that $(t + \delta, x + \delta h) \in B(t,x;\epsilon)$ for all $\delta \in (0,\delta_0)$. According to (7) for $\delta \in (0,\delta_0)$ the equality

$$\begin{aligned} &\omega(t+\delta, x+\delta h) - \omega(t,x) = \min_{i_0} \max_{j_0} [\varphi(t+\delta, x+\delta h,i_0,j_0) - \varphi(t,x,i_0,j_0)] \\ &\qquad \text{for } i_0 \in I_0(t,x),\ j_0 \in J_0(t,x,i_0) \end{aligned} \tag{8}$$

is valid. Functions $(t,x) \to \varphi(t,x,i,j)$ are differentiable, therefore

$$\varphi(t + \delta, x + \delta h, i, j) - \varphi(t,x,i,j) = \delta[\partial\varphi(t,x,i,j)/\partial t + (\partial\varphi(t,x,i,j)/\partial x)h] + \alpha(\delta,i,j)\delta$$

where $\alpha(\delta,i,j) \to 0$ as $\delta \to +0$. Since the sets I and J are finite we obtain that

$$|\alpha(\delta,i,j)| \leqslant \alpha_*(\delta), \quad i \in I, \quad j \in J,$$

where $\alpha_*(\delta) \to 0$ as $\delta \to 0$. Substituting this estimate into (8) we have

$$|\delta^{-1}[\omega(t + \delta, x + \delta h) - \omega(t,x)] - \min_{i_0} \max_{j_0} [\partial\varphi(t,x,i_0,j_0)/\partial t + (\partial\varphi(t,x,i_0,j_0)/\partial x)h] \leqslant \alpha_*(\delta).$$

We have thus shown that a piecewise-smooth function is differentiable in the direction of $(1,h)$ and its derivative in that direction exists and is defined by the equality

$$D\omega(t,x)|(1,h) = \min_{i_0} \max_{j_0} [\partial\varphi(t,x,i_0,j_0)/\partial t + (\partial\varphi(t,x,i_0,j_0)/\partial x)h] \tag{9}$$

where $i_0 \in I_0(t,x)$, $j_0 \in J_0(t,x,i_0)$.

The notion of a piecewise-smooth function can be generalized as follows. For any point $(t_*,x_*) \in T^0 \times R^n$ let us find a neighborhood $B(t_*,x_*;\epsilon)$ and a set of differentiable functions $(t,x) \to \varphi(t,x,i,j)$ ($i \in I$, $j \in J$, I and J are finite sets) so that for $(t,x) \in B(t_*,x_*;\epsilon)$ the equality (4) is valid. Such a function ω will then be called *locally piecewise-smooth*. It follows from the proof presented above that locally piecewise-smooth functions are directionally differentiable and their derivatives in the direction of $(1,h)$ are given by equality (9).

Example 2. *The upper envelope of a family of smooth functions.*

Consider a function $\omega : T \times R^n \to R$ defined by equality

$$\omega(t,x) \overset{\text{def}}{=} \sup_{\ell \in R^m} \varphi(t,x,\ell). \tag{10}$$

We shall assume that the following conditions are fulfilled:

(a) $\varphi(t,x,\ell) = \varphi_1(t,x,\ell) - \varphi_2(\ell)$;

(b) the function $\varphi_1 : T \times R^n \times R^m \to R$ is continuous. The derivatives $(t,x,\ell) \to \partial\varphi_1(t,x,\ell)/\partial t$ and $(t,x,\ell) \to \partial\varphi_1(t,x,\ell)/\partial x$ are defined and continuous on $T^0 \times R^n \times R^m$. The following estimates are valid

$$\sup\{|\partial\varphi_1(t,x,\ell)/\partial t| : (t,x) \in \mathfrak{D}, \ \ell \in \mathfrak{L}\} < \infty,$$

$$\sup\{\|\partial\varphi_1(t,x,\ell)/\partial x\| : (t,x) \in \mathfrak{D}, \ \ell \in \mathfrak{L}\} < \infty,$$

where $\mathfrak{D}$ and $\mathfrak{L}$ are arbitrary bounded sets in $T^0 \times R^n$ and R^m respectively.

(c) the function φ_2: $R^m \to [a,\infty]$ is lower semi-continuous; here a is a finite number. The effective domain of function φ_2, i.e.

$$\mathrm{dom}\varphi_2 \overset{\mathrm{def}}{=} \{\ell \in R^m : \varphi_2(\ell) < \infty\},$$

is non-empty and bounded.

Observe that for all $(t,x) \in T \times R^n$ and $\ell \in \mathrm{dom}\ \varphi_2$ we have the inequality $\varphi(t,x,\ell) \leqslant \varphi_1(t,x,\ell) - a$. Therefore $\omega(t,x)$ is a finite number. Define the set

$$\mathfrak{L}_0(t,x) \overset{\mathrm{def}}{=} \{\ell_0 \in R^m : \varphi(t,x,\ell_0) = \omega(t,x)\}. \tag{11}$$

We show that $\mathfrak{L}_0(t,x) \neq \phi$. Let $\ell_i \in R^m$ $(i = 1, 2, \ldots)$ $\lim\limits_{i\to\infty} \varphi(t,x,\ell_i) = \omega(t,x)$. In can be assumed that $\ell_i \in \mathrm{dom}\ \varphi_2$ $(i = 1,2, \ldots)$ and $\lim\limits_{i\to\infty} \ell_i = \ell_*$. Since function φ_2 is lower semi-continuous, the function $\ell \to \varphi(t,x,\ell)$ is upper semi-continuous. Hence

$$\omega(t,x) = \lim_{i\to\infty} \varphi(t,x,\ell_i) \leqslant \varphi(t,x,\ell_*).$$

The strict inequality $\omega(t,x) < \varphi(t,x,\ell_*)$ contradicts equality (10). We thus obtain that $\varphi(t,x,\ell_*) = \omega(t,x)$, i.e. $\ell_* \in \mathfrak{L}_0(t,x)$.

The following lemma is valid.

Lemma 4.3.1. *Let conditions* (a) – (c) *be fulfilled. Then the function* $\omega : T \times R^n \to R$ *defined by equality* (10) *satisfies the Lipschitz condition* (1) *in every bounded domain* $\mathfrak{D} \subset T \times R^n$. *At each point* $(t,x) \in T^0 \times R^n$ *the function* ω *is directionally differentiable . Its derivative in direction* $(1,h)$ *is determined by the equality*

$$D\omega(t,x)|(1,h) = \max_{\ell\in\mathfrak{L}_0(t,x)} \left[\frac{\partial\varphi_1(t,x,\ell)}{\partial t} + \frac{\partial\varphi_1(t,x,\ell)}{\partial x}h\right]. \tag{12}$$

Proof. Let $\mathfrak{D}$ be a bounded convex set in $T^0 \times R^n$. In accordance with condition (b) we have

$$|\partial\varphi_1(t,x,\ell)/\partial t| \leqslant \lambda < \infty\ ,$$

$$\|\partial\varphi_1(t,x,\ell)/\partial x\| \leqslant \lambda < \infty\ ,$$

for all $(t,x) \in \mathfrak{D}$, $\ell \in \mathrm{dom}\ \varphi_2$. Therefore utilizing the mean value theorem [17, 67] we obtain

$$|\varphi(t_1,x_1,\ell) - \varphi(t_2,x_2,\ell)| \leqslant \lambda[|t_1 - t_2| + \|x_1 - x_2\|]$$

$$\text{for all } (t_i,x_i) \in cl\ \mathcal{D},\quad \ell \in \text{dom}\ \varphi_2,\ i = 1, 2.$$

Let $\ell_i \in \mathfrak{L}_0(t_i,x_i)$ $(i = 1, 2)$. From (10), (11) we have

$$\omega(t_1,x_1) \geqslant \omega(t_2,x_2) + \varphi(t_1,x_1,\ell_2) - \varphi(t_2,x_2,\ell_2) \geqslant \omega(t_2,x_2)$$
$$- \lambda[|t_1 - t_2| + \|x_1 - x_2\|].$$

Analogously we obtain $\omega(t_2,x_2) \geqslant \omega(t_1,x_1) - \lambda[|t_1 - t_2| + \|x_1 - x_2\|]$. Thus the Lipschitz condition is verified.

Consider the multi-valued mapping $(t,x) \to \mathfrak{L}_0(t,x)$. Clearly $\mathfrak{L}_0(t,x) \subset \text{dom}\ \varphi_2$ for all $(t,x) \in T \times R^n$. Let $(t_i,x_i) \in T \times R^n$, $\ell_i \in \mathfrak{L}_0(t_i,x_i)$ $(i = 1, 2, \ldots)$. $(t_i,x_i) \to (t_*,x_*)$ and $\ell_i \to \ell_*$ for $i \to \infty$. We shall show that $\ell_* \in \mathfrak{L}_0(t_*,x_*)$. From (10) and (11) we have

$$\varphi(t_i,x_i,\ell_i) \geqslant \varphi(t_i,x_i,\ell)\quad (i = 1, 2, \ldots),\quad \ell \in \text{dom}\ \varphi_2.$$

The function $(t,x,\ell) \to \varphi(t,x,\ell)$ is upper semi-continuous, therefore

$$\overline{\lim_{i\to\infty}}\ \varphi(t_i,x_i,\ell_i) \leqslant \varphi(t_*,x_*,\ell_*).$$

The function $(t,x) \to \varphi(t,x,\ell)$ is continuous for any $\ell \in \text{dom}\ \varphi_2$, therefore

$$\lim_{i\to\infty}\ \varphi(t_i,x_i,\ell) = \varphi(t_*,x_*,\ell).$$

We thus obtain that

$$\varphi(t_*,x_*,\ell_*) \geqslant \varphi(t_*,x_*,\ell)\quad (\ell \in \text{dom}\ \varphi_2),$$

i.e. $\ell_* \in \mathfrak{L}_0(t_*,x_*)$. It follows from the proposition proved for $(t_i,x_i) = (t_*,x_*)$ $(i = 1, 2, \ldots)$ that $\mathfrak{L}_0(t_*,x_*)$ is a closed subset of the bounded set dom φ_2, i.e. $\mathfrak{L}_0(t_*,x_*)$ is a compact set in R^m.

Let $(t,x) \in T^0 \times R^n$, $h \in R^n$. Consider the sequence $\alpha_i > 0$ $(i = 1, 2, \ldots)$ such that $\alpha_i \to 0$ as $i \to \infty$, $t + \alpha_i < \theta$ and

$$\underline{\lim_{\delta\to+0}}\ \delta^{-1}[\omega(t + \delta,x + \delta h) - \omega(t,x)] =$$
$$\lim_{i\to\infty}\ \alpha_i^{-1}[\omega(t + \alpha_i,x + \alpha_i h) - \omega(t,x)]\ .$$

It follows from (10) and (11) that

$$\omega(t + \alpha_i,x + \alpha_i h) - \omega(t,x) \geqslant \varphi_1(t + \alpha_i,\ x + \alpha_i h,\ell) - \varphi_1(t,x,\ell)$$

$$\text{where } \ell \in \mathfrak{L}_0(t,x).$$

The mean value theorem implies that

$$\varphi_1(t + \alpha_i, x + \alpha_i h, \ell) - \varphi_1(t,x,\ell)$$
$$= \alpha_i \left[\frac{\partial\varphi_1(t + \alpha_i^*, x + \alpha_i^* h, \ell)}{\partial t} + \frac{\partial\varphi_1(t + \alpha_i^*, x + \alpha_i^* h, \ell)}{\partial x} h \right]$$
$$\text{where} \quad \alpha_i^* \in (0, \alpha_i)$$

Taking condition (b) into account we obtain the inequality

$$\underline{\lim}_{\delta \to +0} \delta^{-1}[\omega(t + \delta, x + \delta h) - \omega(t,x)]$$
$$\geqslant \lim_{i\to\infty} \left[\frac{\partial\varphi_1(t + \alpha_i^*, x + \alpha_i^* h, \ell)}{\partial t} + \frac{\partial\varphi_1(t + \alpha_i^*, x + \alpha_i^* h, \ell)}{\partial x} h \right]$$
$$= \frac{\partial\varphi_1(t,x,\ell)}{\partial t} + \frac{\partial\varphi_1(t,x,\ell)}{\partial x} h.$$

This inequality is valid for all $\ell \in \mathfrak{L}_0(t,x)$, therefore

$$\underline{\lim}_{\delta \to +0} \delta^{-1}[\omega(t + \delta, x + \delta h) - \omega(t,x)]$$
$$\geqslant \max_{\ell \in \mathfrak{L}_0(t,x)} \left[\frac{\partial\varphi(t,x,\ell)}{\partial t} + \frac{\partial\varphi(t,x,\ell)}{\partial x} h \right]. \tag{13}$$

Consider a sequence $\beta_i = (0, \theta - t)$ $(i = 1, 2, ...)$ such that $\beta \to 0$ as $i \to \infty$ and

$$\overline{\lim}_{\delta \to +0} \delta^{-1}[\omega(t + \delta, x + \delta h) - \omega(t,x)]$$
$$= \lim_{i\to\infty} \beta_i^{-1}[\omega(t + \beta_i, x + \beta_i h) - \omega(t,x)].$$

It follows from (10) and (11) that

$$\omega(t + \beta_i, x + \beta_i h) - \omega(t,x) \leqslant \varphi_1(t + \beta_i, x + \beta_i h, \ell_i) - \varphi_1(t,x,\ell_i)$$
$$\text{for } \ell_i \in \mathfrak{L}_0(t + \beta_i, x + \beta_i h).$$

Since $\ell_i \in \operatorname{dom} \varphi_2 (i = 1, 2, ...)$ and $\operatorname{dom} \varphi_2$ is bounded we may assume that $\ell_i \to \ell_*$ as $i \to \infty$. As it has been shown above, $\ell_* \in \mathfrak{L}_0(t,x)$. Using the mean-value theorem and taking condition (b) into account we arrive at the inequality

$$\overline{\lim}_{\delta \to +0} \delta^{-1}[\omega(t + \delta, x + \delta h) - \omega(t,x)]$$
$$\leqslant \lim_{i\to\infty} \left[\frac{\partial\varphi_1(t + \beta_i^*, x + \beta_i^* h, \ell_i)}{\partial t} + \frac{\partial\varphi_1}{\partial x} (t + \beta_i^*, x + \beta_i^* h, \ell_i) h \right]$$
$$= \frac{\partial\varphi_1(t,x,\ell_*)}{\partial t} + \frac{\partial\varphi_1(t,x,\ell_*)}{\partial x} h \leqslant \max_{\ell \in \mathfrak{L}_0(t,x)} \left[\frac{\partial\varphi_1(t,x,\ell)}{\partial t} + \frac{\partial\varphi_1(t,x,\ell)}{\partial x} h \right]. \tag{14}$$

Here $\beta_i^* \in (0,\beta_i)$. The inequalities (13) and (14) imply (12). The lemma is proved.

Generally a function ω satisfying the Lipschitz condition is not differentiable directionally. One can easily construct an example of such a function. However, for any locally Lipschitzian function we may define upper and lower directional derivatives, which are the finite numbers.

Let $(t,x,) \in T^0 \times R^n$ and $h \in R^n$ as above. Introduce the following notation:

$$(15)\qquad \begin{aligned} D^+\omega(t,x)|\,(1,h) &\overset{\text{def}}{=} \overline{\lim_{\delta\to+0}}\ \delta^{-1}[\omega(t+\delta,x+\delta h)-\omega(t,x)],\\ D^-\,\omega(t,x)|\,(1,h) &\overset{\text{def}}{=} \underline{\lim_{\delta\to+0}}\ \delta^{-1}[\omega(t+\delta,x+\delta h)-\omega(t,x)]. \end{aligned}$$

Numbers $D^+\omega(t,x)|(1,h)$ and $D^-\omega(t,x)|(1,h)$ are called, respectively, the *upper and lower derivatives of the function* ω *at point* (t,x) *in the direction* $(1,h)$. If function ω is differentiable at point (t,x) in direction $(1,h)$ then evidently the equalities

$$D^+\omega(t,x)|(1,h) = D\omega(t,x)|(1,h) = D^-(t,x)|(1,h)$$

are valid.

We shall prove the following assertion.

Lemma 4.3.2. *Let a function* ω *satisfy condition* (1) *in each bounded region* $\mathfrak{D} \subset T \times R^n$. *Then for any choice of point* $(t,x) \in T^0 \times R^n$ *the functions*

$$h \to g^+(h) \overset{\text{def}}{=} D^+\omega(t,x)|(1,h), \quad h \to g^-(h) \overset{\text{def}}{=} D^-\omega(t,x)|(1,h)$$

will satisfy the Lipschitz condition in the whole space R^n.

Proof. Let $\epsilon > 0$ and let λ be the Lipschitz constant for the function ω in the domain $\mathfrak{D} \overset{\text{def}}{=} B(t,x;\epsilon)$. We select two vectors $h_* \in R^n$ and $h^* \in R^n$. According to (15) we have

$$g^+(h_*) = \lim_{i\to\infty} [\omega(t+\alpha_i,x+\alpha_i h_*) - \omega(t,x)]\alpha_i^{-1},$$

$$g^+(h^*) = \lim_{i\to\infty} [\omega(t+\beta_i,x+\beta_i h^*) - \omega(t,x)]\beta_i^{-1},$$

where α_i, β_i $(i = 1, 2, \ldots)$ are sequences of positive numbers which converge to zero. Without loss of generality one can assume that

$$(t+\alpha_i,x+\alpha_i h) \in B(t,x;\epsilon),\ (t+\beta_i,x+\beta_i h) \in B(t,x;\epsilon)\ (i = 1, 2, \ldots)$$

with $h = h_*$ and $h = h^*$. It follows from the definition of an upper derivative that when the numbers α_i are replaced by β_i the following estimate is valid.

$$g^+(h_*) \geq \overline{\lim_{i\to\infty}} [\omega(t + \beta_i, x + \beta_i h_*) - \omega(t,x)]\beta_i^{-1}$$
$$= \overline{\lim_{i\to\infty}} [\omega(t + \beta_i, x + \beta_i h^*) - \omega(t,x) + r_i(h_*,h^*)]\beta_i^{-1},$$

where

$$r_i (h_*,h^*) \overset{\text{def}}{=} \omega(t + \beta_i, x + \beta_i h_*) - \omega(t + \beta_i, x + \beta_i h^*).$$

From (1) we have $r_i(h_*,h^*) \geq - \lambda\beta_i \| h_* - h^* \|$. Hence

$$g^+(h_*) \geq \overline{\lim_{i\to\infty}} [\omega(t + \beta_i,\ x + \beta_i h^*) - \omega(t,x)]\ \beta_i^{-1}$$
$$- \lambda \| h_* - h^* \| = g^+(h^*) - \lambda \| h_* - h^* \|.$$

Interchanging the positions of vectors h_* and h^* we arrive at the Lipschitz condition

$$|g^+(h_*) - g^+(h^*)| \leq \lambda \| h_* - h^* \|.$$

For the function $h \to \overline{g}(h)$ the assertion of Lemma 4.3.2 is proved analogously.

4.4. Generalization of the Main Equation

The conditions of u-stability and v-stability introduced in §4.2 can be defined in various ways. In particular, they can be defined in infinitesimal form. In this case the conditions of stability become inequalities for directional derivatives. Therefore the necessary and sufficient conditions which should be satisfied by the value function will consist of boundary conditions and two inequalities, one of which expresses the condition of u-stability while the other, the condition of v-stability. These inequalities are of the same form for those points at which the function is differentiable as well as for the points of the singular set. At points where the function is differentiable the inequalities obtained become the main equation. Thus these inequalities may be considered as generalizations of the main equation. On singular surfaces where several smooth functions join together, the inequalities indicate how the gradients of these smooth components should be related.

4.4.1. In this section we shall consider differential games, either with payoff functionals of the form

(1) $$\gamma(x(\cdot)) \overset{\text{def}}{=} \sigma(x(\theta))$$

or with payoff functionals of the form

(2) $$\gamma(x(\cdot)) \overset{\text{def}}{=} \min_{t_0 \leq t \leq \theta} \rho(t,x(t)).$$

We shall assume that functions $\sigma: R^n \to R$ and $\rho: T \times R^n \to R$ are locally Lipschitz continuous. By Theorems 3.4.1, 3.4.2 the value functions of differential games with the payoff functionals (1) and (2) satisfy the Lipschitz condition in any bounded region $\mathcal{D} \subset T \times R^n$.

In Section 4.2 properties of u-stability and v-stability were defined for a continuous function $\omega: T \times R^n \to R$. It was shown that the property of u-stability may be defined by any one of the equivalent inequalities (4.2.3), (4.2.5), (4.2.6) while the property of v-stability may be expressed by any one of the equivalent inequalities (4.2.4), (4.2.7), (4.2.8). If one assumes that the function ω satisfies the Lipschitz condition (4.3.1), then yet another two inequalities may be obtained – one which expresses the property of u-stability and the other, that of v-stability.

Below we shall often utilize the sets

$$(3)\qquad \begin{aligned} \mathcal{F}_1(t,x,u) &\overset{\text{def}}{=} co\ \{f(t,x,u,v) : v \in \mathcal{Q}\}, \\ \mathcal{F}_2(t,x,v) &\overset{\text{def}}{=} co\ \{f(t,x,u,v) : u \in \mathcal{P}\} \end{aligned}$$

introduced in §2.2.1. Recall that the sets of motions $X_1(t_*,x_*,u)$ and $X_2(t_*,x_*,v)$ utilized in the definitions of stable bridges and functions (cf. §2.2 and §4.2) were defined as the totality of solutions of differential inclusions

$$\dot{x}(t) \in \mathcal{F}_1(t,x(t),u) \quad \text{and} \quad \dot{x}(t) \in \mathcal{F}_2(t,x(t),v),$$

satisfying the initial condition $x(t_*) = x_*$.

Let the function $\omega: T \times R^n \to R$ satisfy the Lipschitz condition (4.3.1) in any bounded region $\mathcal{D} \subset T \times R^n$. Consider the following inequalities:

$$(4)\qquad \sup_{(t_*,x_*)\in\mathcal{E}} \max_{v\in\mathcal{Q}} \min_{h\in\mathcal{F}_2(t_*,x_*,v)} D^-\omega(t_*,x_*)|(1,h) \leqslant 0,$$

$$(5)\qquad \inf_{(t_*,x_*)\in\mathcal{E}} \min_{u\in\mathcal{P}} \max_{h\in\mathcal{F}_1(t_*,x_*,u)} D^+\omega(t_*,x_*)|(1,h) \geqslant 0.$$

Here, as in §4.2, $\mathcal{E}$ is an open region in $T^0 \times R^n$.

Note that according to Lemma 4.3.2 the function $h \to g^-(h) \overset{\text{def}}{=} D^-\omega(t_*,x_*)|(1,h)$ satisfies the Lipschitz condition. The set $\mathcal{F}_2(t_*,x_*,v)$ is compact in R^n and the multi-valued mapping $v \to \mathcal{F}_2(t_*,x_*,v)$ is continuous. Therefore the maximum over v and the minimum over h are attained in inequality (4). The same considerations show that in (5) the minimum over u and the maximum over h are attained. The following lemma is valid.

Lemma 4.4.1. *For a locally Lipschitzian function* $\omega: T \times R^n \to R$ *inequality* (4) *is equivalent to any one of the inequalities* (4.2.3), (4.2.5), *and* (4.2.6); *inequality* (5) *is equivalent to any one of the inequalities*

(4.2.4), (4.2.7), *and* (4.2.8).

Proof. Let $(t_*,x_*) \in \mathcal{E}$, $v \in \mathcal{Q}$. We shall show that the following inequalities are valid:

$$(6) \quad \inf_{x(\cdot)} \varliminf_{\delta\to 0} [\omega(t_* + \delta, x(t_* + \delta)) - \omega(t_*,x_*)]\delta^{-1} \leqslant \min_h D^-\omega(t_*,x_*)|(1,h),$$

$$(7) \quad \inf_{x(\cdot)} \varlimsup_{\delta\to +0} [\omega(t_* + \delta, x(t_* + \delta)) - \omega(t_*,x_*)]\delta^{-1} \geqslant \min_h D^-\omega(t_*,x_*)|(1,h),$$

where $x(\cdot) \in X_2(t_*,x_*,v)$, $h \in \mathcal{F}_2(t_*,x_*,v)$.

We shall first prove inequality (6). Select an $h_* \in \mathcal{F}_2(t_*,x_*,v)$ and a sequence δ_k $(k = 1, 2, ...)$, $\delta_k \to +0$ as $k \to \infty$, such that

$$(8) \quad \begin{aligned} \min_{h\in\mathcal{F}_2(t_*,x_*,v)} D^-\omega(t_*,x_*)|(1,h) &= D^-\omega(t_*,x_*)|(1,h_*) \\ &= \lim_{k\to\infty} [\omega(t_* + \delta_k, x_* + \delta_k h_*) - \omega(t_*,x_*)]\delta_k^{-1}. \end{aligned}$$

We show that a solution $x_*(\cdot) \in X_2(t_*,x_*,v)$ exists such that

$$(9) \quad x_*(t_* + \delta_k) = x_* + \delta_k h_* + \delta_k g_k, \quad \|g_k\| \to 0 \text{ as } k \to \infty.$$

This solution can be determined as follows. Select a function $(t,x) \to f^0(t,x)$ from the condition

$$\operatorname{dist}(h_*,\mathcal{F}_2(t,x,v)) = \|h_* - f^0(t,x)\|, \quad f^0(t,x) \in \mathcal{F}_2(t,x,v).$$

Since for all (t,x) the set $\mathcal{F}_2(t,x,v)$ is a convex compact set and the multi-valued mapping $(t,x) \to \mathcal{F}_2(t,x,v)$ is continuous, the function $(t,x) \to f^0(t,x)$ will also be continuous. Consider the differential equation

$$\dot{x}_*(t) = f^0(t,x_*(t)), \quad x(t_*) = x_*.$$

Observe that the equality $f^0(t_*,x_*) = h_*$ follows from the construction of the function f^0. Therefore a solution $x_*(\cdot)$ of the equation under consideration satisfies relations (9).

We set

$$(10) \quad r_k \overset{\text{def}}{=} [\omega(t_* + \delta_k, x_*(t_* + \delta_k)) - \omega(t_* + \delta_k, x_* + \delta_k h_*)].$$

It follows from (9) and the Lipschitz condition (4.3.1) that

$$(11) \quad r_k\delta_k^{-1} \to 0 \text{ as } k \to \infty.$$

Therefore

$$\inf_{x(\cdot)} \underline{\lim}_{\delta\to+0} [\omega(t_* + \delta, x(t_* + \delta)) - \omega(t_*, x_*)]\delta^{-1}$$
$$\leq \underline{\lim}_{\delta\to+0} [\omega(t_* + \delta, x_*(t_* + \delta)) - \omega(t_*, x_*)]\delta^{-1}$$
$$\leq \overline{\lim}_{k\to\infty} [\omega(t_* + \delta_k, x_* + \delta_k h_*) - \omega(t_*, x_*) + r_k]\delta_k^{-1}$$
$$= D^-\omega(t_*, x_*)|(1, h_*) = \min_{h\in F_2(t_*,x_*,v)} D^-\omega(t_*, x_*)|(1, h).$$

Inequality (6) is thus proved.

We now prove inequality (7). Let $\epsilon > 0$. Select a motion $x_*(\cdot) \in X_2(t_*, x_*, v)$ and a sequence δ_k ($k = 1, 2$), $\delta_k \to 0$ as $k \to \infty$ such that

$$\inf_{x(\cdot)} \overline{\lim}_{\delta\to+0} [\omega(t_* + \delta, x(t_* + \delta)) - \omega(t_*, x_*)]\delta^{-1}$$
$$\geq \lim_{k\to\infty} [\omega(t_* + \delta_k, x_*(t_* + \delta_k)) - \omega(t_*, x_*)]\delta_k^{-1} - \epsilon. \tag{12}$$

From the continuity of the mapping $(t,x) \notin F_2(t,x,v)$ it follows that

$$x_*(t_* + \delta_k) = x_* + \delta_k h_k + \delta_k \overset{*}{g}_k, \quad h_k \in F_2(t_*, x_*, v), \quad \|\overset{*}{g}_k\| \to 0$$
$$\text{as } k \to \infty.$$

The set $F_2(t_*, x_*, v)$ is compact, therefore one can assume that $h_k \to h_* \in F_2(t_*, x_*, v)$ as $k \to \infty$. We thus arrive at (9) where $g_k = \overset{*}{g}_k + h_k - h_*$. From (10) – (12) we obtain

$$\inf_{x(\cdot)} \overline{\lim}_{\delta\to+0} [\omega(t_* + \delta, x(t_* + \delta)) - \omega(t_*, x_*)]\delta^{-1} + \epsilon$$
$$\geq \lim_{k\to\infty} [\omega(t_* + \delta_k, x_* + \delta_k h_*) - \omega(t_*, x_*) + r_k]\delta_k^{-1}$$
$$\geq D^-\omega(t_*, x_*)|(1, h_*) \geq \min_{h\in F_2(t_*,x_*,v)} D^-\omega(t_*, x_*)|(1, h).$$

Inequality (7) is thus proved.

Inequalities (6) and (7) are valid for all $(t_*, x_*) \in \mathcal{G}$ and $v \in \mathcal{Q}$. Therefore, utilizing (6) we obtain the implication (4) $\Rightarrow$ (4.2.6) and utilizing (7) we have (4.2.5) $\Rightarrow$ (4). In accordance with Lemma 4.2.1 inequalities (4.2.3), (4.2.5), and (4.2.6) are equivalent. Hence inequality (4) will be equivalent to any one of the conditions (4.2.3), (4.2.5), and (4.2.6). Analogously one can prove that inequality (5) will be equivalent to any one of the inequalities (4.2.4), (4.2.7), and (4.2.8).

We now state the basic results of this chapter. Below, two theorems are presented which immediately follow from Theorems 4.2.1, 4.2.2, and Lemma 4.4.1.

Theorem 4.4.1. *Let function* $\sigma : R^n \to R$ *be locally Lipschitzian. In order that a function* $\omega : T \times R^n \to R$ *be the value function of a differential game with payoff of the form* (1), *it is necessary and sufficient that the following set of conditions be fulfilled:*

(a) ω *is locally Lipschitzian;*

(13) (b) $\omega(\theta,x) = \sigma(x), \quad x \in R^n;$

(14) (c) $\max_{v \in Q} \min_{h \in F_2(t,x,v)} D^-\omega(t,x)|(1,h) \leq 0;$

(15) (d) $\min_{u \in P} \max_{h \in F_1(t,x,u)} D^+\omega(t,x)|(1,h) \geq 0$

for all $t < \theta$, $x \in R^n$.

Theorem 4.4.2. *Let the function* $\rho : T \times R^n \to R$ *be locally Lipschitzian. In order that a function* $\omega : T \times R^n \to R$ *be the value function of a differential game with payoff of the form* (2), *it is necessary and sufficient that the following set of conditions be fulfilled:*

(a) ω *is locally Lipschitzian;*

(16) (b) $\omega(t,x) \leq \rho(t,x)$, $\omega(\theta,x) = \rho(\theta,x)$ for $t \in T$, $x \in R^n$;

(c) *in the region* $\mathcal{E} \overset{\text{def}}{=} \{(t,x) \in T^0 \times R^n : \omega(t,x) < \rho(t,x)\}$ *inequality* (14) *will be fulfilled;*

(d) *for all* $(t,x) \in T^0 \times R^n$ *the inequality* (15) *be fulfilled.*

We note that a function ω which satisfies the inequalities (14) and (15) with a small error will be close to the value function. The following proposition is valid.

Let $\omega : T \times R^n \to R$ be a locally Lipschitzian function satisfying equality (13) and the inequality

(17) $$\max_{v \in Q} \min_{h \in F_2(t,x,v)} D^-\omega(t,x)|(1,h) \leq \epsilon \quad (t < \theta, \ x \in R^n).$$

Then

$$\omega(t,x) + \epsilon(\theta - t) \geq \omega^0(t,x), \quad t \in T, \ x \in R^n,$$

where ω^0 is the value function of a differential game with a payoff of the form (1).

Proof. Consider the function $\omega_*(t,x) = \omega(t,x) + \epsilon(\theta - t)$, $(t,x) \in T \times R^n$. This function satisfies inequality (14), i.e. it is u-stable in the region $T^0 \times R^n$. According to Lemma 4.2.3 the estimate $\omega^0 \leqslant \omega_*$ is valid, q.e.d.

The following propositions are also true.

If a locally Lipschitzian function $\omega : T \times R^n \to R$ satisfies equality (13) and the inequality

$$(18) \qquad \min_{u \in P} \max_{h \in F_1(t,x,u)} D^+\omega(t,x)|(1,h) \geqslant -\epsilon \quad (t < \theta,\ x \in R^n),$$

then the estimate

$$\omega(t,x) - \epsilon(\theta - t) \leqslant \omega^0(t,x), \quad t \in T, \quad x \in R^n$$

is valid.

If a locally Lipschitzian function $\omega : T \times R^n \to R$ satisfies equality (13) and inequalities (17), (18) then the estimate

$$|\omega^0(t,x) - \omega(t,x)| \leqslant \epsilon(\theta - t), \quad t \in T, \quad x \in R^n$$

is valid.

Analogous estimates may be obtained for the value function of a differential game with payoff of the form (2).

4.4.2. We now state and prove several corollaries to Theorems 4.4.1 and 4.4.2.

Corollary 4.4.1. *Let $\omega^0 : T \times R^n \to R$ be the value function of a differential game with payoff of the form* (1), *where σ is a locally Lipschitzian function. Then ω^0 is a generalized solution of the main equation* (4.1.1) *with the boundary condition* (4.1.2).

Proof. According to the definition of a generalized solution (cf. (§4.1.2) it is required to show that the function ω satisfies the main equation (4.1.1) at all points $(t,x) \in T^0 \times R^n$ at which it is differentiable. Note that in view of Rademacher's theorem [27, 65] a locally Lipschitzian function ω^0 is non-differentiable on a set of measure zero. Let the function ω^0 be differentiable at the point (t,x), then for any $h \in R^n$

$$(19) \qquad \begin{aligned} D^-\omega(t,x)|\,(1,h) &= D\omega(t,x)|\,(1,h) = D^+\omega(t,x)|\,(1,h) \\ &= \frac{\partial\omega^0(t,x)}{\partial t} + \frac{\partial\omega^0(t,x)}{\partial x}h\,. \end{aligned}$$

Therefore inequalities (14) and (15) can be rewritten as

$$\frac{\partial \omega^0(t,x)}{\partial t} + \max_{v \in Q} \min_{h \in F_2(t,x,v)} \frac{\partial \omega^0(t,x)}{\partial x} h = \frac{\partial \omega^0(t,x)}{\partial t}$$

$$+ \max_{v \in Q} \min_{u \in P} \frac{\partial \omega^0(t,x)}{\partial x} f(t,x,u,v) \leqslant 0,$$

$$\frac{\partial \omega^0(t,x)}{\partial t} + \min_{u \in P} \max_{v \in Q} \frac{\partial \omega^0(t,x)}{\partial x} f(t,x,u,v) \geqslant 0.$$

Taking condition (1.2.8) into account we arrive at equality (4.1.1). Corollary 4.4.1 is thus proved.

Theorem 4.4.2 yields

Corollary 4.4.2. *Let ω^0 be the value function of a differential game with payoff of the form* (2), *where ρ is a locally Lipschitzian function. Then the function ω^0 satisfies relations* (16) *and at the points of differentiability relations*

$$(20) \qquad \frac{\partial \omega^0(t,x)}{\partial t} + \min_{u \in P} \max_{v \in Q} \frac{\partial \omega^0(t,x)}{\partial x} f(t,x,u,v) = 0 \quad \text{for } \omega^0(t,x) < \rho(t,x),$$

and

$$(21) \qquad \frac{\partial \omega^0(t,x)}{\partial t} + \min_{u \in P} \max_{v \in Q} \frac{\partial \omega^0(t,x)}{\partial x} f(t,x,u,v) \geqslant 0 \quad \text{for } (t,x) \in T^0 \times R^n$$

are fulfilled.

Corollaries 4.4.1 and 4.4.2 establish the necessary conditions to be satisfied by the value function. Utilizing Theorems 4.4.1 and 4.4.2 we obtain the following sufficient conditions from equality (19).

Corollary 4.4.3. *Let a continuous function $\omega^0 : T \times R^n \to R$ be differentiable on $T^0 \times R^n$ and satisfy the main equation* (4.1.1) *for all $(t,x) \in T^0 \times R^n$ and the boundary condition* (4.1.2). *Then this function is the value function of a differential game with payoff* (1).

Corollary 4.4.4. *Let a continuous function $\omega^0 : T \times R^n \to R$ be differentiable on $T^0 \times R^n$, satisfy relations* (16), *equality* (20) *in the region $\mathcal{E}^0 \overset{\text{def}}{=} \{(t,x) \in T^0 \times R^n : \omega^0(t,x) < \rho(t,x)\}$, and the inequality* (21) *in the region $T^0 \times R^n$. Then this function is the value function of a differential game with pay-off* (2).

We note once more that it is quite seldom that value functions turn out to be differentiable on the whole $T^0 \times R^n$. Very often value functions can be represented as a splicing of a finite number of smooth functions. In these cases, value functions can be represented in the form (4.3.4), either on the whole space $T \times R^n$ or locally. A piecewise-smooth function defined by the relation (4.3.4) is directionally differentiable, and for its derivative in the direction $(1,h)$ the formula (4.3.9) is valid. Taking this into account we immediately obtain from Theorems 4.4.1 and 4.4.2 the following assertions.

Theorem 4.4.3. *In order that a piecewise-smooth function* $\omega : T \times R^n \to R$ *be the value function of a differential game with payoff* (1) *it is necessary and sufficient that it satisfy the boundary condition* (13) *and for* $(t,x) \in T^0 \times R^n$ *satisfy the inequalities*

$$\text{(22)} \qquad \max_{v} \min_{h_2} \min_{i} \max_{j} \left[\frac{\partial \varphi(t,x,i,j)}{\partial t} + \frac{\partial \varphi(t,x,i,j)}{\partial x} h_2 \right] \leqslant 0,$$

$$\text{(23)} \qquad \min_{u} \max_{h_1} \min_{i} \max_{j} \left[\frac{\partial \varphi(t,x,i,j)}{\partial t} + \frac{\partial \varphi(t,x,i,j)}{\partial x} h_1 \right] \geqslant 0,$$

where $u \in P$, $v \in Q$, $h_2 \in F_2(t,x,v)$, $h_1 \in F_1(t,x,u)$, $i \in I_0(t,x)$, $j \in J_0(t,x,i)$, *and the sets* I_0 *and* J_0 *are defined in accordance with* (4.3.5).

Theorem 4.4.4. *In order that a piecewise-smooth function* $\omega : T \times R^n \to R$ *be the value function of a differential game with payoff* (2), *it is necessary and sufficient that it satisfy relations* (16), *the inequality* (22) *in the region* $\mathcal{E} \stackrel{\text{def}}{=} \{(t,x) \in T^0 \times R^n : \omega(t,x) < \rho(t,x)\}$, *and the inequality* (23) *in the region* $T^0 \times R^n$.

In the case when the value function is piecewise-smooth one can select in the space of positions $T \times R^n$ regions $\mathfrak{D}_{ij}$ in which $\omega^0(t,x) = \varphi(t,x,i,j)$ and the function $\varphi(\cdot,i,j)$ either satisfies the main equation provided the payoff is determined by equality (1), or satisfies the relations of the form (20), (21) provided the payoff is defined by equality (2). On singular surfaces where several smooth functions $\varphi(\cdot,i,j)$ join together, inequalities (22) and (23) indicate how the gradients of functions $\varphi(\cdot,i,j)$, $i \in I_0(t,x)$, $j \in J_0(t,x,i)$ should be related.

Theorems 4.4.3 and 4.4.4 establish necessary and sufficient conditions which should be satisfied by the value function under the additional assumption that this function is piecewise-smooth. It should, however, be admitted that we don't have a sufficiently general result which will show under what assumptions on function σ (or ρ) and f the value function ω will be piecewise-smooth.

Remark. The assertions presented above can be modified to obtain the conditions under which the local-Lipschitzian function ω coincides with the value function ω^0 on the set $cl\,\mathfrak{D}$, where

$$\mathfrak{D} \stackrel{\text{def}}{=} \{(t,x) \in T^0 \times R^n : a < \omega(t,x) < b\}, \quad (-\infty \leqslant a < b \leqslant \infty)$$

(rather than on the whole space $T \times R^n$). To do this it is sufficient to replace the region $T^0 \times R^n$ by the region $\mathcal{D}$ in the statements of the theorems and corollaries.

4.4.3. If desirable, one can interpret the problem of optimal control as a particular case of a differential game. One can obtain assertions analogous to Theorems 4.4.1 and 4.4.2 for the value function in a problem of optimal control.

Consider the following problem of optimal control. Let the motion of the controlled system be described by equation

$$\text{(24)} \qquad \dot{x}(t) = f(t,x(t),u(t)).$$

Here, as before, $x(t)$ is an n-dimensional phase vector, $u(t)$ is a control which satisfies the constraint $u(t) \in \mathcal{P}$, and $\mathcal{P}$ is a compact set in R^p. We assume that the function $f : T \times R^n \times \mathcal{P} \to R^n$ is continuous and possesses continuous partial derivatives $\partial f/\partial x$ and $\partial f/\partial t$ on $T^0 \times R^n \times \mathcal{P}$. Also for an arbitrary $(t,x) \in T \times R^n$ the set $\{f(t,x,u) : u \in \mathcal{P}\}$ is convex and for any initial position (t_0,x_0) and an arbitrary measurable control $u(\cdot) : [t_0,\theta] \to \mathcal{P}$ the motion $x(\cdot,t_0,x_0,u(\cdot))$ of the system (24) can be extended up to time $t = \theta$. It is also assumed that a continuously differentiable function $\sigma : R^n \to R$ is given. The control problem under consideration consists of determining an optimal control $u^0(\cdot)$ which yields the minimum

$$\text{(25)} \qquad \sigma(x(\theta,t_0,x_0,u^0(\cdot)) = \min_{u(\cdot)} \sigma(x(\theta,t_0,x_0,u(\cdot)) \overset{\text{def}}{=} \omega^0(t_0,x_0).$$

Here the minimum is computed over all measurable functions $u(\cdot) : [t_0,\theta] \to \mathcal{P}$. The function $(t_0,x_0) \to \omega^0(t_0,x_0)$ will be called the value function as before.

Observe that the value function defined for a problem of optimal control will coincide with the value function for a differential game in which the motion of the controlled system is described by the equation $\dot{x}(t) = f(t,x(t),u(t)) + v(t)$, $v(t) \in \mathcal{Q} = \{0\}$, i.e. the control of the second player is fictitious. The payoff in this game is given by the functional $\gamma(x(\cdot)) = \sigma(x(\theta))$. This fact allows us to transfer the results obtained for the value function of a differential game to the case of a problem of optimal control.

In the statement of the theorem presented below, a class Ω of functions $\omega : T \times R^n \to R$ is considered. This class is defined as follows: we say that a function ω belongs to the class Ω if

$$\omega(t,x) = \min_{\alpha \in A} \varphi(t,x,\alpha), \ (t,x) \in T \times R^n .$$

Here A is a metric compact set and $\varphi : T \times R^n \times A \to R$ is a continuous function which possesses continuous derivatives $\partial\varphi/\partial x$ and $\partial\varphi/\partial t$ on $T \times R^n \times A$.

The following theorem is valid.

Theorem 4.4.5. *In order that a function* $\omega : T \times R^n \to R$ *be the value function in the optimal control problem under consideration, it is necessary and sufficient that one of the following two sets of conditions be fulfilled.*

(1.a) *function* ω *is directionally differentiable*;

(1.b) $\omega(\theta,x) = \sigma(x), \ x \in R^n$;

(1.c) $\min\limits_{u \in P} D\omega(t,x)|(1,f(t,x,u)) = 0, \ \ (t,x) \in T^0 \times R^n$;

(2.a) *function* ω *is contained in the class* Ω;

(2.b) $\omega(\theta,x) = \sigma(x), \ \ x \in R^n$;

(2.c) $$\frac{\partial\omega(t,x)}{\partial t} + \min_{u \in P} \frac{\partial\omega(t,x)}{\partial x} f(t,x,u) = 0$$

at all points (t,x) *at which function* ω *is differentiable.*

Proofs of this and of the next theorem are omitted. These can be carried out based on Theorems 4.4.1 and 4.4.2 and utilizing the above mentioned fact that the value function in a control problem coincides with the value function in a differential game with a fictitious second player.

For a control problem where the target functional is defined by equality (2), the following assertion is valid.

Theorem 4.4.6. *In order that a function* $\omega : T \times R^n \to R$ *be the value function in a control problem where the target functional is defined by equality* (2), *it is necessary and sufficient that one of the following two groups of conditions be fulfilled:*

(1.a) *function* ω *is directionally differentiable*;

(1.b) $\omega(\theta,x) = \rho(\theta,x), \ \omega(t,x) \leqslant \rho(t,x)$ *for* $(t,x) \in T \times R^n$;

(1.c) $\min\limits_{u \in P} D\omega(t,x)| \, (1,f(t,x,u)) = 0$ *for* $(t,x) \in T^0 \times R^n$

and $\omega(t,x) < \rho(t,x)$;

(1.d) $\min\limits_{u \in P} D\omega(t,x)| \, (1,f(t,x,u)) \geqslant 0$ *for* $(t,x) \in T^0 \times R^n$

and $\omega(t,x) = \rho(t,x)$;

(2.a) *function* ω *is contained in the class* Ω;

(2.b) $\omega(\theta,x) = \rho(\theta,x), \ \omega(t,x) \leqslant \rho(t,x)$ *for* $(t,x) \in T \times R^n$;

(2.c) $$\frac{\partial\omega(t,x)}{\partial t} + \min_{u \in P} \frac{\partial\omega(t,x)}{\partial x} f(t,x,u) = 0$$

at all points (t,x) *where the function* ω *is differentiable and* $\omega(t,x) < \rho(t,x)$;

$$\text{(2.d)} \quad \frac{\partial\omega(t,x)}{\partial t} + \min_{u \in P} \frac{\partial\omega(t,x)}{\partial x} f(t,x,u) \geqslant 0$$

at all points (t,x) *where the function* ω *is differentiable and* $\omega(t,x) = \rho(t,x)$.

4.4.4. Theorems 4.4.5 and 4.4.6 can be obtained in the same manner as stated above, i.e. utilizing the fact that the value function in a control problem coincides with the value function of the corresponding differential game. Such an approach is quite natural here since we have prepared the appropriate tools for its realization. However, if we are interested in control problems (24) only (and not differential games), Theorems 4.4.5 and 4.4.6 can be obtained without using the notion of differential games. Consider, for example, Theorem 4.4.5. Conditions (1.c) and (2.c) play the dominant role in this theorem. Both of these conditions express (in an infinitesimal form) the following property of the value function in a control problem: for any $(t_*,x_*) \in T^0 \times R^n$ and $t^* \in (t_*,\theta]$ there exists a measurable control $u(\cdot) : [t_*,t^*] \to P$ such that for the solution $x(\cdot) = x(\cdot,t_0,x_0,u(\cdot))$ $(x(t_*) = x_*)$ of equation (24) the equality $\omega^0(t^*,x(t^*)) = \omega^0(t_*,x_*)$ is valid.

This property is just a modification of a well known *optimality principle* in control theory. Thus, equations (1.c) and (2.c) can be obtained from this well known principle.

We now return to differential games. To develop methods of dynamic programming for differential games, it is first required to formulate the corresponding optimality principle so that it is universal and efficient. However, this turns out to be much more difficult than the formulation of such a principle in problems of optimal (programmed) control or in multistage (discrete) games.

There are several contenders for such a principle. We mention, in particular, *the transition principle* and the *principle of semipermeability* (cf. (Isaacs [32]). To clarify their meaning consider the case when

$$f(t,x,u,v) = f^{(1)}(t,x,u) + f^{(2)}(t,x,v)$$

and we shall assume that vectograms $f^{(1)}(t,x,P) \overset{\text{def}}{=} \{f^{(1)}(t,x,u) : u \in P\}$ and $f^{(2)}(t,x,Q) \overset{\text{def}}{=} \{f^{(2)}(t,x,v) : v \in Q\}$ are convex for all $(t,x) \in T \times R^n$. Assume also that functions ω considered below are differentiable in any direction $(1,h)$, $h \in R^n$, i.e.

$$D^+\omega(t,x)|(1,h) = D^-\omega(t,x)|(1,h) = D\omega(t,x)|(1,h).$$

Consider first the transition principle. Expressed in an infinitesimal form the function ω satisfying this principle should satisfy the equality

$$(26)\qquad \max_{v\in Q}\min_{u\in P} D\omega(t,x)|\,(1,f^{(1)}(t,x,u) + f^{(2)}(t,x,v))$$
$$= \min_{u\in P}\max_{v\in Q} D\omega(t,x)|\,(1,f^{(1)}(t,x,u) + f^{(2)}(t,x,v)) = 0.$$

The inequalities

$$(27)\qquad \begin{aligned} &\min_{u\in P}\max_{v\in Q} D\omega(t,x)|\,(1,f^{(1)}(t,x,u) + f^{(2)}(t,x,v)) \leqslant 0,\\ &\max_{v\in Q}\min_{u\in P} D\omega(t,x)|\,(1,f^{(1)}(t,x,u) + f^{(2)}(t,x,v)) \geqslant 0 \end{aligned}$$

correspond to the condition of semipermeability.

If the function ω is differentiable at the point (t,x) then equality (26) and inequalities (27) are equivalent to the equality (the main equation)

$$\frac{\partial\omega(t,x)}{\partial t} + \min_{u\in P}\frac{\partial\omega(t,x)}{\partial\omega} f^{(1)}(t,x,u)$$
$$+ \max_{v\in Q}\frac{\partial\omega(t,x)}{\partial x} f^{(2)}(t,x,v) = 0.$$

Consider now the case when the point (t,x) belongs to the singular set, i.e. at this point the value function ω^0 is not differentiable. According to Theorem 4.4.1 (or 4.4.2) the value function satisfies inequalities (14) or (15). One can give examples where strict inequality is valid in (14) or (15) on a singular set, i.e.

$$\max_{v\in Q}\min_{u\in P} D\omega(t,x)|\,(1,f^{(1)}(t,x,u) + f^{(2)}(t,x,v)) < 0,$$
$$\min_{u\in P}\max_{v\in Q} D\omega(t,x)|\,(1,f^{(1)}(t,x,u) + f^{(2)}(t,x,v)) > 0.$$

As is seen from these inequalities, relations (26) and (27) may be violated on singular sets. Hence the principles on which these inequalities are based are not universal. On the other hand, properties of u-stability and v-stability are universal. They are fulfilled where the value function is differentiable as well as on singular surfaces.

4.5. Examples

In this section, we consider several examples of differential games in which Theorems 4.4.1 and 4.4.2 are used.

4.5.1. Let the motion of the controlled system be described by the equation

$$\dot{x}(t) = tu(t) + v(t)$$

where x, u, and v are two-dimensional vectors, $t \leqslant 0$, $\theta = 0$.

The controls of the players satisfy the constraints $\|u(t)\| \leqslant 1$, $\|v(t)\| \leqslant 1$. The payoff is given by the equality

$$\gamma_1(x(\cdot)) \overset{\text{def}}{=} \|x(0)\|.$$

In §4.1.2 the following formula is presented to define the value function of a given game:

$$\omega_1(t,x) = \max\{a_1(t,x),\ a_2(t)\},$$

$$(1)\qquad a_1(t,x) \overset{\text{def}}{=} \|x\| - t^2/2 - t,$$

$$a_2(t) \overset{\text{def}}{=} -t^2/2 - t \ \text{ for } \ -1 \leqslant t \leqslant 0, \quad a_2(t) \overset{\text{def}}{=} 1/2 \text{ for } t \leqslant -1.$$

We show that function ω_1 satisfies conditions (a) – (c) indicated in Theorem 4.4.1. It follows immediately from (1) that the function ω_1 is locally Lipschitzian and satisfies the boundary condition $\omega(0,x) = \|x\|$. We verify that inequalities (4.4.14) and (4.4.15) are fulfilled. Three cases are possible:

(j) $\omega_1(t,x) = a_1(t,x) > a_2(t)$,

(jj) $\omega_1(t,x) = a_2(t) > a_1(t,x)$,

(jjj) $\omega_1(t,x) = a_1(t,x) = a_2(t)$.

Case (j). Inequality $a_1(t,x) > a_2(t)$ implies that $\|x\| > 0$. Therefore in a sufficiently small neighborhood of the point (t,x) the function $(t,x) \to \omega_1(t,x) = a_1(t,x) = \|x\| - t^2/2 - t$ is differentiable. We thus obtain

$$\frac{\partial \omega_1(t,x)}{\partial t} + \min_u \max_v \frac{\partial \omega(t,x)}{\partial x}(tu + v)$$
$$= -t - 1 + \min_u \max_v \langle x/\|x\|,\ tu + v\rangle = 0.$$

(Recall that $t < 0$, therefore $\min_{\|u\| \leqslant 1} \langle x/\|x\|,\ tu\rangle = t$.) Thus, in case (j) the main equation is fulfilled at the point (t,x), which yields the inequalities (4.4.14) and (4.4.15).

Case (jj). Inequality $a_2(t) > a_1(t,x)$ can be valid only for $t < -1$. In this case, in a sufficiently small neighborhood of the point (t,x) the equality $\omega_1(\tau,y) = 1/2$ holds. Therefore the function satisfies the main equation at the point (t,x).

Case (jjj). Assume first that $t \in [-1,0)$. Then $\|x\| = 0$ and for $\tau \geqslant t$ and $y \in R^2$ we have $\omega_1(\tau,y) = a_1(\tau,y) = \|y\| - \tau^2/2 - \tau$. Therefore $D\omega_1(t,x)|(1,h) = -t - 1 + \|h\|$. For $t \in [-1,0)$ the equalities

$$\max_{\|v\|\leq 1} \min_{\|u\|\leq 1} \|tu + v\| = 1 + t$$

$$\min_{\|u\|\leq 1} \max_{\|v\|\leq 1} \|tu + v\| = 1$$

are valid. From these equalities we obtain

$$\max_{v} \min_{u} D\omega(t,x)| (1,tu + v) = 0,$$

$$\min_{u} \max_{v} D\omega_1(t,x)| (1,tu + v) = -t > 0,$$

i.e. inequalities (4.4.14) and (4.4.15) are fulfilled.

Now let $t < -1$. Then in a sufficiently small neighborhood of the point (t,x) the equality

$$\omega_1(t,x) = \max\{1/2, \|x\| - t^2/2 - t\}, \|x\| > 0$$

is valid. According to equality (4.3.9), we obtain

$$D\omega_1(t,x)|(1,h) = \max\{0, -t - 1 - \langle x/\|x\|,h\rangle\}.$$

Therefore

$$\max_{v} \min_{u} D\omega_1(t,x)| (1,tu + v) = 0,$$

$$\min_{u} \max_{v} D\omega_1(t,x)| (1,tu + v) = 0.$$

It is thus established that function ω_1 satisfies all the conditions (a) – (c) stated in Theorem 4.4.1. Hence this function is the value function of the game under consideration.

4.5.2.. Consider a differentiable game in which the motion of the controlled system is described by the equation

(2) $$\dot{x}(t) = u(t) + v(t), \quad u(t) \in P, \ v(t) \in Q$$

where x, u, and v are n-dimensional vectors. We assume that P and Q are convex compact sets in R^n. Let the payoff be defined by the equality

(3) $$\gamma_2(x(\cdot)) \overset{\text{def}}{=} \sigma(x(\theta)).$$

We assume that function $\sigma : R^n \to R$ is convex and satisfies the Lipschitz condition

(4) $$|\sigma(x_1) - \sigma(x_2)| \leq \lambda \|x_1 - x_2\|, \ x_i \in R^n, \ i = 1, 2.$$

To write down the formula which defines the value function, we shall utilize certain concepts of convex analysis (cf. §A.2).

Let $\sigma^* : R^n \to (-\infty,\infty]$ be a function conjugate to function σ. The function σ^* is defined by equality

$$\sigma^*(\ell) \stackrel{\text{def}}{=} \sup_{x \in R^n} [\langle \ell, x \rangle - \sigma(x)], \ \ell \in R^n.$$

The function σ^* is convex and lower semi-continuous. The equality

(5) $$\sigma(x) = \sup_{\ell \in R^n} [\langle \ell, x \rangle - \sigma^*(\ell)]$$

is valid. Observe also that when the Lipschitz condition (4) is satisfied the set

(6) $$\mathfrak{L} \stackrel{\text{def}}{=} \operatorname{dom} \sigma^* \stackrel{\text{def}}{=} \{\ell \in R^n : \sigma^*(\ell) < \infty\}$$

will be bounded.

Consider the function

(7) $$\omega_2(t,x) \stackrel{\text{def}}{=} \sup_{\ell \in R^n} [\langle \ell, x \rangle + (\theta - t)\kappa(\ell) - \sigma^*(\ell)]$$

where

(8) $$\kappa(\ell) \stackrel{\text{def}}{=} \min_{u \in P} \langle \ell, u \rangle + \max_{v \in Q} \langle \ell, v \rangle.$$

We shall prove that the function $\omega_2 : T \times R^n \to R$ is the value function of the differential game (2), (3).

Set

(9) $$\mathfrak{L}_0(t,x) \stackrel{\text{def}}{=} \{\ell_0 \in R^n : \langle \ell_0, x \rangle + (\theta - t)\kappa(\ell_0) - \sigma^*(\ell_0) = \omega_2(t,x)\}.$$

Utilizing the lower semi-continuity of the function σ^* one can prove that for any $(t,x) \in T \times R^n$ the set $\mathfrak{L}_0(t,x)$ is a non-empty compact set contained in the bounded set $\mathfrak{L}$ and the multi-valued mapping $(t,x) \to \mathfrak{L}_0(t,x)$ is upper semi-continuous (cf. Section 4.3.2). Equality (7) can be rewritten as

(10) $$\omega_2(t,x) = \max_{\ell \in \mathfrak{L}} [\langle \ell, x \rangle + (\theta - t)\kappa(\ell) - \sigma^*(\ell)].$$

We show that the function ω_2 satisfies conditions (a) – (c) stated in Theorem 4.4.1. It can be seen from (10) – where the set $\mathfrak{L}$ is bounded – that the function ω_2 is locally Lipschitzian. It follows from (5) and (7) that function ω_2 satisfies the boundary condition $\omega_2(\theta,x) = \sigma(x)$. It remains to verify that inequalities (4.4.14) and (4.4.15) are fulfilled.

Function ω_2 is an upper envelope of a family of smooth functions. According to Lemma 4.3.1 this function is directionally differentiable and the formula

(11) $$D\omega_2(t,x) | (1,h) = \max_{\ell \in \mathfrak{L}_0(t,x)} [\langle \ell, h \rangle - \kappa(\ell)]$$

is valid. We shall show that the inequality

(12) $$\max_{v\in Q}\min_{u\in P} D\omega_2(t,x)|(1,u+v) \leqslant 0 \quad \text{for} \quad (t,x) \in T^0 \times R^n$$

is valid. We fix the point $(t,x) \in T^0 \times R^n$ and introduce the notation

$$\mathfrak{L}_0 \overset{\text{def}}{=} \mathfrak{L}_0(t,x), \quad \Pi \overset{\text{def}}{=} rpm(\mathfrak{L}_0),$$

i.e. Π is the set of probability measures normed on a compact set $\mathfrak{L}_0$. Let

$$\pi \in \Pi, \quad \ell_\pi = \int_{\mathfrak{L}_0} \ell\pi(d\ell).$$

The set $\mathfrak{L}$ defined in accordance with (6) is convex, therefore $\ell_\pi \in \mathfrak{L}$. From the definition of the set $\mathfrak{L}_0 = \mathfrak{L}_0(t,x)$ (cf.(9)) follows the inequality

$$\langle \ell,x\rangle + (\theta - t)\kappa(\ell) - \sigma^*(\ell) \geqslant \langle \ell_\pi,x\rangle + (\theta - t)\kappa(\ell_\pi) - \sigma^*(\ell_\pi)$$

$$\text{for all } \ell \in \mathfrak{L}_0.$$

Integrating the l.h.s. of this inequality with respect to measure π we obtain

(13) $$(\theta - t)\int_{\mathfrak{L}_0} \kappa(\ell)\pi(d\ell) - \int_{\mathfrak{L}_0} \sigma^*(\ell)\pi(d\ell) \geqslant (\theta - t)\kappa(\ell_\pi) - \sigma^*(\ell_\pi).$$

The function σ^* is convex, therefore

(14) $$\int_{\mathfrak{L}_0} \sigma^*(\ell)\pi(d\ell) \geqslant \sigma^*(\ell_\pi).$$

From (13), (14), and inequality $\theta - t > 0$ we have

(15) $$\int_{\mathfrak{L}_0} \kappa(\ell)\pi(d\ell) \geqslant \kappa(\ell_\pi).$$

Since the equality

$$D\omega_2(t,x)|(1,h) = \max_{\ell\in\mathfrak{L}_0} [\langle \ell,h\rangle - \kappa(\ell)] = \max_{\pi\in\Pi} [\langle \ell_\pi,h\rangle - \int_{\mathfrak{L}_0} \kappa(\ell)\,\pi(d\ell)]$$

is valid we have in accordance with (11) that

(16) $$\max_{v\in Q}\min_{u\in P} D\omega_2(t,x)|(1,u+v) = \max_{v\in Q}\min_{u\in P}\max_{\pi\in\Pi} [\langle \ell_\pi,u+v\rangle - \int_{\mathfrak{L}_0} \kappa(\ell)\pi(d\ell)].$$

The expression appearing in square brackets depends linearly on u and π, the sets P and Π are convex. Therefore in accordance with the minimax theorem [21, 34, 55], operations *min* and *max* can be interchanged. It follows from (15), (16), and (8) that

$$\max_{v\in Q}\min_{u\in P} D\omega_2(t,x)|(1,u+v) = \max_{\pi\in\Pi}\max_{v\in Q}\min_{u\in P} [\langle \ell_\pi,u+v\rangle - \int_{\mathfrak{L}_0}\kappa(\ell)\pi(d\ell)]$$
$$= \max_{\pi\in\Pi} [\kappa(\ell_\pi) - \int_{\mathfrak{L}_0} \kappa(\ell)\pi(d\ell)] \leqslant 0.$$

Inequality (12) is thus proved.

We now show that the inequality

(17) $$\min_{u \in P} \max_{v \in Q} D\omega_2(t,x)|(1,u+v) \geqslant 0$$

is also valid. According to (11), (8), and in view of the inequality *min max* $\geqslant$ *max min* we obtain

$$\min_{u \in P} \max_{v \in Q} D\omega_2(t,x)|(1,u+v)$$

$$\geqslant \max_{\ell \in \mathfrak{L}_0(t,x)} \min_{u \in P} \max_{v \in Q} [\langle \ell, u+v \rangle - \kappa(\ell)] = 0.$$

Inequality (17) is thus proved.

We have thus shown that function ω_2 satisfies all the conditions (a) – (c) of Theorem 4.4.1. Accordingly, ω_2 is the value function of the differential game (2), (3).

4.5.3. Let the motion of a controlled system be described by equation (2). Consider a differential game in which the payoff is defined by the equality

(18) $$\gamma_3(x(\cdot)) \overset{\text{def}}{=} \min_{t_0 \leqslant t \leqslant \theta} \rho(x(t)).$$

Here $\rho : R^n \to R$ is a convex function satisfying the Lipschitz condition on the whole space R^n, i.e. function ρ satisfies the same conditions as function σ in the preceding example. Denote by ρ^* the function conjugate to ρ. Set

(19) $$\chi(\ell) \overset{\text{def}}{=} \min \{\kappa(\ell), 0\}$$

where the quantity $\kappa(\ell)$ is defined in accordance with (6).

We show that the value function of the differential game (2), (18) is determined by the relation

(20) $$\omega_3(t,x) = \sup_{\ell \in R^n} [\langle \ell, x \rangle + (\theta - t)\,\chi(\ell) - \rho^*(\ell)].$$

Since the set $\mathfrak{L} \overset{\text{def}}{=} dom \rho^*$ is bounded, the function ω_3 is locally Lipschitzian. It follows from the equality

$$\rho(x) = \sup_{\ell \in R^n} [\langle \ell, x \rangle - \rho^*(\ell)]$$

that $\omega_3(\theta,x) = \rho(x)$. Taking into account that $\chi(\ell) \leqslant 0$ we arrive at $\omega_3(t,x) \leqslant \rho(x)$. Hence the function ω_3 satisfies the conditions (a), (b) indicated in Theorem 4.4.2. We shall prove that function ω_3 satisfies conditions (c), (d) of this theorem also.

Introduce the following notation:

$$(21)\qquad \begin{aligned} &\varphi(\ell,t) \overset{\text{def}}{=} \rho^*(\ell) - (\theta - t)\chi(\ell),\\ &\mathcal{L}_0(t,x) \overset{\text{def}}{=} \{\ell_0 \in R^n : \langle \ell_0, x\rangle - \varphi(\ell_0,t) = \omega_3(t,x)\}. \end{aligned}$$

As in the previous example, $\mathcal{L}_0(t,x)$ is a non-empty compact set contained in a bounded set $\mathcal{L}$. The multi-valued mapping $(t,x) \to \mathcal{L}_0(t,x)$ is upper semi-continuous. Function ω_3 is directionally differentiable and the formula

$$(22)\qquad D\omega_3(t,x)|\,(1,h) = \max_{\ell \in \mathcal{L}_0(t,x)} [\langle \ell,h\rangle - \chi(\ell)],\ (t,x) \in T^0 \times R^n,\ h \in R^n$$

is valid (cf. Lemma 4.3.1). It follows from (8), (19), and the inequality *min max* ⩾ *max min* that

$$\begin{aligned} \min_{u\in P} \max_{v\in Q} D\omega_3(t,x)|\,(1,u+v) &\geqslant \max_{\ell\in\mathcal{L}_0(t,x)} \min_{u\in P} \max_{v\in Q} [\langle \ell, u+v\rangle - \chi(\ell)] \\ &= \max_{\ell\in\mathcal{L}_0(t,x)} [\kappa(\ell) - \chi(\ell)] \geqslant 0. \end{aligned}$$

We have thus shown that function ω_3 satisfies condition (d) of Theorem 4.4.2. It remains to verify that function ω_3 satisfies condition (c) of Theorem 4.4.2, i.e. the inequality

$$(23)\qquad \max_{v\in Q} \min_{u\in P} D\omega_3(t,x)|\,(1,u+v) \leqslant 0$$

is valid for any position $(t,x) \in T^0 \times R^n$ where $\omega_3(t,x) < \rho(x)$.

We introduce additional notation. Let $\varphi^{**}(\cdot,t)$ be the second conjugate function to function $\varphi(\cdot,t)$ (21). We observe that $\varphi^{**}(\cdot,t)$ is a convex, lower semi-continuous function. Function $\varphi^{**}(\cdot,t)$ is a convex hull of the function $\varphi(\cdot,t)$, i.e. $\varphi^{**}(\cdot,t) = conv\varphi(\cdot,t)$ is the largest convex minorant of the function $\varphi(\cdot,t)$. The following inequality and equality are valid:

$$(24)\qquad \varphi^{**}(\ell,t) \leqslant \varphi(\ell,t) \quad (\ell \in R^n,\ t \in T),$$

$$(25)\qquad \begin{aligned} &\omega_3(t,x) = \max_{\ell\in R^n} [\langle \ell,x\rangle - \varphi(\ell,t)] = \max_{\ell\in R^n} [\langle \ell,x\rangle - \varphi^{**}(\ell,t)],\\ &\qquad (t,x) \in T \times R^n. \end{aligned}$$

We set

$$(26)\qquad \mathcal{L}_*(t,x) \overset{\text{def}}{=} \{\ell_* \in R^n : \langle \ell_*, x\rangle - \varphi^{**}(\ell_*,t) = \omega_3(t,x)\}.$$

For any $(t,x) \in T \times R^n$ the set $\mathcal{L}_*(t,x)$ is a non-empty convex compact set contained in $\mathcal{L} = dom\rho^*$. The inclusion

$$(27)\qquad \mathcal{L}_0(t,x) \subset \mathcal{L}_*(t,x),\quad (t,x) \in T \times R^n,$$

is valid where $\mathcal{L}_0(t,x)$ is a set defined in accordance with (21).

We choose a position $(t,x) \in T^0 \times R^n$. Two cases are possible.

(j) there exist $\ell_* \in \mathcal{L}_*(t,x)$ such that $\kappa(\ell_*) > 0$;

(jj) $\kappa(\ell) \leq 0$ for all $\ell \in \mathcal{L}_*(t,x)$.

Consider the case (j). The function $\kappa(\ell)$ is continuous, therefore

$$\kappa(\ell) > 0 \quad \text{for } \ell \in B(\ell_*,\epsilon) \overset{\text{def}}{=} \{\ell \in R^n : \|\ell - \ell_*\| \leq \epsilon\},$$

where ϵ is a positive number. It follows from (19) and (21) that

$$\chi(\ell) = 0, \quad \varphi(\ell,t) = \rho^*(\ell), \; \ell \in B(\ell_*;\epsilon).$$

Taking (24) into account we obtain

$$\varphi^{**}(\ell,t) \leq \rho^*(\ell), \quad \ell \in B(\ell_*;\epsilon).$$

From (19) and (21) we have

$$\varphi(\ell,t) \geq \rho^*(\ell), \quad \ell \in R^n,$$

i.e. ρ^* is a convex minorant of the function $\varphi(\cdot,t)$. By the definition $\varphi^{**}(\cdot,t)$ is the largest convex minorant of the function $\varphi(\cdot,t)$. Therefore

$$\varphi^{**}(\ell,t) \geq \rho^*(\ell), \quad \ell \in R^n.$$

Thus we obtain the equality

$$\varphi^{**}(\ell,t) = \rho^*(\ell), \quad \ell \in B(\ell_*;\epsilon). \tag{28}$$

Taking into account that $\ell_* \in \mathcal{L}_*(t,x)$ we have from (26) and (28) that

$$\begin{aligned} \omega_3(t,x) &= \max_{\ell \in R^n} [\langle \ell,x\rangle - \varphi^{**}(\ell,t)] = \max_{\ell \in B(\ell_*;\epsilon)} [\langle \ell,x\rangle - \varphi^{**}(\ell,t)] \\ &= \max_{\ell \in B(\ell_*;\epsilon)} [\langle \ell,x\rangle - \rho^*(\ell)] = \langle \ell_*,x\rangle - \rho^*(\ell_*). \end{aligned}$$

Function $\ell \to \langle \ell,x\rangle - \rho^*(\ell)$ is concave and attains its maximum at the center of a ball $B(\ell_*;\epsilon)$, therefore this maximum will be global, i.e.

$$\langle \ell_*,x\rangle - \rho^*(\ell_*) = \max_{\ell \in R^n} [\langle \ell,x\rangle - \rho^*(\ell)] = \rho(x).$$

Thus in the case (j) we have an equality $\omega_3(t,x) = \rho(x)$, hence in this case inequality (23) need not be verified.

Consider the case (jj). Relation (20) differs from relation (7) only by replacing functions κ and σ^* by functions χ and ρ^*. Therefore in the same manner as in the preceding example we obtain

(29) $$D\omega_3(t,x)|(1,h) = \max_{\pi\in\Pi} [\langle \ell_\pi, h\rangle - \int_{\mathfrak{L}_0} \chi(\ell)\pi(d\ell)],$$

(30) $$\int_{\mathfrak{L}_0} \chi(\ell)\pi(d\ell) \geqslant \chi(\ell_\pi).$$

Here $\mathfrak{L}_0 \overset{\text{def}}{=} \mathfrak{L}_0(t,x)$ (cf. (21)), $\Pi \overset{\text{def}}{=} rpm(\mathfrak{L}_0)$. According to (27) we have $\mathfrak{L}_0 \subset \mathfrak{L}_*(t,x)$. Recall that $\mathfrak{L}_*(t,x)$ is convex and $\ell_\pi = \int_{\mathfrak{L}_0} \ell\pi(d\ell)$, therefore $\ell_\pi \in \mathfrak{L}_*(t,x)$. In the case under consideration, the inequality $\kappa(\ell) \leqslant 0$ is valid for $\ell \in \mathfrak{L}_*(t,x)$, hence in accordance with (19) we have

$$\chi(\ell) = \kappa(\ell), \quad \int_{\mathfrak{L}_0} \chi(\ell)\pi(d\ell) = \int_{\mathfrak{L}_0} \kappa(\ell)\pi(d\ell), \quad \chi(\ell_\pi) = \kappa(\ell_\pi).$$

Relations (29) and (30) can be rewritten as

$$D\omega_3(t,x)|(1,h) = \max_{\pi\in\Pi}[\langle \ell_\pi, h\rangle - \int_{\mathfrak{L}_0} \kappa(\ell)\, \pi(d\ell)],$$

$$\int_{\mathfrak{L}_0} \kappa(\ell)\pi(d\ell) \geqslant \kappa(\ell_\pi).$$

We thus obtain

$$\max_{v\in Q} \min_{u\in P} D\omega_3(t,x)|(1,u+v) =$$
$$\max_{v\in Q} \min_{u\in P} \max_{\pi\in\Pi} [\langle \ell_\pi, u+v\rangle - \int_{\mathfrak{L}_0} \kappa(\ell)\pi(d\ell)].$$

This equality is analogous to equality (16). Utilizing this equality we arrive, as in the previous example, at inequality (23).

It has thus been established that function ω_3 (20) satisfies all the conditions (a) – (d) stated in Theorem 4.4.2. Hence this function is the value function of the differential game (2), (18).

4.5.4. In the examples presented above, we have been able to obtain sufficiently simple formulas to define the value function. It should, however, be admitted that the possibilities for an analytic construction of the value function are quite limited. Even for a control system with a very simple motion equation and a simple payoff function it is, as a rule, either not possible to determine the value function explicitly or this function is given by a very complicated expression. To illustrate this situation, we shall present the following example whose solution was obtained by A. M. Taras'ev.

Let the motion of the controlled system be described by the equation

(31) $$\dot{x}_1 = x_2 + v, \quad \dot{x}_2 = u$$

where the scalar controls of the players satisfy the restrictions $|u| \leqslant$

1, $|v| \leqslant 1$. We set $T \stackrel{\text{def}}{=} [0,2]$, i.e. here $\theta = 2$. The payoff functional is defined by equality

$$\gamma_1(x(\cdot)) \stackrel{\text{def}}{=} \max\{|x_1(2)|, |x_2(2)|\}.$$

In this example the value function $\omega_4 : [0, 2] \times R^2 \to R$ is a piecewise-smooth function "pasted together" from 12 smooth functions φ_i ($i = 1, 2, ..., 12$). Functions φ_1, φ_3 and φ_5 are defined by the equalities

$$\varphi_1(t,x) \stackrel{\text{def}}{=} x_1 + (2 - t)x_2 + t - t^2/2$$

$$\varphi_3(t,x) \stackrel{\text{def}}{=} x_2 - (2 - t)$$

$$\varphi_5(t,x) \stackrel{\text{def}}{=} 2 - \varphi_3(t,x) - 2[1 - \varphi_1(t,x) - \varphi_3(t,x)]^{1/2} .$$

Functions φ_7, φ_9, and φ_{11} are defined implicitly

$$2\ln(1 + [1 - \varphi_1(t,x) - \varphi_7]^{1/2}) - 2[1 - \varphi_1(t,x) - \varphi_7]^{1/2}$$
$$- \varphi_3(t,x) - \varphi_7 - 2(2\varphi_7 - 1)^{1/2} = 0,$$

$$2\ln(1 + [1 - \varphi_1(t,x) - \varphi_9]^{1/2}) - 2[1 - \varphi_1(t,x) - \varphi_9]$$
$$- \varphi_3(t,x) - \varphi_9 - 2(1 + \ln\varphi_9) = 0,$$

$$\tfrac{1}{4}(\varphi_{11} - \varphi_3(t,x) - 2)^2 - 1 + \varphi_1(t,x) + \varphi_{11}$$
$$+(\varphi_{11} - \varphi_3(t,x) - 2)[\ln(\varphi_{11} - \varphi_3(t,x) - 2) - \ln(2\psi(\varphi_{11}) - 1)] = 0,$$

where $\psi(\varphi) = \exp[(\varphi - 1) + (2\varphi - 1)^{1/2}]$. The remaining 6 functions are defined by the equalities $\varphi_{i+1}(t,-x) \stackrel{\text{def}}{=} \varphi_i(t,x)$ ($i = 1, 3, ..., 11$).

The space of positions $[0, 2] \times R^2$ is subdivided into subsets $\mathfrak{D}_i$ ($i = 1, 2, ..., 12$) in each one of which the value function ω_4 coincides with the corresponding function φ_i. In the regions $\mathfrak{D}_i$ ($i = 7, 8, ..., 12$) the implicit functions $\varphi_7, \varphi_8, ..., \varphi_{12}$ are uniquely defined and are continuously differentiable. All the functions $\varphi_i : \mathfrak{D}_i \to R$ satisfy the main equation which here is of the form

$$\frac{\partial\varphi(t,x)}{\partial t} + \frac{\partial\varphi(t,x)}{\partial x_1}x_2 + \left|\frac{\partial\varphi(t,x)}{\partial x_1}\right| - \left|\frac{\partial\varphi(t,x)}{\partial x_2}\right| = 0.$$

On singular surfaces where different sets $\mathfrak{D}_i$ and $\mathfrak{D}_j$ are joined together, the inequalities (4.4.14) and (4.4.15) are fulfilled.

A complete description of the manner in which the space $[0, 2] \times R^2$ is subdivided into regions $\mathfrak{D}_i$ turns out to be quite cumbersome. We shall omit this description. We shall only describe the construction of the value function in the region $[1, 2] \times R^2$. In this region the function ω_4 is "pasted together" from the first six functions as follows:

$$\omega_4 = \max\{\varphi_1, \varphi_3, \varphi_6\} \quad \text{for} \quad \varphi_1 + \varphi_3 \geqslant 0,$$

$$\omega_4 = \max\{\varphi_2, \varphi_4, \varphi_5\} \quad \text{for} \quad \varphi_2 + \varphi_4 \geqslant 0,$$

$$\omega_4 = \max\{\varphi_5, \varphi_6\} \quad \text{for} \quad \varphi_1 + \varphi_3 \leqslant 0, \quad \varphi_2 + \varphi_4 \leqslant 0.$$

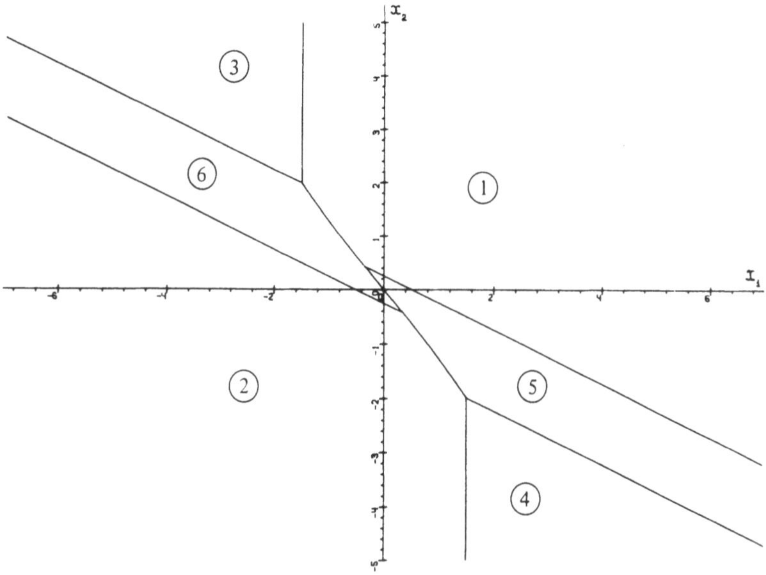

Fig. 4.5.1

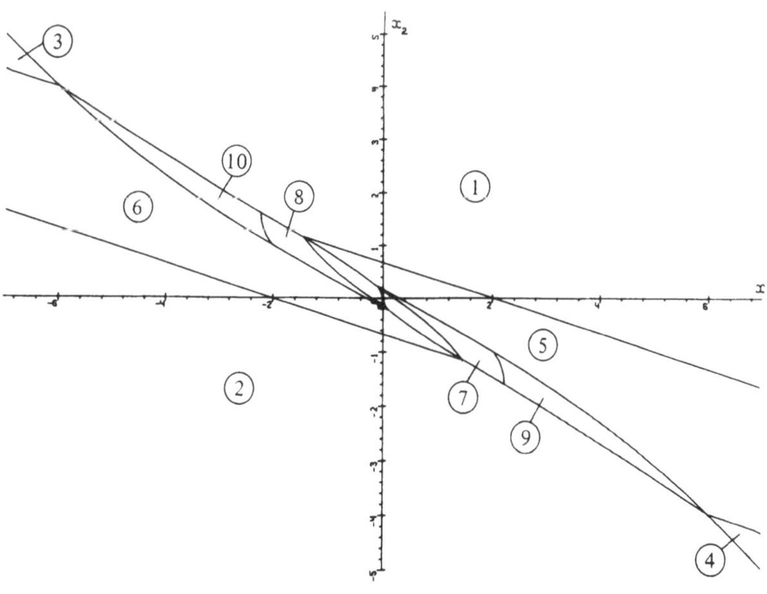

Fig. 4.5.2

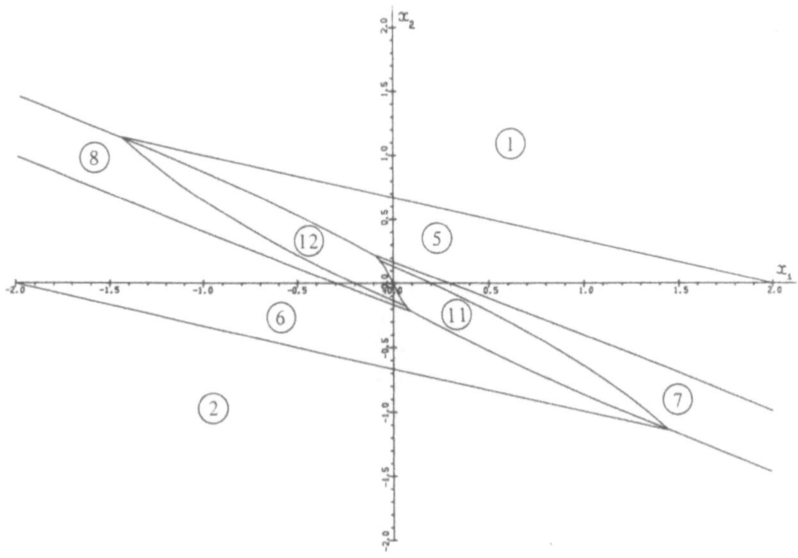

Fig. 4.5.3

Observe that not all the surfaces which separate different regions $\mathfrak{D}_i$ and $\mathfrak{D}_j$ are singular. Some of these functions φ_i and φ_j are combined smoothly. For example, this is the case for functions φ_1 and φ_5, and φ_2 and φ_6.

In Fig. 4.5.1 cross-sections of the regions $\mathfrak{D}_i$ by the hyperplane t = 1 are presented. The numbers in the small circles are indices of these regions. In Fig. 4.5.2 the cross-sections of regions $\mathfrak{D}_i$ by the hyperplane $t = 0$ are shown. A part of these sections is presented in Fig. 4.5.3. on a larger scale.

In the example under consideration the payoff function is defined by equality $\sigma(x) = \max\{|x_1|,|x_2|\}$. This function is piecewise smooth. One could select a payoff function to be a smooth one, for example, the function $\sigma(x) = x_1^2 + x_2^2$. However, in this case, the construction of the value function turns out to be even more difficult.

Chapter 5
EXTREMAL AIMING

5.1. Programmed Maximin and Programmed Absorption

Various procedures for solving minimax and maximin problems of positional control are known in the theory of differential games. These procedures are based on a sound theoretical foundation for a sufficiently wide class of problems. Some of these were mentioned in preceding chapters. However, such universal procedures cannot be implemented in practice since the memory and computational requirements exceed the capacities of existing computers. Therefore it is important to select a type of problem for which it is possible to develop a specific efficient method of solution. This and the next two chapters are devoted to an investigation of these problems.

In this chapter we shall consider linear controlled systems. We shall study differential games with a fixed termination time and a convex pay-off function as well as approach problems with a convex target set. Regularity conditions will be formulated under which the programmed maximin will coincide with the value of a positional differential game. An efficient method for solving the problems under consideration will be constructed, namely the method of extremal aiming based on the solution and synthesis of auxiliary problems of programmed control.

In this section an informal discussion on the method of extremal aiming is presented. Next, for a linear differential game with a fixed termination time, a programmed maximin of the pay-off is determined and various formulas for calculating a programmed maximin will be obtained.

5.1.1. In §5.2 a positional strategy will be defined which is called a *strategy of extremal aiming* (not to be confused with the strategy extremal to a stable bridge). In order to give an intuitive idea about this strategy we consider the following example.

Let two mass points Y and Z be in motion in a plane. Their motion is described by the equations

$$(1) \qquad \ddot{y}_1 = u_1, \; \ddot{y}_2 = u_2, \; \ddot{z}_1 = v_1, \; \ddot{z}_2 = v_2.$$

Here (y_1,y_2) are coordinates of the point Y and (z_1,z_2) are coordinates of the point Z. One may assume that the mass of each point is equal to unity; in this case the vectors $(u_1,u_2)^T$ and $(v_1,v_2)^T$ can be interpreted as forces being applied by the first and second players to points Y and Z, respectively. It is assumed that the controlled actions satisfy the constraints

$$(2) \qquad \|u\| \leqslant \beta_1, \quad \|v\| \leqslant \beta_2 .$$

Consider a differential game with a fixed termination time $t = \theta$. We shall assume that the payoff in this game is the distance between points Y and Z at the time $t = \theta$. The first player attempts to minimize this distance while the second player seeks to maximize it.

To write equation (1) in a normal form we introduce the phase variables $y_3 = \dot{y}_1$, $y_4 = \dot{y}_2$, $z_3 = \dot{z}_1$, $z_4 = \dot{z}_2$. The equation (1) can then be rewritten as

$$(3) \qquad \dot{y}_1 = y_3, \; \dot{y}_2 = y_4, \; \dot{y}_3 = u_1, \; \dot{y}_4 = u_2,$$

$$(4) \qquad \dot{z}_1 = z_3, \; \dot{z}_2 = z_4, \; \dot{z}_3 = v_1, \; \dot{z}_4 = v_2.$$

We set $y = (y_1, ..., y_4)^T$, $z = (z_1, ..., z_4)^T$. The payoff functional will be determined by the equality

$$(5) \qquad \gamma(y(\cdot),z(\cdot)) \stackrel{\text{def}}{=} [(y_1(\theta) - z_1(\theta))^2 + (y_2(\theta) - z_2(\theta))^2]^{1/2}.$$

The strategies of the players will be the functions $(t,y,z) \to U(t,y,z)$ and $(t,y,z) \to V(t,y,z)$ which are determined for $t \leqslant \theta$, $y \in R^4$, $z \in R^4$ and satisfy the constraints $\|U(t,y,z)\| \leqslant \beta_1$, $\|V(t,y,z)\| \leqslant \beta_2$. We shall introduce auxiliary motions which will be utilized in the construction of optimal strategies U^0 and V^0.

Let a time $t_* < \theta$ and a vector $y_* \in R^4$ be given. The set $G_1(t_*,y_*)$ can be defined as a collection of points $(\overset{*}{y}_1,\overset{*}{y}_2)^T \in R^2$ for each of which there exists an admissible control $u(\cdot)$ which transfers the system (3) from the initial state $y(t_*) = y_*$ into the final state $y(\theta)$ such that $y_1(\theta) = \overset{*}{y}_1$, $y_2(\theta) = \overset{*}{y}_2$. The set $G_1(t_*,y_*)$ is called the domain of accessibility of the controlled system. Simply speaking, the domain of accessibility is the set of all possible positions of the point Y at time $t = \theta$. We note that we have defined the domain of accessibility in the subspace of the first two coordinates of the phase vector y. This is explained by the fact that the payoff function depends only on these components of the phase vector.

It is easy to write down explicit expressions that define the domain of accessibility $G_1(t_*,x_*)$. If y_* is a zero vector then $G_1(t_*,y_*)$ is a circle of radius $(\theta - t_*)^2\beta_1/2$ with center at the origin. One can hit the point g^* which is located on the boundary of this circle utilizing the constant control $u(t) = \beta_1 g^* \|g^*\|^{-1}$ $(t^* \leqslant t \leqslant \theta)$. If y_* is an arbitrary vector in R^4 then $G_1(t_*,y_*)$ will be a circle of the same radius but with center at point $c^{(1)}(t_*,y_*)$ with coordinates

$$c_1^{(1)}(t_*,y_*) = y_{*1} + (\theta - t_*)y_{*3}, \quad c_2^{(1)}(t_*,y_*) = y_{*2} + (\theta - t_*)y_{*4}$$

Note that $c^{(1)}(t_*,y_*)$ is the position in which the point Y finds itself at time $t = \theta$ under the zero control $u(t) = 0$ $(t_* \leqslant t \leqslant \theta)$. We thus obtain

$$G_1(t_*,y_*) = \{g \in R^2 : \|g - c^{(1)}(t_*,y_*)\| \leqslant (\theta - t_*)^2\, \beta_1/2\}.$$

Analogously we introduce the notion of the domain of accessibility $G_2(t_*,z_*)$ for the system (4) and obtain

$$G_2(t_*,z_*) = \{g \in R^2 : \|g - c^{(2)}(t_*,z_*)\| \leqslant (\theta - t_*)^2\beta_2/2\},$$

$$c_1^{(2)}(t_*,z_*) = z_{*1} + (\theta - t_*)z_{*3}, \quad c_2^{(2)}(t_*,z_*) = z_{*2} + (\theta - t_*)z_{*4}\,.$$

We now define the quantity

$$\epsilon_0(t,y,z) \overset{\text{def}}{=} \min\{\epsilon \geqslant 0 : \; G_2(t,z) \subset \bar{G}_1^{\epsilon}(t,y)\}. \tag{6}$$

Recall that the symbol $\bar{G}^{\epsilon}$ denotes the closed ϵ-neighborhood of the set G. In our case $\bar{G}_1^{\epsilon}(t,y)$ is a circle of radius $2^{-1}(\theta - t)^2\beta_1 + \epsilon$. Set

$$\begin{aligned} s(t,y,z) &\overset{\text{def}}{=} c^{(1)}(t,y) - c^{(2)}(t,z) \\ &= (y_1 - z_1 + (\theta - t)(y_3 - z_3),\; y_2 - z_2 + (\theta - t)(y_4 - z_4))^{\mathrm{T}}, \\ \epsilon_*(t,y,z) &\overset{\text{def}}{=} \|s(t,y,z\| + (\beta_2 - \beta_1)(\theta - t)^2/2. \end{aligned} \tag{7}$$

It follows from a simple geometric consideration that

$$\epsilon_0(t,y,z) = \begin{cases} \epsilon_*(t,y,z) & \text{for } \epsilon_*(t,y,z) \geqslant 0 \\ 0 & \text{for } \epsilon_*(t,y,z) \leqslant 0. \end{cases} \tag{8}$$

Consider strategies U^0 and V^0 defined by the equalities

$$U^0(t,y,z) \overset{\text{def}}{=} \begin{cases} -\beta_1 s(t,y,z)\;\|s(t,y,z)\|^{-1} & \text{for } \epsilon_0(t,y,z) > 0, \\ (0,0)^{\mathrm{T}} & \text{for } \epsilon_0(t,y,z) = 0. \end{cases} \tag{9}$$

$$(10) \qquad V^0(t,y,z) \overset{\text{def}}{=} \begin{cases} -\beta_2 s(t,y,z)\, \| s(t,y,z) \|^{-1} & \text{for } \epsilon_0(t,y,z) > 0, \\ (0,0)^{\mathrm{T}} & \text{for } \epsilon_0(t,y,z) = 0. \end{cases}$$

In the example under consideration U^0 and V^0 are strategies of extremal aiming. A pair of these strategies forms a saddle point of the differential game (3) – (5).

We shall describe geometrical considerations which explain the choice of these strategies. Assume that $\epsilon_0 = \epsilon_0(t_*,y_*,z_*) > 0$. In Fig. 5.1.1 the full circle is $G_2 = G_2(t_*,z_*)$ and parts of the circumference surrounding the circle $G_1 = G_1(t_*,y_*)$ are presented, the dashed and solid curves representing parts of the circles $G_1^{\bar{\epsilon}_0}$ and G_1, respectively. The boundaries of the circles $G_1^{\bar{\epsilon}_0}$ and G_2 have a common point q^0. Among the points of the circle G_2, the point q^0 is the farthest from the region $_1$. The distance from point q^0 to the domain G_1 equals ϵ_0. If the second player selects a constant control $v(t) = V^0(t_*,y_*,z_*)$ $(t_* \leqslant t \leqslant \theta)$ then at time $t = \theta$ the point Z will be in position q^0. Under this control the second player assures that the inequality $\gamma(y(\cdot),z(\cdot)) \geqslant \epsilon_0$ is fulfilled. Indeed, different positions of point Y at the last instant of time $t = \theta$ correspond to different controls or strategies of the first player. However, all these positions are located in the circle G_1. Therefore the distance between points $Y \in G_1$ and $Z = q^0$ cannot be less than the number ϵ_0.

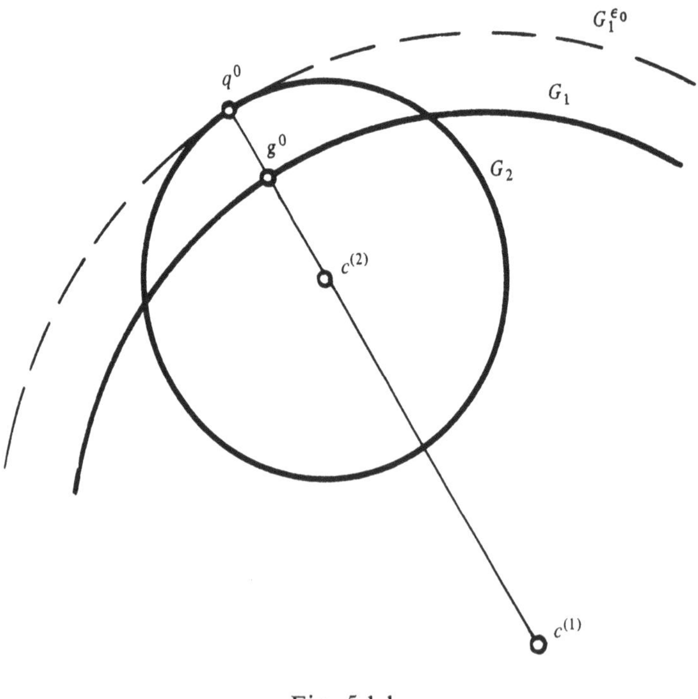

Fig. 5.1.1.

We shall show that the strategy U^0 (9) assures that the opposite bound $\gamma(y(\cdot), z(\cdot)) \leqslant \epsilon_0$ is fulfilled. It will thus be shown that the quantity $\epsilon_0(t_*, y_*, z_*)$ is the value of the differential game (3) – (5) defined for the initial position (t_*, y_*, z_*). We shall first examine the situation when the first player chooses the constant control $u(t) = U^0(t_*, y_*, z_*)$ $(t_* \leqslant t \leqslant \theta)$. It is easy to verify that at time $t = \theta$ the point Y will be in the position g^0 (cf. Fig. 5.1.1). Then the point g^0 satisfies the condition $\|g^0 - q^0\| = \mathrm{dist}(q^0, G_1)$. We also note that $\mathrm{dist}(q^0, G_1) = \max \mathrm{dist}(q, G_1)$ for $q \in G_2$. Thus at time $t = \theta$ under the control $u(t)$ indicated above, point Y will approach the point q^0 within the minimal possible distance. Therefore one can say that the strategy U^0 aims the motion of system (3) into the point q^0 -- the farthest one in region G_2 from the region G_1.

If the second player chooses the constant control $v(t) = V^0(t_*, y_*, z_*)$ $(t_* \leqslant t \leqslant \theta)$ and the first player aims at point q^0, then at time $t = \theta$ the points Z and Y will be at locations q^0 and g^0 and the equality $\gamma(y(\cdot), z(\cdot)) = \|q^0 - g^0\| = \epsilon_0$ will be valid. However, in a minimax problem posed to the first player one should provide for any possible behavior of the opponent -- the second player. We therefore assume that the first player acts using the feedback principle and implements the strategy of extremal aiming in a step-by-step procedure. Assume that the first player uses the control $u(t) = U^0(t_*, y_*, z_*)$ on the time interval $[t_*, t_* + \delta]$. Assume that the second player can select any measurable control $t \to v(t)$, $\|v(t)\| \leqslant \beta_2$. At time $t = t_* + \delta$ the phase states $y(t_* + \delta)$ and $z(t_* + \delta)$ of systems (3) and (4) are implemented. We define the corresponding regions of accessibility $G_1(t_* + \delta, y(t_* + \delta))$ and $G_2(t_* + \delta, z(t_* + \delta))$. It follows from the definitions of the regions of accessibility that

$$G_1(t_* + \delta, y(t_* + \delta)) \subset G_1(t_*, y_*),$$

$$G_2(t_* + \delta, z(t_* + \delta)) \subset G_2(t_*, y_*).$$

Moreover under the above indicated choice of control $u(t)$ the region of accessibility $G_1(t_* + \delta, y(t_* + \delta))$ will, as before, contain the point g^0. Utilizing this fact it can be shown that

(11) $$\epsilon_0(t_* + \delta, y(t_* + \delta), z(t_* + \delta)) - \epsilon_0(t_*, y_*, z_*) \leqslant \mu(\delta),$$

where $\mu(\delta)$ is a quantity of a higher order of smallness relative to δ.

At time $t = t_* + \delta$ the first player chooses a new control $u(t) = U^0(t_* + \delta, y(t_* + \delta), z(t_* + \delta))$ $(t_* + \delta \leqslant t < t_* + 2\delta)$ which aims the motion of the system into a new point q^0 defined for the regions of accessibility $G_1(t_* + \delta, y(t_* + \delta))$ and $G_2(t_* + \delta, z(t_* + \delta))$. This control assures that at the next step the estimate analogous to the estimate (11) is fulfilled and so on. As a result we have at time $t = \theta$ that $\epsilon_0(\theta, y(\theta), z(\theta)) \leqslant \epsilon_0(t_*, y_*, z_*) + \alpha(\delta)$, where $\alpha(\delta) \to 0$ as $\delta \to 0$. Observe that for $t = \theta$ the regions $G_1(\theta, y)$ and $G_2(\theta, z)$ contain only one point, (y_1, y_2) and (z_1, z_2) respectively. Therefore $\epsilon_0(\theta, y, z) = [(y_1 -$

$z_1)^2 + (y_2 - z_2)^2]^{1/2}$. Approaching the limit as $\delta \to 0$ we obtain that the strategy of extremal aiming U^0 assures that the estimate

(12) $$\gamma(y(\cdot),z(\cdot)) = \epsilon_0(\theta,y(\theta),z(\theta)) \leqslant \epsilon_0(t_*,y_*,z_*)$$

is fulfilled. It was assumed in the above arguments that function ϵ_0 (6) takes on positive values. If, however, $\epsilon_0(t,y,z) = 0$ then $U^0(t,y,z)$ can be chosen arbitrarily (in (9) it was assumed that $U^0(t,y,z) = (0,0)^T$) which does not affect the final conclusions.

As mentioned before, the constant control $v(t) = V^0(t_*,y_*,z_*)$ $(t_* \leqslant t \leqslant \theta)$ assures that the estimate $\gamma(y(\cdot), z(\cdot)) \geqslant \epsilon_0(t_*,y_*,z_*)$ is fulfilled. It is easy to verify that this inequality remains valid when the positional strategy V^0 (10) is implemented in a step-by-step procedure. We thus conclude that the pair of strategies U^0 (9) and V^0 (10) forms a saddle point of the differential game (3) – (5) under consideration and the quantity ϵ_0 (6) is the value of this game.

Strategies of extremal aiming U^0 and V^0 (9), (10) were defined above based on geometric considerations. We note that these strategies can also be defined from the following analytic considerations. In the region where $\epsilon_0(t,y,z) > 0$ the function ϵ_0 is continuously differentiable and satisfies the main equation. We define for $\epsilon_0(t,y,z) > 0$ strategies U^0 and V^0 in accordance with (4.1.3) and (4.1.4). In the region where $\epsilon_0(t,y,z) = 0$ functions U^0 and V^0 can be selected arbitrarily; in particular it can be assumed that $U^0(t,y,z) = (0,0)^T$, $V^0(t,y,z) = (0,0)^T$. These constructions yield formulas (9), (10).

It is important to keep in mind that the quantity $\epsilon_0(t_*,y_*,z_*)$ is a *programmed maximin* of the payoff, i.e.

(13) $$\epsilon_0(t_*,y_*,z_*) = \max_{v(\cdot)} \min_{u(\cdot)} r(y(\theta,t_*,y_*,u(\cdot)),z(\theta,t_*,z_*,v(\cdot))$$

where $r(y,z) = [(y_1 - z_1)^2 + (y_2 - z_2)^2]^{1/2}$.
Here the maximum and the minimum are computed over all programmed controls $u(\cdot)$ and $v(\cdot)$ which are identified with measurable functions satisfying constraints (2). The fact that formulas (6) and (13) coincide follows from the definition of regions of accessibility. The quantity $\epsilon_0(t_*,y_*,z_*)$ defined by relation (6) or equivalently by relation (13) plays the role of the forecasted error. This forecast is formed based on a solution of the auxiliary problems of programmed control. In accordance with the rule of extremal aiming described above, the control for the first player is chosen in such a manner that the forecast does not become worse, more precisely, an increase in the forecasted error will be minimal.

In the general case, the definition of strategies of extremal aiming will be presented in Section 5.2. Therein conditions are stated under which extremal aiming can be used. In subsections 5.1.2 – 5.1.5 a programmed maximin of a convex payoff function is defined and some of its properties are considered.

5.1.2. Consider a linear differential game in which the motion of the controlled system is described by the equation

$$(14) \qquad \dot{x}(t) = A(t)x(t) + B(t)u(t) + C(t)v(t).$$

As usual $x \in R^n$, and $A(t)$, $B(t)$, and $C(t)$ are $n \times n$, $n \times p$, and $n \times q$ matrices. It is assumed that the entries of these matrices depend continuously on parameter t. Controls of players satisfy the bounds $u(t) \in \mathcal{P}$, $v(t) \in \mathcal{Q}$ where $\mathcal{P}$ and $\mathcal{Q}$ are convex compact sets in R^p and R^q respectively. We note that the convexity of the sets $\mathcal{P}$ and $\mathcal{Q}$ is an inessential assumption introduced in order to simplify writing down certain relations. The payoff in the game under consideration is defined by the equality

$$(15) \qquad \gamma(x(\cdot)) \stackrel{\text{def}}{=} \sigma(x(\theta)).$$

Here θ is a fixed time and $\sigma : R^n \to R$ is a convex function which satisfies the Lipschitz condition

$$(16) \qquad |\sigma(x^{(1)}) - \sigma(x^{(2)})| \leqslant \lambda \| x^{(1)} - x^{(2)} \|, \quad x^{(i)} \in R^n, \ i = 1, 2.$$

Let $\sigma^* : R^n \to (-\infty, +\infty]$ be a function conjugate to the function σ. Function σ^* is defined by equality (A.3). It is known (cf. §A.2) that function σ^* is convex lower semi-continuous, the set $\mathfrak{L} \stackrel{\text{def}}{=} \operatorname{dom}\sigma^* \stackrel{\text{def}}{=} \{\ell \in R^n : \sigma^*(\ell) < \infty\}$ is bounded and equality (A.4) is valid.

We define a programmed maximin

$$(17) \qquad \omega_p(t_*,x_*) \stackrel{\text{def}}{=} \max_{v(\cdot)\in \mathbf{V}_p} \ \min_{u(\cdot)\in \mathbf{U}_p} \sigma(x(\theta,t_*,x_*, u(\cdot),v(\cdot))).$$

Here $\mathbf{U}_p$ and $\mathbf{V}_p$ are sets of all the measurable functions $u(\cdot) : T \to \mathcal{P}$ and $v(\cdot) : T \to \mathcal{Q}$. These functions are called *programmed controls* (*or open-loop controls*). The symbol $x(\cdot,t_*,x_*,u(\cdot),v(\cdot))$ denotes the solution $x(\cdot) : [t_*,\theta] \to R^n$ of equation (14) satisfying the initial condition $x(t_*) = x_*$. It is assumed that $t_* \in T$; in the calculations and formulation of results that follow, one can set either $T = (-\infty,\theta]$ or $T = [t_\#,\theta]$. As will be shown below both the maximum and the minimum are attained in (17).

We now define another more convenient relationship for calculating the programmed maximin (17). In accordance with formula (1.2.14) we have

$$x(\theta,t_*,x_*,u(\cdot),v(\cdot)) = \Phi(\theta,t_*)x_* + \int_{t_*}^{\theta} \Phi(\theta,t)B(t)u(t)dt$$
$$+ \int_{t_*}^{\theta} \Phi(\theta,t)C(t)v(t)dt,$$

where the matrix-function $t \to \Phi(\theta,t)$ satisfies equation (1.2.15). We set

(18)
$$G_1(t_*,\theta) \overset{\text{def}}{=} \{g = \int_{t_*}^{\theta} \Phi(\theta,t)B(t)u(t)dt : u(\cdot) \in \mathbf{U}_p\},$$
$$G_2(t_*,\theta) \overset{\text{def}}{=} \{g = \int_{t_*}^{\theta} \Phi(\theta,t)C(t)v(t)dt : v(\cdot) \in \mathbf{V}_p\}.$$

Note that $G_1(t_*,\theta)$ is the region of accessibility of the controlled system (14) for $v(t) = 0$ $(t \in T)$ and $x(t_*) = 0$ while $G_2(t_*,\theta)$ is the region of accessibility of the system (14) for $u(t) = 0$ $(t \in T)$ and $x(t_*) = 0$. It is known that $G_1(t_*,\theta)$ and $G_2(t_*,\theta)$ are convex compact sets in R^n (cf. for example, [39, 45]). Utilizing the notation introduced above, relation (17) can be rewritten as

$$\omega_p(t,x) = \max_{g_2 \in G_2(t,\theta)} \min_{g_1 \in G_1(t,\theta)} \sigma(\Phi(\theta,t)x + g_1 + g_2).$$

Function $(g_1,g_2) \to (\sigma(\Phi(\theta,t)x + g_1 + g_2)$ is continuous (it even satisfies the Lipschitz condition), therefore the maximum and the minimum are attained here. Consequently the maximum and the minimum are also attained in relation (17).

According to (A.4) we have

$$\sigma(x) = \sup_{\ell}[\langle \ell,x\rangle - \sigma^*(\ell)].$$

Therefore

(19)
$$\omega_p(t,x) = \max_{g_2} \min_{g_1} \sup_{\ell} [\langle \ell,\Phi(\theta,t)x\rangle + \langle \ell,g_1 + g_2\rangle - \sigma^*(\ell)]$$
$$\text{for } g_2 \in G_2(t,\theta),\ g_1 \in G_1(t,\theta),\ \ell \in \operatorname{dom}\sigma^*.$$

The expression which appears in (19) in square brackets will be denoted by $\chi(t,x,\ell,g_1,g_2)$. The function $\ell \to \chi(t,x,\ell,g_1,g_2)$ is concave on the convex set $\operatorname{dom}\sigma^*$ and the function $g_1 \to \chi(t,x,\ell,g_1,g_2)$ is linear on the convex compact set $G_1(t,\theta)$. By the minimax theorem (cf. for example [21, 34, 55]) the operations of *min* and *sup* in (19) can be interchanged. We thus obtain

(20)
$$\omega_p(t,x) = \max_{g_2} \sup_{\ell} \min_{g_1} \chi(t,x,\ell,g_1,g_2)$$
$$= \sup_{\ell} [\langle \ell,\Phi(\theta,t)x\rangle + \min_{g_1} \langle \ell,g_1\rangle + \max_{g_2} \langle \ell,g_2\rangle - \sigma^*(\ell)]$$

for $\ell \in \operatorname{dom}\sigma^*$, $g_1 \in G_1(t,\theta)$, $g_2 \in G_2(t,\theta)$.

Set

(21)
$$\kappa_1(\ell,\tau,\theta) = \min_{u \in P} \langle \ell,\Phi(\theta,\tau)B(\tau)u\rangle,$$
$$\kappa_2(\ell,\tau,\theta) = \max_{v \in Q} \langle \ell,\Phi(\theta,\tau)C(\tau)v\rangle,$$
$$\kappa(\ell,\tau,\theta) = \kappa_1(\ell,\tau,\theta) + \kappa_2(\ell,\tau,\theta).$$

Utilizing Theorem A.3.2 (the theorem on measurable selection) it can be shown that there exist controls $u_\ell(\cdot) \in \mathbf{U}_p$ and $v_\ell(\cdot) \in \mathbf{V}_p$ which satisfy the equalities

$$\langle \ell, \Phi(\theta,\tau)B(\tau)u_\ell(\tau)\rangle = \kappa_1(\ell,\tau,\theta),$$

$$\langle \ell, \Phi(\theta,\tau)C(\tau)v_\ell(\tau)\rangle = \kappa_2(\ell,\tau,\theta), \quad \tau \in T.$$

Therefore in accordance with (18) and (21) we obtain

$$\min_{g_1 \in G_1(t,\theta)} \langle \ell, g_1\rangle + \max_{g_2 \in G_2(t,\theta)} \langle \ell, g_2\rangle = \int_t^\theta \kappa(\ell,\tau,\theta)d\tau.$$

We introduce the following notation:

(22) $$\psi(\ell,t) \overset{\text{def}}{=} \sigma^*(\ell) - \int_t^\theta \kappa(\ell,\tau,\theta)d\tau.$$

Now relation (20) can be rewritten as

(23) $$\omega_p(t,x) = \max_{\ell \in \text{dom}\sigma^*} [\langle \ell, \Phi(\theta,t)x\rangle - \psi(\ell,t)].$$

Note that the maximum is actually attained in (23). To verify this consider the function $\varphi : T \times R^n \times R^n \to (-\infty,\infty]$ defined by the equality $\varphi(t,x,\ell) = \langle \ell, \Phi(\theta,t)x\rangle - \psi(\ell,t)$. Taking into account that the function φ^* in (22) is lower semicontinuous and the set $\text{dom}\sigma^*$ is bounded we obtain that the function φ satisfies all the conditions (a) – (c) stipulated in §4.3.2. Therefore, as shown in §4.3.2 the maximum in (23) is attained (cf. also, an analogous remark in §A.1).

We shall note several properties of functions of a programmed maximin. The following lemma is valid.

Lemma 5.1.1. *Function $\omega_p : T \times R^n \to R$ satisfies the Lipschitz condition in every bounded region $\mathcal{D} \subset T \times R^n$. For any $t \in T$ the function $\omega_p(t,\cdot)$ is convex.*

Proof. Set

$$\varphi_1(t,x,\ell) \overset{\text{def}}{=} \langle \ell, \Phi(\theta,t),x\rangle + \int_t^\theta \kappa(\ell,\tau,\theta)d\tau, \quad \varphi_2(\ell) = \sigma^*(\ell).$$

It is easy to verify that functions φ_1 and φ_2 satisfy the conditions (a) – (c) of Lemma 4.3.1. Therefore the function ω_p will be locally Lipschitzian.

In accordance with (22) and (23), $\omega_p(t,x) = \max\varphi(t,x,\ell)$ for $\ell \in \text{dom}\sigma^*$; here $x \to \varphi(t,x,\ell)$ is a linear function. We obtain that $\omega_p(t,\cdot)$ is an upper envelope of the family of linear functions $\varphi(t,\cdot,\ell)$ ($\ell \in \text{dom}\sigma^*$). Therefore $\omega_p(t,\cdot)$ will be a convex function, q.e.d.

5.1.3. Consider a particular case of the differential game (14), (15). Assume as above that the motion of the controlled system is described by equation (14) and the payoff functional is of the form

(24) $\gamma(x(\cdot)) \overset{\text{def}}{=} \operatorname{dist}(x(\theta), M)$

where M is a convex closed set in R^n. The function $x \to \operatorname{dist}(x, M)$ is convex and satisfies the Lipschitz condition (with the Lipschitz constant equal to 1), i.e. this function possesses the properties stated in the differential game (14), (15) for the function σ. Let $\varphi_M : R^n \to (-\infty,\infty]$ be the support function of the set M, i.e.

(25) $\varphi_M(\ell) \overset{\text{def}}{=} \sup_{y \in M} \langle \ell, y \rangle .$

It is known (cf., for example, [63]) that the distance to a convex set can be expressed by means of a support function as follows:

(26) $\operatorname{dist}(x, M) = \sup\{\langle \ell, x \rangle - \varphi_M(\ell) : \|\ell\| \leqslant 1\}.$

Define a programmed maximin

(27) $\epsilon_0(t_*, x_*) \overset{\text{def}}{=} \max_{v(\cdot) \in \mathbf{V}_p} \min_{u(\cdot) \in \mathbf{U}_p} \operatorname{dist}(x(\theta, t_*, x_*, u(\cdot), v(\cdot)), M).$

In order to obtain a more convenient relation for computing the quantity $\epsilon_0(t_*, x_*)$ we shall use equality (26). Set $\sigma(x) = \operatorname{dist}(x, M)$, $\sigma^*(\ell) = \varphi_M(\ell)$ for $\|\ell\| \leqslant 1, \sigma^*(\ell) = \infty$ for $\|\ell\| > 1$. It is easy to observe that using our notation the equality (26) becomes equality (A.4), which was utilized in the derivation of formula (23). Hence in the case under consideration equality (23) is valid with $\omega_p(t,x) = \epsilon_0(t,x)$ and the function σ^* defined in the manner indicated above. Observe here that

$$\operatorname{dom}\sigma^* = \operatorname{dom}\varphi_M \cap B \quad \text{where} \quad B \overset{\text{def}}{=} \{\ell \in R^n : \|\ell\| \leqslant 1\}.$$

Reverting to the initial notation we obtain

(28) $\epsilon_0(t,x) = \max\{\langle \ell, \Phi(\theta,t)x \rangle - \varphi(\ell,t) : \ell \in \operatorname{dom}\varphi_M \cap B\},$

(29) $\varphi(\ell,t) \overset{\text{def}}{=} \varphi_M(\ell) - \int_t^\theta \kappa(\ell,\tau,\theta)d\tau.$

It follows from (21), (25), and (29) that the function $\varphi(\cdot,t)$ is positive homogeneous, i.e. $\varphi(\alpha\ell,t) = \alpha\varphi(\ell,t)$ for $\alpha > 0$, $\ell \in R^n$. Therefore the set $\operatorname{dom}\varphi_M$ is convex. If $\epsilon_0(t,x) > 0$ then the maximum in (28) is attained for vectors with unit norm.

Assume now that the set M is of the form

(30) $M = \{y \in R^n : [y]_m \in M_*\},$

where $[y]_m$ is a vector consisting of the first m components of an n-dimensional vector y, M_* is a convex compact set in R^m. It follows from (25) and (30) that in this case $\operatorname{dom}\varphi_M = \{\ell \in R^n; \ell_i = 0, i = n - m + 1, ..., n\}$. Utilizing this fact, the equalities (28) and (29) can be rewritten as

(31)
$$\epsilon_0(t,x) = \max_{|s|\leq 1} [\langle s,[\Phi(\theta,t)]_m x\rangle - \varphi^{(m)}(s,t)],$$
$$\varphi^{(m)}(s,t) \overset{\text{def}}{=} \varphi_{M}^{(m)}(s) - \int_t^{\theta} \kappa^{(m)}(s,\tau,\theta)d\tau,$$

(32)
$$\varphi_M^{(m)}(s) \overset{\text{def}}{=} \max_{z\in M_*} \langle s,z\rangle,\ \kappa^{(m)}(s,\tau,\theta) \overset{\text{def}}{=} \kappa_1^{(m)}(s,\tau,\theta) + \kappa_2^{(m)}(s,\tau,\theta),$$
$$\kappa_1^{(m)}(s,\tau,\theta) \overset{\text{def}}{=} \min_{u\in P} \langle s,[\Phi(\theta,\tau)]_m B(\tau)u\rangle$$
$$\kappa_2^{(m)}(s,\tau,\theta) \overset{\text{def}}{=} \max_{v\in Q} \langle s,[\Phi(\theta,\tau)]_m C(\tau)v\rangle,$$

where s is an m-dimensional vector, $[\Phi(\theta,\tau)]_m$ is an $(m \times n)$-matrix consisting of the first m rows of the matrix $\Phi(\theta,\tau)$.

5.1.4. We now state two problems whose solutions are considered in the following sections.

Problem 5.1.1. *Let an initial position (t_0,x_0), a closed convex set $M \subset R^n$, and the instant of time $\theta > t_0$ be given. It is required to determine a positional strategy $U^0 : T \times R^n \to P$ such that for any motion $x(\cdot) \in X_1(t_0,x_0,U^0)$ the condition $x(\theta) \in M$ is fulfilled.*

Problem 5.1.2. *Let an initial position (t_0,x_0) and a closed convex set $M \subset R^n$ be given. It is required to determine a time $\theta \geq t_0$ and a positional strategy $U^0 : R \times R^n \to P$ such that $x(\theta) \in M$ for any motion $x(\cdot) \in X_1(t_0,x_0,U^0)$.*

The concepts of a set of programmed absorption and the time of programmed absorption defined below will be utilized in the course of our investigation of these problems.

Definition. *Let a convex, closed set M and an instant of time $t = \theta$ be given. The set of programmed absorption is, by definition, a collection of positions $(t_*,x_*) \subset T \times R^n$ possessing the following property: for any programmed control $v(\cdot) \in \mathbf{V}_p$ there exists a programmed control $u(\cdot) \in \mathbf{U}_p$ such that $x(\theta,t_*,x_*,u(\cdot),v(\cdot)) \in M$.*

Comparing this definition with the definition of a programmed maximin (27) it is easy to observe that the set

(33)
$$W_p \overset{\text{def}}{=} \{(t_*,x_*) \in T \times R^n : \epsilon_0(t_*,x_*) = 0\}$$

will be the set of programmed absorption.

Since the function $(t,x) \to \epsilon_0(t,x)$ is continuous (even locally-Lipschitzian) and the function $x \to \epsilon_0(t,x)$ is convex for any $t \in T$ the set W_p will be closed in $T \times R^n$ and for any $t \in T$ the set $W_p[t] \overset{\text{def}}{=} \{w \in R^n : (t,w) \in W_p\}$ will be convex. Observe that these properties can also be obtained directly from the initial definition of the set of programmed absorption.

Assume now that the number θ is not fixed but is a variable parameter. We introduce this parameter into the notation of the programmed maximin (27), i.e. in place of $\epsilon_0(t_*,x_*)$ we shall write $\epsilon_0(t_*,x_*,\theta)$.

Definition. *For a given initial position* (t_0,x_0) *the number*

$$(34) \qquad \theta_p(t_0,x_0) \overset{\text{def}}{=} \min\{\theta \geqslant t_0 : \epsilon_0(t_0,x_0,\theta) = 0\}$$

is called the time of programmed absorption.

5.1.5. Formulas for the programmed maximin obtained in §§5.1.2 and 5.1.3 for a linear system (14) can be carried over to the case of a linear system of the form

$$(35) \qquad \dot{x} = A(t)x + g(t,u,v)$$

where $g: T \times \mathcal{P} \times \mathcal{Q} \to R^n$ is a continuous function. Here the sets $\mathcal{P}$ and $\mathcal{Q}$ are not assumed to be convex. Instead it is required that the saddle point condition in a small game be fulfilled, i.e.

$$(36) \qquad \min_{u\in\mathcal{P}} \max_{v\in\mathcal{Q}} \langle \ell,g(t,u,v)\rangle = \max_{v\in\mathcal{Q}} \min_{v\in\mathcal{P}} \langle \ell,g(t,u,v)\rangle$$
$$(\ell \in R^n, \quad t \in T).$$

For the system (35) the programmed maximin ω_p and ϵ_0 can be computed using formulas (22), (23), (28), (29), and (31), (32) in which one should set

$$(37) \qquad \begin{aligned} \kappa(\ell,\tau,\theta) &= \min_{u\in\mathcal{P}} \max_{v\in\mathcal{Q}} \langle \ell,\Phi(\theta,\tau)g(\tau,u,v)\rangle \\ &= \max_{v\in\mathcal{Q}} \min_{u\in\mathcal{P}} \langle \ell,\Phi(\theta,\tau)g(\tau,u,v)\rangle, \end{aligned}$$

$$(38) \qquad \begin{aligned} \kappa^{(m)}(s,\tau,\theta) &= \min_{u\in\mathcal{P}} \max_{v\in\mathcal{Q}} \langle s,[\Phi(\theta,\tau)]_m g(\tau,u,v)\rangle \\ &= \max_{v\in\mathcal{Q}} \min_{u\in\mathcal{P}} \langle s,[\Phi(\theta,\tau)]_m g(\tau,u,v)\rangle. \end{aligned}$$

The interchangeability of the operations *min* and *max* in (37) and (38) follows from condition (36) and the well known equalities $\langle \ell,\Phi g\rangle = \langle \Phi^T\ell,g\rangle$, $\langle s,[\Phi]_m g\rangle = \langle [\Phi]_m^T s,g\rangle$ (recall that ^{T} is the transposition index).

For the system (35) the above mentioned formulas for computing the quantities $\omega_p(t,x)$ and $\epsilon_0(t,x)$ can be deduced in the same manner as in §§5.1.2 and 5.1.3, starting with the definitions of programmed maximins. We shall omit these derivations. Equalities (22), (23),

(28), (29), (31), and (32) can be utilized formally without attaching the actual meaning to the quantities $\omega_p(t,x)$ and $\epsilon_0(t,x)$.

5.2. Extremal Aiming in Linear Controlled Systems

Here we define strategies of extremal aiming and state conditions under which these strategies can be used. Whole classes of quite interesting problems satisfy these conditions. An important advantage of the method of extremal aiming is its efficiency, i.e., relatively few computations are required to implement this method.

5.2.1. We first describe strategies of extremal aiming for the differential game (5.1.14), (5.1.15). Let ω_p be the function of a programmed maximin defined by equality (5.1.17) or equivalently by formulas (5.1.27) and (5.1.23). We state conditions under which the function ω_p will coincide with the value function of the differential game (5.1.14) and (5.1.15).

Define the set

$$(1) \qquad \mathfrak{L}_0(t,x) \overset{\text{def}}{=} \{\ell_0 \in \operatorname{dom}\sigma^* : \langle \ell_0, \Phi(\theta,t)x\rangle - \psi(\ell_0,t) = \omega_p(t,x)\}.$$

This set consists of vectors which yield a maximum in (5.1.23). As it was shown in §4.3.2 the set $\mathfrak{L}_0(t,x)$ is a non-empty compact set. We introduce the condition:

Condition 5.2.1. *For any $t \in T^0$ the function $\psi(\cdot,t)$ is strictly convex on the set $\mathfrak{L} = \operatorname{dom}\sigma^*$.*

Recall that the function $\varphi : \mathfrak{L} \to R$ is called strictly convex if for any $\ell_i \in \mathfrak{L}$ $(i = 1, 2)$, $\ell_1 \neq \ell_2$ and $\alpha \in (0,1)$ the inequality

$$\varphi((1-\alpha)\ell_1 + \alpha\ell_2) < (1-\alpha)\varphi(\ell_1) + \alpha\varphi(\ell_2)$$

is valid. As is well known, when condition 5.2.1 is fulfilled the set $\mathfrak{L}_0(t,x)$ will contain a unique element which we shall denote by $\ell_0(t,x)$. This proposition is valid for any choice of the point $(t,x) \in T^0 \times R^n$ (recall that either $T^0 = (-\infty, \theta)$ or $T^0 = (t_{\#}, \theta))$.

Definition. *Functions $U^0 : T \times R^n \to \mathcal{P}$ and $V^0 : T \times R^n \to \mathcal{Q}$ selected from the conditions*

$$(2) \qquad \begin{aligned} \langle \ell_0(t,x), \Phi(\theta,t)B(t)U^0(t,x)\rangle &= \min_{u\in\mathcal{P}} \langle \ell_0(t,x), \Phi(\theta,t)B(t)u\rangle \\ &\overset{\text{def}}{=} \kappa_1(\ell_0(t,x),t,\theta), \end{aligned}$$

$$(3)\qquad \langle \ell_0(t,x),\Phi(\theta,t)C(t)V^0(t,x)\rangle = \max_{v\in Q} \langle \ell_0(t,x),\Phi(\theta,t)C(t)v\rangle \overset{\text{def}}{=} \kappa_2(\ell_0(t,x),t,\theta)$$

are called strategies of extremal aiming.

Note that the vector $\ell_0(t,x)$ is defined for $t < \theta$; for $t = \theta$ the functions U^0 and V^0 can be selected arbitrarily. For $t \in T^0$ conditions (2) and (3) may define the functions U^0 and V^0 nonuniquely. The theorem presented below is valid for any functions U^0 and V^0 satisfying the conditions (2) and (3).

Before proceeding to the statement and proof of the above theorem, we shall discuss briefly how extremal aiming is implemented in a step-by-step procedure. Let a certain partition $\Delta \overset{\text{def}}{=} \{t_0 = \tau_0 < \tau_1 < \ldots < \tau_{k+1} = \theta\}$ with a constant step $\delta = \tau_{i+1} - \tau_i$ ($i = 0, 1, \ldots, k$) be given. At time $t = \tau_i$ the phase state $x = x(\tau_i)$ is measured and the solution of problem (5.1.23) starts. As a result of solving this problem the maximizing vector $\ell_0(\tau_i,x(\tau_i))$ is determined. Next the control $U^0(\tau_i,x(\tau_i))$ is determined from condition (2). These problems do not require large amounts of computation. Note that for computing the vector $\ell_0(\tau_i,x(\tau_i))$, the vector $\ell_0(\tau_{i-1},x(\tau_{i-1}))$ determined on the preceding step can be used as an initial approximation. Let the above indicated calculations be implemented during time δ_*. At time $\tau_i + \delta_*$ the control $U^0(\tau_i,x(\tau_i))$ will be determined which is then implemented in the system (5.1.14) on the interval $[\tau_i + \delta_*,\tau_{i+1} + \delta_*)$. Observe that on the time interval $[\tau_{i+1},\tau_{i+1} + \delta_*)$ the calculation of the new control $U^0(\tau_{i+1},x(\tau_{i+1}))$ is carried out.

Clearly this procedure can be implemented provided the time required for computing these problems does not exceed the step $\delta = \tau_{i+1} - \tau_i$. If, however, $\delta_* > \delta$ then one can calculate the function $(t,x) \to \ell_0(t,x)$ or construct the synthesis $(t,x) \to U^0(t,x)$ in advance, in the region of relevance to our problem.

This method of control turns out to be stable relative to the delay δ_* due to the time spent on computing and to the errors in the definition of the vector $\ell_0(\tau_i,x(\tau_i))$ and the control $U^0(\tau_i,x(\tau_i))$. It is important to emphasize that in order to determine control it is not necessary to compute the value $\omega_p(\tau_i,x(\tau_i))$ and moreso it is not required to reconstruct the gradient of the function ω_p based on its values in the neighborhood of the position $(\tau_i,x(\tau_i))$; it is sufficient to determine vector $\ell_\epsilon(\tau_i,x(\tau_i))$ which gives an approximate solution of problem (5.1.23).

Theorem 5.2.1. *Let conditions* 5.2.1 *be fulfilled. Then the function of the programmed maximin* ω_p *defined by the equalities* (5.1.22) *and* (5.1.23) *coincides with the value function of the differential game* (5.14), (5.15) *on the whole space* $T \times R^n$. *For any initial position* $(t_0,x_0) \in T \times R^n$ *the pair of strategies of extremal aiming* U^0 *and* V^0 *forms a saddle point of the differential game* (5.1.14), (5.1.15).

Proof. Under the condition 5.2.1 the function $(t,x) \to \ell_0(t,x)$ is continuous in the region $T^0 \times R^n$. This assertion is easily proved by contradiction. Assuming that the function ℓ_0 is not continuous we obtain that the set $\mathfrak{L}_0$ may contain more than one element.

Utilizing the continuity of the function ℓ_0 we prove that the function ω_p is continuously differentiable in the region $T^0 \times R^n$. Let $(t,x) \in T^0 \times R^n$, $(t+\delta,x) \in T^0 \times R^n$ for all $\delta \in (-\delta_0,\delta_0)$. Set $\ell_* \overset{\text{def}}{=} \ell_0(t,x)$, $\ell_\delta \overset{\text{def}}{=} \ell_0(t+\delta,x)$. According to (5.1.22) and (5.1.23) we have

$$\omega_p(t,x) = \langle \ell_*,\Phi(\theta,t)x\rangle + \int_t^\theta \kappa(\ell_*,\tau,\theta)d\tau - \sigma^*(\ell_*)$$

$$\geqslant \langle \ell_\delta,\Phi(\theta,t)x\rangle + \int_t^\theta \kappa(\ell_\delta,\tau,\theta)d\tau - \sigma^*(\ell_\delta),$$

$$\omega_p(t+\delta,x) = \langle \ell_\delta,\Phi(\theta,t+\delta)x\rangle + \int_{t+\delta}^\theta \kappa(\ell_\delta,\tau,\theta)d\tau - \sigma^*(\ell_\delta)$$

$$\geqslant \langle \ell_*,\Phi(\theta,t+\delta)x\rangle + \int_{t+\delta}^\theta \kappa(\ell_*,\tau,\theta)d\tau - \sigma^*(\ell_*).$$

Therefore

$$\omega_p(t+\delta,x) - \omega_p(t,x) \geqslant \langle \ell_*,\Phi(\theta,t+\delta)x\rangle - \langle \ell_*,\Phi(\theta,t)x\rangle - \int_t^{t+\delta} \kappa(\ell_*,\tau,\theta)d\tau,$$

$$\omega_p(t+\delta,x) - \omega_p(t,x) \leqslant \langle \ell_\delta,\Phi(\theta,t+\delta)\rangle - \langle \ell_\delta,\Phi(\theta,t)x\rangle - \int_t^{t+\delta} \kappa(\ell_\delta,\tau,\theta)d\tau.$$

The matrix-function $t \to \Phi(\theta,t)$ satisfies the equation (1.2.15), therefore

$$\langle \ell,\Phi(\theta,t+\delta)x\rangle - \langle \ell,\Phi(\theta,t)x\rangle = -\delta\langle \ell,\Phi(\theta,t)A(t)x\rangle + \delta\alpha_1(\ell,\delta).$$

The function $\kappa(\ell,\cdot,\theta)$ is continuous, therefore

$$\int_t^{t+\delta} \kappa(\ell,\tau,\theta)d\tau = \delta\kappa(\ell,t,\theta) + \delta\alpha_2(\ell,\delta).$$

Here $|\alpha_i(\ell,\delta)| \leqslant \alpha(\delta)$ $(i = 1, 2)$ for any ℓ belonging to the bounded set $\mathfrak{L} = \operatorname{dom}\sigma^*$ and $\alpha(\delta) \to 0$ as $\delta \to 0$. Taking into account that $\ell_\delta \to \ell_* = \ell_0(t,x)$ for $\delta \to 0$ we obtain

$$\frac{\partial\omega_p(t,x)}{\partial t} = -\langle \ell_0(t,x),\Phi(\theta,t)A(t)x\rangle - \kappa(\ell_0(t,x),t,\theta). \tag{4}$$

We define the derivative $\partial\omega_p(t,x)/\partial x$. Let

$$(t,x) \in T^0 \times R^n,\ h \in R^n,\quad \ell_* \overset{\text{def}}{=} \ell_0(t,x), \ell_h \overset{\text{def}}{=} \ell_0(t,x+h).$$

In accordance with (5.1.23) we have

$$\omega_p(t,x+h) - \omega_p(t,x) \geqslant \langle \ell_*, \Phi(\theta,t)h \rangle,$$

$$\omega_p(t,x+h) - \omega_p(t,x) \leqslant \langle \ell_h, \Phi(\theta,t)h \rangle.$$

Taking into account that $\lim \ell_h = \ell_*$ for $\|h\| \to 0$ we obtain that a linear form $h \to \langle \ell_*, \Phi(\theta,t)h \rangle$ satisfies the relation

$$\lim_{\|h\|\to 0} |\omega_p(t,x+h) - \omega_p(t,x) - \langle \ell_*, \Phi(\theta,t)h \rangle| \, \|h\|^{-1} = 0.$$

Consequently,

(5) $$\frac{\partial \omega_p(t,x)}{\partial x} = \ell_0^T(t,x)\Phi(\theta,t).$$

Here we utilize the equality

(6) $$\langle \ell, \Phi h \rangle = \ell^T \Phi h \quad (\ell \in R^n, \ h \in R^n),$$

where as before, the symbol ^{T} denotes transposition (recall that in accordance with the convention accepted above the derivative $\partial\omega_p(t,x)/\partial x$ should be a row-vector).

It follows from equalities (4) and (5) that in the region $T^0 \times R^n$ the derivatives of the function ω_p are continuous. From (5), (6), and (5.1.21) we have

$$\min_{u \in P} \frac{\partial \omega_p(t,x)}{\partial x} B(t)u = \kappa_1(\ell_0(t,x),t,\theta),$$

$$\max_{v \in Q} \frac{\partial \omega_p(t,x)}{\partial x} C(t)v = \kappa_2(\ell_0(t,x),t,\theta),$$

$$\kappa_1(\ell,t,\theta) + \kappa_2(\ell,t,\theta) = \kappa(\ell,t,\theta).$$

Therefore it follows from (4) – (6) and (5.1.14) that in the region $T^0 \times R^n$ the function ω_p satisfies the main equation

(7) $$\frac{\partial \omega_p(t,x)}{\partial t} + \frac{\partial \omega_p(t,x)}{\partial x} A(t)x + \min_{u \in P} \frac{\partial \omega_p(t,x)}{\partial x} B(t)u + \max_{v \in Q} \frac{\partial \omega_p(t,x)}{\partial x} C(t)v = 0.$$

Directly from (5.1.17) we have the boundary condition $\omega_p(\theta,x) = \sigma(x)$.

Consequently, in accordance with Theorem 4.1.1, the function ω_p coincides with the value function of the differential game (5.1.14) and (5.1.15) on the whole space $T \times R^n$. Next it follows from (5) that for the system (5.1.14) under consideration conditions (4.1.3) and (4.1.4) become conditions (2) and (3). Therefore, in accordance with Theorem 4.1.1 for any initial position $(t_0,x_0) \in T \times R^n$ the pair of strategies of extremal aiming forms a saddle point of the differential game (5.1.14) and (5.1.15). Theorem 5.2.1 is thus proved.

Remark. In the case under consideration the value of the positional differential game coincides with the programmed maximin (5.1.17). Hence for any initial position $(t_0,x_0) \in T \times R^n$ the saddle point is formed not only by the pair of positional strategies U^0 and V^0, but also by the pair $(U^0, v^0(\cdot))$, where U^0 is a positional strategy of extremal aiming, $v^0(\cdot) \in \mathbf{V}_p$ is a programmed control selected from the condition

$$(8) \quad \min_{u(\cdot)\in \mathbf{U}_p} \sigma(x(\theta,t_0,x_0,u(\cdot),v^0(\cdot))) = \max_{v(\cdot)\in \mathbf{V}_p} \min_{u(\cdot)\in \mathbf{U}_p} \sigma(x(\theta,t_0,x_0,u(\cdot),v(\cdot))).$$

It follows from the proof of Theorem 5.2.1 that Condition 5.2.1 can be relaxed and replaced by the following condition.

Condition 5.2.2. *For any position $(t,x) \in \mathcal{E}$ the set $\mathfrak{L}_0(t,x)$ contains the unique element $\mathfrak{l}_0(t,x)$. Here*

$$(9) \qquad \mathcal{E} \overset{\text{def}}{=} \{(t,x) \in T^0 \times R^n : a < \omega_p(t,x) < b\}$$

where $-\infty \leqslant a < b \leqslant \infty$.

When Condition 5.2.1 is replaced by 5.2.2 the definition of a strategy of extremal aiming should be defined more exactly as follows. Assume that conditions (2) and (3) define functions U^0 and V^0 in the region $\mathcal{E}$. Outside of this region functions U^0 and V^0 can be chosen arbitrarily.

In the statement of the Theorem 5.2.2 presented below, the following notation is utilized:

$$\mathcal{E}_* \overset{\text{def}}{=} \{(t,x) \in T \times R^n : a \leqslant \omega_p(t,x) < b\},$$

$$\mathcal{E}^* \overset{\text{def}}{=} \{(t,x) \in T \times R^n : a < \omega_p(t,x) \leqslant b\}.$$

Theorem 5.2.2. *Let Condition 5.2.2 be fulfilled. If $(t_0,x_0) \in \mathcal{E}_*$, then*

$$(10) \qquad \sigma(x(\theta)) \leqslant \omega_p(t_0,x_0) \quad \text{for all} \quad x(\cdot) \in X_1(t_0,x_0,U^0).$$

If $(t_0,x_0) \in \mathcal{E}^*$, then

$$(11) \qquad \sigma(x(\theta)) \geqslant \omega_p(t_0,x_0) \quad \text{for all} \quad x(\cdot) \in X_2(t_0,x_0,V^0).$$

Here U^0 and V^0 are strategies of extremal aiming. In the region cl$\mathcal{E}$ *the function of the programmed maximin ω_p coincides with the value function of the differential game* (5.1.14) *and* (5.1.15). *For the initial position $(t_0,x_0) \in \mathcal{E}$ the pair of strategies U^0,V^0 forms a saddle point of the differential game* (5.1.14) and (5.1.15).

It follows from the proof of Theorem 5.2.1 that when the condition 5.2.2 is fulfilled, the function ω_p is continuously differentiable in the region $\mathcal{E}$ and satisfies the main equation in this region. Therefore the validity of Theorem 5.2.2 follows from Theorem 4.1.1.

5.2.2. Let the payoff function be of the form $\sigma(x) = \text{dist}(x, M)$ -- the distance from the point x to the convex closed set $M \subset R^n$. In this case the programmed maximin is defined by equalities (5.1.28) and (5.1.29) and is denoted by $\epsilon_0(t,x)$. Now set $\mathcal{L}_0(t,x)$ to be the set of maximizing vectors in (5.1.28), i.e.

$$(12) \qquad \mathcal{L}_0(t,x) \overset{\text{def}}{=} \{\ell_0 \in \text{dom}\varphi_M \cap B : \langle \ell_0, \Phi(\theta,t)x\rangle - \varphi(\ell_0,t) = \epsilon_0(t,x)\},$$

and recall that $B \overset{\text{def}}{=} \{\ell \in R^n : \|\ell\| \leqslant 1\}$. According to (5.1.21), (5.1.25), and (5.1.29) for any $t \in T$ the function $\varphi(\cdot,t)$ is positive homogeneous and $\text{dom}\varphi(\cdot,t) = \text{dom}\varphi_M$ is a convex cone. Introduce the following:

Condition 5.2.3. *For any $t \in T$ the function $\varphi(\cdot,t)$ (5.1.29) is convex.*

The role of this condition is clarified by the following assertion.

Lemma 5.2.1. *Let Condition 5.2.3 be fulfilled. Then for any position $(t,x) \in \mathcal{E}^+$ the set $\mathcal{L}_0(t,x)$ defined in (12) contains a unique element $\ell_0(t,x)$. Here*

$$(13) \qquad \mathcal{E}^+ \overset{\text{def}}{=} \{(t,x) \in T^0 \times R^n : \epsilon_0(t,x) > 0\}.$$

Proof. Assume the contrary. Let $(t,x) \in T^0 \times R^n$, $\epsilon_0(t,x) > 0$, $\ell_i \in \mathcal{L}_0(t,x)$ $(i = 1,2)$, $\ell_1 \neq \ell_2$. Consider the vector $\ell_* = (1/2)\ell_1 + (1/2)\ell_2$. In accordance with (12) we have $\ell_i \in \text{dom}\varphi_M \cap B$ $(i = 1, 2)$, therefore $\ell_* \in \text{dom}\varphi_M \cap B$. It follows from the definition of the set $\mathcal{L}_0(t,x)$ (cf. (12)) and Condition 5.2.3 that

$$(14) \qquad \begin{aligned} \langle \ell_*, \Phi(\theta,t)x\rangle - \varphi(\ell_*,t) \geqslant 1/2[\langle \ell_1, \Phi(\theta,t)x\rangle \\ - \varphi(\ell_1,t) + \langle \ell_2, \Phi(\theta,t)x\rangle - \varphi(\ell_2,t)] = \epsilon_0(t,x) > 0. \end{aligned}$$

Since the function $\varphi(\cdot,t)$ is positive homogeneous, this inequality implies that $\|\ell_*\| > 0$. Taking into account that $\|\ell_i\| \leqslant 1$ $(i = 1, 2)$ we obtain $\|\ell_*\| \in (0,1)$. Define the vector $\ell^* = \lambda\ell_*$, where $\lambda = \|\ell_*\|^{-1} > 1$. It follows from positive homogeneity of the function $\varphi(\cdot,t)$ and inequality (14) that

$$\langle \ell^*, \Phi(\theta,t)x\rangle - \varphi(\ell^*,t) = \lambda[\langle \ell_*, \Phi(\theta,t)x\rangle - \varphi(\ell_*,t)] \geqslant \lambda\epsilon_0(t,x) > \epsilon_0(t,x).$$

Here $\ell^* \in \text{dom}\varphi_M \cap B$. We thus obtain a contradiction to relation (5.1.28). Lemma 5.2.1 is thus proved.

We shall note without proof that the converse assertion is also valid: for any position $(t,x) \in \mathcal{E}^+$ let the set $\mathcal{L}_0(t,x)$ consist of a single element, then Condition 5.2.3 is fulfilled. We now utilize Theorem 5.2.2. Assume that Condition 5.2.3 is fulfilled. Then in accordance with

Lemma 5.2.1, Condition 5.2.2 will be fulfilled with $\mathcal{E} = \mathcal{E}^+$ and $\omega_p = \epsilon_0$. Note the dist(x, $\mathcal{M}$) ≥ 0 and $\epsilon_0(t,x) \geq 0$ therefore $\mathcal{E}_* \overset{def}{=} \{(t,x) \in T^p \times R^n : \epsilon_0(t,x) \geq 0\} = T \times R^n$. We obtain the following assertion.

Theorem 5.2.3. *Let the condition 5.2.3 be fulfilled. Then, for any initial position $(t_0,x_0) \in T \times R^n$, the programmed maximin $\epsilon_0(t_0,x_0)$* (5.1.28) *coincides with the value of the positional differential game* (5.1.14), (5.1.24). *The pair of strategies of extremal aiming U^0 and V^0 defined by relations* (2) *and* (3) *forms a saddle point of the differential game* (5.1.14), (5.1.24).

We present a modification of Condition 5.2.3 and Theorem 5.2.3 for the case when the set $\mathcal{M}$ is of the form (5.1.30). As remarked in §5.1.3 in this case the dom$\varphi_{\mathcal{M}}$ coincides with the linear subspace $\{\ell \in R^n : \ell_i = 0,\ i = n - m + 1, \ldots, n\}$ and therefore Condition 5.2.3 becomes:

Condition 5.2.4. *For any $t \in T$ the function $\varphi^{(m)}(\cdot,t)\colon R^m \to R$ is convex (recall that the function $\varphi^{(m)}$ is defined in accordance with* (5.1.32)).

Define the set

$$(15)\quad S_0(t,x) \overset{def}{=} \{s_0 \in R^m : \|s_0\| \leq 1,\ \langle s_0, [\Phi(\theta,t)]_m x\rangle - \varphi^{(m)}(s_0,t) = \epsilon_0(t,x)$$

(this set consists of vectors which yield the maximum in (5.1.31)). If Condition 5.2.4 is fulfilled, then for any position $(t,x) \in \mathcal{E}^+$ (13) the set $S_0(t,x)$ contains a unique vector which will be denoted by $s_0(t,x)$. This assertion is proved in the same manner as Lemma 5.2.1. In the case under consideration, strategies of extremal aiming may be defined from the conditions

$$(16)\quad \begin{aligned}&\langle s_0(t,x),[\Phi(\theta,t)]_m B(t)U^0(t,x)\rangle\\ &= \min_{u\in\mathcal{P}} \langle s_0(t,x),[\Phi(\theta,t)]_m B(t)u\rangle \overset{def}{=} k_1^{(m)}(s_0(t,x),t,\theta),\end{aligned}$$

$$(17)\quad \begin{aligned}&\langle s_0(t,x),[\Phi(\theta,t)]_m C(t)V^0(t,x)\rangle\\ &= \max_{v\in\mathcal{Q}} \langle s_0(t,x),[\Phi(\theta,t)]_m C(t)v\rangle \overset{def}{=} \kappa_1^{(m)}(s_0(t,x),t,\theta).\end{aligned}$$

Recall that $[\Phi(\theta,t)]_m$ is an $(m \times n)$-matrix formed by the first m rows of the matrix $\Phi(\theta,t)$. Thus the following modification of Theorem 5.2.3 is valid.

Theorem 5.2.4. *Let the set $\mathcal{M}$ be of the form* (5.1.30) *and let condition* 5.2.4 *be fulfilled. Then, for any initial position $(t_0,x_0) \in T \times R^n$, the programmed maximin $\epsilon_0(t_0,x_0)$* (5.1.31) *coincides with the value of the positional differential game* (5.1.14), (5.1.24). *The pair of strategies of extremal* *aiming U^0 and V^0 defined by the relations* (16) *and* (17) *form a saddle point of the differential game under consideration.*

Under Condition 5.2.3 (or Condition 5.2.4) the function ϵ_0 is continuously differentiable in the region $\mathcal{E}^+$ (13). Its derivatives are defined by relations (4) and (5) in which $\partial\omega_p/\partial t$ and $\partial\omega_p/\partial x$ must be replaced by $\partial\epsilon_0/\partial t$ and $\partial\epsilon_0/\partial x$. These formulas are valid also in the particular case when the set M is of the form (5.1.30). However, taking into account that in this case $\ell_0(t,x) = s_{01}(t,x), ..., s_{0m}(t,x), 0, ..., 0)^T$, expressions for the derivatives of the function ϵ_0 (5.1.31) can be rewritten as

$$(18)\qquad \begin{aligned} \frac{\partial\epsilon_0(t,x)}{\partial t} &= -\langle s_0(t,x),[\Phi(\theta,t)]_m A(t)x\rangle - \kappa^{(m)}(s_0(t,x),t,\theta), \\ \frac{\partial\epsilon_0(t,x)}{\partial x} &= s_0^T(t,x)[\Phi(\theta,t)]_m \,. \end{aligned}$$

These formulas will be used in the next section.

Utilizing Theorems 5.2.3 and 5.2.4 one may arrive at sufficient conditions under which Problem 5.1.1 possesses a solution. Let Condition 5.2.3 (or Condition 5.2.4) be fulfilled. Then, in accordance with Theorem 5.2.3 (or Theorem 5.2.4), the strategy of extremal aiming U^0 assures that the estimate

$$\mathrm{dist}(x(\theta), M) \leqslant \epsilon_0(t_0,x_0) \quad \text{for all} \quad x(\cdot) \in X_1(t_0,x_0,U^0)$$

is fulfilled. If we assume that $\epsilon_0(t_0,x_0) = 0$, we then obtain that $\mathrm{dist}(x(\theta), M) = 0$ or equivalently $x(\theta) \in M$ for all $x(\cdot) \in X_1(t_0,x_0,U^0)$. Recall that equality $\epsilon_0(t_0,x_0,) = 0$ is equivalent to the inclusion $(t_0,x_0) \in W_p$ where W_p is the set of programmed absorption introduced in §5.1.4. We can thus state the following assertion.

Corollary 5.2.1. *Let Condition* 5.2.3 (*or Condition* 5.2.4 *provided the set* M *is of the form* (5.1.30)) *be fulfilled. Assume that the initial position* (t_0,x_0) *is contained in the set of programmed absorption* W_p (5.1.33). *Then the strategy of extremal aiming* U^0 *defined by relation* (2) (*or relation* (16)) *yields a solution of Problem* 5.1.1.

5.2.3. We note some particular cases when Conditions 5.2.3 and 5.2.4 are fulfilled.

Assume that for all $t \in T$ the function

$$\ell \to \kappa(\ell,t,\theta) \overset{\text{def}}{=} \min_{u\in P} \langle \ell,\Phi(\theta,t)B(t)u\rangle + \max_{v\in Q} \langle \ell,\Phi(\theta,t)C(t)v\rangle$$

is concave on a convex cone $\mathrm{dom}\varphi_M$. Then for all $t \in T$ function $\varphi(\cdot,t) : R^n \to (-\infty,\infty]$ will be convex, i.e. the condition 5.2.3 will be fulfilled (recall that the function φ is determined by relation (5.1.29)).

To verify this assertion we note that $\mathrm{dom}\varphi(\cdot,t) = \mathrm{dom}\varphi_M$. Therefore it is required to show that function $\varphi(\cdot,t)$ is convex on the cone $\mathrm{dom}\varphi_M$. Under the assumption stated above the function

$$\ell \to \int_t^\theta \kappa(\ell,\tau,\theta)d\tau \tag{19}$$

will be concave on $\mathrm{dom}\varphi_M$. The supporting function φ_M (5.1.25) is convex. Therefore the function $\varphi(\cdot,t) : \mathrm{dom}\varphi_M \to R$ which is defined in accordance with (5.1.29) as the difference of the convex support function φ_M and the concave function (19) will be convex.

Analogously the following assertion can be verified.

For all $t \in T$ let the function

$$s \to \kappa^{(m)}(s,t,\theta) \overset{\mathrm{def}}{=} \min_{u\in P} \max_{v\in Q} \langle s,[\Phi(\theta,t)]_m(B(t)u + C(t)v)\rangle \tag{20}$$

be concave on R^m. Then Condition 5.2.4 is fulfilled.

The condition of concavity of function $\kappa^{(m)}(\cdot,t,\theta)$ can be expresed geometrically. In order to do this, we introduce the following notation and concepts.

Let G_1 and G_2 be compact sets in R^m. Set

$$G_1 + G_2 \overset{\mathrm{def}}{=} \{g + g_2 \ : g_1 \in G_1, \ g_2 \in G_2\},$$

this set is called the algebraic sum of sets G_1 and G_2. Let $G = G_1 + G_2$; it is easy to verify that

$$\max_{g\in G} \langle s,g\rangle = \max_{g_1\in G_1} \langle s,g_1\rangle + \max_{g_2\in G_2} \langle s,g_2\rangle. \tag{21}$$

Define the sets

$$\begin{aligned} P^{(m)}(t,\theta) &\overset{\mathrm{def}}{=} \{ [\Phi(\theta,t)]_m B(t)u : u \in P\}, \\ Q^{(m)}(t,\theta) &\overset{\mathrm{def}}{=} \{ [\Phi(\theta,t)]_m C(t)v : v \in Q\}. \end{aligned} \tag{22}$$

Since the sets P and Q are convex compact sets in R^p and R^q respectively, the sets $P^{(m)}(t,\theta)$ and $Q^{(m)}(t,\theta)$ will be convex compact sets in R^m. Assume that for any $t \in T$ the equality

$$P^{(m)}(t,\theta) = Q^{(m)}(t,\theta) + \mathcal{H}^{(m)}(t,\theta) \tag{23}$$

is valid where $\mathcal{H}^{(m)}(t,\theta)$ is a set in R^m. One can verify that equality (23) defines the set $\mathcal{H}^{(m)}(t,\theta)$ uniquely and the set $\mathcal{H}^{(m)}(t,\theta)$ turns out to be a convex compact set. We shall show that (23) implies that function $\kappa^{(m)}(\cdot,t,\theta)$ (20) is concave.

According to (21), (23) we have

$$\begin{aligned} &\max \{\langle s,g\rangle : g \in P^{(m)}(t,\theta)\} \\ &= \max \{\langle s,g\rangle : g \in Q^{(m)}(t,\theta)\} + \max \{\langle s,g\rangle : g \in \mathcal{H}^{(m)}(t,\theta)\}. \end{aligned}$$

It follows from (22) and (5.1.32) that

$$\max\{\langle s,g\rangle : g \in \mathcal{P}^{(m)}(t,\theta)\} = \max_{u\in\mathcal{P}} \langle s, -[\Phi(\theta,t)]_m B(t)u\rangle$$

$$= -\min_{u\in\mathcal{P}} \langle s,[\Phi(\theta,t)]_m B(t)u\rangle = -\kappa_1^{(m)}(s,t,\theta),$$

$$\max\{\langle s,g\rangle : g \in \mathcal{Q}^{(m)}(t,\theta)\} = \max_{v\in\mathcal{Q}} \langle s, [\Phi(\theta,t)]_m C(t)v\rangle$$

$$\overset{\text{def}}{=} \kappa_2^{(m)}(s,t,\theta).$$

We obtain

$$\kappa_1^{(m)}(s,t,\theta) + \kappa_2^{(m)}(s,t,\theta) = -\max\{\langle s,g\rangle : g \in \mathcal{H}^{(m)}(t,\theta)\}. \tag{24}$$

It is obvious from this equality that the function $\kappa^{(m)}(\cdot,t,\theta) = \kappa_1^{(m)}(\cdot,t,\theta) + \kappa_2^{(m)}(\cdot,t,\theta)$ is concave. Note that the converse is also valid: if function $\kappa^{(m)}(\cdot,t,\theta)$ is concave then equality (23) is valid.

We thus prove that Condition 5.2.4 will be fulfilled if for all $t \in T$ the equality (23) is valid. Note that equality (23) is a generalization of the *"unitype" condition* of pursuer and evader to be discussed in the next section.

5.2.4. Results obtained for the controlled system (5.1.14) carry over to the case of a system of the form (5.1.35). For the controlled system (5.1.35) strategies of extremal aiming are determined from the conditions

$$\begin{aligned}
&\max_{v\in\mathcal{Q}} \langle \ell_0(t,x),\Phi(\theta,t)g(t,U^0(t,x),v)\rangle \\
&\quad = \min_{u\in\mathcal{P}} \max_{v\in\mathcal{Q}} \langle \ell_0(t,x),\Phi(\theta,t)g(t,u,v)\rangle, \\
&\min_{u\in\mathcal{P}} \langle \ell_0(t,x),\Phi(\theta,t)g(t,u,V^0(t,x,))\rangle \\
&\quad = \max_{v\in\mathcal{Q}} \min_{u\in\mathcal{P}} \langle \ell_0(t,x),\Phi(\theta,t)g(t,u,v)\rangle.
\end{aligned} \tag{25}$$

Here the vector $\ell_0(t,x)$ has the same meaning as in relations (2), (3), i.e. $\ell_0(t,x)$ is the unique maximizing vector in (5.1.23) or (5.1.28) where $\kappa(\ell,t,\theta)$ is defined in accordance with (5.1.37). Relation (25) is a modification of relations (2) and (3). Relations (16) and (17) are modified analogously. We are not going to present statements of theorems for a system of the form (5.1.35) since they are a verbatim repetition of the statements of Theorems 5.2.1 – 5.2.4.

5.2.5. We consider two simple examples which illustrate the content of this section. First we shall consider the differential game discussed in Section 5.1.1. In this game the motion of the controlled systems is described by equations (5.1.3) and (5.1.4) and the payoff functional is given by equality (5.1.5). We introduce a new phase vector $x = y - z \in R^4$, its motion in accordance with (5.1.3) and (5.1.4) satisfies the equation

(26) $\dot{x}_1 = x_3,\ \dot{x}_2 = x_4,\ \dot{x}_3 = u_1 - v_1,\ \dot{x}_4 = u_2 - v_2.$

It is assumed that the controls of the players satisfy the bounds (5.1.2). For the controlled system (26) the payoff functional is defined by equality

(27) $\gamma(x(\cdot)) = (x_1^2(\theta) + x_2^2(\theta))^{1/2}.$

Let $\omega^0(t_0,x_0)$ be the value of the differential game (26), (27) and let strategies $(t,x) \to U^0(t,x)$ and $(t,x) \to V^0(t,x)$ form a saddle point of the differential game (26), (27). It is easy to show that the pair of strategies $(t,y,z) \to U^0(t,y - z)$ and $(t,y,z) \to V^0(t,y - z)$ forms a saddle point of the differential game (5.1.3) – (5.1.5) and the quantity $\omega^0(t_0,y_0 - z_0)$ is the value of the differential game (5.1.3) – (5.1.5) defined for the initial position (t_0,y_0,z_0).

Note that $(x_1^2 + x_2^2)^{1/2} = \mathrm{dist}(x, M_*)$ where $M_* = \{x \in R^4 : [x]_2 = (x_1,x_2)^T = (0,0)\}^T$. We shall therefore try to utilize Theorem 5.2.4. In this example, $m = 2$,

$$B(t) = -C(t) = \begin{bmatrix} 0 \\ E \end{bmatrix}, \quad [\Phi(\theta,t)]_2 = (E, (\theta - t)E), \tag{28}$$

where $0 = \begin{bmatrix} 0 & 0 \\ 0 & 0 \end{bmatrix}, \qquad E = \begin{bmatrix} 1 & 0 \\ 0 & 1 \end{bmatrix},$

$$P = \{u \in R^2 : \|u\| \leqslant \beta_1\}, \quad Q = \{v \in R^2 : \|v\| \leqslant \beta_2\}.$$

In accordance with (5.1.32) we have

$$\begin{aligned} \kappa_1^{(2)}(s,\tau,\theta) &= -\beta_1(\theta - \tau)(s_1^2 + s_2^2)^{1/2}, \\ \kappa_2^{(2)}(s,\tau,\theta) &= \beta_2(\theta - \tau)(s_1^2 + s_2^2)^{1/2}, \\ \kappa^{(2)}(s,\tau,\theta) &= -(\beta_1 - \beta_2)(\theta - \tau)(s_1^2 + s_2^2)^{1/2}. \end{aligned} \tag{29}$$

Since $\beta_1 > \beta_2$ the function $\kappa(\cdot,\tau,\theta)$ is concave for all $\tau \leqslant \theta$. Therefore as shown in Section 5.2.3, Condition 5.2.4 is fulfilled.

We shall write the formula which defines the quantity $\epsilon_0(t,x)$. Taking (28) into account we have

$$\begin{aligned} &\langle s,[\Phi(\theta,t)]_2 x\rangle = s_1 g_1(t,x) + s_2 g_2(t,x), \\ &g_1(t,x) = x_1 + (\theta - t)x_3,\ \ g_2(t,x) = x_2 + (\theta - t)x_4. \end{aligned} \tag{30}$$

In our example $\varphi^{(2)}(s) = 0$ and

$$\int_t^\theta \kappa(s,\tau,\theta)d\tau = -(1/2)(\beta_1 - \beta_2)(\theta - t)^2(s_1^2 + s_2^2)^{1/2}.$$

Therefore according to (5.1.31) we obtain

(31) $\epsilon_0(t,x) = \max_{\|s\| \leqslant 1} [s_1 g_1(t,x) + s_2 g_2(t,x) - (1/2)(\beta_1 - \beta_2)(\theta - t)^2\|s\|].$

It is easy to observe that for $\epsilon_0(t,x) > 0$ the maximum in (31) is attained at the unique vector

(32)
$$s_0(t,x) = (g_1(t,x)/\|g(t,x)\|,\ g_2(t,x)/\|g(t,x)\|)^T,$$
$$\|g(t,x)\| = (g_1^2(t,x) + g_2^2(t,x))^{1/2}.$$

Therefore

(33)
$$\epsilon_0(t,x) = \begin{cases} \epsilon_*(t,x) & \text{for } \epsilon_*(t,x) \geqslant 0 \\ 0 & \text{for } \epsilon_*(t,x) \leqslant 0 \end{cases}$$

where $\epsilon_*(t,x) = \|g(t,x)\| - (1/2)(\beta_1 - \beta_2)(\theta - t)^2$.

For $\epsilon_0(t,x) > 0$ we define the optimal strategies of the players from the conditions (16), (17); for $\epsilon_0(t,x) = 0$ the values of $U^0(t,x)$ and $V^0(t,x)$ can be selected arbitrarily on the compact sets and We obtain

(34)
$$U^0(t,x) = \begin{cases} -\beta_1 s_0(t,x) & \text{for } \epsilon_0(t,x) > 0 \\ (0,0)^T & \text{for } \epsilon_0(t,x) = 0. \end{cases}$$

(35)
$$V^0(t,x) = \begin{cases} -\beta_2 s_0(t,x) & \text{for } \epsilon_0(t,x) > 0 \\ (0,0)^T & \text{for } \epsilon_0(t,x) = 0. \end{cases}$$

If in formulas (30), (32) – (35) we substitute $x = y - z$ then the quantity $\epsilon_0(t,x)$ (33) coincides with the quantity $\epsilon_0(t,y,z)$ (5.1.7), (5.1.8) and the strategies U^0 and V^0 (34), (35) coincide with the strategies U^0 and V^0 (5.1.9), (5.1.10).

Consider the second example. Let two controlled systems be given. The motion of the first system (a mass point) is described by equation (5.1.3) and the motion of the second system (an inertia-free point) is described by the equation

(36) $\dot{z}_1 = v_1,\ \dot{z}_2 = v_2.$

Controls of the players satisfy the constraints (5.1.2) as before, but it is not required that the inequality $\beta_1 > \beta_2$ be fulfilled. We assume that the payoff is defined by the equality

(37) $\gamma(y(\cdot),z(\cdot)) = \max\{[(y_1(\theta) - z_1(\theta))^2 + (y_2(\theta) - z_2(\theta))^2]^{1/2} - r, 0\}$

where r is a positive number.

We introduce new phase variables $x_1 = y_1 - z_1$, $x_2 = y_2 - z_2$, $x_3 = y_3$, $x_4 = y_4$. For these variables we have the following equations:

(38) $\dot{x}_1 = x_3 - v_1,\ \dot{x}_2 = x_4 - v_2,\ \dot{x}_3 = u_1,\ \dot{x}_4 = u_2.$

We assume that the new payoff functional is given by the equality

(39) $\gamma(x(\cdot)) = \mathrm{dist}(x(\theta), M),$

where

$$M = \{x \in R^4 : [x]_2 \in M_*\}, \quad M_* = \{w \in R^2 : \|w\| \leqslant r\}.$$

Analogous to the preceding example, the initial differential game (5.1.3), (36), (37) is equivalent to the transformed game (38), (39).

In our example

$$B(t) = \begin{bmatrix} 0 \\ E \end{bmatrix}, \quad C(t) = \begin{bmatrix} -E \\ 0 \end{bmatrix},$$

$$[\Phi(\theta,t)]_2 = (E, (\theta - t)E),\ t \leqslant \theta.$$

Therefore

$$\kappa_1^{(2)}(s,\tau,\theta) = -\beta_1(\theta - \tau)\|s\|, \quad \kappa_2^{(2)}(s,\tau,\theta) = \beta_2\|s\|,$$

$$\kappa^{(2)}(s,\tau,\theta) = [\beta_2 - \beta_1(\theta - \tau)]\|s\|.$$

Next, in accordance with (5.1.32), (39) we have $\varphi_M^{(2)}(s) = r\|s\|$,

(40) $\varphi^{(2)}(s,t) = [r + (1/2)\beta_1(\theta - t)^2 - \beta_2(\theta - t)]\|s\|.$

We shall assume that

(41) $r \geqslant \beta_2^2\,(2\beta_1)^{-1}.$

Under the condition (41) the estimate $r + (1/2)\beta_1(\theta - t)^2 - \beta_2(\theta - t) \geqslant 0$ is valid for all $t \leqslant \theta$, therefore the function $\varphi^{(2)}(\cdot,t)$ (40) will be convex, i.e. Condition 5.2.4 will be fulfilled.

According to (5.1.31) we obtain the formula

(42) $$\epsilon_0(t,x) = \max_{\|s\| \leqslant 1} [s_1 g_1(t,x) + s_2 g_2(t,x) - (r + (1/2)\beta_1(\theta - t)^2 - \beta_2(\theta - t))\|s\|]$$

for computing the quantity $\epsilon_0(t,x)$. Here $g_1(t,x)$ and $g_2(t,x)$ are defined by equalities (30). For $\epsilon_0(t,x) > 0$ the maximum in (42) is attained at the unique vector $s_0(t,x)$ (32). Optimal strategies U^0 and V^0 in the differential game (38), (39) are defined as in the previous example by relations (34), (35).

Note that in the second example the function $s \to \kappa^{(2)}(s,\tau,\theta) = [\beta_2 - \beta_1(\theta - \tau)]\|s\|$ for $\tau \in (\theta - \beta_2/\beta_1,\theta]$ is not concave, i.e. here the condition of uni-typedness stated in Section 5.2.3 is not fulfilled. Nevertheless, in this example Condition (5.2.4) is valid when inequality (41) is satisfied.

5.3. Extremal Aiming in Problems of Pursuit and Evasion

In the preceding section a differential game with a fixed termination time and a convex payoff function was considered. Optimal strategies in this game were determined within the framework of the method of extremal aiming. As is shown below, extremal aiming can be used for solving problems of pursuit and evasion.

5.3.1. First consider Problem 5.1.2 (cf. Section 5.1.4). We assume that in the phase space R^n the set $M = \{x \in R^n : [x]_m \in M_*\}$ is given, where M_* is a convex compact set in R^m. In the theorem presented below, conditions are stated under which Problem 5.1.2 can be solved utilizing the strategy of extremal aiming. First, before we state this theorem we note that in Section 5.2 the time of termination of the game $t = \theta$ was fixed. Here we assume that θ is a variable parameter. Introducing this parameter into the quantities (5.1.31) and (5.1.32) we write

$$(1)\qquad \varphi^{(m)}(s,t,\theta) \overset{\text{def}}{=} \varphi^{(m)}_M(s) - \int_t^\theta \kappa^{(m)}(s,\tau,\theta)d\tau,$$

$$(2)\qquad \epsilon_0(t,x,\theta) \overset{\text{def}}{=} \max_{\|s\|\leqslant 1} [\langle s,[\Phi(\theta,t)]_m x\rangle - \varphi^{(m)}(s,t,\theta)].$$

The following theorem is valid.

Theorem 5.3.1. *Let a number $\theta \geqslant t_0$ exist for a given initial position (t_0,x_0) such that $\epsilon_0(t_0,x_0,\theta) = 0$ and for any $t \in [t_0,\theta]$ the function $\varphi^{(m)}(\cdot,t,\theta) : R^m \to R$ is convex. Then the strategy of extremal aiming U^0 defined by equality* (5.2.6) *assures the fulfillment of inclusion $x(\theta) \in M$ for all $x(\cdot) \in X_1(t_0,x_0,U^0)$, i.e. the number θ and strategy U^0 will be a solution of Problem* 5.1.2.

This theorem follows directly from Theorem 5.2.4.

Problem 5.1.2 is somewhere between Problems 5.1.1, in which the encounter time with the set M is preassigned, and the pursuit problem which in the case under consideration is stated as follows:

Let $x(\cdot):[t_0,\infty) \to R^n$ be a continuous function. We set

$$T_0(x(\cdot)) \overset{\text{def}}{=} \{t \geqslant t_0 : x(t) \in M\},$$

$$\tau_0(x(\cdot)) \overset{\text{def}}{=} \begin{cases} \min\{t : t \in T_0(x(\cdot))\} & \text{for } T_0(x(\cdot)) \neq \phi \\ \infty & \text{for } T_0(x(\cdot)) = \phi. \end{cases}$$

The pursuit problem considered here consists of determining the positional strategy $U_0 : R \times R^n \to P$ such that

$$\sup\{\tau_0(x(\cdot)) : x(\cdot) \in X_1(t_0,x_0,U^0)\}$$

$$= \min_U \sup\{\tau^0(x(\cdot)) : x(\cdot) \in X_1(t_0,x_0,U)\}.$$

We shall discuss the problem of optimality of the strategy of extremal aiming in a pursuit problem. Assume that for any $\theta \geqslant t_0$ and all $t \in [t_0,\theta]$ the function $\varphi^{(m)}(\cdot,t,\theta)$ is convex. Assume also that $\theta \geqslant t_0$ exist such that $\epsilon_0(t_0,x_0,\theta) = 0$. Under these assumptions it is natural to select the smallest root of the equation

$$\epsilon_0(t_0,x_0,\theta) = 0 \qquad (\theta \geqslant t_0).$$

This root is denoted by $\theta_p(t_0,x_0)$ and a special term -- the *time of programmed absorption* -- was introduced for it in Section 5.1.4. In accordance with Theorem 5.3.1 where we set $\theta = \theta_p(t_0,x_0)$, the strategy U^0 assures the encounter of motions $x(\cdot) \in X_1(t_0,x_0,U^0)$ with the set M at time $t = \theta_p(t_0,x_0)$.

For any $\theta_* < \theta_p(t_0,x_0)$ we have $\epsilon_0(t_0,x_0,\theta_*) > 0$. Therefore the second player can assure the fulfillment of inequality

$$\text{dist}(x(\theta_*),M) \geqslant \epsilon_0(t_0,x_0,\theta_*) > 0 \text{ for all } x(\cdot) \in X_2(t_0,x_0,V^0).$$

Here V^0 is the strategy of extremal aiming defined according to (5.2.17) where we set $\theta = \theta_*$. Note that the positional strategy V^0 can be replaced by the maximin programmed control $v^0(\cdot)$ (cf. the analogous remark in Section 5.2.1). Positional strategy V^0 excludes the possibility of point $x(\theta)$ falling into the set M at a *fixed time* $t = \theta_*$. However, in general there may exist a motion $x(\cdot) \in X_2(t_0,x_0,V^0)$ and a time $t_* \in (t_0,\theta_*)$ such that $x(t_*) \in M$. In other words the strategy V^0 does not guarantee evasion from an encounter in the time interval $[t_0,\theta_*]$. Thus the following problem (the evasion problem) arises.

Let a number $\theta_* < \theta_p(t_0,x_0)$ be given. It is required to construct a strategy V_* such that

$$\text{(3)}\qquad \begin{aligned}&\operatorname{dist}(x(t),M) > 0 \quad \text{for all } x(\cdot) \in X_2(t_0,x_0,V_*)\\ &\qquad\qquad \text{and for all } \ t \in [t_0,\theta_*].\end{aligned}$$

This problem is of interest on its own and moreover allows us to answer the question concerning the optimality of a strategy of extremal aiming U^0 in the pursuit problem. Indeed, assume that for any $\theta_* < \theta_p(t_0,x_0)$ one can construct a strategy V_* which assures that condition (3) is fulfilled. We then obtain

$$\sup_V \inf_{x(\cdot)} \tau^0(x(\cdot)) \geqslant \theta_p(t_0,x_0) \quad (\text{here } \ x(\cdot) \in X_2(t_0,x_0,V_*)).$$

Consequently, the strategy U^0 such that

$$\sup_{x(\cdot)} \tau^0(x(\cdot)) \leqslant \theta_p(t_0,x_0) \quad \text{for} \quad x(\cdot) \in X_1(t_0,x_0,U^0)$$

will be optimal in the pursuit problem and the time of programmed absorption $\theta_p(t_0,x_0)$ will be the value of a differential game of pursuit-evasion.

5.3.2. Let (t_0,x_0) be a given initial position, $\theta_p(t_0,x_0)$ be the corresponding time of programmed absorption, and θ_* be an arbitrary number in the interval $(t_0,\theta_p(t_0,x_0))$. Consider the evasion problem on the time interval $[t_0,\theta_*)$. A solution of this problem will be constructed in the form of a modified strategy of extremal aiming.

Define the function

$$\text{(4)}\qquad \lambda(t,x) \overset{\text{def}}{=} \int_t^{\theta_*} \epsilon_0^{-1}(t,x,\theta)d\theta.$$

It follows from the discussion below that this function plays the role of a Lyapunov function of a certain kind. Note that this function is defined in the region

$$\text{(5)}\qquad G \overset{\text{def}}{=} \{(t,x) : t < \theta_*, \ \min_{t\leqslant\theta\leqslant\theta_*} \epsilon_0(t,x,\theta) > 0\}.$$

We state the following properties of the region G and function λ.

(a) $(t_0,x_0) \in G$;

(b) if $(t,x) \in G$, then $\operatorname{dist}(x, M) > 0$;

(c) if $(t_k,x_k) \in G$ $(k = 1, 2, \ldots)$, $(t_k,x_k) \to (t_*,x_*)$ as $k \to \infty$, $t_* < \theta_*$, $(t_*,x_*) \notin G$, then $\lambda(t_k,x_k) \to \infty$ as $k \to \infty$.

Property (a) follows from the definition of the number $\theta_p(t_0,x_0)$ (5.1.34) and the inequalities $t_0 < \theta_* < \theta_p(t_0,x_0)$.

It follows from the definition of the programmed maximin (5.1.27) that $\epsilon_0(t,x,t) = \operatorname{dist}(x,M)$. This equality yields property (b).

We shall prove property (c). It can be verified that the function ϵ_0 (2) satisfies the condition

$$|\epsilon_0(t,x,\theta_1) - \epsilon_0(t,x,\theta_2)| \leqslant \beta|\theta_1 - \theta_2| \tag{6}$$

$$\theta_i \in [t,\theta_*],\ i = 1, 2,\ t \in [t_0,\theta_*],\ x \in \mathfrak{D}\ \ (\mathfrak{D} \text{ is a compactum in } R^n).$$

Let $\epsilon_k \overset{\text{def}}{=} \min_{t_k \leqslant \theta \leqslant \theta_*} \epsilon_0(t_k,x_k,\theta) = \epsilon_0(t_k,x_k,\theta_k)$.

It follows from (6) that $\epsilon_0(t_k,x_k,\theta) \leqslant \epsilon_k + \beta|\theta - \theta_k|$. Therefore

$$\lambda(t_k,x_k) \geqslant \int_{t_k}^{\theta_*} [\epsilon_k + \beta|\theta - \theta_k|]^{-1} d\theta.$$

Since $\lim_{k \to \infty} (t_k,x_k) = (t_*,x_*) \notin G$, we have that $\epsilon_k \to 0$ as $k \to \infty$.

Hence $\lambda(t_k,x_k) \to \infty$ as $k \to \infty$.

Property (c) is thus proved.

Let $x(\cdot) : [t_0,\theta_*] \to R^n$ be a continuous function satisfying the condition $x(t_0) = x_0$. For $t = t_0$ we have $(t_0,x_0) \in G$. Let t_* be the time at which the point $(t,x(t))$ departs from the region G for the first time. Assume that

$$\lambda(t,x(t)) \leqslant \lambda_* < \infty \quad \text{for all} \quad t \in [t_0,t_*). \tag{7}$$

Then $t_* = \theta_*$.

Suppose the contrary. Let $(t_*,x(t_*)) \notin G$ and $t_* < \theta_*$. Consider the sequence $(t_k,x(t_k))$ where $t_k < t_*$ $(k = 1, 2, ...)$ and $t_k \to t_*$ as $k \to \infty$. For this sequence we have

$$(t_k,x(t_k)) \in G, \quad \lambda(t_k,x(t_k)) \leqslant \lambda_* \ (k = 1, 2, ...)$$

which contradicts property (c).

Thus the graph of the function $x(\cdot)$ satisfying the bound (7) remains in the region G up to time $t = \theta_*$. Therefore in accordance with property (b) we have $x(t) \notin M$ for $t_0 \leqslant t < \theta_*$.

Now we can assert the following: if the evader possesses a strategy which assures that the quantity $\lambda(t,x,(t))$ is nonincreasing in the region G, then this strategy guarantees evasion from hitting M on the time interval $[t_0,\theta_*)$.

We now proceed directly to the construction of such a strategy. We shall assume that the following conditions are fulfilled.

Condition 5.3.1. *For all* $t \in [t_0,\theta_*]$ *and for all* $\theta \in [t,\theta_*]$ *the function* $\varphi^{(m)}(\cdot,t,\theta) : R^m \to R$ *is convex (function* $\varphi^{(m)}$ *is defined by equality* (1)).

Condition 5.3.2. *Let*

$$(8) \qquad \xi(s,t,\theta,u,v) \overset{\text{def}}{=} \langle s,[\Phi(\theta,t)]_m B(t)u\rangle + \langle s,[\Phi(\theta,t)]_m C(t)v\rangle - \kappa^{(m)}(s,t,\theta).$$

(Recall that $\kappa^{(m)}$ *is defined by equality* (5.1.32)). *It is assumed that for all* $t \in [t_0,\theta_*]$ *and* $u \in P$ *the inequality*

$$\max_{v\in Q} \min_{t\leqslant\theta\leqslant\theta_*} \min_{\|s\|\leqslant 1} \xi(s,t,\theta,u,v) \geqslant 0$$

is fulfilled.

Note that in Section 5.3.3 a more convenient condition for checking the Condition 5.3.3 is stated which implies Conditions 5.3.1 and 5.3.2.

We introduce the following notation. Let $(t,x) \in G$ and $\theta \in [t,\theta_*]$. Condition 5.3.1 implies that the vector $s_0(t,x,\theta)$ which yields the maximum in (2) will be unique. The function $(t,x,\theta) \to s_0(t,x,\theta)$ will be continuous for $(t,x) \in G$ and $\theta \in [t,\theta_*]$. Set

$$(9) \qquad \ell_*(t,x) \overset{\text{def}}{=} \int_t^{\theta_*} \epsilon_0^{-2}(t,x,\theta)[\Phi(\theta,t)]_m^T s_0(t,x,\theta)d\theta.$$

Recall that $[\Phi]_m$ is the $(m \times n)$-matrix constructed from the first m rows of the $(n \times n)$-matrix Φ and the upper index ^{T} indicates transpose. Hence $\ell_*(t,x)$ is an n-dimensional column-vector. This vector is defined for the positions $(t,x) \in G$. Denote by $V_0(t,x)$ the set of vectors $v_0 \in Q$ such that the condition for the maximum

$$(10) \qquad \langle \ell_*(t,x),C(t)v_0\rangle = \max_{v\in Q} \langle \ell_*(t,x),C(t)v\rangle$$

will be fulfilled. This condition defines the set $V_0(t,x)$ for $(t,x) \in G$. If, however, $(t,x) \notin G$ we then set $V_0(t,x) = Q$.

Consider the differential inclusion

$$(11) \qquad \dot{x}(t) \in \{A(t)x(t) + B(t)u + C(t)v : u \in P, \ v \in V_0(t,x(t))\}.$$

The set of solutions $x(\cdot) : [t_0,\theta_*] \to R^n$ of this differential inclusion satisfying the initial condition $x(t_0) = x_0$ will be denoted by $X(t_0,x_0,V_0)$. Note that for any position (t,x) the set $V_0(t,x)$ is convex and the multi-valued mapping $(t,x) \to V_0(t,x)$ is upper semicontinuous. Therefore in view of Theorem A.4.1 the set $X(t_0,x_0,V_0)$ is nonempty. The following lemma is valid.

Lemma 5.3.1. *Let Conditions* 5.3.1 *and* 5.3.2 *be fulfilled. Then*

$$\operatorname{dist}(x(t),M) > 0 \ \text{ for all } \ t \in [t_0,\theta_*) \ \text{ and } \ x(\cdot) \in X(t_0,x_0,V_0).$$

Proof. If $(t,x) \in G$, then Condition 5.3.1 implies that for any $\theta \in [t,\theta_*]$ the derivatives $\partial\epsilon_0(t,x,\theta)/\partial t$ and $\partial\epsilon_0(t,x,\theta)/\partial x$ exist. These derivatives are defined by the following relations (cf. (5.2.18)):

$$\frac{\partial \epsilon_0(t,x,\theta)}{\partial t} = -\langle s_0(t,x,\theta),[\Phi(\theta,t)]_m A(t)x\rangle - \kappa^{(m)}(s_0(t,x,\theta),t,\theta),$$

$$\frac{\partial \epsilon_0(t,x,\theta)}{\partial x} = s_0^T(t,x,\theta)[\Phi(\theta,t)]_m.$$

According to (4) and (9) we obtain that for $(t,x) \in G$ there exist derivatives

$$\frac{\partial \lambda(t,x)}{\partial t} = -\epsilon_0^{-1}(t,x,\theta)$$

$$+\int_t^{\theta_*} \epsilon_0^{-2}(t,x,\theta)[\langle s_0(t,x,\theta),[\Phi(\theta,t)]_m A(t)x\rangle + \kappa^{(m)}(s_0(t,x,\theta),t,\theta)]d\theta, \quad (12)$$

$$\frac{\partial \lambda(t,x)}{\partial x} = -\epsilon_*^T(t,x) = -\int_t^{\theta_*} \ell_0^{-2}(t,x,\theta)s_0^T(t,x,\theta)[\Phi(\theta,t)]_m d\theta.$$

Choose an arbitrary solution $x(\cdot)$ belonging to the set $X(t_0,x_0,V_0)$. Let t_* be the instant of time at which $(t,x(t)) \notin G$ for the first time. Consider the function $t \to \lambda(t,x(t))$. This function is defined on the half-interval $[t_0,t_*)$ and is absolutely continuous on any segment $[t_0,t_* - \delta]$, where $\delta \in (0,t_* - t_0]$. For almost all $t \in [t_0, t_* - \delta]$ the derivative $d\lambda(t,x(t))/dt$ exists. From (11) and (12) we have

$$\frac{d\lambda(t,x(t))}{dt} = -\epsilon_0^{-1}(t,x(t),t)$$

$$-\int_t^{\theta_*} \epsilon_0^{-2}(t,x(t),\theta)\xi(s_0(t,x(t),\theta),t,\theta,u(t),v_0(t))d\theta. \quad (13)$$

Here $u(t) \in P$, $v_0(t) \in V_0(t,x(t))$, and the quantity $\xi(s,t,\theta,u,v)$ is defined by equality (8).

Select a vector $v_* \in Q$ from the condition

$$\min_{(s,\theta)} \xi(s,t,\theta,u(t),v_*) = \max_v \min_{(s,\theta)} \xi(s,t,\theta,u(t),v)$$

$$\text{where } \|s\| \leqslant 1, \quad \theta \in [t,\theta_*], \quad v \in Q.$$

Replace vector $v_0(t)$ in (13) by vector v_*. Taking (10), (12) into account we obtain

$$\frac{d\lambda(t,x(t))}{dt} < -\int_t^{\theta_*} \epsilon_0^{-2}(t,x(t),\theta)\ \xi(s_0(t,x(t),\theta),t,\theta,u(t),v_*)d\theta.$$

From the definition of vector v_* and Condition 5.3.2 we have

$$\xi(s_0(t,x(t),\theta),t,\theta,u(t),v_*) \geqslant \min_{(s,\theta)} \xi(s,t,\theta,u(t),v_*) \geqslant 0.$$

Therefore $d\lambda(t,x(t))/dt < 0$. This estimate is valid for almost all $t \in [t_0,t_* - \delta]$. Hence $\lambda(t_* - \delta,\ x(t_* - \delta)) < \lambda(t_0,x_0)$; here δ is an arbitrary number in $(0,t_* - t_0]$. We thus arrive at relation (7) from which -- as shown above -- it follows that $x(t) \notin \mathcal{M}$ for all $t \in [t_0,\theta_*)$. Lemma 5.3.1 is thus proved.

Consider the strategy $V_0 : T \times R^n \to \mathcal{Q}$ such that

$$(14) \qquad V_0(t,x) \in \mathcal{V}_0(t,x) \ \text{ for all } \ (t,x) \in T \times R^n$$

i.e. function V_0 is a selector of a multi-valued mapping $\mathcal{V}_0$. Let $\mathcal{X}_2(t_0,x_0,V_0)$ be the set of constructive motions generated by the strategy V_0. This set is embedded in the set $\mathcal{X}(t_0,x_0,\mathcal{V}_0)$ of generalized motions generated by a multi-valued strategy $\mathcal{V}_0$ (cf. Section A.4). Therefore from Lemma 5.3.1 follows

Theorem 5.3.2. *Let Conditions* 5.3.1 *and* 5.3.2 *be fulfilled. Then the positional strategy* V_0 (14) *guarantees an evasion on the interval* $[t_0,\theta_*)$, *i.e.* $x(t) \notin \mathcal{M}$ *for all* $t \in [t_0,\theta_*)$ *and* $x(\cdot) \in \mathcal{X}_2(t_0,x_0,V_0)$.

Note that the proof of Theorem 5.3.2 can be carried out using bounds on the variation of function $\lambda(t,x(t))$ along step-by-step motions generated by the strategy V_0. However, such a proof turns out to be more cumbersome than the proof of Lemma 5.3.1.

We shall briefly discuss the meaning of condition (10). The selection of control v_0 satisfying this condition signifies aiming in the direction of the vector $\ell_*(t,x)$ which is the average of all possible directions of an extremal aiming $\ell^*(t,x,\theta) = [\Phi(\theta,t)]_m^T s_0(t,x,\theta)$, corresponding to different values of the parameter $\theta \in [t,\theta_*]$. According to (9), when computing the averaged vector $\ell_*(t,x)$ the weight function $\theta \to \epsilon_0^{-2}(t,x,\theta)$ is used. This function is such that the larger the weight of the direction $\ell^*(t,x,\theta)$, the smaller the corresponding value of $\epsilon_0(t,x,\theta)$. The averaged direction of extremal aiming selected in this manner assures that the inequality $\epsilon_0(t,x(t),\theta) > 0$ remains valid for all $\theta \in [t,\theta_*]$ and for all $t \in [t_0,\theta_*)$. Thus the strategy is a modified strategy of extremal aiming.

Combining Theorems 5.3.1 and 5.3.2 we obtain the following assertion.

Theorem 5.3.3. *Let Conditions* 5.3.1 *and* 5.3.2 *be fulfilled for all* $\theta \geqslant t_0$. *Let a time of programmed absorption* $\theta_p(t_0,x_0)$ *exist. Then* $\theta_p(t_0,x_0)$ *is the value of the differential game of pursuit-evasion under consideration. The pair of positional strategies* U^0 *and* V_ϵ *forms the* ϵ*-equilibrium situation in this game. Here* U^0 *is a strategy of extremal aiming defined by equality* (5.2.16) *where* $\theta = \theta_p(t_0,x_0)$ *and* V_ϵ *is a modified strategy of extremal aiming defined in accordance with* (9), (10), *and* (14) *where* $\theta_* = \theta_p(t_0,x_0) - \epsilon$.

5.3.3. Before proceeding to a discussion of examples, we state the conditions which are more convenient for verification than Conditions 5.3.1 and 5.3.2. These conditions were originally given in [28].

Condition 5.3.3. *Let sets* $\mathcal{P}^{(m)}(t,\theta)$ *and* $\mathcal{Q}^{(m)}(t,\theta)$ *be defined by the equalities* (5.2.22). *Let the equality*

$$(15) \qquad \mathcal{P}^{(m)}(t,\theta) = \mathcal{Q}^{(m)}(t,\theta) + \mathcal{H}^{(m)}(t,\theta)$$

be fulfilled for all $t \in [t_0,\theta_*]$ *and all* $\theta \in [t,\theta_*]$. *Assume also that for any* $t \in [t_0,\theta_*]$ *and* $u \in \mathcal{P}$ *there exists a vector* $v_*(t,u) \in \mathcal{Q}$ *satisfying the condition*

$$(16) \qquad -[\Phi(\theta,t)]_m(B(t)u + C(t)v_*(t,u)) \overset{\text{def}}{=} g(t,\theta,u) \in \mathcal{H}^{(m)}(t,\theta)$$

for all $\theta \in [t,\theta_*]$.

As shown in Section 5.2.3 equality (15) implies that function $\varphi^{(m)}(\cdot,t,\theta)$ (cf. (1)) is convex, i.e. the Condition 5.3.1 is fulfilled.

We shall show that Condition 5.3.2 will follow from relation (16). We utilize equalities (5.1.32) and (5.2.24) which we rewrite as

$$\kappa^{(m)}(s,t,\theta) \overset{\text{def}}{=} \kappa_1^{(m)}(s,t,\theta) + \kappa_2^{(m)}(s,t,\theta) = \min\{\langle s, -g\rangle : g \in \mathcal{H}^{(m)}(t,\theta)\}.$$

Therefore

$$\langle s, -g\rangle - \kappa_1^{(m)}(s,t,\theta) - \kappa_2^{(m)}(s,t,\theta) \geqslant 0$$

if $g \in \mathcal{H}^{(m)}(t,\theta)$. In particular, this inequality will be valid for all $t \in [t_0,\theta_*]$ and $\theta \in [t,\theta_*]$ for the vector $g(t,\theta,u)$ introduced in (16). We obtain that Condition 5.3.2 is fulfilled. Thus it is shown that Conditions 5.3.3 imply Conditions 5.3.1 and 5.3.2.

5.3.4. As the first example, consider the differential game of pursuit-evasion for *uni-type objects.*

Let the motions of the pursuer and evader be described by uni-type equations

$$(17) \qquad \dot{y} = A(t)y + B(t)u,$$

$$(18) \qquad \dot{z} = A(t)z + B(t)v.$$

Here y and z are n-dimensional phase vectors, $A(t)$ and $B(t)$ are $(n \times n)$ and $(n \times p)$-matrices. As usual, the controls of the players satisfy the constraints $u \in \mathcal{P}$ and $v \in \mathcal{Q}$, where $\mathcal{P}$ and $\mathcal{Q}$ are convex compact sets. We shall assume that

$$(19) \qquad \mathcal{P} = \alpha\mathcal{Q}, \quad \alpha > 1.$$

Here $\alpha\mathcal{Q} \stackrel{\text{def}}{=} \{\alpha v : v \in \mathcal{Q}\}$. When the equality (19) is fulfilled the systems (17) and (18) are called uni-type systems.

Let m components forming m-dimensional vectors $[y]_m$ and $[z]_m$ be selected out of the phase vectors y and z respectively. Without loss of generality we assume that $[y]_m$ and $[z]_m$ consist of the first m components of the vectors y and z. The pursuer attempts to bring the point $[y(t)]_m$ nearer to the point $[z(t)]_m$ so that the distance between them will be at most a given number. The time $t = \tau$ at which this approach occurs for the first time is called the encounter time or the capture time. The encounter condition may be defined as

$$(20) \qquad [y(\tau)]_m - [z(\tau)]_m \in \mathcal{M} .$$

If the distance between the points $[y]_m$ and $[z]_m$ is measured in the Euclidean metric then $\mathcal{M}_* \stackrel{\text{def}}{=} \{w \in R^m : \|w\| \leq r\}$, where r is a given number, the ultimate distance between the pursuer and the evader. In the general case we shall assume that $\mathcal{M}_*$ is a convex compact set in R^m.

Set $x = y - z$. The motion of the vector x is described by the equation

$$(21) \qquad \dot{x} = A(t)x + B(t)u - B(t)v.$$

Consider the differential game of pursuit-evasion for the system (21), assuming that the encounter time is determined by the condition $[x(t)]_m \in \mathcal{M}_*$. Let the strategies $(t,x) \to U^0(t,x)$ and $(t,x) \to V^\epsilon(t,x)$ form an ϵ-saddle point in this game. Then the strategies $(t,y,z) \to U^0(t,y - z)$ and $(t,y,z) \to V^\epsilon(t,y - z)$ form an ϵ-saddle point in the initial differential game (17), (18), and (20).

Thus, in this example it is sufficient to determine the solutions of problems of pursuit and evasion for the system (21). This system is a particular case of the system (5.1.14) with $C(t) = - B(t)$ and equality (19) fulfilled. We shall show that in the case under consideration Condition 5.3.3 is fulfilled.

We determine the sets $\mathcal{P}^{(m)}(t,\theta)$ and $\mathcal{Q}^{(m)}(t,\theta)$. From the equalities $B(t) = - C(t)$ and $\mathcal{P} = \alpha\mathcal{Q}$ we have

$$\mathcal{P}^{(m)}(t,\theta) = \alpha\, \mathcal{Q}^{(m)}(t,\theta).$$

Taking into account that the set $\mathcal{Q}^{(m)}(t,\theta)$ is convex we obtain

$$(22) \qquad \mathcal{P}^{(m)}(t,\theta) = \mathcal{Q}^{(m)}(t,\theta) + (\alpha - 1)\, \mathcal{Q}^{(m)}(t,\theta).$$

Hence the equality (15) is fulfilled with $\mathcal{H}^{(m)}(t,\theta) = (\alpha - 1)\, \mathcal{Q}^{(m)}(t,\theta)$. It is easy to verify that for the vector $v_*(u) = \alpha^{-1}u$ the inclusion (16) is valid. We have established that Condition 5.3.3 is fulfilled for the system (21), (19). Thus, in this case, one can use the results of this section when solving problems of pursuit and evasion.

5.3.5. Consider in somewhat more detail a differential game of pursuit-evasion in which the motions of the pursuing and evading systems are described by equations (1.2.18) and (1.2.19), respectively. In these equations we shall assume that $m_1 = m_2 = 1$. Let the encounter condition be determined by the equalities $y_1(\tau) = z_1(\tau)$, $y_2(\tau) = z_2(\tau)$, i.e. the pursuer is aiming so that the geometric positions (coordinates) of the pursued and evading points coincide.

We now introduce new phase variables $x_1 = y_1 - z_1$, $x_2 = y_2 - z_2$, $x_3 = y_3$, $x_4 = y_4$, $x_5 = z_3$, and $x_6 = z_4$. The variation in time of the new phase vector $x \in R^6$ is described by the equations

(23)
$$\begin{gathered} \dot{x}_1 = x_3 - x_5, \quad \dot{x}_2 = x_4 - x_6, \\ \dot{x}_3 = -\alpha_1 x_3 + u_1, \quad \dot{x}_4 = -\alpha_1 x_4 + u_2, \\ \dot{x}_5 = -\alpha_2 x_5 + v_1, \quad \dot{x}_6 = -\alpha_2 x_6 + v_2. \end{gathered}$$

Controls of the players satisfy the constraints

(24)
$$(u_1^2 + u_2^2)^{1/2} \leqslant \beta_1, \quad (v_1^2 + v_2^2)^{1/2} \leqslant \beta_2.$$

For system (23) the encounter condition becomes the equality $x_1(\tau) = x_2(\tau) = 0$. Hence here $M_* = \{0\}$ is a set in R^2 consisting of the single zero vector. It is easy to verify that the initial differential game posed for the system (1.2.18) and (1.2.19) will be equivalent to the differential game for the tranformed system (23).

We shall assume that

(25)
$$\beta_1/\alpha_1 \geqslant \beta_2/\alpha_2, \quad \beta_1 \geqslant \beta_2 .$$

The first two rows of the 6×6 matrix $\Phi(\theta,t)$ defined for the system (23), in accordance with (1.2.15) and (1.2.16), will be needed in the calculation below. These two rows form a 2×6 matrix

(26)
$$[\Phi(\theta,t)]_2 = (E, r_1(\theta,t)E, - r_2(\theta,t)E).$$

Here E is the (2×2)-unit matrix

(27)
$$\begin{gathered} r_1(\theta,t) = [1 - \exp\alpha_1(t - \theta)]\alpha_1^{-1}, \\ r_2(\theta,t) = [1 - \exp\alpha_2(t - \theta)]\alpha_2^{-1}. \end{gathered}$$

Therefore

$$[\Phi(\theta,t)]_2 Bu = r_1(\theta,t)u, \quad [\Phi(\theta,t)]_2 Cv = r_2(\theta,t)v.$$

In accordance with (5.2.22) we obtain

$$P^{(2)}(t,\theta) = \{g \in R^2 : \|g\| \leqslant \beta_1 r_1(\theta,t)\},$$

$$Q^{(2)}(t,\theta) = \{g \in R^2 : \|g\| \leqslant \beta_2 r_2(\theta,t)\}.$$

One can prove that if conditions (25) are satisfied then the inequalities

$$\beta_1 r_1(\theta,t) \geqslant \beta_2 r_2(\theta,t) \text{ for all } -\infty < t \leqslant \theta < \infty$$

are valid. Therefore, in this example equality (15) is fulfilled. It is easy to verify that for the vector $v_*(u) = \beta_2 u \beta_1^{-1}$, the inclusion (16) will be valid. Hence, for the example under consideration Condition 5.3.3 will be fulfilled.

To write down the formula for the programmed maximin $\epsilon_0(t,x,\theta)$ (7) we note that here

$$[\Phi(\theta,t)]_2 x = (g_1(t,x,\theta), g_2(t,x,\theta))^T,$$

$$g_1(t,x,\theta) = x_1 + r_1(t,\theta)x_3 - r_2(t,\theta)x_5,$$

(28) $$g_2(t,x,\theta) = x_2 + r_1(t,\theta)x_4 - r_2(t,\theta)x_6,$$

$$\kappa_1^{(2)}(s,t,\theta) = -\beta_1 r_1(t,\theta)(s_1^2 + s_2^2)^{1/2},$$

$$\kappa_2^{(2)}(s,t,\theta) = \beta_2 r_2(t,\theta)(s_1^2 + s_2^2)^{1/2}.$$

Therefore

$$\epsilon_0(t,x,\theta) = \max_{\|s\| \leqslant 1} \{s_1 g_1(t,x,\theta) + s_2 g_2(t,x,\theta) - \|s\|[(\beta_1 \alpha_1^{-1} - \beta_2 \alpha_2^{-1})(\theta - t) + \beta_1 \alpha_1^{-2}(\exp\alpha_1(t-\theta) - 1) - \beta_2 \alpha_2^{-2}(\exp\alpha_2(t-\theta) - 1)]\}.$$

If $\epsilon_0(t,x,\theta) > 0$, the maximum is then obtained here at the unique vector

(29) $$s_0(t,x,\theta) = (g_1(t,x,\theta)/\|g(t,x,\theta)\|, \; g_2(t,x,\theta)/\|g(t,x,\theta)\|)^T.$$

Hence

(30) $$\epsilon_0(t,x,\theta) = \max\{\epsilon_*(t,x,\theta), 0\}$$

where

$$\epsilon_*(t,x,\theta) = \|g(t,x,\theta)\| - \beta_1[(\theta - t)\alpha_1^{-1} + (\exp\alpha_1(t - \theta) - 1)\alpha_1^{-2}]$$
$$+ \beta_2[(\theta - t)\alpha_2^{-1} + (\exp\alpha_2(t - \theta) - 1)\alpha_2^{-2}].$$

It can be shown that for an arbitrary initial position the equation $\epsilon_0(t_0,x_0,\theta) = 0$ $(\theta \geqslant t_0)$ possesses a root, i.e. the time of programmed absorption $\theta_p(t_0,x_0)$ exists.

Relation (5.2.16) determines the strategy

(31) $$U^0(t,x) = \begin{cases} -\beta_1 g(t,x,\theta)/\|g(t,x,\theta)\| & \text{for } \|g(t,x,\theta)\| > 0 \\ (0,0)^T & \text{for } \|g(t,x,\theta)\| = 0 \end{cases}$$

in this example. According to Theorem 5.3.3 for $\theta = \theta_p(t_0,x_0)$ this strategy will be an optimal strategy in a pursuit problem.

We now turn to an evasion problem. From relations (9), (10), and (14) we obtain that an evasion strategy is determined by the equality

(32) $$V^0(t,x) = \beta_2 s^*(t,x)/\|s^*(t,x)\|,$$

where

(33) $$s^*(t,x) = \int_t^{\theta_*} \epsilon_0(t,x,\theta)^{-2} r_2(t,\theta) s_0(t,x,\theta) d\theta$$

where the quantities $\epsilon_0(t,x,\theta)$, $r_2(t,\theta)$ and the vector $s_0(t,x,\theta)$ are defined by equalities (30), (27), (29), $\theta_* < \theta_p(t_0,x_0)$. Note that strategy V_0 is determined by equality (32) in the region G (5). Outside this region we complete the definition by setting $V_0(t,x) = (0,0)^T$.

In this example the pair of strategies U^0 (31) and V_0 (32), (33), where $\theta_* = \theta_p(t_0,x_0) - \epsilon$, forms an ϵ-equilibrium situation.

5.4. Regularity Conditions for a Programmed Maximin

Conditions for which a programmed maximin coincides with the value of the positional differential game will be called regularity conditions for a programmed maximin. Relations which determine a programmed maximin are quite simple and easily implemented numerically. It is therefore important to study regularity conditions in more detail, relax the sufficient conditions stated in Section 5.2, and obtain the necessary and sufficient regularity conditions for a programmed maximin. This section is devoted to these problems.

5.4.1. We shall again consider a differential game with a fixed termination time. Assume that the motion of the controlled system is described by equation (5.1.14) and the payoff is given by equality (5.1.15) where $\sigma : R^n \to R$ is a convex function satisfying the Lipschitz condition.

It follows from the results of Section 3.1 that in the differential game under consideration (5.1.14) and (5.1.15) for any initial position $(t_0,x_0) \in T \times R^n$, the value $\omega^0(t_0,x_0)$ exists. Recall that

$$\omega^0(t_0,x_0) \overset{\text{def}}{=} \min_U \max_{x(\cdot)\in X_1(t_0,x_0,U)} \sigma(x(\theta)) = \max_V \min_{x(\cdot)\in X_2(t_0,x_0,V)} \sigma(x(\theta)).$$

Here U and V are the positional strategies of the first and the second players, respectively.

For the differential game (5.1.14), (5.1.15) in Section 5.1, the programmed maximin $\omega_p(t_0,x_0)$ was introduced. It is easy to show that the value of a positional differential game and programmed maximin are related by the inequality

$$(1) \qquad \omega^0(t_0,x_0) \geqslant \omega_p(t_0,x_0), \quad (t_0,x_0) \in T \times R^n.$$

This inequality is valid not only for the differential game (5.1.14) and (5.1.15) but actually for all linear and nonlinear differential games with different types of payoff functionals. Inequality (1) expresses the sufficiently obvious fact: the result guaranteed to the second player in the class of positional strategies cannot be worse (i.e. less) than the result guaranteed to him/her in the class of programmed controls. Utilizing this consideration one can construct a formal proof of inequality (1).

However, another approach seems to be more convenient for future discussion. First we shall show that the function ω_p is ν-stable. This fact will be required in subsequent arguments; moreover, it easily implies inequality (1).

Thus we verify that function ω_p satisfies the condition of ν-stability in the region $T^0 \times R^n$. According to the definition presented in Section 4.2 it is required to show that for any $(t_*,x_*) \in T^0 \times R^n$, $t^* \in (t_*,\theta]$ and $u_* \in P$ a programmed control $v(\cdot) \in \mathbf{V}_p$ exists for which $\omega_p(t^*,x(t^*,t_*,x_*,u_*,v(\cdot)) \geqslant \omega_p(t_*,x_*)$.

Let $(\tau,y) \in T \times R^n$, $v_*(\cdot) \in \mathbf{V}_p$. We define the accessibility region

$$G(\tau,y,v_*(\cdot)) \overset{\text{def}}{=} \{x(\theta,\tau,y,u(\cdot),v_*(\cdot)) : u(\cdot) \in \mathbf{U}_p\}.$$

Let $(t_*,x_*) \in T^0 \times R^n$, $t^* \in (t_*,\theta]$, $u_*(\cdot) \in \mathbf{U}_p$ and $x^* = x(t^*,t_*,x_*,u_*(\cdot),v_*(\cdot))$. It is easy to verify that

$$(2) \qquad G(t^*,x^*,v_*(\cdot)) \subset G(t_*,x_*,v_*(\cdot)).$$

We select the control $v_*(\cdot) \in \mathbf{V}_p$ from the condition

$$(3) \qquad \min_{u(\cdot)\in\mathbf{U}_p} \sigma(x(\theta,t_*,x_*,u(\cdot),v_*(\cdot))) = \max_{v(\cdot)\in\mathbf{V}_p} \min_{u(\cdot)\in\mathbf{U}_p} \sigma(x(\theta,t_*,x_*,u(\cdot),v(\cdot))) = \omega_p(t_*,x_*).$$

Set $x^* = x(t^*,t_*,x_*,u_*,v_*(\cdot))$. Relations (2) and (3) imply that

$$\omega_p(t^*,x^*) = \max_{v(\cdot)} \min_{u(\cdot)} \sigma(x(\theta,t^*,x^*,u(\cdot),v(\cdot)))$$
$$\geq \min_{u(\cdot)} \sigma(x(\theta,t^*,x^*,u(\cdot),v_*(\cdot))) = \min\{\sigma(x) : x \in G(t^*,x^*,v_*(\cdot))\}$$
$$\geq \min \{\sigma(x) : x \in G(t_*,x_*,v_*(\cdot))\}$$
$$= \min_{u(\cdot)} \sigma(x(\theta,t_*,x_*,u(\cdot),v_*(\cdot))) = \omega_p(t_*,x_*) ,$$

q.e.d.

We have thus shown that the function of programmed maximin ω_p satisfies the condition of v-stability in the region $T^0 \times R^n$. This function is continuous (moreover it is locally-Lipschitzian (cf. Lemma 5.1.1)) and satisfies the boundary condition $\omega_p(\theta,x) = \sigma(x)(x \in R^n)$. Therefore Lemma (4.2.5) implies inequality (1).

Examples can be presented for which the strict inequality $\omega^0(t_0,x_0) > \omega_p(t_0,x_0)$ holds. In particular, such an inequality is valid in the following example.

Let x, u, and v be 2-dimensional vectors, $\theta = 0$, $\sigma(x) = \|x\|$, $P = Q = \{w \in R^2 : \|w\| \leq 1\}$. The motion equation is of the form $\dot{x} = tu + v$. In this example, the payoff function is determined by equalities (4.1.12) and (4.1.13) according to which for $t_0 = -2$, $x_0 = (0,0)^T$ we obtain $\omega^0(t_0,x_0) = 1/2$. We compute for this initial position (t_0,x_0) the value of the programmed maximin. By definition we have

$$\omega_p(t_0,x_0) = \max_{v(\cdot)} \min_{u(\cdot)} \left\| \int_{-2}^{0} tu(t)dt + \int_{-2}^{0} v(t)dt \right\|.$$

We introduce the following notation:

$$G_1 \overset{\text{def}}{=} \{\int_{-2}^{0} tu(t)dt : u(\cdot) \in \mathbf{U}_p\} = \{g \in R^2 : \|g\| \leq 2\},$$
$$G_2 \overset{\text{def}}{=} \{\int_{-2}^{0} v(t)dt : v(\cdot) \in \mathbf{V}_p\} = \{g \in R^2 : \|g\| \leq 2\}.$$

Utilizing this notation we have

$$\omega_p(t_0,x_0) = \max_{g_2 \in G_2} \min_{g_1 \in G_1} \|g_1 + g_2\| = 0.$$

In this example we obtain $\omega^0(t_0,x_0) = 1/2 > \omega_p(t_0,x_0) = 0$ for $t_0 = -2$ and $x_0 = (0,0)^T$.

Thus, without additional assumptions a programmed maximin and the value of a positional differential game may not coincide. Conditions under which the function of the programmed maximin ω_p coincides with the value of the function on the whole space $T \times R^n$ or in the region $\mathcal{E}$ of the form (5.2.9) are called *regularity conditions of programmed maximin*. As was shown in Theorems 5.2.1 and 5.2.2, Conditions 5.2.1 and 5.2.2 are sufficient conditions for regularity. Note that under these conditions the maximum in relation (5.1.23) is attained at the unique vector $\ell_0(t,x)$.

The purpose of this section is to state and prove necessary and sufficient regularity conditions for the programmed maximin. We note that under the conditions stipulated below there may be several vectors yielding the maximum in (5.1.23)

5.4.2. We first study the conditions under which the function of the programmed maximin ω_p (5.1.23) coincides with the value function ω^0 in the whole space $T \times R^n$. As it was established in Lemma 5.1.1 the function ω_p: $T \times R^n \to R$ is locally Lipschitzian. It follows directly from (5.1.17) that $\omega_p(\theta,x) = \sigma(x)$ $(x \in R^n)$. Moreover, it was shown above that function ω_p is v-stable in the region $T^0 \times R^n$. Therefore according to Theorem 4.2.1 we obtain that functions ω_p and ω^0 coincide if and only if the function of the programmed maximin ω_p satisfies the condition of u-stability in the region $T^0 \times R^n$. We thus arrive at the following conclusion: a necessary and sufficient regularity condition for the function of the programmed maximin should express nothing else but the condition of u-stability of this function in the region $T^0 \times R^n$.

We now turn to the formula (5.1.23) which can be rewritten as

$$\omega_p(t,x) = \sup_{\ell \in R^n} [\varphi_1(t,x,\ell) - \varphi_2(\ell)].$$

Here

$$\varphi_1(t,x,\ell) \overset{\text{def}}{=} \langle \ell,\Phi(\theta,t)x \rangle + \int_t^\theta \kappa(\ell,\tau,\theta)d\tau, \quad \varphi_2(\ell) = \sigma^*(\ell).$$

The function φ_1 is continuous on $T \times R^n \times R^n$. In accordance with (1.2.15), its derivatives are determined by the equalities

$$\frac{\partial \varphi_1(t,x,\ell)}{\partial t} = - \langle \ell,\Phi(\theta,t)A(t)x \rangle - \kappa(\ell,t,\theta),$$

$$\frac{\partial \varphi_1(t,x,\ell)}{\partial t} = \ell^T\Phi(\theta,t).$$

Functions $(t,x,\ell) \to \partial\varphi_1(t,x,\ell)/\partial t$ and $(t,x,\ell) \to \partial\varphi_1(t,x,\ell)/\partial x$ are continuous on $T^0 \times R^n \times R^n$. The function $\varphi_2 = \sigma^* : R^n \to [a,\infty]$ is lower semicontinuous. Since $\sigma^*(\ell) = \sup\{\langle \ell,x \rangle - \sigma(x) : x \in R^n\}$ it follows that $\sigma^*(\ell) \geqslant - \sigma(0)$. It therefore could be assumed that $a = -\sigma(0)$. As it was noted in Section 5.1.2 the set $\mathrm{dom}\varphi_2 = \mathrm{dom}\sigma^*$ is bounded.

Thus all the conditions of Lemma 4.3.1 are fulfilled and we obtain from this lemma that

$$(4) \quad D\omega_p(t,x)|(1,h) = \max_{\ell \in \mathfrak{L}_0(t,x)} [- \langle \ell,\Phi(\theta,t)A(t)x \rangle - \kappa(\ell,t,\theta) + \langle \ell,\Phi(\theta,t)h \rangle].$$

Here $(t,x) \in T^0 \times R^n$, $h \in R^n$ and the set $\mathfrak{L}_0(t,x)$ is defined by the equality

$$\mathfrak{L}_0(t,x) \overset{\text{def}}{=} \{\ell_0 \in R^n : \varphi_1(t,x,\ell_0) - \varphi_2(\ell_0) = \omega_p(t,x)\}$$

(5)
$$= \{\ell_0 \in R^n : \langle \ell_0, \Phi(\theta,t)x\rangle + \int_t^\theta [\kappa_1(\ell_0,\tau,\theta) + \kappa_2(\ell_0,\tau,\theta)]d\tau - \sigma^*(\ell_0)$$

$$= \max_{\ell \in R^n} [\langle \ell, \Phi(\theta,t)x\rangle + \int_t^\theta [\kappa_1(\ell,\tau,\theta) + \kappa_2(\ell,\tau,\theta)]\; d\tau - \sigma^*(\ell)]\}.$$

Note that $\mathfrak{L}_0(t,x)$ is a compact set in $\mathrm{dom}\sigma^*$.

Since ω_p is a locally Lipschitzian function in accordance with the results obtained in Section 4.4, the condition of the u-stability of function ω_p can be expressed by the inequality

(6)
$$\max_{v \in Q} \min_{u \in P} D\omega_p(t,x)|\,(1,f(t,x,u,v)) \leqslant 0.$$

In our case $f(t,x,u,v) = A(t)x + B(t)u + C(t)v$. After substituting this equality in formula (4), inequality (6) can be written as

$$\max_{v \in Q} \min_{u \in P} \max_{\ell \in \mathfrak{L}_0(t,x)} [\langle \ell, \Phi(\theta,t)B(t)u\rangle + \langle \ell, \Phi(\theta,t)C(t)v\rangle - \kappa(\ell,t,\theta)] \leqslant 0.$$

It is easy to show that a strict inequality cannot be attained here. Indeed let $\ell_0 \in \mathfrak{L}_0(t,x)$, then according to (5.1.21) we have

$$\max_{v \in Q} \min_{u \in P} \max_{\ell \in \mathfrak{L}_0(t,x)} [\langle \ell, \Phi(\theta,t)B(t)u\rangle + \langle \ell, \Phi(\theta,t)C(t)v\rangle - \kappa(\ell,t,\theta)]$$

$$\geqslant \max_{v \in Q} \min_{u \in P} [\langle \ell_0, \Phi(\theta,t)B(t)u\rangle + \langle \ell_0, \Phi(\theta,t)C(t)v\rangle - \kappa(\ell_0,t,\theta)] = 0.$$

We thus arrive at the following condition.

Condition 5.4.1. *For any position $(t,x) \in T^0 \times R^n$ the equality*

(7)
$$\max_{v \in Q} \min_{u \in P} \max_{\ell \in \mathfrak{L}_0(t,x)} [\langle \ell, \Phi(\theta,t)B(t)u\rangle + \langle \ell, \Phi(\theta,t)C(t)v\rangle - \kappa(\ell,t,\theta)] = 0$$

is fulfilled.

This condition is a condition of u-stability of function ω_p in the domain $T^0 \times R^n$. Before proceeding to the final statement of the result we consider yet another relationship which is equivalent to equality (7).

We introduce the following notation: $\mathfrak{L}_0 \overset{\text{def}}{=} \mathfrak{L}_0(t,x), \Pi \overset{\text{def}}{=} \mathrm{rpm}(\mathfrak{L}_0)$ i.e. Π is a set of probability measures normalized on the compact set $\mathfrak{L}_0$. Observe that for any continuous function $\ell \to r(\ell)$ the equality

$$\max_{\ell \in \mathfrak{L}_0} r(\ell) = \max_{\pi \in \Pi} \int_{\mathfrak{L}_0} r(\ell)\pi(d\ell)$$

is valid. Set

$$\chi(t,\pi,u,v) \overset{\text{def}}{=} \langle \ell_\pi, \Phi(\theta,t)B(t)u\rangle + \langle \ell_\pi, \Phi(\theta,t)C(t)v\rangle$$

$$- \int_{\mathcal{L}_0} \kappa(\ell t,\theta)\pi(d\ell), \quad \ell_\pi \overset{\text{def}}{=} \int_{\mathcal{L}_0} \ell\pi(d\ell).$$

Utilizing this notation, equality (7) can be rewritten as

$$\max_{v\in Q} \min_{v\in P} \max_{\pi\in\Pi} \chi(t,\pi,u,v) = 0.$$

Function $u \to \chi(t,\pi,u,v) : P \to R$ and functional $\pi \to \chi(t,\pi,u,v) : \Pi \to R$ are linear, the sets P and Π are convex. We can therefore utilize the minimax theorem [21, 34, 55] here and interchange the operations of $\max_\pi$ and $\min_u$. Taking (5.1.21) into account we obtain

$$\max_{\pi\in\Pi} \max_{v\in Q} \min_{u\in P} \chi(t,\pi,u,v) = \max_{\pi\in\Pi} [\kappa(\ell_\pi,t,\theta) - \int_{\mathcal{L}_0} \kappa(\ell,t,\theta)\pi(d\ell)] = 0.$$

Thus Condition 5.4.1 is equivalent to the following condition.

Condition 5.4.2. *For any position* $(t,x) \in T^0 \times R^n$ *the equality*

$$\max_{\pi\in\Pi} [\kappa(\ell_\pi,t,\theta) - \int_{\mathcal{L}_0(t,x)} \kappa(\ell,t,\theta)\pi(d\ell)] = 0 \tag{8}$$

is fulfilled, where $\ell_\pi = \int_{\mathcal{L}_0(t,x)} \ell\pi(d\ell)$.

The meaning of condition (8) can be clarified by the following example. Assume that $\mathcal{L}_0 = \{\ell^{(1)},\ell^{(2)}\}$. In this case equality (8) means that the linear function

$$\alpha \to \alpha\kappa(\ell^{(1)},t,\theta) + (1-\alpha)\kappa(\ell^{(2)},t,\theta) : [0,1] \to R$$

majorizes the function $\alpha \to \kappa(\ell_\alpha,t,\theta)$ where $\ell_\alpha = \alpha\ell^{(1)} + (1-\alpha)\ell^{(2)}$.

In the arguments presented above it was shown that the locally Lipschitzian function ω_p satisfies the equality $\omega_p(\theta,x) = \sigma(x)$ and the condition of v-stability in the region $T^0 \times R^n$. The condition of u-stability of this function is expressed by Condition 5.4.1 or Condition 5.4.2. Therefore, according to Theorem 4.2.1 we obtain the following assertion.

Theorem 5.4.1. *In order that the function of the programmed maximin* ω_p (5.1.23) *coincide with the value function* ω^0 *of a positional differential game* (5.1.14), (5.1.15) *on the whole space* $T \times R^n$, *it is necessary and sufficient that one of the conditions* 5.4.1 *or* 5.4.2 *be fulfilled.*

Thus Conditions 5.4.1 and 5.4.2 are necessary and sufficient regularity conditions of a programmed maximin on the whole space $T \times R^n$. Consider a modification of Theorem 5.4.1 which follows from the remark presented in Section 4.4.2.

Theorem 5.4.2. *Let the region $\mathcal{E}$ be defined according to* (5.2.9). *The equality*

$$\omega_p(t,x) = \omega^0(t,x) \quad \text{for} \quad (t,x) \in \text{cl}\ \mathcal{E}$$

is valid if and only if equality (7) *or equality* (8) *are fulfilled for all* $(t,x) \in \mathcal{E}$.

5.4.3. We state several corollaries of Theorems 5.4.1 and 5.4.2.

Corollary 5.4.1. *Let the function $\psi(\cdot,t) : R^n \to (-\infty,\infty]$ be convex for all $t \in T$. Then the function of the programmed maximin ω_p and the value function ω^0 coincide on the whole space $T \times R^n$ (recall that function ψ is defined by the equality* (5.1.22)).

Proof. If the function $\psi(\cdot,t)$ is convex, then the set $\mathcal{L}_0(t,x)$ defined according to (5) will be a convex compact set. We shall show that the function $\kappa(\cdot,t,\theta)$ is convex on $\mathcal{L}_0(t,x)$. Let $\delta > 0$. Consider the functions $\psi(\cdot,t + \delta)$ and $\psi(\cdot,t)$. These functions are convex. From (5) we have

$$\psi(\ell,t) = \langle \ell,\Phi(\theta,t)x \rangle - \omega_p(t,x), \ell \in \mathcal{L}_0(t,x),$$

i.e. the function $\psi(\cdot,t)$ is linear on $\mathcal{L}_0(t,x)$. Therefore the function $\psi(\cdot,t + \delta) - \psi(\cdot,t)$ is convex on $\mathcal{L}_0(t,x)$. According to (5.1.22)

$$\psi(\ell,t + \delta) - \psi(\ell,t) = \int_t^{t+\delta} \kappa(\ell,\tau,\theta)d\tau.$$

Thus we obtain that for any $\delta > 0$ the function

$$\ell \to \int_t^{t+\delta} \kappa(\ell,\tau,\theta)d\tau$$

is convex on $\mathcal{L}_0(t,x)$. This implies that $\kappa(\cdot,t,\theta)$ is convex on $\mathcal{L}_0(t,x)$.

For a convex function $\kappa(\cdot,t,\theta) : \mathcal{L}_0(t,x) \to R$ the inequality

$$\kappa(\ell_\pi,t,\theta) \leqslant \int_{\mathcal{L}_0(t,x)} \kappa(\ell,t,\theta)\pi(d\ell) \quad \text{for all} \quad \pi \in \Pi = \text{rpm}(\mathcal{L}_0(t,x))$$

is valid.
Hence

$$\max_{\pi\in\Pi} [\kappa(\ell_\pi,t,\theta) - \int_{\mathcal{L}_0(t,x)} \kappa(\ell,t,\theta)\pi(d\ell)] \leqslant 0.$$

Strict inequality is impossible here. We thus arrive at equality (8). From Thoerem 5.4.1 we obtain $\omega_p(t,x) = \omega^0(t,x)$, $(t,x) \in T \times R^n$. Corollary 5.4.1 is thus proved.

Consider the case when $\sigma(x) = \mathrm{dist}(x,\mathcal{M})$, i.e. the distance from point x up to the set $\mathcal{M}$. We assume that $\mathcal{M} = \{y \in R^n : [y]_m = (y_1, ..., y_m)^T \in \mathcal{M}_*\}$, where $\mathcal{M}_*$ is a convex compact set in R^m. Recall that in this case the programmed maximin is determined by relations (5.1.31) and (5.1.32) and is denoted by $\epsilon_0(t,x)$. Define the set

(9)
$$S_0(t,x) \overset{\text{def}}{=} \{s_0 \in R^m : \|s_0\| \leqslant 1, \ \langle s_0, [\Phi(\theta,t)]_m x\rangle + \int_t^\theta \kappa^{(m)}(s_0,\tau,\theta)d\tau - \varphi_{\mathcal{M}}^{(m)}(s_0) = \epsilon_0(t,x)\}.$$

If $(t,x) \in \mathcal{E}^+$ then, as it was noted in Section 5.2.3, $\|s_0\| = 1$ for all $s_0 \in S_0(t,x)$. Recall that $\mathcal{E}^+ \overset{\text{def}}{=} \{(t,x) \in T^0 \times R^n : \epsilon_0(t,x) > 0\}$.

Condition 5.4.3. *For all* $(t,x) \in \mathcal{E}^+$ *the equality*

(10)
$$\max_{v\in \mathcal{Q}} \min_{u\in \mathcal{P}} \max_{s\in S_0(t,x)} [\langle s,[\Phi(\theta,t)]_m B(t)u\rangle + \langle s,[\Phi(\theta,t)]_m C(t)v\rangle - \kappa^{(m)}(s,t,\theta)] = 0$$

is valid.

Condition 5.4.4. *For all* $(t,x) \in \mathcal{E}^+$ *the equality*

(11)
$$\max_{\pi\in\Pi} [\kappa^{(m)}(s_\pi,t,\theta) - \int_{S_0(t,x)} \kappa^{(m)}(s,t,\theta)\pi(ds)] = 0$$

is valid where

$$\Pi \overset{\text{def}}{=} \mathrm{rpm}(S_0(t,x)), \quad s_\pi \overset{\text{def}}{=} \int_{S_0(t,x)} s\pi(ds).$$

We note that Conditions 5.4.3 and 5.4.4 are equivalent. The case under consideration differs from the general case only in notation. Therefore taking into account that $\epsilon_0(t,x) \geqslant 0$ for all $(t,x) \in T \times R^n$ we arrive at the following modification of Theorem 5.4.2.

Corollary 5.4.2. *In order that the programmed maximin* ω_p (5.1.31) *coincide with the value of a positional differential game* (5.1.14), (5.1.24) *in the whole space* $T \times R^n$, *it is necessary and sufficient that one of the conditions* 5.4.3 *or* 5.4.4 *be fulfilled.*

5.4.4. We now turn to Problem 5.1.1 which was posed in Section 5.1.4. Recall the statement of this problem. Let an initial position (t_0,x_0), a termination time $t = \theta$, and the target set $\mathcal{M} \subset R^n$ be given. It is required to determine the positional strategy U^0 such that $x(\theta) \in \mathcal{M}$ for all $x(\cdot) \in X_1(t_0,x_0,U^0)$. Using the terminology adopted in Chapter 2 we can assert that this problem is an $(\mathcal{M}^*, \mathcal{N})$-approach problem, where $\mathcal{M}^* \overset{\text{def}}{=} \{(\theta,x) : x \in \mathcal{M}\}$, $\mathcal{N} \overset{\text{def}}{=} T \times R^n$. For the $(\mathcal{M}^*, \mathcal{N})$-

approach problem under consideration one can define the maximal u-stable bridge W_0 and the set of programmed absorption W_p. It is useful to compare these two sets.

According to Theorem 2.4.3 the maximal u-stable bridge W_0 can be defined as follows: A point $(t_0,x_0) \in W_0$ if and only if for any choice of positional strategy V there exists a motion $x(\cdot) \in X_2(t_0,x_0,V)$ which satisfies the condition $x(\theta) \in M$. This definition is the reason that the maximal u-stable bridge W_0 is called a set of positional absorption.

We now present the definition of a set of programmed absorption W_p (cf. Section 5.1.4). A point $(t_0,x_0) \in W_p$ if and only if for any programmed control $v(\cdot) \in \mathbf{V}_p$ there exists a motion $x(\cdot) \in X_2(t_0,x_0,v(\cdot))$ such that $x(\theta) \in M$. Here

$$X_2(t_0,x_0,v(\cdot)) \stackrel{\text{def}}{=} \{x(\cdot,t_0,x_0,u(\cdot),v(\cdot)) : u(\cdot) \in \mathbf{U}_p\}.$$

The class of positional strategies V is wider than the class of programmed controls $v(\cdot)$. Utilizing this fact one can prove that

$$(12) \qquad W_0 \subset W_p.$$

Observe that this inclusion is an analog of inequality (1). Moreover, (12) follows from (1) if one sets $\sigma(x) = \text{dist}(x,M)$.

Let ω^0 be the value function of a differential game (5.1.14), (5.1.24). It is known that (cf. Section 3.1)

$$W_0 = \{(t,x) \in T \times R^n : \omega^0(t,x) = 0\}.$$

It is also known (cf. Section 5.1.4) that

$$W_p = \{(t,x) \in T \times R^n : \epsilon_0(t,x) = 0\}.$$

Therefore from Corollary 5.4.2 we obtain

Corollary 5.4.3. *Let* $M \stackrel{\text{def}}{=} \{y \in R^n : [y]_m \in M_*\}$, *where* M_* *is a convex compact set in* R^m. *Let one of the conditions 5.4.3 or 5.4.4 be fulfilled. Then the sets of positional and programmed absorptions coincide, i.e.*

$$W_0 = W_p.$$

Conditions 5.4.3 and 5.4.4 are weaker than Condition 5.2.4. Note that in Section 5.2 a strategy of extremal aiming U^0 (5.2.16) was constructed. Under the conditions stipulated in Corollary 5.2.1 (cf. Section 5.2.2) the strategy U^0 yields a solution to the approach problem. In Section 5.2 some advantages of the strategy of extremal aiming were stated. When replacing the regularity conditions stated in Section 5.2 by the weaker conditions 5.4.3 and 5.4.4, one can try to modify the strategy of extremal aiming in the following manner, for example. Replace the unique maximizing vector $s_0(t,x)$ in conditions (5.2.31) by an arbitrary vector $s_0 \in S_0(t,x)$ and then define vector $U^0(t,x)$ in accordance with (5.2.16). Such a strategy would have the advantages described in Section 5.2.1. However, this modification of extremal aiming is not suitable for solving the problem under consideration since it does not assure that the point $x(\theta)$ will fall into the set $\mathcal{M}$.

Therefore, under Conditions 5.4.3 and 5.4.4 a positional strategy of the first player which yields a solution of Problem 5.1.1 is usually defined as a strategy extremal to the stable bridge $\mathcal{W}_0 = \mathcal{W}_p$. In accordance with the results of Chapter 2, such a strategy would guarantee the encounter $x(\theta) \in \mathcal{M}$ from the initial position $(t_0,x_0) \in \mathcal{W}_0 = \mathcal{W}_p$, i.e. from the initial position in which $\epsilon_0(t_0,x_0) = 0$. The advantages furnished in our situation by auxiliary programmed constructions are that the stable bridge coincides here with the set of a programmed absorption and is defined by the relatively simple relations (5.1.31) and (5.1.32). An analogous remark is also valid for the differential game (5.1.14) and (5.1.15). Under Conditions 5.4.1 or 5.4.2 it is not possible to construct an extremal aiming which guarantees an optimal result. A minimax strategy should be defined as a strategy extremal to a u-stable bridge $\mathcal{W} = \{(t,x) \in T \times R^n : \omega^0(t,x) \leqslant \omega^0(t_0,x_0)\}$. According to Theorem 5.4.1 we have $\omega^0 = \omega_p$, thus relations (5.1.23) facilitate the construction of the bridge $\mathcal{W}$.

5.4.5. We present examples which will illustrate the results of Sections 5.4.2 and 5.4.3.

Consider first the differential game in which the motion of the controlled system is described by the equation

$$\dot{x} = u + v \tag{14}$$

and the payoff is determined by the equality $\gamma(x(\cdot)) = \sigma(x(\theta))$, where σ is convex and satisfies the Lipschitz condition. We assume that the controls of players satisfy the constraints $u \in \mathcal{P}$, $v \in \mathcal{Q}$, $\mathcal{P}$ and $\mathcal{Q}$ are convex compact sets in R^n. Recall that such a differential game was considered in Section 4.5.2.

In this example the value of the positional differential game coincides with the programmed maximin. Indeed, since $\Phi(\theta,t) = E$ (the unit matrix) for the system (14), according to (5.1.21) – (5.1.23) we obtain

$$\begin{aligned} &\kappa(\ell,t,\theta) \overset{\text{def}}{=} (\theta - t)\,\kappa(\ell), \\ &\kappa(\ell) \overset{\text{def}}{=} \min_{u\in\mathcal{P}} \langle \ell,u\rangle + \max_{v\in\mathcal{Q}} \langle \ell,v\rangle, \qquad (15) \\ &\omega_p(t,x) = \max_{\ell\in R^n} [\langle \ell,x\rangle + (\theta - t)\kappa(\ell) - \sigma^*(\ell)]. \end{aligned}$$

It can be verified that Condition 5.4.1 is fulfilled in the game under consideration. This condition was actually verified in Section 4.5.2 and it is therefore not repeated; we directly utilize Theorem 5.4.1 which implies that the value of the programmed maximin ω_p (15) will be the value of the differential game under consideration. Formula (15) coincides with formula (4.5.10) given in Section 4.5.

Consider another more specific example. Let the motion of the controlled system be described by equations (5.2.26) and the payoff be given by equality (5.2.27). It is assumed that the control of the second player, in the same manner as in Section 5.2.5, is subject to the constraints

$$(v_1^2 + v_2^2)^{1/2} = \beta_2 \tag{16}$$

and the control of the first player now satisfies the constraint

$$|u_1| \leqslant \beta_1, \quad |u_2| \leqslant \beta_1. \tag{17}$$

It is assumed that $\beta_1 > \beta_2$. Using the calculations presented in Section 5.2.5 and taking into account that in the case under consideration

$$\kappa_1^{(2)}(s,t,\theta) = -\beta_1(\theta - t)(|s_1| + |s_2|),$$

we obtain the following formula which determines the programmed maximin:

$$\begin{aligned}\epsilon_0(t,x) = \max_{\|s\| \leqslant 1} [&s_1 g_1(t,x,\theta) + s_2\, g_2(t,x,\theta) \\ &- (\theta - t)^2/2(\beta_1(|s_1| + |s_2|) - \beta_2(s_1^2 + s_2^2)^{1/2})],\end{aligned} \tag{18}$$

where $g_i(t,x,\theta)$ $(i = 1,2)$ are defined by the equalities (5.2.30). Note that for $\epsilon_0(t,x) > 0$ the maximum in (18) is attained at the unit vectors s_0, i.e. $s_1^2 + s_2^2 = 1$ for $s \in S_0(t,x)$. Taking this into account one can show that for $\epsilon_0(t,x) > 0$ the maximum (18) is attained at either the unique vector $s_0(t,x)$ or $S_0(t,x) = \{s_0^{(1)}, s_0^{(2)}\}$, where $s_0^{(1)}$ and $s_0^{(2)}$ are vectors of the form

$$s_0^{(1)} = ((-1)^i,0)^T, \quad s_0^{(2)} = (0,(-1)^j)^T \quad (i,j = 1, 2).$$

We thus obtain four different pairs of vectors $\{s_0^{(1)}, s_0^{(2)}\}$.

We show that in this example Condition 5.4.3 is fulfilled. If the set $S_0(t,x)$ consists of a single vector then equality (10) follows immediately from relations (5.1.32).

Let the set $S_0(t,x)$ consist of a pair of vectors of the type indicated above. For definiteness set $s_0^{(1)} = (1,0)^T$, $s_0^{(2)} = (0,1)^T$. We write the expression for the quantity appearing on the l.h.s. of equality (10). We obtain

complicate the symbolism of concepts and formulas.

We define two more types of strategies.

Multivalued strategies. Consider a multivalued mapping $(t,x) \to \mathcal{V}(t,x)$ defined on the space of positions $T \times R^n$. We assume that $\mathcal{V}(t,x) \subset \mathcal{Q}$ and that $\mathcal{V}(t,x)$ is a convex compact set for all $(t,x) \in T \times R^n$. It is supposed that the mapping $\mathcal{V}$ is upper semicontinuous. The totality of such mappings $\mathcal{V}$ will be denoted by $\mathbf{V}_{mv}$ (mv-multivalued). For a multivalued strategy $\mathcal{V} \in \mathbf{V}_{mv}$ the set $\mathcal{X}(t_0,x_0,\mathcal{V})$ is defined as a collection of absolutely continuous solutions $x(\cdot) : [t_0,\theta] \to R^n$ of the differential inclusion

$$\dot{x}(t) \in \{A(t)x(t) + B(t)u + C(t)v : u \in \mathcal{P},\ v \in \mathcal{V}(t,x,(t))\} \tag{3}$$

satisfying the initial condition $x(t_0) = x_0$.

Programmed strategies. In Section 5.1. the set $\mathbf{V}_p$ was introduced consisting of all measurable functions $v(\cdot) : T \to \mathcal{Q}$. The set of motions $\mathcal{X}(t_0,x_0,v(\cdot))$ generated by a programmed strategy $v(\cdot) \in \mathbf{V}_p$ is defined by the equality

$$\mathcal{X}(t_0,x_0,v(\cdot)) \overset{\text{def}}{=} \{x(\cdot,t_0,x_0,u(\cdot),v(\cdot)) : u(\cdot) \in \mathbf{U}_p\}.$$

For each one of these types of strategies for the second player we shall determine an optimal result. Thus consider the following quantities:

$$\omega^0(t_0,x_0) \overset{\text{def}}{=} \omega_{fb}(t_0,x_0) \overset{\text{def}}{=} \max_{V}\min_{x(\cdot)} \sigma(x(\theta)) \quad \text{for } V \in \mathbf{V}_{fb},\ x(\cdot) \in \mathcal{X}_2(t_0,x_0,V), \tag{4}$$

$$\omega_{cfb}(t_0,x_0) \overset{\text{def}}{=} \sup_{V}\min_{x(\cdot)} \sigma(x(\theta)) \text{ for } V \in \mathbf{V}_{cfb},\ x(\cdot) \in \mathcal{X}_2(t_0,x_0,V), \tag{5}$$

$$\omega_c(t_0,x_0) \overset{\text{def}}{=} \sup_{V}\min_{x(\cdot)} \sigma(x(\theta)) \text{ for } V \in \mathbf{V}_c,\ x(\cdot) \in \mathcal{X}^c(t_0,x_0,V), \tag{6}$$

$$\omega_{mv}(t_0,x_0) \overset{\text{def}}{=} \sup_{V}\min_{x(\cdot)} \sigma(x(\theta)) \text{ for } V \in \mathbf{V}_{mv},\ x(\cdot) \in \mathcal{X}(t_0,x_0,\mathcal{V}), \tag{7}$$

$$\omega_p(t_0,x_0) \overset{\text{def}}{=} \max_{v(\cdot)}\min_{x(\cdot)} \sigma(x(\theta)) = \max_{v(\cdot)}\min_{u(\cdot)} \sigma(x(\theta,t_0,x_0,u(\cdot),v(\cdot))) \quad \text{for } v(\cdot) \in \mathbf{V}_p,\ x(\cdot) \in \mathcal{X}(t_0,x_0,v(\cdot)),\ u(\cdot) \in \mathbf{U}_p. \tag{8}$$

As was shown in Section 5.4.1 the inequality $\omega^0(t_0,x_0) \geqslant \omega_p(t_0,x_0)$ is valid. Functions ω^0 and ω_p coincide provided the regularity conditions are fulfilled. These conditions single out a whole class of quite interesting differential games. However, in the whole collection of linear differential games of the form (5.1.14), (5.1.15) those which satisfy the regularity condition apparently constitute only a small fraction. In other words the fulfillment of the strict inequality $\omega^0(t_0,x_0) > \omega_p(t_0,x_0)$ is a more typical state of affairs. This is clear since $\omega^0(t_0,x_0)$ is an optimal result for the second player in the class of positional strategies which implement a feedback control depending on the position of the game that was realized. Under this type of control the second player enjoys more possibilities than under a programmed control.

Strategies $V \in \mathbf{V}_{cfb}$, $v \in \mathbf{V}_c$ and $\mathcal{V} \in \mathbf{V}_{mv}$ also implement a control for the second player depending on the position that was realized. At first glance, they would have to be hardly different from the positional strategies $V \in \mathbf{V}_{fb}$ in this respect. Therefore one would expect that except for relatively rare occasions the strict inequalities $\omega_{cfb}(t_0,x_0) > \omega_p(t_0,x_0)$, $\omega_c(t_0,x_0) > \omega_p(t_0,x_0)$, $\omega_{mv}(t_0,x_0) > \omega_p(t_0,x_0)$ must be valid. However, this hypothesis is not true as the following theorem indicates.

Theorem 5.5.1. *For the linear differential game* (5.1.14), (5.1.15) *where* $\sigma : R^n \to R$ *is a convex function for each initial position* $(t_0,x_0) \in T \times R^n$, *the equalities*

$$\omega_{cfb}(t_0,x_0) = \omega_c(t_0,x_0) = \omega_{mv}(t_0,x_0) = \omega_p(t_0,x_0) \tag{9}$$

are valid.

Before proceeding to the proof of the theorem we shall make a few comments concerning equality (9).

Consider first the equality

$$\omega_{mv}(t_0,x_0) = \omega_p(t_0,x_0). \tag{10}$$

Let $V : T \times R^n \to Q$ be an arbitrary function. For this function we define two sets of motions $X_2(t_0,x_0,V)$ and $X^0(t_0,x_0,V)$. The first set is the set of constructive motions determined by a limiting transition of sequences of step-by-step motions. The second set $X^0(t_0,x_0,V)$ is a set of generalized motions. It is defined as a collection of solutions of differential inclusion (3), where

$$\mathcal{V}(t,x) \overset{\text{def}}{=} co[\bigcap_{\epsilon>0} cl\{V(\tau,y) : |t - \tau| + \|x - y\| \leqslant \epsilon\}]$$

(recall that such a construction was described in Section 1.3.4 for strategies of the first player). Under our definition of a multivalued strategy $\mathcal{V}$ we have $X^0(t_0,x_0,V) = X(t_0,x_0,\ \ )$. Therefore equality (10) implies that for any function $V : T \times R^n \to Q$ the bound

$$\min_{x(\cdot)\in X^0(t_0,x_0,V)} \sigma(x(\theta)) \leqslant \omega_p(t_0,x_0) \tag{11}$$

is valid. This bound asserts that by choosing an arbitrary function V and working with the whole set of generalized motions

$X^0(t_0,x_0,V)$ the second player cannot improve on the optimal result attained in the class of programmed controls. On the other hand, as was mentioned above in many examples, a strict inequality $\omega^0(t_0,x_0) > \omega_p(t_0,x_0)$ is valid, i.e. there exists a function $V^0 : T \times R^n \to$ such that

$$\min_{x(\cdot)\in X_2(t_0,x_0,V^0)} \sigma(x(\theta)) > \omega_p(t_0,x_0). \tag{12}$$

We emphasize that here $X_2(t_0,x_0,V^0)$ is the set of constructive motions. Thus relations (11) and (12) demonstrate the difference between two formalizations of strategies, one is based on the utilization of constructive motions and the other is based on the utilization of generalized motions. As seen from (11) and (12), the formalization utilizing the set of generalized motions does not permit the second player to make full use of all the information available to him/her concerning the current position of the game.

Consider the relations

$$\omega^0(t_0,x_0) = \omega_{fb}(t_0,x_0) > \omega_p(t_0,x_0) = \omega_{cfb}(t_0,x_0).$$

These relations show that the result guaranteed by a discontinuous positional strategy $V^0 \in \mathbf{V}_{fb}$ *cannot be approximated* by choosing continuous positional strategies $V \in \mathbf{V}_{cfb}$. The conclusions obtained herein supplement the arguments which were presented in §1.3.4 and §1.4.4 where the formalization of positional strategies suggested in Chapter 1 was discussed.

5.5.2. We sketch the main points of the proof of Theorem 5.5.1.

We shall show that

$$\omega_p(t_0,x_0) \leqslant \omega_{cfb}(t_0,x_0), \quad \omega_p(t_0,x_0) \leqslant \omega_c(t_0,x_0),$$
$$\omega_p(t_0,x_0) \leqslant \omega_{mv}(t_0,x_0). \tag{13}$$

Let the programmed control $v^0(\cdot) \in \mathbf{V}_p$ be selected from the condition

$$\min_{x(\cdot)\in X(t_0,x_0,v^0(\cdot))} \sigma(x(\theta)) = \max_{v(\cdot)\in \mathbf{V}_p} \min_{x(\cdot)\in X(t_0,x_0,v(\cdot))} \sigma(x(\theta)) = \omega_p(t_0,x_0).$$

The measurable function $v^0(\cdot)$ can be approximated by a continuous function $v_\epsilon(\cdot)$ such that

$$\min\{\sigma(x(\theta)) : x(\cdot) \in X(t_0,x_0,v_\epsilon(\cdot))\} \geqslant$$
$$\geqslant \min\{\sigma(x(\theta)) : x(\cdot) \in X(t_0,x_0,v^0(\cdot))\} - \epsilon. \tag{14}$$

The function $v_\epsilon(\cdot)$ can be determined, for example, by using Lusin's theorem [36, 50, 74].

Define a continuous function

$$V_\epsilon(t,x) = v_\epsilon(t) \quad \text{for} \quad (t,x) \in T \times R^n.$$

Consider also the mapping

$$\mathcal{V}_\epsilon(t,x) = \{v_\epsilon(t)\} \quad \text{for } (t,x) \in T \times R^n.$$

The equalities

$$(15) \qquad X_2(t_0,x_0,V_\epsilon) = X^0(t_0,x_0,V_\epsilon) = X(t_0,x_0,\mathcal{V}_\epsilon) = X(t_0,x_0,v_\epsilon(\cdot))$$

are valid. Here the first set is a set of constructive motions generated by a continuous positional strategy $V_\epsilon \in \mathbf{V}_{cfb}$. The second set is the set of motions generated by a continuous strategy $V^\epsilon \in \mathbf{V}_c$ and defined in accordance with (1), (2). The third set is defined in accordance with (3). In view of (5) – (7) we have from (14), (15) that

$$\omega_{cfb}(t_0,x_0) \geqslant \omega_p(t_0,x_0) - \epsilon, \quad \omega_0(t_0,x_0) \geqslant \omega_p(t_0,x_0) - \epsilon,$$

$$\omega_{mv}(t_0,x_0) \geqslant \omega_p(t_0,x_0) - \epsilon.$$

Here $\epsilon > 0$ is an arbitrarily small number, thus the bound (13) is verified.

Assume that

$$(16) \qquad \omega_{mv}(t_0,x_0) > \omega_p(t_0,x_0).$$

Then there exists a multivalued strategy $\mathcal{V}_* \in \mathbf{V}_{mv}$ such that

$$(17) \qquad \min\{\sigma(x(\theta)) : x(\cdot) \in X(t_0,x_0,\mathcal{V}_*)\} > \omega_p(t_0,x_0).$$

We now present constructions which refute the inequalities (16) and (17).

Define the set

$$(18) \qquad \begin{aligned} \mathbb{R} \overset{\text{def}}{=} \{(u(\cdot),v(\cdot)) \in \mathbf{U}_p \times \mathbf{V}_p : \\ \sigma(x(\theta,t_0,x_0,u(\cdot),v(\cdot)) \leqslant \omega_p(t_0,x_0)\}. \end{aligned}$$

Here, as well as in the succeeding constructions, we shall assume that functions $u(\cdot)$ and $v(\cdot)$ are defined on the interval $[t_0,\theta]$.

We construct a multivalued mapping which assigns to the pair $(u(\cdot),v(\cdot)) \in \mathbb{R}$ a nonempty subset $\mathbb{R}_*(u(\cdot),v(\cdot)) \subset \mathbb{R}$.

Let a pair $(u(\cdot),v(\cdot)) \in \mathbb{R}$ be chosen. The motion $x(\cdot,t_0,x_0,u(\cdot),v(\cdot))$ corresponds to this pair. Set

$$V_*[t,u(\cdot),v(\cdot)] \overset{\text{def}}{=} V_*(t,x(t,t_0,x_0,u(\cdot),v(\cdot))).$$

Denote by $\mathbf{V}_*(u(\cdot),v(\cdot))$ the set of measurable functions $v_*(\cdot) : [t_0,\theta] \to Q$ which for almost all $t \in [t_0,\theta]$ satisfies the inclusion

$$v_*(t) \in V_*[t,u(\cdot),v(\cdot)].$$

It follows from Theorem A.3.2 (the theorem on measurable selection) that $\mathbf{V}_*(u(\cdot),v(\cdot)) \neq \phi$. Note that $\mathbf{V}_*(u(\cdot),v(\cdot)) \subset \mathbf{V}_p$.

We now form the set $\mathbb{R}_*(u(\cdot),v(\cdot))$. Set

$$\mathbb{R}_*(u(\cdot),v(\cdot)) \overset{\text{def}}{=} \{(u_*(\cdot),v_*(\cdot)) \in \mathbb{R} : v_*(\cdot) \in \mathbf{V}_*(u(\cdot),v(\cdot))\}.$$

Observe that it follows from the definition of the programmed maximin that for any $v_*(\cdot) \in \mathbf{V}_p$ there exists $u_*(\cdot) \in \mathbf{U}_p$ satisfying the inequality

$$(19) \qquad \sigma(x(\theta,t_0,x_0,u_*(\cdot),v_*(\cdot))) \leqslant \omega_p(t_0,x_0).$$

We have thus constructed the mapping $(u(\cdot),v(\cdot)) \to \mathbb{R}_*(u(\cdot)),v(\cdot))$. It turns out that the constructed mapping possesses a fixed point, i.e. there exists a pair $(u^0(\cdot),v^0(\cdot)) \in \mathbb{R}$ which satisfies the condition

$$(20) \qquad (u^0(\cdot),v^0(\cdot)) \in \mathbb{R}_*(u^0(\cdot),v^0(\cdot)).$$

(A proof of this assertion reduces to the verification of conditions for the well known fixed point theorem (cf. [7]). We shall omit this inspection.)

We shall show that the existence of a fixed point $(u^0(\cdot),v^0(\cdot))$(20) contradicts inequality (17). Indeed, the inclusion $(u^0(\cdot),v^0(\cdot)) \in \mathbb{R}$ means that the inequality

$$(21) \qquad \sigma(x(\theta,t_0,x_0,u^0(\cdot),v^0(\cdot))) \leqslant \omega_p(t_0,x_0)$$

is fulfilled. On the other hand, from condition (20) using the construction of the set $\mathbb{R}_*(u^0(\cdot),v^0(\cdot))$, we obtain that the motion $x(\cdot,t_0,x_0,u^0(\cdot),v^0(\cdot))$ satisfies the differential inclusion

$$\dot{x}(t) \in \{A(t)x(t) + B(t)u + C(t)v : u \in P,\ \ v \in V_*(t,x,(t))\},$$

since $u^0(t) \in P$, $v^0(t) \in V_*(t,x(t,t_0,x_0,u^0(\cdot),v^0(\cdot)))$ for almost all $t \in [t_0,\theta]$. We have thus shown that $x(\cdot,t_0,x_0,u^0(\cdot),v^0(\cdot)) \in X(t_0,x_0,V_*)$ and this motion satisfies the inequality (21). This yields a contradiction to (17).

The contradiction obtained proves the inequality $\omega_{mv}(t_0,x_0) \leqslant \omega_p(t_0,x_0)$. From this inequality and the inequality $\omega_{mv}(t_0,x_0) \geqslant \omega_p(t_0,x_0)$ (cf. (13)) follows the equality $\omega_{mv}(t_0,x_0) = \omega_p(t_0,x_0)$.

Now let V be an arbitrary continuous strategy, i.e. $V \in \mathbf{V}_c$. Set

$$V_V(t,x) \stackrel{\text{def}}{=} \{V(t,x)\}, \quad (t,x) \in T \times R^n.$$

It is easy to observe that the mapping V_V defined herein is contained in the class $\mathbf{V}_{mv}$ and the equality $X^c(t_0,x_0,V) = X(t_0,x_0,V_V)$ is valid. Therefore, the inequality $\omega_c(t_0,x_0) \leqslant \omega_p(t_0,x_0)$ follows from the constructions presented above. Taking the second inequality in (13) into account we arrive at the equality $\omega_c(t_0,x_0) = \omega_p(t_0,x_0)$.

It remains to verify the inequality $\omega_{cfb}(t_0,x_0) \leqslant \omega_p(t_0,x_0)$. Assume the contrary. Then there exists a continuous positional strategy $V_* \in \mathbf{V}_{cfb}$ and a partition $\Delta = \{t_0 = \tau_0 < \tau_1 < ... < \tau_{k+1} = \theta\}$ such that

$$\text{(22)} \qquad \min\{\sigma(x(\theta,t_0,x_0,u(\cdot),V_*,\Delta)) : u(\cdot) \in \mathbf{U}_p\} < \omega_p(t_0,x_0).$$

We construct the mapping $(u(\cdot),v(\cdot)) \to \mathbf{R}^*(u(\cdot),v(\cdot))$ which assigns to the pair $(u(\cdot),v(\cdot)) \in \mathbf{R}$ a nonempty subset $\mathbf{R}^*(u(\cdot),v(\cdot)) \subset \mathbf{R}$ The only difference here from the constructions considered above is that in place of the set $\mathbf{V}_*(u(\cdot),v(\cdot))$ a unique control

$$v_*(t) \stackrel{\text{def}}{=} V_*(\tau_i,x(\tau_i,t_0,x_0,u(\cdot),v(\cdot))), \quad \tau_i \leqslant t < \tau_{i+1}, \quad i = 0, 1, ..., k$$

is introduced. As in the preceding case, the constructed mapping possesses a fixed point which contradicts inequality (22). This contradiction proves the inequality $\omega_{cfb} \leqslant \omega_p(t_0,x_0)$.

5.5.3. Theorem 5.5.1 can be utilized for an investigation of the necessary conditions under which the value function of a differential game (5.1.14), (5.1.15) will be continuously differentiable.

In the theorem presented below the following notation is used:

$$\mathcal{E}^0 \stackrel{\text{def}}{=} \{(t,x) \in T^0 \times R^n : a < \omega^0(t,x) < b\},$$

$$\mathcal{E}_p \stackrel{\text{def}}{=} \{(t,x) \in T^0 \times R^n : a < \omega_p(t,x) < b\}.$$

Here $-\infty \leqslant a < b \leqslant \infty$.

Theorem 5.5.2. *Let the value function ω^0 of the differential game* (5.1.14), (5.1.15) *be continuously differentiable in the region $\mathcal{E}^0$. Then* $\omega^0(t,x) = \omega_p(t,x)$ for $(t,x) \in cl\ \mathcal{E}^0$ *and for any position $(t,x) \in \mathcal{E}^0$, the set $\mathfrak{L}_0(t,x)$* (5.2.1) *contains the unique element $\mathfrak{l}_0(t,x)$. For any position (t,x) in the region $\mathcal{E}_p$ let the set $\mathfrak{L}_0(t,x)$* (5.2.1) *contain a unique element $\mathfrak{l}_0(t,x)$. Then $\omega^0(t,x) = \omega_p(t,x)$ for $(t,x) \in cl\ \mathcal{E}_p$ and the function ω^0 is continuously differentiable in the region $\mathcal{E}_p$.*

Proof. Consider first the second part of Theorem 5.5.2. Let $\mathfrak{L}_0(t,x) = \{\mathfrak{l}_0(t,x)\}$ for $(t,x) \in \mathcal{E}_p$ i.e. Condition 5.2.2 is fulfilled. Then according to Theorem 5.2.2, $\omega^0(t,x) = \omega_p(t,x)$ for $(t,x) \in cl\ \mathcal{E}_p$. In the proof of Theorems 5.2.1 and 5.2.2 it was shown that the function ω_p

is continuously differentiable in the region $\mathcal{E}_p$. Thus the second part of Theorem 5.5.2 is proved.

Let the function ω^0 be continuously differentiable in the region $\mathcal{E}_p^0$. Define a multivalued strategy $\mathcal{V}^0$ as follows. Set

$$\mathcal{V}^0(t,x) \overset{\text{def}}{=} \{v^0 \in \mathcal{Q} : (\partial\omega^0(t,x)/\partial x)C(t)v^0$$

$$= \max_{v\in\mathcal{Q}} (\partial\omega^0(t,x)/\partial x)C(t)v\} \quad \text{for} \quad (t,x) \in \mathcal{E}^0$$

$$\mathcal{V}^0(t,x) \overset{\text{def}}{=} \mathcal{Q} \quad \text{for} \quad (t,x) \in (T \times R^n)\setminus\mathcal{E}^0.$$

For an arbitrary initial position $(t_0,x_0) \in \mathcal{E}^0$ the inequality

$$\min\{\sigma(x(\theta)) : x(\cdot) \in X(t_0,x_0, \mathcal{V}^0)\} \geqslant \omega^0(t_0,x_0) \tag{23}$$

is valid (cf. the proof of Theorem 4.4.1).

From (23), in accordance with (7), we have $\omega_{mv}(t_0,x_0) \geqslant \omega^0(t_0,x_0)$. From (9) we obtain $\omega_p(t_0,x_0) \geqslant \omega^0(t_0,x_0)$. On the other hand it is known (cf. 5.4.1)) that $\omega_p(t_0,x_0) \leqslant \omega^0(t_0,x_0)$. Consequently

$$\omega^0(t_0,x_0) = \omega_p(t_0,x_0) \quad \text{for all} \quad (t_0,x_0) \in \mathcal{E}^0.$$

In particular it follows from this equality that one cannot have a strict inequality in (23).

We have thus obtained that the function ω_p is continuously differentiable in the region $\mathcal{E}^0$.

Let $(t,x) \in \mathcal{E}^0$, $h \in R^n$. We shall compute the directional derivative $D\omega_p(t,x)|(0,h)$. Since the function ω_p is differentiable at point (t,x) we have

$$D\omega_p(t,x)|(0,h) = (\partial\omega_p(t,x)/\partial x)h\,. \tag{24}$$

Consequently $h \to D\omega_p(t,x)|(0,h)$ is a linear function. On the other hand

$$D\omega_p(t,x)|(0,h) = \max_{\ell\in\mathcal{L}_0(t,x)} \langle \ell,\Phi(\theta,t)h\rangle, \quad h \in R^n. \tag{25}$$

is valid. This formula is an analog of equality (5.4.4). The validity of formula (25) follows easily from the proof of Lemma 4.3.1. From (24), (25) we obtain

$$(\partial\omega_p(t,x)/\partial x)h = \max_{\ell\in\mathcal{L}_0(t,x)} \langle \ell,\Phi(\theta,t)h\rangle, \quad h \in R^n.$$

This equality implies that the set $\mathcal{L}_0(t,x)$ must contain a single element. Theorem 5.5.2 is thus proved.

Setting $a = -\infty$, $b = \infty$ in the statement of Theorem 5.5.2 we obtain

Corollary 5.5.1. *In order that the value function* ω^0 *of the differential game* (5.1.14), (5.1.15) *be continuously differentiable in the region* $T^0 \times R^n$ *it is necessary and sufficient that* $\omega^0(t,x) = \omega_p(t,x)$ *for all* $(t,x) \in T \times R^n$ *and that condition* 5.5.2 *be fulfilled in the region* $T^0 \times R^n$.

In conclusion consider the differential game (5.1.14), (5.1.24). Set

$$\mathcal{E}^+ \overset{\text{def}}{=} \{(t,x) \in T^0 \times R^n : \epsilon_0(t,x) > 0\}.$$

where $\mathcal{E}_0(t,x)$ is a programmed maximin defined in accordance with (5.1.27), (5.1.28). Recall that Condition 5.2.2 for the function ϵ_0 in the region $\mathcal{E}^+$ is equivalent to Condition 5.2.3. Let ω^0 be the value function of the differential game (5.1.14), (5.1.24). Recall also that the equality of the functions ω^0 and ϵ_0 in the region $\mathcal{E}^+$ yields the equality of these functions in the whole space $T \times R^n$. Therefore the following corollary is valid.

Corollary 5.5.2. *In order that the value function* ω^0 *of the differential game* (5.1.14), (5.1.24) *be continuously differentiable in the region* $\mathcal{E}^+$ *it is necessary and sufficient that* $\epsilon_0(t,x) = \omega^0(t,x)$ *for all* $(t,x) \in T \times R^n$ *and that condition* 5.2.3 *be fulfilled.*

In the case when the set M is of the form (5.1.30), Corollary 5.5.2 remains valid, moreover Condition 5.2.3 can in this case be replaced by a more convenient Condition 5.2.4.

Chapter 6
EXTREMAL AIMING FOR NONLINEAR DIFFERENTIAL GAMES

6.1. Generalized Programmed Controls

The method of extremal aiming can be developed for the solution of nonlinear differential games. As in the linear case, it is useful to consider auxiliary problems of programmed control. The purpose of these investigations is to construct a positional strategy from a synthesis of the solutions of the problems of programmed control. As a rule, we consider problems in which it is required to determine the maximin of a target functional in the class of programmed controls. Under specific conditions, to be called regularity conditions, the programmed maximin coincides with the value of a positional differential game. In the preceding chapter some auxiliary problems for programmed controls were considered and regularity conditions on linear differential games were formulated. We now proceed in the same fashion for nonlinear differential games. Obviously when one proceeds to the nonlinear case new problems arise which require solution. The first such problem is choosing a class of programmed controls. Recall that up until now programmed controls were identified with measurable time functions satisfying given constraints. For linear controlled systems in a class of such programmed controls, a solution of the auxiliary problem under consideration is attained (i.e. optimal controls do exist.) However, as is known in the case of a nonlinear controlled system, optimal programmed controls may not exist. In such a case one can either try to develop programmed constructions working with ϵ-optimal solutions or confine himself to the consideration of systems for which there exist optimal programmed controls. In the course of implementation of the first possibility very cumbersome constructions are encountered. The second route leads to an unjustifiable narrowing of the class of problems under investigation.

There is, however, a third possibility. This involves extending the class of programmed controls and transition to generalized programmed controls which take values on a set of probability measures. Such a device is well known for problems of optimal control. In this section it will be developed for maximin problems of programmed control. First we present in Section 6.1.1 some necessary facts from functional analysis. In Section 6.1.2 formal definitions of generalized programmed controls are given. Next in Section 6.1.3 motivation for these definitions is presented, and it is shown how the ordinary and generalized controls are interconnected. In Section 6.1.4 properties of generalized programmed controls are considered.

6.1.1. First we present some basic results of functional analysis. A detailed exposition of these facts can be obtained, for example, in the text [74].

Let $\mathcal{Z}$ be a metric compact set. Assume for definiteness, that $\mathcal{Z}$ is a compact subset of a Euclidean space R^m. The symbol $\mathcal{C}(\mathcal{Z})$ denotes the space of continuous functions $g : \mathcal{Z} \to R$. On this space the norm

$$|g|_{\mathcal{C}} \overset{\text{def}}{=} \max_{z \in \mathcal{Z}} |g(z)|$$

is defined.

Let $T = [t_*, \theta]$, assume that on the interval T the Lebesgue measure is defined. Consider the space $\mathcal{B}(T, \mathcal{Z})$, whose elements are the functions $(t,z) \to h(t,z) : T \times \mathcal{Z} \to R$ satisfying the following conditions:

(a) $h(t,\cdot) \in \mathcal{C}(\mathcal{Z})$ for all $t \in T$;

(b) $h(\cdot,z)$ is measurable for all $z \in \mathcal{Z}$;

(c) there exists an integrable function $t \to r(t) : T \to R$ such that

$$|h(t,\cdot)|_{\mathcal{C}} \overset{\text{def}}{=} \max_{z \in \mathcal{Z}} |h(t,z)| \leqslant r(t)$$

for $t \in T$.

More precisely, the elements of the space $\mathcal{B}(T, \mathcal{Z})$ are equivalence classes. Two functions $h^{(1)}$ and $h^{(2)}$ are equivalent provided $h^{(1)}(t,\cdot) = h^{(2)}(t,\cdot)$ for almost all $t \in T$. On the space $\mathcal{B}(T, \mathcal{Z})$ the norm

$$|h|_{\mathcal{B}} = \int_T |h(t,\cdot)|_{\mathcal{C}} dt$$

is defined. It is known that the space $\mathcal{B}(T, \mathcal{Z})$ is separable. The set of functions of the form

$$h(t,z) = \sum_{i=1}^{k} g_i^{(1)}(t) g_i^{(2)}(z) \qquad g_i^{(1)} \in \mathcal{C}(T) \quad g_i^{(2)} \in \mathcal{C}(\mathcal{Z})$$

is everywhere dense in $\mathcal{B}(T, \mathcal{C})$.

We shall describe a general form of linear continuous functionals defined on spaces $\mathcal{C}(\mathcal{Z})$ and $\mathcal{B}(T, \mathcal{Z})$.

Let $frm(\mathcal{Z})$ be the set whose elements are countably-additive regular functions α defined on a σ-algebra of Borel sets in Z. These functions take on values on the real line R. It is known that any function $\alpha \in frm(Z)$ possesses a finite total variation

$$|\alpha|(Z) \overset{\text{def}}{=} \sup \sum_{i=1}^{k} |\alpha(G_i)| < \infty,$$

where $G_i \subset Z$ are Borel sets, $G_i \cap G_j = \phi$ $(i \neq j)$, k is an integer. The elements of the set $frm(Z)$ are called *finite regular measures* and also *Radon measures*. The set $frm(Z)$ is a vector space. On this space a (strong) norm can be introduced which is defined by the equality

$$\|\alpha\|_s \overset{\text{def}}{=} |\alpha|(Z).$$

It is known that the space $frm(Z)$ is isomorphic to the space $C^*(Z)$, i.e. the space of linear continuous functionals defined on $C(Z)$. For any $\ell \in C^*(Z)$ there exists a unique $\alpha \in frm(Z)$ such that

$$\ell(g) = \int_Z g(z)\alpha(dz), \quad g \in C(Z).$$

On the other hand, for any $\alpha \in frm(Z)$ this equality defines a linear continuous functional $\ell : C(Z) \to R$.

In addition to the above mentioned strong norm in the space $frm(Z)$ (and hence in the space $C^*(Z)$) one can define the so-called weak norm. For this purpose a countable everywhere dense subset $\{g_i\}$ $(i = 1, 2, ...)$ is selected in the space $C(Z)$. Set

$$\varphi_i(\alpha) \overset{\text{def}}{=} \int_Z g_i(z)\alpha(dz), \quad i = 1, 2, ...,$$
$$\|\alpha\|_w \overset{\text{def}}{=} \sum_{i=1}^{\infty} 2^{-i}|\varphi_i(\alpha)|(1 + |g_i|_C)^{-1}. \tag{1}$$

The symbol $(frm(Z), \|\cdot\|_w)$ denotes a normed space $frm(Z)$ in which the norm is defined by equality (1).

In the space $frm(Z)$ we select a subset $rpm(Z)$ whose elements are regular probability measures, i.e. for any $\alpha \in rpm(Z)$ the relations $\alpha(Z) = 1$, $\alpha(G) \geqslant 0$, where $G \subset Z$ an arbitrary Borel set is valid. It is known that $rpm(Z)$ is a compact set in the space $(frm(Z), \|\cdot\|_w)$.

We now proceed to a description of a linear continuous functional of a general form defined on the space $B(T, Z)$. Consider the functions

$$\beta : T \to (\mathrm{frm}(Z), \|\cdot\|_w). \tag{2}$$

We shall assume that these functions satisfy the following two conditions. First, functions β are essentially bounded, i.e. for any function β there exists a number $b < \infty$ such that the inequality $\|\beta(t)\|_s = |\beta(t)|(Z) \leqslant b$ is fulfilled almost everywhere on T (here

$\|\beta(t)\|_s$ is a strong norm). Secondly, the functions β should be measurable. Recall that on the interval T a Lebesque measure is defined and $(frm(\mathcal{Z}), \|\cdot\|_w)$ is a normed space, therefore, the measurability property of functions β can be introduced in the usual manner. It is also known that essentially bounded functions will be measurable if and only if for any $g \in C(\mathcal{Z})$ the function

$$t \to \varphi(t) \overset{\text{def}}{=} \int_{\mathcal{Z}} g(z)\beta(t)(dz) : T \to R$$

is Lebesgue measurable.

Let $h \in B(T, \mathcal{Z})$. Introduce the notation

$$h(t,\beta(t)) \overset{\text{def}}{=} \int_{\mathcal{Z}} h(t,z)\beta(t)(dz). \tag{3}$$

It can be shown that for any $h \in B(T, \mathcal{Z})$ and for any function β (2) with the properties stated above the mapping

$$t \to \psi(t) \overset{\text{def}}{=} h(t,\beta(t) : T \to R$$

will be Lebesgue measurable.

Let $B^*(T, \mathcal{Z})$ be the space conjugate to the space $B(T, \mathcal{Z})$. It is known that the space whose elements are the functions β is isomorphic to the space $B^*(T, \mathcal{Z})$. For any $\ell \in B^*(T, \mathcal{Z})$ there exists a unique function β of the form (2) with the properties indicated above such that

$$\ell(h) = \int_T h(t,\beta(t))dt, \quad h \in B(T, \mathcal{Z}). \tag{4}$$

On the other hand, for any essentially bounded and measurable function β the equality (4) defines a linear continuous functional $\ell : B(T, \mathcal{Z}) \to R$. It should be clarified that functions $\beta^{(1)}$ and $\beta^{(2)}$ are assumed to be equivalent if $\beta^{(1)}(t) = \beta^{(2)}(t)$ almost everywhere on T. Below we shall identify the elements of the space $B^*(T, \mathcal{Z})$ with the corresponding essentially bounded and measurable functions β of the form (2) (more precisely with equivalence classes of such functions).

On the space $B^*(T, \mathcal{Z})$ a weak norm can be defined. Let $\{h_i\}$, $i = 1, 2, \ldots$, be a countable everywhere dense set in $B(T, \mathcal{Z})$. Set

$$\begin{aligned} \psi_i(\beta) &\overset{\text{def}}{=} \int_T h_i(t,\beta(t))dt \quad (i = 1, 2, \ldots) \\ \|\beta\|_w &\overset{\text{def}}{=} \sum_{i=1}^{\infty} 2^{-i} \, |\psi_i(\beta)|(1 + |h_i|_B)^{-1}. \end{aligned} \tag{5}$$

The symbol $(B^*(T, \mathcal{Z}), \|\cdot\|_w)$ will denote the normed vector space $B^*(T, \mathcal{Z})$ in which the weak norm is defined by equality (5).

Select the set

$$\mathfrak{L} \overset{\text{def}}{=} \{\beta \in B^*(T, \mathcal{Z}) : \beta(t) \in rpm(\mathcal{Z}) \quad \text{a.e. on } T\} \tag{6}$$

in the space $B^*(T, Z)$. Thus one can assert that the measurable functions $\beta : T \to rpm(Z)$ are the elements of the set $\mathfrak{L}$. These functions are essentially bounded since $|\beta(t)|(Z)_* = 1$ $(t \in T)$. It is known that the set $\mathfrak{L}$ is compact in the space $(B^*(T, Z), \|\cdot\|_w)$. Note also that the topology of the set $\mathfrak{L}$ generated by the topology of the space $(B^*(T, Z), \|\cdot\|_w)$ coincides with the topology in $\mathfrak{L}$ induced by the weak star topology in $B^*(T, Z)$. In other words the following assertion is valid. Let $\beta_* \in \mathfrak{L}$ and $\beta_i \in \mathfrak{L}$ $(i = 1, 2, ...)$. Then $\lim_{i\to\infty}\beta_i = \beta_*$ (i.e. $\lim_{i\to\infty}\|\beta_i - \beta_*\|_w = 0$) if and only if

$$\lim_{i\to\infty} \int_T h(t,\beta_i(t))dt = \int_T h(t,\beta_*(t))dt$$

for all $h \in B(T, Z)$.

6.1.2. We now proceed to the definition of generalized programmed controls. Let P and Q be compact sets in R^p and R^q respectively. The set $\mathfrak{L}$ defined by relation (6) in which we set $Z = P \times Q$ will be denoted by the symbol H. The elements of the set H are denoted by the letter η and are called the generalized programmed controls. We thus obtain that the set H consists of measurable functions $\eta : T \to rpm(P \times Q)$.

Define the motions generated by generalized controls $\eta \in H$. We assume that the motion of a controlled system is described by an equation of the form (1.2.1). We shall assume that the function f satisfies conditions (i), (iii), and (iv) indicated in Section 1.2.2. Condition (ii) in this chapter is replaced by a stronger condition.

(ii)* The derivative $\partial f/\partial x$ exists and is continuous on $T \times R^n \times P \times Q$.

Let $\eta \in H$. We introduce the following notation:

$$f(t,x,\eta(t)) \overset{\text{def}}{=} \int_{P\times Q} f(t,x,u,v)\eta(t)(d(u,v)). \tag{7}$$

Observe that this notation corresponds to notation (3). We shall state certain properties of the function

$$(t,x) \to f(t,x,\eta(t)): \ T \times R^n \to R^n.$$

It can be shown that this function possesses the following properties:

(a) for any $x \in R^n$ the function $t \to f(t,x,\eta(t)): T \to R^n$ is measurable and bounded:

(b) for all $(t,x) \in T \times R^n$ the derivative

$$\frac{\partial f(t,x,\eta(t))}{\partial x} = \int_{P\times Q} \frac{\partial f(t,x,u,v)}{\partial x} \eta(t)(d(u,v))$$

exists for any $x \in R^n$, the function $t \to \partial f(t,x,\eta(t))/\partial x$ is measurable and bounded on T, and for any $t \in T$ the function $x \to \partial f(t,x,\eta(t))/\partial x$ is continuous on R^n.

(c) the bound

$$\langle x, f(t,x,\eta(t))\rangle \leq \xi(1 + \|x\|^2), \quad (t,x) \in T \times R^n$$

is valid.

It is known from the theory of ordinary differential equations (cf., for example, [13, 34, 74]) that there exists a unique absolutely continuous solution $x(\cdot) : T \to R^n$ of the differential equation

$$\dot{x}(t) = f(t,x(t),\eta(t)), \tag{8}$$

satisfying the initial condition $x(t_0) = x_0$, where (t_0,x_0) is a point in $T \times R^n$. This solution will be denoted by the symbol $x(\cdot,t_0,x_0,\eta)$.

Along with the set H defined above we shall consider the set N whose elements are measurable functions $\nu : T \to rpm(\mathcal{Q})$. Thus the set N is nothing else than the set $\mathfrak{L}$ defined by the relation (6) in which we set $\mathcal{Z} = \mathcal{Q}$. The elements of the set N as well as the elements of the set H will be called generalized programmed controls.

Before introducing the next notion we note the following fact. The space $\mathcal{B}(T, \mathcal{Q})$ can naturally be imbedded into the space $\mathcal{B}(T, \mathcal{P} \times \mathcal{Q})$. For $h \in \mathcal{B}(T, \mathcal{Q})$ and $\eta \in H$ we set

$$\int_T dt \int_{\mathcal{P}\times\mathcal{Q}} h(t,v)\eta(t)(d(u,v)) \stackrel{\text{def}}{=} \int_T dt \int_{\mathcal{P}\times\mathcal{Q}} h_*(t,u,v)\eta(t)(d(u,v)) = \int_T h_*(t,\eta(t))dt,$$

where

$$h_*(t,u,v) = h(t,v), \quad (t,u,v) \in T \times \mathcal{P} \times \mathcal{Q}.$$

Definition. *Let $\nu \in N$ and $\eta \in H$. We say that the generalized control η is consistent with a generalized control ν if for all $h \in \mathcal{B}(T, \mathcal{Q})$ the equality*

$$\int_T dt \int_{\mathcal{P}\times\mathcal{Q}} h(t,v)\eta(t)(d(u,v)) \stackrel{\text{def}}{=} \int_T dt \int_{\mathcal{Q}} h_*(t,u,v)\eta(t)(d(u,v)) = \int_T h(t,\nu(t))dt \tag{9}$$

is valid. The set of generalized controls $\eta \in H$ consistent with the control $\nu \in N$ will be denoted by $H(\nu)$.

Note that the consistency condition of controls η and ν may be defined in various equivalent ways. For example, the equality (9) can be replaced by the equality

$$\int_{\mathcal{P}\times\mathcal{Q}} g(v)\eta(t)(d(u,v)) = \int_{\mathcal{Q}} g(v)\nu(t)(d\tau) \tag{10}$$

which should be satisfied for any choice of continuous function $g : \mathcal{Q} \to R$ and for almost all $t \in T$. One can also show that consistency conditions (9) and (10) are equivalent to the condition

$$(11) \qquad \eta(t)(\mathcal{P} \times \mathcal{G}) = \nu(t)(\mathcal{G})$$

which should be fulfilled for almost all $t \in T$ for any Borel set $\mathcal{G} \subset \mathcal{Q}$.

Concluding our first encounter with generalized programmed controls we note that in what follows the sets H and N will be considered as compact metric spaces in which the metric topology is induced by the topology of normed vector spaces $(\mathcal{B}^*(T, \mathcal{P}\times \mathcal{Q}), \|\cdot\|_w)$ and $(\mathcal{B}^*(T, \mathcal{Q}), \|\cdot\|_w)$. Below in Section 6.1.4 properties of generalized controls are considered, especially the consistency property of controls η and ν.

6.1.3. Generalized controls η and ν will be used for the definition of the programmed maximin

$$(12) \qquad \omega_p(t_0,x_0) = \max_{\nu\in N} \min_{\eta\in H(\nu)} \sigma(x(\theta,t_0,x_0,\eta)).$$

Here $\sigma : R^n \to R$ is a continuous function (it is assumed that the payoff is determined by the equality (3.1.1)), $x(\theta,t_0,x_0,\eta)$ is the right-hand end of the solution $x(\cdot,t_0,x_0,\eta)$ of equation (8). It will be shown that in equality (12) both maximum and minimum are attained. Properties of a function of the programmed maximin will be considered. These problems will be studied in the next sections; first, however, we shall clarify the meaning of our formal constructions.

Consider generalized controls $\eta \in H$ and compare them with the ordinary controls $(u(\cdot),v(\cdot)) \in \mathbf{U}_p \times \mathbf{V}_p$. Recall that symbols $\mathbf{U}_p$ and $\mathbf{V}_p$ designate measurable functions $u(\cdot) : T \to \mathcal{P}$ and $v(\cdot) : T \to \mathcal{Q}$. The adjective "ordinary" is added here in order to distinguish the controls $(u(\cdot),v(\cdot)) \in \mathbf{U}_p \times \mathbf{V}_p$ from the generalized controls $\eta \in H$.

We choose an arbitrary pair of ordinary controls $(u(\cdot), v(\cdot)) \in \mathbf{U}_p \times \mathbf{V}_p$. To this pair we correspond a generalized control $\eta \in H$ such that

$$(13) \qquad \eta(t) = \delta(u(t), v(t)), \quad t \in T.$$

Here $\delta(u,v)$ is the Dirac measure concentrated at the point $(u,v) \in \mathcal{P} \times \mathcal{Q}$. Note that in accordance with the notation (7) for the control η of the form (13) the equality

$$f(t,x,\eta(t)) = f(t,x,u(t),v(t)), \quad (t,x) \in T \times R^n,$$

will be valid. Therefore the solution $x(\cdot,t_0,x_0,\eta)$ of equation (8) generated by controls of form (13) coincides with the solution $x(\cdot,t_0,x_0,u(\cdot),v(\cdot))$ of equation (1.2.1). Thus it can be asserted that the set of ordinary controls $(\mathbf{U}_p \times \mathbf{V}_p)$ is imbedded in the set of generalized controls H. It is known that in the case of such imbedding the set $(\mathbf{U}_p \times \mathbf{V}_p)$ is everywhere dense in the metric space H.

Now compare the sets of solutions of differential equations (1.2.1) and (8). Set

$$X^{(1)} \overset{\text{def}}{=} \{x(\cdot,t_0,x_0,u(\cdot),v(\cdot)) : (u(\cdot),v(\cdot)) \in \mathbf{U}_p \times \mathbf{V}_p\},$$

$$X^{(2)} \overset{\text{def}}{=} \{x(\cdot,t_0,x_0,\eta) : \eta \in H\}.$$

It is known that $X^{(2)} = cl\ X^{(1)}$, here $cl\ X^{(1)}$ is the closure of the set $X^{(1)}$ in the metric of the space $C_n(T)$. We note also that the sets $X^{(1)}$ and $X^{(2)}$ coincide if for all $(t,x) \in T \times R^n$ the vectogram $\{f(t,x,u,v) : u \in P,\ v \in Q\}$ is convex. These facts are presented here without proofs since they are required only to clarify the meaning of formal constructions.

Thus it follows from the above that the set of ordinary controls $(u(\cdot),v(\cdot))$ can be supplemented and closed so that a closed set of motions will be obtained. Such a closure is attained by means of introducing generalized controls. As a result of these constructions the set of generalized controls becomes a compact metric space.

Analogously, the set N which consists of generalized controls ν can be interpreted as a closure of the set of ordinary controls $v(\cdot) \in \mathbf{V}_p$. Next, one can clarify the meaning of the set $H(\nu)$. Assume that the generalized control $\nu \in N$ is of the form $\nu(t) = \delta(v(t))$ $(t \in T)$, $v(\cdot) \in \mathbf{V}_p$ and $\delta(v)$ is a Dirac measure concentrated at point v (i.e. the generalized control μ actually coincides with the ordinary control $v(\cdot) \in \mathbf{V}_p$). Define the set

$$(14) \qquad \mathbb{R}(v(\cdot)) \overset{\text{def}}{=} \{(u(\cdot),v(\cdot)) : u(\cdot) \in \mathbf{U}_p\} \subset \mathbf{U}_p \times \mathbf{V}_p .$$

The set $\mathbb{R}(v(\cdot))$ described in this manner can be imbedded in the metric space H. The closure of the set $\mathbb{R}(v(\cdot))$ will be exactly the set $H(\nu)$. If we define the set of motions $\{x(\cdot,t_0,x_0,u(\cdot),v)(\cdot)) : u(\cdot) \in \mathbf{U}_p\}$, then in the case under consideration, the closure of this set in the metric of the space $C_n(T)$ will coincide with the set $\{x(\cdot,t_0,x_0,\eta) : \eta \in H(\nu)\}$.

The following example may serve as a convenient illustration of the relationship between ordinary and generalized controls. Let the motions of two controlled systems be described by the equations

$$(15) \qquad \dot{y}(t) = f^{(1)}(t,y(t),u(t)), \quad \dot{z}(t) = f^{(2)}(t,z(t),v(t)).$$

Consider a differential game in which the payoff is the quantity $\|y(\theta) - z(\theta)\|$. For this game define the programmed maximin. We shall accomplish this utilizing ordinary programmed controls. Denote by $y(\cdot,t_0,y_0,u(\cdot))$ and $z(\cdot,t_0,z_0,v(\cdot))$ solutions of differential equations (15) satisfying the initial conditions $y(t_0) = y_0$, $z(t_0) = z_0$. In accordance with the constructions described in Section 5.1, the programmed maximin in the case under consideration should be defined by the equality

$$\omega_p(t_0,x_0) = \sup_{v(\cdot)\in V_p} \inf_{u(\cdot)\in U_p} \|y(\theta,t_0,y_0,u(\cdot)) - z(\theta,t_0,z_0,v(\cdot))\|.$$

Utilizing notation (14) this relation can be rewritten as

$$\omega_p(t_0,x_0) = \sup_{v(\cdot)\in V_p} \inf_{(u(\cdot),v(\cdot))\in R(v(\cdot))} \|y(\theta,t_0,y_0,u(\cdot)) - z(\theta,t_0,z_0,v(\cdot))\|.$$

We now turn to generalized controls. It follows from the discussion presented above that the maximin (more precisely, sup inf) defined in the class of ordinary programmed controls will coincide with the maximin defined in the class of generalized programmed controls, i.e.

$$\sup_{v(\cdot)\in V_p} \inf_{(u(\cdot),v(\cdot))\in R(v(\cdot))} \|y(\theta,t_0,y_0,u(\cdot)) - z(\theta,t_0,z_0,v(\cdot))\|$$

$$= \max_{\nu\in N} \min_{\eta\in H(\nu)} \|y(\theta,t_0,y_0,\eta) - z(\theta,t_0,z_0,\eta)\|.$$

In the r.h.s. of this equality the maximum and minimum are actually attained.

In conclusion we note the following. Generalized controls η and ν are defined as functions $t \to \eta(t)$ and $t \to \nu(t)$ which take on values on sets of probability measures $rpm(\mathcal{P} \times \mathcal{Q})$ and $rpm(\mathcal{Q})$. One should not, however, expect that such a definition will result in the appearance of certain probabilistic constructions. As in the case of ordinary controls all constructions remain deterministic, in particular the motions generated by generalized controls $\eta \in H$. To estimate the quality of these motions we shall continue as above to utilize a deterministic index (it would not be replaced by a probabilistic characteristic such as mathematical expectation). The introduction of generalized programmed controls is a purely technical device which allows us to close appropriately the set of programmed controls.

6.1.4. Consider sets of generalized programmed controls H, N, and $H(\nu)$. We note once again that H and N are compact metric spaces in which the metric topology is induced by the topology of normed vector spaces $(\mathcal{B}^*(T, \mathcal{P}\times\mathcal{Q}), \|\cdot\|_w)$ and $(\mathcal{B}^*(T,\mathcal{Q}), \|\cdot\|_w)$. We prove the following assertion.

Theorem 6.1.1. *For any $\nu \in N$ the set $H(\nu)$ is nonempty and compact in H. The multivalued mapping $\nu \to H(\nu)$ from N into H is continuous.*

Proof. Let $\nu \in N$, $u_* \in \mathcal{P}$, $\delta(u_*) \in rpm(\mathcal{P})$ be the Dirac measure concentrated at point u_*. Set

(16) $\qquad \eta(t) = \delta(u_*) \times \nu(t) \quad (t \in T),$

i.e. the measure $\eta(t) \in rpm(\mathcal{P} \times \mathcal{Q})$ defined as a product of measures $\delta(u_*) \in rpm(\mathcal{P})$ and $\nu(t) \in rpm(\mathcal{Q})$. It follows from (16) that for any $g \in C(\mathcal{P} \times \mathcal{Q})$

$$\int_{\mathcal{P}\times\mathcal{Q}} g(u,v)\eta(t)(d(u,v)) \overset{\text{def}}{=} g(\eta(t)) = \int_{\mathcal{Q}} g(u_*,v)\nu(t)(dv) \overset{\text{def}}{=} g(u_*,\nu(t)).$$

Since $\nu \in N$, the function $t \to g(u_*,\nu(t))$ is measurable. We obtain that for any $g \in C(\mathcal{P} \times \mathcal{Q})$ the function $t \to g(\eta(t))$ is measurable, consequently equality (16) defines a generalized control $\eta \in H$. Next, it follows from (16) that for any $h \in \mathcal{B}(T, \mathcal{Q})$ equality (9) is valid, i.e. $\eta \in H(\nu)$. We have thus shown that $H(\nu) \neq \phi$.

Let $\nu_i \in N$, $\eta_i \in H(\nu_i)$ $(i = 1, 2, \ldots)$, $\nu_i \to \nu$, $\eta_i \to \eta$ as $i \to \infty$. We show that $\eta \in H(\nu)$.

Choose $h \in \mathcal{B}(T, \mathcal{Q})$. It follows from the condition $\eta_i \in H(\nu_i)$ that

$$\int_T dt \int_{\mathcal{P}\times\mathcal{Q}} h(t,v)\eta_i(t)(d(u,v)) = \int_T dt \int_{\mathcal{Q}} h(t,v)\nu_i(t)(dv).$$

From the convergence $\eta_i \to \eta$ and $\nu_i \to \nu$ as $i \to \infty$ we obtain (cf. Section 6.1.1) that the left-hand and right-hand sides of this equality converge to the left-hand and right-hand sides of equality (9). We have thus established that $\eta \in H(\nu)$.

Hence we have proved that the multivalued mapping $\nu \to H(\nu)$ is upper semicontinuous (cf. Section A.3). Note that from this fact, for $\nu_i = \nu$ $(i = 1, 2, \ldots)$ follows the closure of the set $H(\nu)$. Since this set is contained in the compact metric space H, we have that $H(\nu)$ is compact.

It remains to show that the multivalued mapping $\nu \to H(\nu)$ is lower semi-continuous. Let $\nu \in N$, $\eta_* \in H(\nu)$, $\nu_k \to \nu_*$ as $k \to \infty$. It is required to establish the existence of a sequence $\eta_k \in H(\nu_k)$ $(k = 1, 2, \ldots)$ which converges to η_*.

Assume the contrary. Let a number $r > 0$ exist such that

$$\{\eta \in H : \|\eta - \eta_*\|_w < r\} \cap H(\nu_k) = \phi, \quad (k = 1, 2, \ldots). \tag{17}$$

For an arbitrary $\nu \in N$ in the set $H(\nu)$ we single out a subset $H_\#(\nu)$ whose elements are functions $\eta : T \to rmp(\mathcal{P} \times \mathcal{Q})$ of the form

$$\eta(t) = \delta(u(t)) \times \nu(t) \qquad (t \in T).$$

Here $u(\cdot) : T \to \mathcal{P}$ is a piecewise constant control, i.e. $u(t) = u^{(i)}$ as $t \in T_i$ $(i = 1, 2, \ldots, m)$, where m is a positive integer, $u^{(i)} \in \mathcal{P}$, $T_i \subset T$ are measurable sets such that

$$T_i \cap T_j = \phi \text{ for } i \neq j, \quad \bigcup_{i=1}^m T_i = T.$$

Observe that for any $\eta \in H_\#(\nu)$ and $h \in \mathcal{B}(T, \mathcal{P} \times \mathcal{Q})$ the equality

(18)
$$\int_T h(t,\eta(t))dt \overset{\text{def}}{=} \int_T dt \int_{P\times Q} h(t,u,v)\eta(t)(d(u,v))$$
$$= \sum_{i=1}^{m} \int_{T_i} dt \int_Q h(t,u^{(i)},v)\nu(t)(dv) \overset{\text{def}}{=} \sum_{i=1}^{m} \int_{T_i} h(t,u^{(i)},\nu(t))dt$$

is valid.

The set $H_{\#}(\nu)$ is everywhere dense in the set $H(\nu)$. The proof of this fact is analogous to the proof of Theorem IV.2.1 in the text [74].

Let

$$\eta_{\#}(t) = \delta(u_*(t)) \times \nu_*(t),\ \eta_k(t) = \delta(u_*(t)) \times \nu_k(t),\quad t \in T. \tag{19}$$

Here a piecewise-constant control $u_*(\cdot)$ is chosen in such a manner that the estimate

$$\|\eta_{\#} - \eta_*\|_w < r/2 \tag{20}$$

is valid. For any function $h \in B(T, P\times Q)$ the function $h^{(i)}$ defined by the equality

$$h^{(i)}(t,v) = \begin{cases} h(t,u^{(i)},v) & \text{for}\quad t \in T_i,\quad v \in Q, \\ 0 & \text{for}\quad t \in T\backslash T_i,\ v \in Q \end{cases}$$

is contained in the space $B(T, Q)$. Therefore the convergence $\nu_k \to \nu_*$ implies the convergence

$$\int_T h^{(i)}(t,\nu_k(t))dt = \int_{T_i} h(t,u^{(i)},\nu_k(t))dt \to \int_T h^{(i)}(t,\nu_*(t))dt$$
$$= \int_{T_i} h(t,u^{(i)},\nu_*(t))dt \quad \text{for}\quad k \to \infty.$$

Next we obtain from (18) and (19) that

$$\int_T h(t,\eta_k(t)dt \to \int_T h(t,\eta_{\#}(t))dt \quad \text{for}\quad k \to \infty$$

This relation is valid for any function $h \in B(T, P \times Q)$; consequently,

$$\|\eta_k - \eta\ \|_w \to 0 \quad \text{as}\quad k \to \infty. \tag{21}$$

It is evident from (20) and (21) that for k sufficiently large the estimate $\|\eta_k - \eta_*\| < r$ is valid. By construction of the function η_k (cf. (19)) we have the inclusion $\eta_k \in H(\nu_k)$. We thus arrive at a contradiction with the relation (17). Theorem 6.1.1 is proved.

6.2. Programmed Maximin

We have shown in the preceding section how the programmed maximin is defined in the class of generalized programmed controls.

Here this definition will be considered in more detail. In particular, it will be proved in Section 6.2.2 that in a problem with a fixed termination time and continuous quality index, a programmed maximin depends continuously on the initial position. The proof of this fact, as well as the proofs of some other assertions, is based on Lemma 6.2.1 and Theorem 6.2.1. Lemma 6.2.1 establishes the continuous dependence of solutions of differential equations on the initial conditions and parameters, and Theorem 6.2.1 deals with the differentiability of solutions with respect to the initial states. These assertions are presented in Section 6.2.1.

6.2.1. Various versions of theorems on continuous dependence of solutions on the initial conditions and parameters are known in the theory of ordinary differential equations. For subsequent constructions, Lemma 6.2.1 presented on the next page will be sufficient.

Consider the function

$$(t,y,u,v,\alpha) \to g(t,y,u,v,\alpha) : T \times R^m \times \mathcal{P} \times \mathcal{Q} \times A \to R^m.$$

Here $T = [t_\#,\theta]$, $\mathcal{P}$ and $\mathcal{Q}$ are compact sets in R^p and R^q, A is a metric compact set. We shall assume that the function g satisfies the following conditions:

(a) g is continuous

(b) $\partial g/\partial y$ exists and is continuous on $T \times R^m \times \mathcal{P} \times \mathcal{Q} \times A$.

Yet another condition will be stated below. Meanwhile consider the following construction. Let $\eta : t \to \eta(t) : T \to rpm(\mathcal{P} \times \mathcal{Q})$ be a measurable function (cf. the definitions presented in Sections 6.1.1 and 6.1.2). Recall that function η is an element of the set H and it is called a generalized programmed control. In accordance with (6.1.7) we introduce the following notation:

$$(1) \qquad g(t,y,\eta(t),\alpha) \overset{\text{def}}{=} \int_{\mathcal{P}\times\mathcal{Q}} g(t,y,u,v,\alpha)\eta(t)(d(u,v)).$$

We note several properties of the function

$$(t,y,\alpha) \to g(t,y,\eta(t),\alpha).$$

For any $(y,\alpha) \in R^m \times A$ the function $t \to g(t,y,\eta(t),\alpha)$ is measurable and bounded on T. For any $(t,y,\alpha) \in T \times R^m \times A$ the derivative $\partial g(t,y,\eta(t),\alpha)/\partial y$ exists. For any $(y,\alpha) \in R^m \times A$ the function $t \to \partial g(t,y,\eta(t),\alpha)/\partial y$ is measurable and bounded on T. For any $t \in T$ the function $(y,\alpha) \to \partial g(t,y,\eta(t),\alpha)/\partial y$ is continuous on $R^m \times A$.

It is known from the theory of ordinary differential equations (cf., for example, [13, 34, 74]) that there exists a unique solution of the equation

$$(2) \qquad \dot{y}(t) = g(t,y(t),\eta(t),\alpha)$$

satisfying the initial condition $y(t_0) = y_0$. Under the conditions stipulated above, this solution will be defined on some interval $(t_*,t^*) \subset T$ which contains point t_0. We shall now add to the conditions (a) and (b) the following assumption.

(c) for any $(t_0,y_0,\eta,\alpha) \in T \times R^m \times H \times A$ a solution of equation (2) with the initial condition $y(t_0) = y_0$ can be extended over the whole interval T. This solution is denoted by the symbol $y(\cdot,t_0,y_0,\eta,\alpha)$. Assume also that for any bounded region $\mathfrak{D} \subset R^m$ the set

$$\{y(t,t_0,y_0,\eta,\alpha) \in R^m : (t,t_0,y_0,\eta,\alpha) \in T \times T \times \mathfrak{D} \times H \times A\}$$

is bounded.

Note also that condition (c) is fulfilled if the bound

$$\langle y,g(t,y,u,v,\alpha)\rangle \leqslant \kappa(1 + \|y\|^2)$$

is satisfied for all $(t,y,u,v,\alpha) \in T \times R^m \times \mathcal{P} \times \mathcal{Q} \times A$; here κ is a constant.

Thus for any $(t_0,y_0,\eta,\alpha) \in T \times R^m \times H \times A$ there exists a unique solution $y(\cdot,t_0,y_0,\eta,\alpha)$ of equation (2) with the initial condition $y(t_0) = y_0$. This solution is defined on the whole interval T, and generates the mapping

$$(3) \qquad (t_0,y_0,\eta,\alpha) \to y(\cdot,t_0,y_0,\eta,\alpha) : T \times R^m \times H \times A \to C_m(T).$$

Recall that $C_m(T)$ is the space of continuous function $y(\cdot) : T \to R^m$ in which the norm is defined by equality $|y(\cdot)| \overset{\text{def}}{=} \max\|y(t)\|$ for $t \in T$. We present the proof of the following assertion.

Lemma 6.2.1. *Let the function g satisfy conditions* (a) – (c). *Then the mapping* (3) *is continuous.*

Proof. Let

$$(t_0^{(k)},y_0^{(k)},\eta^{(k)},\alpha^{(k)}) \in T \times R^m \times H \times A, \quad y^{(k)}(\cdot) = y(\cdot,t_0^{(k)},y_0^{(k)},\eta^{(k)},\alpha^{(k)}).$$

$$k = 1, 2, \ldots, \quad (t_0^{(k)},y_0^{(k)},\eta^{(k)},\alpha^{(k)}) \to (t_0,y_0,\eta_0,\alpha_0) \text{ as } k \to \infty.$$

Functions $y^{(k)}(\cdot)$ are uniformly bounded and equicontinuous. Therefore in view of Arzela's theorem (cf., for example, [18, 36, 74]) there exists a limit point of the sequence $y^{(k)}(\cdot)$. To simplify the notation we shall assume that $y^{(k)}(\cdot) \to y_*(\cdot)$ as $k \to \infty$. We represent the solution $y^{(k)}(\cdot)$ in the integral form

$$(4) \qquad y^{(k)}(t) = y_0^{(k)} + \int_{t_0^{(k)}}^{t} g(\tau,y^{(k)}(\tau),\eta^{(k)}(\tau),\alpha^{(k)})d\tau \qquad (t \in T).$$

Set

$$g^{(k)}(\tau,u,v) \overset{\text{def}}{=} g(\tau,y^{(k)}(\tau),u,v,\alpha^{(k)}),$$

$$g_*(\tau,u,v) \overset{\text{def}}{=} g(\tau,y_*(\tau),u,v,\alpha_0).$$

Functions $g^{(k)}$ and g_* are continuous on $T \times P \times Q$. A sequence of functions $g^{(k)}$ $(k = 1, 2, ...)$ converges to the function g_* uniformly on $T \times P \times Q$. Equality (4) can be rewritten as follows:

$$y^{(k)}(t) = y_0^{(k)} + I_1^{(k)}(t) + I_2^{(k)}(t),$$

$$I_1^{(k)}(t) = \int_{t_0}^{t} \int_{P\times Q} g_*(\tau,u,v)\eta^{(k)}(\tau)(d(u,v)),$$

$$I_2^{(k)}(t) = \int_{t_0}^{t} \int_{P\times Q} (g^{(k)}(\tau,u,v) - g_*(\tau,u,v)\eta^{(k)}(\tau)(d(u,v))$$

$$+ \int_{t_0^{(k)}}^{t_0} d\tau \int_{P\times Q} (g^{(k)}(\tau,u,v)\,\eta^{(k)}(\tau)(d(u,v)).$$

It is easy to show that

$$\lim_{k\to\infty} I_1^{(k)}(t) = \int_{t_0}^{t} g_*(\tau,\eta_0(\tau)d\tau \overset{\text{def}}{=} \int_{t_0}^{t} d\tau \int_{P\times Q} g_*(\tau,u,v)\eta_0(\tau)(d(u,v)),$$

$$\lim_{k\to\infty} I_2^{(k)}(t) = 0.$$

Therefore, approaching the limit in equation (4) we obtain

$$y_*(t) = y_0 + \int_{t_0}^{t} g_*(\tau,\eta_0(\tau))d\tau = y_0 + \int_{t_0}^{t} g_*(\tau,y_*(\tau),\eta_0(\tau),\alpha_0)d\tau.$$

Hence the function $y_*(\cdot)$ coincides with the solution $y(\cdot,t_0,y_0,\eta_0,\alpha_0)$. Since this fact is valid for any limit member of the sequence $y(\cdot,t_0^{(k)}, y_0^{(k)},\eta^{(k)},\alpha^{(k)})$, the continuity of mapping (3) is proved.

Note that Lemma 6.2.1 remains valid if condition (b) is replaced by a weaker assumption

$$\| g(t,y^{(1)},u,v,\alpha) - g(t,y^{(2)},u,v,\alpha) \| \in \lambda(\mathfrak{D}) \| y^{(1)} - y^{(2)} \| ,$$

$$(t,y^{(i)},u,v,\alpha) \in T \times \mathfrak{D} \times P \times Q \times A, \quad i = 1, 2,$$

where $\mathfrak{D}$ is a bounded set in R^m.

We consider yet another two lemmas which are actually corollaries of Lemma 6.2.1. First we introduce the following condition.

(c)* Let a collection $(t_0,y_0,\eta) \in T \times R^m \times H$ be selected and fixed. Assume that for this element and for any $\alpha \in A$ the solution of equation (2) satisfying the condition $y(t_0) = y_0$ is defined on the whole interval T. This solution will be denoted by $y(\cdot,\alpha)$. Assume also that $\sup \| y(t,\alpha) \| < \infty$ for $t \in T$, $\alpha \in A$.

Lemma 6.2.2. *Let the function g satisfy conditions* (a), (b), *and* (c)*. *Then the function* $\alpha \to y(\cdot,\alpha) : A \to C_m(T)$ *is continuous.*

It is easy to observe that the proof of this assertion will be a simplified version of the proof of Lemma 6.2.1.

Next we set in the conditions of Lemma 6.2.1 $m = n$, $y = x$, $g(t,y,u,v,\alpha) = f(t,x,u,v)$ (function f satisfies the conditions (i), (iii) given in Section 1.2.2 and condition (ii)* given in Section 6.1.2; this function does not depend on the parameter α). Let $x(\cdot,t_0,x_0,\eta)$ be a solution of the differential equation (6.1.8) satisfying the initial condition $x(t_0) = x_0$. Then according to Lemma 6.2.1 we have

Lemma 6.2.3. *The mapping*

$$(t_0,x_0,\eta) \to x(\cdot,t_0,x_0,\eta) : T \times R^n \times H \to C_n(T)$$

is continuous. In particular, the function

$$(t_0,x_0,\eta) \to x(\theta,t_0,x_0,\eta) : T \times R^n \times H \to R^n$$

is continuous.

Moreover, we shall show that the function $x_0 \to x(\cdot,t_0,x_0,\eta)$ satisfies the Lipschitz condition.

Let G be a bounded set in R^n. It follows from condition (1.2.7) that the set

$$\mathcal{D}(G) \stackrel{\text{def}}{=} \{(t,x(t\cdot t_0,x_0,\eta)) \in T \times R^n : (t_0,x_0,\eta) \in T \times G \times H\}$$

is bounded (cf. the corresponding arguments in the proof of Lemma 1.3.1). It follows from condition (ii)* (cf. Section 6.1.2) that the function f satisfies the Lipschitz condition

$$(5) \quad \begin{gathered} \|f(t,x^{(1)},u,v) - f(t,x^{(2)},u,v)\| \leqslant \lambda \|x^{(1)} - x^{(2)}\| \\ (t,x^{(i)}) \in \mathcal{D}(G), \ (u,v) \in P \times Q, \ \lambda = \lambda(\mathcal{D}(G)). \end{gathered}$$

Let

$$(t_0,x_0^{(i)},\eta) \in T \times G \times H, \ x^{(i)}(t) = x(t,t_0,x_0^{(i)},\eta) \ (i = 1, 2),$$

$$s(t) = x^{(1)}(t) - x^{(2)}(t).$$

Using (5) one can obtain

$$\frac{d\|s(t)\|^2}{dt} = 2\langle s(t),f(t,x^{(1)}(t),\eta(t)) - f(t,x^{(2)}(t),\eta(t))\rangle \leqslant 2\lambda\|s(t)\|^2.$$

It follows from this estimate that

$$\max_{t \in T} \|s(t)\| = \max_{t_\# \leqslant t \leqslant \theta} \|x(t,t_0,x_0^{(1)},\eta) - x(t,t_0,x_0^{(2)},\eta)\|$$

(6)

$$\leqslant \|x_0^{(1)} - x_0^{(2)}\| \exp\lambda(\theta - t_\#), \ (t_0,x_0^{(i)},\eta) \in T \times G \times H, \ i = 1, 2.$$

We have thus shown that the mapping $x_0 \to x(\cdot,t_0,x_0,\eta)$ satisfies the Lipschitz condition.

In addition to the above properties the function $x_0 \to x(\theta,t_0,x_0,\eta)$ is also differentiable. This fact is used quite substantially below. To state the appropriate assertion we introduce some notation.

It follows from condition (ii)* (cf. Section 6.1.2) that the function $x \to f(t,x,\eta(t))$ is differentiable and the equality

$$\frac{\partial f(t,x,\eta(t))}{\partial x} = \int_{P \times Q} \frac{\partial f(t,x,u,v)}{\partial x} \eta(t)(d(u,v)) \quad ((t,x,\eta) \in T \times R^n \times H) \tag{7}$$

is valid. Select a point $(t_0,x_0,\eta) \in T \times R^n \times H$ and define the corresponding solution $x(\cdot,t_0,x_0,\eta)$ of the equation (6.1.8). Next we consider the $n \times n$ matrix

$$L(t) \overset{\text{def}}{=} \frac{\partial f(t,x(t,t_0,x_0,\eta),\eta(t))}{\partial x} = \int_{P \times Q} \frac{\partial f(t,x(t,t_0,x_0,\eta),u,v)}{\partial x} \eta(t)(d(u,v)). \tag{8}$$

Note that the generalized control η appears twice in the definition of the matrix $L(t)$: firstly this control generates the motion $x(\cdot,t_0,x_0,\eta)$, secondly, the integration in (8) is taken over the measure $\eta(t)$. Each component function

$$t \to \frac{\partial f_i(t,x(t,t_0,x_0,\eta),\eta(t))}{\partial x_j} : T \to R \qquad (i,j = 1, 2, ..., n) \tag{9}$$

is measurable and bounded. Therefore a unique solution of the linear differential equation

$$\frac{d\Phi(t)}{dt} = L(t)\,\Phi(t) \qquad (t \in T) \tag{10}$$

exists, satisfying the condition $\Phi(t_0) = E$. Here $\Phi(t)$ and E are $n \times n$ matrices, E is the unit matrix. This solution is denoted by the symbol $\Phi(\cdot,t_0,x_0,\eta)$.

The following theorem is valid.

Theorem 6.2.1. *For any $(t_0,\eta) \in T \times R^n$ the function $x_0 \to x(\theta,t_0,x_0,\eta)$: $R^n \to R^n$ is differentiable and the equality*

$$\frac{\partial x(\theta,t_0,x_0,\eta)}{\partial x_0} = \Phi(\theta,t_0,x_0,\eta) \tag{11}$$

is valid. The mapping

$$(t_0,x_0,\eta) \to \Phi(\theta,t_0,x_0,\eta)$$

is continuous on $T \times R^n \times H$, *i.e. the components of the matrix* $\Phi(\theta,t_0,x_0,\eta)$ *depend continuously on* (t_0,x_0,η).

Proof. Select and fix a collection $(t_0,x_0,\eta,y_0) \in T \times R^n \times H \times R^n$. Let $A = [-1,1]$. For $\alpha \in A$, $\alpha \neq 0$, we set

$$(12)\qquad \begin{gathered} z(t) \stackrel{\text{def}}{=} x(t,t_0,x_0,\eta), \quad z(t,\alpha) \stackrel{\text{def}}{=} x(t,t_0,x_0 + \alpha y_0,\eta), \\ y(t,\alpha) \stackrel{\text{def}}{=} [z(t,\alpha) - z(t)]\alpha^{-1}. \end{gathered}$$

According to (6.1.8) we have

$$(13)\qquad \begin{gathered} \dot z(t) = f(t,z(t),\eta(t)), \quad z(t_0) = x_0, \\ \dot z(t,\alpha) = f(t,z(t,\alpha),\eta(t)), \quad z(t_0,\alpha) = x_0 + \alpha y_0, \\ \dot y(t,\alpha) = [f(t,z(t) + \alpha y(t,\alpha), \eta(t)) - f(t,z(t),\eta(t))]\alpha^{-1}, \quad y(t_0,\alpha) = y_0. \end{gathered}$$

Introduce the function

$$(t,y,u,v,\alpha) \to g(t,y,u,v,\alpha) : T \times R^n \times P \times Q \times A \to R^n$$

defined by the relations

$$(14)\qquad \begin{gathered} g(t,y,u,v,\alpha) = [f(t,z(t) + \alpha y,u,v) - f(t,z(t),u,v)]\alpha^{-1} \\ \text{for } \alpha \in [-1,0) \cup (0,1], \\ g(t,y,u,v,0) = \frac{\partial f(t,z(t),u,v)}{\partial x} y. \end{gathered}$$

This function is continuous and continuously differentiable with respect to y, i.e. the function g (14) satisfies the conditions (a) and (b). Moreover, for the function g (14) condition (c)* is fulfilled. Indeed for $\alpha \neq 0$ the function $y(\cdot,\alpha)$ (12) is defined on T and satisfies equation (13), i.e. $\dot y(t,\alpha) = g(t,y(t,\alpha),\eta(t),\alpha)$. For $\alpha = 0$ there exists a solution of the linear differential equation

$$(15)\qquad \dot y(t,0) = \frac{\partial f(t,z(t),\eta(t))}{\partial x} y(t,0), \quad y(t_0,0) = y_0.$$

This solution is defined and is unique on the whole interval T. To complete the verification of condition (c)* it is yet required to show that $\sup_{t,\alpha}\|y(t,\alpha)\| < \infty$ for $t \in T$, $\alpha \in A$. Indeed, we have from (6) and (12) that for $\alpha \neq 0$

$$\|y(t,\alpha)\| \leqslant \|y_0\| \exp \lambda(\theta - t_\#).$$

Here the quantity λ is defined by condition (5) in which $G = \{\alpha y_0, \alpha \in [-1,1]\}$. For $\alpha = 0$ the function $y(t,0)$ satisfies equation (15), and is therefore bounded.

Thus the function g (14) satisfies all the conditions of Lemma 6.2.2, hence the function $\alpha \to y(\cdot,\alpha)$: $[-1,1] \to C_n(T)$ is continuous. In particular we obtain that

$$\lim_{\alpha\to 0} y(\theta,\alpha) = y(\theta,0).$$

As was shown above (cf. (12), (13)), in the case under consideration we have

$$y(t,\alpha) = [x(t,t_0,x_0 + \alpha y_0,\eta) - x(t,t_0,x_0,\eta)]\alpha^{-1}.$$

Hence,

$$\lim_{\alpha\to 0}[x(\theta,t_0,x_0 + \alpha y_0,\eta) - x(\theta,t_0,x_0,\eta)]\alpha^{-1} = y(\theta,0). \tag{16}$$

Recall that the function $t \to y(t,0)$ satisfies equation (15). Taking (8) and (12) into account, (15) can be written as

$$\dot{y}(t,0) = L(t)y(t,0), \quad y(t_0,0) = y_0.$$

Therefore the equality $y(t,0) = \Phi(t,t_0,x_0,\eta)y_0$ is valid. For $t = \theta$ it follows from (16) that

$$\lim_{\alpha\to 0}[x(\theta,t_0,x_0+\alpha y_0,\eta) - x(\theta,t_0,x_0,\eta)]\alpha^{-1} = \Phi(\theta,t_0,x_0,\eta)y_0. \tag{17}$$

Let $e^{(i)}$ be an n-dimensional vector whose i-th component equals 1 and all the other components equal 0. Setting $y_0 = e^{(i)}$ in equality (17) we obtain that the function $x_0 \to x(\theta,t_0,x_0,\eta)$ possesses the partial derivative

$$\frac{\partial x(\theta,t_0,x_0,\eta)}{\partial x_{0,i}} = \Phi(\theta,t_0,x_0,\eta)e^{(i)}.$$

To complete the proof of Theorem 6.2.1 it remains to show that the function $(t_0,x_0,\eta) \to \Phi(\theta,t_0,x_0,\eta)$ is continuous.

Now let

$$\alpha \overset{\text{def}}{=} (t_0,x_0,\zeta) \in T \times G \times H \overset{\text{def}}{=} A$$

where G is a compact set in R^n. Let Y be an $n \times n$-matrix (it can be viewed as a vector of dimensionality n^2). Set

$$g(t,Y,u,v,\alpha) = \frac{\partial f(t,x(t,t_0,x_0,\zeta),u,v)}{\partial x}Y. \tag{18}$$

This equality defines a function g which satisfies conditions (a), (b), (c). Let $\tau_0 \in T$, $\eta \in H$. Consider the differential equation

(19) $\dot{Y}(t) = g(t,Y(t),\eta(t),\alpha).$

A solution of this equation which satisfies the condition $Y(\tau_0) = E$ will be denoted by $Y(\cdot,\tau_0,\eta,\alpha)$. In accordance with Lemma 6.2.1 the mapping

(20) $(\tau_0,\eta,\alpha) \to Y(\cdot,\tau_0,\eta,\alpha)$

will be continuous on $T \times H \times A$. Observe that here $\alpha = (t_0,x_0,\zeta) \in T \times G \times H$. It follows from (8), (10) and (18), (19) that for $t_0 = \tau_0$ and $\zeta = \eta$ the equality

$$\Phi(\cdot,t_0,x_0,\eta) = Y(\cdot,\tau_0,\eta,\alpha)$$

is valid. Therefore the continuity of the mapping (20) implies the continuity of the mapping $(t_0,x_0,\eta) \to \Phi(\cdot,t_0,x_0,\eta)$. In particular we obtain that the mapping $(t_0,x_0,\eta) \to \Phi(\theta,t_0,x_0,\eta)$ is continuous on $T \times G \times H$. Since here G is an arbitrary compact set in R^n, this mapping is continuous on $T \times R^n \times H$. Thus the Theorem 6.2.1 is proved.

6.2.2. We now proceed to the definition of a programmed maximin. We shall assume that the payoff of the differential game is of the form (3.1.1). Moreover, we shall assume that the function $\sigma : R^n \to R$ can be represented in the form

(21) $\sigma(x) = \min_{\alpha \in A} \xi(x,\alpha).$

Here A is a metric compact set. The function $\xi: R^n \times A \to R$ is continuous. It is also assumed that in the region $\mathcal{E} \times A$ the derivative $\partial\xi/\partial x$ is defined and continuous. The region $\mathcal{E}$ is given by the relation

(22) $\mathcal{E} \overset{\text{def}}{=} \{x \in R^n: \sigma(x) \in (a,b)\}$

where a and b are numbers (possibly improper), i.e. $-\infty \leqslant a < b \leqslant \infty$.

In particular, one can simply assume that $\sigma : R^n \to R$ is a continuously differentiable function. Clearly, in this case the function σ belong to the type indicated above. As another example, consider the function

(23) $\sigma(x) \overset{\text{def}}{=} \operatorname{dist}(x,M) \quad (x \in R^n),$

where M is a compact set in R^n. In this case we have

$$\sigma(x) = \min_{y \in M} \|x - y\|,$$

i.e. here $\alpha = y$, $A = M$, $\xi(x,\alpha) = \|x - y\|$, $\mathcal{E} = \{x \in R^n: \operatorname{dist}(x,M) > 0\}$.

As mentioned above, the main topic of investigation in this chapter are functions of a programmed maximin. For a given initial

initial position $(t_0,x_0) \in T \times R^n$ a *programmed maximin* is defined by the relation

$$\begin{aligned}\omega_p(t_0,x_0) &\overset{\text{def}}{=} \max_{\nu\in N} \min_{\eta\in H(\nu)} \sigma(x(\theta,t_0,x_0,\eta)) \\ &= \max_{\nu\in N} \min_{\eta\in H(\nu)} \min_{\alpha\in A} \xi(x(\theta,t_0,x_0,\eta),\alpha).\end{aligned} \tag{24}$$

Recall that the sets of generalized programmed controls N and $H(\nu)$ were defined in Section 6.1.2; $x(\theta,t_0,x_0,\eta)$ is a point at which the solution of the differential equation (6.1.8) arrives at time $t = \theta$ from the initial position (t_0,x_0). Introduce the notation

$$\varphi(t_0,x_0,\alpha,\eta) \overset{\text{def}}{=} \xi(x(\theta,t_0,x_0,\eta),\alpha). \tag{25}$$

Utilizing this notation the quantity $\omega_p(t_0,x_0)$ can be defined as follows:

$$\omega_p(t_0,x_0) \overset{\text{def}}{=} \max_{\nu\in N} \min_{(\alpha,\eta)\in A\times H(\nu)} \varphi(t_0,x_0,\alpha,\eta). \tag{26}$$

As will be shown in Theorem 6.2.2 presented below, the maxima and minima in (24) and (26) are actually obtained. To state this theorem we introduce the following mappings:

$$\psi(t_0,x_0,\nu) \overset{\text{def}}{=} \inf_{(\alpha,\eta)\in A\times H(\nu)} \varphi(t_0,x_0,\alpha,\eta), \tag{27}$$

$$\omega_p(t_0,x_0) \overset{\text{def}}{=} \sup_{\nu\in N} \psi(t_0,x_0,\nu), \tag{28}$$

$$\Pi_0(t_0,x_0,\nu) \overset{\text{def}}{=} \{(\alpha_0,\eta_0) \in A \times H(\nu)\colon \varphi(t_0,x_0,\alpha_0,\eta_0) = \psi(t_0,x_0,\nu)\}, \tag{29}$$

$$N_0(t_0,x_0) \overset{\text{def}}{=} \{\nu_0 \in N\colon \psi(t_0,x_0,\nu_0) = \omega_p(t_0,x_0)\}. \tag{30}$$

Note that equality (28) should be considered as the initial definition of the quantity $\omega_p(t_0,x_0)$. The equivalence of definitions (28), (26), and (24) will follow from the fact that the sets $\Pi_0(t_0,x_0,\nu)$ and $N_0(t_0,x_0)$ are nonempty. Recall also that in accordance with the constructions in Sections 6.1.1 and 6.1.2, the sets of generalized programmed controls H and N are viewed as compact metric spaces.

The following theorem is valid.

Theorem 6.2.2. *For any* $(t_0,x_0,\nu) \in T \times R^n \times N$, $\Pi_0(t_0,x_0,\nu)$ *is a nonempty compact set in* $A \times H$. *For any* $(t_0,x_0) \in T \times R^n$ *the set* $N_0(t_0,x_0)$ *is a nonempty compact set in* N. *The functions*

$$\psi : (t_0,x_0,\nu) \to \psi(t_0,x_0,\nu): T \times R^n \times N \to R,$$

$$\omega_p : (t_0,x_0) \to \omega_p(t_0,x_0): T \times R^n \to R$$

are continuous. The multivalued mappings $(t_0,x_0,\nu) \to \Pi_0(t_0,x_0,\nu)$ *and* $(t_0,x_0) \to N_0(t_0,x_0)$ *are upper semicontinuous.*

Proof. Let $(t_0,x_0,\nu) \in T \times R^n \times N$. We shall show that the set $\Pi_0(t_0,x_0,\nu)$ is nonempty. In accordance with Theorem 6.1.1 the set $H(\nu)$ is compact in H. The function φ (25) is continuous on $T \times R^n \times A \times H$. This fact follows from the equality (25) where $\xi: R^n \times A \to R$ is a continuous function and $(t_0,x_0,\eta) \to x(\theta,t_0,x_0,\eta): T \times R^n \times H \to R^n$ is also a continuous function (cf. Lemma 6.2.3). We therefore obtain that the continuous function $(\alpha,\eta) \to \varphi(t_0,x_0,\alpha,\eta)$ attains its minimum on the compact set $A \times H(\nu)$. We have established that $\Pi_0(t_0,x_0,\nu) \neq \phi$.

We shall show that the function ψ (27) is continuous on $T \times R^n \times N$. Let $(t_k,x_k,\nu_k) \in T \times R^n \times N$ $(k = 1,2, ...)$, $\lim(t_k,x_k,\nu_k) = (t_*,x_*,\nu_*)$ as $k \to \infty$. It is required to verify that

$$\lim_{k\to\infty} \psi(t_k,x_k,\nu_k) = \psi(t_*,x_*,\nu_*). \tag{31}$$

One can suppose without loss of generality that the limit in (31) exists.

Consider the sequence

$$(\alpha_k,\eta_k) \in \Pi_0(t_k,x_k,\nu_k) \qquad (k = 1,2, ...).$$

Since $\Pi_0(t_k,x_k,\nu_k) \subset A \times H(\nu_k) \subset A \times H$, where A and H are metric compact sets (hence A and H are sequentially compact), one can select from the sequence (α_k,η_k) $(k = 1,2, ...)$ a subsequence $(\alpha_{k_\ell},\eta_{k_\ell})$ $(\ell = 1,2, ...)$ which converges to the limit $(\alpha_*,\eta_*) \in A \times H(\nu_*)$. Note that the inclusion $\eta_* \in H(\nu_*)$ follows from Theorem 6.1.1. From (27) and (29) we have

$$\psi(t_{k_\ell},x_{k_\ell},\nu_{k_\ell}) = \varphi(t_{k_\ell},x_{k_\ell},\alpha_{k_\ell},\eta_{k_\ell}) \qquad (\ell = 1,2, ...).$$

Since the function φ is continuous, approaching the limit we obtain

$$\begin{aligned} \lim_{k\to\infty} \psi(t_k,x_k,\nu_k) &= \lim_{\ell\to\infty} \psi(t_{k_\ell},x_{k_\ell},\nu_{k_\ell}) \\ &= \lim_{\ell\to\infty} \varphi(t_{k_\ell},x_{k_\ell},\alpha_{k_\ell},\eta_{k_\ell}) = \varphi(t_*,x_*,\alpha_*,\eta_*) \\ &\geqslant \min_{(\alpha,\eta)\in A\times H(\nu_*)} \varphi(t_*,x_*,\alpha,\eta) = \psi(t_*,x_*,\nu_*). \end{aligned} \tag{32}$$

We now prove the opposite bound. Choose a point $(\alpha_*,\eta_*) \in \Pi_0(t_*,x_*,\nu_*)$. According to (27), (29) we have $\psi(t_*,x_*,\nu_*) = \varphi(t_*,x_*,\alpha_*,\eta_*)$. Since $\eta_* \in H(\nu_*)$ and $\nu_k \to \nu_*$ as $k \to \infty$, it follows from Theorem 6.1.1

that one can select a sequence $\eta^{(k)} \in H(\nu_k)$ $(k = 1,2, ...)$ satisfying the condition $\eta^{(k)} \to \eta_*$ as $k \to \infty$. From (27) we obtain

$$\psi(t_k,x_k,\nu_k) \leqslant \varphi(t_k,x_k,\alpha_*,\eta^{(k)}) \qquad (k = 1,2, ...).$$

Approaching the limit we have

$$\lim_{k\to\infty} \psi(t_k,x_k,\nu_k) \leqslant \lim_{k\to\infty} \varphi(t_k,x_k,\alpha_*,\eta^{(k)}) = \varphi(t_*,x_*,\alpha_*,\eta_*)$$
$$= \psi(t_*,x_*,\nu_*).$$

From this inequality and inequality (32) the equality (31) follows. We thus have proved that the function ψ (27) is continuous.

We verify that the multivalued mapping $(t,x,\nu) \to \Pi_0(t,x,\nu)$ is upper semicontinuous. Let $(t_k,x_k,\nu_k) \to (t_*,x_*,\nu_*)$ and $(\alpha_k,\eta_k) \to (\alpha_*,\eta_*)$ as $k \to \infty$, where $(\alpha_k,\eta_k) \in \Pi_0(t_k,x_k,\nu_k)$ $(k = 1,2, ...)$. It is required to show that $(\alpha_*,\eta_*) \in \Pi_0(t_*,x_*,\nu_*)$. From (29) we have that

$$\varphi(t_k,x_k,\alpha_k,\eta_k) = \psi(t_k,x_k,\nu_k) \qquad (k = 1,2, ...).$$

Using the continuity of functions φ and ψ and approaching the limit as $k \to \infty$ we obtain

$$\varphi(t_*,x_*,\alpha_*,\eta_*) = \psi(t_*,x_*,\nu_*).$$

Note that the compactness of the set A and Theorem 6.1.1 imply the inclusion $(\alpha_*,\eta_*) \in A \times H(\nu_*)$. Thus $(\alpha_*,\eta_*) \in \Pi_0(t_*,x_*,\nu_*)$, and the upper semicontinuity of the multivalued mapping $(t,x,\nu) \to \Pi_0(t,x,\nu)$ is proved.

The continuity of the function $\nu \to \psi(t_0,x_0,\nu)$ implies that this function attains its maximum on a compact set N. Hence $N_0(t_0,x_0) \neq \phi$.

We now verify that the function $(t_0,x_0) \to \omega_p(t_0,x_0)$ is continuous. Let $(t_k,x_k) \to (t_*,x_*)$ as $k \to \infty$. It is required to show that $\lim \omega_p(t_k,x_k) = \omega_p(t_*,x_*)$ as $k \to \infty$ (without loss of generality it can be assumed that the limit exists in this case). Let $\nu_k \in N_0(t_k,x_k)$ $(k = 1,2, ...)$. To simplify the notation we suppose that $\nu_k \to \nu_*$ as $k \to \infty$. Continuity of the function ψ implies that

$$\lim_{k\to\infty} \omega_p(t_k,x_k) = \lim_{k\to\infty} \psi(t_k,x_k,\nu_k) = \psi(t_*,x_*,\nu_*)$$
$$\leqslant \max_{\nu\in N} \psi(t_*,x_*,\nu) = \omega_p(t_*,x_*).$$

Let $\nu^* \in N_0(t_*,x_*)$, i.e. $\psi(t_*,x_*,\nu^*) = \omega_p(t_*,x_*)$. From (28) we have

$$\psi(t_k,x_k,\nu^*) \leqslant \omega_p(t_k,x_k) \qquad (k = 1,2, ...).$$

Approaching the limit we obtain

$$\lim_{k\to\infty} \omega_p(t_k,x_k) \geqslant \lim_{k\to\infty} \psi(t_k,x_k,\nu^*) = \psi(t_*,x_*,\nu^*) = \omega_p(t_*,x_*)$$

which proves the continuity of ω_p.

The continuity of functions ψ (27) and ω_p (28) implies that the multivalued mapping $(t_0,x_0) \to N_0(t_0,x_0)$ (30) is upper semicontinuous. (This is verified in the same manner as the upper semicontinuity of the mapping $(t,x,\nu) \to \Pi_0(t,x,\nu)$.) Theorem 6.2.2 is thus proved.

In conclusion we note the following. Let an initial position $(t_0,x_0) \in T \times R^n$ be given and let $(\alpha_0,\eta_0) \in \Pi_0(t_0,x_0,\nu)$. Recall that in accordance with the discussion in Section 6.1.2, the generalized programmed control η_0 is defined on the whole interval $T = [t_\#,\theta]$. It is easy, however, to observe that as far as the minimizing control η_0 is concerned, only the interval $[t_0,\theta]$ is important; how the control is defined on the interval $[t_\#,t_0)$ is irrelevant. An analogous remark is valid concerning the maximizing control $\nu_0 \in N_0(t_0,x_0)$. This fact will be utilized in succeeding constructions. For this purpose the following notation will be required.

Let η: $T \to rpm(\mathcal{P} \times \mathcal{Q})$ be a generalized programmed control. Let $t_0 \in T = [t_\#,\theta]$. The restriction of the function η to the intervals $[t_\#,t_0)$ and $[t_0,\theta]$ will be denoted by $\eta\|[t_\#,t_0)$ and $\eta\|[t_0,\theta]$ respectively. It follows directly from the definition of the solution $x(\cdot,t_0,x_0,\eta)$ that for any $(t_0,x_0,\eta^{(i)}) \in T \times R^n \times H$, $i = 1,2$, such that $\eta^{(1)}\|[t_0,\theta] = \eta^{(2)}\|[t_0,\theta]$ the equality

$$x(t,t_0,x_0,\eta^{(1)}) = x(t,t_0,x_0,\eta^{(2)}) \quad \text{for} \quad t_0 \leqslant t \leqslant \theta \tag{33}$$

is valid. Next we introduce the notation

$$\Pi^*(t_0,x_0,\nu) \overset{\text{def}}{=} \{(\alpha,\eta\|[t_0,\theta]) \colon (\alpha,\eta) \in \Pi_0(t_0,x_0,\nu)\}. \tag{34}$$

Let $\eta_* \in H(\nu)$ and $(\alpha^*,\eta^*\|[t_0,\theta]) \in \Pi^*(t_0,x_0,\nu)$. Set $\eta_0(t) = \eta_*(t)$ for $t \in [t_\#,t_0)$ and $\eta_0(t) = \eta^*(t)$ for $t \in [t_0,\theta]$. Equality (33) implies that $(\alpha^*,\eta_0) \in \Pi_0(t_0,x_0,\nu)$.

Analogously, let $\nu\|[t_\#,t_0)$ and $\nu\|[t_0,\theta]$ be the restrictions of the generalized programmed control ν: $T \to rpm(\mathcal{Q})$ to the intervals $[t_\#,t_0)$ and $[t_0,\theta]$ respectively. We set

$$N^*(t_0,x_0) \overset{\text{def}}{=} \{\nu\|[t_0,\theta] \colon \nu \in N_0(t_0,x_0)\}. \tag{35}$$

Let

$$\nu_* \in N, \quad \nu^*\|[t_0,\theta] \in N^*(t_0,x_0), \quad \nu_0(t) = \nu_*(t)$$
$$\text{for } t_\# \leqslant t < t_0, \; \nu_0(t) = \nu^*(t) \text{ for } t_0 \leqslant t \leqslant \theta.$$

Then $\nu_0 \in N_0(t_0,x_0)$. This proposition also follows from equality (33).

Finally observe that

$$\Pi^*(t_0,x_0,\nu_1) = \Pi^*(t_0,x_0,\nu_2) \tag{36}$$

provided that $\nu_1\|[t_0,\theta] = \nu_2\|[t_0,\theta]$.

6.3. The Minimum Principle and Maximin Rule

To determine programmed maximin two problems must be solved: the first, that of optimal minimizing control, and the second, that of optimal maximizing control. Below we state the necessary conditions to be satisfied by solutions of these problems. The necessary condition satisfied by minimizing control is called the minimum principle. Actually, this condition is the same as the well known Pontryagin's maximum principle. It is, however, more convenient for our purposes to consider this principle in the form of a minimum condition. The necessary condition to be satisfied by the maximizing control is called here the maximin rule. This rule is justified when a special regularity condition is fulfilled.

6.3.1. In the proofs of the minimum principle and the maximin rule certain facts will be utilized which are related to generalized programmed controls. We shall formulate and substantiate these propositions.

Consider a mapping $\mu: v \to \mu(v): \mathcal{Q} \to rpm(\mathcal{P})$. Let $g: \mathcal{P} \to R$ be a continuous function. Set

$$r_g(v) \overset{\text{def}}{=} \int g(u)\mu(v)(du) \qquad (v \in \mathcal{Q}).$$

Assume that for any continuous function g the function $v \to r_g(v): \mathcal{Q} \to R$ is Borel measurable. In this case we shall call the mapping $v \to \mu(v)$ a measurable one. It is known that this definition of measurability is equivalent to the condition that the function $v \to \mu(v)(G)$ is measurable for any Borel set $G \subset \mathcal{P}$.

Let a measurable mapping $\mu: v \to \mu(v)$ and a generalized programmed control $\nu \in N$ be given. One can associate with the pair (μ,ν) a generalized programmed control $\eta \in H(\nu)$ which is defined as follows. Let $t \in T$, $G_1 \subset \mathcal{P}$, and $G_2 \subset \mathcal{Q}$ be Borel sets. Set

$$\eta(t)(G_1 \times G_2) = \int_{G_2} \mu(v)(G_1)\nu(t)(dv). \tag{1}$$

It is known from measure theory (cf. for example [52]) that relation (1) uniquely determines a measure $\eta(t) \in \text{rpm}(\mathcal{P} \times \mathcal{Q})$. Note that for this measure

$$\int_{\mathcal{P}\times\mathcal{Q}} g(u,v)\eta(t)(d(u,v)) = \int_{\mathcal{Q}} \nu(t)(dv)\int_{\mathcal{P}} g(u,v)\mu(v)(du) \tag{2}$$

for any continuous function $g: \mathcal{P} \times \mathcal{Q} \to R$. Note that equality (2) can also be considered as a definition of the measure $\eta(t)$. This definition is equivalent to definition (1).

Thus for each $t \in T$ one can define by either one of these two method the measure $\eta(t) \in rpm(\mathcal{P} \times \mathcal{Q})$. It can be verified that this results in a measurable function $\eta: t \to \eta(t): T \to rpm(\mathcal{P} \times \mathcal{Q})$, i.e. $\eta \in H$. Moreover, it follows from relations (1) and (2) that control η is consistent with control ν (cf. the definition in Section 6.1.2), i.e. $\eta \in H(\nu)$.

Consider the following particular case. Let $\mu^*(v) = \delta(u(v))$, where $\delta(u)$ is the Dirac measure concentrated at point $u \in P$ and let $v \to u(v)$: $Q \to P$ be a Borel measurable function. The mapping $v \to \delta(u(v))$: $Q \to rpm(P)$ will be measurable in the sense indicated above. Therefore a generalized control $\eta^* \in H(\nu)$ can be associated with the pair (μ^*,ν). Note that in this case the equality (2) becomes

$$\int_{P \times Q} g(u,v)\eta^*(t)(d(u,v)) = \int_{Q} g(u(v),v)\nu(t)(dv). \tag{3}$$

Let a continuous function g_0: $P \times Q \to R$ be given. Select a Borel measurable function $v \to u_0(v)$: $Q \times P$ from the condition

$$g_0(u_0(v),v) = \min_{u \in P} g_0(u,v). \tag{4}$$

According to the theorem on a measurable choice (cf. Section A.3), such a selection is indeed possible. We set in the preceding constructions $u(v) = u_0(v)$. Since equality (3) is valid for $g = g_0$ we have from (4) that

$$\int_{P\times Q} g_0(u,v)\eta^*(t)(d(u,v)) = \int_Q \min_{u\in P} g_0(u,v)\nu(t)(dv) \quad (t \in T). \tag{5}$$

We thus obtain that for any continuous function $g_0 : P \times Q \to R$ and any generalized control $\nu \in N$, one can define a generalized control $\eta^* \in H(\nu)$ such that equality (5) will be fulfilled.

Note also that for any $h \in B(T, P \times Q)$, $\nu \in N$ and $\eta \in H(\nu)$ the inequality

$$\int_T dt \int_{P\times Q} h(t,u,v)\eta(t)(d(u,v)) \geq \int_T dt \int_Q [\min_{u\in P} h(t,u,v)]\nu(t)(dv) \tag{6}$$

is valid. Indeed $h(t,u,v) - \min_{u\in P} h(t,u,v) \geq 0$ for $(t,u,v) \in T \times P \times Q$. Therefore

$$\int_T dt \int_{P\times Q} [h(t,u,v) - \min_{u\in P} h(t,u,v)]\eta(t)(d(u,v)) \geq 0.$$

Using the condition (6.1.9) we obtain inequality (6).

We state yet another assertion. Let $T_i \subset T$ $(i = 1,2)$ be measurable sets. Assume that $T_1 \cap T_2 = \phi$ and $T_1 \cup T_2 = T$. Let $\nu_i \in N$, $\eta_i \in H(\nu_i)$ $(i = 1,2)$. Set $\nu_*(t) = \nu_i(t)$, $\eta_*(t) = \eta_i(t)$ for $t \in T_i$. It follows from the definitions of generalized programmed controls and the definition of the set $H(\nu)$ (cf. Section 6.1.2) that $\nu_* \in N$ and $\eta_* \in H(\nu_*)$. In particular, consider the sets $T_1 = T\backslash[t_*, t_*+\epsilon)$, $T_2 = [t_*, t_*+\epsilon)$, where t_* is a number in the half interval $[t_\#,\theta)$. Let ϵ be a parameter varying in the interval $(0, \theta - t_*)$. In this case, for controls ν_* and η_* defined above, we shall use the notation ν_ϵ and η_ϵ. According to the definition of convergence in spaces N and H (cf. Sections 6.1.1, 6.1.2) we have $\lim_{\epsilon\to0}\nu_\epsilon = \nu_1$, $\lim_{\epsilon\to0}\eta_\epsilon = \eta_1$.

6.3.2. Consider the following problem of optimal control.

Problem 6.3.1. *Let the initial position* $(t_0,x_0) \in T \times R^n$ *and a generalized programmed control* $\nu \in N$ *be given. It is required to obtain a pair* $(\alpha^0,\eta^0) \in A \times H(\nu)$ *satisfying the condition*

$$\varphi(t_0,x_0,\alpha^0,\eta^0) = \min_{(\alpha,\eta)\in A\times H(\nu)} \varphi(t_0,x_0,\alpha,\eta) \overset{\text{def}}{=} \psi(t_0,x_0,\nu). \tag{7}$$

Here the notation (6.2.25) and (6.2.27) is utilized. Note also that for the set of solutions to the problem under consideration, the designation $\Pi_0(t_0,x_0,\nu)$ was introduced (cf. (6.2.29)). It was established in Theorem 6.2.2 that $\Pi_0(t_0,x_0,\nu) \neq \phi$, i.e. there exists a solution for Problem 6.3.1. We introduce the following notation:

$$\begin{aligned} \mathcal{D}_0 &\overset{\text{def}}{=} \{(t,x) \in T \times R^n: \omega_p(t,x) \in (a,b)\}, \\ \mathcal{D}_1 &\overset{\text{def}}{=} \{(t,x,\nu) \in T \times R^n \times N: \psi(t,x,\nu) \in (a,b)\}, \\ \mathcal{D}_2 &\overset{\text{def}}{=} \{(t,x,\alpha,\eta) \in T \times R^n \times A \times H: \varphi(t,x,\alpha,\eta) \in (a,b)\}. \end{aligned} \tag{8}$$

Here (a,b) is the interval which appears in the definition of the domain $\mathcal{E}$. Recall (cf. (6.2.25)) that $\varphi(t_0,x_0,\alpha,\eta) = \xi(x(\theta,t_0,x_0,\eta),\alpha)$. In the domain $\mathcal{E} \times A$ there exists a continuous derivative $\partial\xi/\partial x$ (cf. Section 6.2.2). According to Theorem 6.2.1, the derivative $\partial x(\theta,t_0,x_0,\eta)/\partial x_0$ is defined and continuous on the set $T \times R^n \times H$. We thus obtain that in the domain $\mathcal{D}_2$ there exists a continuous derivative $\partial\varphi(t,x,\alpha,\eta)/\partial x$.

We now state and prove a necessary condition satisfied by the solution of problem 6.3.1.

Theorem 6.3.1. *Let* $(t_0,x_0,\nu) \in \mathcal{D}_1$ *and let* (α^0,η^0) *be the solution of Problem* 6.3.1, i.e. $(\alpha^0,\eta^0) \in \Pi_0(t_0,x_0,\nu)$. *Then for almost all* $t \in [t_0,\theta]$ *the condition*

$$\begin{aligned} &\frac{\partial\varphi(t,x^0(t),\alpha^0,\eta^0)}{\partial x} f(t,x^0(t),\eta^0(t)) \\ &= \int_Q \left[\min_{u\in P} \frac{\partial\varphi(t,x^0(t),\alpha^0,\eta^0)}{\partial x} f(t,x^0(t),u,v)\right] \nu(t)(dv) \end{aligned} \tag{9}$$

is fulfilled where $x^0(t) \overset{\text{def}}{=} x(t,t_0,x_0,\eta^0)$.

Before proceeding to the proof of this theorem we present certain observations. According to (7) and (8), condition $(t_0,x_0,\nu) \in \mathcal{D}_1$ means that $\varphi(t_0,x_0,\alpha^0,\eta^0) \in (a,b)$. It follows from (6.2.25) that $\varphi(t,x^0(t),\alpha^0,\eta^0) = \varphi(t_0,x_0,\alpha^0,\eta^0)$ for $t_0 \leqslant t \leqslant \theta$. Consequently, $(t,x^0(t),\alpha^0,\eta^0) \in \mathcal{D}_2$ and the derivative $\partial\varphi(t,x^0(t),\alpha^0,\eta^0)/\partial x$ is defined for all $t \in [t_0,\theta]$. (Recall, incidentally, that this derivative is a row-vector.)

Consider transformations by which the optimality principle (9) is reduced to the traditional form. We introduce the notation

$$s^{\mathrm{T}}(\theta) \overset{\mathrm{def}}{=} \frac{\partial \xi}{\partial x}(x^0(\theta),\alpha^0)$$

($s^{\mathrm{T}}(\theta)$ is a row-vector). Since $x(\theta,t,x^0(t),\eta^0) = x(\theta,\theta,x^0(\theta),\eta^0) = x^0(\theta)$, it follows that $\partial\xi(x(\theta,t,x^0(t),\eta^0)/\partial x = s^{\mathrm{T}}(\theta)$ for $t_0 \leqslant t \leqslant \theta$. Therefore in accordance with (6.2.11) and (6.2.25) we have

(10) $$\frac{\partial\varphi(t,x^0(t),\alpha^0,\eta^0)}{\partial x} = s^{\mathrm{T}}(\theta)\Phi(\theta,t,x^0(t),\eta^0).$$

From the definition of the matrix $\Phi(\theta,t_0,x_0,\eta)$ (cf. (6.2.8) and (6.2.10)) the equality

(11) $$\Phi(\theta,t,x^0(t),\eta^0) = Z(\theta)\cdot Z^{-1}(t)$$

follows. Here $t \to Z(t)$ is a solution of the matrix differential equation

(12) $$\frac{dZ(t)}{dt} = L(t)Z(t)$$

satisfying the condition $\det Z(t) \neq 0$ $(t \in T)$ (as is known, it is sufficient to assume that $\det Z(t) \neq 0$ for at least one $t \in T$). Set

(13) $$Y(\zeta,\tau) \overset{\mathrm{def}}{=} Z(\zeta)\cdot Z^{-1}(\tau) \qquad (\zeta,\tau \in T).$$

From here it follows that

$$Z^{\mathrm{T}}(\tau)Y^{\mathrm{T}}(\zeta,\tau) = Z^{\mathrm{T}}(\zeta).$$

Differentiating this equality with respect to τ and utilizing the fact that according to (12)

$$\frac{dZ^{\mathrm{T}}(\tau)}{d\tau} = Z^{\mathrm{T}}(\tau)L^{\mathrm{T}}(\tau)$$

we obtain

$$Z^{\mathrm{T}}(\tau)[L^{\mathrm{T}}(\tau)Y^{\mathrm{T}}(\zeta,\tau) + dY^{\mathrm{T}}(\zeta,\tau)/d\tau] = 0.$$

We thus arrive at the equation

$$\frac{dY^{\mathrm{T}}(\zeta,\tau)}{d\tau} = -L^{\mathrm{T}}(\tau)Y^{\mathrm{T}}(\zeta,\tau).$$

From (11) and (13) we have

(14) $$\Phi(\theta,t,x^0(t),\eta^0) = Y(\theta,t), \qquad Y(\theta,\theta) = E,$$
$$\frac{dY^{\mathrm{T}}(\theta,t)}{dt} = -L^{\mathrm{T}}(t)Y^{\mathrm{T}}(\theta,t), \quad t_0 \leqslant t \leqslant \theta.$$

Set

$$s(t) \overset{\text{def}}{=} (\partial\varphi(t,x^0(t),\alpha^0,\eta^0)/\partial x)^{\mathrm{T}}.$$

It follows from (10) and (14) that

$$s(t) = \Phi^{\mathrm{T}}(\theta,t,x^0(t),\eta^0)s(\theta) = Y^{\mathrm{T}}(\theta,t)s(\theta)$$

$$\dot{s}(t) = -L^{\mathrm{T}}(t)s(t), \quad s(\theta) = (\partial\xi(x^0(\theta),\alpha^0)/\partial x)^{\mathrm{T}}.$$

Taking (6.2.8) into account these relations can be rewritten as follows:

$$(15)\qquad \begin{aligned} \dot{s}(t) &= -\left[\frac{\partial f(t,x^0(t),\eta^0(t))}{\partial x}\right]^{\mathrm{T}} s(t),\\ s(\theta) &= (\partial\xi(x^0(\theta),\alpha^0)/\partial x)^{\mathrm{T}}. \end{aligned}$$

We thus obtain that the function $t \to s(t)$ is a solution of the so-called *conjugate equation.* This solution satisfies the indicated condition at the right-hand endpoint. Using the solution of the conjugate equation, condition (9) can be written as

$$(16)\qquad \begin{aligned} &\langle s(t),\ f(t,x^0(t),\eta^0(t))\rangle\\ &= \int_{Q} [\min_{u\in P}\langle s(t),f(t,x^0(t),u,v)\rangle]\nu(t)(dv). \end{aligned}$$

Proof of Theorem 6.3.1. Assume the contrary. Then it follows from inequality (6) that a number $r > 0$ and a set $T_* \subset [t_0,\theta]$ exist such that $\mathrm{mes}(T_*) > 0$ and

$$(17)\qquad \begin{aligned} &\frac{\partial\varphi(t,x^0(t),\alpha^0,\eta^0)}{\partial x} f(t,x^0(t),\eta^0(t))\\ &> \int_{Q} \min_{u\in P}\left[\frac{\partial\varphi(t,x^0(t),\alpha^0,\eta^0)}{\partial x} f(t,x^0(t),u,v)\right]\nu(t)(dv) + r \\ &\qquad\qquad \text{for } t \in T_*. \end{aligned}$$

It is known (cf. e.g. [50]) that the set T_* contains a Lebesgue point, i.e. a point t_* such that

$$(18)\qquad \epsilon^{-1}\mathrm{mes}(T_\epsilon^{(1)}) \to 1, \quad \epsilon^{-1}\mathrm{mes}(T_\epsilon^{(2)}) \to 0 \quad \text{for} \quad \epsilon \to +0$$

where

$$(19)\qquad T_\epsilon^{(1)} \overset{\text{def}}{=} T_* \cap T_\epsilon, \quad T_\epsilon^{(2)} \overset{\text{def}}{=} T_\epsilon \backslash T_\epsilon^{(1)}, \quad T_\epsilon \overset{\text{def}}{=} [t_*, t_*+\epsilon].$$

Set $x_* = x^0(t_*)$. We select a generalized control $\eta^* \in H(\nu)$ from the condition

$$\frac{\partial\varphi(t_*,x_*,\alpha^0,\eta^0)}{\partial x} f(t_*,x_*,\eta^*(t))$$

(20)

$$= \int_Q \left[\min_{u\in P} \frac{\partial\varphi(t_*,x_*,\alpha^0,\eta^0)}{\partial x} f(t_*,x_*,u,v)\right]\nu(t)(dv) \quad (t \in T).$$

Feasibility of such a choice is demonstrated in Section 6.3.1 (cf. relation (5)). We define a generalized control $\eta_\epsilon \in H(\nu)$ as follows

(21) $$\eta_\epsilon(t) = \begin{cases} \eta^0(t) & \text{for } t \in T\backslash T_\epsilon \\ \eta^*(t) & \text{for } t \in T_\epsilon \ . \end{cases}$$

Set $x_\epsilon(t) = x(t,t_0,x_0,\eta_\epsilon)$. Observe that $x_\epsilon(t) = x^0(t)$ for $t_0 \leqslant t \leqslant t_*$, in particular $x_\epsilon(t_*) = x^0(t_*) = x_*$. From the definition of the quantity $\varphi(t,x,\alpha,\eta)$ (6.2.25), we have the following equalities:

$$\varphi(t_0,x_0,\alpha^0,\eta^0) = \varphi(t,x^0(t),\alpha^0,\eta^0),$$

$$\varphi(t_0,x_0,\alpha^0,\eta_\epsilon) = \varphi(t,x_\epsilon(t),\alpha^0,\eta_\epsilon), \qquad t \in [t_0,\theta].$$

Moreover it follows from (21) that

$$\varphi(t,x_\epsilon(t),\alpha^0,\eta_\epsilon) = \varphi(t,x_\epsilon(t),\alpha^0,\eta^0) \quad \text{for} \quad t \in [t_\epsilon,\theta].$$

Here $t_\epsilon = t_* + \epsilon$. Therefore setting $t = t_\epsilon$ in these equalities we obtain

(22)

$$I_\epsilon \overset{\text{def}}{=} \varphi(t_0,x_0,\alpha^0,\eta^0) - \varphi(t_0,x_0,\alpha^0,\eta_\epsilon)$$
$$= \varphi(t_\epsilon,x^0(t_\epsilon),\alpha^0,\eta^0) - \varphi(t_\epsilon,x_\epsilon(t_\epsilon),\alpha^0,\eta^0).$$

We estimate the quantity I_ϵ. Utilizing the mean value theorem [17, 67], we have

(23) $$I_\epsilon = \frac{\partial\varphi(t_\epsilon,y(\lambda(\epsilon),\epsilon),\alpha^0,\eta^0)}{\partial x}(x^0(t_\epsilon) - x_\epsilon(t_\epsilon)).$$

Here

(24) $$y(\lambda,\epsilon) \overset{\text{def}}{=} \lambda x^0(t_\epsilon) + (1-\lambda)x_\epsilon(t_\epsilon),$$

where $\lambda(\epsilon)$ is a number in the interval [0,1]. Recall that the point $(t_*,x_*,\alpha^0,\eta^0)$ is contained in the open domain $\mathfrak{D}_2$ (8) in which the derivative $\partial\varphi(t,x,\alpha,\eta)/\partial x$ is defined and continuous. For $\epsilon > 0$ sufficiently small we have the imbedding

$$\{(t_\epsilon,\ y(\lambda,\epsilon),\alpha^0,\eta^0)\colon \lambda \in [0,1]\} \subset \mathfrak{D}_2.$$

Thus the application of the mean value theorem is justified in this case.

In accord with the notation introduced above we have

$$x^0(t_\epsilon) = x_* + \int_{T_\epsilon} f(t,x^0(t),\eta^0(t))dt,$$

$$x_\epsilon(t_\epsilon) = x_* + \int_{T_\epsilon} f(t,x_\epsilon(t),\eta^*(t))dt.$$

Next, taking into account that the functions $\partial\varphi/\partial x$ and f are continuous, we obtain from (23) that

$$I_\epsilon \geqslant \int_{T_\epsilon} \frac{\partial\varphi(t_*,x_*,\alpha^0,\eta^0)}{\partial x}[f(t_*,x_*,\eta^0(t)) - f(t_*,x_*,\eta^*(t)]dt - r\cdot\epsilon/4. \tag{25}$$

This inequality will be valid for $\epsilon > 0$ sufficiently small. Taking again the continuity of functions $\partial\varphi/\partial x$ and f into account we obtain from relations (17), (19), and (20) that for $t \in T_\epsilon^{(1)}$ the inequalities

$$\frac{\partial\varphi(t_*,x_*,\alpha^0,\eta^0)}{\partial x}f(t_*,x_*,\eta^0(t))$$

$$\geqslant \frac{\partial\varphi(t,x^0(t),\alpha^0,\eta^0)}{\partial x}f(t,x^0(t),\eta^0(t)) - r/4$$

$$\geqslant \int_Q \left[\min_{u\in P} \frac{\partial\varphi(t,x^0(t),\alpha^0,\eta^0)}{\partial x}f(t,x^0(t),u,v)\right]\nu(t)(dv) + \frac{3r}{4}$$

$$\geqslant \int_Q \left[\min_{u\in P} \frac{\partial\varphi(t_*,x_*,\alpha^0,\eta^0)}{\partial x}f(t_*,x_*,u,v)\right]\nu(t)(dv) + \frac{2r}{4}$$

$$= \frac{\partial\varphi(t_*,x_*,\alpha^0,\eta^0)}{\partial x}f(t_*,x_*,\eta^*(t)) + \frac{2r}{4}$$

are valid. Substituting this estimate into inequality (25) we obtain

$$I_\epsilon \geqslant 2r\ \mathrm{mes}(T_\epsilon^{(1)})/4 - \Lambda\ \mathrm{mes}(T_\epsilon^{(2)}) - r\epsilon/4.$$

Here the number Λ is chosen in such a manner that for $t \in [t_*,\theta]$ the inequality

$$\frac{\partial\varphi(t_*,x_*,\alpha^0,\eta^0)}{\partial x}[f(t_*,x_*,\eta^0(t)) - f(t_*,x_*,\eta^*(t))] \geqslant -\Lambda$$

is fulfilled.

Finally, it follows from relation (18) that for $\epsilon > 0$ sufficiently small, the strict inequality $I_\epsilon > 0$ is fulfilled. Returning to the definition of the quantity I_ϵ (22) we obtain

$$\varphi(t_0,x_0,\alpha^0,\eta^0) > \varphi(t_0,x_0,\alpha^0,\eta_\epsilon)$$

where $\eta_\epsilon \in H(\nu)$. However this inequality contradicts relation (7). Theorem 6.3.1. is thus proved.

6.3.3. Consider now the problem of optimal maximizing control $\nu^0 \in N$. In accord with the definition of the programmed maximin this problem is stated as follows.

Problem 6.3.2. *For a given initial position* $(t_0,x_0) \in T \times R^n$ *it is required to define a generalized programmed control* $\nu^0 \in N$ *such that*

$$\psi(t_0,x_0,\nu^0) = \max_{\nu\in N} \psi(t_0,x_0,\nu) \overset{\text{def}}{=} \max_{\nu\in N} \min_{\eta\in H(\nu)} \sigma(x(\theta,t_0,x_0,\eta)) \overset{\text{def}}{=} \omega_p(t_0,x_0).$$

Recall that the quantity $\psi(t_0,x_0,\nu)$ is defined by equality (6.2.27). For the set of solutions of Problem 6.3.2 the notation $N_0(t_0,x_0)$ was introduced (cf. (6.2.30)). It was established in Theorem 6.2.2 that $N_0(t_0,x_0) \neq \phi$, i.e. a solution of Problem 6.3.2 exists.

Below we state and prove a necessary condition to be satisfied by a maximizing control ν^0. This condition is called the maximin rule. It should be noted that unlike the minimum principle which is of a quite general nature the maximin rule is valid only when the following condition is fulfilled.

Condition 6.3.1. *For any* $(t_0,x_0) \in \mathfrak{D}_0$ *and* $\nu^0 \in N_0(t_0,x_0)$ *the set* $\Pi^*(t_0,x_0,\nu^0)$ *contains the unique element* $(\alpha^0,\eta^0\|[t_0,\theta])$.

Recall that the sets $\mathfrak{D}_0$ and $\Pi^*(t_0,x_0,\nu^0)$ are defined in accordance with (8) and (6.2.34) respectively. This condition will be called the *first regularity condition.*

Theorem 6.3.2. *Let Condition* 6.3.1 *be fulfilled and let* $(t_0,x_0) \in \mathfrak{D}_0$. *Let* $\nu^0 \in N_0(t_0,x_0)$. *Then for almost all* $t \in [t_0,\theta]$ *the following condition should be fulfilled* (*the maximin rule*):

$$\int_Q \left[\min_{u\in P} \frac{\partial\varphi(t,x^0(t),\alpha^0,\eta^0)}{\partial x} f(t,x^0(t),u,v)\right] \nu^0(t)(dv) = \max_{v\in Q} \min_{u\in P} \frac{\partial\varphi(t,x^0(t),\alpha^0,\eta^0)}{\partial x} f(t,x^0(t),u,v). \tag{27}$$

Here $(\alpha^0,\eta^0) \in \Pi_0(t_0,x_0,\nu^0)$, $x^0(t) = x(t,t_0,x_0,\eta^0)$.

We present certain clarifications. It follows from Condition 6.3.1 that a unique pair $(\alpha^0,\eta^0\|[t_0,\theta]) \in \Pi^*(t_0,x_0,\nu^0)$ can be associated with the control $\nu^0 \in N_0(t_0,x_0)$. Although the control $\eta^0\|[t_\#,t_0]$ can be arbitrary the motion $x^0(t) = x(t,t_0,x_0,\eta^0)$ for $t_0 \leqslant t \leqslant \theta$ is uniquely defined. It follows from the condition $(t_0,x_0) \in \mathfrak{D}_0$ that $(t,x^0(t),\alpha^0,\eta^0) \in \mathfrak{D}_2$ for $t_0 \leqslant t \leqslant \theta$. Hence for each $\nu^0 \in N_0(t_0,x_0)$ on the interval $[t_0,\theta]$ a unique function $t \to \partial\varphi(t,x^0(t),\alpha^0,\eta^0)/\partial x$ is defined.

We also note that utilizing the solution of the conjugate system (15) the maximin rule can be expressed by

$$(28)\qquad \int_{\mathcal{Q}} \Big[\min_{u\in\mathcal{P}} \langle s(t), f(t,x^0(t),u,v)\rangle\Big]\nu^0(t)(dv) = \max_{v\in\mathcal{Q}} \min_{u\in\mathcal{P}} \langle s(t), f(t,x^0(t),u,v)\rangle,$$

which should be satisfied for almost all $t \in [t_0,\theta]$.

Proof of Theorem 6.3.2. Assume the contrary. Let a number $r > 0$ and a set $T_* \subset [t_0,\theta]$ exist such that $\mathrm{mes}(T_*) > 0$ and for $t \in T_*$ the inequality

$$(29)\qquad \int_{\mathcal{Q}} \Big[\min_{u\in\mathcal{P}} \frac{\partial\varphi(t,x^0(t),\alpha^0,\eta^0)}{\partial x} f(t,x^0(t),u,v)\Big]\nu^0(t)(dv) < \max_{v\in\mathcal{Q}} \min_{u\in\mathcal{P}} \frac{\partial\varphi(t,x^0(t),\alpha^0,\eta^0)}{\partial x} f(t,x^0(t),u,v) - r$$

is fulfilled.

Select a measurable function $t \to v^0(t)\colon T \to \mathcal{Q}$ which satisfies the condition

$$\min_{u\in\mathcal{P}} \frac{\partial\varphi(t,x^0(t),\alpha^0,\eta^0)}{\partial x} f(t,x^0(t),u,v^0(t)) = \max_{v\in\mathcal{Q}} \min_{u\in\mathcal{P}} \frac{\partial\varphi(t,x^0(t),\alpha^0,\eta^0)}{\partial x} f(t,x^0(t),u,v).$$

The possibility of such a choice is established in Theorem A.3.2. We set $\nu^*(t) = \delta(v^0(t))$ $(t \in T)$, where $\delta(v)$ is a Dirac measure. We obtain a generalized programmed control $\nu^* \in N$ for which the equality

$$(30)\qquad \int_{\mathcal{Q}} \Big[\min_{u\in\mathcal{P}} \frac{\partial\varphi(t,x^0(t),\alpha^0,\eta^0)}{\partial x} f(t,x^0(t),u,v)\Big]\nu^*(t)(dv) = \max_{v\in\mathcal{Q}} \min_{u\in\mathcal{P}} \frac{\partial\varphi(t,x^0(t),\alpha^0,\eta^0)}{\partial x} f(t,x^0(t),u,v), \qquad t \in T$$

is valid.

Let t_* be a Lebesgue point of the set T_* (cf. (18) and (19)). Consider the control $\nu_\epsilon \in N$ defined as follows:

$$(31)\qquad \nu_\epsilon(t) = \begin{cases} \nu^0(t) & \text{for} \quad t \in T\setminus T_\epsilon \\ \nu^*(t) & \text{for} \quad t \in T_\epsilon \end{cases}$$

According to (26) we have

$$(32)\qquad \psi(t_0,x_0,\nu^0) \geqslant \psi(t_0,x_0,\nu_\epsilon).$$

Let $(\alpha_\epsilon,\eta_\epsilon) \in \Pi_0(t_0,x_0,\nu_\epsilon)$. It follows from (6.2.29) that

$$(33)\qquad \psi(t_0,x_0,\nu_\epsilon) = \varphi(t_0,x_0,\alpha_\epsilon,\eta_\epsilon).$$

Define the control η_ϵ^* such that

$$(34) \qquad \eta_\epsilon^*(t) = \begin{cases} \eta_\epsilon(t) & \text{for} \quad t \in T\backslash T_\epsilon \\ \eta^0(t) & \text{for} \quad t \in T_\epsilon \ . \end{cases}$$

Note that $\eta_\epsilon^* \in H(\nu^0)$ (cf. Section 6.3.1). Therefore according to (6.2.27)

$$(35) \qquad \psi(t_0,x_0,\nu^0) \leqslant \varphi(t_0,x_0,\alpha_\epsilon,\eta_\epsilon^*).$$

From (32), (33), and (35) we obtain

$$(36) \qquad \varphi(t_0,x_0,\alpha_\epsilon,\eta_\epsilon^*) \geqslant \varphi(t_0,x_0,\alpha_\epsilon,\eta_\epsilon).$$

To simplify the notation we shall assume that

$$(37) \qquad \eta_\epsilon|[t_\#,t_0) = \eta_\epsilon^*|[t_\#,t_0) = \eta^0|[t_\#,t_0),$$

where $\eta^0 \in H(\nu^0)$ is the control which appears in conditions (29) and (30). As was shown at the end of Section 6.2.2, selection of control on the interval $[t_\#,t_0)$ cannot affect constructions which are carried out on the interval $[t_0,\theta]$ (t_0 is the initial time in the control problems under consideration.)

According to (31) the control ν_ϵ differs from the control ν^0 only on the interval T_ϵ. Therefore, as was noted in Section 6.3.1 we have $\lim \nu_\epsilon = \nu^0$ as $\epsilon \to +0$. Let a sequence $(\epsilon_k,\alpha_{\epsilon_k},\eta_{\epsilon_k})$ $(k = 1,2, \ldots)$ be selected such that $\epsilon_k \to +0$, $(\alpha_{\epsilon_k},\eta_{\epsilon_k}) \to (\alpha_*,\eta_*)$ as $k \to \infty$. Such a sequence may indeed be selected since the set $A \times H$ is sequentially compact. As was shown in Theorem 6.2.2 the multivalued mapping $\nu \to \Pi_0(t_0,x_0,\nu)$ is upper semicontinuous, therefore the condition $(\alpha_{\epsilon_k},\eta_{\epsilon_k}) \in \Pi_0(t_0,x_0,\nu_{\epsilon_k})$ $(k = 1,2, \ldots)$ implies that $(\alpha_*,\eta_*) \in \Pi_0(t_0,x_0,\nu^0)$. From equality (37) we have $\eta_*|[t_\#,t_0) = \eta^0|[t_\#,t_0)$. On the other hand, Condition 6.3.1 implies that

$$\Pi^*(t_0,x_0,\nu^0) = \{(\alpha^0,\eta^0|[t_0,\theta])\} = \{(\alpha_*,\eta_*|[t_0,\theta])\}.$$

We have thus established that

$$(38) \qquad \lim \alpha_\epsilon = \alpha^0, \quad \lim \eta_\epsilon = \eta^0, \quad \lim \eta_\epsilon^* = \eta^0 \quad \text{as} \quad \epsilon \to +0.$$

The third relation is a consequence of the fact that the control η_ϵ^* differs from the control η_ϵ only on the set T_ϵ (cf. (34)).

Set

$$x_\epsilon(t) = x(t,t_0,x_0,\eta_\epsilon), \quad x_\epsilon^*(t) = x(t,t_0,x_0,\eta_\epsilon^*).$$

From (38) we obtain in accordance with Lemma 6.2.3 that

$$(39)\qquad \lim_{\epsilon\to+0} x_\epsilon(t) = \lim_{\epsilon\to+0} x^*_\epsilon(t) = x^0(t) \overset{\text{def}}{=} x(t,t_0,x_0,\eta^0)$$

uniformly on $[t_0,\theta]$. We estimate the quantity

$$(40)\qquad \begin{aligned} I_\epsilon &\overset{\text{def}}{=} \varphi(t_0,x_0,\alpha_\epsilon,\eta^*_\epsilon) - \varphi(t_0,x_0,\alpha_\epsilon,\eta_\epsilon) \\ &= \varphi(t_\epsilon,x^*_\epsilon(t_\epsilon),\alpha_\epsilon,\eta_\epsilon) - \varphi(t_\epsilon,x_\epsilon(t_\epsilon),\alpha_\epsilon,\eta_\epsilon), \end{aligned}$$

where $t_\epsilon = t_* + \epsilon$. The second equality is a corollary of (6.2.25) and (34). We shall utilize the mean value theorem (cf. [17, 67]) which yields

$$I_\epsilon = \frac{\partial\varphi(t_\epsilon,y(\lambda(\epsilon),\epsilon),\alpha_\epsilon,\eta_\epsilon)}{\partial x}(x^*_\epsilon(t_\epsilon) - x_\epsilon(t_\epsilon)),$$

$$y(\lambda,\epsilon) \overset{\text{def}}{=} \lambda x^*_\epsilon(t_\epsilon) + (1-\lambda)x_\epsilon(t_\epsilon),$$

where $\lambda(\epsilon)$ is a number in the interval $[0,1]$. For $\epsilon > 0$ sufficiently small the mean value theorem can be used since relations (38) and (39) are valid and the point $(t_*,x^0(t_*),\alpha^0,\eta^0)$ is located in the open region $\mathfrak{D}_2$ (8) in which the derivative $\partial\varphi/\partial x$ is defined and continuous. Taking (34) into account we have

$$x^*_\epsilon(t_\epsilon) = x_\epsilon(t_*) + \int_{T_\epsilon} f(t,x^*_\epsilon(t),\eta^0(t))dt,$$

$$x_\epsilon(t_\epsilon) = x_\epsilon(t_*) + \int_{T_\epsilon} f(t,x_\epsilon(t),\eta_\epsilon(t))dt.$$

We thus obtain

$$I_\epsilon = I^{(1)}_\epsilon - I^{(2)}_\epsilon$$

$$I^{(1)}_\epsilon \overset{\text{def}}{=} \frac{\partial\varphi(t_\epsilon,y(\lambda(\epsilon),\epsilon),\alpha_\epsilon,\eta_\epsilon)}{\partial x}\int_{T_\epsilon} f(t,x^*_\epsilon(t),\eta^0(t))dt$$

$$I^{(2)}_\epsilon \overset{\text{def}}{=} \frac{\partial\varphi(t_\epsilon,y(\lambda(\epsilon),\epsilon),\alpha_\epsilon,\eta_\epsilon)}{\partial x}\int_{T_\epsilon} f(t,x_\epsilon(t),\eta_\epsilon(t))dt.$$

In the calculations below, relations (38), (39) and the continuity of functions $\partial\varphi/\partial x$ and f are used without further explanations. The relations presented below will be valid for $\epsilon > 0$ sufficiently small.

Taking into account that $(\alpha^0,\eta^0) \in \Pi_0(t_0,x_0,\nu^0)$, $(\alpha_\epsilon,\eta_\epsilon) \in \Pi_0(t_0,x_0,\nu_\epsilon)$ and utilizing the minimum principle we obtain

$$I_\epsilon^{(1)} \leqslant \int_{T_\epsilon} \frac{\partial\varphi(t,x^0(t),\alpha^0,\eta^0)}{\partial x} f(t,x^0(t),\eta^0(t))dt + \frac{\epsilon r}{4}$$

$$= \int_{T_\epsilon} dt \int_{\mathcal{Q}} \left[\min_{u\in\mathcal{P}} \frac{\partial\varphi(t,x^0(t),\alpha^0,\eta^0)}{\partial x} f(t,x^0(t),u,v)\right] \nu^0(t)(dv) + \frac{\epsilon r}{4}$$

$$I_\epsilon^{(2)} \geqslant \int_{T_\epsilon} \frac{\partial\varphi(t,x_\epsilon(t),\alpha_\epsilon,\eta_\epsilon)}{\partial x} f(t,x_\epsilon(t),\eta_\epsilon(t))dt - \frac{\epsilon r}{4}$$

$$= \int_{T_\epsilon} dt \int_{\mathcal{Q}} \left[\min_{u\in\mathcal{P}} \frac{\partial\varphi(t,x_\epsilon(t),\alpha_\epsilon,\eta_\epsilon)}{\partial x} f(t,x_\epsilon(t),u,v)\right] \nu_\epsilon(t)(dv) - \frac{\epsilon r}{4}$$

$$\geqslant \int_{T_\epsilon} dt \int_{\mathcal{Q}} \left[\min_{u\in\mathcal{P}} \frac{\partial\varphi(t,x^0(t),\alpha^0,\eta^0)}{\partial x} f(t,x^0(t),u,v)\right] \nu^*(t)(dv) - \frac{2\epsilon r}{4}$$

$$= \int_{T_\epsilon} \left[\max_{v\in\mathcal{Q}} \min_{u\in\mathcal{P}} \frac{\partial\varphi(t,x^0(t),\alpha^0,\eta^0)}{\partial x} f(t,x^0(t),u,v)\right] dt - \frac{2\epsilon r}{4}.$$

(In the last relations we have used the facts that $\nu_\epsilon(t) = \nu^*(t)$ for $t \in T_\epsilon$ and equality (30) is fulfilled.)

We now utilize relations (19) and (29) which imply that

$$I_\epsilon = I_\epsilon^{(1)} - I_\epsilon^{(2)} \leqslant -r \text{ mes}(T_\epsilon^{(1)}) + \frac{3\epsilon r}{4} + 2\Lambda \text{ mes}(T_\epsilon^{(2)})$$

where Λ is a constant such that

$$\max_{(u,v,t)} \left|\frac{\partial\varphi(t,x^0(t),\alpha^0,\eta^0)}{\partial x} f(t,x^0(t),u,v)\right| \leqslant \Lambda$$

for $(u,v,t) \in \mathcal{P} \times \mathcal{Q} \times [t_*,\theta]$. Finally, we obtain from (18) that for $\epsilon > 0$ sufficiently small the strict inequality $I_\epsilon < 0$ will be fulfilled. It follows from the definition of the quantity I_ϵ (cf. (40)) that the inequality $I_\epsilon < 0$ contradicts the inequality (36). Theorem 6.3.2 is thus proved.

In conclusion we present a simple example which shows that the maximin rule does not hold if the first regularity condition is not fulfilled (Condition 6.3.1).

Let a function σ be defined in accordance with (6.2.21) by the equality

$$\sigma(x) = \min_{i=1,2} \xi(x,\alpha_i)$$

where

$$x \in R^1, \quad \xi(x,\alpha_1) = \max\{1 + x,0\}, \quad \xi(x,\alpha_2) = \max\{1 - x,0\}.$$

Thus here $A = \{\alpha_1,\alpha_2\}$, $(a,b) \in (0,\infty)$, $\mathcal{E} = (-1,1)$ and

$$\sigma(x) = \begin{cases} 1 + x & \text{for } x \in [-1,0] \\ 1 - x & \text{for } x \in [0,1] \\ 0 & \text{for } x \leqslant -1, \ x \geqslant 1. \end{cases}$$

We shall assume that the motion of the controlled system is described by the equation

$$\dot{x} = u + v, \quad |u| \leqslant 1, \quad |v| \leqslant 1.$$

We select the initial position $t_0 = 0$, $x_0 = 0$. Let $\theta = 1/2$. Then the control $v^0(t) = 0$, $t \in [0, 1/2]$ is contained among the maximizing controls. The set $\Pi_0(t_0,x_0,v^0(\cdot))$ contains two minimizing pairs ($\alpha_1^0 = \alpha_1$, $u_1^0(t) = -1$, and $\alpha_2^0 = \alpha_2$, $u_2^0(t) = 1$). If for these pairs we define solutions of the conjugate system (15), we obtain $s_1(t) = 1$, $s_2(t) = -1$ for $t \in [0, 1/2]$. Clearly the zero control $v^0(t) = 0$ *does not* satisfy the maximin rule (28) for neither $s(t) = s_1(t)$, nor for neither $s(t) = s_2(t)$. Moreover, this control does not satisfy the maximin rule for any nonzero solution $s(t)$.

6.4. Directional Derivatives of Functions of a Programmed Maximin

With the minimum principle and the maximin rule at our disposal one can prove that a function of the programmed maximin is directionally differentiable. Below we shall present a proof of this fact and obtain expressions for directional derivatives. Next, we state conditions under which a function of the programmed maximin is continuously differentiable.

6.4.1. We shall assume that condition 6.3.1 (the first regularity condition) is fulfilled. Let $(t_*,x_*) \in \mathfrak{D}_0$, $h \in R^n$, $0 < \epsilon < \theta - t_*$. Recall that the region $\mathfrak{D}_0$ is defined in accordance with (6.3.8). The purpose of derivations below is to show that the limit

$$\lim_{\epsilon \to +0} [\omega_p(t_\epsilon,x_* + \epsilon h) - \omega_p(t_*,x_*)]\epsilon^{-1}$$

exists. Here $t_\epsilon = t_* + \epsilon$ and ω_p: $T \times R^n \to R$ is a function of the programmed maximin defined by equality (6.2.26). This limit is the derivative $D\omega_p(t_*,x_*)|(1,h)$ of the function ω_p evaluated at point (t_*,x_*) in the direction $(1,h)$ (cf. Section 4.3).

We utilize the notation (6.2.27) – (6.2.30). Note that in this notation operations of inf and sup can be replaced by operations of min and max. From (6.2.28) and (6.2.30) we have

$$\begin{aligned} &\omega_p(t_*,x_*) = \psi(t_*,x_*,\nu^0), \\ (1) \qquad &\omega_p(t_\epsilon, x_* + \epsilon h) \geqslant \psi(t_\epsilon, x_* + \epsilon h,\nu^0), \end{aligned}$$

where $\nu^0 \in N_0(t_*,x_*)$. Let $(\alpha_\epsilon,\eta_\epsilon) \in \Pi_0(t_\epsilon,x_* + \epsilon h,\nu^0)$. According to (6.2.27) and (6.2.29) we obtain

$$\begin{aligned}&\psi(t_\epsilon,x_* + \epsilon h,\nu^0) = \varphi(t_\epsilon,x_* + \epsilon h,\alpha_\epsilon,\eta_\epsilon),\\ &\psi(t_*,x_*,\nu^0) \leqslant \varphi(t_*,x_*,\alpha_\epsilon,\eta_\epsilon).\end{aligned} \tag{2}$$

It follows from (1) and (2) that

$$\omega_p(t_\epsilon,x_* + \epsilon h) - \omega_p(t_*,x_*) \geqslant \varphi(t_\epsilon,x_* + \epsilon h,\alpha_\epsilon,\eta_\epsilon) - \varphi(t_*,x_*,\alpha_\epsilon,\eta_\epsilon). \tag{3}$$

Now we specify the choice of the control η_ϵ more precisely. Let $(\alpha^0,\eta^0) \in \Pi_0(t_*,x_*,\nu^0)$. Set

$$\eta_\epsilon(t) = \eta^0(t) \quad \text{for} \quad t \in [t_\#,t_\epsilon). \tag{4}$$

In accordance with the remark presented at the end of Section 6.2.2 the control η_ϵ can be selected in such a manner that it will satisfy the condition $(\alpha_\epsilon,\eta_\epsilon) \in \Pi_0(t_\epsilon,x_* + \epsilon h,\nu^0)$ and the additional condition (4).

Let $x^0(t) = x(t,t_*,x_*,\eta^0)$. It follows from (4) and (6.2.25) that

$$\varphi(t_*,x_*,\alpha_\epsilon,\eta_\epsilon) = \varphi(t_\epsilon,x^0(t_\epsilon),\alpha_\epsilon,\eta_\epsilon).$$

Therefore in accordance with (3) we have

$$\begin{aligned}&\omega_p(t_\epsilon,x_* + \epsilon h) - \omega_p(t_*,x_*)\\ &\qquad \geqslant d(\epsilon) \overset{\text{def}}{=} \varphi(t_\epsilon,x_* + \epsilon h,\alpha_\epsilon,\eta_\epsilon) - \varphi(t_\epsilon,x^0(t_\epsilon),\alpha_\epsilon,\eta_\epsilon).\end{aligned} \tag{5}$$

Note that

$$x_* + \epsilon h - x^0(t_\epsilon) = \epsilon h - \int_{t_*}^{t_\epsilon} f(t,x^0(t),\eta^0(t))dt.$$

For $\epsilon > 0$ sufficiently small, one can utilize the mean value theorem [17, 67] to estimate the quantity $d(\epsilon)$ (an appropriate justification is presented in the same manner as in the proofs of Theorems 6.3.1 and 6.3.2). According to the mean value theorem we have

$$d(\epsilon) = \frac{\partial\varphi(t_\epsilon,y(\lambda(\epsilon),\epsilon),\alpha_\epsilon,\eta_\epsilon)}{\partial x}\left[\epsilon h - \int_{t_*}^{t_\epsilon} f(t,x^0(t),\eta^0(t))dt\right] \tag{6}$$

where $y(\lambda,\epsilon) = \lambda(x_* + \epsilon h) + (1 - \lambda)x^0(t_\epsilon)$, $\lambda(\epsilon)$ is a number in the interval [0,1].

It can be shown that

$$\lim_{\epsilon\to+0} (\alpha_\epsilon,\eta_\epsilon) = (\alpha^0,\eta^0) \tag{7}$$

where $(\alpha^0,\eta^0) \in \Pi_0(t_*,x_*,\nu^0)$. Recall that this pair was mentioned in condition (4). A verification of relation (7) is carried out in the same manner as the relations (6.3.38) were proved in Theorem 6.3.2 and is based on the following facts: (i) the inclusion $(\alpha_\epsilon,\eta_\epsilon) \in \Pi_0(t_\epsilon,x_* + \epsilon h,\nu^0)$; (ii) upper semicontinuity of the multivalued mapping $(t,x) \to \Pi_0(t,x,\nu)$ (cf. Theorem 6.2.2); (iii) Condition (4); (iv) condition 6.3.1. It is also clear that

$$\lim_{\epsilon\to+0} t_\epsilon = t_*, \qquad \lim_{\epsilon\to+0} y(\lambda(\epsilon),\epsilon) = x_*.$$

Utilizing the continuity of functions $\partial\varphi/\partial x$ and f we obtain

$$\text{(8)} \qquad \frac{\partial\varphi(t_\epsilon,y(\lambda(\epsilon),\epsilon),\alpha_\epsilon,\eta_\epsilon)}{\partial x}\epsilon h = \frac{\partial\varphi(t_*,x_*,\alpha^0,\eta^0)}{\partial x}\epsilon h + \epsilon r_1(\epsilon)$$

$$\text{(9)} \qquad \frac{\partial\varphi(t_\epsilon,y(\lambda(\epsilon),\epsilon),\alpha_\epsilon,\eta_\epsilon)}{\partial x}\int_{t_*}^{t_\epsilon} f(t,x^0(t),\eta^0(t))dt$$

$$= \int_{t_*}^{t_\epsilon} \frac{\partial\varphi(t,x^0(t),\alpha^0,\eta^0)}{\partial x} f(t,x^0(t),\eta^0(t))dt + \epsilon r_2(\epsilon)$$

$$= \int_{t_*}^{t_\epsilon} dt \int_{\mathcal{Q}} \left[\min_{u\in P} \frac{\partial\varphi(t,x^0(t),\alpha^0,\eta^0)}{\partial x} f(t,x^0(t),u,v)\right]\nu^0(t)(dv) + \epsilon r_2(\epsilon)$$

$$= \int_{t_*}^{t_\epsilon} \left[\max_{v\in\mathcal{Q}}\min_{u\in P} \frac{\partial\varphi(t,x^0(t),\alpha^0,\eta^0)}{\partial x} f(t,x^0(t),u,v)\right]dt + \epsilon r_2(\epsilon)$$

$$= \epsilon \max_{v\in\mathcal{Q}}\min_{u\in P} \frac{\partial\varphi(t_*,x_*,\alpha^0,\eta^0)}{\partial x} f(t_*,x_*,u,v) + \epsilon(r_2(\epsilon) + r_3(\epsilon)).$$

Here $r_i(\epsilon) \to 0$ as $\epsilon \to 0$. In relations (9) first we use the fact that $(\alpha^0,\eta^0) \in \Pi_0(t_*,x_*,\nu^0)$; this implies that control η^0 satisfies the minimum principle (6.3.9); next we use the fact that the control $\nu^0 \in N_0(t_*,x_*)$ satisfies the maximin rule (6.3.27).

Before substituting relations (8) and (9) into the estimates (5) and (6) we introduce yet another notation. According to Condition 6.3.1 for any $(t_*,x_*) \in \mathfrak{D}_0$ and $\nu^0 \in N_0(t_*,x_*)$ the set $\Pi^*(t_*,x_*,\nu^0)$ (cf. (6.2.34)) contains a unique element. Therefore it follows from the remark presented at the end of Section 6.2.2 that

$$\text{(10)} \qquad \left\{\frac{\partial\varphi^{\mathrm{T}}(t_*,x_*,\alpha^0,\eta^0)}{\partial x} : (\alpha^0,\eta^0) \in \Pi_0(t_*,x_*,\nu^0)\right\} = \{s^0(t_*,x_*,\nu^0)\}.$$

Here $s^0(t_*,x_*,\nu^0)$ is the unique element (column-vector) contained in the set appearing on the l.h.s. of (10). Therefore equality (10) defines a column-vector $s^0(t_*,x_*,\nu^0)$ for all $(t_*,x_*) \in \mathfrak{D}_0$ and $\nu^0 \in N_0(t_*,x_*)$. Observe that the function

$$\text{(11)} \qquad \nu \to s^0(t_*,x_*,\nu)\colon N_0(t_*,x_*) \to R^n$$

is continuous. This fact follows from the continuity of the mapping $(\alpha,\eta) \to \partial\varphi(t_*,x_*,\alpha,\eta)/\partial x$, the upper semicontinuity of the multivalued mapping $\nu \to \Pi_0(t_*,x_*,\nu)$, and the uniqueness of the vector $s^0(t_*,x_*,\nu^0)$ defined by equality (10).

Utilizing notation (10) and relations (6), (8), and (9) we obtain

$$d(\epsilon) = [\langle s^0(t_*,x_*,\nu^0),h\rangle - \max_{v\in\mathcal{Q}} \min_{u\in\mathcal{P}} \langle s^0(t_*,x_*,\nu^0),f(t_*,x_*,u,v)\rangle]\epsilon$$
$$+ \epsilon(r_1(\epsilon) - r_2(\epsilon) - r_3(\epsilon)).$$

Recall here that $r_i(\epsilon) \to 0$ as $\epsilon \to +0$. Therefore it follows from (5) that

$$\underline{\lim}_{\epsilon\to+0} [\omega_p(t_\epsilon,x_* + \epsilon h) - \omega_p(t_*,x_*)]\epsilon^{-1}$$
$$\geqslant \langle s^0(t_*,x_*,\nu^0),h\rangle - \max_{v\in\mathcal{Q}} \min_{u\in\mathcal{P}} \langle s^0(t_*,x_*,\nu^0),f(t_*,x_*,u,v)\rangle.$$

This bound is valid for any choice of control $\nu^0 \in N_0(t_*,x_*)$. Hence

$$\underline{\lim}_{\epsilon\to+0} [\omega_p(t_\epsilon,x_* + \epsilon h) - \omega_p(t_*,x_*)]\epsilon^{-1}$$
$$\geqslant \max_{\nu\in N_0(t_*,x_*)} [\langle s^0(t_*,x_*,\nu^0),h\rangle - \tag{12}$$
$$- \max_{v\in\mathcal{Q}} \min_{u\in\mathcal{P}} \langle s^0(t_*,x_*,\nu^0),f(t_*,x_*,u,v)\rangle].$$

The use of the max operation is justified here since $N_0(t_*,x_*)$ is a compact set in N (cf. Theorem 6.2.2) and the function $\nu \to s^0(t_*,x_*,\nu)$ (11) is continuous.

Thus the inequality (12) bounds the derivative $D\omega_p(t_*,x_*)|(1,h)$ from below. Consider an analogous upper bound. Let $\epsilon_k > 0$ ($k = 1,2, ...$) be a sequence such that $\epsilon_k \to 0$ as $k \to \infty$ and

$$\lim_{k\to\infty} [\omega(t^{(k)},x_* + \epsilon_k h) - \omega(t_*,x_*)]\epsilon_k^{-1}$$
$$= \overline{\lim}_{\epsilon\to+0} [\omega(t_\epsilon,x_* + \epsilon h) - \omega(t_*,x_*)]\epsilon^{-1}. \tag{13}$$

Here $t^{(k)} = t_* + \epsilon_k$. Let $\nu_k^* \in N_0(t^{(k)},x_* + \epsilon_k h)$ ($k = 1,2, ...$). Since the set N is a compact metric space (cf. Section 6.1.2) we can assume without loss of generality that the sequence ν_k^* ($k = 1,2, ...$) is convergent, i.e. we set $\lim \nu_k^* = \nu^0 \in N_0(t_*,x_*)$ as $k \to \infty$. The inclusion $\nu^0 \in N_0(t_*,x_*)$ follows from the semicontinuity of the multivalued mapping $(t,x) \to N_0(t,x)$ (cf. Theorem 6.2.2).

Let the generalized control ν_k be defined as follows:

$$\nu_k|[t_\#,t^{(k)}) = \nu^0|[t_\#,t^{(k)}), \quad \nu_k|[t^{(k)},\theta] = \nu_k^*|[t^{(k)},\theta]. \tag{14}$$

It is easy to verify that as above

$$\nu_k \in N_0(t^{(k)}, x_* + \epsilon_k h) \quad (k = 1,2, ...) \tag{15}$$

$$\lim_{k\to\infty} \nu_k = \nu^0 \in N_0(t_*,x_*).$$

Next let

$$(16) \qquad (\alpha_k,\eta_k) \in \Pi_0(t_*,x_*,\nu_k).$$

Once more without loss of generality it can be assumed that

$$(17) \qquad \lim_{k\to\infty} (\alpha_k,\eta_k) = (\alpha^0,\eta^0) \in \Pi_0(t_*,x_*,\nu^0).$$

This inclusion is a corollary of the upper semicontinuity of the multivalued mapping $\nu \to \Pi_0(t_*,x_*,\nu)$ (cf. Theorem 6.2.2).

From (15), (6.2.28), and (6.2.30) we have

$$(18) \qquad \begin{aligned} &\omega_p(t^{(k)}, x_* + \epsilon_k h) = \psi(t^{(k)}, x_* + \epsilon_k h, \nu_k), \\ &\omega_p(t_*,x_*) \geqslant \psi(t_*,x_*,\nu_k). \end{aligned}$$

It follows from (16), (6.2.27), and (6.2.29) that

$$(19) \qquad \begin{aligned} &\psi(t^{(k)}, x_* + \epsilon_k h,\nu_k) \leqslant \varphi(t^{(k)}, x_* + \epsilon_k h,\alpha_k,\eta_k), \\ &\psi(t_*,x_*,\nu_k) = \varphi(t_*,x_*,\alpha_k,\eta_k). \end{aligned}$$

Relations (18) and (19) yield

$$(20) \qquad \begin{aligned} \omega_p(t^{(k)}, x_* + \epsilon_k h) - \omega_p(t_*,x_*) &\leqslant d_k \\ &\overset{\text{def}}{=} \varphi(t^{(k)}, x_* + \epsilon_k h,\alpha_k,\eta_k) - \varphi(t_*,x_*,\alpha_k,\eta_k). \end{aligned}$$

Set $x^{(k)}(t) = x(t,t_*,x_*,\eta_k)$. Observe that in view of Lemma 6.2.3 we have from (17) that $x^{(k)}(t) \to x^0(t) = x(t,t_0,x,\eta^0)$ as $k \to \infty$. According to (6.2.25) the equality $\varphi(t_*,x_*,\alpha_k,\eta_k) = \varphi(t^{(k)},x^{(k)}(t^{(k)}), \alpha_k,\eta_k)$ is valid. Therefore

$$d_k = \varphi(t^{(k)}, x_* + \epsilon_k h,\alpha_k,\eta_k) - \varphi(t^{(k)},x^{(k)}(t^{(k)}),\alpha_k,\eta_k).$$

Note that

$$x^{(k)}(t^{(k)}) = x_* + \int_{t_*}^{t_k} f(t,x^{(k)}(t),\eta_k(t))dt.$$

As above, the mean value theorem can be used here (for k sufficiently large) which yields

$$(21) \qquad d_k = \frac{\partial\varphi(t^{(k)},y^{(k)}(\lambda_k),\alpha_k,\eta_k)}{\partial x} \left[\epsilon_k h - \int_{t_*}^{t^{(k)}} f(t,x^{(k)}(t),\eta_k(t))dt\right]$$

where $y^k(\lambda) = \lambda(x_* + \epsilon_k h) + (1 - \lambda)x^{(k)}(t^{(k)})$ and λ_k is a number in the interval [0,1].

Since the functions $\partial\varphi/\partial x$ and f are continuous, utilizing notation (10) we have from (17) that

$$(22)\qquad \frac{\partial\varphi(t^{(k)},y^{(k)}(\lambda_k),\alpha_k,\eta_k)}{\partial x}\,\epsilon_k h = \langle s^0(t_*,x_*,\nu^0,h\rangle\epsilon_k + \epsilon_k r_k^{(1)},$$

$$(23)\qquad \frac{\partial\varphi(t^{(k)},y^{(k)}(\lambda_k),\alpha_k,\eta_k)}{\partial x}\int_{t_*}^{t^{(k)}} f(t,x^{(k)}(t),\eta_k(t))dt$$

$$= \int_{t_*}^{t^{(k)}} \frac{\partial\varphi(t,x^{(k)}(t),\alpha_k,\eta_k)}{\partial x} f(t,x^{(k)}(t),\eta_k(t))dt + \epsilon_k r_k^{(2)}$$

$$= \int_{t_*}^{t^{(k)}} dt\int_{Q}\left[\min_{u\in P}\frac{\partial\varphi(t,x^{(k)}(t),\alpha_k,\eta_k)}{\partial x} f(t,x^{(k)}(t),u,v)\right]\nu_k(t)(dv)+\epsilon_k r_k^{(2)}$$

$$= \int_{t_*}^{t^{(k)}} dt\int_{Q}\left[\min_{u\in P}\frac{\partial\varphi(t,x^0(t),\alpha^0,\eta^0)}{\partial x} f(t,x^0(t),u,v)\right]\nu^0(t)(dv) + \epsilon_k(r_k^{(2)} + r_k^{(3)})$$

$$= \int_{t_*}^{t^{(k)}} \left[\max_{v\in Q}\min_{u\in P}\frac{\partial\varphi(t,x^0(t),\alpha^0,\eta^0)}{\partial x} f(t,x^0(t),u,v)\right]dt + \epsilon_k(r_k^{(2)} + r_k^{(3)})$$

$$= \epsilon_k \max_{v\in Q}\min_{u\in P}\langle s^0(t_*,x_*,\nu^0),f(t_*,x_*,u,v)\rangle + \epsilon_k(r_k^{(2)} + r_k^{(3)} + r_k^{(4)}).$$

Here $r_k^{(i)} \to 0$ as $k \to \infty$ $(i = 1,2,3,4)$. In deriving the estimate (23) we first use the fact that $(\alpha_k,\eta_k) \in \Pi_0(t_*,x_*,\nu_k)$, therefore the control η_k satisfies the minimum principle (6.3.9). Next we take into account the equality $\nu_k(t) = \nu^0(t)$ for $t \in [t_*,t^{(k)})$ (cf. (14)); finally we use the inclusion $\nu^0 \in N_0(t_*,x_*)$ and the fact that control ν^0 satisfies the maximin rule.

Substituting the estimates (22) and (23) into (20) and (21) we have

$$[\omega_p(t^{(k)}, x_* + \epsilon_k h) - \omega_p(t_*,x_*)]\epsilon_k^{-1}$$
$$\leqslant \langle s^0(t_*,x_*,\nu^0,h\rangle - \max_{v\in Q}\min_{u\in P}\langle s^0(t_*,x_*,\nu^0),\ f(t_*,x_*,u,v)\rangle + r_k^{(1)} - r_k^{(2)} - r_k^{(3)} - r_k^{(4)}.$$

Recall here that ν^0 is an element in the set $N_0(t_*,x_*)$. From equality (13) we conclude that

$$(24)\qquad \overline{\lim_{\epsilon\to+0}}\ [\omega(t_\epsilon,\ x_* + \epsilon h) - \omega(t_*,x_*)]\epsilon^{-1}$$
$$\leqslant \max_{\nu^0\in N_0(t_*,x_*)}[\langle s^0(t_*,x_*,\nu^0,h\rangle - \max_{v\in Q}\min_{u\in P}\langle s^0(t_*,x_*,\nu^0),f(t_*,x_*,u,v)\rangle].$$

The estimates (12) and (24) imply that the directional derivative

$$
(25)\quad \begin{aligned} D\omega(t_*,x_*)|(1,h) &\overset{\text{def}}{=} \lim_{\epsilon\to+0}[\omega\ (t_* + \epsilon,\ x_* + \epsilon h) - \omega(t_*,x_*)]\epsilon^{-1} \\ &= \max_{\nu^0\in N_0(t_*,x_*)}[\langle s^0(t_*,x_*,\nu^0),h\rangle - \max_{v\in\mathcal{Q}}\min_{u\in\mathcal{P}}\langle s^0(t_*,x_*,\nu^0),f(t_*,x_*,u,v)\rangle] \end{aligned}
$$

exists.

Thus the following assertion is proved.

Theorem 6.4.1. *Let Condition* 6.3.1 *be fulfilled. Then for any* $(t_*,x_*) \in \mathfrak{D}_0$ (6.3.8) $(t_* < \theta)$, $h \in R^n$ *the function of the programmed maximin* ω_p *is differentiable at the point* (t_*,x_*) *in the direction* $(1,h)$. *The directional derivative* $D\omega_p(t_*,x_*)|(1,h)$ *is defined by equality* (25) *where the sets* $N_0(t_*,x_*)$ *and the vector* $s^0(t_*,x_*,\nu^0)$ *are defined by relations* (6.2.30) *and* (10).

6.4.2. Below we shall state the conditions under which a function of the programmed maximin is continuously differentiable (in the ordinary sense). However, before stating these conditions we again discuss briefly directional derivatives. It was established in Theorem 6.4.1 that the function ω_p is differentiable in the direction $(1,h)$, where $h \in R^n$ is an arbitrary vector. This assertion may be amplified and one can show that a function of the programmed maximin is differentiable at the point $(t_*,x_*) \in \mathfrak{D}_0$ in the direction $(0,h)$ and the corresponding directional derivative is defined by the equation

$$
(26)\quad \begin{aligned} D\omega_p(t_*,x_*)|(0,h) &\overset{\text{def}}{=} \lim_{\epsilon\to+0}[\omega_p(t_*,x_* + \epsilon h) - \omega_p(t_*,x_*)]\epsilon^{-1} \\ &= \max_{\nu^0\in N_0(t_*,x_*)}\langle s^0(t_*,x_*,\nu^0),h\rangle. \end{aligned}
$$

To verify this proposition it is sufficient to replace the numbers $t_\epsilon = t_* + \epsilon$ and $t^{(k)} = t_* + \epsilon_k$ in the proof of Theorem 6.4.1 by the number t_*. Following this replacement the constructions described in Section 6.4.1 yield the equality (26).

We now proceed to the statement of the second regularity condition.

Condition 6.4.1. *Let condition* 6.3.1 *be fulfilled and moreover, let the following assertion be valid: for any point* $(t_0,x_0) \in \mathfrak{D}_0$ *the set* $N^*(t_0,x_0)$ *contains a single element.*

Recall that the sets $\mathfrak{D}_0$ and N^* are defined by relations (6.3.8) and (6.2.35). Consider the vector $s^0(t_*,x_*,\nu^0)$ defined by equality (10). According to the remark presented at the end of Section 6.2.2 for any $\nu_1^0 \in N_0(t_*,x_*)$ and $\nu_2^0 \in N_0(t_*,x_*)$ such that $\nu_1^0|[t_*,\theta] = \nu_2^0|[t_*,\theta]$ we have $s^0(t_*,x_*,\nu_1^0) = s^0(t_*,x_*,\nu_2^0)$. Therefore under Condition 6.4.1 we have

$$(27) \qquad \{s^0(t_*,x_*,\nu^0)\colon \nu^0 \in N_0(t_*,x_*)\} = \{s^\#(t_*,x_*)\}$$

where $s^\#(t_*,x_*)$ is the single element of the set appearing on the l.h.s. of relation (27). Note that the function

$$(28) \qquad (t_*,x_*) \to s^\#(t_*,x_*)\colon \mathcal{D}_0 \to R^n$$

defined by equality (27) is continuous.

We shall prove this assertion. Let $(t_k,x_k) \in \mathcal{D}_0$ $(k = 1,2, \ldots)$ and $(t_k,x_k) \to (t_*,x_*)$ as $k \to \infty$. Let $\nu_* \in N_0(t_*,x_*)$ and $(\alpha_*,\eta_*) \in \Pi_0(t_*,x_*,\nu_*)$. Consider the sequence $\nu_k \in N_0(t_k,x_k)$ and $(\alpha_k,\eta_k) \in \Pi_0(t_k,x_k,\nu_k)$ $(k = 1,2, \ldots)$. According to the remark presented at the end of Section 6.2.2 one can assume that

$$(29) \qquad \nu_k|[t_\#,t_k) = \nu_*|[t_\#,t_k), \qquad \eta_k|[t_\#,t_k) = \eta_*|[t_\#,t_k).$$

We shall prove that $\nu_k \to \nu_*$ as $k \to \infty$. Let ν^* be a limit point of the sequence ν_k $(k = 1,2, \ldots)$. It follows from the upper continuity of the multivalued mapping $(t,x) \to N_0(t,x)$ (cf. Theorem 6.2.2) that $\nu^* \in N_0(t_*,x_*)$. Inclusions $\nu^* \in N_0(t_*,x_*)$ and $\nu_* \in N_0(t_*,x_*)$ and Condition 6.4.1 yield that $\nu^*|[t_*,\theta] = \nu_*|[t_*,\theta]$. From (29) we also have that $\nu^*|[t_\#,t_*) = \nu_*|[t_\#,t_*)$. Hence $\nu_* = \nu^*$ which proves that $\nu_k \to \nu_*$ as $k \to \infty$.

Next it is verified analogously that $(\alpha_k,\eta_k) \to (\alpha_*,\eta_*)$. According to definitions (10) and (27) the equalities

$$s^\#(t_*,x_*) = (\partial\varphi(t_*,x_*,\alpha_*,\eta_*)/\partial x)^T$$

$$s^\#(t_k,x_k) = (\partial\varphi(t_k,x_k,\alpha_k,\eta_k)/\partial x)^T$$

are valid. Taking into account that the function $\partial\varphi/\partial x$ is continuous in the region $\mathcal{D}_2$ (6.3.8) and $(t_k,x_k,\alpha_k,\eta_k) \to (t_*,x_*,\alpha_*,\eta_*)$ as $k \to \infty$ we obtain that $s^\#(t_k,x_k) \to s^\#(t_*,x_*)$. We thus have verified the continuity of the function $s^\#$ (28).

If Condition 6.4.1 is fulfilled then utilizing the notation (27), relation (26) can be rewritten as

$$D\omega_p(t_*,x_*)|(0,h) = \langle s^\#(t_*,x_*),h\rangle.$$

This implies that

$$(30) \qquad \frac{\partial\omega_p(t_*,x_*)}{\partial x_i} = s_i^\#(t_*,x_*)$$

$$(\partial\omega_p(t_*,x_*)/\partial x)^T = s^\#(t_*,x_*).$$

Hence the vector $s^\#(t_*,x_*)$ is the gradient of the function $x \to \omega_p(t_*,x)$ at the point $x = x_*$.

Next, if Condition 6.4.1 is fulfilled, formula (25) with $h = 0$ yields that

$$D\omega_{\mathrm{p}}(t_*,x_*)|(1,0) = -\max_{v\in\mathcal{Q}}\min_{u\in\mathcal{P}}\langle s^{\#}(t_*,x_*),f(t_*,x_*,u,v)\rangle.$$

Note that $D\omega_{\mathrm{p}}(t_*,x_*)|(1,0)$ is just the partial derivative from the right with respect to t of the function ω_{p}. This derivative depends continuously on (t_*,x_*), therefore the partial derivative from the left with respect to t also exists and moreover, these two derivatives coincide. Hence we obtain

$$(31)\qquad \frac{\partial\omega_{\mathrm{p}}(t_*,x_*)}{\partial t} = -\max_{v\in\mathcal{Q}}\min_{u\in\mathcal{P}}\langle s^{\#}(t_*,x_*),f(t_*,x_*,u,v)\rangle.$$

The partial derivatives $\partial\omega_{\mathrm{p}}(t_*,x_*)/\partial x$ and $\partial\omega_{\mathrm{p}}(t_*,x_*)/\partial t$ are continuous in the region $\mathfrak{D}_0$ (6.3.8) and therefore the function ω_{p} is continuously differentiable in this region.

Thus the following assertion has been established.

Theorem 6.4.2. *Let Condition* 6.4.1 *be fulfilled. Then the function of the programmed maximin* ω_{p} *will be continuously differentiable in the region* $\mathfrak{D}_0$ (6.3.8) *and its partial derivatives are defined by equalities* (30) *and* (31) *where the vector* $s^{\#}$ *is defined in accordance with* (27).

6.5. Stability Conditions for the Programmed Maximin

In this section, strategies of optimal aiming are defined for a nonlinear differential game. The construction of strategies of extremal aiming leads to a synthesis of solutions of auxiliary problems of programmed control (Problems 6.3.1 and 6.3.2). The proposed control method may be substantiated under the second regularity condition (Condition 6.4.1). In this case, the functions of the programmed maximin and the value function of a differential game coincide and are continuously differentiable. Below we shall consider also another stability condition of the programmed maximin which is somewhat weaker than Condition 6.4.1.

6.5.1. First we shall assume that the second regularity condition (Condition 6.4.1) is fulfilled. According to Theorem 6.4.2 in this case the function of the programmed maximin is continuously differentiable in the region $\mathfrak{D}_0$ (6.3.8) and its partial derivatives are expressed by relations (6.4.30) and (6.4.31). Now we define *strategies of extremal aiming* (these should not be confused with strategies extremal to stable bridges). We identify these strategies with functions U^0: $T \times R^n \to \mathcal{P}$ and V^0: $T \times R^n \to \mathcal{Q}$ which assign to each position $(t,x) \in \mathfrak{D}_0$, the points $u^0 = U^0(t,x)$ and $v^0 = V^0(t,x)$ satisfying the conditions

$$(1)\qquad \max_{v\in\mathcal{Q}}\langle s^{\#}(t,x),f(t,x,u^0,v)\rangle = \min_{u\in\mathcal{P}}\max_{v\in\mathcal{Q}}\langle s^{\#}(t,x),f(t,x,u,v)\rangle,$$

$$(2)\qquad \min_{u\in P} \langle s^{\#}(t,x),f(t,x,u,v^0)\rangle = \max_{v\in Q}\min_{u\in P} \langle s^{\#}(t,x),f(t,x,u,v)\rangle.$$

Recall that the vector $s^{\#}(t,x)$ is the gradient of the function $x \to \omega_p(t,x)$. In the region $(T \times R^n)\setminus\mathfrak{D}_0$ functions U^0 and V^0 may be defined arbitrarily.

Before stating Theorem 6.5.1 we recall that in the differential game considered herein the motion is described by equation (1.2.1) and the r.h.s. of this equation satisfies the saddle point condition in a small game (cf. equality (1.2.8)). Recall also that the payoff of this game is of the form (3.1.1) where the representation (6.2.21) is valid for the function σ.

Theorem 6.5.1. *Let the second regularity condition be fulfilled (Condition* 6.4.1). *Then in the region* cl $\mathfrak{D}_0$ *the function of the programmed maximin coincides with the value function of the differential game under consideration. If* $\omega_p(t_0,x_0) \in [a,b)$, *then the strategy* U^0 *defined by equality* (1) *will be an optimal strategy for the first player. If* $\omega_p(t_0,x_0) \in (a,b]$ *then* V^0 (2) *is an optimal strategy for the second player. For any initial position* $(t_0,x_0) \in \mathfrak{D}_0$ *the pair of strategies* (U^0,V^0) *forms a saddle point of the game under consideration.*

Proof. It follows from (6.4.30) and (6.4.31) that in the region $\mathfrak{D}_0$ the function of the programmed maximin satisfies the main equation (4.1.1). Also it follows directly from the definition of a programmed maximin that $\omega_p(\theta,x) = \sigma(x)$ $(x \in R^n)$. Taking equality $s^{\#}(t,x) = (\partial\omega_p(t,x)/\partial x)^T$ into account we obtain that relations (1), (2) coincide with relations (4.1.3), (4.1.4). Comparing conditions and assertions of Theorems 4.1.1 and 6.5.1 we conclude that all the assertions of Theorem 6.5.1 follow from Theorem 4.1.1.

We shall now discuss the result obtained. Observe firstly that under the second regularity condition the function of a programmed maximin becomes a solution of the main equation (4.1.1). Programmed constructions developed in this chapter are actually one of the ways of integrating equation (4.1.1) by the *method of characteristics*, the role of which is played here by the optimal programmed motions $x(\cdot,t_0,x_0,\eta^0)$ obtained as a result of solving Problems 6.3.1 and 6.3.2.

Next we note the following. To determine the strategies of extremal aiming it is necessary to know the gradient with respect to x of the function of the programmed maximin. However, for this purpose, it is not necessary to compute the value of this function at the given position or in its neigborhood. This gradient may be defined as follows. For a given position (t_*,x_*) we obtain solutions of auxiliary problems of programmed control (Problems 6.3.1 and 6.3.2) and consider the corresponding conjugate system (6.3.15). Let $s(\cdot)$ be a solution of the conjugate system. The value of the function $s(\cdot)$ at point $t = t_*$ is the required gradient $(\partial\omega_p(t_*,x_*)/\partial x)^T = s^{\#}(t_*,x_*)$. Naturally one can utilize the optimality criteria -- the minimum principle and the maximin rule -- when solving Problems 6.3.1 and 6.3.2.

We also note that the method of extremal aiming can be developed without using generalized programmed controls. In this case, auxiliary problems are considered in the class of ordinary programmed controls. Optimal solutions of these problems may not be available, therefore it may be necessary to deal with ϵ-optimal controls and carry out rather cumbersome limiting transitions.

6.5.2. In Theorem 6.5.1 a sufficient condition is stated in which the programmed maximin coincides with the value of the differential game under consideration. This condition may be relaxed. To state a third regularity condition we introduce the following notation:

$$\kappa(s,t,x) \overset{\text{def}}{=} \max_{v\in Q} \min_{u\in P} \langle s, f(t,x,u,v)\rangle = \min_{u\in P} \max_{v\in Q} \langle s, f(t,x,u,v)\rangle \tag{3}$$

(here condition (1.2.8) is taken into account).

Condition 6.5.1. *Let the first regularity condition (Condition* (6.3.1) *be fulfilled and let for any point* $(t,x) \in \mathcal{D}_0$ (6.3.8) *the equality*

$$\max_{v\in Q} \min_{h\in F_2(t,x,v)} \max_{\nu^0\in N_0(t,x)} [\langle s^0(t,x,\nu^0),h\rangle - \kappa(s^0(t,x,\nu^0),t,x)] = 0 \tag{4}$$

be fulfilled.

Recall that the set $F_2(t,x,v)$ is defined by relation (2.2.1) and that the set $N_0(t,x)$ is a collection of maximizing controls (cf. (6.2.30)).

Note that the quantity appearing on the l.h.s. of equality (4) is always non-negative. Indeed, let $\nu_* \in N_0(t,x)$. Then

$$\max_v \min_h \max_{\nu^0} [\langle s^0(t,x,\nu^0),h\rangle - \kappa(s^0(t,x,\nu^0),t,x)] \geqslant \max_v \min_h \langle s^0(t,x,\nu_*),h\rangle - \kappa(s^0(t,x,\nu_*),t,x) = 0.$$

Here we are using the fact that

$$\max_{v\in Q} \min_{h\in F_2(t,x,v)} \langle s,h\rangle = \kappa(s,t,x). \tag{5}$$

If Condition 6.4.1 is fulfilled, i.e. $s^0(t,x,\nu^0) = s^{\#}(t,x)$ for all $\nu^0 \in N_0(t,x)$, then (5) implies (4). Thus Condition 6.4.1 yields Condition 6.5.1.

Note also that along with equality (5) we have

$$\min_{u\in P} \max_{h\in F_1(t,x,u)} \langle s,h\rangle = \kappa(s,t,x),$$

where $\mathcal{F}_1(t,x,u)$ is the set defined by relation (2.2.1). Therefore the inequality

$$(6) \qquad \min_{u\in P} \max_{h\in\mathcal{F}_1(t,x,u)} \max_{\nu^0\in N_0(t,x)} [\langle s^0(t,x,\nu^0),h\rangle - \kappa(s^0(t,x,\nu^0),t,x)] \geqslant 0$$

holds.

The following assertion is valid.

Theorem 6.5.2. *Let the third regularity condition be fulfilled (Condition 6.5.1). Then in the region cl $\mathcal{D}_0$ the programmed maximin coincides with the value of the differential game under consideration.*

Proof. Taking Theorem 6.4.1 and equality (6.4.25) into account, relations (4) and (6) can be expressed as follows:

$$\max_{v\in Q} \min_{h\in\mathcal{F}_2(t,x,v)} D\omega_p(t,x|(1,h) = 0,$$

$$\min_{u\in P} \max_{h\in\mathcal{F}_1(t,x,u)} D\omega_p(t,x)|(1,h) \geqslant 0.$$

These relations are valid in the region $\mathcal{D}_0$. Recall also that $\omega_p(\theta,x) = \sigma(x)$. Therefore the assertion of Theorem 6.5.2 follows from the results of Section 4.4 (cf. Theorem 4.4.1 and the remark at the end of Section 4.4.2). Theorem 6.5.2 is proved.

Since the function of the programmed maximin in the region *cl* $\mathcal{D}_0$ coincides with the value function of a differential game, the set

$$W_0 \overset{\text{def}}{=} \{(t,x) \in T \times R^n: \omega_p(t,x) \leqslant \omega_p(t_0,x_0)\}$$

$$(\text{where } \omega_p(t_0,x_0) \in [a,b))$$

is a u-stable bridge. If the initial position is such that $\omega_p(t_0,x_0) \in [a,b)$, then the strategy U^e extremal to the set W_0 will be an optimal strategy for the first player. Although under the third regularity condition it is not possible to construct optimal strategies in the form of synthesis of problems of programmed control as in the preceding case, nevertheless the fact that the programmed maximin and the value coincide facilitates the construction of a stable bridge and in the final analysis the construction of an optimal strategy as well.

It follows from Theorems 4.4.1 and 6.4.1 that when the first regularity condition (Condition 6.3.1) is satisfied, the fulfillment of equality (4) is a necessary and sufficient condition for the coincidence of a programmed maximin and the value of the game under consideration in the region *cl* $\mathcal{D}_0$.

6.5.3. In conclusion we define the set of programmed absorption and state the condition under which it coincides with the maximal

u-stable bridge.

Consider an approach problem in which it is required to assure that a phase point falls into a given compact set $\mathcal{M}$ at a fixed time $t = \theta$. Denote by $\mathcal{W}_p$ the totality of positions $(t_0,x_0) \in T \times R^n$ such that for any generalized control $\nu \in N$ there exists a generalized control $\eta \in H(\nu)$ for which the corresponding motion $x(\cdot,t_0,x_0,\eta)$ will satisfy the condition $x(\theta,t_0,x_0,\eta) \in \mathcal{M}$. The set $\mathcal{W}_p$ is called the *set of programmed absorption.* Let the function σ be defined by equality (6.2.23). Then the set of programmed absorption can be defined as follows:

$$\mathcal{W}_p = \{(t_0,x_0) \in T \times R^n\colon \max_{\nu \in N} \min_{\eta \in H(\nu)} \sigma(x(\theta,t_0,x_0,\eta)) = 0\}.$$

As a corollary to Theorem 6.5.2 we shall obtain the following assertion.

Let the function σ be defined by equality (6.2.23). Let Condition 6.5.1 be fulfilled in the region $\mathcal{D}_0$ (6.3.8) (where $a = 0$, b is a positive number). Then the set of programmed absorption $\mathcal{W}_p$ coincides with the maximal u-stable bridge in the approach problem under consideration.

Chapter 7
PRIOR STABLE SETS

7.1. Prior Stable Path

It was shown in Chapter 2 that the solution of the problem of approach and evasion can be reduced to the definition of corresponding u-stable and v-stable bridges. In the general case however, the numerical construction of such bridges is quite a difficult problem. Indeed, the implementation of well known constructive definitions of stable bridges is impractical due to the excessively large number of calculations required. Therefore it becomes necessary -- when studying individual types of approach and evasion problems -- to develop special, less laborious methods of constructing stable bridges. One such method was considered in the preceding two chapters where it was shown that under specified conditions (regularity conditions) construction of a stable bridge in an approach problem may be reduced to construction of auxiliary problems of programmed absorption. In other words, under regularity conditions the set of programmed absorption possesses the property of u-stability. Recall that the maximal u-stable bridge is always contained in the set of programmed absorption. Therefore, in cases when regularity conditions are fulfilled, the set of programmed absorption turns out to be not just a u-stable bridge but the maximal u-stable bridge in the approach problem under consideration.

However, when constructing the required stable bridge, one may proceed in a different way, namely by constructing bridges which would be contained in the maximal stable bridge and for which the stability property is inherent in the very method of construction. The simplest starting point is to attempt to construct an a priori stable path $\{(t, w(t)):\ t_0 \leqslant t \leqslant \theta\}$ which would associate the initial position $(t_0, x_0 = w(t_0))$ with the target set M . Such paths are considered in this section. In the next subsection prior stable manifolds are defined formed by solutions of differential inclusions.

In Section 7.2 conditions are stated under which, utilizing prior stable sets, one can construct optimal strategies for the players. In the conclusion of this chapter the connection between the method of a priori stable sets and the method of programmed absorption is discussed.

7.1.1. Before proceeding to the definition of a prior stable path we recall the definition of a u-stable bridge presented in Section 2.2.1. According to this definition the set $\mathcal{W} \subset T \times R^n$ is called a u-stable bridge in an $(\mathcal{M}, \mathcal{N})$-approach problem provided

$$(1) \qquad \mathcal{W} \subset \mathcal{N}, \qquad \mathcal{W}(\theta) \subset \mathcal{M}$$

and the following condition (the condition of u-stability) is fulfilled: for any $(t_*, x_*) \in \mathcal{W}$, $t^* \in (t_*, \theta]$ and $v \in \mathcal{Q}$ there exist $x(\cdot) \in X_2(t_*, x_*, v)$ such that

$$(2) \qquad \{(t, x(t)) \colon t_* \leqslant t \leqslant t^*\} \cap \{\mathcal{W}(t^*) \cup \mathcal{M}) \neq \phi .$$

Here $\mathcal{W}(\tau) = \{(\tau, x) \colon (\tau, x) \in \mathcal{W}\}$ and $X_2(t_*, x_*, v)$ is the set of absolutely-continuous solutions $x(\cdot)\colon [t_*, \theta] \to R^n$ of the differential inclusion

$$(3) \qquad \dot{x}(t) \in \mathcal{F}_2(t, x(t), v) \overset{\text{def}}{=} \mathrm{co}\{f(t, x(t), u, v) \colon u \in \mathcal{P}\}$$

satisfying the initial condition $x(t_*) = x_*$.

Recall that $\mathcal{M}$ and $\mathcal{N}$ are closed sets in $T \times R^n$ and $\mathcal{M} \subset \mathcal{N}$. If $\mathcal{W}$ is a u-stable bridge in an $(\mathcal{M}, \mathcal{N})$-approach problem, then according to Lemma 2.2.5, the closure of the set $\mathcal{W}$ will also be a u-stable bridge in an $(\mathcal{M}, \mathcal{N})$-approach problem. We therefore assume that the u-stable bridges considered below are closed sets. Let $x(\cdot)\colon [t_*, \theta] \to R^n$ be a continuous function and, as usual,

$$T_0(x(\cdot)) = \{t \geqslant t_* \colon (t, x(t)) \in \mathcal{M}\},$$

$$\tau^0(x(\cdot)) = \begin{cases} \min\{t \colon t \in T_0(x(\cdot))\} & \text{for} \quad T_0(x(\cdot)) \neq \phi \\ \theta + 1 & \text{for} \quad T_0(x(\cdot)) = \phi . \end{cases}$$

We shall show that in the definition of a u-stable bridge, relation (2) may be replaced by the following condition:

$$(4) \qquad (t, x(t)) \in \mathcal{W} \quad \text{for all} \quad t_* \leqslant t \leqslant \tau^0(x(\cdot)).$$

Thus we shall prove that the following lemma is valid.

Lemma 7.1.1. *In order that a closed set* $\mathcal{W} \subset T \times R^n$ *be a u-stable bridge in an* $(\mathcal{M}, \mathcal{N})$*-approach problem it is necessary and sufficient that it satisfy relations* (1) *and that the following condition be fulfilled: for* *any* $(t_*,x_*) \in \mathcal{W}$ *and* $v \in \mathcal{Q}$ *there exists a solution* $x(\cdot) \in X_2(t_*,x_*,v)$ *for which the inclusion* (4) *is valid.*

Proof. (4) implies (2), therefore the sufficiency of conditions (1) and (4) is obvious.

Let $\mathcal{W}$ be a u-stable bridge in an $(\mathcal{M},\mathcal{N})$-approach problem. Let $(t_0,x_0) \in \mathcal{W}$ and $v \in \mathcal{Q}$. Select a partition $\Delta = \{t_0 = \tau_0 < \tau_1 < \cdots < \tau_{k+1} = \theta\}$. It follows from the condition of u-stability that a solution $x(\cdot) \in X_2(t_0,x_0,v)$ exists, satisfying the condition

$$(\tau_i,x(\tau_i)) \in \mathcal{W} \quad \text{for all} \quad \tau_i \in [t_0,\tau^0(x(\cdot))].$$

Note that the estimate $\tau^0(x(\cdot)) \leqslant \theta$ follows from the condition $\mathcal{W}(\theta) \subset \mathcal{M}$.

We select a sequence of partitions $\Delta_\ell = \{t_0 = \tau_{\ell,0} < \tau_{\ell,1} < \cdots < \tau_{\ell,k(\ell)+1} = \theta\}$ $(\ell = 1,2, \ldots)$. Assume that diam $\Delta_\ell \to 0$ as $\ell \to \infty$. For each one of these partitions Δ_ℓ we define a solution $x_\ell(\cdot) \in X_2(t_0,x_0,v)$ satisfying the conditions

$$(5) \qquad \begin{aligned} &(\tau_{\ell,i},\, x_\ell(\tau_{\ell,i})) \in \mathcal{W} \quad \text{for all} \quad \tau_{\ell,i} \in [t_0,\tau^0(x_\ell(\cdot))),\\ &\tau^0(x_\ell(\cdot)) \leqslant \theta. \end{aligned}$$

As is known, $X_2(t_0,x_0,v)$ is a compact set in $C_n([t_0,\theta])$ (cf. Section A.4), therefore it can be assumed without loss of generality that $x_\ell(\cdot) \to x_*(\cdot) \in X_2(t_0,x_0,v)$. We also assume that $\tau^0(x_\ell(\cdot)) \to \tau_* \leqslant \theta$ as $\ell \to \infty$.

Let $t \in [t_0,\tau_*]$. Since diam $\Delta_\ell \to 0$ as $\ell \to \infty$ it follows from (5) that a sequence $(t_m,x_m) \in \mathcal{W}$ $(m = 1,2, \ldots)$ exists which converges to the point $(t,x_*(t))$. Since $\mathcal{W}$ is a closed set we obtain that $(t,x_*(t)) \in \mathcal{W}$. It follows from the definition of number τ_* and of the functional $\tau^0(x(\cdot))$ that the inequality $\tau^0(x_*(\cdot)) \leqslant \tau_*$ is valid. We have thus shown that a solution $x_*(\cdot) \in X_2(t_0,x_0,v)$ exists satisfying the condition $(t,x_*(t)) \in \mathcal{W}$ for all $t \in [t_0,\tau^0(x_*(\cdot))]$. Lemma 7.1.1. is thus proved.

7.1.2. Let $\tau \leqslant \theta$, $w(\cdot)$: $[t_0,\tau] \to R^n$ be a continuous function. Consider the set

$$(6) \qquad \mathcal{W} = \{(t,\, w(t))\colon t_0 \leqslant t \leqslant \tau\}.$$

We shall assume that

$$(7) \qquad \mathcal{W} \subset \mathcal{N},\ \ (t,w(t)) \notin \mathcal{M} \ \text{ for } \ t_0 \leqslant t < \tau,\ \ (\tau,w(\tau)) \in \mathcal{M}.$$

The set $\mathcal{W}$ considered herein is the graph of function $w(\cdot)$. This set represents a path which passes through the initial position $(t_0,x_0) = w(t_0))$ and connects it with the target set $\mathcal{M}$. We shall state the conditions under which this path will be a u-stable bridge in an

$(\mathcal{M}, \mathcal{N})$-approach problem.

We introduce the notation

$$(8) \qquad \mathcal{F}_2(t,w) \overset{\text{def}}{=} \bigcap_{v \in \mathcal{Q}} \mathcal{F}_2(t,w,v)$$

where the sets $\mathcal{F}_2(t,w,v)$ are defined according to (2.2.1), i.e. $\mathcal{F}_2(t,w,v) = \text{co}\{f(t,w,u,v)\colon u \in \mathcal{P}\}$. The following theorem is valid.

Theorem 7.1.1. *Let $w(\cdot)\colon [t_0,\vartheta] \to R^n$ be a continuous function satisfying conditions (7). In order that the set W (6) be a u-stable bridge in an $(\mathcal{M}, \mathcal{N})$-approach problem it is necessary and sufficient that the function $w(\cdot)$ be absolutely continuous and satisfy the differential inclusion*

$$(9) \qquad \dot{w}(t) \in \mathcal{F}_2(t, w(t))$$

for almost all $t \in [t_0,\vartheta]$.

Proof. Let $\mathcal{Q}_*$ be a countable set everywhere dense in $\mathcal{Q}$. Since the multivalued mapping $v \to \mathcal{F}_2(t,w,v)$ is continuous, one can prove that

$$\mathcal{F}_2(t,w) = \bigcap_{v \in \mathcal{Q}_*} \mathcal{F}_2(t,w,v).$$

Let the set W defined by relation (6) be a u-stable bridge in an $(\mathcal{M}, \mathcal{N})$-approach problem. Since $(t,w(t)) \notin \mathcal{M}$ for $t_0 \leqslant t < \vartheta$, it follows from Lemma 7.1.1 that for any $v \in \mathcal{Q}$ a solution $x(\cdot) \in X_2(t_0,x_0,v)$ exists satisfying the inclusion $(t,x(t)) \in W$ for all $t_0 \leqslant t \leqslant \vartheta$. Here $x_0 = w(t_0)$. In the case under consideration the inclusion $(t,x(t)) \in W$ signifies the equality $x(t) = w(t)$. We thus obtain that $w(\cdot)$ is an absolutely-continuous function which for each $v \in \mathcal{Q}$ satisfies the differential inclusion

$$(10) \qquad \dot{w}(t) \in \mathcal{F}_2(t,w(t),v) \quad \text{for almost all} \quad t \in [t_0,\vartheta].$$

Let $T(v)$ be the set of points $t \in [t_0,\vartheta]$ for which the inclusion (10) is violated. Since $\text{mes}(T(v)) = 0$ for any $v \in \mathcal{Q}$ and the set $\mathcal{Q}_*$ is countable we have $\text{mes}(\cup_{v \in \mathcal{Q}_*} T(v)) = 0$. Hence for almost all $t \in [t_0,\vartheta]$ the inclusion

$$\dot{w}(t) \in \bigcap_{v \in \mathcal{Q}_*} \mathcal{F}_2(t,w(t),v) = \mathcal{F}_2(t,w(t))$$

is valid. Thus the necessity of condition (9) is proved.

Let an absolutely-continuous function $w(\cdot)\colon [t_0,\vartheta] \to R^n$ satisfy conditions (7) and the differential inclusion (9). Then for any $v \in \mathcal{Q}$ the function $w(\cdot)$ satisfies the differential inclusion (10). Therefore Lemma 7.1.1 implies that the set W (6) will be a u-stable bridge in an $(\mathcal{M}, \mathcal{N})$-approach problem. Theorem 7.1.1 is thus proved.

Now we define a strategy U_e extremal to the path $\mathcal{W}$ (6). Clearly in the case under consideration

$$\mathcal{W}[t] \overset{\text{def}}{=} \{w \in R^n: (t,w) \in \mathcal{W}\} = \{w(t)\} \quad \text{for} \quad t_0 \leqslant t \leqslant \tau$$

$$\mathcal{W}[t] = \phi \quad \text{for} \quad t \in T \setminus [t_0,\tau].$$

Therefore according to a general definition of extremal strategy (cf. Section 2.4.1) we obtain that the strategy U_e extremal to the path $\mathcal{W}$ (6) must satisfy the condition

$$\max_{v \in \mathcal{Q}} \langle x - w(t), f(t,x,U_e(t,x),v)\rangle \tag{11}$$
$$= \min_{u \in \mathcal{P}} \max_{v \in \mathcal{Q}} \langle x - w(t), f(t,x,u,v)\rangle \quad \text{for} \quad (t,x) \in [t_0,\tau] \times R^n.$$

For $t \in T \setminus [t_0,\tau]$, $x \in R^n$, the value of the function U_e may be selected in the compact set $\mathcal{P}$ arbitrarily. If an absolutely-continuous function $w(\cdot)$: $[t_0,\tau] \to R^n$ satisfies relations (7) and differential inclusion (9), then in view of Theorems 2.4.5 and 7.1.1 the strategy U_e extremal to the path $\mathcal{W}$ (6) will retain the position $(t,x(t))$ on the set $\mathcal{W}$ up to its encounter with the set $\mathcal{M}$. Note that in the case under consideration this assertion implies that $x(t) = w(t)$ for all $x(\cdot) \in \mathcal{X}_1(t_0,x_0,U_e)$ and $t_0 \leqslant t \leqslant \tau$. We thus arrive at the following assertion.

Theorem 7.1.2. *Let $w(\cdot)$: $[t_0,\tau] \to R^n$ be an absolutely continuous function satisfying relations* (7) *and the differential inclusion* (9) *and the initial condition $w(t_0) = x_0$. Let U_e be a strategy extremal to the path $\mathcal{W}$(6), i.e. this strategy is chosen in accordance with condition* (11). *Then for any motion $x(\cdot) \in \mathcal{X}_1(t_0,x_0,U_e)$ the condition of $(\mathcal{M}, \mathcal{N})$-approach holds. Moreover, for all $x(\cdot) \in \mathcal{X}_1(t_0,x_0,U_e)$ and all $t \in [t_0,\tau]$ the equality $x(t) = w(t)$ is valid.*

7.1.3. Analogous results can be obtained for the problem of $(\mathcal{M}, \mathcal{N})^\epsilon$-evasion. Recall that in this problem it is required to determine a positional strategy V^ϵ: $T \times R^n \to \mathcal{Q}$ so that for all motions $x(\cdot) \in \mathcal{X}_2(t_0,x_0,V^\epsilon)$ the condition of $(\mathcal{M}, \mathcal{N})^\epsilon$-evasion will be fulfilled, i.e. relations (2.1.8) will be satisfied. As noted in Section 2.1.1, the condition of $(\mathcal{M}, \mathcal{N})^\epsilon$-evasion is equivalent to the condition of $(\mathcal{K}_\epsilon, \mathcal{L}_\epsilon)$-approach where the sets $\mathcal{K}_\epsilon$ and $\mathcal{L}_\epsilon$ are defined in accordance with (2.1.9).

The following theorem is valid.

Theorem 7.1.3. *Let $t_0 \in T$, $t_0 < \tau \leqslant \theta$, $w(\cdot)$: $[t_0,\tau] \to R^n$ be a continuous function satisfying the conditions*

$$\mathcal{W} \overset{\text{def}}{=} \{(t,w(t)): t_0 \leqslant t \leqslant \tau\} \subset \mathcal{L}_\epsilon \tag{12}$$

$$(t,w(t)) \notin \mathcal{K}_\epsilon \quad \textit{for} \quad t_0 \leqslant t < \tau, \quad (\tau,w(\tau)) \in \mathcal{K}_\epsilon. \tag{13}$$

In order that the set $\mathcal{W}$ *be a v-stable bridge in an* $(K_\epsilon, \mathfrak{L}_\epsilon)$*-approach problem it is necessary and sufficient that the function* $w(\cdot)$ *be absolutely continuous and satisfy the differential inclusion*

$$\dot{w}(t) \in \mathcal{F}_1(t,w(t)) \quad \textit{for almost all} \quad t \in [t_0,\tau], \tag{14}$$

where

$$\mathcal{F}_1(t,w) \overset{\text{def}}{=} \bigcap_{u \in \mathcal{P}} \mathcal{F}_1(t,w,u). \tag{15}$$

This theorem is an analog of Theorem 7.1.1 and is therefore presented without proof. Note that the graph of function $w(\cdot)$, i.e. the set $\mathcal{W}$ which represents the path that emanates from the initial position $(t_0, x_0 = w(t_0))$, escapes the ϵ-neighborhood of the set M and is cut short either at time $\tau = \theta$ or at time $\tau \leqslant \theta$ when the point $(t,w(t))$ leaves the ϵ-neighborhood of the set N for the first time.

Define the strategy V_e: $T \times R^n \to \mathcal{Q}$ extremal to the path $\mathcal{W}$ (12). According to the general definition given in Section 2.4.3 we obtain that in the case under consideration the strategy V_e must satisfy the condition

$$\begin{aligned} &\min_{u \in \mathcal{P}} \langle w(t) - x,\ f(t,x,u,V_e(t,x)) \rangle \\ &\quad = \max_{v \in \mathcal{Q}} \min_{u \in \mathcal{P}} \langle w(t) - x,\ f(t,x,u,v) \rangle \ \text{ for } \ (t,x) \in [t_0,\tau] \times R^n. \end{aligned} \tag{16}$$

From Theorems 2.4.6 and 7.1.3 follows

Theorem 7.1.4. *Let* $w(\cdot)$: $[t_0,\tau] \to R^n$ *be an absolutely continuous function satisfying conditions* (12), (13), *differential inclusion* (14), *and the initial condition* $w(t_0) = x_0$. *Let* V_e *be a strategy extremal to the path* $\mathcal{W}$ (12) *and defined in accordance with condition* (16). *Then for any motion* $x(\cdot) \in X_2(t_0,x_0,V_e)$ *the condition of* $(K_\epsilon, \mathfrak{L}_\epsilon)$*-approach is valid, or equivalently, the condition of* $(M,N)^\epsilon$*-evasion is valid. Moreover, for all* $x(\cdot) \in X_2(t_0,x_0,V_e)$ *and all* $t \in [t_0,\tau]$ *the equality* $x(t) = w(t)$ *holds.*

7.1.4. The differential inclusions (9) and (14) are pivotal for the proposed method of solving problems of approach and evasion. Sets $\mathcal{F}_2(t,w)$ and $\mathcal{F}_1(t,w)$ appearing on the r.h.s. of differential inclusions (9) and (14) were defined by relations (8) and (15). Sometimes it is more convenient to utilize the equivalent analytical descriptions of these sets rather than these geometric definitions.

We introduce the following notation. Let $(t,w) \in T \times R^n$, $\ell \in R^n$. Set

$$\begin{aligned} \kappa_1(\ell,t,w) &\overset{\text{def}}{=} \min_{u \in \mathcal{P}} \max_{v \in \mathcal{Q}} \langle \ell, f(t,w,u,v) \rangle \\ \kappa_2(\ell,t,w) &\overset{\text{def}}{=} \min_{v \in \mathcal{Q}} \max_{u \in \mathcal{P}} \langle \ell, f(t,w,u,v) \rangle . \end{aligned} \tag{17}$$

Lemma 7.1.2. *The vector* h *belongs to the set* $\mathcal{F}_i(t,w)$, ($i = 1,2$) *if and only if the inequality*

$$(18) \qquad \max_{\|\ell\|=1} [\langle \ell,h \rangle - \kappa_i(\ell,t,w)] \leqslant 0$$

is fulfilled.

Proof. Let $h \in \mathcal{F}_2(t,w)$. According to (8) we have $h \in \mathcal{F}_2(t,w,v)$ for any $v \in \mathcal{Q}$. It is easy to verify that

$$(19) \qquad \begin{aligned} &\max\{\langle \ell,g \rangle : g \in \mathcal{F}_2(t,w,v)\} \\ &= \max_{u \in \mathcal{P}} \langle \ell, f(t,w,u,v) \rangle \text{ for all } \ell \in R^n, \ v \in \mathcal{Q}. \end{aligned}$$

Hence

$$\langle \ell,h \rangle \leqslant \min_{v \in \mathcal{Q}} \max_{u \in \mathcal{P}} \langle \ell, f(t,w,u,v) \rangle \text{ for all } \ell \in R^n .$$

Thus inequality (18) is proved for $i = 2$.

Assume now that the vector $h \in R^n$ satisfies inequality (18) for $i = 2$. Select a point $v \in \mathcal{Q}$. Set

$$\mathcal{E}(\ell) = \{x \in R^n: \langle \ell,x \rangle \leqslant \max_{u \in \mathcal{P}} \langle \ell,f(t,w,u,v) \rangle$$

where $\ell \in R^n$, $\|\ell\| = 1$. According to (19) the set $\mathcal{E}(\ell)$ is a support half-space to the set $\mathcal{F}_2(t,w,v)$. It follows from (18) that

$$h \in \mathcal{E}(\ell) \quad (\ell \in R^n, \ \|\ell\| = 1),$$

i.e. the vector h belongs to the intersection of these half-spaces. Since the set $\mathcal{F}_2(t,w,v)$ is convex and closed it coincides with the intersection of its support half-spaces. We thus obtain $h \in \mathcal{F}_2(t,w,v)$. This inclusion is valid for arbitrary $v \in \mathcal{Q}$. Hence $h \in \mathcal{F}_2(t,w)$. Lemma 7.1.2 is proved.

It follows from (17) that the function $\kappa_i(\cdot,t,w)$: $R^n \to R^1$ is positive homogeneous, i.e. $\kappa_i(\alpha\ell,t,w) = \alpha\kappa_i(\ell,t,w)$ ($\alpha > 0$, $\ell \in R^n$). Assume that the function $\kappa_i(\cdot,t,w)$ is convex, then this function will be a support function for the set $\mathcal{F}_i(t,w)$ (cf. Section A.2). In this case the equality

$$(20) \qquad \kappa_i(\ell,t,w) = \max\{\langle \ell,h \rangle : h \in \mathcal{F}_i(t,w)\}$$

is valid.

7.2. Stable Integral Manifolds

In this section we shall consider stable bridges which are obtained from combining graphs of all solutions of certain differential inclusions. Such bridges are called stable integral manifolds. Under the conditions stipulated below, the method of stable integral manifolds and the method of prior stable paths complement each other in the following sense: from any initial position either a stable path emanates in an approach (evasion) problem or a stable integral manifold in an evasion (approach) problem.

7.2.1. In the preceding section construction of stable bridges in the form of stable paths was described. According to this description, in order to construct a u-stable path in an $(\mathcal{M}, \mathcal{N})$- approach problem it is required to define a solution of the differential inclusion (7.1.9) satisfying the conditions (7.1.7). Such a path is the most narrow u-stable bridge. The first player is capable of holding the motion of the controlled system on his/her chosen path. This means that the opportunities for the first player heavily prevail over the opportunities of the second player. Formally the advantage of the first player over the second is expressed by the relation $\mathcal{F}_2(t,w(t)) \neq \phi$ for $t_0 \leqslant t \leqslant T$.

Analogously, one can say that a v-stable path in an $(\mathcal{M}, \mathcal{N})^\epsilon$-evasion problem can be constructed if the second player has a large advantage over the first, which is formally expressed by the condition $\mathcal{F}_1(t,w(t)) \neq \phi$ for $t_0 \leqslant t \leqslant T$.

The method of prior stable paths is very convenient, since it allows us to reduce construction of a u-stable or v-stable bridges to the determination of an appropriate solution of differential inclusion (7.1.9) or (7.1.14). However, application of this method is constrained by the requirement that the sets $\mathcal{F}_2(t,w(t))$ or $\mathcal{F}_1(t,w(t))$ are nonempty; this condition is only rarely valid. The method presented below provides far wider possibilities for construction of sets possessing stability properties.

Let $(t,x) \to \mathcal{F}(t,x)$ be a multivalued mapping satisfying the following conditions:

(i_1) $\mathcal{F}(t,x) \cap \mathcal{F}_2(t,x,v) \neq \phi$ for all $v \in \mathcal{Q}$ and $(t,x) \in \mathcal{D}$;

(ii) $\mathcal{F}(t,x)$ is convex and $\mathcal{F}(t,x) \subset \mathrm{co}\{f(t,x,u,v)\colon u \in P,\ v \in \mathcal{Q}\}$ for all $(t,x) \in \mathcal{D}$;

(iii) the mapping $(t,x) \to \mathcal{F}(t,x)$ is upper semicontinuous on $\mathcal{D}$.

Here $\mathcal{D}$ is an open set in the space $T \times R^n$. Recall that the set $\mathcal{F}_2(t,x,v)$ is defined in accordance with (2.2.1).

We complete the definition of the mapping $(t,x) \to \mathcal{F}(t,x)$ on the whole space $T \times R^n$ setting

(1) $\mathcal{F}(t,x) = \text{co}\{f(t,x,u,v)\colon u \in \mathcal{P},\ v \in \mathcal{Q}\}$ for $(t,x) \in (T \times R^n)\backslash\mathcal{D}$.

Note that with this definition the upper semicontinuity of the mapping $\mathcal{F}$ is retained.

Consider the differential inclusion

(2) $\dot{x}(t) \in \mathcal{F}(t,x(t))$.

Let $(t_0,x_0) \in T \times R^n$. Denote by $X(t_0,x_0,\mathcal{F})$ the set of absolutely continuous solutions $x(\cdot)\colon [t_0,\theta] \to R^n$ of the differential inclusion (2) satisfying the initial condition $x(t_0) = x_0$. Under these assumptions the set $X(t_0,x_0,\mathcal{F})$ is a nonempty compact set in $C_n([t_0,\theta])$.

Next we shall assume that for any solution $x(\cdot) \in X(t_0,x_0,\mathcal{F})$ the condition of $(\mathcal{M}, \mathcal{N})$-approach is fulfilled, i.e.

$$\tau^0(x(\cdot)) \leqslant \theta, \quad (t,x(t)) \in \mathcal{N} \quad \text{for} \quad t_0 \leqslant t \leqslant \tau^0(x(\cdot)),$$

where the functional τ^0 is defined according to (2.1.1).

Define the set

(3) $\mathcal{W} = \{(t,x(t))\colon\ t_0 \leqslant t \leqslant \tau^0(x(\cdot)),\quad x(\cdot) \in X(t_0,x_0,\mathcal{F})\}$.

It will be shown below that the set $cl\,\mathcal{W}$ is a u-stable bridge in an $(\mathcal{M}, \mathcal{N})$-approach problem. This set will be called a *u-stable integral manifold*.

Before proceeding to the proof of the above stated fact we note the following: In the construction considered above, it can be assumed at once that the mapping $(t,x) \to \mathcal{F}(t,x)$ is defined and satisfies conditions (i_1)-(iii) on the whole space $T \times R^n$. However, we have preferred to define the mapping $\mathcal{F}$ first in a certain efficient region $\mathcal{D}$ and then extend the definition over the whole space $T \times R^n$. The fact of the matter is that in the course of implementing the method of stable integral manifolds, situations arise when certain considerations (by means of which the sets $\mathcal{F}(t,x)$ and $\mathcal{W}$ are defined) are effective not on the whole space $T \times R^n$ but only in the region $\mathcal{D}$. In these cases a stable integral manifold can be defined in the following modified manner. Let $(t_0,x_0) \in \mathcal{D}$. Consider a solution of the differential inclusion (2) which we shall continue up to the time of the first exit of the point $(t,x(t))$ to the boundary of the region $\mathcal{D}$. Note that the set of solutions will now consist of functions $t \to x(t)$ defined, generally speaking, on different time intervals. This set will be denoted by $X_*(t_0,x_0,\mathcal{F})$. As above, we shall assume that for any solution $x(\cdot) \in X_*(t_0,x_0,\mathcal{F})$ the condition of an $(\mathcal{M}, \mathcal{N})$-approach is fulfilled. This assumption means, in particular, that for any solution $x(\cdot) \in X_*(t_0,x_0,\mathcal{F})$ the point $(t,x(t))$ encounters the set $\mathcal{M}$ not later than it hits the boundary of the region $\mathcal{D}$. We set

(4) $\mathcal{W}_* = \{(t,x(t))\colon t_0 \leqslant t \leqslant \tau^0(x(\cdot)),\ x(\cdot) \in X_*(t_0,x_0,\mathcal{F})\}$.

It is easy to verify that $\mathcal{W}_* = \mathcal{W}$ provided $\mathcal{D}$ is an open set and $\mathcal{W}_* \subset \mathcal{D}$ (here $\mathcal{W}$ is a set defined according to (3)). The assertion stated herein is not of a fundamental nature and is presented only to clarify the role of the set $\mathcal{D}$.

Theorem 7.2.1. *Let the mapping* $(t,x) \to \mathcal{F}(t,x)$ *satisfy the conditions* (i_1), (ii), (iii) *and equality* (1). *Assume that all the solutions* $x(\cdot) \in \mathcal{X}(t_0,x_0,\mathcal{F})$ *of differential inclusion* (2) *satisfy the condition of an* $(\mathcal{M}, \mathcal{N})$*-approach and that the condition* $\mathcal{W} \subset \mathcal{D}$ *is fulfilled, where* $\mathcal{W}$ *is the set defined by equality* (3). *Then the set cl* $\mathcal{W}$ *will be a u-stable bridge in an* $(\mathcal{M}, \mathcal{N})$*-approach problem.*

Proof. We shall show that the set $\mathcal{W}$ is a u-stable bridge in an $(\mathcal{M}, \mathcal{N})$-approach problem.

Let $(t,w) \in \mathcal{W}$. According to (3) a solution $x(\cdot) \in \mathcal{X}(t_0,x_0,\mathcal{F})$ exists such that $x(t) = w$, $\tau^0(x(\cdot)) \geq t$. This solution satisfies the condition of an $(\mathcal{M}, \mathcal{N})$-approach, therefore $(t,x(t)) = (t,w) \in \mathcal{N}$. We have thus shown that $\mathcal{W} \subset \mathcal{N}$.

Let $(\theta,w) \in \mathcal{W}$. Then a solution $x(\cdot) \in \mathcal{X}(t_0,x_0,\mathcal{F})$ exists such that $x(\theta) = w$, $\tau^0(x(\cdot)) = \theta$. It follows from the definition of functional τ^0 (2.1.1) that $(\theta,x(\theta)) = (\theta,w) \in \mathcal{M}$. We have thus shown that $\mathcal{W}(\theta) \subset \mathcal{M}$.

We now show that the set $\mathcal{W}$ satisfies the condition of u-stability. Let $(t_*,x_*) \in \mathcal{W}$, $t^* \in (t_*,\theta]$, $v \in \mathcal{Q}$. According to (3) a solution $x(\cdot) \in \mathcal{X}(t_0,x_0,\mathcal{F})$ exists such that $x(t_*) = x_*$, $\tau^0(x(\cdot)) \geq t_*$. Moreover, the function $x(\cdot)$: $[t_0,\theta] \to R^n$ belonging to the set $\mathcal{X}(t_0,x_0,\mathcal{F})$ and satisfying the above-stated conditions can be chosen so that it will satisfy the differential inclusion

$$\dot{x}(t) \in \mathcal{F}(t,x(t)) \cap \mathcal{F}_2(t,x(t),v) \tag{5}$$

almost everywhere on the interval $[t_*,\theta]$. Indeed, let a function $x^{(1)}(\cdot)$: $[t_0,\theta] \to R^n$ belong to the set $\mathcal{X}(t_0,x_0,\mathcal{F})$ and satisfy the conditions $x^{(1)}(t_*) = x_*$, $\tau^0(x^{(1)}(\cdot)) \geq t_*$. We now define the solution $x^{(2)}(\cdot)$: $[t_*,\theta] \to R^n$ of the differential inclusion (5) satisfying the initial condition $x^{(2)}(t_*) = x^{(1)}(t_*) = x_*$. Such a solution exists since the multivalued mapping $(t,x) \to \mathcal{F}(t,x) \cap \mathcal{F}_2(t,x,v)$ satisfies all the requirements stated in Section A.4. Set $x(t) = x^{(1)}(t)$ for $t_0 \leq t \leq t_*$ and $x(t) = x^{(2)}(t)$ for $t_* \leq t \leq \theta$. Clearly this function is an element of the set $\mathcal{X}(t_0,x_0,\mathcal{F})$ and its restriction to the interval $[t_*,\theta]$ is an element of the set $\mathcal{X}_2(t_*,x_*,v)$ which consists of all the solutions of the differential inclusion (7.1.3) satisfying the initial condition $x(t_*) = x_*$. Two cases are possible:

(j) $\{(t,x(t))\colon t_* \leq t \leq t^*\} \cap \mathcal{M} \neq \phi$;
(jj) $\{(t,x(t))\colon t_* \leq t \leq t^*\} \cap \mathcal{M} \neq \phi$.

In case (j) we obtain that a solution of the differential inclusion (7.1.3) whose graph intersects the set $\mathcal{M}$ emanates from the position $(t_*,x_*) \in \mathcal{W}$. In case (jj) the inequality $\tau^0(x(\cdot)) \geq t^*$ is valid.

Therefore in case (jj) according to (3) the inclusion $(t^*,x(t^*)) \in \mathcal{W}$ is valid. Thus the condition of u-stability of the set $\mathcal{W}$ is proved.

To complete the proof we shall use Lemma 2.2.5. According to this lemma the set $cl\ \mathcal{W}$ will be a u-stable bridge in an $(\mathcal{M}, \mathcal{N})$-approach problem. Theorem 7.2.1 is proved.

Now we present the formulation of an analogous result for an $(\mathcal{M}, \mathcal{N})^{\epsilon}$-evasion problem. Let $(t,x) \to \mathcal{F}(t,x)$ be a multivalued mapping defined in the open region $\mathcal{D} \subset T \times R^n$ which satisfies the condition

$$(\mathrm{i}_2) \qquad \mathcal{F}(t,x) \cap \mathcal{F}_1(t,x,u) \neq \phi$$

for all $(t,x) \in \mathcal{D}$ and $u \in \mathcal{P}$, as well as conditions (ii) and (iii) stated above. We complete the definition of the mapping $(t,x) \to \mathcal{F}(t,x)$ on the whole space $T \times R^n$ by means of the equality (1).

The following theorem is valid.

Theorem 7.2.2. *Let a multivalued mapping* $(t,x) \to \mathcal{F}(t,x)$ *satisfy the conditions* (i_2), (ii), (iii) *and equality* (1). *Assume that all the solutions* $x(\cdot) \in X(t_0,x_0,\mathcal{F})$ *of the differential inclusion* (2) *satisfy the condition of an* $(\mathcal{K}_{\epsilon},\mathcal{L}_{\epsilon})$*-approach (or equivalently the condition of an* $(\mathcal{M}, \mathcal{N})^{\epsilon}$*-evasion). Assume also that* $\mathcal{W} \subset \mathcal{D}$ *where*

$$(6) \qquad \mathcal{W} \overset{\mathrm{def}}{=} \{(t,x(t)) \colon t_0 \leqslant t \leqslant t^{\epsilon}(x(\cdot)),\ \ x(\cdot) \in X(t_0,x_0,\mathcal{F})\}.$$

Then the set $cl\ \mathcal{W}$ *will be a* v*-stable bridge in a problem of* an $(\mathcal{K}_{\epsilon},\mathcal{L}_{\epsilon})$*-approach.*

Recall that $\mathcal{K}_{\epsilon}$ and $\mathcal{L}_{\epsilon}$ are closed sets defined according to (2.1.9). Functional $t^{\epsilon}(x(\cdot))$ is defined by relations (2.1.7).

7.2.2. We compare the method of prior stable paths with the method of stable integral manifolds. In both methods stable bridges are defined using solutions of certain differential inclusions. In the $(\mathcal{M}, \mathcal{N})$-approach problem a stable path is identified with the graph of one of the solutions of differential inclusion (7.1.9), satisfying conditions (7.1.7). To construct such a path it is required that at least the sets $\mathcal{F}_2(t,w)$ be nonempty in a certain region of the space of positions. Moreover one could say that the wider the sets $\mathcal{F}_2(t,w)$ the more likely it is possible to construct the required path.

In the second method, a stable integral manifold in an $(\mathcal{M}, \mathcal{N})$-approach problem is defined by means of the totality of solutions of differential inclusion (2) where the mapping $\mathcal{F}$ satisfies the conditions (i_1), (ii), and (iii). It is required in this method that all the solutions of the differential inclusion (2) satisfy the condition of an $(\mathcal{M}, \mathcal{N})$-approach. Therefore, here -- unlike the case for the method of prior stable paths -- the wider the sets $\mathcal{F}(t,w)$, the harder it is to construct the required stable integral manifold.

In the nonrigorous discussion presented above, possibilities for solving the problem of an $(\mathcal{M}, \mathcal{N})$-approach using the two methods of prior stable sets were considered. We shall discuss the situation where an approach problem is solved by the method of prior stable paths while an evasion problem is solved by the method of stable integral manifolds. We shall show that under certain conditions these two methods complement each other.

First we prove the following assertion.

Lemma 7.2.1. *Let the equality (the condition of saddle point in a small game)*

$$(7)\qquad \min_{u\in\mathcal{P}} \max_{v\in\mathcal{Q}} \langle \ell, f(t,w,u,v)\rangle = \max_{v\in\mathcal{Q}} \min_{u\in\mathcal{P}} \langle \ell, f(t,w,u,v)\rangle \qquad (\ell \in R^n)$$

be satisfied. If the function $\kappa_1(\cdot,t,w)\colon R^n \to R^1$ *is convex, then*

$$(8)\qquad \mathcal{F}_1(t,w) \cap \mathcal{F}_2(t,w,v) \neq \phi$$

for any $v \in \mathcal{Q}$. *If the function* $\kappa_2(\cdot,t,w)\colon R^n \to R^1$ *is convex, then*

$$(9)\qquad \mathcal{F}_2(t,w) \cap \mathcal{F}_1(t,w,u) \neq \phi$$

for all $u \in \mathcal{P}$.

Here the sets $\mathcal{F}_2(t,w)$, $\mathcal{F}_1(t,w)$, and functions κ_i are defined by the relations (7.1.8), (7.1.15), and (7.1.17), respectively.

Proof. Let the function $\kappa_1(\cdot,t,w)$ be convex. Suppose the contrary, i.e. a point $v_* \in \mathcal{Q}$ exists such that $\mathcal{F}_1(t,w) \cap \mathcal{F}_2(t,w,v_*) = \phi$ where

$$\mathcal{F}_2(t,w,v) \overset{\text{def}}{=} \text{co}\{f(t,w,u,v)\colon u \in \mathcal{P}\}.$$

According to the Hahn-Banach theorem [34, 36] a hyperplane exists which subdivides the convex compact sets $\mathcal{F}_1(t,w)$ and $\mathcal{F}_2(t,w,v_*)$, i.e. a vector $\ell_* \in R^n$ exists such that $\langle \ell_*,h\rangle < \langle \ell_*,g\rangle$ for all $h \in \mathcal{F}_1(t,w)$ and all $g \in \mathcal{F}_2(t,w,v_*)$. This inequality can be rewritten as

$$\max\{\langle \ell_*,h\rangle\colon h \in \mathcal{F}_1(t,w)\} < \min\{\langle \ell_*,g\rangle : g \in \mathcal{F}_2(t,w,v_*)\}$$
$$= \min_{u\in\mathcal{P}} \langle \ell_*, f(t,w,u,v_*)\rangle \leqslant \max_{v\in\mathcal{Q}} \min_{u\in\mathcal{P}} \langle \ell_*, f(t,w,u,v)\rangle.$$

Taking equalities (7) and (7.1.17) into account we obtain

$$(10)\qquad \max\{\langle \ell_*,h\rangle : h \in \mathcal{F}_1(t,w)\} < \kappa_1(\ell_*,t,w).$$

It was mentioned in the preceding subsection that for a convex function $\kappa_1(\cdot,t,w)$, equality (7.1.20) is valid for $i = 1$. This equality contradicts inequality (10) and this contradiction proves relation (8).

Analogously one can prove that relation (9) is valid provided the function $\kappa_2(\cdot,t,w)$ is convex. Note that in this proof we use the equality

$$\max_{u\in P}\min_{v\in Q} \langle \ell, f(t,w,u,v)\rangle = \min_{v\in Q}\max_{u\in P} \langle \ell, f(t,w,u,v)\rangle,$$

which follows from (7) and the well known identity

$$\min_{u\in P}\max_{v\in Q} r(u,v) = -\max_{u\in P}\min_{v\in Q} -(r(u,v)).$$

Thus Lemma 7.2.1 is proved.

Assume that the function $\ell \to \kappa_2(\ell,t,w)$ is convex for all $(t,w) \in T \times R^n$. It then follows from Lemma 7.2.1 that in particular, $\mathcal{F}_2(t,w) \neq \phi$ for all $(t,w) \in T \times R^n$. We select an initial position $(t_0,x_0) \in T \times R^n$. Denote by $X(t_0,x_0,\mathcal{F}_2)$ the set of absolutely continuous solutions $w(\cdot)$: $[t_0,\theta] \to R^n$ of the differential inclusion

$$\dot{w}(t) \in \mathcal{F}_2(t,w(t))$$

satisfying the initial condition $w(t_0) = x_0$. Note that $X(t_0,x_0,\mathcal{F}_2)$ is a nonempty compact set in $C_n([t_0,\theta])$. Recall also that the sets M and N are closed in $T \times R^n$. Therefore one and only one of the following two cases are possible:

(i) a solution $w_*(\cdot) \in X(t_0,x_0,\mathcal{F}_2)$ exists such that $\tau_* = \tau^0(w_*(\cdot)) \leqslant \theta$, $(t,w_*(t)) \in N$ for $t_0 \leqslant t \leqslant \tau_*$, i.e. this solution satisfies the condition of an (M, N)-approach;

(ii) an $\epsilon > 0$ exists such that for all the solutions $w(\cdot) \in X(t_0,x_0,\mathcal{F}_2)$ the condition of $(M, N)^\epsilon$-evasion or equivalently the condition of $(K_\epsilon,\mathcal{L}_\epsilon)$-approach is valid.

In case (i) we define the set

$$(11) \qquad W_1 = \{(t,w_*(t)): t_0 \leqslant t \leqslant \tau^0(w_*(\cdot))\}.$$

According to Theorem 7.1.1 this set is a u-stable bridge in a problem of an (M, N)-approach (a priori stable path). Next, as shown in Theorem 7.1.2 the strategy U_e extremal to the set W_1 yields the solution of an (M, N)-approach problem.

In case (ii) we consider the set

$$(12) \qquad W_2 = cl\{(t,w(t)): t_0 \leqslant t \leqslant t^\epsilon(w(\cdot)),\ w(\cdot) \in X(t_0,x_0,\mathcal{F}_2)\}.$$

We verify that the set W_2 is a v-stable bridge in a problem of an $(M, N)^\epsilon$-evasion. We shall use the convexity condition of the function $\kappa_2(\cdot,t,w)$ and Lemma 7.2.1. We obtain that the mapping $(t,w) \to \mathcal{F}_2(t,w)$ satisfies condition (i_2) for all $(t,w) \in T \times R^n$. Clearly the mapping $\mathcal{F}_2$ satisfies conditions (ii) and (iii) also. Therefore it follows from Theorem 7.2.2 that the set W_2 will be a v-stable integral manifold. Strategy V_e extremal to the set W_2 yields a solution of an

$(\mathcal{M}, \mathcal{N})^{\epsilon}$-evasion problem.

We have thus proved the following assertion.

Theorem 7.2.3. *Let the function $\kappa_2(\cdot,t,w)$: $R^n \to R^1$ be convex for all $(t,w) \in T \times R^n$. Then for any initial position $(t_0,x_0) \in T \times R^n$ the following alternative is valid: either a u-stable path $\mathcal{W}_1$ (11) emanates from this initial position and the strategy U_e extremal to the set $\mathcal{W}_1$ yields a solution of an $(\mathcal{M}, \mathcal{N})$-approach problem; or for some $\epsilon > 0$ the position (t_0,x_0) is located on a v-stable integral manifold $\mathcal{W}_2$ (12) and the strategy V_e extremal to the set $\mathcal{W}_2$ yields a solution of an $(\mathcal{M}, \mathcal{N})^{\epsilon}$-evasion problem.*

Assume now that for all $(t,w) \in T \times R^n$ function $\kappa_1(\cdot,t,w)$ is convex. Let $(t_0,x_0) \in T \times R^n$ and let $\mathcal{X}(t_0,x_0,\mathcal{F}_1)$ be the set of all absolutely continuous solutions of the differential inclusion $\dot{w}(t) \in \mathcal{F}_1(t,w(t))$ satisfying the initial condition $w(t_0) = x_0$. Two cases are possible:

(j) for any solution $w(\cdot) \in \mathcal{X}(t_0,x_0,\mathcal{F}_1)$ the condition of an $(\mathcal{M}, \mathcal{N})$-approach is fulfilled;

(jj) there exists an $\epsilon > 0$ and a solution $w_*(\cdot) \in \mathcal{X}(t_0,x_0,\mathcal{F}_1)$ such that the condition of an $(\mathcal{M}, \mathcal{N})^{\epsilon}$-evasion is satisfied.

In case (j) we define the set

$$(13) \qquad \mathcal{W}^{(1)} = cl\{(t,w(t)): t_0 \leqslant t \leqslant \tau^0(w(\cdot)),\ w(\cdot) \in \mathcal{X}(t_0,x_0,\mathcal{F}_1)\}.$$

According to Theorem 7.2.1 and Lemma 7.2.1 this set will be a u-stable integral manifold in an $(\mathcal{M}, \mathcal{N})$-approach problem.

In case (jj) we consider the set

$$(14) \qquad \mathcal{W}^{(2)} = \{(t,w_*(t)): t_0 \leqslant t \leqslant t^{\epsilon}(w(\cdot))\}.$$

It follows from Theorem 7.1.3 that the set $\mathcal{W}^{(2)}$ is a v-stable path in an $(\mathcal{M}, \mathcal{N})^{\epsilon}$-evasion problem.

We thus obtain the following assertion.

Theorem 7.2.4. *Let the function $\kappa_1(\cdot,t,w)$ be convex for all $(t,w) \in T \times R^n$. Then for any initial position $(t_0,x_0) \in T \times R^n$ the following alternative is valid: either the position (t_0,x_0) is located on a u-stable integral manifold $\mathcal{W}^{(1)}$ (13) and the strategy U_e extremal to the set $\mathcal{W}^{(1)}$ yields a solution of an $(\mathcal{M}, \mathcal{N})$-approach problem or a v-stable path $\mathcal{W}^{(2)}$ (14) emanates from the position (t_0,x_0) and the strategy V_e extremal to the set $\mathcal{W}^{(2)}$ yields a solution of an $(\mathcal{M}, \mathcal{N})^{\epsilon}$-evasion problem.*

The conditions of Theorems 7.2.3 and 7.2.4 can be somewhat relaxed. We shall, for example, assume that the function $\kappa_2(\cdot,t,w)$ is convex not for all $(t,w) \in T \times R^n$ but in some region $\mathcal{D}$. In this case the constructions stated in Theorem 7.2.3 are modified as follows.

Let int $D \neq \phi$. The solution $w(\cdot) \in \mathcal{X}(t_0,x_0,\mathcal{F}_2)$ will now be considered on the interval $[t_0, \theta_*(w(\cdot))]$ where $\theta_*(w(\cdot))$ is the time at which the point $(t,w(t))$ hits the boundary of the region $\mathcal{D}$ for the first time. If the solution $w_*(\cdot) \in X(t_0,x_0,\mathcal{F}_2)$ exists satisfying the condition of an $(\mathcal{M},\mathcal{N})$-approach then the set $\mathcal{W}_1$ (11) corresponding to this solution will be a u-stable path in an $(\mathcal{M},\mathcal{N})$-approach problem. If, however, for all solutions $w(\cdot) \in X(t_0,x_0,\mathcal{F}_2)$ the condition of an $(\mathcal{M},\mathcal{N})^\epsilon$-evasion is fulfilled and $(t,w(t)) \in \operatorname{int} \mathcal{D}$ for all $t_0 \leqslant t \leqslant t^\epsilon(w(\cdot))$ then the set $\mathcal{W}_2$ (12) will be a v-stable integral manifold.

Recall that in Sections 5.2 and 5.3 the so-called condition of "uni-typedness" was considered. Assumptions concerning the convexity of functions $\kappa_1(\cdot,t,w)$ or $\kappa_2(\cdot,t,w)$ can be interpreted as a further generalization of this condition.

In Chapter 3 several types of differential games were considered in which optimal strategies for players forming saddle points can be defined in the form of strategies solving the corresponding problems of $(\mathcal{M},\mathcal{N})$-approach and $(\mathcal{M},\mathcal{N})$-evasion. Under the conditions of Theorems 7.2.3 and 7.2.4, optimal strategies may be developed within the framework of constructions presented in these theorems.

Under certain conditions one can construct u-stable and v-stable bridges combining the methods of prior stable paths and stable integral manifolds. Under these conditions the space of positions is subdivided into regions in which one can utilize either the method of prior stable paths or the method of stable integral manifolds. A description of these conditions and subsequent constructions is rather cumbersome and is therefore not presented here.

7.2.3. As an example, consider the problem of damping a pendulum. It is known that the motion of a pendulum can be described by the equations

$$(15) \qquad \dot{x}_1 = x_2, \quad \dot{x}_2 = -\lambda \sin x_1 + (u + v).$$

Here u is the control and v is the disturbance. It is assumed that

$$(16) \qquad |u| \leqslant \alpha, \quad |v| \leqslant \beta, \quad \alpha > \beta.$$

The target set is defined by the relations

$$(17) \qquad \mathcal{M} = \{(t,x_1,x_2)\colon t \in R^1, \;\; x_2 = 0, \;\; x_1 = 2k\pi, \;\; k = 0,\pm 1, \ldots\}.$$

It is required to select a strategy which will assure the damping of the pendulum during the shortest time. Formally this means the following.

Let $x = (x_1,x_2)^{\mathrm{T}} \in R^2$ and $x(\cdot)\colon \; [t_0,\theta] \to R^2$ be a continuous function. Define a functional $x(\cdot) \to \tau^0(x(\cdot))$ setting $\tau^0(x(\cdot)) = \infty$ if $(t,x(t)) \notin \mathcal{M}$ for all $t \geqslant t_0$, and $\tau^0(x(\cdot)) = \min\{t \geqslant t_0\colon (t,x(t)) \in \mathcal{M}\}$ otherwise. Let $U\colon \; R^1 \times R^2 \to [-\alpha,\alpha]$ be a strategy and let $\Gamma_1(U) = \sup\{\tau^0(x(\cdot))\colon \; x(\cdot) \in X_1(t_0,x_0,U)\}$ where $X_1(t_0,x_0,U)$ is a set of motions generated by this strategy. It is required to determine a strategy U_0 such that

$$\Gamma_1(U_0) = \min_U \Gamma_1(U)$$

where the minimum is taken over all the positional strategies U.

We shall attempt to construct an optimal strategy U_0 in the form of a strategy extremal to a stable path. Define for the example under consideration the set $\mathcal{F}_2(t,x)$ (7.1.8) and a function $\kappa_2(t,x)$ (7.1.17). We have

$$\mathcal{F}_2(t,x) = \{(x_2, -\lambda \sin x_1 + h)^T: |h| \leqslant \alpha - \beta\}$$

$$\kappa_2(\ell,t,x) = \ell_1 x_2 - \lambda \ell_2 \sin x_1 + (\alpha - \beta)|\ell_2|.$$

Note that here the function $\kappa_2(\cdot,t,x)$ is convex for all $(t,x) \in R^1 \times R^2$ (the quantity $\kappa_2(\ell,t,x)$ does not depend on the variable t).

In our example the differential inclusion (7.1.9) can be written as follows:

$$(18) \qquad \dot{w}_1(t) = w_2(t), \quad \dot{w}_2(t) = -\lambda \sin w_1(t) + h(t), \quad h(t) \in [\beta-\alpha, \alpha-\beta].$$

Let $w(\cdot,t_0,x_0,h(\cdot))$ be a solution of equation (18).

Consider an auxiliary problem of optimal control. In this problem it is required to obtain an optimal programmed control $h_0(\cdot)$: $[t_0,\infty) \to [\beta-\alpha, \alpha-\beta]$ such that

$$\tau^0(w(\cdot,t_0,x_0,h_0(\cdot))) = \min_{h(\cdot)} \tau^0(w(\cdot,t_0,x_0,h(\cdot)))$$

where the minimum is defined on the set of all possible measurable functions $h(\cdot)$: $[t_0,\infty) \to [\beta-\alpha, \alpha-\beta]$. In simple words, it is required to determine a control $h_0(\cdot)$ which will transfer the system (18) onto set $\mathcal{M}$ (17) during a minimal of time.

Let this problem be solved. Set

$$w_*(t) = w(t,t_0,x_0,h_0(\cdot)) \quad \text{and} \quad \tau_* = \tau^0(w_*(\cdot)).$$

According to Theorem 7.2.3, the strategy U_e extremal to the path $\{(t,w_*(t))\colon\ t_0 \leqslant t \leqslant \tau_*\}$ assures that the system (15) will be brought into the set $\mathcal{M}$ (17) at time $t = \tau_*$. Note that in accordance with (7.1.11) this strategy is defined by the equality

$$U_e(t,x) = \begin{cases} +\alpha & \text{for } w_{*2}(t) - x_2 \geqslant 0 \\ -\alpha & \text{for } w_{*2}(t) - x_2 < 0 \end{cases}.$$

The constructed strategy U_e is optimal. Indeed, for any $\tau < \tau_*$ we have

$$\{(t,w(t,t_0,x_0,h(\cdot)))\colon\ t_0 \leqslant t \leqslant \tau\} \cap \mathcal{M} = \phi.$$

This relation is valid for all admissible controls $h(\cdot)$. Therefore, according to Theorem 7.2.3, the set of solutions of the differential equation (18) obtained in the course of sorting all the admissible controls $h(\cdot)$ generates a v-stable integral manifold which does not intersect the set $\mathcal{M}$. Therefore no strategy U can assure that the system (15) will be brought into the set $\mathcal{M}$ during a period of time smaller than $\tau_* - t_0$.

This example shows that in those cases when the method of prior stable paths can be applied, a construction of positional strategy reduces in essence to a solution of an auxiliary problem of programmed control.

7.3. Prior Stable Sets for a Linear System

In this section we shall define prior stable paths and stable integral manifolds in the particular case when the motion of the controlled system is described by a linear differential equation. We shall consider the relation between the method of extremal aiming described in Chapter 5 and the method of prior stable sets.

7.3.1. Let the motion of the controlled system be described by a linear differential equation

$$(1) \qquad \dot{x}(t) = A(t)x(t) + B(t)u(t) + C(t)v(t).$$

Here, as usual, x is an n-dimensional phase vector: $A(t)$, $B(t)$, and $C(t)$ are $n \times n$, $n \times p$, and $n \times q$ matrices whose elements depend continuously on variable t. It is assumed that the players' controls satisfy the constraints $u(t) \in \mathcal{P}$ and $v(t) \in \mathcal{Q}$ where $\mathcal{P}$ and $\mathcal{Q}$ are convex compact sets in R^p and R^q respectively.

We introduce the following notation:

$$(2) \qquad \begin{aligned} B(t)\mathcal{P} + C(t)v &\stackrel{\text{def}}{=} \{B(t)u + C(t)v\colon\ u \in \mathcal{P}\} \\ \mathcal{H}(t) &\stackrel{\text{def}}{=} \bigcap_{v \in \mathcal{Q}} (B(t)\mathcal{P} + C(t)v). \end{aligned}$$

To begin with we shall assume that

$$(3) \qquad \mathcal{H}(t) \neq \phi, \quad t \in T.$$

Before proceeding to the statement of the problem and the formulation of the result we shall briefly comment on the definition of the set $\mathcal{H}(t)$ and condition (3). First, note that in the case under consideration $f(t,x,u,v) = A(t)x + B(t)u + C(t)v$. Therefore according to (7.1.8) and (2) we obtain $\mathcal{F}_2(t,x) = \{A(t)x + h\colon h \in \mathcal{H}(t)\}$. Hence condition (3) is equivalent to the condition $\mathcal{F}_2(t,x) \neq \phi$ for $(t,x) \in T \times R^n$.

The set $\mathcal{H}(t)$ (2) can be defined using the operation of geometric subtraction (cf., e.g. [29]). This operation is introduced as follows: Let $\mathcal{A}$ and $\mathcal{B}$ be sets in R^n. Set

$$\mathcal{A} \overset{*}{-} \mathcal{B} \overset{\text{def}}{=} \{c \in R^n \colon \mathcal{B} + c \subset \mathcal{A}\}.$$

The set $\mathcal{A} \overset{*}{-} \mathcal{B}$ is called the geometric difference of the sets $\mathcal{A}$ and $\mathcal{B}$. In the theory of differential games this operation is widely used in L. S. Pontryagin's works (see, e.g., [59]).

It is easy to verify that

$$\mathcal{A} \overset{*}{-} \mathcal{B} = \bigcap_{b \in \mathcal{B}} (\mathcal{A} - b).$$

Therefore the set $\mathcal{H}(t)$ (2) can be defined as follows:

$$\mathcal{H}(t) = B(t)\mathcal{P} \overset{*}{-} (-C(t)\mathcal{Q}).$$

If $\mathcal{H}(t) \neq \phi$, then the set $(-C(t)\mathcal{Q}) \overset{\text{def}}{=} \{-C(t)v \colon v \in \mathcal{Q}\}$ can be translated inside the set $B(t)\mathcal{P}$, i.e. one can say in this case that the set $(-C(t)\mathcal{Q})$ is "smaller" than the set $B(t)\mathcal{P}$. Thus condition (3) expresses the advantage of the first player over the second and has the geometric meaning indicated above.

Clearly the set $\mathcal{H}(t)$ is a convex compact set in R^n for all $t \in T$. It can be shown that the multivalued mapping $t \to \mathcal{H}(t)$ is upper semi-continuous.

Finally, a remark concerning the assumption that the sets $\mathcal{P}$ and $\mathcal{Q}$ are convex is in order. This is not a fundamental supposition. If $\mathcal{P}$ is a compact set in R^p (not necessarily convex) then in the definition of the set $\mathcal{H}(t)$, $\mathcal{P}$ must be replaced by $\mathrm{co}\,\mathcal{P}$. It is easy to verify that the equality

$$\mathcal{H}(t) \overset{\text{def}}{=} \bigcap_{v \in \mathcal{Q}} (B(t)\,\mathrm{co}\,\mathcal{P} + C(t),v) = \bigcap_{v \in \mathrm{co}\,\mathcal{Q}} (B(t)\,\mathrm{co}\,\mathcal{P} + C(t)v)$$

is valid. Hence it is immaterial whether $\mathcal{Q}$ is convex or not.

We now proceed to the construction of a prior stable path. Consider for system (1) the problem of an $(\mathcal{M}, \mathcal{N})$-approach (cf. Section 2.1.1). Recall that $\mathcal{M} \subset \mathcal{N}$ and $\mathcal{M}, \mathcal{N}$ are closed sets in $T \times R^n$ (it is not required here that the sets $\mathcal{M}$ or $\mathcal{N}$ satisfy any additional requirements of convexity, boundedness, etc.). To construct the desired path we turn to the following problem of programmed control.

Let an initial position $(t_0,x_0) \in T \times R^n$ be given. Consider the controlled system

$$\dot{w}(t) = A(t)w(t) + h(t). \tag{4}$$

Here $t \to h(t)$ is a measurable function satisfying the condition $h(t) \in \mathcal{H}(t)$ for almost all $t \geq t_0$. Such functions are called admissible controls. Let $w(\cdot,t_0,x_0,h(\cdot))$ be a solution of equation (4) satisfying the initial condition $w(t_0) = w_0$. It is required to determine an admissible control $h_*(\cdot)$ for which the corresponding solution $w(\cdot,t_0,x_0,h_*(\cdot))$ satisfies the condition of an $(\mathcal{M}, \mathcal{N})$-approach.

Assume that the problem posed possesses a solution. Define the set

$$\mathcal{W} = \{(t,w(t,t_0,x_0,h_*(\cdot)): \; t_0 \leqslant t \leqslant \tau\}.$$

Here τ is the time at which the point $(t,w(t,t_0,x_0,h_*(\cdot))$ hits the set $\mathcal{M}$ for the first time. This set $\mathcal{W}$ is the desired prior stable path.

The fact that this set is a u-stable bridge in the $(\mathcal{M}, \mathcal{N})$-approach problem under consideration follows from Theorem 7.1.1 which was proven in the general case. Note that in the case under consideration the positional strategy U extremal to the path $\mathcal{W}$ is defined by the following relation:

$$\langle (x - w_*(t)), \; B(t)U_e(t,x)\rangle = \min_{u \in P} \langle (x - w_*(t)), \; B(t)u\rangle,$$

where

$$w_*(t) = w(t,t_0,x_0,h_*(\cdot)), \quad t_0 \leqslant t \leqslant \tau.$$

The strategy U_e constructed herein yields a solution of an $(\mathcal{M}, \mathcal{N})$-approach problem.

We thus obtain that an under condition (3) the solution of the game-theoretic problem of an $(\mathcal{M}, \mathcal{N})$-approach can be reduced to the solution of an auxiliary problem of programmed control of the system (4). Note that for the constructions presented herein, it is sufficient to have that $\mathcal{H}(t) \neq \phi$ for $t_0 \leqslant t \leqslant \tau$.

7.3.2. Consider the differential game of pursuit-evasion. As above we shall assume that the motion of the controlled system is described by equation (1). The payoff functional is determined as follows. Let a closed set $\mathcal{M}$ be defined in the space $R^1 \times R^n$. Let $x(\cdot)\colon [t_0,\infty) \to R^n$ be a continuous function. Set

$$T_0(x(\cdot)) = \{t \geqslant t_0\colon \; (t,x(t)) \in \mathcal{M}\}$$

(5)
$$\tau^0(x(\cdot)) = \begin{cases} \min\{t\colon t \in T_0(x(\cdot))\} & \text{for} \quad T_0(x(\cdot)) \neq \phi \\ \infty & \text{for} \quad T_0(x(\cdot)) = \phi. \end{cases}$$

Functional $x(\cdot) \to \tau^0(x(\cdot))$ is the payoff functional in the game under consideration. Recall that a functional of this type was introduced in Section 3.2.3. The particular case considered here is the case in which the phase constraint is absent, i.e. here $\mathcal{N} = R^1 \times R^n$. The differential game of pursuit-evasion consists of a pursuit problem in which the first player attempts to assure the encounter of the position $(t,x(t))$ with the target set $\mathcal{M}$ during the smallest guaranteed time and of an evasion game in which the second player aims to either prevent the falling of point $(t,x(t))$ into the set M or tries to delay as much as possible the instant of such an encounter.

In order to construct strategies of players forming a saddle point in the differential game under consideration we shall utilize Theorem 7.2.3. Define the quantity $\kappa_2(\ell,t,x)$ (7.1.17) for the system (1). We obtain

$$\kappa_2(\ell,t,x) = \langle \ell, A(t)x\rangle + \kappa(\ell,t),$$

where

$$\kappa(\ell,t) = \max_{u\in \mathcal{P}} \langle \ell, B(t)u\rangle + \min_{v\in \mathcal{Q}} \langle \ell, C(t)v\rangle. \tag{6}$$

We shall assume that for all $t \in R^1$ function $\kappa(\cdot,t)$: $R^n \to R^1$ is convex. Clearly, in the given case the condition of convexity of the function $\kappa(\cdot,t)$ is equivalent to the assumption that the function $\kappa_2(\cdot,t,x)$ is convex. According to Lemma 7.2.1 we have $\mathcal{H}(t) \neq \phi$ for all $t \in R^1$. Furthermore one can show that the condition of convexity of function $\kappa(\cdot,t)$ is equivalent to the following equality:

$$B(t) \quad = \mathcal{H}(t) - C(t)\mathcal{Q} \ ;$$

here the sets $\{B(t)u\colon u \in \mathcal{P}\}$ and $\{h - C(t)v\colon h \in \mathcal{H}(t),\ v \in \mathcal{Q}\}$ appear on the left-hand and right-hand sides respectively.

We turn now to the following auxiliary problem of programmed control. It is required to determine an admissible control $h_0(\cdot)$: $[t_0,\infty) \to R^n$ (satisfying the constraint $h_0(t) \in \mathcal{H}(t)$) such that

$$\tau^0(w(\cdot,t_0,x_0,h_0(\cdot))) = \min_{h(\cdot)} \tau^0(w(\cdot,t_0,x_0,h(\cdot)) = \theta_0(t_0,x_0). \tag{7}$$

Here (t_0,x_0) is the initial position, $w(\cdot,t_0,x_0,h(\cdot))$ is the solution of equation (1), and the minimum in (7) is computed over all admissible controls $h(\cdot)$. We shall denote this minimum by $\theta_0(t_0,x_0)$. If an admissible control $h(\cdot)$ exists such that $\tau_0(w(\cdot,t_0,x_0,h(\cdot)) < \infty$ then one can prove that a control $h_0(\cdot)$ exists which is optimal, i.e. the minimum in (7) is actually attained. The quantity $\theta_0(t_0,x_0) - t_0$ is the optimal time in the problem of transferring the system (1) from the initial position (t_0,x_0) into the target set M.

We are assuming that the initial position (t_0,x_0) is fixed. Introduce the following notation:

$$\tau^0(h(\cdot)) \overset{\text{def}}{=} \tau^0(w(\cdot,t_0,x_0,h(\cdot))),$$

$$w(\cdot,h(\cdot)) \overset{\text{def}}{=} w(\cdot,t_0,x_0,h(\cdot)),$$

$$T(h(\cdot)) \overset{\text{def}}{=} [t_0,\tau^0(h(\cdot))] \quad \text{for} \quad \tau^0(h(\cdot)) < \infty,$$

$$T(h(\cdot)) \overset{\text{def}}{=} [t_0,\infty) \quad \text{for} \quad \tau^0(h(\cdot)) = \infty,$$

$$\mathcal{W}(h(\cdot)) \overset{\text{def}}{=} \{(t,w(t,h(\cdot))): \ t \in T(h(\cdot))\}.$$

Here $h(\cdot)$ is an arbitrary admissible control.

Define now a u-stable path $\mathcal{W}_1$ and a v-stable integral manifold $\mathcal{W}_2$. Set

$$\mathcal{W}_1 \overset{\text{def}}{=} \mathcal{W}(h_0(\cdot)) \tag{8}$$

$$\mathcal{W}_2 \overset{\text{def}}{=} cl\left[\bigcup_{h(\cdot)} \mathcal{W}(h(\cdot))\right]. \tag{9}$$

In (8) $h_0(\cdot)$ is an optimal control in problem (7) and in (9) the union is taken over all admissible controls $h(\cdot)$. If problem (7) does not possess a solution, i.e. $\tau^0(w(\cdot,t_0,x_0,h(\cdot)) = \infty$ for all admissible controls $h(\cdot)$ then the path $\mathcal{W}_1$ does not exist and only a v-stable manifold is considered.

The following assertion is valid.

For all $t \in R^1$ let the function $\kappa(\cdot,t)$: $R^n \to R^1$ be convex. Let $\omega^0(t_0,x_0)$ be the value of the differential game of pursuit-evasion under consideration. Then $\omega^0(t_0,x_0) = \theta_0(t_0,x_0)$ where $\theta_0(t_0,x_0)$ is an optimal result in problem (7). If $\theta_0(t_0,x_0) < \infty$ then the strategy U_e extremal to the path $\mathcal{W}_1$ (8) guarantees that the pursuer will encounter the point $(t,x(t))$ with the set M at time $t = \theta_0(t_0,x_0)$. For $\theta_0(t_0,x_0) < \infty$ and for $\theta_0(t_0,x_0) = \infty$ the strategy V_e extremal to the set $\mathcal{W}_2$ (9) guarantees the evader an evasion of the point $(t,x(t))$ from an encounter with the set M on the interval $[t_0,\theta_0(t_0,x_0))$.

This assertion actually follows from Theorem 7.2.3. We note only the following fact. It follows from the construction of the set $\mathcal{W}_2$ (9) and from Theorem 7.2.3 that the set $\mathcal{W}_2$ is v-stable and for all $(t,w) \in \mathcal{W}_2$, $t < \theta_0(t_0,x_0)$ we have that $(t,w) \notin M$ for all constructive motions $x(\cdot) \in X_2(t_0,x_0,V_e)$. Here both cases $\theta_0(t_0,x_0) < \infty$ and $\theta_0(t_0,x_0) = \infty$ are admissible. Thus the strategy V_e yields the solution of the evasion problem. Note that here constructive motions are considered. When considering the corresponding step-by-step motions it is important to distinguish between the case $\theta_0(t_0,x_0) < \infty$ and the case $\theta_0(t_0,x_0) = \infty$. In the first case, for any $\epsilon > 0$ one can find a δ such that for a partition satisfying the bound diam $\Delta \leqslant \delta$ condition $(t,x(t)) \notin M$ for $t_0 \leqslant t \leqslant \theta_0(t_0,x_0) - \epsilon$ will be valid for any step-by-step motion $x(\cdot) \in X(t_0,x_0,V_e,\Delta)$. In the second case one cannot assert that for step-by-step motions the evasion from M holds on the whole semi-axis $[t_0,\infty)$; however, the following proposition is valid: for any arbitrarily large number $\theta > t_0$ one can find $\delta > 0$ such that for any step-by-step motion $x(\cdot) \in X_2(t_0,x_0,V_e,\Delta)$ where diam $\Delta \leqslant \delta$, the evasion condition $(t,x(t)) \notin M$ is valid for all $t_0 \leqslant t \leqslant \theta$. Some additional aspects of approximating constructive motions will be discussed in Chapter 8.

7.3.3. Now consider the differential game in which the payoff is given by the equality

(10) $\gamma(x(\cdot)) = \mathrm{dist}([x(\theta)]_m, \mathcal{M})$.

Here θ is a fixed time, $[x]_m = (x_1, ..., x_m)^T$ is an m-dimensional vector consisting of the first m components of the n-dimensional phase vector x, $\mathcal{M}$ is a closed set in R^m, $\mathrm{dist}([x]_m, \mathcal{M})$ is the distance from the point $[x]_m$ to the set $\mathcal{M}$. As above, we shall assume that the motion of the controlled system is described by equation (1). The construction of optimal strategies to be suggested below in the game under consideration is based on Theorems 7.1.1 and 7.2.3. However, before using these theorems we shall proceed to new phase variables which will allow us to relax somewhat the rather stringent assumption (3) and the convexity condition of the function $\kappa(\cdot,t)$.

We introduce the transformation

$$(t,y) = (t, [\Phi(\theta,t)]_m x).$$

Here y is an m-dimensional vector, $\Phi(\theta,t)$ is an $n \times n$ matrix, the function $t \to \Phi(\theta,t)$ satisfies equation (1.2.15), $[\Phi(\theta,t)]_m$ is an $m \times n$ matrix consisting of the first m rows of matrix $\Phi(\theta,t)$. The motion of the new phase vector y will be described by the equation (cf. Section 2.5.4)

(11) $\dot{y}(t) = B_*(t)u(t) + C_*(t)v(t)$.

Here $B_*(t) = [\Phi(\theta,t)]_m B(t), C_*(t) = [\Phi(\theta,t)]_m C(t)$. We define the payoff functional

(12) $\gamma_*(y(\cdot)) = \mathrm{dist}(y(\theta), \mathcal{M})$

on the motions $y(\cdot)\colon [t_0,\theta] \to R^n$ of the controlled system (11). The initial differential game (1), (10) is equivalent to the transformed game (11), (12). (The equivalence is in the sense defined in Section 2.5.4.)

Let

(13) $\kappa_*(s,t) \overset{\mathrm{def}}{=} \max\limits_{u\in\mathcal{P}} \langle s,B_*(t)u\rangle + \min\limits_{v\in\mathcal{Q}} \langle s,C_*(t)v\rangle$.

When considering the differential game (11), (12) we shall assume that for all $t \in [t_0,\theta]$ the function $\kappa_*(\cdot,t)\colon R^m \to R^1$ is convex. According to Lemma 7.2.1 we obtain that under this condition the set

(14) $\mathcal{H}_*(t) \overset{\mathrm{def}}{=} \bigcap\limits_{v\in\mathcal{Q}} (B_*(t)\mathcal{P} + C_*(t)v)$

is nonempty for all $t \in [t_0,\theta]$. A measurable function $h(\cdot)\colon [t_0,\theta] \to R^n$ satisfying the condition $h(t) \in \mathcal{H}_*(t)$ for almost all $t \in [t_0,\theta]$ will be called an admissible control.

We note the following. If for the initial differential game (11), (12) the function $\kappa(\cdot,t)\colon R^n \to R^1$ defined according to (6) is convex, then the function $\kappa_*(\cdot,t)$: $R^m \to R^1$ will also be convex. The converse assertion is, however, generally not valid. Namely, it is possible to have for $m < n$ situations in which the function $\kappa_*(\cdot,t)$ is convex and the set $\mathcal{H}_*(t)$ is nonempty while for the initial differential game the corresponding conditions of convexity of function $\kappa(\cdot,t)$ (6) and the nonemptiness of the set $\mathcal{H}(t)$ (2) are not fulfilled. Thus the transition to the transformed differential game (11), (12) allows us to relax the conditions under which it is possible to construct prior stable sets.

We introduce the following notation:

$$w(t,t_0,y_0,h(\cdot)) \overset{\text{def}}{=} y_0 + \int_{t_0}^{t} h(\tau)d\tau ,$$

$$\mathcal{W}(h(\cdot)) \overset{\text{def}}{=} \{(t,w(t,t_0,y_0,h(\cdot))): t_0 \leqslant t \leqslant \theta\},$$

(15) $$w_*(t_0,y_0) \overset{\text{def}}{=} \min_{h(\cdot)} \operatorname{dist}(w(\theta,t_0,y_0,h(\cdot)), M)$$
$$= \operatorname{dist}(w(\theta,t_0,y_0,h_0(\cdot)), M),$$

(16) $$\mathcal{W}_1 \overset{\text{def}}{=} \mathcal{W}(h_0(\cdot)),$$

(17) $$\mathcal{W}_2 \overset{\text{def}}{=} \bigcup_{h(\cdot)} \mathcal{W}(h(\cdot)).$$

The minimum in (15) and the union in (17) are taken over the set of all admissible controls $h(\cdot)$.

Observe that here the u-stable path $\mathcal{W}_1$ is defined as a result of solving the problem for programmed control of the system $\dot{w}(t) = h(t)$, $w(t_0) = y_0$. In this problem it is required to obtain an optimal admissible control $h_0(\cdot)$ for which the distance from point $w(\theta)$ to the set M will be minimal. The graph of the solution $w(\cdot,t_0,y_0,h_0(\cdot))$ generated by the optimal control $h_0(\cdot)$ forms a path $\mathcal{W}_1$.

The set $\mathcal{W}_2$ (17) is a v-stable integral manifold. It is obtained as a result of combining graphs of solutions $w(\cdot,t_0,y_0,h(\cdot))$ corresponding to all possible admissible controls $h(\cdot)$. It can be shown that the set $\mathcal{W}_2$ (17) is closed in $T \times R^n$.

Utilizing Theorem 7.2.3 we obtain the following assertion.

For all $t \in [t_0,\theta]$ let the function $\kappa_*(\cdot,t)$: $R^m \to R^1$ defined according to (13) be convex. Then the value of the differential game (11), (12) coincides with the quantity $w_*(t_0,y_0)$ (15), which is the optimal result in the auxiliary problem of programmed control. The saddle point of the differential game (11), (12) is formed by the strategy U_e extremal to the path $\mathcal{W}_1$ (16) and the strategy V_e extremal to the stable integral manifold $\mathcal{W}_2$ (17).

With this assertion at our disposal it is easy to define the value and the saddle point for the initial differential game (1), (10).

Recall that the differential game (1), (10) was studied in Chapter 5 in the framework of the method of extremal aiming. Note that in Chapter 5, the differential game (1), (10) was considered for the case when the set $\mathcal{M}$ is convex. Here the solution is obtained without this assumption. In this sense the method of a priori stable sets proposed in this chapter has a certain advantage over the method of extremal aiming.

However, in this section optimal strategies in the differential game (11), (12) (or equivalently in the differential game (1), (10)) are constructed under the condition that the function $s \to \kappa_*(s,t)$ is convex.

We compare this condition with the regularity conditions introduced in Chapter 5. We shall now assume that the set $\mathcal{M}$ is convex. It is easy to observe that quantities $\kappa_*(s,t)$ (6) and $\kappa^{(m)}(s,t,\theta)$ (5.1.32) are related as $\kappa_*(s,t) = -\kappa^{(m)}(-s,t,\theta)$. Consider the function $s \to \varphi^{(m)}(s,t)$ defined by equality (cf. (5.1.32))

$$\text{(18)}\qquad \begin{aligned} \varphi^{(m)}(s,t) &\overset{\text{def}}{=} \varphi_{\mathcal{M}}^{(m)}(s) - \int_t^\theta \kappa^{(m)}(s,\tau,\theta)d\tau \\ &= \varphi_{\mathcal{M}}^{(m)}(s) + \int_t^\theta \kappa_*(-s,\tau)d\tau. \end{aligned}$$

Here $\varphi_{\mathcal{M}}^{(m)}(s) = \sup\{\langle s,y\rangle : y \in \mathcal{M}\}$. The function $\varphi_{\mathcal{M}}^{(m)}(\cdot)$ is convex. If for all $\tau \in [t_0,\theta]$ the function $\kappa_*(\cdot,\tau)$ is convex then, as one can observe from (18), the function $\varphi^{(m)}(\cdot,t)$ will be convex for all $t \in [t_0,\theta]$, i.e. in this case Condition 5.2.4 is fulfilled. This is sufficient for optimal strategies of the players to be defined in the framework of the method of extremal aiming. The converse is generally not true, i.e. situations are possible when the function $\varphi^{(m)}(\cdot,t)$ is convex and therefore the method of extremal aiming may be used, while the functions $\kappa_*(\cdot,\tau)$ are not and hence the method of stable sets cannot be utilized. In this sense one can assert that the method of extremal aiming has wider applications.

In conclusion, we note that for linear differential games, construction of a stable path was described in L. S. Pontryagin's paper [59] which was followed by a number of investigations (cf. in particular [30, 53]). For nonlinear systems the method of prior stable paths and stable integral manifolds was introduced in paper [71].

7.4. Comparison of Controlled Systems

Situations are possible when two controlled systems are considered -- the initial system and a certain simplified auxiliary system. The following question arises: under what conditions will a stable bridge defined for an auxiliary system remain stable with respect to the initial system? This section is devoted to a discussion of this problem.

7.4.1. Consider two controlled systems

$$\text{(1)}\qquad \dot{x}(t) = f^{(1)}(t,x(t),u^{(1)}(t),v^{(1)}(t)),$$

(2) $\dot{x}(t) = f^{(2)}(t,x(t),u^{(2)}(t),v^{(2)}(t))$.

Here the functions $f^{(i)}$: $T \times R^n \times P^{(i)} \times Q^{(i)} \to R^n$ $(i = 1,2)$ satisfy all the conditions stated in Section 1.2.2 including the saddle point condition in a small game (cf. (1.2.8)). We assume that $P^{(i)}$ and $Q^{(i)}$ are compact sets in R^{p_i} and R^{q_i} respectively. Controls $u^{(i)}(\cdot)$ and $v^{(i)}(\cdot)$ satisfy the constraints $u^{(i)}(t) \in P^{(i)}$, $v^{(i)}(t) \in Q^{(i)}$. Assume that functions $f^{(1)}$ and $f^{(2)}$ are related in the following manner:

(3)
$$\min_{u \in P^{(1)}} \max_{v \in Q^{(1)}} \langle \ell, f^{(1)}(t,x,u,v) \rangle$$
$$\geqslant \max_{v \in Q^{(2)}} \min_{u \in P^{(2)}} \langle \ell, f^{(2)}(t,x,u,v) \rangle \quad (t,x,\ell) \in T \times R^n \times R^n.$$

Note that this first inequality can be interpreted as follows: The opportunities available to the first player when he/she uses system (2) are not worse than when he/she uses system (1). The exact meaning of inequality (3) will be expressed in Theorem 7.4.1 presented below. In the proof of this theorem the following fact will be used.

Let $(t,x) \in T \times R^n$, $u^{(1)} \in P^{(1)}$, $v^{(2)} \in Q^{(2)}$. Set

(4) $$\mathcal{F}_1^{(1)}(t,x,u^{(1)}) \overset{\text{def}}{=} \mathrm{co}\{f^{(1)}(t,x,u^{(1)},v^{(1)}): \ v^{(1)} \in Q^{(1)}\},$$

(5) $$\mathcal{F}_2^{(2)}(t,x,v^{(2)}) \overset{\text{def}}{=} \mathrm{co}\{f^{(2)}(t,x,u^{(2)},v^{(2)}): \ u^{(2)} \in P^{(2)}\}.$$

Then

(6) $$\mathcal{F}_1^{(1)}(t,x,u^{(1)}) \cap \mathcal{F}_2^{(2)}(t,x,v^{(2)}) \neq \phi .$$

Assume the contrary. Then in view of the Hahn-Banach theorem (cf. [34, 36]) there exists a vector $\ell \in R^n$ such that $\langle \ell,h^{(1)} \rangle < \langle \ell,h^{(2)} \rangle$ for all $h^{(1)} \in \mathcal{F}_1^{(1)}(t,x,u^{(1)})$ and for all $h^{(2)} \in \mathcal{F}_2^{(2)}(t,x,v^{(2)})$. This inequality can be rewritten as

$$\max_{v^{(1)} \in Q^{(1)}} \langle \ell, f^{(1)}(t,x,u^{(1)},v^{(1)}) \rangle < \min_{u^{(2)} \in P^{(2)}} \langle \ell, f^{(2)}(t,x,u^{(2)},v^{(2)}) \rangle.$$

Therefore we obtain

$$\min_{u^{(1)} \in P^{(1)}} \max_{v^{(1)} \in Q^{(1)}} \langle \ell, f^{(1)}(t,x,u^{(1)},v^{(1)}) \rangle$$
$$< \max_{v^{(2)} \in Q^{(2)}} \min_{u^{(2)} \in P^{(2)}} \langle \ell, f^{(2)}(t,x,u^{(2)},v^{(2)}) \rangle.$$

We thus arrive at a contradiction to condition (3) and assertion (6) is proved.

We shall use relation (6) in the proof of the following assertion.

Theorem 7.4.1. *Let* $\mathcal{W}$ *be a closed u-stable bridge in an* $(\mathcal{M}, \mathcal{N})$*-approach problem posed for the controlled system* (1). *If condition* (3) *is fulfilled, then the set* $\mathcal{W}$ *will be a u-stable bridge in a problem of an* $(\mathcal{M}, \mathcal{N})$*-approach posed for the controlled system* (2).

Proof. Let $(t_*,x_*) \in \mathcal{W}$, $t^* \in (t_*,\theta]$, $v^{(2)} \in \mathcal{Q}^{(2)}$. Denote by $X_2(t_*,x_*,v^{(2)})$ the set of solutions $x(\cdot)$: $[t_*,\theta] \to R^n$ of the differential inclusion

$$\dot{x}(t) \in \mathcal{F}_2^{(2)}(t,x(t),v^{(2)}) \tag{7}$$

satisfying the initial condition $x(t_*) = x_*$. Set

$$\mathcal{Z} \overset{\text{def}}{=} \{(t,x(t)): t_* \leqslant t \leqslant t^*,\ \ x(\cdot) \in X_2(t_*,x_*,v^{(2)}\}. \tag{8}$$

It will be proved below that

$$\mathcal{Z} \cap (\mathcal{W}(t^*) \cup \mathcal{M}) \neq \phi \tag{9}$$

(recall that $\mathcal{W}(t^*) \overset{\text{def}}{=} \{(t^*,w) \in \mathcal{W}\}$).

Relation (9) means that the set $\mathcal{W}$ possesses the property of u-stability in relation to system (2). Since $\mathcal{W} \subset \mathcal{N}$ and $\mathcal{W}(\theta) \subset \mathcal{M}$ the proof of the theorem reduces to verification of relation (9).

We shall first show that the set possesses the following property: for any $(\tau_*,z_*) \in \mathcal{Z}$ and $u^{(1)} \in \mathcal{P}^{(1)}$ a solution $w(\cdot)$: $[\tau_*,t^*] \to R^n$ ($w(\tau_*) = z_*$) of the differential inclusion

$$\dot{w}(t) \in \mathcal{F}_1^{(1)}(t,w(t),u^{(1)}) \tag{10}$$

exists satisfying the condition $(t,w(t)) \in \mathcal{Z}$ for $\tau_* \leqslant t \leqslant t^*$. This property is a condition of v-stability of the set $\mathcal{Z}$ relative to system (1).

Let $(\tau_*,z_*) \in \mathcal{Z}$. According to the definition of the set $\mathcal{Z}$ a solution $x_*(\cdot) \in X_2(t_*,x_*,v^{(2)})$ exists such that $x_*(\tau_*) = z_*$. Consider the solution $w(\cdot)$: $[\tau_*,t^*] \to R^n$ of the differential inclusion

$$\dot{w}(t) \in \mathcal{F}_1^{(1)}(t,w(t),u^{(1)}) \cap \mathcal{F}_2^{(2)}(t,w(t),v^{(2)}) \tag{11}$$

satisfying the initial condition $w(\tau_*) = z_*$. Here $(t,x) \to \mathcal{F}_1^{(1)}(t,x,u^{(1)})$ and $(t,x) \to \mathcal{F}_2^{(2)}(t,x,v^{(2)})$ are multivalued continuous mappings, the sets $\mathcal{F}_1^{(1)}(t,x,u^{(1)})$ and $\mathcal{F}_2^{(2)}(t,x,v^{(2)})$ are convex compact sets; moreover, relation (6) is fulfilled. Therefore the solution $w(\cdot)$ under consideration does exist (cf. Section A.4). Define the function $x^*(\cdot)$: $[t_*,t^*] \to R^n$ setting $x^*(t) = x_*(t)$ for $t_* \leqslant t \leqslant \tau_*$, $x^*(t) = w(t)$ for $\tau_* \leqslant t \leqslant t^*$. The function $x^*(\cdot)$ is contained in the set $X_2(t_*,x_*,v^{(2)})$. Therefore, according to the definition of the set $\mathcal{Z}$ we obtain $(t,x^*(t)) = (t,w(t)) \in \mathcal{Z}$ for $\tau_* \leqslant t \leqslant t^*$. Since the function $w(\cdot)$ satisfies the inclusion (11) it is also a solution of the differential inclusion (10).

Thus the property of v-stability of the set Z (8) is proved. Note that the set Z can be viewed as a u-stable integral manifold generated by solutions of the differential inclusion (7).

For the controlled system (1) we define a pair of strategies $U_e^{(1)}: T \times R^n \to P^{(1)}$ and $V_e^{(1)}: T \times R^n \to Q^{(1)}$. We assume that $U_e^{(1)}$ is a strategy extremal to the set W and $V_e^{(1)}$ is a strategy extremal to the set Z (note that the set Z is closed). Let $X_1(t_*,x_*,U_e^{(1)})$, $X_2(t_*,x_*,V_e^{(1)})$, and $X(t_*,x_*,U_e^{(1)},V_e^{(1)})$ be sets of constructive motions of the system (1) generated by the strategy $U_e^{(1)}$, strategy $V_e^{(1)}$, and the pair of strategies $U_e^{(1)}$, $V_e^{(1)}$ respectively. Recall (cf. Section 1.5.2) that $X(t_*,x_*,U_e^{(1)},V_e^{(1)}) \subset X_1(t_*,x_*,U_e^{(1)}) \cap X_2(t_*,x_*,V_e^{(1)}) \neq \phi$. We select a motion $x(\cdot) \in$

$X(t_*,x_*,U_e^{(1)},V_e^{(1)})$.

Since W is a u-stable bridge for the controlled system (1), $(t_*,x_*) \in W$, and $x(\cdot) \in X_1(t_*,x_*,U_e^{(1)})$ it follows from Theorem 2.4.5 that

$$(12) \qquad (t,x(t)) \in W \quad \text{for} \quad t_* \leqslant t \leqslant \tau,$$

where τ is the time at which $(t,x(t)) \in M$ for the first time.

On the other hand $(t_*,x_*) \in Z$. Z is a v-stable set for the system (1), $x(\cdot) \in X_2(t_*,x_*,V_e^{(1)})$. Therefore, according to Theorem 2.4.6

$$(13) \qquad (t,x(t)) \in Z \quad \text{for} \quad t_* \leqslant t \leqslant t^*.$$

Two cases are possible: $\tau < t^*$ and $\tau \geqslant t^*$. In the first case we obtain $(\tau,x(\tau)) \in M \cap Z$, from the definition of the number τ and from (13), i.e. in this case relation (9) is valid. In the second case, it follows from (12) and (13) that $(t^*,x(t^*)) \in W(t^*) \cap Z$, i.e. in this case, relation (9) is also fulfilled. Theorem 7.4.1 is thus proved.

If for any closed sets $M \subset T \times R^n$ and $N \subset T \times R^n$ $(M \subset N)$ a u-stable bridge in a problem of an (M, N)-approach posed for system (1) is a u-stable bridge in a problem of an (M, N)-approach posed for system (2), then we shall say that *system* (2) *u-majorizes system* (1). It was established in Theorem 7.4.1 that inequality (3) is a sufficient condition under which the system (2) u-majorizes system (1). One can show that inequality (3) is also a necessary condition for system (2) to u-majorize system (1).

7.4.2. We shall introduce analogous notions for the problem of an $(K, \mathfrak{L})$-approach. Recall that this problem is posed to the second player. Usually $K = K_\epsilon$, $\mathfrak{L} = \mathfrak{L}_\epsilon$ where the sets K_ϵ and $\mathfrak{L}_\epsilon$ are defined by equalities (2.1.9). In this case the problem of an $(K_\epsilon, \mathfrak{L}_\epsilon)$-approach is equivalent to the problem of an $(M, N)^\epsilon$-evasion.

If for any closed sets K and $\mathfrak{L}$ $(K \subset \mathfrak{L} \subset T \times R^n)$ a v-stable bridge in a problem of an $(K, \mathfrak{L})$-approach posed for system (1) is a v-stable bridge in a problem of an $(K, \mathfrak{L})$-approach posed for system (2), we then say that *system* (2) *v-majorizes system* (1). The following theorem is valid.

Theorem 7.4.2. *In order that system* (2) *v-majorizes system* (1) *it is sufficient that*

$$\max_{v\in \mathcal{Q}^{(1)}} \min_{u\in \mathcal{P}^{(1)}} \langle \ell, f^{(1)}(t,x,u,v)\rangle \leq \min_{u\in \mathcal{P}^{(2)}} \max_{v\in \mathcal{Q}^{(2)}} \langle \ell, f^{(2)}(t,x,u,v)\rangle \tag{14}$$

for all $(t,x,\ell) \in T \times R^n \times R^n$.

This theorem is proved in the same manner as Theorem 7.4.1. We note without proof that inequality (14) is a necessary condition for system (2) to v-majorize system (1).

We introduce yet another notion. We say that controlled systems (1) and (2) are equivalent if

$$\max_{v\in \mathcal{Q}^{(1)}} \min_{u\in \mathcal{P}^{(1)}} \langle \ell, f^{(1)}(t,x,u,v)\rangle = \min_{u\in \mathcal{P}^{(2)}} \max_{v\in \mathcal{Q}^{(2)}} \langle \ell, f^{(2)}(t,x,u,v)\rangle \tag{15}$$

for all $(t,x,\ell) \in T \times R^n \times R^n$. Recall that here the operations *min* and *max* are interchangeable. Theorems 7.4.1 and 7.4.2 yield the following assertion.

Theorem 7.4.3. *If the systems* (1) *and* (2) *are equivalent then system* (1) *u-majorizes and v-majorizes system* (2) *and system* (2) *u-majorizes and v-majorizes system* (1) *simultaneously.*

In this section only part of the results related to problems of comparisons and equivalence of controlled systems is considered. In Section 10.5 we shall return to these problems.

Chapter 8
QUALITATIVE PROBLEMS IN THE THEORY OF DIFFERENTIAL GAMES

8.1. The Problem of Stability of Solutions

Formalization of game-theoretical control problems adopted in Chapter 1 admits a transition to practical control procedures. This transition is performed by constructing step-by-step motions and it is assumed that when controls are formed in a step-by-step procedure, the exact information about the realized position of the controlled process is available. However, in reality informational noise of one kind or another is unavoidable. For example, it is possible to have either imprecise measurement of the phase states of the system or a delay of information about the realized state. It turns out that in differential games, situations are possible when even small informational errors can disrupt the proposed positional method of control. In these cases, solutions of game-theoretical problems of control turn out to be incorrect, i.e. unstable with respect to the informational errors, and the problem of its regularization arises. In this and in succeeding sections of this chapter the problem of stability of solutions of game-theoretical problems of control is investigated.

8.1.1. Example. We present an example of a game-theoretical problem of control in which informational noise destroys the ideal solution.

Let three points Y, $Z^{(1)}$, and $Z^{(2)}$ move on the plane. For clarity we shall assume that point Y characterizes the position of a hound and $Z^{(1)}$ and $Z^{(2)}$ the positions of two hares. Assume also that the velocities of points Y, $Z^{(1)}$, and $Z^{(2)}$ can change instantaneously in their magnitude and direction. We thus assume that the motion equations are of the form

$$\dot{y}(t) = u(t), \quad \dot{z}^{(1)}(t) = v^{(1)}(t), \quad \dot{z}^{(2)}(t) = v^{(2)}(t),$$

where y, $z^{(1)}$, $z^{(2)}$, u, $v^{(1)}$, and $v^{(2)}$ are two-dimensional vectors. We assume that the velocities of points Y, $Z^{(1)}$, and $Z^{(2)}$ are bounded in their absolute value, i.e. the inequalities

$$\|u(t)\| \leqslant \alpha, \quad \|v^{(i)}(t)\| \leqslant \beta, \quad (i = 1,2), \quad \alpha > \beta$$

are fulfilled.

We shall consider the following pursuit problem: it is required to determine a strategy U^0 which assures that point $y(t)$ encounters at least one of the points $z^{(1)}(t)$, $z^{(2)}(t)$. In other words, in this problem the hound must catch at least one of the hares.

It is easy to verify that a solution of this problem is provided by the following simple method of control: point $y(t)$ must move with maximal speed to that point $z^{(i)}(t)$ $(i = 1,2)$ which, at given time t is closer to the point $y(t)$. Thus the required strategy $(t,x) \to U^0(t,x)$

$$(x = (y_1, y_2, z_1^{(1)}, z_2^{(1)}, z_1^{(2)}, z_2^{(2)})^T \in R^6)$$

can be given by the relations

$$U^0(t,x) = \begin{cases} \alpha(z^{(1)} - y)/\|z^{(1)} - y\|, & \text{if } \|z^{(1)} - y\| \leqslant \|z^{(2)} - y\|, \\ \alpha(z^{(2)} - y)/\|z^{(2)} - y\|, & \text{if } \|z^{(2)} - y\| < \|z^{(1)} - y\|. \end{cases} \tag{1}$$

If for the target functional we take the time elapsed before the point $y(t)$ encounters one of the points $z^{(i)}(t)$ $(i = 1,2)$, then the strategy U^0 will be an optimal minimax strategy for the pursuer.

In a step-by-step procedure the strategy U^0 forms the control $u(t)$ as follows. Let a partition $\Delta = \{t_0 < \tau_0 < \tau_1 < \cdots\}$, $\tau_i \to \infty$ as $i \to \infty$ be selected. At time $t = \tau_i$ the pursuer determines the closest of the two points $z^{(1)}(\tau_i)$, $z^{(2)}(\tau_i)$ and it is towards this point that the velocity vector $u(\tau_i)$ of the pursuit point is directed and this velocity vector remains unchanged on the half-interval $[\tau_i,\tau_{i+1})$. For any $\epsilon > 0$ for $\delta = \operatorname{diam}\Delta$ sufficiently small, the method of control assures the ϵ-approach of the pursuing point $y(t)$ towards one of the evading points not later than at time

$$\tau^0 = t_0 + \min\left\{\frac{\|y_0 - z_0^{(1)}\|}{\alpha - \beta}, \frac{\|y_0 - z_0^{(2)}\|}{\alpha - \beta}\right\}.$$

However, this conclusion is valid under the condition that the phase vectors $y(\tau_i)$, $z^{(1)}(\tau_i)$, and $z^{(2)}(\tau_i)$ are all measured precisely by the pursuer. If, however, the phase states are measured with errors, then this informational noise may destroy the proposed method of pursuit.

Indeed, let the control of the pursuer be formed in the following manner:

$$u_*(t) = u_*(\tau_i) = U^0(\tau_i, x_*(\tau_i)), \quad \tau_i \leqslant t < \tau_{i+1} = \tau_i + \delta, \tag{2}$$

in accordance with the chosen strategy U^0. Here $x_*(\tau_i)$ is the result of an imprecise measurement of the phase vector $x(\tau_i)$, i.e. it is supposed that the positions of points $y(\tau_i)$, $z^{(1)}(\tau_i)$, and $z^{(2)}(\tau_i)$ are measured imprecisely.

Let the measurement errors satisfy the bound

(3) $\|x_*(\tau_i) - x(\tau_i)\| \leqslant \xi.$

According to the game-theoretical nature of the pursuit problem we shall assume that the measurement errors are under the control of the opponent. Therefore according to the concept of "guaranteed result" one should take into account the situation when this informational noise is realized in the most unfavorable manner for the pursuer. Consider the following situation: let the point $y(t_*)$ be in a small neighborhood of the line AB equally distant from points $z^{(1)}(t_*)$ and $z^{(2)}(t_*)$. Starting from this instant of time let points $z^{(1)}(t_*)$ and $z^{(2)}(t_*)$ move along the rays A_1B_1 and A_2B_2 respectively, remaining at an equal distance from the axis AB at all times. Moreover, in view of the incorrect determination at each one of these times $t = \tau_i$ of the closest of the points $z^{(1)}(\tau_i)$, $z^{(2)}(\tau_i)$, the motion of the point $y(t)$ will take place in the vicinity of the ray AB. This will occur as a result of an alternate aiming at either false point $z_*^{(1)}(\tau_i)$ or false point $z_*^{(2)}(\tau_i)$. Such a motion will not approach either one of the pursued points.

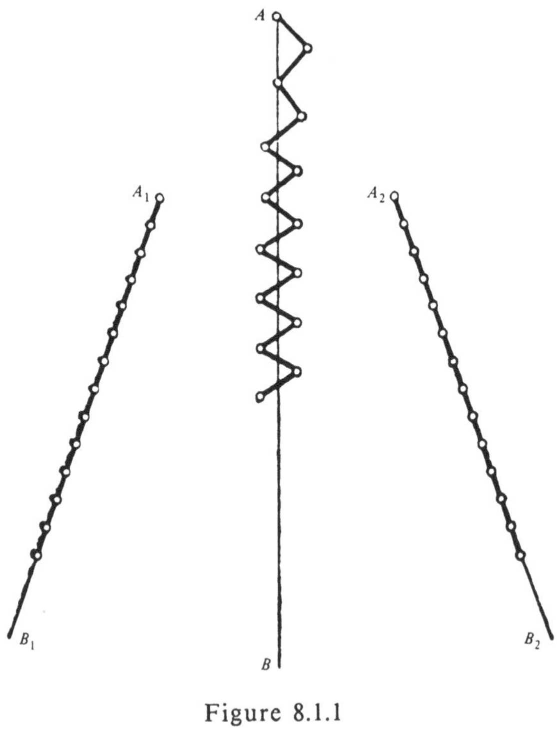

Figure 8.1.1

This cyclic motion of $y(t)$ when choosing the control $u_*(t)$ (2) occurs in the vicinity of the surface of discontinuity of the function U^0 (1), i.e. in the vicinity of the positions for which the equality $\|y - z^{(1)}\| = \|y - z^{(2)}\|$ is valid. Note that such cyclic behavior may occur in the case when the number δ in relation (2) and the number ξ in the bound (3) are related by the inequality $\delta \leqslant \delta_0(\xi)$ where the quantity $\delta_0(\xi)$ is of the same order of smallness as the quantity ξ. Therefore one could recommend the following: in order that errors should not destroy the proposed control procedure one should bound from below the distance between the times τ_i. However, such a recommendation suffers from the defect that the condition $\tau_{i+1} - \tau_i > \delta_0(\xi_i)$ $(i = 0,1, \ldots)$ may result in an unnecessary coarseness of the step-by-step procedure. As a result the solution of the problem will again be destroyed since the discrepancy between step-by-step and constructive motions on which the required outcome of the game is achieved grows substantially. Thus we are compelled to reject this recommendation.

We could try to avoid the destructive action of informational noise by replacing the discontinuous strategy U^0 (1) by some continuous strategy U_0. In the example under consideration we could proceed as follows: at the initial time $t = t_0$ we select out of the two points $z_0^{(1)} = z^{(1)}(t_0)$ and $z_0^{(2)} = z^{(2)}(t_0)$ the point $z_0^{(i)}$ which is closer to the point $y_0 = y(t_i)$. For definiteness let us assume that $z_0^{(1)}$ is such a point. Next, during the time $t \geqslant t_0$ the pursuit problem is considered only for the point $z^{(1)}(t)$. The strategy U_0 which yields a solution of this problem is determined by the function $U_0(t,x) = \alpha(z^{(1)} - y)/\|z^{(1)} - y\|$. The function $(t,x) \to U_0(t,x)$ is continuous at all positions (t,x) except those for which the equality $y = z^{(1)}$ is fulfilled. It is easy to verify that the step-by-step control procedure generated by strategy U_0 turns out to be stable relative to the informational noise.

8.1.2. In the example considered above, two solutions of the pursuit problem were indicated. One of these was defined by a discontinuous strategy U^0 and this solution turned out to be incorrect, i.e. unstable relative to small informational noise. Another regularized solution was defined by the continuous strategy U_0. Realization of this strategy in a step-by-step control procedure is stable relative to informational noise. However, such a regularization based on the transition to continuous strategies may not be possible to implement in a general case. The point is that situations are possible in differential games when a solution of the game-theoretical control problem is being furnished by a discontinuous strategy which cannot be approximated or replaced appropriately by a continuous strategy. This fundamental fact has been emphasized in Sections 1.4.4 and 5.5.1. Therefore it turns out that in the general case a stable solution of game-theoretical control problems may not be possible to obtain based on ordinary procedures of positional control of type (2), and to implement a correct solution it is necessary to develop some improved control methods. The next two sections are devoted to an investigation of this problem.

However, before proceeding to the actual construction of stable control procedures we shall discuss further the motivation for this construction. The example considered above clearly shows the reason why the required pursuit process is destroyed. But this single example may be an insufficiently convincing reason since one may form the impression that unstable strategies occur only in exotic cases due to artificial constructions. Therefore we shall briefly discuss the problem of stability of strategies extremal to stable bridges. Recall that these strategies actually form the equilibrium situations in the positional differential games under consideration.

Consider a controlled system of the general form (1.2.1). Let $U^e : T \times R^n \to \mathcal{P}$ be a strategy extremal to some u-stable bridge $\mathcal{W}$. According to the definition of this strategy given in Section 2.4.1 the function U^e is defined by the condition

$$\max_{v \in \mathcal{Q}} \langle x - e(t,x),\ f(t,x,U^e(t,x),v) \rangle = \min_{u \in \mathcal{P}} \max_{v \in \mathcal{Q}} \langle x - e(t,x),\ f(t,x,u,v) \rangle, \tag{4}$$

where $e(t,x)$ is the point in the set $\mathcal{W}[t] = \{w \in R^n\colon (t,w) \in \mathcal{W}\}$ closest to the point x. It was mentioned in Section 2.4.1 that the function $(t,x) \to U^e(t,x)$ in general is not continuous. The appearance of discontinuities in the function U^e may be for two reasons. Firstly, it may turn out that the function $(t,x) \to e(t,x)$ has discontinuities. The presence of these discontinuities is associated with the fact that the point in the set $\mathcal{W}[t]$ which is closest to the point x may not be unique. Secondly the minimum in (4) may be attained on a non-unique element $U^e(t,x)$. Thus, if the function $(t,x) \to e(t,x)$ is continuous then the function U^e may have discontinuities for the second reason. However, in this case, informational errors do not destroy the step-by-step procedure of control corresponding to strategy U^e, i.e. in the presence of a small amount of informational noise the strategy U^e will retain step-by-step motions in the vicinity of a u-stable bridge.

The situation is quite different if the function $(t,x) \to e(t,x)$ is discontinuous. In this case small errors in the measurement of a phase point x will correspond to a substantial variation in the corresponding values of the function e. Such sliding conditions may occur in the presence of informational noise when there is often a substantial variation in the direction $s_*(\tau_i) = (x_*(\tau_i) - e(\tau_i,x_*(\tau_i)))$ in which the control $u_*(\tau_i) = U^e(\tau_i,x_*(\tau_i))$ aims at the motion of the system (1.2.1). The resulting action of such a sliding control $u_*(\cdot)$ does not assure the required motion in the vicinity of a stable bridge. As in the previous example, two hares may escape from the dog that is rushing between them -- so also in this situation the boundary of the stable bridge may move away from the motion generated by the sliding control $u_*(\cdot)$.

8.1.3. Thus in general, a strategy extremal to the stable bridge may turn out to be unstable relative to informational noise. Also, the arguments presented above indicate that the informational noise cannot destroy the action of an extremal strategy if for all $t \in T$ the function $x \to e(t,x)$ is continuous. Such a situation is valid if for all $t \in T$ the sets $W[t] = \{w \in R^n: (t,w) \in W\}$ are convex.

To formulate the appropriate assertion we introduce certain notation. As usual, let $\Delta = \{t_0 = \tau_0 < \tau_1 < \dots \tau_{k+1} = \theta\}$ be a partition of the time interval $[t_0,\theta]$. We shall assume that at time $t = \tau_i$ imprecise measurements of the phase position of the system are made. These measurements result in vectors $x_*(\tau_i) = x(\tau_i) + y(\tau_i)$. Here $x(\tau_i)$ is the actual phase vector of the system, $y(\tau_i)$ is the error of a measurement conducted at time $t = \tau_i$. The totality of errors $y(\tau_i)$, $(i = 0,1, \dots, k)$ will be called informational noise and will be denoted by $y(\cdot)$. We shall assume that in a step-by-step procedure the control of the first player is formed according to the following rule:

$$u_*(t) = u_*(\tau_i) = U(\tau_i,x_*(\tau_i)) = U(\tau_i,x(\tau_i) + y(\tau_i)),$$
$$\tau_i \leqslant t < \tau_{i+1}, \quad i = 0,1, \dots, k. \tag{5}$$

Denote by $x(\cdot,t_0,x_0,U,v(\cdot),\Delta,y(\cdot))$ the step-by-step motion of the controlled system satisfying the initial condition $x(t_0) = x_0$ and generated by the control $u_*(\cdot)$ (5) coupled with a certain measurable control $v(\cdot)$: $[t_0,\theta] \to Q$. Note that this collection $(t_0,x_0,U,v(\cdot),\Delta,y(\cdot))$ uniquely determines the step-by-step motion.

The following assertion is valid.

Theorem 8.1.1. *Let W be a closed u-stable bridge in an (M, N)-approach problem and let $(t_0,x_0) \in W$. Assume that for all $t \in T$ the sets $W[t] = \{w \in R^n: (t,w) \in W\}$ are convex. Let U^e be a strategy extremal to the set W. Then for any $\epsilon > 0$ one can find a $\delta > 0$ and a $\xi > 0$ such that the step-by-step motion $x(\cdot,t_0,x_0,U^e,v(\cdot),\Delta,y(\cdot))$ satisfies the condition of an $(\bar{M}^\epsilon, \bar{N}^\epsilon)$-approach. Here $\bar{M}^\epsilon$ and $\bar{N}^\epsilon$ are closed ϵ-neighborhoods of the sets M and N, $\operatorname{diam}\Delta \leqslant \delta$, $v(\cdot)$: $[t_0,\theta] \to Q$ is a measurable control of the second player, $y(\cdot)$ is an arbitrary informational noise satisfying the bound $\|y(\tau_i)\| \leqslant \xi$ $(i = 0,1, \dots, k)$.*

In the proof of this theorem as well as in the proof of Theorem 8.2.1 presented below, a local estimate is used - which is an analog of the estimate (2.3.23). We shall introduce the notation that will be required for the statement of the above mentioned estimate.

Let $X(t_0,x_0)$ be the set of solutions $x(\cdot)$: $[t_0,\theta] \to R^n$ of the differential inclusion $\dot{x}(t) \in \operatorname{co}\{f(t,x(t),u,v): u \in P, v \in Q\}$ satisfying the initial condition $x(t_0) = x_0$. We set

$$\mathcal{D}(t_0,x_0) \stackrel{\text{def}}{=} \{(t,x(t)): t_0 \leq t \leq \theta,\ x(\cdot) \in X(t_0,x_0)\},$$

(6) $$\mathcal{D}_0 \stackrel{\text{def}}{=} \bigcup_{(t_0,x_0)\in\mathcal{D}} \mathcal{D}(t_0,x_0),$$

(7) $$\mathcal{D}_1 \stackrel{\text{def}}{=} \{(t,x+y) \in T \times R^n: (t,x) \in \mathcal{D}_0,\ \|y\| \leq 1\}.$$

Here $\mathcal{D}$ is a compact set in $[t_{\#},\theta] \times R^n \subset T \times R^n$. Note that according to Corollary A.4.1 the sets $\mathcal{D}_0$ and $\mathcal{D}_1$ are compact in $T \times R^n$.

We replace the sets $\mathcal{D}_0$ (2.3.9) and $\mathcal{D}_1$ (2.3.10) by the sets $\mathcal{D}_0$ (6) and $\mathcal{D}_1$ (7) respectively. Next, we define the quantity $\alpha(\delta)$ by the equality (2.3.15). Let $\xi \geq 0$, set

(8) $$\alpha_{\#}(\delta,\xi) \stackrel{\text{def}}{=} \alpha(\delta) + 2(\Lambda + \lambda)\xi,$$

(9) $$\beta_{\#}(t,\delta,\xi) \stackrel{\text{def}}{=} (\xi^2 + \alpha_{\#}(\delta,\xi)\lambda^{-1})\exp 2\lambda(t - t_{\#}) - \alpha_{\#}(\delta,\xi)\lambda^{-1}$$

(the numbers λ and Λ are defined here in the same manner as in Section 2.3.2). Note that the function $\beta(\cdot) = \beta_{\#}(\cdot,\delta,\xi)$ satisfies the differential equation

$$\frac{d\beta(t)}{dt} = 2\lambda\beta(t) + 2\alpha_{\#}(\delta,\xi)$$

and the initial condition $\beta(t_{\#}) = \xi^2$.

We select numbers $\delta_0 > 0$ and $\xi_0 > 0$ so that the inequality

(10) $$\beta_{\#}(t,\delta,\xi) < 1 \quad \text{for} \quad t \in [t_{\#},\theta],\ \ \delta \in (0,\delta_0),\ \ \xi \in (0,\xi_0)$$

is fulfilled. We determine points $u_0 \in P$ and $v_0 \in Q$ from the conditions

(11) $$\max_{v\in Q} \langle s^*, f(t_*,x^*,u_0,v)\rangle = \min_{u\in P}\max_{v\in Q} \langle s^*, f(t_*,x^*,u,v)\rangle$$

(12) $$\min_{u\in P} \langle s^*, f(t_*,x^*,u,v_0)\rangle = \max_{v\in Q}\min_{u\in P} \langle s^*, f(t_*,x^*,u,v)\rangle.$$

The following estimate is valid.

Lemma 8.1.1. *Let the set $\mathcal{D}_0$ and quantities $\beta_{\#}$, δ_0, and ξ_0 be defined according to* (6), (9), (10). *Let the points u_0 and v_0 be selected from conditions* (11) *and* (12). *Assume that*

$$(t_*,x_*) \in \mathcal{D}_0,\quad \|x_* - w_*\|^2 \leq \beta_{\#}(t_*,\delta,\xi),$$

$$\|x_* - x^*\| \leq \xi,\qquad \|s^* - (x_* - w_*)\| \leq \xi,$$

$$\delta \in (0,\delta_0),\quad \xi \in (0,\xi_0).$$

Let $x(\cdot)$: $[t_*,\theta] \to R^n$ and $w(\cdot)$: $[t_*,\theta] \to R^n$ be solutions of the *differential inclusions*

$$\dot{x}(t) \in \mathcal{F}_1(t,x(t),u_0) \stackrel{\text{def}}{=} \text{co}\{f(t,x(t),u_0,v)\colon v \in \mathcal{Q}\},$$

$$\dot{w}(t) \in \mathcal{F}_2(t,w(t),v_0) \stackrel{\text{def}}{=} \text{co}\{f(t,w(t),u,v_0)\colon u \in P\},$$

satisfying the initial conditions $x(t_*) = x_*$ *and* $w(t_*) = w_*$. *Then the estimate*

$$(13) \qquad \|x(t) - w(t)\|^2 \leqslant \beta_{\#}(t,\delta,\xi) \quad \text{for} \quad t_* \leqslant t \leqslant t_* + \delta$$

is valid.

Proof. Replacing $\beta(t,\delta)$ by $\beta_{\#}(t,\delta,\xi)$ and repeating verbatim the arguments presented in the proof of Lemma 2.3.1 we arrive at the estimates (2.3.28) and (2.3.29). Next, utilizing the inequalities (2.3.14) and the bounds

$$\|s(t_*)\| = \|x_* - w_*\| \leqslant \sqrt{\beta_{\#}(t_*,\delta,\xi)} < 1,$$

$$\|s^* - s(t_*)\| = \|s^* - (x_* - w_*)\| \leqslant \xi,$$

$$\|x^* - x_*\| \leqslant \xi$$

we arrive at

$$\psi_1(t_*,x_*,s(t_*),u_0) \leqslant \psi_1(t_*,x^*,s^*,u_0) + (\lambda + \Lambda)\xi,$$

$$\psi_2(t_*,x_*,s(t_*),v_0) \geqslant \psi_2(t_*,x^*,s^*,v_0) - (\lambda + \Lambda)\xi.$$

Thus from (2.3.28), (2.3.29), and (8) we have

$$\psi_1(t,x(t),s(t),u_0) \leqslant \psi_1(t_*,x^*,s^*,u_0) + \tfrac{1}{2}\alpha_{\#}(\delta,\xi),$$

$$\psi_2(t,w(t),s(t),v_0) \geqslant \psi_2(t_*,x^*,s^*,v_0) - \lambda\|s(t)\|^2 - \tfrac{1}{2}\alpha_{\#}(\delta,\xi).$$

According to (1.2.8), (2.3.13), (11), and (12) the equality

$$\psi_1(t_*,x^*,s^*,u_0) = \psi_2(t_*,x^*,s^*,v_0)$$

is valid.

Next the proof of estimate (13) again coincides with the proof of Lemma 2.3.1 in which $\alpha(\delta)$ and $\beta(t,\delta)$ must be replaced by $\alpha_{\#}(\delta,\xi)$ and $\beta_{\#}(t,\delta,\xi)$. Thus Lemma 8.1.1 is proved.

We note yet another fact which will be utilized in the proof of Theorem 8.1.1.

Let $\mathcal{E}$ be a convex closed set in R^n. Denote by $e(x)$ the point of set $\mathcal{E}$ closest to the point x, i.e.

(14) $$\|e(x) - x\| = \min_{e\in\mathcal{E}} \|e - x\| \overset{\text{def}}{=} \text{dist}(x,\mathcal{E}).$$

Consider the function

$$x \to s(x) \overset{\text{def}}{=} (x - e(x)): R^n \to R^n.$$

We shall show that this function satisfies the Lipschitz condition with constant equal to 1, i.e.

(15) $$\|s(x^{(1)}) - s(x^{(2)})\| \leqslant \|x^{(1)} - x^{(2)}\|.$$

Let $x^{(i)} \in R^n$, $e^{(i)} = e(x^{(i)})$, $i = 1,2$. Since the set $\mathcal{E}$ is convex the interval connecting points $e^{(1)}$ and $e^{(2)}$ is contained in $\mathcal{E}$. Furthermore, from (14) the inequalities

$$\langle x^{(1)} - e^{(1)}, e^{(2)} - e^{(1)}\rangle \leqslant 0, \quad \langle x^{(2)} - e^{(2)}, e^{(1)} - e^{(2)}\rangle \leqslant 0$$

follow. Combining these inequalities we have

$$\langle x^{(1)} - x^{(2)}, e^{(2)} - e^{(1)}\rangle + \langle e^{(2)} - e^{(1)}, e^{(2)} - e^{(1)}\rangle \leqslant 0.$$

Therefore

$$\|s(x^{(1)}) - s(x^{(2)})\|^2 = \|x^{(1)} - e^{(1)} - x^{(2)} + e^{(2)}\|^2$$

$$= \|x^{(1)} - x^{(2)}\|^2 + 2\langle x^{(1)} - x^{(2)}, e^{(2)} - e^{(1)}\rangle + \|e^{(2)} - e^{(1)}\|^2$$

$$\leqslant \|x^{(1)} - x^{(2)}\|^2 - \|e^{(2)} - e^{(1)}\|^2 \leqslant \|x^{(1)} - x^{(2)}\|^2.$$

Thus the inequality (15) is proved.

Proof of Theorem 8.1.1. Let the numbers $\delta_0 > 0$ and $\xi_0 > 0$ be chosen in such a manner that the bound (10) is fulfilled. Let the point (t_0,x_0) be contained in the compact set $\mathcal{D}_0$ (6). It is evident from (8), (9) that $\beta_{\#}(\theta,\delta,\xi) \to 0$ as $\delta \to 0$, $\xi \to 0$. Therefore to prove Theorem 8.1.1 it is sufficient to establish the following assertion: for $\xi \in (0,\xi_0)$ and $\delta = \text{diam}\,\Delta \in (0,\delta_0)$, any step-by-step motion $x(\cdot) = x(t_0,x_0,U^e,v(\cdot),\Delta,y(\cdot))$ satisfies the condition of an $(\bar{M}^\epsilon, \bar{N}^\epsilon)$-approach where

$$\epsilon = \sqrt{\beta_{\#}(\theta,\delta,\xi)} + (\Lambda + 1)\delta.$$

Note that this assertion differs from the fact established in the proof of Theorem 2.4.1 as the quantity $\beta(\theta,\delta)$ is replaced by the quantity $\beta_{\#}(\theta,\delta,\xi)$.

We shall now indicate how the proof of Theorem 2.4.1 must be modified in order to obtain the above stated assertion.

Firstly, in relation (2.4.10) the points $x(\tau_i)$ and $w_{(i)}$ should be replaced by the points $x_*(\tau_i) = x(\tau_i) + y(\tau_i)$ and $w^*_{(i)} = e(\tau_i, x_*(\tau_i))$. Note that

$$\|x(\tau_i) - x_*(\tau_i)\| \leqslant \xi$$
(16)
$$\|x(\tau_i) - w_{(i)} - (x_*(\tau_i) - w^*_{(i)})\| \leqslant \xi.$$

The second estimate follows from inequality (15). In the differential inclusion (2.4.11) the initial condition is not changed, i.e.

$$w^{(i)}(\tau_i) = w_{(i)} = e(\tau_i,\ x(\tau_i)).$$

Next, estimate (2.4.16) is replaced by the estimate

$$\mathrm{dist}^2(x(\tau_i),\ \mathcal{W}[\tau_i]) \leqslant \beta_{\#}(\tau_i, \delta, \xi).$$

In the proof of this estimate the reference to Lemma 2.3.1 is replaced by a reference to Lemma 8.1.1. Note from estimate (16) we infer that inequalities $\|x_* - x^*\| \leqslant \xi$ and $\|s^* - (x_* - x^*)\| \leqslant \xi$ are fulfilled. These inequalities are present in the conditions of Lemma 8.1.1.

After these replacements, one can repeat the arguments in full in the proof of Theorem 2.4.1 and obtain the assertion stated above. In this manner Theorem 8.1.1 will be proved.

8.2. Guidance Control

We now proceed to a description of regularization of an extremal construction which yields the solution of the game-theoretical control problems that are stable relative to informational noise. The essence of the proposed regularization consists of introducing an auxiliary guidance point which moves on a stable bridge. The motion of this auxiliary point and the motion of the initial system are formed so that they will mutually track each other. An introduction of the guidance may be viewed as an inclusion into the control system of a certain standard model not subject to the effects of informational noise. Information concerning the evolution of the guidance is furnished by computations carried out on a computer.

8.2.1. We shall first describe a control procedure with guidance for an $(\mathcal{M}, \mathcal{N})$-approach problem.

Let the motion of the controlled system be described by the differential equation

(1) $$\dot{x}(t) = f(t, x(t), u(t), v(t)).$$

We shall assume that the function $f\colon T \times R^n \times \mathcal{P} \to \mathcal{Q} \to R^n$ satisfies all the conditions (i) - (iv) stipulated in Section 1.2.2. As usual it is assumed that the controls of the first and second players satisfy the constraints $u(t) \in P$ and $v(t) \in Q$ where P and Q are compact sets in R^p and R^q respectively. When describing the control procedure suggested herein we shall adhere in principal to the terminology used in the theory of differential games, i.e. we shall assume that the motion of the system (1) is controlled by two competing players. Moreover, this description is also fully acceptable in minimax control problems where there is not direct competition between two sides and the function $t \to v(t)$ characterizes certain undetermined disturbance.

We shall consider a step-by-step procedure in which the control $u(t)$ is formed by the rule

$$(2) \qquad u(t) = U(\tau_i, x_*(\tau_i), w(\tau_i)), \quad \tau_i \leq t < \tau_{i+1}, \quad i = 0,1, ..., k.$$

Here U: $T \times R^n \times R^n \to P$ is a function and $\Delta = \{t_0 = \tau_0 < \tau_1 < ... < \tau_{k+1} = \theta\}$ is a partition of the interval $[t_0,\theta]$. Function U and partition Δ are selected by the first player. It is assumed that the first player measures the phase states $x(\tau_i)$ imprecisely and the n-dimensional vector $x_*(\tau_i)$ appearing in (2) is a result of this error. We are assuming that the measurement error satisfies the bound

$$(3) \qquad \|x_*(\tau_i) - x(\tau_i)\| \leq \xi \qquad (i = 0,1, ..., k).$$

According to (2) the selection of control $u(t)$ ($\tau_i \leq t < \tau_{i+1}$) depends also on the n-dimensional vector $w(\tau_i)$ which will be called the *phase position of the guidance* at time $t = \tau_i$.

Before describing the evolution of a guide we shall note the basic distinction between the controlled system and the guidance. This will be done by means of the example described in Section 1.1.3. Suppose that a problem controlling the movement of an aircraft is being considered. Let a differential equation of the type (1) be formed which describes with sufficient precision the motion of the aircraft in accordance with appropriate physical laws and technical possibilities. The phase variables $x_1(t)$, ..., $x_n(t)$ in this example will be geometrical coordinates of the aircraft, the components of the velocity vector, and several other quantities. Controlling parameters $(u_1(t), ..., u_p(t))$ characterize the aileron angles, the mode of the operation of the engines, and so on. Finally parameters $(v_1(t), ..., v_q(t))$ characterize the effect of the disturbance - the force and the direction of the wind. Let the automatic device assign the control according to rule (2) where $x_*(\tau_i) = (x_{*1}(\tau_i), ..., x_{*n}(\tau_i))^T$ is the result of an imprecise measurement of phase variables $x_1(\tau_i)$, ..., $x_n(\tau_i)$. We emphasize that these measurements characterize the state of a real world controlled object, namely that of an aircraft. These measurements are carried out by appropriate technical devices such as data sensors of height, velocity, etc.

Simultaneously, calculations are carried out on a computer. These calculations consist of cycles; each cycle is implemented on the time interval $[\tau_i, \tau_{i+1})$ $(i = 0,1, ..., k)$. The initial data for calculations carried out on the time interval $[\tau_i, \tau_{i+1})$ are the vectors $x_*(\tau_i)$ and $w(\tau_i)$. As a result of these calculations we obtain the vector $w(\tau_{i+1})$ which is the phase position of the guidance at time $t = \tau_{i+1}$. The algorithm which evaluates the vector $w(\tau_{i+1})$ is described below. For the time being we shall note that the mathematical model of the guidance implemented on a computer as described above is based on a differential equation which delineates the motion of a real-world controlled object.

Thus the controlled action $u(t)$ (2) is selected depending on the vectors $x_*(\tau_i)$ and $w(\tau_i)$. The first of these vectors is the result of measurements of the position of a real-world controlled object. The second vector is obtained as a result of modelling on a computer an auxiliary system -- that of the guidance. It turns out that this additional information allows us to overcome the destructive action of the informational noise.

We now turn to a formal construction of a guidance control procedure in an $(\mathcal{M}, \mathcal{N})$-approach problem. First define the functions $U^0: T \times R^n \times R^n \to \mathcal{P}$ and $V_*: T \times R^n \times R^n \to \mathcal{Q}$, satisfying the conditions

$$
\max_{v \in \mathcal{Q}} \langle x - w, f(t,x,U^0(t,x,w),v) \rangle = \min_{u \in \mathcal{P}} \max_{v \in \mathcal{Q}} \langle x - w, f(t,x,u,v) \rangle, \tag{4}
$$

$$
\min_{u \in \mathcal{P}} \langle x - w, f(t,x,u,V_*(t,x,w)) \rangle = \max_{v \in \mathcal{Q}} \min_{u \in \mathcal{P}} \langle x - w, f(t,x,u,v) \rangle. \tag{5}
$$

Observe that conditions (4) and (5) define functions U^0 and V_* in general, albeit not uniquely. In the succeeding constructions one can utilize any pair of functions U^0 and V_* satisfying the above stated conditions.

Let $\mathcal{W}$ be a u-stable bridge in an $(\mathcal{M}, \mathcal{N})$-approach problem. We shall assume that the initial position (t_0, x_0) of the controlled system (1) is located on the set $\mathcal{W}$. Select a partition $\Delta = \{t_0 = \tau_0 < \tau_1 < ... < \tau_{k+1} = \theta\}$. At the initial time $t = t_0 = \tau_0$ the first player is carrying out an imprecise measurement of the initial phase vector $x_0 = x(t_0)$. Let $x_*(\tau_0)$ be the result of this measurement. We determine the point $w(\tau_0) \in R^n$ from the condition

$$
\|x_*(\tau_0) - w(\tau_0)\| = \min\{\|x_*(\tau_0) - w\|: w \in \mathcal{W}[t_0]\} \tag{6}
$$

where $\mathcal{W}[t_0] = \{w \in R^n: (t_0, w) \in \mathcal{W}\}$ and it is assumed that this set is closed. Thus $w(\tau_0)$ is a point on the set $\mathcal{W}[t_0]$ closest to the point

$x_*(\tau_0)$ (if there are several such points $w(\tau_0)$ any one of them can be selected).

On the first time interval $[\tau_0,\tau_1)$ we select a constant control for the first player $u^0(t) = U^0(\tau_0,x_*(\tau_0),w(\tau_0))$, $\tau_0 \leqslant t < \tau_1$. The selected control coupled with a certain measurable control for the second player generates a motion of the controlled system on the time interval $[\tau_0,\tau_1]$. This motion satisfies the differential equation

$$\text{(7)} \qquad \dot{x}(t) = f(t,x(t),u^0(t),v(t))$$

and the initial condition $x(t_0) = x_0$. On the first interval we also define the guidance motion $w(t)$. For this purpose we select a constant control $v_*^{(0)} = V_*(\tau_0,x_*(\tau_0),w(\tau_0))$ and consider the differential inclusion (cf. (2.2.1), (2.2.3))

$$\dot{w}(t) = \mathcal{F}_2(t,w(t),v_*^{(0)}).$$

Recall that $(\tau_0,w(\tau_0)) \in \mathcal{W}$. Therefore according to the condition of u-stability of the set $\mathcal{W}$ among the solutions of the differential inclusion there exists a solution such that either

$$\text{(8)} \qquad (\tau_1,\ w(\tau_1)) \in \mathcal{W}$$

or

$$\text{(9)} \qquad \{(t,w(t)):\ t_0 \leqslant t \leqslant \tau_1\} \cap \mathcal{M} \neq \phi\ .$$

We shall select as the motion of the guidance on the interval $[\tau_0,\tau_1]$ just such a solution of the differential inclusion. Thus on the first time interval the motion of the guidance will either encounter the set $\mathcal{M}$, i.e. condition (9) will be fulfilled or this motion will terminate on the set $\mathcal{W}$, i.e. the inclusion (8) is valid.

Let $x(\tau_i)$ and $w(\tau_i)$ be the phase positions of the controlled system and the guidance system at time $t = \tau_i$. Let $x_*(\tau_i)$ be the result of an imprecise measurement of the phase position $x(\tau_i)$. On the next time interval $[\tau_i,\tau_{i+1})$ the first player chooses the constant control

$$\text{(10)} \qquad u^0(t) = U^0(\tau_i,x_*(\tau_i),w(\tau_i)),\ \ \tau_i \leqslant t < \tau_{i+1}.$$

This control, coupled with a certain measurable control of the second player $v(t)$, generates a motion $x(t)$ which satisfies equation (7). In order to determine the motion of the guidance $w(t)$ at $\tau_i \leqslant t \leqslant \tau_{i+1}$ we select the constant control

$$\text{(11)} \qquad v_*^{(i)} = V_*(\tau_i,x_*(\tau_i),w(\tau_i))$$

and consider the differential inclusion

(12) $\dot{w}(t) \in \mathcal{F}_2(t,w(t),v_*^{(i)})$.

Assume that as a result of constructions on preceding time intervals we obtain that $(\tau_i,w(\tau_i)) \in W$. Then according to the condition of u-stability of the set W among the solutions of the differential inclusion (12) there exists a solution such that either

(13) $(\tau_{i+1},w(\tau_{i+1})) \in W$

or

(14) $\{(t,w(t)): \tau_i \leqslant t \leqslant \tau_{i+1}\} \cap M \neq \phi$.

We choose such a solution as the motion of the guidance on the interval $[\tau_i,\tau_{i+1}]$. If there are several solutions of the differential inclusion (12) satisfying either (13) or (14), any one of them is selected.

It will be shown below that these constructions assure the mutual tracking of the motions of the controlled system and of the guidance. These constructions are defined until condition (14) is fulfilled on a certain time interval $[\tau_i,\tau_{i+1}]$, i.e. the guidance falls into the target set. It follows from the definition of a u-stable bridge W that the encounter $(\tau,w(\tau)) \in M$ will necessarily occur at $\tau = \tau^0(w(\cdot)) \leqslant \theta$. By the very nature of the problem of an (M,N)-approach it is sufficient to carry out the construction of motions $x(\cdot)$ and $w(\cdot)$ up to the time $\tau = \tau^0(w(\cdot))$. However, in further arguments it will be convenient to continue the motions $w(t)$ and $x(t)$ up to the time $t = \theta$, assuming formally that on the intervals $[\tau_i,\tau_{i+1}]$ where $\tau_i \geqslant \tau^0(w(\cdot))$ any solution of the differential inclusion (12), with the control $v_*^{(i)}$ defined by equality (11) as before, is chosen as the motion of the guidance. For $\tau_i \geqslant \tau^0(w(\cdot))$ the control $u^0(t)$ (10) is selected in system (7) as before.

In order to explain more meaningfully how the motion of the guidance is formed we shall assume that for all $(t,x,v) \in T \times R^n \times Q$ the set $\{f(t,x,u,v): u \in P\}$ is convex. Under this assumption the differential inclusion (12) can be replaced by the differential equation

(15) $\dot{w}(t) = f(t,w(t),u_*(t),v_*^{(i)})$.

Consider the following problem. Let $(\tau_i,w(\tau_i)) \in W$. It is required to determine the control $u_*(\cdot): [\tau_i,\tau_{i+1}) \to P$ so that the corresponding motion of the system (15) satisfies either condition (13) or condition (14). The following should be noted. Although in the notation of equation (15) two controlling vectors $v_*^{(i)}$ and $u_*(t)$ appear, their choice is subordinated only to the first player. Control $v_*^{(i)}$, which is constant on the interval $[\tau_i,\tau_{i+1})$, is assigned by the first player in accordance with (11), and under the choice of the control $u_*(\cdot)$: $[\tau_i,\tau_{i+1}) \to P$ the first player is solving the above stated problem.

Recall that the motion of the guidance is simulated on a computer. The input to the computer at time $t = \tau_i$ is the vectors $w(\tau_i)$ and $v_*^{(i)}$. After the required calculations are carried out, the vector $w(\tau_{i+1})$ appears at the output at time $t = \tau_{i+1}$. The guidance is associated with the controlled system by the fact that piecewise-constant controls $v_*(t) = v_*^{(i)}$ $(\tau_i \leqslant t < \tau_{i+1},\ i = 0, 1, ..., k)$ in (12) or (15) which are selected in accordance with (11) depend on the results of the measurements $x_*(\tau_i)$. It is assumed that the motion of guidance is simulated on a computer with sufficient precision, i.e. we shall assume that the phase positions of the guidance are computed without errors.

The meaning of relations (10) and (11) is as follows: the control of the real motion $x(t)$ and the computer simulated motion of the guidance $w(t)$ are implemented in such a way that the motions mutually track each other. Indeed, control $u^0(t)$ (10) is chosen so that the maximal shift of the motion $x(t)$ on the point $w(\tau_i)$ will be assured while control $v_*^{(i)}$ (11) is selected from the aiming condition of the motion $w(t)$ on the point $x_*(\tau_i)$.

Finally we note that the process of formation of the motion $w(t)$ can be interpreted as follows. Consider a differential game in which the first player possesses an informational advantage, i.e. at time $t = \tau_i$ he/she knows not only the realized position $(\tau_i,w(\tau_i))$ but also the constant control $v_*^{(i)}$ which will be realized by the opponent on the interval $[\tau_i,\tau_{i+1})$ (recall that a similar game was discussed in Section 2.1.3). It can be assumed that in the procedure described above the motion of the guidance is selected in the course of the game. In reality the first player's opponent will be a fictitious player. Actually the first player plays with himself/herself and assigns control $v_*^{(i)}$ according to rule (11).

8.2.2. The procedure of the control with guidance constructed above yields a solution of an $(\mathcal{M}, \mathcal{N})$-approach problem which is stable relative to informational noise. The following theorem is valid.

Theorem 8.2.1. *Let $(t_0,x_0) \in \mathcal{W}$, where $\mathcal{W}$ is a u-stable bridge in the problem of an $(\mathcal{M}, \mathcal{N})$-approach. Then for any $\epsilon > 0$ a $\delta > 0$ and a $\xi > 0$ can be found such that for any control of the opponent $v(\cdot)$: $[t_0,\theta] \to \mathcal{Q}$ and any partition Δ and any informational noise satisfying the bound* $\operatorname{diam}\Delta \leqslant \delta$ *and bound* (3) *the motion of the system* (7) *in which the control $u^0(\cdot)$ is constructed according to the procedure indicated above will satisfy the condition of an $(\overline{\mathcal{M}^{\epsilon}}, \overline{\mathcal{N}^{\epsilon}})$-approach.*

Proof. According to constructions described in Section 8.2.1, the pair of motions $(x(\cdot),w(\cdot))$ corresponds to the collection $(t_0,x_0,\Delta,v(\cdot),y(\cdot))$. The first motion is the motion of the controlled system, the second is the guidance motion. Here (t_0,x_0) is the initial position, Δ is a partition selected by the first player, $v(\cdot)$: $[t_0,\theta] \to \mathcal{Q}$ is the control of the second player, $y(\cdot) = \{y(\tau_i) = x_*(\tau_i) - x(\tau_i)$: $i = 0,1, ..., k\}$ is the informational noise. We shall assume that $\delta = \operatorname{diam} \Delta < \delta_0$,

$$\|y(\tau_i)\| \leqslant \xi < \xi_0 \quad (i = 0,1, \ldots, k),$$

where the numbers $\delta_0 > 0$ and $\xi_0 > 0$ are selected in such a manner that the inequality (8.1.10) is fulfilled. It is also assumed that the point (t_0,x_0) is contained in the compact set $\mathfrak{D}_0$ (8.1.6).

We shall show that the pair of motions $(x(\cdot),w(\cdot))$ satisfies the estimate

$$(16) \qquad \|x(t) - w(t)\|^2 \leqslant \beta_\#(t,\delta,\xi), \quad t_0 \leqslant t \leqslant \theta,$$

where the quantity $\beta_\#(t,\delta,\xi)$ is defined by the equality (8.1.9). It is evident from (8.1.9) and (6) that at the initial time $t = t_0$ the inequality

$$\|x(t_0) - w(t_0)\|^2 \leqslant \xi^2 \leqslant \beta_\#(t_0,\delta,\xi)$$

is valid. Controls $u^0(t)$ $(t_0 \leqslant t < \tau_1)$ (4), (10) and $v_*^{(0)}$ (5), (11) are selected in accordance with conditions (8.1.11) and (8.1.12). The bounds

$$\|x_*(\tau_0) - x(\tau_0)\| \leqslant \xi,$$

$$\|(x_*(\tau_0) - w(\tau_0)) - (x(\tau_0) - w(\tau_0))\| \leqslant \xi$$

are valid. This means that the inequalities

$$\|x^* - x_*\| \leqslant \xi, \quad \|s^* - (x_* - w_*)\| \leqslant \xi$$

which are stipulated in Lemma 8.1.1 are fulfilled. Thus, all the conditions of Lemma 8.1.1 are satisfied on the first interval $[\tau_0,\tau_1)$ which implies that the estimate (16) is valid for $t \in [\tau_0,\tau_1]$. Applying Lemma 8.1.1 successively on the next intervals, we obtain the estimate (16) for all $t \in [t_0,\theta]$.

It follows from the construction of the motion of the guidance that a time $\tau^0 \in [t_0,\theta]$ exists such that

$$(17) \qquad (\tau^0,w(\tau^0)) \in M \text{ and } (\tau_i,w(\tau_i)) \in W \subset N \text{ for } \tau_0 \leqslant \tau_i < \tau^0.$$

Utilizing the arguments presented at the end of the proof of Theorem 2.4.1 one can obtain from (16) and (17) that the motion $x(\cdot)$ satisfies the condition of an $(M^{\bar{r}}, N^{\bar{r}})$-approach where $r = \sqrt{\beta_\#(\theta, \delta, \xi)} + (\Lambda+1)\delta$. Since $\beta_\#(\theta,\delta,\xi) \to 0$ as $\delta \to 0$, $\xi \to 0$, Theorem 8.2.1 is proved.

8.2.3. In Section 8.2.1 a specific type of control procedure with guidance was defined. The control procedure constructed therein assures a stable motion in the vicinity of a u-stable bridge. We shall now introduce a more general definition of control procedure with guidance.

The triple (U,ψ,χ) will be called *a control procedure with guidance for the first player.*

Here U: $T \times R^n \times R^n \to P$ is a function which forms a control in the system (1). The function ψ : $T^2_+ \times R^n \times R^n \to R^n$ is called the *guidance transition function.* Symbol T^2_+ denotes the set $\{(t_1,t_2) \in T \times T$: $t_1 \leqslant t_2\}$. The value of the transition function $\psi(t_1,t_2,x_*,w_*)$ is the position w^* in which the guidance will be at time $t = t_2$ under the condition that at time $t = t_1$ the guidance is in position w_* and as the result of measuring the phase state of the controlled system the vector x_* has been obtained. The third function χ: $T \times R^n \to R^n$ is utilized for the determination of the *initial state of the guidance.*

Under a given initial time and partition $\Delta = \{t_0 = \tau_0 < \tau_1 < ... < \tau_{k+1} = \theta\}$ selected by the first player, the control procedure with guidance (U,ψ,χ) forms a *step-by-step motion* of the system (1) in the following manner.

Let $x_*(\tau_0)$ be the result of an imperfect measurement of the initial position $x(t_0) = x_0$ of the controlled system (1). The initial position of the guidance is defined by the equality

$$w(\tau_0) = \chi(\tau_0,x_*(\tau_0)).$$

We select a constant control $u(t) = U(\tau_0,x_*(\tau_0),w(\tau_0))$ on the interval $[\tau_0,\tau_1)$. This control, together with some measurable control $v(\cdot)$: $[\tau_0,\tau_1) \to Q$ determines the motion of the system (1) on the interval $[\tau_0,\tau_1)$.

At time $t = \tau_1$ the location of the guidance is determined by means of its transition function ψ. Set

$$w(\tau_1) = \psi(\tau_0, \tau_1, x_*(\tau_0), w(\tau_0)).$$

An inaccurate measurement of the realized position of system (1) is carried out at time $t = \tau_1$. As a result of this measurement the first player obtains vector $x_*(\tau_1)$. Next we select a constant control $u(t) = U(\tau_1, x_*(\tau_1), w(\tau_1))$, $\tau_1 \leqslant t < \tau_2$ which, coupled with some measurable control for the second player $v(\cdot)$: $[\tau_1,\tau_0) \to Q$ generates the motion of the system (1) on the interval $[\tau_1,\tau_2)$.

Let these constructions be carried out up to the time $t = \tau_{i-1}$. At time $t = \tau_i$ the location of the guidance will be defined by the equality

$$w(\tau_i) = \psi(\tau_{i-1}, \tau_i, x_*(\tau_{i-1}), w(\tau_{i-1})),$$

where $x_*(\tau_{i-1})$ is the result of the measurement of the phase state $x(\tau_{i-1})$ of the system (1) carried out by the first player at time $t = \tau_{i-1}$ and $w(\tau_{i-1})$ is the phase position of the guidance at time $t = \tau_{i-1}$. Let $x_*(\tau_i)$ be the result of the measurement of the phase state $x(\tau_i)$

of system (1) carried out by the first player. We shall assume that the control $u(t)$ in the system (1) on the interval $[\tau_i,\tau_{i+1})$ is determined by the equality

$$(18) \qquad u(t) = U(\tau_i, x_*(\tau_i), w(\tau_i)), \quad \tau_i \leqslant t < \tau_{i+1}.$$

This control and some other control $v(\cdot)$: $[\tau_i,\tau_{i+1}) \to \mathcal{Q}$ generate the motion of the system (1) on the interval $[\tau_i,\tau_{i+1}]$. Such constructions are continued until the last instant of time $t = \theta$.

A procedure for a control with guidance for the second player will also be defined as a triple of functions (V,ψ,χ). Here V: $T \times R^n \times R^n \to \mathcal{Q}$ is a function; the general definition of functions ψ and χ is the same as in a control procedure for the first player.

It can be shown that the procedure with guidance constructed in Section 8.2.1 fits within the framework of the general definition presented herein. We are not going to verify this. Instead we shall consider within the framework of a general definition a control procedure with guidance for the second player which will result in a stable solution of an $(\mathcal{M},\mathcal{N})^\epsilon$-evasion problem. Note that the definition presented below will be an analog of constructions described in Section 8.2.1.

Let the sets K_ϵ and $\mathcal{L}_\epsilon$ be defined by the equalities (2.1.9). Then, as is known, the condition of an $(\mathcal{M},\mathcal{N})^\epsilon$-evasion will be equivalent to the condition of a $(K_\epsilon,\mathcal{L}_\epsilon)$-approach. Let $\mathcal{W}$ be a closed v-stable bridge in a problem of a $(K_\epsilon,\mathcal{L}_\epsilon)$-approach.

We define a control procedure with guidance (V^0,ψ^0,e). The function V^0: $T \times R^n \times R^n \to \mathcal{Q}$ is selected from the condition

$$\min_{u\in\mathcal{P}} \langle w - x,\ f(t,x,u,V^0(t,x,w))\rangle$$
$$= \max_{v\in\mathcal{Q}}\ \min_{u\in\mathcal{P}} \langle w - x,\ f(t,x,u,v)\rangle.$$

The function e: $T \times R^n \to R^n$ will be defined according to (2.4.2), (2.4.4). Note that here the function $e = \chi$ defines the initial position of the guidance $w(\tau_0) = e(t_0,x_*(\tau_0))$ (cf. the analogous relation (6)).

It remains to define the transition function ψ^0. Let the function U_*: $T \times R^n \times R^n \to \mathcal{P}$ be selected via the condition

$$\max_{v\in\mathcal{Q}} \langle w - x,\ f(t,x,U_*(t,x,w),v)\rangle$$
$$= \min_{u\in\mathcal{P}}\ \max_{v\in\mathcal{Q}} \langle w - x,\ f(t,x,u,v)\rangle.$$

Let $(t_*,w_*,x_*) \in T \times R^n \times R^n$. Consider the differential inclusion (cf. (2.2.1))

$$(19) \qquad \dot{w}(t) \in \mathcal{F}_1(t,w(t),\ U_*(t_*,x_*,w_*)).$$

Assume that $(t_*,w_*) \in \mathcal{W}$, $t^* \in [t_*,\theta]$. Since $\mathcal{W}$ is a v-stable bridge in a $(K_\epsilon,\mathcal{L}_\epsilon)$-approach problem there exists a solution $w(\cdot)$: $[t_*,t^*] \to R^n$ of the differential inclusion (19) with the initial condition $w(t_*) = w_*$ which satisfies either the inclusion $(t^*, w(t^*)) \in W$ or the condition $\{(t,w(t)): t_* \leqslant t \leqslant t^*\} \cap K_\epsilon \neq \phi$. We shall assume that in the case under consideration the transition function of the guidance ψ^0 assigns to the collection (t_*, t^*, x_*, w_*) a point $w^* = w(t^*)$; the solution $w(\cdot)$ of the differential inclusion (19) arrives at this point at time $t = t^*$. Let $(t_*, w_*) \notin W$, $t^* \in [t_*, \theta]$, $x_* \in R^n$. In this case we select an arbitrary solution $w(\cdot): [t_*, t^*] \to R^n$ of the differential inclusion (19) satisfying the initial condition $w(t_*) = w_*$ and set $\psi^0(t_*, t^*, x_*, w_*) = w(t^*)$. Observe that the function ψ^0 is in general nonuniquely defined (as are functions V^0 and e).

We have thus defined a control procedure with guidance for the second player. It can be shown that in the case when $(t_0, x_0) \in W$, the step-by-step motion generated by this procedure (V^0, ψ^0, e) for a sufficiently refined partition Δ and sufficiently small informational errors (3) will remain in a small neighborhood of a v-stable bridge W. In this manner the control procedure will result in the solution of an evasion problem which will be stable relative to informational errors. The proof of this fact almost completely coincides with the proof of Theorem 8.2.1 and is thus not presented here.

8.2.4. In Sections 8.2.1 - 8.2.3 control procedures with guidance were considered which yield a stable solution of the problem of an (M, N)-approach and an $(M, N)^\epsilon$-evasion. In these procedures the evolution of the guidance was defined by means of the differential equation $\dot{w}(t) = f(t, w(t), u_*(t), v_*(t))$ which is identical to equation (1) describing the motion of the initial controlled system. More precisely, when constructing the motion of the guidance, the differential inclusion (12) or (19) was considered.

We show that the coincidence of equations describing the motions of the controlled system and of the guidance is not a necessary condition. This will be done based on an example of an (M, N)- approach problem.

Assume that the motion of the guidance is defined by means of the differential equation

$$\dot{w}(t) = g(t,\ w(t),\ u_*(t),\ v_*(t)). \tag{20}$$

Here the function $g: T \times R^n \times P_* \times Q_* \to R^n$ satisfies all the conditions (i) - (iv) stated in Section 1.2.2. The sets P_* and Q_* are compact sets in finite-dimensional spaces. We shall assume that functions f and g are related by the inequality

$$\begin{aligned} &\min_{u \in P} \max_{v \in Q} \langle \ell, f(t,x,u,v) \rangle \\ &\qquad \leqslant \max_{v_* \in Q_*} \min_{u_* \in P_*} \langle \ell, g(t,x,u_*,v_*) \rangle \end{aligned} \tag{21}$$

which should be fulfilled for all $(t, x, \ell) \in T \times R^n \times R^n$.

Let $\mathcal{W}$ be a closed u-stable bridge in a problem of an $(\mathcal{M},\mathcal{N})$-approach defined for the system (20). The transition function of the guidance ψ_*: $T_+^2 \times R^n \times R^n \to R^n$ will be defined as follows. Assume that the function V_*: $T \times R^n \times R^n \to \mathcal{Q}_*$ is selected here from the condition

$$\min_{u_* \in \mathcal{P}_*} \langle x - w,\ g(t,x,u_*,V_*(t,x,w))\rangle$$
$$= \max_{v_* \in \mathcal{Q}_*} \min_{u_* \in \mathcal{P}_*} \langle x - w,\ g(t,x,u_*,v_*)\rangle.$$

Consider the differential inclusion

$$\dot{w}(t) \in \text{co}\{g(t,w(t),u_*,V_*(t_*,x_*,w_*)) \colon u_* \in \mathcal{P}_*\}. \tag{22}$$

If $(t_*,w_*) \in \mathcal{W}$, $t^* \in [t_*,\theta]$, $x_* \in R^n$ then the function ψ_* assigns to the collection (t_*,t^*,x_*,w_*) the point $w^* = w(t^*)$ into which the solution $w(\cdot)$: $[t_*,\theta] \to R^n$ is transferred. This solution $w(\cdot)$ is selected in such a manner that it satisfies either condition $(t^*,w(t^*)) \in \mathcal{W}$ or $\{(t,w(t)\colon t_* \leqslant t \leqslant t^*\} \cap \mathcal{M} \neq \phi$. If, however, $(t_*,w_*) \notin \mathcal{W}$, $t^* \in [t_*,\theta]$, $x \in R^n$, then $\psi_*(t_*,t^*,x_*,w_*) = w(t^*)$, where $w(\cdot)$: $[t_*,\theta] \to R^n$ $(w(t_*) = w_*)$ is an arbitrary solution of the differential inclusion (22).

According to the general definition, the control procedure with guidance for the first player is defined by the triple of functions (U_0,ψ_*,e). Here ψ_* is the transition function for the guidance defined in the manner indicated above. Functions U^0 and e are defined by equalities (4) and (2.4.2), (2.4.4) respectively. For the procedure of a guidance control (U_0,ψ_*,e) the assertion of Theorem 8.2.1 remains valid. The only revision which should be made in the statement of this assertion is that now $\mathcal{W}$ is a u-stable bridge in a problem of an $(\mathcal{M},\mathcal{N})$-approach posed for the system (20). We shall omit the verification of this assertion for the control procedure proposed herein since it basically coincides with the proof of Theorem 8.2.1.

We now compare the control procedure defined in Section 8.2.1 with the procedure (U_0,ψ_*,e) described herein. Assume that $\mathcal{W}$ is a maximal u-stable bridge in an $(\mathcal{M},\mathcal{N})$-approach problem posed for the system (20). Under the condition (21), in view of Theorem 7.4.1, the set $\mathcal{W}$ will be a u-stable bridge in an $(\mathcal{M},\mathcal{N})$-approach problem for the controlled system (1). However, if the controlled systems (1) and (20) are not equivalent, i.e. for some $(t,x,\ell) \in T \times R^n \times R^n$ a strict inequality is possible in condition (21), then it is feasible that the set $\mathcal{W}$ will not coincide with the maximal u-stable bridge defined for the initial system (1). Therefore initial positions may exist for which the control procedure described in Section 8.2.1 will assure the solution of an $(\mathcal{M},\mathcal{N})$-approach problem, but for the solution of this problem one cannot utilize procedure (U_0,ψ_*,e). In this respect the procedure considered herein is inferior to the procedure in Section 8.2.1. However, situations are possible when either system (20) turns out to be more convenient for constructing the required transfer function for the guidance ψ_* or when utilizing system (20) the estimate of the discrepancy between the motions of the controlled system and the guidance turn out to be substantially better than in the case discussed in Sections 8.2.1 and 8.2.2. In this respect the proposed modification of the control procedure with guidance has advantages over the procedure defined in Section 8.2.1.

Concluding the general discussion of control procedures with guidance we note the following. The definition of a guidance transition function plays the central role in the construction of control procedures with guidance. It can be asserted that the whole burden of solving problems of positional control is transferred here to the construction of the guidance's evolution.

Along with the fact that a control procedure with guidance is stable with respect to informational noise, it is also very important that when working with these procedures one enjoys the freedom of choosing auxiliary methods. For example, as was noted above, when forming the motion of a guidance one can consider a game with informational discrimination of an opponent. In certain cases it becomes convenient to define the motion of a guidance as a result of solving auxiliary game-theoretical problems posed in the class of countercontrols or quasi-strategies. We shall not dwell on formal definitions of these notions. We shall only note that a direct utilization of countercontrols or quasi-strategies (outside the framework of control procedures with guidance) is possible in the presence of additional information about the opponent's control.

8.2.5. Example. Consider the following evasion problem. Let the motion of a controlled system be described by the linear differential equation

$$\dot{x}(t) = A(t)x(t) + B(t)u(t) + C(t)v(t).$$

Suppose that this system satisfies the conditions stipulated in Section 7.3. We shall thus assume that for all $t \in T$

$$(23)\qquad \begin{aligned} \mathcal{H}(t) &\overset{\text{def}}{=} \bigcap_{v \in \mathcal{Q}} (B(t)\mathcal{P} + C(t)v) \neq \phi, \\ B(t)\mathcal{P} &= \mathcal{H}(t) - C(t)\mathcal{Q}\,. \end{aligned}$$

Recall that

$$B(t)\mathcal{P} = \{B(t)u\colon u \in \mathcal{P}\} \quad \text{and}$$

$$\mathcal{H}(t) - C(t)\mathcal{Q} = \{h - C(t)v\colon h \in \mathcal{H}(t),\ v \in \mathcal{Q}\}.$$

Let a closed set $\mathcal{M}$ be defined in the space $R^1 \times R^n$. Consider a differential game of pursuit-evasion in which the payoff functional is defined by equality (7.3.5). It was shown in Section 7.3.2 that under stipulated conditions the value of this differential game of pursuit-evasion coincides with the optimal result in the problem

(7.3.7). In Section 7.3 optimal strategies of players were defined which form a saddle point for this game. In particular a solution of the evasion problem was obtained in the form of a strategy extremal to a v-stable integral manifold $\mathcal{W}_2$ (7.3.9).

Now consider another solution of the evasion problem. This solution will be constructed in the form of a control procedure with guidance. To simplify the exposition we shall present the basic derivations under the assumption that informational errors are absent.

According to a general definition of a control procedure with guidance it is required to define a triple of functions (V_0, ψ_0, χ_*).

Assume that the function $(t,x,w) \to V_0(t,x,w)$ is selected from the condition

(24)
$$\langle w - x,\ C(t)V_0(t,x,w)\rangle = \max_{v \in \mathcal{Q}} \langle w - x,\ C(t)v\rangle$$
$$(V_0(t,x,w) \in \mathcal{Q},\quad (t,x,w) \in R^1 \times R^n \times R^n).$$

The function χ_*: $R^1 \times R^n \to R^n$ is defined by the equality $\chi_*(t,x) = x$ for all $(t,x) \in R^1 \times R^n$.

In order to construct a guidance transition function ψ_0 we shall first define the function h_0: $R^1 \times R^n \to R^n$ satisfying the following conditions: for any $s \in R^n$ the function $h_0(\cdot,s)$: $R^1 \to R^n$ is measurable; for all $(t,s) \in R^1 \times R^n$ the inclusion $h_0(t,s) \in \mathcal{H}(t)$ is fulfilled; for all $(t,s) \in R^1 \times R^n$ the equality

(25)
$$\langle s, h_0(t,s)\rangle = \max\{\langle s,h\rangle : h \in \mathcal{H}(t)\}$$

is fulfilled. Observe that the multivalued mapping $t \to \mathcal{H}(t)$ is continuous. The existence of such a function h_0 thus follows from the theorem on measurable selection (cf. Section A.3).

Let $x_* \in R^n$, $w_* \in R^n$, $t_* \leqslant t^*$. Define the solution $w(\cdot)$: $[t_*,t^*] \to R^n$ of the differential equation

(26)
$$\dot{w}(t) = A(t)w(t) + h_0(t, x_* - w_*),\quad w(t_*) = w_*.$$

We suppose that the mapping ψ_0 assigns to the collection (t_*,t^*,x_*,w_*) a point $\psi_0(t_*,t^*,x_*,w_*) = w^* = w(t^*)$; the solution $w(\cdot)$ of equation (26) arrives at this point at time $t = t^*$.

Let an initial position $(t_0,x_0) \in R^1 \times R^n$ be given. Assume that for this position in the problem (7.3.7) the quantity $\theta_0 = \theta_0(t_0,x_0)$ is finite. Select a partition $\Delta = \{(t_0 = \tau_0 < \tau_1 < \cdots < \tau_{k+1} = \theta_0(t_0,x_0)\}$. In a step-by-step procedure the motions of the controlled system and the guidance satisfy the equations

(27)
$$\dot{x}(t) = A(t)x(t) + B(t)u(t) + C(t)v_0(t),$$

(28)
$$\dot{w}(t) = A(t)w(t) + h_*(t).$$

Here

$$v_0(t) = V_0(\tau_i, x(\tau_i), w(\tau_i)),$$

(29) $$h_*(t) = h_0(t, x(\tau_i) - w(\tau_i)),$$

$$\tau_i \leqslant t < \tau_{i+1} \quad (i = 0,1,2, \dots, k)$$

and the initial states of motions $x(\cdot)$ and $w(\cdot)$ coincide, i.e. $x(t_0) = w(t_0) = x_0$.

Observe that such a definition of guidance corresponds to the construction of its transition function ψ_0, i.e. the motion $w(\cdot)$ passes at times $t = \tau_i$, $i = 0,1, \dots, k+1$, through the points which are assigned by the transition function according to the rule

$$w(\tau_0) = \chi_*(t_0, x_0) = x_0,$$

$$w(\tau_i) = \psi_0(\tau_{i-1}, \tau_i, x(\tau_{i-1}), w(\tau_{i-1})) \quad (i = 1,2, \dots, k+1).$$

The function $h_*(\cdot)$: $[t_0, \theta_0] \to R^n$ which appears in equation (28) satisfies the constraint $h_*(t) \in \mathcal{H}(t)$ $(t_0 \leqslant t \leqslant \theta_0)$. Therefore according to (7.3.7) for any $\theta < \theta_0 = \theta_0(t_0, x_0)$ an $\epsilon > 0$ exists such that the motion of the guidance $w(\cdot)$ (28) does not fall into an ϵ-neighborhood of the set M in the time interval $[t_0, \theta]$, i.e.

(30) $$(t, w(t)) \notin M^\epsilon \quad \text{for} \quad t_0 \leqslant t \leqslant \theta.$$

We shall prove below that for a sufficiently refined subdivision Δ the estimate

(31) $$\max_{t_0 \leqslant t \leqslant \theta} \|x(t) - w(t)\| \leqslant \epsilon/2$$

is valid. (Here $x(\cdot)$ and $w(\cdot)$ are a pair of motions which satisfy the equations (27) and (28) and are defined within the framework of the control procedure considered herein.) This estimate is valid for any control $u(\cdot)$: $[t_0, \theta] \to P$ chosen by the pursuer. It follows from (30) and (31) that the control procedure with guidance constructed here assures the evader evasion from encountering the sets $M^{\epsilon/2}$ on the interval $[t_0, \theta]$. (The number $\theta < \theta_0 = \theta(t_0, x_0)$ can be selected arbitrarily close to the number $\theta_0(t_0, x_0)$, which is the value of the differential game.)

It thus remains to verify the estimate (31). Set $s(t) = x(t) - w(t)$. We obtain from (27) and (28) that

(32) $$\frac{d\|s(t)\|^2}{dt} = 2\langle s(t), A(t)s(t)\rangle + 2\langle s(t), B(t)u(t) + C(t)v_0(t) - h_*(t)\rangle$$

where $v_0(t)$ and $h_*(t)$ are defined according to (29). In view of (23) the equality $B(t)u(t) = h(t) - C(t)v(t)$ is valid, where $h(t)$ and $v(t)$ are points in $\mathcal{H}(t)$ and $\mathcal{Q}$ respectively. Therefore,

$$\langle s(t),\ B(t)u(t) + C(t)v_0(t) - h_*(t)\rangle$$

$$= \langle s(\tau_i), C(\tau_i)(v_0(t) - v(t))\rangle - \langle s(\tau_i), h_*(t) - h(t)\rangle$$

(33)
$$+ [\langle s(t), C(t)(v_0(t) - v(t))\rangle - \langle s(\tau_i), C(\tau_i)(v_0(t) - v(t))\rangle$$

$$- \langle s(t) - s(\tau_i),\ h_*(t) - h(t)\rangle] \leqslant \alpha(\delta),$$

$$\alpha(\delta) \to 0 \text{ as } \delta = \operatorname{diam} \Delta \to 0.$$

In the derivation of this estimate the inequalities

$$\langle s(\tau_i),\ C(\tau_i)(v_0(t) - v(t))\rangle \leqslant 0$$

and

$$\langle s(\tau_i),\ h_*(t) - h(t)\rangle \geqslant 0, \quad \tau_i \leqslant t < \tau_{i+1},$$

were utilized. These inequalities follow from (24) and (25). The quantity appearing in square brackets in (33) approaches zero as $\delta \to 0$ in view of the continuity of functions $s(\cdot)$ and $C(\cdot)$.

Substituting the estimate (33) into (32) we arrive at the inequality

$$\frac{d\|s(t)\|^2}{dt} \leqslant 2\langle s(t), A(t)s(t)\rangle + 2\alpha(\delta).$$

Therefore taking $\|s(t_0)\| = 0$ and $\alpha(\delta) \to 0$ as $\delta \to 0$ into account we obtain that for $\delta = \operatorname{diam} \Delta$ sufficiently small the estimate (31) is valid.

We have thus shown that the proposed control procedure with guidance yields a solution of the evasion problem. This conclusion is justified in the case when the current position is measured precisely. One can, however, repeat -- with minor variations -- the arguments presented above in the case when informational noise is present and verify that the proposed solution remains stable with respect to this noise.

8.3. Stabilization of Solutions of Game-Theoretical Control Problems

It was shown in the preceding section that a control procedure with guidance yields solutions approach and evasion problems which are stable relative to informational noise. The stability of the proposed solution is expressed in Theorem 8.2.1 in a canonical form, i.e. the statement of this theorem contains the familiar clause: "... for any $\epsilon > 0$, a $\delta > 0$ and a $\xi > 0$ can be found such that ..." In Theorem 8.2.1 an important qualitative result has been established. However, from a practical point of view the problem of stability of solutions requires further investigation. The fact of the matter is that according to the estimates (8.2.16) and (8.1.9) established in the proof of Theorem 8.2.1, the values of parameters δ and ξ which are admissible for a given value of ϵ may turn out to be exceedingly small. The basic reason for this is that estimates (8.2.16) and (8.1.9) contain an exponential factor which may take on enormously large values provided the time interval $[t_0,\theta]$ is sufficiently large. A need therefore arises to improve further control procedures with guidance which would allow us to overcome this defect. This section is devoted to a discussion of this problem.

8.3.1. Return once more to the control procedure with guidance described in Section 8.2.1. We shall characterize this procedure using notions from stability and stabilization theories (cf., e.g., [2, 48]). According to the description presented in Section 8.2.1 two motions $x(\cdot)$ and $w(\cdot)$ were considered. One of them is the motion of the controlled system (8.2.1) which is formed in real-world conditions in the presence of informational noise. One cannot assure for this motion an exact fulfillment of the desired $(\mathcal{M}, \mathcal{N})$-approach condition. This condition is satisfied for $x(\cdot)$ with a certain error. The second motion -- that of the guidance -- is simulated on a computer under ideal conditions and for this motion the $(\mathcal{M}, \mathcal{N})$-approach is fulfilled precisely. The motion of guidance $w(\cdot)$ and the motion of the controlled system $x(\cdot)$ may be called unperturbed and perturbed motions respectively. However, problems considered in stability and stabilization theories are usually reduced to a form in which the unperturbed motion vanishes. Following this tradition we shall define the difference $s(t) = x(t) - w(t)$ which will be called perturbation and the corresponding function $t \to s(t)$ the perturbed motion, while the function $t \to s_0(t) \equiv 0$ will be called unperturbed motion (this is a preliminary terminology which will be stipulated below more precisely).

To estimate perturbations in stability and stabilization theories, Lyapunov functions are utilized. According to a well known approach, to investigate a stability problem or to solve a stabilization problem it is sufficient to construct a Lyapunov function in such a manner that this function and its derivative will possess specific properties governed by the motion equation. In the control procedure under consideration we have actually adhered to this approach. Indeed, to estimate perturbations we have used the function $s \to L(s) = \|s\|^2$ which plays the role of Lyapunov's function. In the proof of Lemma 8.1.1 the derivative $dL(s(t))/dt = ds^2(t)/dt$ was considered. According to (8.1.11) and (8.1.12) controls u_0 and v_0 were in reality chosen from the condition which would assure the least possible value of the derivative $dL(s(t))/dt$. The same applies to controls $u^0(t)$ (8.2.10) and $v_*(t)$ (8.2.11) which are selected in a control procedure with a guidance described in Section 8.2.1. This control procedure can be viewed as one of the methods of stabilizing the perturbed motion $s(\cdot) = x(\cdot) - w(\cdot)$. The function $L(s) = \|s\|^2$ is perhaps the simplest Lyapunov function that can be used for this purpose. An improvement of a control with guidance may be achieved by constructing a more suitable -- for a given specific case -- Lyapunov function. And this indeed is one of the problems of the theory of stabilizing motions.

To achieve greater freedom of action in the succeeding constructions we shall assume that the guidance's motion is described by the differential equation (8.2.20) where the function g satisfies the inequality (8.2.21). We shall also assume that for all $(t,w,v_*) \in T \times R^n \times \mathcal{Q}_*$ the set $\{g(t,w,u_*,v_*): u_* \in \mathcal{P}_*\}$ is convex. This nonessential assumption allows us to simplify somewhat the exposition and in place of differential inclusion (8.2.22) consider the differential equation (8.2.20).

Note that in a control procedure with guidance the controlled systems (8.2.1) and (8.2.20) are connected to each other. The phase state of the combined system (8.2.1) and (8.2.20) is characterized by a $2n$- dimensional vector $(x_1, ..., x_n, w_1, ..., w_n)^T$. We introduce a new phase variable $(s_1, ..., s_n, w_1, ..., w_n)^T$ where $s = x - w$. According to (8.2.1) and (8.2.20) we obtain that the motion of a new phase vector is described by a system of differential equations

$$
\begin{aligned}
\dot{s}(t) &= f(t,w(t) + s(t),u_0(t),v(t)) - g(t,w(t),u_*(t),v_*(t)), \\
\dot{w}(t) &= g(t,w(t),u_*(t),v_*(t)).
\end{aligned} \tag{1}
$$

The motion $(s^0(\cdot), w^0(\cdot))$ of the system (1) such that $s^0(t) = 0$ for $t_0 \leqslant t \leqslant \theta$ will be called an unperturbed motion. We shall be interested in the estimates of perturbations along the first n components of a $2n$-dimensional phase vector $(s_1, ..., s_n, w_1, ..., w_n)$.

Choose a continuously differentiable function $(t,s,w) \to L(t,s,w)$: $T \times R^n \times R^n \to [0,\infty)$. Assume that $L(t,s,w) = 0$ if and only if $s = 0$. The function L plays the role of a Lyapunov function. We shall (formally) evaluate its derivative for the system (1). We obtain

$$
\begin{aligned}
\left[\frac{dL(t,s,w)}{dt}\right]_{(1)} &= \frac{\partial L(t,s,w)}{\partial t} + \frac{\partial L(t,s,w)}{\partial s}[f(t,w + s,u_0,v) \\
&\quad - g(t,w,u_*,v_*)] + \frac{\partial L(t,s,w)}{\partial w} g(t,w,u_*,v_*).
\end{aligned} \tag{2}
$$

Here $\partial L/\partial s$ and $\partial L/\partial w$ are row-vectors $(\partial L/\partial s_1, ..., \partial L/\partial s_n)$ and $(\partial L/\partial w_1, ..., \partial L/\partial w_n)$. Define the functions $(t,x,w) \to U^0(t,x,w)$: $T \times R^n \times R^n \to \mathcal{P}$ and $(t,x,w) \to V^*(t,x,w)$: $T \times R^n \times R^n \to \mathcal{Q}_*$ by the conditions

$$
\begin{aligned}
&\max_{v \in \mathcal{Q}} \frac{\partial L(t,x - w,w)}{\partial s} f(t,x,U_0(t,x,w),v) \\
&= \min_{u \in \mathcal{P}} \max_{v \in \mathcal{Q}} \frac{\partial L(t,x - w,w)}{\partial s} f(t,x,u,v),
\end{aligned} \tag{3}
$$

$$
\begin{aligned}
&\min_{u_* \in \mathcal{P}_*} \left[\frac{\partial L(t,x-w,w)}{\partial s} - \frac{\partial L(t,x-w,w)}{\partial w}\right] g(t,w,u_*,V_*(t,x,w)) \\
&= \max_{v_* \in \mathcal{Q}_*} \min_{u_* \in \mathcal{P}_*} \left[\frac{\partial L(t,x-w,w)}{\partial s} - \frac{\partial L(t,x-w,w)}{\partial w}\right] g(t,w,u_*,v_*).
\end{aligned} \tag{4}
$$

It follows from (2) - (4) that for $u_0 = U_0(t,x,w) = U_0(t,w+s,w)$, $v_* = V_*(t,x,w) = V_*(t,w+s,w)$ and for all $v \in \mathcal{Q}$, $u_* \in \mathcal{P}_*$ the inequality

$$(5) \qquad \left[\frac{dL(t,s,w)}{dt}\right]_{(1)} \leqslant K(t,s,w)$$

is valid where

$$(6) \qquad \begin{aligned} K(t,s,w) = {} & \frac{\partial L(t,s,w)}{\partial t} + \min_{u \in \mathcal{P}} \max_{v \in \mathcal{Q}} \frac{\partial L(t,s,w)}{\partial s} f(t,w+s,u,v) \\ & - \max_{v_* \in \mathcal{Q}_*} \min_{u_* \in \mathcal{P}_*} \left[\frac{\partial L(t,s,w)}{\partial s} - \frac{\partial L(t,s,w)}{\partial w}\right] g(t,w,u_*,v_*). \end{aligned}$$

Assume that one succeeds in finding function L such that the quantity K corresponding to it via (6) satisfies the bound

$$(7) \qquad K(t,s,w) \leqslant 0.$$

This inequality should be satisfied for $\|s\| \leqslant a$ and for $(t,w) \in \mathcal{D}_a$, where a is a positive number,

$$\mathcal{D}_a = \{(t,x + y): (t,x) \in \mathcal{D}_0, \ \|y\| \leqslant a\}$$

and the set $\mathcal{D}_0$ is defined by relations of the form (8.1.6). Moreover we shall assume that in the region under consideration the functions

$$(t,s,w) \to \partial L(t,s,w)/\partial t, \quad (t,s,w) \to \partial L(t,s,w)/\partial s,$$

$$(t,s,w) \to \partial L(t,s,w)/\partial w, \quad (t,x) \to f(t,x,u,v), \text{ and}$$

$$(t,w) \to g(t,w,u_*,v_*)$$

satisfy the Lipschitz condition.

Assuming that the function V_* was selected from condition (4) we define the transition function of the guidance in the manner as was done in Section 8.2.4. We thus obtain a control procedure with guidance (U_0,ψ_*,e) where the functions U_0 and e are defined by relations (3) and (2.4.2), (2.4.4).

Let the first player select a partition $\Delta = \{t_0 = \tau_0 < \tau_1 < \cdots < \tau_{k+1} = \theta\}$. Assume that the initial position (t_0,x_0) is located on a u-stable bridge W in a problem of an (M, N)-approach posed for the system (8.2.20). Let $(x(\cdot),w(\cdot))$ be a pair of step-by-step motions which are realized when the first player selects the above indicated control procedure with guidance. This motion was realized when controls $u_0(t) = U_0(\tau_i,x_*(\tau_i),w(\tau_i))$ and $v_*(t) = V_*(\tau_i,x_*(\tau_i),w(\tau_i))$ $(\tau_i \leqslant t < \tau_{i+1},\ i = 0,1, \ldots, k)$ were selected. Control $u_0(\cdot)$ was realized in the controlled system (8.2.1), control $v_*(\cdot)$ was utilized in system (8.2.20) when the motion of the guidance was formed. Here $x_*(\tau_i)$ are results of imprecise measurements of the position of the system

(8.2.1) realized at time $t = \tau_i$. Errors in these measurements satisfy the bound (8.2.3).

Along these motions we shall define the function $L(t) = L(t,s(t),w(t))$, where $s(t) = x(t) - w(t)$. We calculate the derivative $dL(t)/dt$. This derivative is defined by the equality (2) in which one should substitute $s = s(t)$, $w = w(t)$, $u_0 = u_0(t)$, $v = v(t)$, $u_* = u_*(t)$, $v_* = v_*(t)$; here $t \to v(t)$ is the control selected by the second player, $t \to u_*(t)$ is the control selected in the guidance system (8.2.20) in such a manner that the guidance moves along the bridge $\mathcal{W}$. After some simple calculations we obtain

$$\frac{dL(t)}{dt} \leqslant K(\tau_i,s_*(\tau_i),w(\tau_i)) + \Bigg[\frac{\partial L(t,s(t),w(t))}{\partial t} - \frac{\partial L(\tau_i,s_*(\tau_i),w(\tau_i))}{\partial t}$$

$$+ \frac{\partial L(t,s(t),w(t))}{\partial s} f(t,x(t),u_0(t),v(t)) - \frac{\partial L(\tau_i,s_*(\tau_i),w(\tau_i))}{\partial s} f(\tau_i,x_*(\tau_i),u_0(t),v(t))$$

$$+ \left[\frac{\partial L(t,s(t),w(t))}{\partial w} - \frac{\partial L(t,s(t),w(t))}{\partial s}\right] g(t,w(t),u_*(t),v_*(t))$$

$$- \left[\frac{\partial L(\tau_i,s_*(\tau_i),w(\tau_i))}{\partial w} - \frac{\partial L(\tau_i,s_*(\tau_i),w(\tau_i))}{\partial s}\right] g(\tau_i,w(\tau_i),u_*(t),v_*(t))\Bigg]$$

$$(\tau_i \leqslant t < \tau_{i+1}, \quad i = 0,1, \ldots, k).$$

Here $s_*(\tau_i) = x_*(\tau_i) - w(\tau_i)$. The expression appearing in square brackets can be bounded from above by the quantity $(\delta + \xi)r$ where r is a positive number, $\delta = \operatorname{diam} \Delta$, ξ is a bound on the errors of measurements (cf. (8.2.3)).

We thus obtain

$$\text{(8)} \qquad \frac{dL(t,s(t),w(t))}{dt} \leqslant K(\tau_i,s_*(\tau_i),w(\tau_i)) + (\delta + \xi)r.$$

Taking inequality (7) into account we arrive at

$$\frac{dL(t,s(t),w(t))}{dt} \leqslant (\delta + \xi)r \qquad (t_0 \leqslant t \leqslant \theta).$$

Hence

$$\text{(9)} \qquad L(t,s(t),w(t)) \leqslant L(t_0,s(t_0),w(t_0)) + (\delta + \xi)(t - t_0)r \quad (t_0 \leqslant t \leqslant \theta).$$

Assume that

$$L(t,s,w) \geqslant \rho(s) \quad \text{for all} \quad (t,s,w) \in T \times R^n \times R^n.$$

We are particularly interested in the case when $\rho : R^n \to [0,\infty)$ is a positive definite continuous function, i.e. such that $\rho^{-1}(0) \overset{\text{def}}{=} \{s \in R^n: \rho(s) = 0\} = \{0\}$. In particular, the above requirements are fulfilled if $\rho(s) = \|s\|^b$ $(b > 0)$.

Thus let $L(t,s,w) \geqslant \rho(s)$. Then the inequality

(10) $\rho(s(t)) \leqslant \rho(s(t_0)) + (\delta + \xi)(t - t_0)r + \alpha(t_0,x_0,\xi)$

follows from (9). Here $\alpha(t_0,x_0,\xi) = \max\{L(t_0,s,w) - \rho(s)\colon \|s\| \leqslant \xi, \|w-x_0\| \leqslant \xi\}$. Observe that $\alpha(t_0,x_0,\xi) \to 0$ as $\xi \to +0$.

Thus we arrive at an estimate of perturbations which does not contain an exponential factor and depends linearly on t. One can continue to improve procedures with guidance and make sure that the final estimate will be of the form

(11) $\rho(s(t)) \leqslant \alpha(\xi,\delta)$ for all $t \geqslant t_0$

where $\alpha(\xi,\delta) \to 0$ as $\xi \to +0$ and $\delta \to +0$. To achieve this, the Lyapunov function L must be constructed in such a manner that the function K (6) will satisfy the appropriate estimate of the form $K(t,s,w) \leqslant -\rho_*(s)$, where ρ_* is a positive definite function.

The control procedure with guidance (U_0,ψ_*,e) defined above which assures that estimates of the form (10) or form (11) are satisfied, will be called a stabilized control procedure. An additional problem which arises in the course of construction of stabilized control procedures involves the determination of an appropriate Lyapunov function. When choosing this function, which in particular must satisfy conditions (5) - (7), one can vary the function g within definite limits.

8.3.2. In the nonlinear case, construction of the Lyapunov function for a stabilized control procedure is a complicated problem. For a linear control system the Lyapunov function may be defined using the well known results of stability and stabilization theories.

Thus let the motion of the controlled system be described by the equation

(12) $\dot{x}(t) = Ax(t) + Bu(t) + Cv(t)$.

Here A, B, and C are fixed $n \times n$, $n \times p$, and $n \times q$ matrices, respectively. Controls of players satisfy the constraints $u(t) \in \mathcal{P}$, $v(t) \in \mathcal{Q}$, where $\mathcal{P}$ and $\mathcal{Q}$ are convex compact sets in R^p and R^q respectively. We consider for the system (12) an $(\mathcal{M}, \mathcal{N})$-approach problem where $\mathcal{M}$ and $\mathcal{N}$ are certain closed sets ($\mathcal{M} \subset \mathcal{N} \subset T \times R^n$).

We shall assume that the motion of the guidance is described by the equation

(13) $\dot{w}(t) = Aw(t) + Bu_*(t) + Cv_*(t)$

where $u_*(t) \in \mathcal{P}_*$, $v_*(t) \in \mathcal{Q}_*$ and the sets $\mathcal{P}_*$ and $\mathcal{Q}_*$ are convex compact sets in R^p and R^q.

We shall assume that

$$(14) \qquad P_*^{\bar{\alpha}} \subset \mathcal{P}, \quad \mathcal{Q}^{\bar{\alpha}} \subset \mathcal{Q}_*.$$

Here $\alpha > 0$, $P_*^{\bar{\alpha}}$ and $Q^{\bar{\alpha}}$ are closed α-neighborhoods of the sets $\mathcal{P}_*$ and $\mathcal{Q}$. It is easy to verify that when conditions (14) are fulfilled the functions

$$(15) \qquad \begin{aligned} &f(t,x,u,v) = Ax + Bu + Cv \quad ((t,x,u,v) \in T \times R^n \times \mathcal{P} \times \mathcal{Q}), \\ &g(t,w,u_*,v_*) = Aw + Bu_* + Cv_* \quad ((t,w,u_*,v_*) \in T \times R^n \times \mathcal{P}_* \times \mathcal{Q}_*) \end{aligned}$$

will satisfy the inequality (8.2.21), i.e. the system (12) will u-majorize the system (13) (cf. Section 7.4).

In order to determine the desired Lyapunov function we shall consider the auxiliary controlled system

$$(16) \qquad \dot{s} = As + Bh^{(1)} - Ch^{(2)}.$$

We shall assume that this system is stabilizable (cf. [48]). This means that one can select a $p \times n$ matrix $H^{(1)}$ and a $q \times n$ matrix $H^{(2)}$ so that the zero solution of equation

$$(17) \qquad \dot{s} = As + BH^{(1)}s - CH^{(2)}s$$

will be asymptotically stable in the Lyapunov sense ([48]). In other words the matrices $H^{(1)}$ and $H^{(2)}$ can be selected in such a manner that all the eigenvalues of the $n \times n$ matrix $A + BH^{(1)} - CH^{(2)}$ possess a negative real part. According to a well known Lyapunov theorem ([48]) for any negative definite quadratic form

$$(18) \qquad M(s) = \langle s, M_* s \rangle$$

a positive definite quadratic form

$$(19) \qquad L(s) = \langle s, L_* s \rangle$$

will be found whose derivative will satisfy the equality

$$(20) \qquad \left[\frac{dL}{dt}\right]_{(17)} \overset{\text{def}}{=} 2\langle L_* s, (A + BH^{(1)} - CH^{(2)})s \rangle = M(s) = \langle s, M_* s \rangle$$

in view of equation (17). Note that M_* is an $n \times n$ negative definite symmetric matrix, and L_* is an $n \times n$ positive definite symmetric matrix.

Thus we shall assume that the Lyapunov function $(t,s,w) \to L(t,s,w)$ which is utilized in the stabilized control procedure is defined by equality (19), i.e. in the case under consideration the function L does not depend on the variables t and w.

We shall indicate how in the case under consideration relations (1) - (6) will look. Taking (15) into account we obtain that equation (1) can be written as follows:

$$(21)\qquad \begin{aligned} \dot{s}(t) &= As(t) + B(u_0(t) - u_*(t)) + C(v(t) - v_*(t)) \\ \dot{w}(t) &= Aw(t) + Bu_*(t) + Cv_*(t). \end{aligned}$$

Next, taking into account that here $\partial L/\partial t = 0$, $\partial L/\partial w = 0$, and $\partial L/\partial s = 2(L_* s)^T$ we arrive at

$$(22)\qquad \left[\frac{dL}{dt}\right]_{(21)} = 2\langle L_* s, As\rangle + 2\langle L_* s, B(u_0 - u_*)\rangle + 2\langle L_* s, C(v - v_*)\rangle.$$

This is the form of expression (2) in the case under consideration. Conditions (3) and (4) which define functions $(t,x,w) \to U_0(t,x,w)$ and $(t,x,w) \to V_*(t,x,w)$ are now expressed as follows:

$$(23)\qquad \begin{aligned} \langle L_*(x - w), BU_0(t,x,w)\rangle &= \min_{u\in P} \langle L_*(x - w), Bu\rangle, \\ \langle L_*(x - w), CV_*(t,x,w)\rangle &= \max_{v\in Q_*} \langle L_*(x - w), Cv\rangle. \end{aligned}$$

It is easy to observe that these funtions do not depend on the variable t and their values are determined by the difference $x - w$.

In view of (6) and (22) we obtain

$$(24)\qquad \begin{aligned} \frac{1}{2} K(s) \overset{\text{def}}{=} \ & \langle L_* s, As\rangle + \min_{u\in P} \langle L_* s, Bu\rangle \\ & + \max_{v\in Q} \langle L_* s, Cv\rangle - \min_{u_*\in P_*} \langle L_* s, Bu_*\rangle \\ & - \max_{v_*\in Q_*} \langle L_* s, Cv_*\rangle. \end{aligned}$$

In the case under consideration the quantity $K(t,s,w)$ (6) does not depend on the variables t and w, therefore in equation (24) this quantity is denoted by $K(s)$. We shall show that the inequality

$$(25)\qquad K(s) \leqslant M(s) = \langle s, M_* s\rangle$$

is valid for $\|s\| \leqslant a$. Here the number $a > 0$ is selected in such a manner that the bounds

$$(26)\qquad \|H^{(1)}s\| \leqslant \alpha, \quad \|H^{(2)}s\| \leqslant \alpha \quad \text{for} \quad \|s\| \leqslant a$$

will be fulfilled where α is a positive number appearing in condition (14).

We define the points $u_s \in P_*$ and $v_s \in Q$ in such a manner that

$$\langle L_* s, Bu_s\rangle = \min_{u_*\in P_*} \langle L_* s, Bu_*\rangle$$

$$\langle L_* s, Cv_s\rangle = \max_{v\in Q} \langle L_* s, Cv\rangle.$$

Set $u^0 = u_s + H^{(1)}s$, $v^* = v_s + H^{(2)}s$. If condition (26) is fulfilled then according to (14) we have $u^0 \in \mathcal{P}$, $v^* \in \mathcal{Q}_*$. Therefore

$$\min_{u \in \mathcal{P}} \langle L_* s, Bu \rangle \leqslant \min_{u_* \in \mathcal{P}_*} \langle L_* s, u_* \rangle + \langle L_* s, H_s^{(1)} \rangle,$$

$$\max_{v_* \in \mathcal{Q}} \langle L_* s, Cv_* \rangle \geqslant \max_{v \in \mathcal{Q}} \langle L_* s, Cv \rangle + \langle L_* s, H^{(2)} s \rangle.$$

Substituting these inequalities in expression (24) we obtain

$$\frac{1}{2} K(s) \leqslant \langle L_* s, As \rangle + \langle L_* s, BH^{(1)} s \rangle - \langle L_* s, CH^{(2)} s \rangle.$$

Taking equality (20) into account we arrive at the estimate (25).

We now approach the completion of construction and justification of a stabilized control procedure. Let a u-stable bridge be defined for the system (13) in a problem of an $(\mathcal{M}, \mathcal{N})$-approach. The transition function of guidance ψ_* will be defined as follows. Let the function V_* be selected from condition (23). Consider the differential equation

$$(27) \qquad \dot{w}(t) = Aw(t) + Bu_*(t) + CV_*(t_*, x_*, w_*)$$

with the initial condition $w(t_*) = w_*$. If $(t_*, w_*) \in \mathcal{W}$, $t^* \in [t_*, \theta]$, $x_* \in R^n$, we then select a function $u_*(\cdot)$: $[t_*, t^*] \to \mathcal{P}_*$ so that the solution $w(\cdot)$ of equation (27) will satisfy one of the conditions $(t^*, w(t^*)) \in \mathcal{W}$ or $\{(t, w(t)): t_* \leqslant t \leqslant t^*\} \cap \mathcal{M} \neq \phi$. In this case we set $\psi_*(t_*, t^*, x_*, w_*) = w(t^*)$. If, however, $(t_*, w_*) \notin \mathcal{W}$, $t^* \in [t_*, \theta]$, $x_* \in R^n$, then the function $u_*(\cdot)$ in (27) will be selected arbitrarily and we shall again set $\psi_*(t_*, t^*, x_*, w_*) = w(t^*)$.

Thus a control procedure with guidance (U_0, ψ_*, e) is defined. Here functions U_0 and e are selected in accordance with (23) and (2.4.4). Let the first player choose a certain partition $\Delta = \{t_0 = \tau_0 < \tau_1 < \cdots < \tau_{k+1} = \theta\}$. Assume that $(t_0, x_0) \in \mathcal{W}$. Consider an arbitrary pair of step-by-step motions $x(\cdot)$ and $w(\cdot)$ which are generated by the above mentioned procedure (U_0, ψ_*, e). Recall that $s(t) = x(t) - w(t)$ is a perturbation of the motion of system (12) with respect to the motion of guidance (13).

According to (5), (8), and (25) we obtain

$$(28) \qquad \frac{dL(s(t))}{dt} \leqslant M(s_*(\tau_i)) + (\delta + \xi) r_1 \quad (\tau_i \leqslant t \leqslant \tau_{i+1},\ i = 0, 1, \ldots, k).$$

Recall that here $s_*(\tau_i) = x_*(\tau_i) - w(\tau_i)$, $x_*(\tau_i)$ is a result of an imprecise measurement of the location of system (12) realized at time $t = \tau_i$. It follows from (8.2.3) that $\|s(\tau_i) - s_*(\tau_i)\| \leqslant \xi$. The estimate (28) can be used for $\|s_*(\tau_i)\| \leqslant a$. In this estimate $L(s)$ and $-M(s)$ are positive definite quadratic forms. Therefore, as is known, $-M(s) \geqslant \kappa_1 L(s)$ $(s \in R^n)$ where κ_1 is a positive number.

Next note that

$$(29) \qquad s_*(\tau_i) = s(t) + y(t), \quad \|y(t)\| \leqslant (\delta + \xi) r_2 \quad \text{for } \tau_i \leqslant t \leqslant \tau_{i+1}.$$

Therefore, the inequality $\|s_*(\tau_i)\| \leqslant a$ will follow from the estimate $\|s(t)\| \leqslant a - (\delta + \xi)r_2$, $t \in [\tau_i, \tau_{i+1}]$. Moreover for $\|s_*(\tau_i)\| \leqslant a$ we obtain from (29) the estimate

$$\kappa_1 L(s_*(\tau_i)) \geqslant \kappa_1 L(s(t)) - (\delta + \xi)r_3 \quad (\tau_i \leqslant t \leqslant \tau_{i+1}).$$

From (28) and the estimates indicated above we obtain for $\|s(t)\| \leqslant a - (\delta + \xi)r_2$ that

$$\begin{aligned} \frac{dL(s(t))}{dt} &\leqslant M(s_*(\tau_i)) + (\delta + \xi)r_1 \leqslant -\kappa_1 L(s_*(\tau_i)) + (\delta + \xi)r_1 \\ &\leqslant -\kappa_1 L(s(t)) + (\delta + \xi)(r_1 + r_3). \end{aligned} \tag{30}$$

Assume that parameters δ and ξ are chosen sufficiently small so that

$$\{s \in R^n\colon \kappa_1 L(s) \leqslant (\delta + \xi)(r_1 + r_3)\} \subset \{s \in R^n\colon \|s\| < a - (\delta + \xi)r_2\}.$$

We shall also assume that

$$\kappa_1 L(s) \leqslant (\delta + \xi)(r_1 + r_3) \quad \text{for} \quad \|s\| \leqslant \xi. \tag{31}$$

Let

$$\begin{aligned} S \overset{\text{def}}{=} \{s \in R^n\colon \|s\| < a - (\delta + \xi)r_2, \\ \kappa_1 L(s) > (\delta + \xi)(r_1 + r_3)\} \neq \phi. \end{aligned} \tag{32}$$

If $s \in S$ then $-\kappa_1 L(s) + (\delta + \xi)(r_1 + r_3) < 0$. Therefore taking into account that $\|s(t_0)\| \leqslant \xi$ we obtain from (30) and (31) that $\kappa_1 L(s(t)) \leqslant (\delta + \xi)(r_1 + r_3)$. Finally note that for a positive definite quadratic form L the estimate $L(s) \geqslant \kappa_2 \|s\|^2$ $(s \in R^n)$ is valid. We thus conclude that

$$\|s(t)\| \leqslant \kappa_3(\delta + \xi)^{1/2} \quad (t_0 \leqslant t \leqslant \theta).$$

Since the motion of the guidance $w(\cdot)$ satisfies the condition of an $(\mathcal{M}, \mathcal{N})$-approach then the motion of the controlled system (12) will satisfy the condition of an $(\overline{\mathcal{M}^\epsilon}, \overline{\mathcal{N}^\epsilon})$-approach where

$$\epsilon \leqslant \kappa_3(\delta + \xi)^{1/2}, \quad t_0 \leqslant t \leqslant \theta. \tag{33}$$

It is important to emphasize that this estimate does not depend on the length of the interval $[t_0, \theta]$.

Thus the control procedure with guidance (U_0, ψ_*, e) defined above is a stabilized control procedure.

8.4. Evasion on an Infinite Time Interval

In this section the problem of evasion on an infinite time interval $[t_0,\infty)$ is considered. Special features of the statement of such a problem are discussed and alternative conditions for its solvability are formulated. In the case of a linear controlled system a stabilized evasion procedure is constructed.

8.4.1. In this section we shall assume that either $T = [t_{\#},\infty)$ or $T = (-\infty,\infty)$. Let closed sets M and N be given ($M \subset N$) in the space $T \times R^n$. Below we shall consider the problem of evasion on an infinite interval $[t_0,\infty)$. However, before stating the central problem we shall consider the problems of an (M, N)-approach and an (M, N)-evasion on a bounded interval of time $[t_0,\theta]$ and discuss certain problems associated with the limit transition as $\theta \to \infty$.

Let an initial position (t_0,x_0) be given and a number $\theta > t_0$ be selected. According to Theorem 2.1.1 the following alternatives are valid: either the problem of an (M, N)-approach is solvable on the chosen time interval $[t_0,\theta]$ or the problem of an $(M, N)^{\epsilon}$-evasion is solvable for some $\epsilon = \epsilon(\theta)$ on this time interval $[t_0,\theta]$. Here and below the sets M and N are defined on the space $T \times R^n$ and do not depend on the parameter θ. If for some finite θ the first of the above-stated alternatives is valid then apparently for any choice of controls the second player will not be able to guarantee evasion on an infinite time interval $[t_0,\infty)$.

Therefore we shall assume that for any $\theta > t_0$ the second alternative is valid. In other words, for any $\theta > t_0$ a positional strategy V_θ: $T \times R^n \to Q$ and a number $\epsilon = \epsilon(\theta) > 0$ exist such that for any motion $x(\cdot) \in X_2(t_0,x_0,V_\theta)$ the following condition of an $(M, N)^{\epsilon}$-evasion will be fulfilled:

$$(t,x(t)) \notin M^{\epsilon(\theta)} \quad \text{for} \quad t_0 \leqslant t \leqslant \min\{t^{\epsilon(\theta)}(x(\cdot)),\theta\}.$$

Recall that $t^{\epsilon}(x(\cdot))$ is the instant of time when the point $(t,x(t))$ exits the set N^{ϵ} (cf. (2.1.7)) for the first time. M^{ϵ} and N^{ϵ} are open neighborhoods of the sets M and N. Thus let the above-stated assumption be fulfilled. Then the following question arises naturally: is it always the case that when the given condition is fulfilled one will be able to find a strategy V_∞ : $T \times R^n \to Q$ which would guarantee the (M, N)-evasion on the whole time interval $[t_0,\infty)$?

We shall state the problem in a precise manner. Let $x(\cdot)$: $[t_0,\infty) \to R^n$ be a continuous function. Set

$$T_0(x(\cdot)) = \{t \geqslant t_0\text{: } (t,x(t)) \in M\},$$

$$\tau^0(x(\cdot)) = \begin{cases} \min\{t \in T_0(x(\cdot))\} & \text{for} \quad T_0(x(\cdot)) \neq \phi \\ \infty & \text{for} \quad T_0(x(\cdot)) = \phi. \end{cases}$$

We say that the function $x(\cdot)$: $[t_0,\infty) \to R^n$ satisfies the condition of an (M, N)-evasion on the interval $[t_0,\infty)$ if one of the two conditions is fulfilled: either $\tau^0(x(\cdot)) = \infty$ or $\{(t,x(t)): t_0 \leqslant t \leqslant \tau^0(x(\cdot))\} \not\subset N$. We say that the strategy V_∞: $T \times R^n \to Q$ guarantees the (M, N)-evasion on the time interval $[t_0,\infty)$ if for any motion $x(\cdot) \in X_2(t_0,x_0,V_\infty)$ the condition of an (M, N)-evasion on $[t_0,\infty)$ is fulfilled. Here $X_2(t_0,x_0,V_\infty)$ is the set of constructive motions generated by the strategy V_∞. Elements of this set are continuous functions $x(\cdot)$: $[t_0,\infty) \to R^n$ defined by limit transition from the corresponding sequences of step-by-step motions. Here limits are considered which are uniform on any finite interval $[t_0,t^*]$ (cf. an analogous construction in Section 3.2). Now the question we are dealing with is posed in a sufficiently precise manner.

One can verify by means of a simple example that the answer to this question will be negative. Indeed, let the motion of a scalar variable x be described by equation

(1) $$\dot{x} = u + v$$

and the controls of the first and second players satisfy the bounds

(2) $$|u| \leqslant 1, \quad |v| \leqslant 2.$$

We are assuming here that $T = [0,\infty)$. The set M will be defined by the relation

$$M = M_1 \cup M_2 ,$$

$$M_1 = \{(t,x) \in T \times R^1: (t-1)^2 - x^2 \geqslant 1\},$$

$$M_2 = \{(2,x): |x| \geqslant 1\}.$$

We are assuming that the set N coincides with the whole space $T \times R^1$.

Let $(t_0,x_0) = (1,0)$. We shall show that for any $\theta > t_0 = 1$ a strategy V_θ exists which guarantees evasion on the interval $[t_0,\theta]$.

For example one can choose this strategy to be the strategy of the second player extremal to the v-stable path $W_\theta = \{(t,w_\theta(t)): 1 \leqslant t \leqslant \theta\}$ where $w_\theta(t) = 0$ for $1 \leqslant t \leqslant 1 + \delta(\theta)$, $w_\theta(t) = t - 1 - \delta(\theta)$ for $t \geqslant 1 + \delta(\theta)$. The parameter $\delta(\theta)$ is selected in such a manner that for $t_0 = 1 \leqslant t \leqslant \theta$ the path W does not intersect the set M.

Thus in the example under consideration for any $\theta > t_0 = 1$, the second of the above alternatives is fulfilled. However, in this example there is no strategy V_∞ which would guarantee an (M, N)-evasion on the infinite interval $[t_0,\infty)$.

Indeed, let V: $T \times R \to [-2,2]$ be an arbitrary strategy and let $X_2(t_0,x_0,V,\Delta)$ be the set of step-by-step motions generated by this strategy for the partition Δ chosen by the second player. If strategy V precludes an encounter with the set M_2 then there exist $\epsilon > 0$ and

$\delta > 0$ such that for all $x(\cdot) \in X_2(t_0,x_0,V,\Delta)$ the inequality $|x(2)| \leq 1-\epsilon$ holds, provided diam $\Delta < \delta$. We shall assume for definiteness that a sequence of step-by-step motions $x_\ell(\cdot) = x(\cdot,t_0,x_0,u_\ell(\cdot),V,\Delta_\ell)$ exists such that $x_\ell(2) \in [-1+\epsilon, 0)$ $(\ell = 1,2, \ldots)$ and diam $\Delta_\ell \to 0$ as $\ell \to \infty$. Define the control $u^{(\ell)}(t) = u_\ell(t)$ for $t_0 = 1 \leq t < 2$, $u^\ell(t) = 1$ for $t \geq 2$. Consider the step-by-step motions $x^\ell(\cdot) = x(\cdot,t_0,x_0,u^{(\ell)}(\cdot),V,\Delta_\ell)$. Note that $x^{(\ell)}(2) = x_\ell(2) \in [-1+\epsilon,0)$. According to (1) and (2) and in view of the choice of the control $u^{(\ell)}(t) = 2$ for $t \geq 1$ we obtain that $x^{(\ell)}(t) \geq 1 - t + \epsilon$ $(t \geq 2, \ell = 1,2, \ldots)$. It follows from here that any step-by-step motion $x^{(\ell)}(\cdot)$ encounters the set M_1 not later than at time $\theta(\epsilon) = (1+\epsilon)^2/2\epsilon$. Therefore the set of constructive motions $_2(t_0,x_0,V)$ generated by the strategy V contains motions which also encounter the set M_1 not later than at time $\theta(\epsilon)$. This example shows that there is no strategy V_∞ which would guarantee an (M, N)-evasion on an infinite time interval $[t_0,\infty)$.

The example discussed above possesses the following property. Let V_θ $(\theta \in [t_0,\infty))$ be a family of positional strategies which preclude the encounter of motions $x(\cdot) \in X_2(t_0,x_0,V_\theta)$ with the set M on the interval $[t_0,\theta]$. Let M^* be an open neighborhood of the set M. Then a $\theta^* \in (t_0,\infty)$ exists such that for any $\theta > \theta^*$ the set $X_2(t_0,x_0,V_\theta)$ will contain motions which encounter M^* on the interval $[t_0,\theta]$. Thus the family of strategies V_θ does not guarantee a *uniform evasion* from the set M on the interval $[t_0,\theta]$.

Here we have used the term "uniform evasion". In the general case this concept is introduced as follows:

Let M^* and N^* be open neighborhoods of the sets M and N. For a continuous function $x(\cdot)\colon [t_0,\infty) \to R^n$ we define the set

$$T^*(x(\cdot)) = \{t \geq t_0\colon (t,x(t)) \notin N^*\} \tag{3}$$

and a number

$$t^*(x(\cdot)) = \begin{cases} \min\{t \in T^*(x(\cdot))\} & \text{for } T^*(x(\cdot)) \neq \phi \\ \infty & \text{for } T^*(x(\cdot)) = \phi\,. \end{cases} \tag{4}$$

We say that the family of strategies $V_\theta\colon T \times R^n \to Q$ $(\theta \in [t_0,\infty))$ guarantees a uniform (M, N)-evasion if open sets M^* and N^* $(M^* \supset M,$ $N^* \supset N)$ exist and for any $\theta \in [t_0,\infty)$ an arbitrary motion $x(\cdot) \in X_2(t_0,x_0,V_\theta)$ satisfies the following condition of an (M^*, N^*)-evasion on the interval $[t_0,\theta]$:

$$(t,x(t)) \notin M^* \quad \text{for} \quad t_0 \leq t \leq \min\{\theta,t^*(x(\cdot))\}. \tag{5}$$

It turns out that the following assertion is valid. If a family of strategies V_θ exists, guaranteeing uniform (M, N)-evasion, then there exists a strategy V_∞ which guarantees an (M, N)-evasion on an infinite time interval $[t_0,\infty)$. Moreover, the strategy V_∞ will assure (M^*, N^*)-evasion on the interval $[t_0,\infty)$, i.e. for any motion $x(\cdot) \in$

$X_2(t_0,x_0,V_\infty)$ the following condition will be fulfilled:

$$(t, x(t)) \notin M^* \quad \text{for} \quad t_0 \leqslant t \leqslant t^*(x(\cdot)).$$

The validity of this assertion follows from the theorem presented below.

We conclude this preliminary discussion and state several conclusions. When the evasion problem on an infinite time interval is posed as was done in the beginning of this section, the resulting problem will be quite cumbersome for both theoretical investigations and a practical solution. Indeed, in this case the transition from evasion on an arbitrarily large time interval to evasion on an infinite interval $[t,\infty)$ breaks down. Moreover a formal solution cannot be approximated by any practically realizable method of control. We have thus introduced the notion of uniform evasion and we shall confine ourselves to a study of those situations in which a uniform (M, N)-evasion takes place, or equivalently we shall consider a problem of (M^*, N^*)-evasion where M^* and N^* are some open neighborhoods of the sets M and N. In this manner we shall regularize the initial improper evasion problem.

8.4.2. The theorem presented herein contains alternative conditions for solvability of the (M^*, N^*)-evasion problem.

Theorem 8.4.1. *Let M^* and N^* be certain open sets ($M^* \subset N^* \subset T \times R^n$). For any initial position $(t_0,x_0) \in T \times R^n$, one and only one of the following two assertions is valid: either there exists a strategy $V_\infty : T \times R^n \to Q$ which yields a solution of the problem of (M^*, N^*)-evasion on an infinite time interval $[t_0,\infty)$ or there exists a number $\theta < \infty$, closed sets $M_* \subset M^*$, $N_* \subset N^*$, and a strategy $U: T \times R^n \to P$ such that for any motion $x(\cdot) \in X_1(t_0,x_0,U)$ the condition of an (M_*, N_*)-approach will be fulfilled on the interval $[t_0,\theta]$.*

We shall present basic elements of the proof of Theorem 8.4.1. Select a time θ. Consider an auxiliary differential game of approach-evasion whose duration is bounded by the time $t = \theta$. Set

$$M^*_{(\theta)} = \{(t,x) \in M^*: t \leqslant \theta\}, \quad N^*_{(\theta)} = \{(t,x) \in N^*: t \leqslant \theta\}.$$

The game under consideration is composed of the problem of an $(M^*_{(\theta)}, N^*_{(\theta)})$-evasion posed for the second player and the problem of an $(M^*_{(\theta)}, N^*_{(\theta)})$-approach posed for the first player. Note that the condition of an $(M^*_{(\theta)}, N^*_{(\theta)})$-evasion is equivalent to the condition of $(K_{(\theta)}, \mathcal{L}_{(\theta)})$-approach where

$$K_{(\theta)} = K^{(1)}_{(\theta)} \cup K^{(2)}_{(\theta)},$$

$$K^{(1)}_{(\theta)} = \{(t,x) \in T \times R^n: t \leqslant \theta, \ (t,x) \notin N^*\},$$

$$K^{(2)}_{(\theta)} = \{(\theta,x): (\theta,x) \notin M^*\},$$

$$\mathfrak{L}_{(\theta)} = \{(t,x) \in T \times R^n\colon t \leqslant \theta,\ \ (t,x) \notin M^*\}.$$

A study of auxiliary problems may be carried out within the framework of constructions presented in Chapter 2. Thus the assertions presented below represent results obtained in Chapter 2 and reformulated appropriately for the problems considered herein.

The sets $K_{(\theta)}$ and $\mathfrak{L}_{(\theta)}$ introduced above are closed. Therefore one can determine the maximal v-stable bridge $W_{(\theta)}$ in a problem of $(K_{(\theta)}, \mathfrak{L}_{(\theta)})$-approach. We shall state several properties of the set $W_{(\theta)}$.

(j). The set $W_{(\theta)}$ is closed and satisfies the condition $W_{(\theta)} \subset \mathfrak{L}_{(\theta)}$ or equivalently the condition $W_{(\theta)} \cap M^* = \phi$.

(jj). The set $W_{(\theta)}$ satisfies the condition of v-stability, i.e. for any $(t_*,x_*) \in W_{(\theta)}$, $t^* \in (t_*,\theta]$ and $u \in P$ there exists a solution $x(\cdot) \in X_1(t_*,x_*,u)$ such that either $(t^*,x(t^*)) \in W_{(\theta)}$ or $\{(t,x(t))\colon t_* \leqslant t \leqslant t^*\} \not\subset N^*$.

Recall that $X_1(t_*,x_*,u)$ is the set of solutions of differential inclusion (2.2.2).

(jjj) If $t_0 < \theta$, $(t_0,x_0) \notin W_{(\theta)}$ then there exist closed sets $M_* \subset M^*$ and $N_* \subset N^*$ ($M_* \subset N_*$) and a strategy $U\colon T \times R^n \to P$ such that for any motion $x(\cdot) \in X_1(t_0,x_0,U)$ the condition of (M_*, N_*)-approach on the time interval $[t_0,\theta]$ is fulfilled.

Now let θ be a parameter taking values $\theta = k = 1,2, \ldots$. Consider the corresponding sets W_k. Define the set

$$W_\infty = \Big\{(t_*,x_*) \in T \times R^n\colon\ \exists (t_{k_\ell}, x_{k_\ell}) \in W_{k_\ell},\ \ \ell = 1,2, \ldots,\ \lim_{\ell\to\infty} (t_{k_\ell}, x_{k_\ell}) = (t_*,x_*)\Big\}.$$

It is easy to verify that the set W_∞ is closed and satisfies the condition $W_\infty \cap M^* = \phi$. One can also show that the set W_∞ satisfies the following condition of v-stability: for any $(t_*,x_*) \in W_\infty$, $t^* \in (t_*,\infty)$, $u \in P$ there exists a solution $x(\cdot) \in X_1(t_*,x_*,u)$ satisfying either the condition $(t^*,x(t^*)) \in W_\infty$ or the condition $\{(t,x(t)\colon t_* \leqslant t \leqslant t^*\} \not\subset N^*$. Verification of this assertion is carried out basically in the same manner as the proof of the property of u-stability of the set W_*^1 in Lemma 2.2.4.

Let V_∞ be a strategy extremal to the set W_∞. If $(t_0,x_0) \in W_\infty$ then $(t,x(t)) \in W_\infty$ for all $t_0 \leqslant t \leqslant t^*(x(\cdot))$ and all $x(\cdot) \in X_2(t_0,x_0,V_\infty)$. Since $W_\infty \cap M^* = \phi$ the strategy V_∞ will yield a solution to the problem of an (M^*, N^*)-evasion on an infinite time interval $[t_0,\infty)$. Thus in the case when $(t_0,x_0) \in W_\infty$ the first alternative formulated in Theorem 8.4.1 is valid.

If, however, $(t_0,x_0) \notin W_\infty$ then it follows from the definition of the set W_∞ that there exists an integer k such that $(t_0,x_0) \notin W_k$. In this case, according to property (jjj) of the set $W_{(\theta)}$, the second alternative is fulfilled. Theorem 8.4.1 is thus proved.

Note that it follows from the proof of this theorem that a family of strategies $V_{(\theta)}$ ($\theta \in [t_0,\infty)$) exists which guarantees a uniform (M, N)-evasion. This is equivalent to the existence of a strategy V_∞ which assures (M^*, N^*)-evasion on the whole time interval $[t_0,\infty)$.

8.4.3. Let $M^* = M^\epsilon$, $N^* = N^\epsilon$ be open ϵ-neighborhoods of closed sets M and N. Assume that for the problem of an (M^*, N^*)-evasion on the interval $[t_0,\infty)$ the second player constructs a stabilized control procedure with guidance which will assure the fulfillment of the estimate of the form (8.3.11). We are assuming that in this procedure the guidance leaves the initial point which lies on the appropriate v-stable bridge and moves along this bridge either for all $t \geq t_0$ or up to the exit of position $(t,w(t))$ from the set N^*. Then for sufficiently small parameters δ and ξ we obtain that step-by-step motions of the controlled system will satisfy the condition of $(M,N)^{\epsilon/2}$-evasion.

We are not going to describe the construction of stabilized control procedure for an evasion problem in the general case of a non-linear system. On the whole, this construction is carried out along the lines presented in Section 8.3.1. We shall confine ourselves to the case when the motion of the controlled system is described by the linear equation (8.3.12). We shall assume that the motion of the guidance is described by equation (8.3.13) in which the control $u_*(t)$ and $v_*(t)$ satisfy the bounds $u_*(t) \in P_*$ and $v_*(t) \in Q_*$. We shall assume that

$$P^{\bar{\alpha}} \subset P_*, \quad Q_*^{\bar{\alpha}} \subset Q \tag{6}$$

where $\alpha > 0$, $P^{\bar{\alpha}}$ and $Q_*^{\bar{\alpha}}$ are closed α-neighborhoods of the sets P and Q_*. When the condition (6) is fulfilled the system (8.3.12) v-majorizes the system (8.3.13) (cf. Section 7.4). We shall assume that the auxiliary controlled system

$$\dot{s}(t) = As(t) - Bh^{(1)} + Ch^{(2)} \tag{7}$$

is stabilizable, i.e. $p \times n$ and $q \times n$ matrices $H^{(1)}$ and $H^{(2)}$ can be selected so that the zero solution of the equation

$$\dot{s} = As - BH^{(1)}s + CH^{(2)}s \tag{8}$$

will be asymptotically stable. In this case, for any negative definite quadratic form $M(s) = \langle s,M_*s\rangle$ there exists a positive definite quadratic form $L(s) = \langle s,L_*s\rangle$ whose derivative, in view of equation (7), will satisfy the equality

$$\left(\frac{dL}{dt}\right)_{(8)} = 2\langle L_*s, (A - BH^{(1)} + CH^{(2)})s\rangle = M(s) = \langle s,M_*s\rangle. \tag{9}$$

Define the functions $(t,x,w) \to V_0(t,x,w)$ and $(t,x,w) \to U_*(t,x,w)$ from the conditions

$$\begin{aligned} &\langle L_*(w - x), CV_0(t,x,w)\rangle = \max_{v\in Q} \langle L_*(w - x), Cv\rangle, \\ &\langle L_*(w - x), BU_*(t,x,w)\rangle = \min_{u\in P_*} \langle L_*(w - x), Bu\rangle . \end{aligned} \tag{10}$$

Observe that in accordance with the general definition of the control procedure with guidance (cf. Section 8.2.4) we are assuming that functions V_0 and U_* are defined on $T \times R^n \times R^n$, although in this case the values of the functions are determined only by the difference $w - x$.

Let $\mathcal{M}^* = \mathcal{M}^\epsilon$ and $\mathcal{N}^* = \mathcal{N}^\epsilon$. Let a v-stable bridge $\mathcal{W}_\infty$ be defined for the system (8.3.13) in a problem of an $(\mathcal{M}^*, \mathcal{N}^*)$-evasion on an infinite time interval. Consider the differential equation

$$\dot{w}(t) = Aw(t) + BU_*(t_*,x_*,w_*) + Cv_*(t) \tag{11}$$

with the initial condition $w(t_*) = w_*$. Assume that $(t_*,x_*) \in \mathcal{W}_\infty$, $t^* \in (t_*,\infty)$, $x_* \in R^n$. Select a function $v_*(\cdot)$: $[t_*,t^*] \to \mathcal{Q}$ in such a manner that a solution $w(\cdot)$ of equation (11) will satisfy one of the conditions $(t^*,w(t^*)) \in \mathcal{W}_\infty$ or $(t,w(t))$: $t_* \leqslant t \leqslant t^*\} \not\subset \mathcal{N}^*$. Such a solution exists since $\mathcal{W}_\infty$ is a v-stable bridge in a problem of an $(\mathcal{M}^*, \mathcal{N}^*)$-evasion. In the case under consideration we are assuming that the value of the transition function of the guidance $\psi_*(t_*,t^*,x_*,w_*)$ coincides with point $w(t^*)$ into which the solution $w(\cdot)$ arrives at time $t = t^*$. If, however $(t_*,w_*) \notin \mathcal{W}_\infty$, $t^* \in (t_*,\infty)$, $x_* \in R^n$ then the function $v_*(\cdot)$: $[t_*,t^*] \to$ in (9) is selected in an arbitrary manner and we are assuming again that $\psi_*(t_*,t^*,x_*,w_*) = w(t^*)$.

Thus a control procedure with guidance (V_0,ψ_*,e) is defined where the function e is selected according to (2.4.4). Assume that $(t_0,x_0) \in \mathcal{W}_\infty$. Let $\delta = \operatorname{diam} \Delta \overset{\text{def}}{=} \sup(\tau_{i+1} - \tau_i)$, where $\Delta = \{t_0 = \tau_0 < \tau_1 < \tau_2 \ldots\}$ ($\tau_i \to \infty$ as $i \to \infty$), be a partition of the half-intervals $[t_0,\infty)$ selected by the second player. It is also assumed that the second player measures the current position of the system (8.3.12) imprecisely and the errors in measurements satisfy the bound (8.2.3). Consider an arbitrary pair of motions $x(\cdot)$ and $w(\cdot)$ which are formed by the above mentioned control procedure with guidance. Here $x(\cdot)$ is the motion of the controlled system (8.3.12), and $w(\cdot)$ is the motion of the guidance which satisfies equation (8.3.13). Using arguments which coincide on the whole with the derivations presented in Section 8.3.2 we arrive at the bound

$$\|x(t) - w(t)\| \leqslant \kappa(\delta + \xi)^{1/2} \quad \text{for} \quad t \geqslant t_0. \tag{12}$$

Since the position of a guidance moves along a v-stable bridge $\mathcal{W}_\infty$ either for all $t \geqslant t_0$ or up to time of its exit from the set $\mathcal{N}^\epsilon$, it follows from (12) that for δ and ξ sufficiently small the motion $x(\cdot)$ will satisfy the condition of an $(\mathcal{M}, \mathcal{N})^{\epsilon/2}$-evasion.

Thus a stabilized solution of an $(\mathcal{M}, \mathcal{N})$-evasion problem on an infinite time interval has been obtained.

8.4.4. Consider the following example. Let the motion of the controlled system be described by the equations

$$\dot{x}_1 = x_2, \quad \dot{x}_2 = -x_1 + x_2 + u + v \tag{13}$$

with the constraints

$$|u| \leqslant 2, \quad |v| \leqslant 1.$$

For this system it is required to construct a control procedure which, for all $t \geqslant t_0$, would exclude the falling of point $x(t)$ into the circle of radius r with center at the origin, i.e. here

$$M = \{(t,x)\colon t \in R^1, \ x_1^2 + x_2^2 \leqslant r^2\}$$

$$N = R^1 \times R^2.$$

Assume that the motion of the guidance is described by the equation

$$(14) \qquad \dot{w}_1 = w_2, \quad \dot{w}_2 = -w_1 + w_2 + u_* + v_*$$

under the constraints

$$|u_*| \leqslant 2 + \alpha, \quad |v_*| \leqslant 1 - \alpha$$

(α is a sufficiently small positive number). Consider also a controlled system

$$(15) \qquad \dot{w}_1 = w_2, \quad \dot{w}_2 = -w_1 + w_2 + h(t)$$

under the constraints

$$|h| \leqslant 1 + 2\alpha .$$

Define the set

$$(16) \qquad W_\infty = \{(t, w(t,t_0,x_0,h(\cdot))\colon \ t \geqslant t_0, \ h(\cdot) \in \mathcal{H}\}.$$

Here $\mathcal{H}$ is a set of measurable functions $h(\cdot)\colon [t_0,\infty) \to [-1 - 2\alpha, 1 + 2\alpha]$ and $w(\cdot,t_0,x_0,h(\cdot))$ is a solution of equation (15) satisfying the initial condition $w(t_0) = x_0$.

Let the initial position (t_0,x_0) and parameters $r > 0$ and $\epsilon > 0$ be such that $W_\infty \cap M^\epsilon = \phi$. Note that such a selection of the initial point and parameters is indeed feasible since the system $\dot{w}_1 = w_2$, $\dot{w}_2 = -w_1 + w_2$ is unstable. Assume also that for the selected initial position there exists a motion of the controlled system (for some measurable $u(\cdot)\colon [t_0,\infty) \to [-2,2]$ and $v(\cdot)\colon [t_0,\infty) \to [-1,1]$) which falls into the set M . The problem posed is meaningful, namely not every control of the second player assures the evasion from M , and as will be shown below, there exists a control method which will assure the required evasion.

We now describe the proposed solution. In the case under consideration the system $\dot{s}_1 = s_2$, $\dot{s}_2 = -s_1 + s_2 - h^{(1)} + h^{(2)}$

corresponds to the system (7). This system can be stabilized by setting $-h^{(1)} + h^{(2)} = -2s_2$. We thus obtain an asymptotically stable system

$$(17) \qquad \dot{s}_1 = s_2, \quad \dot{s}_2 = -s_1 - s_2.$$

Select a negative definite quadratic form

$$M(s) = -(s_1^2 + s_2^2).$$

The positive definite quadratic form corresponding to it is

$$L(s) = \tfrac{3}{2}\, s_1^2 + s_1 s_2 + s_2^2.$$

Indeed it can be verified directly that

$$(dL(s)/dt)_{(17)} = M(s).$$

If in accordance with (10) we define functions V_0 and U_* then in our example we obtain

$$V_0(t,x,w) = \operatorname{sign}(x_1 - w_1 + 2x_2 - 2w_2),$$

$$U_*(t,x,w) = -(2 + \alpha)\operatorname{sign}(x_1 - w_1 + 2x_2 - 2w_2).$$

We define the transition function of the guidance. Note that the set $\mathcal{W}$ (16) represents a v-stable integral manifold. Therefore the evolution of the guidance can be determined in the following manner. Let $w(\cdot,t_*,x_*,w_*)$ be a solution of the differential equation

$$(18) \qquad \begin{aligned} \dot{w}_1(t) &= w_2(t), \\ \dot{w}_2(t) &= -w_1(t) + w_2(t) - (1+2\alpha)\operatorname{sign}(x_{*1} - w_{*1} + 2x_{*2} - 2w_{*2}), \end{aligned}$$

satisfying the initial condition $w(t_*) = w_*$. Assume that in our example $\psi(t_*,t^*,x_*,w_*) = w(t^*,t_*,x_*,w_*)$. Note that equation (18) is obtained as a result of substituting the controls $u_* = U_*(t_*,x_*,w_*)$ and $v_* = [-(1-\alpha)(2+\alpha)^{-1}]U_*(t_*,x_*,w_*)$ into (14).

Suppose that the initial position (t_0,x_0) is measured precisely. Select the point $(t_0,w_0) = (t_0,x_0)$ as the initial position of the guidance. Then according to the constructions presented above, the position of guidance $(t,w(t))$ will remain on the set $\mathcal{W}_\infty$ (16) for all $t \geqslant t_0$. Since $\mathcal{W}_\infty \cap M^\epsilon = \phi$ it follows from the estimate (12) that for $\delta > 0$ and $\xi > 0$ sufficiently small the constructed control procedure assures for the second player that the inequality $(x_1^2(t) + x_2^2(t))^{1/2} \geqslant r + \epsilon/2$ be fulfilled for all $t \geqslant t_0$. Thus this control procedure yields a solution of the evasion problem under consideration.

8.5. Universal Optimal Strategies

In this section we return to a discussion of the positional method of control. We shall deal, for definiteness, with a minimax control problem in which the payoff functional is defined in accordance with (3.1.1). This problem, coupled with the corresponding maximin problem, comprises a differential game with a fixed termination time. It was shown in Section 3.1 that in this game for any initial position (t_0,x_0) there exists a saddle point (U_0,V_0) which is formed by a pair of optimal strategies U_0 and V_0 of the first and second players respectively.

In order to define an optimal strategy U_0 the first player must possess the following information about the problem posed. Firstly, he or she should know the controlled system. Formally this means that the first player knows the function f appearing in the r.h.s. of equation (1.2.1). In particular, he or she has knowledge of the sets P and Q which constrain the choice of controls $u(t)$ and $v(t)$. Secondly, the first player knows the payoff functional, i.e., in this case the function σ and the termination time θ. Thirdly, the first player has knowledge of the initial position (t_0,x_0) of the game. Possessing this initial information, the first player can construct the corresponding u-stable bridge containing the initial position (t_0,x_0) and next he or she can determine the strategy which is extremal to this bridge. As a result of these constructions a strategy U_0 optimal for the given initial position will be defined.

Suppose now that the first player has to select his own strategy before knowing the initial position. As above, it is assumed that the first player has knowledge of the controlled system and of the payoff functional. However, the specific initial position is not indicated. The first player can process the information available to him in a certain manner and in particular he or she can determine the value function of the differential game, its partial derivatives (wherever they exist), and so on. The following question arises. Is it always possible for the first player in this situation to determine a positional strategy $(t,x) \to U^0(t,x)$ such that it will be optimal for any initial position?

In what follows, such a strategy will be called a universal optimal strategy. A counterexample presented below shows that the reply to this problem is negative. Note that when constructing this counterexample the strategies, motions, and guaranteed results were defined according to the formalization presented in Chapter 1. Next in this section new definitions of positional strategies and the corresponding motions are presented. It is shown that within the framework of the modification proposed herein there always exists a universal optimal strategy.

8.5.1. Consider a differential game with a fixed termination time (cf. Section 3.1). Assume that the motion of the controlled system is described by the differential equation

(1) $\dot{x} = f(t,x,u,v)$

where the function f satisfies all the conditions (i) - (iv) stated in Section 1.2.2. The payoff in this game is defined by equality

(2) $\gamma(x(\cdot)) = \sigma(x(\theta))$

where $\sigma: R^n \to R^1$ is a continuous function and θ is a fixed time of termination of the game. According to Theorem 3.1.1 for any initial position $(t_0,x_0) \in T \times R^n$ there exists a saddle point (U_0,V_0). Below we shall confine ourselves to a discussion of the problem posed to the first player. Recall that the optimal strategy U^0 for the first player is defined by the following condition:

$$\Gamma_1(U_0,t_0,x_0) = \min \Gamma_1(U,t_0,x_0)$$

where

$$\Gamma_1(U,t_0,x_0) = \max\{\sigma(x): x(\cdot) \in X_1(t_0,x_0,U)\}.$$

Here the minimum is taken over all possible positional strategies U: $T \times R^n \to P$, the set $X_1(t_0,x_0,U)$ is a totality of all the constructive motions generated by the strategy U. It is important to note the following: strategy U_0 is defined here as an optimal strategy for the given initial position (t_0,x_0). If another initial position (t_0',x_0') were selected it may then be possible that

$$\Gamma_1(U_0,t_0',x_0') > \min_U \Gamma_1(U,t_0',x_0') = \Gamma_1(U_0',t_0',x_0').$$

It is not too difficult to give an appropriate example.

Consider now another definition of an optimal strategy. Let $\mathcal{D}$ be a region in the space of positions $T \times R^n$. In particular $\mathcal{D}$ may be the whole space $T \times R^n$.

Definition. *A positional strategy U^0: $T \times R^n \to P$ is called a universal optimal strategy for the region $\mathcal{D}$ if*

(3) $\Gamma_1(U^0,t_0,x_0) = \min_U \Gamma_1(U,t_0,x_0)$ *for all* $(t_0,x_0) \in \mathcal{D}$.

Let a be a number, ω^0: $T \times R^n \to R$ be the value function of the game under consideration. Define the sets

$$\mathcal{D}_a = \{(t_0,x_0) \in T \times R^n : \omega^0(t_0,x_0) = a\}$$

$$\mathcal{W}_a = \{(t,x) \in T \times R^n : \omega^0(t,x) \leq a\}.$$

Let U^e be a strategy extremal to the set $\mathcal{W}_a$. Following from the results of Section 3.1, strategy U^e will be universally optimal for the region $\mathcal{D}_a$. This fact is presented in order to illustrate the definition of a universal optimal strategy and is of no interest on its own.

Consider another example. Assume that the value function ω^0 is continuously differentiable in the region $T^0 \times R^n$. Then according to Theorem 4.1.1, strategy U^0 defined by the conditions (4.1.3), (4.1.4) will be a universal optimal strategy on the whole space $T \times R^n$.

Consider now the general case. Let the function σ be locally-Lipschitzian. Then, as it was established in Theorem 3.4.1, the value function ω^0 will also be locally Lipschitzian. According to the Rademacher theorem ([27,65]) the function ω^0 will be almost everywhere differentiable. Denote by the so-called singular set consisting of points $(t,x) \in T \times R^n$ at which the function ω^0 is not differentiable. At any point $(t,x) \in (T^0 \times R^n)\backslash S$ the derivative $\partial\omega^0(t,x)/\partial x$ exists and therefore at such a point one can formally use the conditions (4.1.3), (4.1.4) for the definition of strategy U^0. Then one could try to argue as follows. A singular set S on which the strategy U^0 is not defined is of measure zero. It is irrelevant how the strategy U^0 is defined on such a set. One would therefore expect that the strategy U^0 constructed in this manner will be universally optimal on the whole space $T \times R^n$.

However this argument turns out to be wrong. In the counterexample presented below one cannot define a universal optimal strategy using the method indicated above. Moreover in this counterexample there is in general no universal optimal strategy on the whole space of strategies.

8.5.2. We state one simple property of a universal optimal strategy that will be utilized for substantiation of the counterexample considered below.

Lemma 8.5.1. *Let $\mathfrak{D}$ and G be open regions in $T \times R^n$. Assume that in the differential game* (1), (2) *there exists a positional strategy U^0 which is universally optimal in $\mathfrak{D}$. Assume also that the value function ω^0 is continuously differentiable in the region G and $\mathfrak{D} \supset G$. Then for any $\epsilon > 0$ there exists a set $\mathcal{E}_\epsilon$ everywhere dense in G such that at each point $(t,x) \in \mathcal{E}_\epsilon$ the inequality*

$$\max_{v\in Q} (\partial\omega^0(t,x)/\partial x) f(t,x,U^0(t,x),v)$$
$$\leqslant \min_{u\in P} \max_{v\in Q} (\partial\omega^0(t,x)/\partial x) f(t,x,u,v) + \epsilon$$

is fulfilled.

Proof. Assume the contrary. Let there exist numbers $\epsilon > 0$ and $\alpha > 0$ and a point $(t_*,x_*) \in G$ such that at any point

$$(t,x) \in B((t_*,x_*);\ \alpha) \overset{\text{def}}{=} \{(\tau,y) \in G : |t_* - \tau| \leqslant \alpha,\ \|x_* - y\| \leqslant \alpha\}$$

the inequality

$$(4)\qquad \max_{v\in Q} (\partial\omega^0(t,x)/\partial x) f(t,x,U^0(t,x),v) > \min_{u\in P}\max_{v\in Q} (\partial\omega^0(t,x)/\partial x) f(t,x,u,v) + \epsilon$$

is fulfilled. At points $(t,x) \in B(t_*,x_*);\alpha)$ the function ω^0 is differentiable, therefore according to Corollary 4.4.1, at these points the function ω^0 satisfies the main equation

$$(5)\qquad \frac{\partial\omega^0(t,x)}{\partial t} + \min_{u\in P}\max_{v\in Q} \frac{\partial\omega^0(t,x)}{\partial x} f(t,x,u,v) = 0.$$

It follows from (4) and (5) that at all points $(t,x) \in B((t_*,x_*);\ \alpha)$ the inequality

$$(6)\qquad \frac{\partial\omega^0(t,x)}{\partial t} + \max_{v\in Q} \frac{\partial\omega^0(t,x)}{\partial x} f(t,x,U^0(t,x),v) > \epsilon$$

will be valid.

Consider step-by-step motions $x(\cdot,t_*,x_*,U^0,v(\cdot),\Delta)$. Select a time $t^* \in (t_*,\theta]$ so that $(t,x(t,t_*,x_*,U^0,v(\cdot),\Delta)) \in B((t_*,x_*);\alpha)$ for all $t_* \leqslant t \leqslant t^*$, for all admissible $v(\cdot)$ and for any partition Δ. It follows from inequality (6) and from the continuous differentiability of function ω^0 in the region $B((t_*,x_*);\alpha)$ that one can select a sufficiently small $\delta > 0$ so that for any partition Δ satisfying the bound diam $\Delta \leqslant \delta$ one can select a control $v_\Delta(\cdot)$: $[t_*,t^*) \to Q$ such that the step-by-step motion $x_\Delta(\cdot) = x(\cdot,t_*,x_*,U^0,v_\Delta(\cdot),\Delta)$: $[t_*,t^*] \to R^n$ satisfies the inequality

$$\omega^0(t^*,x(t^*)) \geqslant \omega^0(t_*,x_*) + \frac{\epsilon}{2}(t^* - t_*).$$

Next, utilizing the property of v-stability of function ω^0 (cf. Section 4.2) a control $v_\Delta(\cdot)$ can be defined on the interval $[t^*,\theta]$ in such a manner that the step-by-step motion $x_\Delta(\cdot)$ will satisfy the inequality

$$\omega^0(t,x_\Delta(t)) \geqslant \omega^0(t^*,x_\Delta(t^*)) \quad \text{for} \quad t \in [t^*,\theta].$$

In particular it follows from these estimates that for $t = \theta$

$$\sigma(x_\Delta(\theta)) \geqslant \omega^0(t_*,x_*) + \frac{\epsilon}{2}(t^* - t_*).$$

This inequality is valid for any partition Δ satisfying the bound diam $\Delta \leqslant \delta$. Therefore approaching the limit as diam $\Delta \to 0$, we obtain that there exists a constructive motion $x_*(\cdot) \in X_1(t_*,x_*,U^0)$ for which $\sigma(x_*(\theta)) \geqslant \omega^0(t_*,x_*) + (t^* - t_*)\epsilon/2$. Hence, according to (3.1.3) and (3.1.6)

$$\max\{\sigma(x(\theta)) \colon x(\cdot) \in X_1(t_*,x_*,U^0)\}$$
$$\geqslant \omega^0(t_*,x_*) + (t^* - t_*)\frac{\epsilon}{2} > \min_U \Gamma_1(U,t_*,x_*) = \omega^0(t_*,x_*).$$

We have thus obtained a contradiction to the assumption that strategy U^0 is optimal for the point $(t_*,x_*) \in \mathfrak{D}$. Lemma 8.5.1 is proved.

Example. Consider a controlled system

$$(7) \qquad \dot{x}_1 = u_1 + v_1, \quad \dot{x}_2 = u_2 + v_2$$

with the constraints

$$(8) \qquad u_1^2 + u_2^2 \leqslant 2, \quad v_1^2 + v_2^2 \leqslant 2.$$

Suppose that the payoff is defined by the equality

$$(9) \qquad \gamma(x(\cdot)) \overset{\text{def}}{=} \sigma(x(1)) \overset{\text{def}}{=} |x_1(1)| + x_2(1).$$

Here $\theta = 1$ is a fixed time of termination of the game.

It is easy to verify that in the differential game under consideration the value function is expressed by the equality $\omega^0(t,x) = \sigma(x) = |x_1| + x_2$ $(t \leqslant 1,\ x \in R^2)$. One can apply the notion of equivalent systems (cf. Section 7.4) and verify that the system (7), (8) is equivalent to the system $\dot{w}_1 = 0$, $\dot{w}_2 = 0$, whence the equality $\omega_0(t,x) = \sigma(x)$ follows immediately.

Define an open bounded region

$$(10) \qquad \mathfrak{D} \overset{\text{def}}{=} \{(t,x)\colon 0 < t < 1,\ \|x\| < 2\sqrt{2}\, t\}.$$

In subsequent constructions we shall utilize the following self-evident property of the region $\mathfrak{D}$. Let $(t_*,x_*) \in \mathfrak{D}$ and let $x(\cdot)$ be the motion of the system (7), (8) satisfying the initial condition $x(t_*) = x_*$ and generated by an arbitrary pair of admissible controls $u(\cdot)$ and $v(\cdot)$. Then $(t,x(t)) \in \mathfrak{D}$ for all $t \in [t_*,1)$. The purpose of the arguments below is to provide the following assertion.

Assertion 1. *In the differential game* (7) - (9) *a positional strategy for the first player which would be universally optimal in the region* $\mathfrak{D}$ (10) *does not exist.*

Assume the contrary. Let a strategy U^0 exist which is universally optimal in the domain $\mathfrak{D}$. Since $\omega^0(t,x) = |x_1| + x_2$, the function ω will be continuously differentiable in the region

$$G = \{(t,x) \in \mathfrak{D} :\ x_1 \neq 0\}.$$

According to Lemma 8.5.1 for any $\epsilon > 0$ a set $\mathcal{E}_\epsilon$ exists, everywhere dense in G, such that at each point $(t,x) \in \mathcal{E}_\epsilon$ the relations

$$U^0(t,x) = h^0(x) + h^*(t,x),$$

(11) $$h^0(x) = (-1,-1)^T \text{ for } x_1 > 0,\ h^0(x) = (+1,-1)^T \text{ for } x_1 < 0,$$

$$\|h^*(t,x)\| \leqslant \alpha(\epsilon),\quad \alpha(\epsilon) \to 0 \text{ for } \epsilon \to 0$$

will be fulfilled.

Select an initial position (t^0,x^0) such that

(12) $$t^0 \in (0,1),\quad x^0 = (0,x_2^0)^T,\quad |x_2^0| < 2\sqrt{2}\,t^0.$$

It is easy to see that $(t^0,x^0) \in \mathfrak{D}$. Choose a positive number ϵ such that the inequality

(13) $$2\epsilon + \alpha(\epsilon) < \sqrt{2} - 1$$

will be fulfilled. Below we shall construct a partition

$$\Delta = \{t^0 = \tau_0 < \tau_1 < \dots < \tau_{k+1} = 1\},\ \epsilon/2 \leqslant \tau_{i+1} - \tau_i \leqslant \epsilon,\ i = 0,1,\dots,k$$

and a piecewise-constant control

$$v(t) = v(\tau_i),\quad \tau_i \leqslant t < \tau_{i+1},\quad \|v(\tau_i)\| \leqslant \sqrt{2},\quad i = 0,1,\dots,k$$

such that for the step-by-step motion $x(\cdot) = x(\cdot,t^0,x^0,U^0,v(\cdot),\Delta)$ the inequality

(14) $$x_2(1) \geqslant x_2^0 + (1 - t^0)(\sqrt{2} - 1) - \beta(\epsilon),$$

will be valid, where

(15) $$\beta(\epsilon) = (1 - t^0)(2\epsilon + \alpha(\epsilon)) + 3\epsilon\sqrt{2}\,.$$

Since $\beta(\epsilon) \to 0$ as $\epsilon \to 0$, approaching the limit as $\epsilon \to 0$ in equality (14) we obtain the existence of a constructive motion $x^*(\cdot) \in X_1(t^0,x^0,U^0)$ satisfying the inequality $\sigma(x^*(1)) = |x_1^*(1)| + x_2^*(1) \geqslant x_2^*(1) \geqslant x_2^0 + (1-t^0)(\sqrt{2} - 1) > x_2^0$. According to (12) we have $x_1^0 = 0$. Therefore $\omega^0(t^0,x^0) = |x_1^0| + x_2^0 = x_2^0$. Consequently $\sigma(x^*(1)) > \omega^0(t^0,x^0)$. This inequality contradicts the fact that strategy U^0 is optimal for the initial position $(t^0,x^0) \in \mathfrak{D}$.

Thus for the proof of Assertion 1 it is sufficient to construct a partition Δ and control $v(\cdot)$ such that the step-by-step motion $x(\cdot) = x(\cdot,t^0,x^0,U^0,v(\cdot),\Delta)$ satisfies the inequality (14). First we shall prove the following auxiliary assertion.

Assertion 2. *Let* $\epsilon \in (0,1)$, $(t_*,x_*) \in \mathfrak{D}$, $t_* < 1 - \epsilon$, $\|u_*\| \leqslant \sqrt{2}$, $v_\epsilon = (0,\sqrt{2} - \epsilon)^T$. *Then there exist* v^* *and* t^* *such that*

$$\|v^*\| \leqslant \epsilon, \quad t^* \in [t_* + \epsilon/2, t_* + \epsilon], \quad (t^*,x(t^*)) \in \mathcal{E}_\epsilon,$$

where

$$x(t^*) = x_* + (u_* + v_\epsilon + v^*)(t^* - t_*).$$

To prove this assertion we set

$$x(t,v^*) \overset{\text{def}}{=} x_* + (u_* + v_\epsilon + v^*)(t - t_*),$$

$$Z \overset{\text{def}}{=} \{(t,x(t,v^*)) \colon t_* + \epsilon/2 \leqslant t \leqslant t_* + \epsilon, \quad \|v^*\| \leqslant \epsilon\}.$$

Since $(t_*,x_*) \in \mathcal{D}$ and $\|u_* + v_\epsilon + v^*\| \leqslant 2\sqrt{2}$, we have $Z \subset \mathcal{D}$ (cf. (10)). According to Lemma 8.5.1 the set $\mathcal{E}_\epsilon$ is everywhere dense in $G \overset{\text{def}}{=} \{(t,x) \in \mathcal{D} : x_1 \neq 0\}$. Therefore $\mathcal{E}_\epsilon$ is everywhere dense in $\mathcal{D}$. The set Z is imbedded in $\mathcal{D}$ and contains interior points; hence $Z \cap \mathcal{E}_\epsilon \neq \phi$, q.e.d.

We now proceed directly to the construction of the required partition Δ and control $v(\cdot)$. Let an initial position (t^0,x^0) and a number $\epsilon > 0$ be chosen satisfying the relations (12), (13). The times τ_i $(i = 0,1, ..., k)$ and controls $v(t) = v(\tau_i)$ $(\tau_i \leqslant t < \tau_{i+1})$ will be defined by induction. At the first step we shall use Assertion 2 in which we set $(t_*,x_*) = (t^0,x^0)$, $u_* = U^0(t^0,x^0) = u^0(\tau_0)$. According to this assertion the time $t^* = \tau_1 \in [t^0 + \epsilon/2, t^0 + \epsilon]$ and the vector $v^*(\tau_0)$ $(\|v^*(\tau_0)\| \leqslant \epsilon)$ can be selected in such a manner that the controlled system $\dot{x}(t) = u^0(\tau_0) + v_\epsilon + v^*(\tau_0)$, $\tau_0 = t^0 \leqslant t < \tau_1$, would move from the initial position (t^0,x^0) to the position $(\tau_1, x(\tau_1)) \in \mathcal{E}_\epsilon \subset \mathcal{D}$. Thus at the first step the time τ_1 and control $v(t) = v_\epsilon + v^*(\tau_0)$, $\tau_0 \leqslant t < \tau_1$ are determined.

Let the times $\tau_1, ..., \tau_i$ and controls

$$u^0(t) = U^0(\tau_j,x(\tau_j)), \quad v(t) = v_\epsilon + v^*(\tau_j), \quad \|v^*(\tau_j)\| \leqslant \epsilon,$$

$$\tau_j \leqslant t < \tau_{j+1}, \quad j = 0,1, ..., i-1$$

be determined. Here $x(t)$ is the motion of the controlled system

$$\text{(16)} \qquad \dot{x}(t) = u^0(t) + v(t), \quad t^0 \leqslant t < \tau_i, \quad x(t^0) = x^0.$$

Note that $\|u^0(t) + v(t)\| \leqslant 2\sqrt{2}$, therefore $(t,x(t)) \in \mathcal{D}$ for $t^0 \leqslant t \leqslant \tau_i$.

Assume now that $\tau_i < 1 - \epsilon$. Once more we shall use Assertion 2 in which we set $(t_*,x_*) = (\tau_i,x(\tau_i))$, $u_* = U^0(\tau_i,x(\tau_i)) = u^0(\tau_i)$. At the $(i+1)$-th step we choose the time $t^* = \tau_{i+1} \in [\tau_i + \epsilon/2, \tau_i + \epsilon]$ and the vector $v^*(\tau_i)$ $(\|v^*(\tau_i)\| \leqslant \epsilon)$ such that for $u^0(t) = U^0(\tau_i,x(\tau_i))$, $v(t) = v_\epsilon + v^*(\tau_i)$ $(\tau_i \leqslant \tau < \tau_{i+1})$ the motion of the system (16) extended up to the time $t = \tau_{i+1}$ will satisfy the condition $(\tau_{i+1}, x(\tau_{i+1})) \in \mathcal{E}_\epsilon \subset \mathcal{D}$.

These constructions are performed for i such that $\tau_i < 1 - \epsilon$. Let the inequality $\tau_i \geqslant 1 - \epsilon$ be realized for the first time for $i = k$. On the last interval $[\tau_k,1]$ we set in system (16)

$$u^0(t) = U^0(\tau_k, x(\tau_k)), \quad v(t) = v_\epsilon.$$

Consider the properties of the constructed motion $x(\cdot)$. This motion is just the step-by-step motion $x(\cdot,t^0,x^0,U^0,v(\cdot),\Delta)$ generated by the strategy U^0 for the above indicated choice of partition Δ and control $v(t) = v_\epsilon + v^*(\tau_i)$, $\tau_i \leqslant t < \tau_{i+1}$, $i = 0,1, ..., k$. According to the constructions presented above we have $(\tau_i,x(\tau_i)) \in \mathcal{E}_\epsilon$ $(i = 1,2, ..., k)$. Therefore it follows from (11) that $u_2^0(\tau_i) = -1 + h_2^*(\tau_i,x(\tau_i))$, $|h_2^*(\tau_i,x(\tau_i)| \leqslant \alpha(\epsilon)$ for $i = 1,2, ..., k$. We also note that $v_2(\tau_i) = \sqrt{2} - \epsilon + v_2^*(\tau_i)$, $|v_2^*(\tau_i)| \leqslant \epsilon$. These estimates imply that

$$x_2(\tau_{i+1}) = x_2(\tau_i) + (\tau_{i+1} - \tau_i)(u_2^0(\tau_i) + v_2(\tau_i))$$

$$\geqslant x_2(\tau_i) + (\tau_{i+1} - \tau_i)[\sqrt{2} - 1 - 2\epsilon - \alpha(\epsilon)] \quad (i = 1,2, ..., k).$$

Taking into account that $\tau_{k+1} = 1$ we arrive at

$$x_2(1) \geqslant x_2(\tau_1) + (1 - \tau_1)[\sqrt{2} - 1 - 2\epsilon - \alpha(\epsilon)].$$

Note that $0 < \tau_1 - t^0 \leqslant \epsilon$, $x_2(\tau_1) \geqslant x_2^0 - 2\epsilon\sqrt{2}$. Moreover, it follows from inequality (13) that the quantity appearing here in square brackets is positive. Therefore utilizing notation (15) we arrive at the estimate $x_2(1) > x_2^0 + (1 - t^0)(\sqrt{2} - 1) - \beta(\epsilon)$. Thus the estimate (14) as well as Assertion 1 are proved.

In the example considered here the payoff function σ (9) is non-differentiable on the line $\{x \in R^2\colon x_1 = 0\}$. However the absence of universal optimal strategies has no connection with this fact. Other examples can be constructed in which the payoff function is everywhere differentiable. Nevertheless in these examples as well as in the differential game (7) - (9) it is impossible to define a universal optimal strategy for certain bounded regions $\mathfrak{D}$. The controlled system

$$\dot{x}_1 = u_1 + v_1, \quad \dot{x}_2 = u_2 + x_1 v_2$$

with the constraints

$$|v_1| \leqslant 1, \quad |v_2| \leqslant 1, \quad |u_1| \leqslant 1 - u_2, \quad u_2 \in [-1,0]$$

may serve as such an example. The payoff functional is defined here by the equality

$$\gamma(x(\cdot)) = - x_2(\theta).$$

Hence $\sigma(x) = -x_2$ and although the payoff function is smooth, the value function ω^0 is only piecewise smooth here. It is defined by the relations

$$\omega^0(t,x) = \begin{cases} -x_2 + |x_1|(\theta - t) & \text{for } \theta - 1 \leqslant t \leqslant \theta \\ -x_2 + |x_1| + x_1^2/2 & \text{for } t \leqslant \theta-1,\ |x_1| \leqslant \theta-1-t \\ -x_2 + |x_1|(\theta - t) - (\theta - t - 1)^2/2 & \text{for } t \leqslant \theta-1,\ |x_1| > \theta-1-t\,. \end{cases}$$

In this example one can verify that for certain bounded regions $\mathfrak{D} \subset T \times R^2$ a universal optimal strategy for the first player is nonexistent.

8.5.3. The examples presented above show that within the framework of formalization described in Chapter 1 it is not always possible to define universal optimal strategies. However one can extend the framework of this formalization, introduce new definitions of positional strategies, and prove the existence of the universal optimal strategy in a sufficiently general case. This problem will be discussed somewhat later; first we shall describe the definition of a universal optimal control procedure with guidance.

As above we shall restrict our discussion to the differential game (1), (2). Let ω^0: $T \times R^n \to R^1$ be the value function of this game. Define for the first player a control procedure with guidance (U^0,ψ^0,χ^0). Here the function U^0: $T \times R^n \times R^n \to P$ is chosen from the condition (8.2.4). To define the transition function of the guidance ψ^0 we shall consider the function V_*: $T \times R^n \times R^n \to \mathcal{Q}$ which satisfies the condition (8.2.5). Let a collection $(t_*,t^*,x_*,w_*) \in T_+^2 \times R^n \times R^n$ be given i.e. $t_* < t^* \leqslant \theta$, $x_* \in R^n$, $w_* \in R^n$. Consider the differential inclusion

$$\dot{w}(t) \in \mathrm{co}\{f(t,w(t),u,\ V(t_*,x_*,w_*)) \colon u \in P\}.$$

Among the solutions of this differential inclusion satisfying the initial condition $w(t_*) = w_*$ we select one for which the inequality $\omega^0(t^*,w(t^*)) \leqslant \omega^0(t_*,w_*)$ is valid. Recall that the value function ω^0 is u-stable in the region $T^0 \times R^n$ (cf. Section 4.2), therefore such a solution $w(\cdot)$ does exist. Set $\psi^0(t_*,t^*,x_*,w_*) = w(t^*)$. The third function χ^0 is defined by the equality $\chi^0(t,x) = x$ for all $(t,x) \in T \times R^n$. This equality means that the initial position of the guidance should coincide with either the initial position of the controlled system (1), provided it is measured precisely, or with the point x_0^* which is the result of an imprecise measurement of the initial position x_0.

We shall briefly describe the manner in which the control procedure (U^0,ψ^0,χ^0) operates. Assume for simplicity that the current position of the controlled system (1) is measured precisely. Let (t_0,x_0) be the initial position. Let the first player choose a partition $\Delta = \{t_0 = \tau_0 < \tau_1 < \ldots < \tau_{k+1} = \theta\}$. According to the general description of a control procedure with guidance (cf. Sections 8.2.1 and 8.2.3), the control of the first player will be formed by the rule

$u^0(t) = U^0(\tau_i,x(\tau_i),w(\tau_i))$, $\tau_i \leqslant t < \tau_{i+1}$, $i = 0,1, ..., k$. Here $x(\tau_i)$ and $w(\tau_i)$ are the phase positions of the controlled system and guidance which are realized at discrete times $t = \tau_i$. Recall that the states of the guidance are assigned by means of a transition function according to the rule $w(\tau_{i+1}) = \psi^0(\tau_i,\tau_{i+1},x(\tau_i),w(\tau_i))$. It follows from the definition of a transition function ψ^0 that $\omega^0(\tau_0,w(\tau_0)) \geqslant \omega^0(\tau_1,w(\tau_1)) \geqslant \cdots \geqslant \omega^0(\tau_{i+1},w(\tau_{k+1})) = \omega^0(\theta,w(\theta)) = \sigma(w(\theta))$. Note that $w(\tau_0) = \chi^0(t_0,x_0) = x_0$. Therefore we have

(17) $$\sigma(w(\theta)) \leqslant \omega^0(t_0,x_0).$$

It was shown in Section 8.2.2 that under our choice of functions U^0 and V_* the motions of the guidance and of the controlled system mutually track each other. Therefore

(18) $$\|x(\theta) - w(\theta)\| \leqslant \alpha(\delta), \quad \alpha(\delta) \to 0 \quad \text{as} \quad \delta = \operatorname{diam} \Delta \to 0.$$

Estimates (17), (18) in view of the continuity of function σ imply that

(19) $$\sigma(x(\theta)) \leqslant \omega^0(t_0,x_0) + \beta(\delta), \qquad \beta(\delta) \to 0 \quad \text{as} \quad \delta = \operatorname{diam} \Delta \to 0.$$

This inequality is valid for any step-by-step motion of the system (1) generated by the control procedure (U^0,ψ^0,χ^0).

Similarly, as was done for positional strategies in Section 1.3, one can now approach the limit as diam $\Delta \to 0$ and determine the constructive motions generated by a control procedure with guidance. For the control procedure (U^0,ψ^0,χ^0) defined above we obtain from the inequality (19) the bound $\sigma(x(\theta)) \leqslant \omega^0(t_0,x_0)$ which will be valid for any constructive motion generated by the procedure (U^0,ψ^0,χ^0). This bound implies that our method of control assures an optimal result for the first player. This fact is valid for any choice of the initial position $(t_0,x_0) \in T \times R^n$. Therefore the control procedure (U^0,ψ^0,χ^0) may be called a universal optimal on the whole space of positions $T \times R^n$. Note in conclusion that for any constructive motion $x(\cdot)$: $[t_0,\theta] \to R^n$ generated by the procedure (U^0,ψ^0,χ^0) the inequality

(20) $$\omega^0(t_2,x(t_2)) \leqslant \omega^0(t_1,x(t_1)) \quad \text{for all} \quad t_0 \leqslant t_1 \leqslant t_2 \leqslant \theta$$

will be valid.

8.5.4. It was established above that in the class of ordinary positional strategies $(t,x) \to U(t,x)$ universal optimal strategies may not exist. In a control procedure with guidance the function $(t,x,w) \to U(t,x,w)$ is used to form controls $u(t)$. The addition of a new n-dimensional variable w allows us to construct a control method which would be optimal for the whole space of positions. We shall now describe another class of strategies containing universal optimal strategies.

We introduce a new definition of positional strategy. Consider the functions

$$(21)\qquad (t,x,\epsilon) \to U(t,x,\epsilon)\colon T \times R^n \times (0,1) \to \mathcal{P}.$$

In order not to introduce new terminology, strategies for the first player corresponding to such a function will be referred to as positional strategies, as before. This new type of positional strategy will be considered only in this section. Where necessary we shall specify which type of positional strategy $(t,x) \to U(t,x)$ or $(t,x,\epsilon) \to U(t,x,\epsilon)$ is used.

Define motions generated by a positional strategy U of the form (21). Let (t_0,x_0) be an initial position. Let $\Delta = \{t_0 = \tau_0 < \tau_1 < \cdots < \tau_{k+1} = \theta\}$ be a partition selected by the first player. Denote by $X_1(t_0,x_0,U,\Delta,\epsilon)$ the set of step-by-step motions generated by strategy U (21) for the chosen partition Δ and for a fixed value of the parameter ϵ ($\epsilon \in (0,1)$). This set consists of absolutely- continuous functions $x(\cdot)\colon [t_0,\theta] \to R^n$ which satisfy the initial condition $x(t_0) = x_0$ and the differential inclusion

$$\dot{x}(t) \in \mathrm{co}\{f(t,x(t),u(t),v)\colon v \in \mathcal{Q}\},$$

where

$$u(t) = U(\tau_i,x(\tau_i),\epsilon),\quad \tau_i \leqslant t < \tau_{i+1},\quad i = 0, \ldots, k.$$

The definition of step-by-step motions presented here is virtually the same as the definitions of these motions presented in Section 1.2.1. We now proceed to the definition of constructive motions. For a positional strategy U of the form (21) constructive motions are defined by means of a double limit. First, for a fixed $\epsilon > 0$ we take the limit of a sequence of step-by-step motions as diam $\Delta \to 0$. Then we approach the limit as $\epsilon \to 0$.

Thus let $\epsilon \in (0,1)$. Define first the set

$$(22)\qquad \begin{aligned} X_1(t_0,x_0,U,\epsilon) &\stackrel{\text{def}}{=} \{x(\cdot)\colon\ \exists\, x_i(\cdot) \in X_1(t_0,x_0,U,\Delta_i,\epsilon),\\ &(i = 1,2, \ldots),\ \mathrm{diam}\,\Delta_i \to 0,\ x_i(\cdot) \to x(\cdot)\ \text{as}\ i \to \infty\}. \end{aligned}$$

Note that this definition is analogous to the definition (1.3.4). The sequence $x_i(\cdot)$ ($i = 1,2, \ldots$) must converge to the function $x(\cdot)$ uniformly on the interval $[t_0,\theta]$.

Now we perform the second limit transition and define the set

$$X_1(t_0,x_0,U) \overset{\text{def}}{=} \{x(\cdot): \exists\, x_j(\cdot) \in X_1(t_0,x_0,U,\epsilon_j) \quad (j = 1,2,\ldots),\ \epsilon_j \to +0,\ x_j(\cdot) \to x(\cdot) \text{ as } j \to \infty\} \tag{23}$$

(as before the convergence $x_j(\cdot) \to x(\cdot)$ is uniform on the interval $[t_0,\theta]$).

The set $X_1(t_0,x_0,U)$ (23) is called a set of constructive motions generated by positional strategy U of the form (21).

One can show that for the set $X_1(t_0,x_0,U)$ (23) defined herein, Lemma 1.3.1 remains valid. The second property of the set $X_1(t_0,x_0,U)$ which is analogous to Lemma 1.3.2 is formulated here as follows. For any $\alpha > 0$, initial position $(t_0,x_0) \in T \times R^n$, and strategy $(t,x,\epsilon) \to U(t,x,\epsilon)$ one can choose a number $\epsilon_0 \in (0,1)$ and a function $\epsilon \to \delta_0(\epsilon)$ ($\delta_0(\epsilon) > 0$ for $\epsilon > 0$) such that the inclusion

$$X_1(t_0,x_0,U,\Delta,\epsilon) \subset X_1^{\alpha}(t_0,x_0,U)$$

will be fulfilled for all $\epsilon \in (0,\epsilon_0)$ and for any partition Δ satisfying the estimate $\operatorname{diam} \Delta \leqslant \delta_0(\epsilon)$. Recall that $X_1^{\alpha}(t_0,x_0,U)$ is an α-neighborhood of the set $X_1(t_0,x_0,U)$.

Analogously one can introduce into consideration positional strategies for the second player

$$(t,x,\epsilon) \to V(t,x,\epsilon)\colon T \times R^n \times (0,1) \to \mathcal{Q}. \tag{24}$$

Positional strategies $(t,x) \to U(t,x)$ and $(t,x) \to V(t,x)$ can be viewed as a particular case of positional strategies of the form (21) and (24). Therefore each time a certain differential game defined on the class of strategies of the form $(t,x) \to U(t,x)$ and $(t,x) \to V(t,x)$ possesses an equilibrium situation, so also does the corresponding game defined in the class of strategies of the form (21) and (24).

The difference between the two formalizations of positional strategies manifests itself in solutions of the problem of universal optimal strategy. We have shown that problem (3) may not have a solution, provided it is considered in the class of strategies of the form $(t,x) \to U(t,x)$. On the other hand, it is established below that for any bounded domain $\mathfrak{D}$ one can construct a universal optimal strategy $(t,x,\epsilon) \to U^0(t,x,\epsilon)$.

We shall thus describe the construction of a universal optimal strategy. Let $\mathfrak{D}$ be a compact set in $T \times R^n$. Define in accordance with (8.1.6) and (8.1.7) sets $\mathfrak{D}_0$ and $\mathfrak{D}_1$. The function $(t,x,\epsilon) \to U^0(t,x,\epsilon)$ constructed below will be a universal optimal strategy for the region $\mathfrak{D}_0$. Recall that $\mathfrak{D}_0 \supset \mathfrak{D}$ and the compact sets $\mathfrak{D}_0$ and $\mathfrak{D}_1$ are contained in the strip $[t_{\#},\theta] \times R^n$.

Replacing the sets $\mathfrak{D}_0$ (2.3.9) and $\mathfrak{D}_1$ (2.3.10) by the sets $\mathfrak{D}_0$ (8.1.6) and $\mathfrak{D}_1$ (8.1.7) we define quantities $\alpha(\epsilon)$ and $\beta(t,\epsilon)$ by the equalities (2.3.15) and (2.3.16) in which we substitute $\delta = \epsilon$ and $t_0 = t_{\#}$. We thus obtain

(25)
$$\alpha(\epsilon) \overset{\text{def}}{=} 2[2\Lambda\epsilon + 1][\zeta(\epsilon) + \lambda\Lambda\epsilon] + 4\Lambda^2\epsilon$$
$$\beta(t,\epsilon) \overset{\text{def}}{=} [\exp 2\Lambda(t - t_{\#}) - 1]\, \alpha(\epsilon)\lambda^{-1}.$$

Let $(t,x,\epsilon) \in [t_*,\theta] \times R^n \times (0,1)$. Introduce the following notation:

(26)
$$\mathcal{W}(t,x,\epsilon) \overset{\text{def}}{=} \{w \in R^n\colon \|x - w\|^2 \leqslant \beta(t,\epsilon)\},$$
$$\omega_0(t,x,\epsilon) \overset{\text{def}}{=} \min_w \omega^0(t,w) \quad \text{for} \quad w \in \mathcal{W}(t,x,\epsilon),$$
$$\mathcal{W}_0(t,x,\epsilon) \overset{\text{def}}{=} \{w_0 \in \mathcal{W}(t,x,\epsilon)\colon \omega^0(t,w_0) = \omega_0(t,x,\epsilon)\}.$$

Here $(t,w) \to \omega^0(t,w)$ is the value function of the game under consideration; $\omega_0(t,x,\epsilon)$ is the smallest value which this function takes on for a fixed t and for w in the sphere $\mathcal{W}(t,x,\epsilon)$. $\mathcal{W}_0(t,x,\epsilon)$ is the set of points $w_0 \in \mathcal{W}(t,x,\epsilon)$ at which this minimum is attained. Choose the function

$$(t,x,\epsilon) \to w_0(t,x,\epsilon)\colon [t_{\#},\theta] \times R^n \times (0,1) \to R^n$$

satisfying the condition

$$w_0(t,x,\epsilon) \in \mathcal{W}_0(t,x,\epsilon) \quad \text{for all} \quad (t,x,\epsilon) \in [t_{\#},\theta] \times R^n \times (0,1).$$

Finally completing the construction of the function $(t,x,\epsilon) \to U^0(t,x,\epsilon)$ we shall assume that it is selected from the condition

(27)
$$\max_{v\in\mathcal{Q}} \langle x - w_0(t,x,\epsilon),\, f(t,x,U^0(t,x,\epsilon),v)\rangle$$
$$= \min_{u\in\mathcal{P}} \max_{v\in\mathcal{Q}} \langle x - w_0(t,x,\epsilon),\, f(t,x,u,v)\rangle.$$

The function U^0 is defined by this condition in the region $[t_{\#},\theta] \times R^n \times (0,1)$. Outside this region the function $U^0\colon T \times R^n \times (0,1) \to \mathcal{P}$ can be defined arbitrarily.

Theorem 8.5.1. *The strategy $(t,x,\epsilon) \to U^0(t,x,\epsilon)$ defined by the condition* (27) *is an optimal universal strategy for the region $\mathfrak{D}_0$* (8.1.6).

Proof. Determine a number $\epsilon_0 > 0$ such that $\beta(t,\epsilon) < 1$ for all $t \in [t_{\#},\theta]$ and $\epsilon \in (0,\epsilon_0)$. Let (t_0,x_0) be a point in the set $\mathfrak{D}_0$. Select a number $\epsilon \in (0,\epsilon_0)$ and a partition $\Delta = \{t_0 = \tau_0 < \tau_1 < \cdots < \tau_{k+1} = \theta\}$. Assume that diam $\Delta \leqslant \epsilon$. Consider an arbitrary step-by-step motion $x(\cdot) \in X_1(t_0,x_0,U^0,\Delta,\epsilon)$. Below it will be shown that

(28)
$$\omega_0(\tau_{i+1}, x(\tau_{i+1}),\epsilon) \leqslant \omega_0(\tau_i, x(\tau_i),\epsilon) \quad (i = 0,1, \ldots, k)$$

(recall that the quantity $\omega_0(t,x,\epsilon)$ is defined in accordance with (26)).

From (28) we arrive at the inequality

$$\omega_0(\theta,x(\theta),\epsilon) \leqslant \omega_0(t_0,x_0,\epsilon).$$

It follows from (25) that

$$\beta(t,\epsilon) \leqslant \beta(\theta,\epsilon) \text{ and } \beta(\theta,\epsilon) \to 0 \text{ as } \epsilon \to 0.$$

Therefore taking into account that the value function ω^0 is continuous and satisfies the condition $\omega^0(\theta,x) = \sigma(x)$ we obtain in accordance with (26)

$$\text{(29)} \qquad \omega^0(\theta,x(\theta)) = \sigma(x(\theta)) \leqslant \omega^0(t_0,x_0) + r(\epsilon), \quad r(\epsilon) \to 0 \text{ as } \epsilon \to 0.$$

This estimate is valid for all $x(\cdot) \in \mathfrak{X}_1(t_0,x_0,U^0,\Delta,\epsilon)$, $(t_0,x_0) \in \mathcal{D}_0$, $\epsilon \in (0,\epsilon_0)$, diam $\Delta \leqslant \epsilon$. Approaching the limit for a fixed ϵ as diam $\Delta \to 0$ and then approaching the limit as $\epsilon \to 0$ we arrive at the inequality

$$\sigma(x(\theta)) \leqslant \omega^0(t_0,x_0)$$

which is valid for all $x(\cdot) \in \mathfrak{X}_1(t_0,x_0,U^0)$ and $(t_0,x_0) \in \mathcal{D}$. The inequality obtained means that the strategy U^0 is an optimal universal strategy for the region $\mathcal{D}_0$.

Thus to complete the proof of Theorem 8.5.1 it remains to justify the inequality (28). To simplify the notation we set $\tau_i = t_*$, $\tau_{i+1} = t^*$, $x(\tau_i) = x_*$, $w_0(\tau_i,x(\tau_i),\epsilon) = w_*$, $U^0(\tau_i,x(\tau_i),\epsilon) = u_0$. Note that in view of (26) and the condition $w_0(t,x,\epsilon) \in \mathcal{W}_0(t,x,\epsilon)$ we have

$$\text{(30)} \qquad \|x_* - w_*\|^2 \leqslant \beta(t_*,\epsilon), \quad \omega^0(t_*,w_*) = \omega_0(t_*,x_*,\epsilon).$$

Let the point $v_0 \in \mathcal{Q}$ be selected from the condition

$$\text{(31)} \qquad \begin{aligned} &\min_{u\in P} \langle x_* - w_*, f(t_*,x_*,u,v_0)\rangle \\ &= \max_{v\in \mathcal{Q}} \min_{u\in P} \langle x_* - w_*, f(t_*,x_*,u,v)\rangle. \end{aligned}$$

Consider two differential inclusions (cf. (2.2.1))

$$\dot{x}(t) \in \mathcal{F}_1(t,x(t),u_0),$$

$$\dot{w}(t) \in \mathcal{F}_2(t,w(t),v_0).$$

We shall assume that solutions of these differential inclusions satisfy the initial conditions $x(t_*) = x_*$ and $w(t_*) = w_*$. Among the solutions $w(\cdot)$ we select a solution which satisfies the inequality

$$\text{(32)} \qquad \omega^0(t^*,w(t^*)) \leqslant \omega^0(t_*,w_*).$$

Such a solution exists since the function ω^0 is u-stable in the region $T^0 \times R^n$ (cf. Section 4.2).

Let $x(\cdot)$ be an arbitrary solution of the first differential inclusion. It follows from (27), (31) and also from the conditions $(t_*,x_*) \in \mathcal{D}_0$, $\|x_* - w_*\|^2 \leqslant \beta(t_*,\epsilon)$ and $t^* - t_* \leqslant \operatorname{diam} \Delta \leqslant \epsilon < \epsilon_0$, that one can use Lemma 2.3.1 here, according to which we have

$$(33) \qquad \|x(t^*) - w(t^*)\|^2 \leqslant \beta(t^*,\epsilon).$$

From (26) and (33) we obtain

$$\omega_0(t^*,x(t^*),\epsilon) \leqslant \omega^0(t^*,w(t^*)).$$

Utilizing (30) and (32) we arrive at the inequality

$$\omega^0(t^*,x(t^*),\epsilon) \leqslant \omega^0(t_*,x_*,\epsilon).$$

Thus the estimate (28) as well as Theorem 8.5.1 are proved.

Chapter 9
MIXED STRATEGIES IN DIFFERENTIAL GAMES

9.1. Deterministic and Stochastic Controls

In the differential games considered in the preceding chapters, the players possessed "equal informational abilities" in the sense that the position $(t,x(t))$ realized at the current time t was known to both players. If the information balance is violated, i.e. one of the players obtains information about the control realized by the opponent at the current time, then this additional information does not change the equality of optimal results guaranteed to the first and second players. For example, if the additional information is available to the first player, he or she is still unable to construct a control procedure which would assure him or her of a result which is better (i.e. smaller) than the value of a positional differential game; on the other hand, if the second player finds himself or herself in the situation of information discrimination he or she can, by choosing a positional maximin strategy, still guarantee a result which is no worse (i.e. not smaller) than the value of a positional game. Hence availability or absence of information about the opponent's control (disturbance) realized at the current time does not affect the optimal guaranteed result.

The assertion above is valid in the case when the controlled system satisfies the saddle point condition in a small game (cf. equality (1.2.8)). If, however, this condition is violated then statements and solutions of game-theoretical problems depend to a large extent on whether or not a player possesses additional information about the control chosen by the opponent. In the next chapter we shall consider the cases when one of the players possesses an informational advantage. In this chapter, the case of mutually independent selection of players' controls is investigated, i.e. it is assumed here that each one of the players knows the realized position $(t,x(t))$ but does not know which control was implemented at the current time by the opponent. Such a case suggests the use of mixed positional strategies. Below, we shall present the appropriate motivation. Then in subsequent sections it is shown how the basic results obtained in the preceding chapters are carried over to the case of differential games considered in the class of mixed strategies. In the conclusion of this chapter, a physically realizable stochastic control procedure is defined which approximates the formal solution obtained in the class of mixed strategies.

9.1.1. In this and in the next chapters, we shall consider a controlled system whose motion is described by equation (1.2.1). We shall assume that the function f on the r.h.s. of equation (1.2.1) satisfies conditions (i) - (iii) (cf. Section 1.2.2). We emphasize once again that now the condition of saddle point in a small game is not required. It turns out that in this case the solution of game-theoretical control problems may depend substantially on the information available to each one of the players about the control realized by the opponent. We note the following three cases.

Case I. At the current time t, the first player knows the control $v(t)$ selected by the second player, while the second player does not know the control $u(t)$ chosen by the first.

Case II. Neither of the players knows the control realized by the opponent at the current time.

Case III. At the current time t the second player knows the control $u(t)$, while the first player does not know the control $v(t)$ chosen by the second.

In all the cases stated above, each one of the players knows the realized position of the game $(t,x(t))$. Three types of differential games correspond to the above three cases. The first and the third types, which differ by the interchange of places of the first and second players, will be considered in the next chapter. In this chapter, differential games of the second type are discussed.

If the initial object of investigation is not a differential game, but a minimax problem of the controlling system (1.2.1) in which $v(\cdot)$ is an undetermined disturbance, then the statement and solution of this problem also depend on the nature of information about the realized disturbance and its properties. We shall indicate the three basic cases which correspond to the above Cases I - III.

Case I*. The disturbance $v(t)$ realized at the current time t is known.

Case II*. It is known that the disturbance $v(t)$ is realized independently of the control $u(t)$ applied at time t to the controlled system. The specific value of the disturbance $v(t)$ is not known.

Case III*. The disturbance $v(t)$ which affects the controlled system is unknown. It is assumed that the disturbance $v(t)$ may be correlated in some manner with the control $u(t)$ which is applied at time t to the controlled system.

Precise statements of the control problems and differential games corresponding to the Cases I - III and I* - III* stated herein will be given in Sections 9.2 and 10.1.

Besides these three cases, other variants of distribution of information about the controls selected by the players are possible. We state, for example, the following case. Let $f(t,x,u,v) = \Sigma f^{(i)}(t,x,u^{(i)},v^{(i)})$, $i = 1,2,3$. Let $u(t) = (u^{(1)}(t),u^{(2)}(t),u^{(3)}(t)) \in P^{(1)} \times P^{(2)} \times P^{(3)}$ and $v(t) = (v^{(1)}(t),v^{(2)}(t),v^{(3)}(t)) \in Q^{(1)} \times Q^{(2)} \times Q^{(3)}$ be controls chosen by the first and second players at the current time t. It is assumed that the first and second players have knowledge of the realized position $(t,x(t))$. It is also assumed that the first player knows control $v^{(1)}(t)$ and that he or she uses this information in his or her choice of control $u^{(1)}(t)$. The second player knows control $u^{(3)}(t)$ and he uses this information for the choice of control $v^{(3)}(t)$. Control $u^{(2)}(t)$ is unknown to the second player and the choice of this control does not depend on control $v(t)$ selected by the second player. Control $v^{(2)}(t)$ is unknown to the first player and the choice of this control does not depend on control $u(t)$.

Possessing this information one can determine the equilibrium situation in the corresponding differential game. However, we are not going to study this case specifically. We shall only note that the constructions necessary for this investigation will be introduced in the present and the next chapters when the basic Cases I - III stated above are discussed.

9.1.2. To clarify the difference between the three types of differential games we consider the following simple example. Let the controlled motion be described by the equation

(1) $$\dot{x} = u \cdot v, \quad |u| = 1, \quad |v| = 1.$$

Here x, u, and v are scalars. $P = Q = \{-1\} \cup \{1\}$. Let $t_0 = 0$, $\theta = 1$, and $x_0 = 0$. The payoff functional is defined by the equality

(2) $$\gamma(x(\cdot)) = \sigma(x(1)) = x(1).$$

In the example under consideration the condition of the saddle point in a small game (equality (1.2.8)) is not fulfilled. Indeed, we have

$$\min_u \max_v u \cdot v = 1, \quad \max_v \min_u u \cdot v = -1 \quad \text{for } |u| = 1, \ |v| = 1.$$

We shall consider three types of differential games for the given controlled system and the payoff functional stipulated by (2).

Type I. Let the function $\varphi\colon Q \to P$ be defined by the equality $\varphi(v) = -v$. Suppose that the first player forms his control by the rule

(3) $$u_1^0(t) = \varphi(v(t)) = -v(t)$$

where $v(t)$ is the control selected by the second player. In the case under consideration, the first player knows the control $v(t)$ selected by the opponent at the current time t, thus enabling the above method of control. It follows from (3) that $u_1^0(t) \cdot v(t) = -v^2(t) = -1$ for any admissible control of the second player. We obtain $\dot{x}(t) = u_1^0(t)v(t) = -1$, hence

$$\gamma(x(\cdot)) = x(1) = x_0 - (\theta - t_0) = -1.$$

On the other hand, it is clear that if the second player chooses the constant control $v_1^0 = 1$, then for any admissible control of the first player, the inequalities $\dot{x}(t) = u(t)v_1^0 \geqslant -1$ and $\gamma(x(\cdot)) = x(1) \geqslant -1$ are fulfilled. We thus conclude that $\omega_1^0(t_0,x_0) = -1$, where $\omega_1^0(t_0,x_0)$ is the value of the differential game under consideration.

The second type of differential games will be considered below; meanwhile we shall present the solution in the case when the second player has an informational advantage.

Type III. Define the function $\psi(u) = u$ $(u \in P)$. Assume that the second player forms its control according to the rule

(4) $$v_3^0(t) = \psi(u(t)) = u(t).$$

Such a method of control is admissible since the second player knows the control $u(t)$ selected by the first player at time t. From (4) we obtain that the second player can assure the fulfillment of the equalities

$$\dot{x}(t) = v_3^0(t)u(t) = 1, \quad \gamma(x(\cdot)) = x(1) = 1.$$

On the other hand, if the first player chooses the constant control $u_3^0 = 1$, then for any admissible control of the second player, the inequalities $\dot{x}(t) = u_3^0 v(t) \leqslant 1$, $\gamma(x(\cdot)) \leqslant 1$ are fulfilled. Hence $\omega_3^0(t_0,x_0) = 1$, where $\omega_3^0(t_0,x_0)$ is the value against the differential game (1), (2) in the case of informational discrimination against the first player.

The solutions obtained are degenerate in the sense that in order to realize optimal strategies u_1^0 (3) and v_3^0 (4) one needs to know only the control realized by the opponent at the current time, i.e. the information about the position $(t,x(t))$ is not required. Moreover, any admissible control methods can serve as optimal strategies for the second (first) player in the differential game of the first (third) type. Nevertheless, this example demonstrates sufficiently clearly the value of additional information available to either one of the players. In a differential game of the first (third) type an informational advantage is possessed by the first (second) player, since he knows the opponent's control realized at the current time. This advantage is expressed by the inequality $\omega_1^0(t_0,x_0) = -1 < 1 = \omega_3^0(t_0,x_0)$. Possessing additional information the first player who minimizes the payoff of the game can assure the fulfillment of equality $\gamma(x(\cdot)) = -1$. If, however, additional information is available to the second player who maximizes the payoff, one can guarantee the fulfillment of the equality $\gamma(x(\cdot)) = 1$.

We now turn to differential games of the second type. First, we shall present a formal solution of this game in the class of mixed strategies and then define a stochastic procedure which will approximate this solution.

Type II. Consider the set

$$\hat{P} \stackrel{\text{def}}{=} \{\mu \stackrel{\text{def}}{=} (p_1,p_2)\colon\ p_1 \geqslant 0,\ \ p_2 \geqslant 0,\ \ p_1 + p_2 = 1\}.$$

The elements of this set are called mixed controls for the first player. Let $\mu = (p_1,p_2) \in \hat{P}$, the numbers p_1 and p_2 can be interpreted as probabilities with which the first player selects "the pure" controls $u_1 = +1$ and $u_2 = -1$. Analogously we introduce the set

$$\hat{Q} \stackrel{\text{def}}{=} \{\nu \stackrel{\text{def}}{=} (q_1,q_2) :\ q_1 \geqslant 0,\ \ q_2 \geqslant 0,\ \ q_1 + q_2 = 1\},$$

whose elements will be called mixed controls for the second player. The numbers q_1 and q_2 can also be interpreted as probabilities with which the second player mixes the "pure" controls $v_1 = 1$ and $v_2 = -1$.

If the first and second players have chosen the mixed controls $\mu = (p_1,p_2)$ and $\nu = (q_1,q_2)$, then the mathematical expectation of the product uv is

$$(5) \qquad (p_1u_1 + p_2u_2)(q_1v_1 + q_2v_2) = (p_1 - p_2)(q_1 - q_2).$$

In general, the first and second players may choose the mixed controls depending on the realized position $(t,x(t))$. In particular they can select programmed mixed controls: $t \to \mu(t) = (p_1(t),p_2(t))$ and $t \to \nu(t) = (q_1(t),q_2(t))$. In this case, according to (5), for each $t \in [0,1]$ the mathematical expectation of the product uv is defined. Formally we shall assume that the derivative $\dot{x}(t)$ is equal to this mathematical expectation, i.e. it is assumed that the motion of the controlled system generated by the mixed controls $t \to \mu(t)$ and $t \to \nu(t)$ is defined by the ordinary differential equation

$$\dot{x}(t) = (p_1(t) - p_2(t))(q_1(t) - q_2(t)). \tag{6}$$

Now consider a differential game in which the motion of the controlled system is described by equation (6) and the payoff functional is determined by equality (2). In equation (6) the controls of the first and second players satisfy the constraints $\mu(t) = (p_1(t),p_2(t)) \in \tilde{P}$, $\nu(t) = (q_1(t),q_2(t)) \in \tilde{Q}$. Note that system (6) satisfies the saddle point condition in a small game, i.e.

$$\begin{aligned} &\max_{(q_1,q_2)\in \tilde{Q}} \ \min_{(p_1,p_2)\in \tilde{P}} s(p_1 - p_2)(q_1 - q_2) \\ &= \min_{(p_1,p_2)\in \tilde{P}} \ \max_{(q_1,q_2)\in \tilde{Q}} s(p_1 - p_2)(q_1 - q_2) \quad (s \in R^1). \end{aligned} \tag{7}$$

It is easy to verify that the differential game (6), (2) possesses the value $\omega_2^0 = 0$. Indeed, by selecting $p_1 = p_2 = 1/2$ the first player assures that the equalities $\dot{x}(t) = 0$ and $\gamma(x(\cdot)) = x(1) = 0$ are fulfilled for any control of the second player. In the same manner, choosing $q_1 = q_2 = 1/2$, the second player can assure the equality $\gamma(x(\cdot)) = 0$ for any control of the first player. We thus obtain that $\omega_2^0(t_0,x_0) = 0$. It also follows that in the differential game (6), (2) the pair of controls $\mu^0 = (1/2, 1/2)$ and $\nu^0 = (1/2, 1/2)$ forms a saddle point.

We now return to the controlled system (1). Consider the following procedure of controlling this system. Let the first player choose the mixed control $\mu^0 = (1/2, 1/2)$ and the second select an arbitrary mixed control $\nu(t) = (q_1(t),q_2(t))$. Note that the pure control $v(t)$ may be considered as a particular case of the mixed control $\nu(t) = (q_1(t),q_2(t))$ which takes on the two values (1,0) or (0,1). Let a partition $\Delta = \{0 = \tau_0 < \tau_1 < \cdots < \tau_{k+1} = 1\}$ be chosen. At each time $t = \tau_i$ the players perform trials of choosing random controls $u(t) = u(\tau_i)$ and $v(t) = v(\tau_i)$ $(\tau_i \leqslant t < \tau_{i+1})$. The random control $u(\tau_i)$ takes on values +1 and −1 with probabilities $p_1^0 = 1/2$ and $p_2^0 = 1/2$, while the control $v(\tau_i)$ takes on values +1 and −1 with probabilities $q_1(\tau_i)$ and $q_2(\tau_i)$ respectively. It is important to note that the random choices of controls $u(\tau_i)$ and $v(\tau_i)$ are mutually independent. According to the procedure indicated above we obtain a set of motions of the system (1) on which a probability distribution can be defined in a natural way.

We thus obtain that for any $\epsilon > 0$ a $\delta > 0$ can be found such that, for all partitions Δ satisfying diam $\Delta \leqslant \delta$ and for any random or deterministic control of the second player, this stochastic procedure of controlling the first player assures that the inequality $\gamma(x(\cdot)) \leqslant \epsilon$ will be fulfilled with probability at least $1 - \epsilon$. Analogously, one can define a stochastic control procedure for the second player and prove that it assures that the estimate $\gamma(x(\cdot)) \geqslant -\epsilon$ will be fulfilled with probability as close to one as possible.

Thus stochastic control procedures approximate the optimal result which was formally defined for the differential game (6), (2). In

Sections 9.4 and 9.5 the approximating property of stochastic control procedures will be proved for the general case of a nonlinear differential game. We note that this property is a manifestation of the well known law of large numbers in probability theory. We emphasize again that the utilization of mixed strategies and their approximation are possible only when each one of the players chooses its control independently of the control implemented at the given time by his/her opponent.

9.1.3. Mixed strategies are usually used in game theory in those cases when the game defined in the class of pure strategies does not possess an equilibrium situation. In this case probability distributions are introduced on the set of pure strategies and the original payoff is replaced by the corresponding mathematical expectation. Under certain conditions one can prove that there exists a pair of probability distributions which forms a saddle point. In the example presented above as well as in the subsequent exposition, the term "mixed strategy" is used in another sense. In the constructions suggested here we do *not* replace the payoff by some averaging index such as the mathematical expectation. In a formal analysis of a differential game in the class of mixed strategies we are actually dealing with a deterministic controlled system (in the example presented above this system is described by equation (6)) and with a deterministic payoff. Probabilistic concepts are necessary for approximating mixed strategies by stochastic control procedures. However, here also one can avoid averaged bounds and obtain that the inequality $\gamma(x(\cdot)) \leqslant \omega_2^0(t_0,x_0) + \epsilon$ (in a problem posed for the first player) or the inequality $\gamma(x(\cdot)) \geqslant \omega_2^0(t_0,x_0) - \epsilon$ (in a problem for the second player) will be fulfilled with probability as close to 1 as possible.

In the formulation of the approximating property it is assumed that when the stochastic procedure which approximates the optimal mixed strategy for the first player is realized, motions may come about in which the payoff $\gamma(x(\cdot))$ is substantially larger than the quantity $\omega_2^0(t_0,x_0)$. In reality the probability measure of the sets of this kind of motions will be small. Here the following question arises. The first player may select two methods of control. Firstly he or she may solve the differential game of the third type and define a deterministic control method which will, in a step-by-step realization, assure the fulfillment of the inequality $\gamma(x(\cdot)) \leqslant \omega_3^0(t_0,x_0) + \epsilon$ for all motions $x(\cdot)$. Secondly, he or she may select a stochastic control procedure which will assure the fulfillment of the inequality $\gamma(x(\cdot)) \leqslant \omega_2^0(t_0,x_0) + \epsilon$ with probability close to one. The quantities $\omega_2^0(t_0,x_0)$ and $\omega_3^0(t_0,x_0)$ are always related by the inequality $\omega_2^0(t_0,x_0) \leqslant \omega_3^0(t_0,x_0)$. If $\omega_2^0(t_0,x_0) = \omega_3^0(t_0,x_0)$ then it would seem more reasonable to use a deterministic method of control. If, however, $\omega_2^0(t_0,x_0) < \omega_3^0(t_0,x_0)$ then one is confronted with the choice between a deterministic control method, which guarantees the result $\gamma(x(\cdot)) \leqslant \omega_3^0(t_0,x_0) + \epsilon$, and a stochastic control procedure which would assure the inequality $\gamma(x(\cdot)) \leqslant \omega_2^0(t_0,x_0) + \epsilon$ with probability close to 1, but at the same time the inequality $\gamma(x(\cdot)) \leqslant \omega_3^0(t_0,x_0) + \epsilon$ may be violated with a small probability. Obviously in certain problems one would utilize deterministic strategies. In other problems it turns out to be preferable to use mixed strategies and stochastic control procedures.

9.2. Mixed Strategies

We now turn to a formalization of a differential game considered in the class of mixed strategies. In this section, statements of game-theoretical problems of approach and evasion are presented for the ideal case, i.e. we assume the players mix their controlling actions instantaneously. Next, a theorem on alternatives is presented which characterizes the differential game under consideration. The content of the formal constructions presented herein will be clarified in Sections 9.4 and 9.5 where stochastic approximating control procedures are considered. We note that the definitions and results presented below are in essence similar to those given in Chapter 2.

9.2.1. We shall introduce some notation and definitions. As usual, let $\mathcal{P}$ and $\mathcal{Q}$ be compact sets in R^p and R^q respectively. The symbols rpm($\mathcal{P}$) and rpm($\mathcal{Q}$) denote the sets of regular probability measures standardized on the compact sets $\mathcal{P}$ and $\mathcal{Q}$ respectively. The elements of the sets rpm($\mathcal{P}$) and rpm($\mathcal{Q}$) will be denoted by the letters μ and ν. Note that in Chapter 6 the letter ν was used to denote the function $\nu: T \to$ rpm($\mathcal{Q}$); in this chapter ν is simply a member of the set rpm($\mathcal{Q}$). The function mentioned above will be denoted by $\nu(\cdot)$.

Let

$$(t,x,\mu,\nu) \in T \times R^n \times \mathrm{rpm}(\mathcal{P}) \times \mathrm{rpm}(\mathcal{Q}).$$

Set

$$\breve{f}(t,x,\mu,\nu) \stackrel{\text{def}}{=} \int_{\mathcal{P}}\int_{\mathcal{Q}} f(t,x,u,v)\mu(du)\nu(dv),$$

$$(1) \qquad \breve{\mathcal{F}}_1(t,x,\mu) \stackrel{\text{def}}{=} \mathrm{co}\left\{\int_{\mathcal{P}} f(t,x,u,v)\mu(du): \ v \in \mathcal{Q}\right\},$$

$$\breve{\mathcal{F}}_2(t,x,\nu) \stackrel{\text{def}}{=} \mathrm{co}\left\{\int_{\mathcal{Q}} f(t,x,u,v)\nu(dv): u \in \mathcal{P}\right\}.$$

It is easy to verify that

$$(2) \qquad \begin{aligned} \breve{\mathcal{F}}_1(t,x,\mu) &= \{\breve{f}(t,x,\mu,\nu): \nu \in \mathrm{rpm}(\mathcal{Q})\}, \\ \breve{\mathcal{F}}_2(t,x,\nu) &= \{\breve{f}(t,x,\mu,\nu): \mu \in \mathrm{rpm}(\mathcal{P})\}. \end{aligned}$$

Recall that the function f satisfies the conditions (i) - (iii) stated in Section 1.2.2. Therefore, for any $(t,x,\mu,\nu) \in T \times R^n \times \text{rpm}(\mathcal{P}) \times \text{rpm}(\mathcal{Q})$, the sets $\mathcal{F}_1(t,x,\mu)$ and $\mathcal{F}_2(t,x,\nu)$ will be convex compact sets in R^n. Multivalued mappings $(t,x) \to \mathcal{F}_1(t,x,\mu)$ and $(t,x) \to \mathcal{F}_2(t,x,\nu)$ are continuous. Moreover, the bound $\langle x,h\rangle \leq \xi(1 + \|x\|^2)$ is valid for all $(t,x) \in T \times R^n$, $h \in \mathcal{F}_1(t,x,\mu)$, or $h \in \mathcal{F}_2(t,x,\nu)$, $\nu \in \text{rpm}(\mathcal{Q})$, $\mu \in \text{rpm}(\mathcal{P})$.

Consider the differential inclusions

$$\dot{x}(t) \in \breve{\mathcal{F}}_1(t,x(t),\mu), \tag{3}$$

$$\dot{x}(t) \in \breve{\mathcal{F}}_2(t,x(t),\nu). \tag{4}$$

The symbol $\breve{X}_1(t_*,x_*,\mu)$ will denote the set of absolutely continuous functions $x(\cdot)$: $[t_*,\theta] \to R^n$ which satisfy the differential inclusion (3). The set $\breve{X}_2(t_*,x_*,\nu)$ of solutions of the differential inclusion (4) is defined analogously. It follows from the properties of multivalued mappings $(t,x) \to \mathcal{F}_1(t,x,\mu)$ and $(t,x) \to \mathcal{F}_2(t,x,\nu)$ that $X_1(t_*,x_*,\mu)$ and $X_2(t_*,x_*,\nu)$ are nonempty compact sets in $C_n([t_*,\theta])$ (cf. Section A.4).

Let $\mathcal{E}_1$ and $\mathcal{E}_2$ be sets of all measurable functions $\mu(\cdot)$: $T \to \text{rpm}(\mathcal{P})$ and $\nu(\cdot)$: $T \to \text{rpm}(\mathcal{Q})$ respectively. Recall that the notion of a measurable function $\nu(\cdot)$ was presented in Section 6.1.1. The totality of these functions was denoted in Chapter 6 by the symbol N. One can define measurable functions $\mu(\cdot)$: $T \to \text{rpm}(\mathcal{P})$ in the same manner that measurable mappings $T \to \text{rpm}(\mathcal{Q})$ were defined in Section 6.1.1.

One can show that for any pair $(\mu(\cdot),\nu(\cdot)) \in \mathcal{E}_1 \times \mathcal{E}_2$ the function $(t,x) \to f(t,x,\mu(t),\nu(t))$ satisfies the following conditions: (1) for any fixed t this function satisfies the Lipschitz condition with respect to x; (2) for any fixed $x \in R^n$ the function $t \to f(t,x,\mu(t),\nu(t))$ is measurable; (3) the bound $\langle x,f(t,x),\mu(t),\nu(t))\rangle \leq \xi(1 + \|x\|^2)$ $((t,x) \in T \times R^n)$ is valid. Therefore, in view of Caratheodory's theorem (cf. [13, 34, 74]), the differential equation

$$\dot{x}(t) = \breve{f}(t,x(t),\mu(t),\nu(t)) \tag{5}$$

possesses a unique solution $x(\cdot) = x(\cdot,t_*,x_*,\mu(\cdot),\nu(\cdot))$ satisfying the initial condition $x(t_*) = x_*$. If $\mu(t) = \mu$ or $\nu(t) = \nu$ for all $t \in T$ then the corresponding solution of equation (5) is denoted by $x(\cdot,t_*,x_*,\mu,\nu(\cdot))$ or $x(\cdot,t_*,x_*,\mu(\cdot),\nu)$. One can verify that the sets $X_1(t_*,x_*,\mu)$ and $X_2(t_*,x_*,\nu)$ introduced above can also be defined in the following manner:

$$\breve{X}_1(t_*,x_*,\mu) = \{x(\cdot,t_*,x_*,\mu,\nu(\cdot)): \nu(\cdot) \in \mathcal{E}_2\},$$

$$\breve{X}_2(t_*,x_*,\nu) = \{x(\cdot,t_*,x_*,\mu(\cdot),\nu): \mu(\cdot) \in \mathcal{E}_1\}.$$

We shall omit proofs of these relations since in the subsequent arguments it is sufficient to suppose that the sets X_1 and X_2 are defined as totalities of solutions of differential inclusions (3) and (4).

9.2.2. We shall define *mixed positional strategies* and the motions generated by them. Let the function $\tilde{U}$: $T \times R^n \to \text{rpm}(P)$ and the initial position $(t_0,x_0) \in T \times R^n$ be given. Select a partition $\Delta = \{t_0 = \tau_0 < \tau_1 < \cdots < \tau_{k+1} = \theta\}$. Consider the differential inclusion

$$(6) \qquad \dot{x}(t) \in \tilde{\mathcal{F}}_1(t,x(t),\mu(t)),$$

where

$$(7) \qquad \mu(t) = \mu(\tau_i) = \tilde{U}(\tau_i,x(\tau_i)), \quad \tau_i \leqslant t < \tau_{i+1}, \quad i = 0,1, \ldots, k.$$

The set of absolutely-continuous solutions $x(\cdot)$: $[t_0,\theta] \to R^n$ of this differential inclusion satisfying the initial condition $x(t_0) = x_0$ will be denoted by $X_1(t_0,x_0,\tilde{U},\Delta)$. The elements of the set $X_1(t_0,x_0,\tilde{U},\Delta)$ will be called *step-by-step motions*.

We shall introduce another definition of step-by-step motions. Consider the differential equation (5) where $\nu(\cdot) \in \mathcal{E}_2$ and the function $\mu(\cdot)$ is formed according to the rule (7). We shall denote by $x(\cdot,t_0,x_0,\tilde{U},\nu(\cdot),\Delta)$ an absolutely continuous function $x(\cdot)$: $[t_0,\theta] \to R^n$ which satisfies this equation and the initial condition $x(t_0) = x_0$. Such a function exists and is unique. The totality of step-by-step motions can be defined by the relation

$$X_1(t_0,x_0,\tilde{U},\Delta) \overset{\text{def}}{=} \{x(\cdot,t_0,x_0,\tilde{U},\nu(\cdot),\Delta): \ \nu(\cdot) \in \mathcal{E}_2\}.$$

Such a definition is equivalent to the definition of step-by-step motions as solutions of differential inclusion (6) presented above.

We now approach the limit as diam $\Delta \to 0$. In analogy with (1.3.4), we define the set

$$(8) \qquad \begin{aligned} X_1(t_0,x_0,\tilde{U}) &\overset{\text{def}}{=} \{x(\cdot): \ \exists x_\ell(\cdot) \in X_1(t_0,x_0,\tilde{U},\Delta_\ell) \quad (\ell = 1,2, \ldots), \\ &x_\ell(\cdot) \to x(\cdot), \ \text{diam}\ \Delta_\ell \to 0 \ \text{as} \ \ell \to \infty\}. \end{aligned}$$

Definition. *A function* $\tilde{U}$: $T \times R^n \to \text{rpm}(P)$ *is called a mixed positional strategy for the first player. A set* $X_1(t_0,x_0,\tilde{U})$ *defined by relation (8) is called a set of constructive motions generated by the strategy* $\tilde{U}$ *and emanating from the initial position* (t_0,x_0).

It should be noted that no probability distribution was defined on the sets $X_1(t_0,x_0,\tilde{U},\Delta)$ and $X_1(t_0,x_0,\tilde{U})$. In spite of the use of probability measures in constructions presented above, step-by-step and constructive motions are viewed as ordinary *deterministic time functions* which do not depend on any random event. Later, for approximating mixed strategies, we shall introduce stochastic motions for which a probability distribution will be defined (cf. Section 9.4).

Lemma 1.3.1 remains valid for the set of constructive motions $X_1(t_0,x_0,\widetilde{U})$. Note that the sets of constructive and step-by-step motions defined herein also possess the property stated in Lemma 1.3.2. For a pure positional strategy U: $T \times R^n \to P$ the corresponding step-by-step control procedure is physically realizable, therefore, in the preceding chapters Lemma 1.3.2 was used for a meaningful interpretation of results obtained for constructive motions. However, when a pure strategy is replaced by a mixed strategy $\widetilde{U}$: $T \times R^n \to \mathrm{rpm}(P)$, Lemma 1.3.2 cannot be used directly for this purpose. This is because the step-by-step procedure defined in accordance with (6), (7) or (5), (7) cannot be physically implemented. It is based on a mathematical idealization which admits instantaneous mixing of controls. Below in Section 9.4 a physically realized approximation of mixed strategies will be defined in which the mixing of controls is spanned over time.

Mixed positional strategies for the second player and the motions generated by them are defined analogously. We shall omit these definitions to avoid repetition.

For any pair $(\widetilde{U},\widetilde{V})$ consisting of mixed strategies for the first and second players, and for any initial position $(t_0,x_0) \in T \times R^n$, one can define a set of constructive motions $X(t_0,x_0,\widetilde{U},\widetilde{V})$. This is done in the same manner as in the case of pure positional strategies (cf. Section 1.5.2). Let

$$\Delta^{(1)} = \{t_0 = \tau_0^{(1)} < \tau_1^{(1)} < \cdots < \tau_{k+1}^{(1)} = \theta\},$$

$$\Delta^{(2)} = \{t_0 = \tau_0^{(2)} < \tau_1^{(2)} < \cdots < \tau_{\ell+1}^{(2)} = \theta\}$$

be partitions selected by the first and second players respectively. Denote by $x(\cdot,t_0,x_0,\widetilde{U},\widetilde{V},\Delta^{(1)},\Delta^{(2)})$ a step-by-step motion which is defined as a solution of differential equation (5) satisfying the initial condition $x(t_0) = x_0$, where

$$\mu(t) = \widetilde{U}(\tau_i^{(1)},x(\tau_i^{(1)})), \quad \tau_i^{(1)} \leqslant t < \tau_{i+1}^{(1)}, \quad i = 0,1, \ldots, k,$$

$$\nu(t) = \widetilde{V}(\tau_j^{(2)},x(\tau_j^{(2)})), \quad \tau_j^{(2)} \leqslant t < \tau_{j+1}^{(2)}, \quad j = 0,1, \ldots, \ell.$$

Next we define the set

$$X(t_0,x_0,\widetilde{U},\widetilde{V}) \overset{\text{def}}{=} \{x(\cdot): \exists\, \Delta_m^{(1)}, \ \exists\, \Delta_m^{(2)} \quad (m = 1,2, \ldots),$$

$$x(\cdot,t_0,x_0,\widetilde{U},\widetilde{V},\Delta_m^{(1)},\Delta_m^{(2)}) \to x(\cdot), \ \operatorname{diam} \Delta_m^{(1)} \to 0,$$

$$\operatorname{diam} \Delta_m^{(2)} \to 0 \ \text{ for } \ m \to \infty\}.$$

One can verify that $X(t_0,x_0,\widetilde{U},\widetilde{V})$ is a nonempty compact set on $C_n([t_0,\theta])$ and

$$X(t_0,x_0,\widetilde{U},\widetilde{V}) \subset X_1(t_0,x_0,\widetilde{U}) \cap X_2(t_0,x_0,\widetilde{V}).$$

Using the definitions introduced herein one can consider a differential game in the class of mixed positional strategies. Corresponding notions of the value, saddle point, and equilibrium situation are introduced formally in the same manner as in Section 1.5.

9.2.3. Consider a differential game of approach-evasion in the class of mixed positional strategies. We present the statements of approach and evasion problems.

Problem 9.2.1. *For a given initial position* (t_0,x_0) *it is required to obtain a mixed positional strategy* $\tilde{U}^0$ *such that for any motion* $x(\cdot) \in X_1(t_0,x_0,\tilde{U}^0)$ *the condition of an* (M, N)*-approach holds.*

Problem 9.2.2. *Let the initial position* (t_0,x_0) *and a number* $\epsilon > 0$ *be given. It is required to determine a mixed positional strategy* $\tilde{V}^\epsilon$ *such that for any motion* $x(\cdot) \in X_2(t_0,x_0,\tilde{V}^\epsilon)$ *the condition of an* $(M, N)^\epsilon$*-evasion holds.*

Recall that the conditions of (M, N)-approach and $(M, N)^\epsilon$-evasion are defined by relations (2.1.3) and (2.1.8) respectively. Statements of Problems 9.2.1 and 9.2.2 differ from the formulation of Problems 2.1.1 and 2.1.2 by replacing positional (pure) strategies $U: T \times R^n \to P$ and $V: T \times R^n \to Q$ with mixed positional strategies $\tilde{U}: T \times R^n \to \text{rpm}(P)$ and $\tilde{V}: T \times R^n \to \text{rpm}(Q)$.

The following theorem is valid.

Theorem 9.2.1. *For any closed sets* M *and* N $(M \subset N \subset T \times R^n)$ *and for any initial position* $(t_0,x_0) \in T \times R^n$, *one and only one of the following two assertions is valid*: (1) *there exists a mixed positional strategy* $\tilde{U}^0$ *which yields a solution to the problem of an* (M, N)*-approach*; (2) *there exists an* $\epsilon > 0$ *and a mixed positional strategy* $\tilde{V}^\epsilon$ *which yields a solution of an* $(M, N)^\epsilon$*-evasion problem.*

A proof of this theorem may be carried out along the lines of the proof of Theorem 2.1.1. Therefore, below we state only basic assertions and indicate the difference in the proofs of these assertions and the corresponding statements in Chapter 2. First we shall describe the essence of the extremal construction utilized in the proof of Theorem 9.2.1.

Definition. *A set* $W \subset T \times R^n$ *will be called a* μ*-stable bridge in Problem* 9.2.1 *if it satisfies the following conditions*:

(1_μ) $W \subset N$, $W(\theta) \subset M(\theta)$,

(2_μ) *for any* $(t_*,x_*) \in W$, $t^* \in (t_*,\theta]$ *and* $\nu \in \text{rpm}(Q)$ *there exists a motion* $x(\cdot) \in X_2(t_*,x_*,\nu)$ *such that either* $(t^*,x(t^*)) \in W$ *or* $\{(t,x(t)): t_* \leq t \leq t^*\} \cap M \neq \phi$.

Recall that $\mathcal{W}(\theta) = \{(\theta,x): (\theta,x) \in \mathcal{W}\}$. Condition (2_μ) is called the condition of μ-stability of the set $\mathcal{W}$. Note that this condition can be stated in the following equivalent form: for any $(t_*,x_*) \in \mathcal{W}$, $t^* \in (t_*,\theta]$ and $\nu \in \text{rpm}(\mathcal{Q})$ there exists a control $\mu(\cdot) \in \mathcal{E}_1$ such that for the corresponding solution $x(\cdot) = x(\cdot,t_*,x_*,\mu(\cdot),\nu)$ of equation (5) either the condition $(t^*,x(t^*)) \in \mathcal{W}$ or the condition $(\tau,x(\tau)) \in \mathcal{M}$ for some $\tau \in [t_*,t^*]$ will be fulfilled.

It was noted in Section 2.1.1 that the condition of an $(\mathcal{M},\mathcal{N})^\epsilon$-evasion is equivalent to the condition of a $(K_\epsilon,\mathfrak{L}_\epsilon)$-approach. For uniformity it is more convenient to define a ν-stable bridge not for an evasion problem but for the corresponding approach problem. Thus we shall introduce the following

Definition. *Let* $K \subset \mathfrak{L} \subset T \times R^n$. *A set* $\mathcal{W} \subset T \times R^n$ *is called a* ν-*stable bridge in a problem of a* $(K,\mathfrak{L})$-*approach if it satisfies the following two conditions*:

(1_ν) $\quad \mathcal{W} \subset \mathfrak{L}, \quad \mathcal{W}(\theta) \subset K,$

(2_ν) *for any* $(t_*,x_*) \in \mathcal{W}$, $t^* \in (t_*,\theta]$, $\mu \in \text{rpm}(\mathcal{P})$ *there exists a motion* $x(\cdot) \in \mathcal{X}_1(t_*,x_*,\mu)$ *such that either* $(t^*,x(t^*)) \in \mathcal{W}$ *or* $\{(t,x(t)): t_* \leqslant t \leqslant t^*\} \cap K \neq \phi$.

Condition (2_ν) is called the condition of ν-stability of the set $\mathcal{W}$.

Below some basic assertions are presented related to μ-stable and ν-stable bridges. After replacing the symbols u,v, $\mathcal{P}$, and $\mathcal{Q}$ by the symbols μ, ν, $\text{rpm}(\mathcal{P})$, and $\text{rpm}(\mathcal{Q})$ respectively, proofs of these assertions will in toto coincide with the proofs of analogous assertions established in Section 2.2.

Let $\mathcal{W}_0^1$ be a μ-stable bridge in a problem of an $(\mathcal{M},\mathcal{N})$-approach and any μ-stable bridge in a problem of an $(\mathcal{M},\mathcal{N})$-approach be imbedded into the set $\mathcal{W}_0^1$. It was shown in Section 2.2 that such a set is closed and can always be defined. The set $\mathcal{W}_0^1$ is called the *maximal* μ-*stable bridge in a problem of* an $(\mathcal{M},\mathcal{N})$-*approach*. Let the sets K_ϵ and $\mathfrak{L}_\epsilon$ be defined by the equalities (2.1.9). We consider the maximal ν-stable bridge $\mathcal{W}_\epsilon^2$ in a problem of a $(K_\epsilon,\mathfrak{L}_\epsilon)$-approach. This set is also closed. The equality

$$T \times R^n = \mathcal{W}_0^1 \cup \left[\bigcup_{\epsilon>0} \mathcal{W}_\epsilon^2\right] \tag{9}$$

is valid (cf. Theorem 2.2.1).

We now construct a mixed strategy $\tilde{U}^e$ extremal to the closed set $\mathcal{W}$. Let the multivalued mapping $(t,x) \to \mathcal{W}[t,x]$ and a function $(t,x) \to e(t,x)$: $T \times R^n \to R^n$ be defined by the relations (2.4.2) and (2.4.4). Consider the function U^e: $T \times R^n \to \text{rpm}(\mathcal{P})$ selected from the condition

$$\max_{\nu} \langle x - e(t,x),\ \tilde{f}(t,x,\mu^e,\nu)\rangle = \min_{\mu} \max_{\nu} \langle x - e(t,x),\ \tilde{f}(t,x,\mu,\nu)\rangle \tag{10}$$

where $\mu^e = \tilde{U}^e(t,x)$, $\nu \in \mathrm{rpm}(\mathcal{Q})$, and $\mu \in \mathrm{rpm}(\mathcal{P})$. Note that the existence of maxima and minima in (10) is a consequence of the weak compactness of the sets $\mathrm{rpm}(\mathcal{P})$ and $\mathrm{rpm}(\mathcal{Q})$ (cf. [18, 55, 74]).

Definition. *A strategy* $\tilde{U}^e$: $T \times R^n \to \mathrm{rpm}(\mathcal{P})$ *defined by relation* (10) *will be called a mixed strategy for the first player extremal to the set* $\mathcal{W}$.

The following assertion is valid.

Lemma 9.2.1. *Let* $\mathcal{W}$ *be a closed* μ*-stable bridge in a problem of an* $(\mathcal{M}, \mathcal{N})$*-approach and let* $(t_0,x_0) \in \mathcal{W}$. *Then the mixed strategy* $\tilde{U}^e$ *extremal to this bridge yields a solution to Problem* 9.2.1, i.e. *for any constructive motion* $x(\cdot) \in X_1(t_0,x_0,\tilde{U})$ *the condition of an* $(\mathcal{M}, \mathcal{N})$*-approach is valid.*

This assertion is an analog of Theorem 2.4.1. Recall that the proof of Theorem 2.4.1 is based on the estimate obtained in Section 2.3. In this case the validity of Lemma 9.2.1 is derived by exactly the same arguments from an analogous estimate.

It is important to note the following. In the proof of Lemma 2.3.1 condition (1.2.8) was used. Now this condition is replaced by the equality

$$\min_{\mu} \max_{\nu} \langle s,\tilde{f}(t,x,\mu,\nu)\rangle = \max_{\nu} \min_{\mu} \langle s,\tilde{f}(t,x,\mu,\nu)\rangle \tag{11}$$
$$(s \in R^n,\ \mu \in \mathrm{rpm}(\mathcal{P}),\ \nu \in \mathrm{rpm}(\mathcal{Q})).$$

This equality is valid for any continuous function $(u,v) \to f(t,x,u,v)$: $\mathcal{P} \times \mathcal{Q} \to R^n$. A proof of this equality is presented, for example, in the books [34, 55].

We state the second lemma which is required for the proof of Theorem 9.2.1.

Lemma 9.2.2. *Let closed sets* $\mathcal{K}_\epsilon$ *and* $\mathcal{L}_\epsilon$ *be defined by relations* (2.1.9). *Let* $\mathcal{W}$ *be a closed* ν*-stable bridge in a* $(\mathcal{K}_\epsilon,\mathcal{L}_\epsilon)$*-approach problem and let* $(t_0,x_0) \in \mathcal{W}$. *Then a mixed strategy* $\tilde{V}^e$ *extremal to the set yields a solution of the Problem* 9.2.2, *i.e. for any constructive motion* $x(\cdot) \in X_2(t_0,x_0,\tilde{V}^e)$ *the condition of an* $(\mathcal{M}, \mathcal{N})^\epsilon$*-evasion or equivalently the condition of a* $(\mathcal{K}_\epsilon,\mathcal{L}_\epsilon)$*-approach is valid.*

Note that the mixed strategy $\tilde{V}^e$: $T \times R^n \to \mathrm{rpm}(\mathcal{Q})$ extremal to a closed set $\mathcal{W}$ is defined by the condition

$$(12)\qquad \min_{\mu} \langle e(t,x) - x, \tilde{f}(t,x,\mu,\nu^e)\rangle = \max_{\nu} \min_{\mu} \langle e(t,x) - x, \tilde{f}(t,x,\mu,\nu)\rangle$$

where

$$\nu^e = \tilde{V}^e(t,x), \quad \mu \in \mathrm{rpm}(P), \quad \nu \in \mathrm{rpm}(Q).$$

Here, as in condition (10), the function e: $T \times R^n \to R^n$ is defined according to (2.4.4).

Utilizing Lemma 9.2.1 or 9.2.2 as well as equality (9), it is easy to complete the proof of Theorem 9.2.1. According to (9) the initial position (t_0,x_0) is either contained in the set W_0^1 or there exists an $\epsilon > 0$ such that $(t_0,x_0) \in W_\epsilon^2$. Since W_0^1 is a μ-stable bridge in an (M,N)-approach problem, when $(t_0,x_0) \in W_0^1$ it follows from Lemma 9.2.1 that Problem 9.2.1 is solvable. If, however, $(t_0,x_0) \in W_\epsilon^2$, then taking into account that W_ϵ^2 is a ν-stable bridge in a problem of a $(K_\epsilon,\mathfrak{L}_\epsilon)$-approach, it follows from Lemma 9.2.2 that Problem 9.2.2 is solvable. It remains to verify that for a given initial position (t_0,x_0) Problems 9.2.1 and 9.2.2 cannot be solved simultaneously. Indeed, assuming the contrary, we obtain that for the motions

$$x(\cdot) \in X(t_0,x_0,\tilde{U}^0,\tilde{V}^\epsilon) \subset X_1(t_0,x_0,\tilde{U}^0) \cap X_2(t_0,x_0,\tilde{V}^\epsilon),$$

the conditions of an (M,N)-approach and of an $(M,N)^\epsilon$-evasion are satisfied. This is, however, impossible, which proves Theorem 9.2.1.

In Chapter 3 several types of payoff functionals were defined. If a payoff functional belongs to one of these types and a differential game is considered in the class of mixed positional strategies for the first and second players, then Theorem 9.2.1 yields the existence of the value of such a differential game. Statements of corresponding game-theoretical problems and related arguments and derivations directly repeat the material presented in Chapter 3. The only difference is that positional (pure) strategies U: $T \times R^n \rightarrow P$ and V: $T \times R^n \to Q$ are replaced by mixed positional strategies $\tilde{U}$: $T \times R^n \to \mathrm{rpm}(P)$ and $\tilde{V}$: $T \times R^n \to \mathrm{rpm}(Q)$ and Theorem 2.1.1 is replaced by Theorem 9.2.1. Thus we are not going to present the corresponding statements and proofs here.

9.3. Dynamic Programming, Extremal Aiming, and A Priori Stable Sets

In this section we present a brief survey of results obtained from the basic assertions in Chapters 4, 5, and 7 when the condition of saddle point in a small game is omitted and pure positional strategies are replaced by mixed ones.

9.3.1. We indicate the changes that should be carried out in order to make the transition from the statements and proofs presented in Chapters 4, 5, and 7 to the corresponding statements and proofs for the differential games considered in the class of mixed positional strategies. To the left of the sign ÷ we shall write the symbols, conditions, equations, etc., which were used in the preceding chapters and to the right of this sign we write the corresponding symbols, conditions, equations, etc., introduced in this chapter. We thus have

$$u \div \mu, \quad v \div \nu, \quad \mathcal{P} \div \text{rpm}(\mathcal{P}), \quad \mathcal{Q} \div \text{rpm}(\mathcal{Q}),$$

$$f(t,x,u,v) \div \breve{f}(t,x,\mu,\nu),$$

$$\text{equation (1.2.1)} \div \text{equation (9.2.5)},$$

$$\text{equality (1.2.8)} \div \text{equality (9.2.11)},$$

$$\mathcal{F}_1(t,x,u) \div \breve{\mathcal{F}}_1(t,x,\mu), \quad \mathcal{F}_2(t,x,v) \div \breve{\mathcal{F}}_2(t,x,\nu),$$

$$U: T \times R^n \to \mathcal{P} \div \breve{U}: T \times R^n \to \text{rpm}(\mathcal{P}), \tag{1}$$

$$V: T \times R^n \to \mathcal{Q} \div \breve{V}: T \times R^n \to \text{rpm}(\mathcal{Q}).$$

First consider how the results of Chapter 4 are changed. We shall omit the proofs of assertions presented below since once the indicated replacements are carried out they do not differ from the proofs of the corresponding assertions presented in Chapter 4.

Let the payoff in a differential game be defined by the functional (3.1.1). In Section 3.1 a differential game (1.2.1), (3.1.1) was defined which was considered in the class of pure positional strategies. Under conditions (1.2.8) the existence of the value for the differential game (1.2.1), (3.1.1) was proved. Removing condition (1.2.8) and utilizing the alterations indicated above, we obtain the differential game (9.2.5), (3.1.1) in the class of mixed positional strategies; this game possesses the value $\omega^0(t_0,x_0)$ for any initial position (t_0,x_0). The following assertion is valid for this game.

Theorem 9.3.1. *Let a continuous function ω^0: $T \times R^n \to R^1$ satisfy the boundary condition $\omega^0(\vartheta,x) = \sigma(x)$ $(x \in R^n)$. Let this function be continuously differentiable in the region $T^0 \times R^n$ and satisfy in this region the equation*

$$\frac{\partial \omega^0(t,x)}{\partial t} + \min_{\mu} \max_{\nu} \left[\frac{\partial \omega^0(t,x)}{\partial x} \breve{f}(t,x,\mu,\nu)\right] = 0. \tag{2}$$

Then this function will be the value function for the differential game under consideration. For any initial position $(t_0,x_0) \in T \times R^n$ a pair of mixed positional strategies $\breve{U}_0$ and $\breve{V}_0$ satisfying for $(t,x) \in T^0 \times R^n$ the conditions

$$\max_{\nu} (\partial\omega^0(t,x)/\partial x)\breve{f}(t,x,\mu_0,\nu)$$
(3)
$$= \min_{\mu} \max_{\nu} (\partial\omega^0(t,x)/\partial x)\breve{f}(t,x,\mu,\nu),$$

$$\min_{\mu} (\partial\omega^0(t,x)/\partial x)\breve{f}(t,x,\mu,\nu_0)$$
(4)
$$= \max_{\nu} \min_{\mu} (\partial\omega^0(t,x)/\partial x)\breve{f}(t,x,\mu,\nu)$$

forms a saddle point for the game. In (2) - (4)

$$\mu \in \text{rpm}(\mathcal{P}), \quad \nu \in \text{rpm}(\mathcal{Q}), \quad \mu_0 = \breve{U}_0(t,x), \quad \textit{and} \quad \nu_0 = \breve{V}_0(t,x).$$

This assertion is an analog of Theorem 4.1.1. The theorem presented below is an analog of Theorem 4.4.1.

Theorem 9.3.2. *Let the function* $\sigma: R^n \to R^1$ *be locally Lipschitzian. In order that the function* $\omega : T \times R^n \to R^1$ *be the value function of the differential game with payoff* (3.1.1) *considered in the class of mixed positional strategies, it is necessary and sufficient that the following set of conditions be fulfilled*:

(a) ω *is locally Lipschitzian,*

(b) $\omega(\theta,x) = \sigma(x) \quad (x \in R^n)$,

(5) (c) $$\max_{\nu} \min_{\mu} D^-\omega(t,x)|(1,\breve{f}(t,x,\mu,\nu)) \leqslant 0,$$

(6) $$\min_{\mu} \max_{\nu} D^+\omega(t,x)|(1,\breve{f}(t,x,\mu,\nu)) \geqslant 0$$

for all $t < \theta$, $x \in R^n$. *(In condition* (c) $\mu \in \text{rpm}(\mathcal{P})$, $\nu \in \text{rpm}(\mathcal{Q})$.)

Note that inequalities (5) and (6) can be rewritten as follows:

(7) $$\max_{\nu} \min_{h_2} D^-\omega(t,x)|(1,h_2) \leqslant 0,$$

(8) $$\min_{\mu} \max_{h_1} D^+\omega(t,x)|(1,h_1) \geqslant 0$$

where $h_2 \in \breve{\mathcal{F}}_2(t,x,\nu)$, $h_1 \in \breve{\mathcal{F}}_1(t,x,\mu)$. From equalities (9.2.2) it follows that inequalities (5) and (6) are equivalent to inequalities (7) and (8).

Now let the payoff be defined by the functional (3.3.1). The following assertion is valid for the differential game (9.2.5), (3.3.1) considered in the class of mixed positional strategies (analogous to Theorem 4.4.2).

Theorem 9.3.3. *Let the function* $\rho: T \times R^n \to R^1$ *be locally Lipschitzian. In order that the function* $\omega: T \times R^n \to R^1$ *be the value function of the* *differential game* (9.2.5), (3.3.1) *considered in the class of mixed positional strategies, it is necessary and sufficient that the following set of conditions is fulfilled*:

(a) ω *is locally Lipschitzian,*

(b) $\omega(t,x) \leqslant \rho(t,x)$, $\omega(\theta,x) = \rho(\theta,x)$ *for* $t \in T$, $x \in R^n$,

(c) *in the region* $\{(t,x) \in T^0 \times R^n: \omega(t,x) < \rho(t,x)\}$ *the inequality* (5) *is fulfilled,*

(d) *for all* $(t,x) \in T^0 \times R^n$ *the inequality* (6) *is fulfilled.*

Next, one could present assertions for the differential games (9.2.5), (3.1.1) and (9.2.5), (3.3.1) which would be modifications of Corollaries 4.4.1 - 4.4.4 and Theorems 4.4.3 and 4.4.4. We are not going to do this, however, since the statements presented above give a sufficiently clear idea of the manner in which the results of Chapter 4 are carried over to the case of differential games considered in the class of mixed positional strategies.

9.3.2. In Chapter 5 the method of extremal aiming was suggested, based on a solution of auxiliary problems of programmed control. In particular, this method was used for a solution of game-theoretical control problems in which the dynamics of the controlled system are described by the equation

$$\dot{x}(t) = A(t)x(t) + g(t,u(t),v(t)).$$

Recall that in Chapter 5 the condition of a saddle point in a small game was fulfilled (cf. equation (5.1.36)). Removing this condition, one could consider a differential game in the class of mixed positional strategies. As above, we shall assume formally that the players can mix their controls instantaneously. Therefore we replace the pure controls u and v by mixed controls which are the measures μ and ν. In accordance with this understanding, we shall consider a controlled system

(9) $$\dot{x}(t) = A(t)x(t) + \tilde{g}(t,\mu_t,\nu_t),$$

where $\mu_t \in \text{rpm}(\mathcal{P})$, $\nu_t \in \text{rpm}(\mathcal{Q})$, and the function $\tilde{g}$ is defined by the equality

(10) $$\tilde{g}(t,\mu,\nu) \overset{\text{def}}{=} \int_{\mathcal{P}}\int_{\mathcal{Q}} g(t,u,v)\mu(du)\nu(dv).$$

Observe that

$$\min_{\mu}\max_{\nu}\langle \ell,\tilde{g}(t,\mu,\nu)\rangle = \max_{\nu}\min_{\mu}\langle \ell,\tilde{g}(t,\mu,\nu)\rangle \quad \text{for} \quad \mu\in \mathrm{rpm}(\mathcal{P}),\ \ \nu\in\mathrm{rpm}(\mathcal{Q}). \tag{11}$$

This equality is valid for all $t \in T$ and $\ell \in R^n$. It replaces the condition (5.1.36) which was utilized in Chapter 5.

We introduce the following notation:

$$\begin{aligned}\tilde{\kappa}(\ell,\tau,\theta) &\overset{\text{def}}{=} \min_{\mu}\max_{\nu}\langle \ell,\Phi(\theta,\tau)\tilde{g}(\tau,\mu,\nu)\rangle \\ &= \max_{\nu}\min_{\mu}\langle \ell,\Phi(\theta,\tau)\tilde{g}(\tau,\mu,\nu)\rangle,\end{aligned} \tag{12}$$

$$\begin{aligned}\tilde{\kappa}^{(m)}(s,\tau,\theta) &\overset{\text{def}}{=} \min_{\mu}\max_{\nu}\langle s,[\Phi(\theta,\tau)]_m\tilde{g}(\tau,\mu,\nu)\rangle \\ &= \max_{\nu}\min_{\mu}\langle s,[\Phi(\theta,\tau)]_m\tilde{g}(\tau,\mu,\nu)\rangle.\end{aligned} \tag{13}$$

These quantities replace the quantities $\kappa(\ell,\tau,\theta)$ and $\kappa^{(m)}(s,\tau,\theta)$ which were defined in Chapter 5 by the equalities (5.1.37) and (5.1.38). Recall that the matrix-function $\tau \to \Phi(\theta,\tau)$ satisfies the equation (1.2.15), s is an m-dimensional vector, $[\Phi(\theta,\tau)]_m$ is an $m\times n$ matrix consisting of the first m rows of the matrix $\Phi(\theta,\tau)$.

Consider the differential game (9), (3.1.1). Assume that this game is defined in the class of mixed positional strategies. As in Chapter 5 it is assumed that the function $\sigma: R^n \to R^1$ is convex and satisfies the Lipschitz condition. Recall that $\sigma^*: R^n \to (-\infty,\infty]$ is the function conjugate to the function σ, and $\mathrm{dom}\ \sigma^* \overset{\text{def}}{=} \{\ell \in R^n: \sigma^*(\ell) < \infty\}$.

Let

$$\tilde{\psi}(\ell,t) \overset{\text{def}}{=} \sigma^*(\ell) - \int_t^{\theta} \tilde{\kappa}(\ell,\tau,\theta)d\tau, \tag{14}$$

$$\tilde{\omega}_p(t,x) \overset{\text{def}}{=} \max_{\ell\in\mathrm{dom}\ \sigma^*}[\langle \ell,\Phi(\theta,t)x\rangle - \tilde{\psi}(\ell,t)], \tag{15}$$

$$\tilde{\mathfrak{L}}_0(t,x) \overset{\text{def}}{=} \{\ell_0 \in \mathrm{dom}\ \sigma^*: \langle \ell_0,\Phi(\theta,t)x\rangle - \tilde{\psi}(\ell_0,t) = \tilde{\omega}_p(t,x)\}. \tag{16}$$

Using the sign $\div$ introduced above one can write

$$\psi(\ell,t)\ \ (5.1.22) \div \tilde{\psi}(\ell,t)\ \ (14),$$

$$\omega_p(t,x)\ \ (5.1.23) \div \tilde{\omega}_p(t,x)\ \ (15),$$

$$\mathfrak{L}_0(t,x)\ \ (5.2.1) \div \tilde{\mathfrak{L}}_0(t,x)\ \ (16).$$

Recall that in Section 5.1 the quantity ω_p was introduced as a programmed maximin. In this section we shall also refer to the quantity $\tilde{\omega}_p$ as a programmed maximin. However, we shall omit here the description of the corresponding programmed constructions which would justify the use of this term and proceed directly from formula (15).

Assume first that for any $t < \theta$ the function $\breve{\psi}(\cdot,t)$ is strictly convex. Then the maximum in (15) is attained at the unique vector, i.e., the set $\breve{\mathfrak{L}}_0(t,x)$ consists of the single element $\ell_0(t,x)$. In this case one could define mixed strategies of extremal aiming. These strategies are denoted by U_0 and V_0 and are defined by means of the following conditions:

$$
(17)\quad
\begin{aligned}
&\max_{\nu} \langle \ell_0(t,x), \Phi(\theta,t)\tilde{g}(t,\mu_0,\nu)\rangle \\
&\qquad = \min_{\mu} \max_{\nu} \langle \ell_0(t,x), \Phi(\theta,t)\tilde{g}(t,\mu,\nu)\rangle \\
&\min_{\mu} \langle \ell_0(t,x), \Phi(\theta,t)\tilde{g}(t,\mu,\nu_0)\rangle \\
&\qquad = \max_{\nu} \min_{\mu} \langle \ell_0(t,x), \Phi(\theta,t)\tilde{g}(t,\mu,\nu)\rangle
\end{aligned}
$$

where

$$\mu \in \text{rpm}(\mathcal{P}), \quad \nu \in \text{rpm}(\mathcal{Q}), \quad \mu_0 = \breve{U}_0(t,x), \quad \nu_0 = \breve{V}_0(t,x).$$

The following theorem is valid (an analog of Theorem 5.2.1).

Theorem 9.3.4. *Let the function $\breve{\psi}(\cdot,t)$ be strictly convex for any $t < \theta$. Then the function of the programmed maximin $\tilde{\omega}_p$ (15) coincides with the value function of the differential game (9), (3.1.1) considered in the class of mixed positional strategies. For any initial position $(t_0,x_0) \in T \times R^n$ the pair of mixed strategies of extremal aiming U_0 and V_0 (17) forms a saddle point of the game under consideration.*

We present yet another theorem which is a carryover of Theorem 5.4.1 to the differential game considered here in the class of mixed strategies.

Theorem 9.3.5. *In order that the function of the programmed maximin $\tilde{\omega}_p$ (15) coincide on the whole space $T \times R^n$ with the value function of the differential game (9), (3.1.1) considered in the class of mixed strategies, it is necessary and sufficient that the equality*

$$\max_{\nu} \min_{\mu} \max_{\ell \in \breve{\mathfrak{L}}_0(t,x)} [\langle \ell, \Phi(\theta,t)\tilde{g}(t,\mu,\nu)\rangle - \tilde{\kappa}(\ell,t,\theta)] = 0$$

be fulfilled for all $t < \theta$ and $x \in R^n$.

Assume now that the payoff is defined by

$$(18)\qquad \gamma(x(\cdot)) = \text{dist}([x(\theta)]_m, \mathcal{M})$$

where $\mathcal{M}$ is a convex compact set in R^m.

Consider the differential game (9), (18) in the class of mixed strategies. For this game, the programmed maximin is determined by the equality

$$(19)\qquad \tilde{\epsilon}_0(t,x) \overset{\text{def}}{=} \max_{\|s\|\leqslant 1} [\langle s,[\Phi(\theta,t)]_m x\rangle - \tilde{\varphi}^{(m)}(s,t)]$$

where s is an m-dimensional vector,

$$(20)\qquad \tilde{\varphi}^{(m)}(s,t) \overset{\text{def}}{=} \max_{\kappa\in M} \langle s,z\rangle - \int_t^\theta \tilde{\kappa}^{(m)}(s,\tau,\theta)d\tau,$$

the quantity $\tilde{\kappa}^{(m)}$ is defined according to (13). We note that $\epsilon_0(t,x)$ (5.1.31) $\div$ $\tilde{\epsilon}_0(t,x)$ (19).

Assume that for any $t < \theta$ the function $\tilde{\varphi}^{(m)}(\cdot,t)$ is convex. Then in the region

$$(21)\qquad \mathcal{E}^+ \overset{\text{def}}{=} \{(t,x) \in T^0 \times R^n\colon \tilde{\epsilon}_0(t,x) > 0\},$$

the maximum in (19) is attained at the unique vector $s_0(t,x)$. (This fact is proved in the same manner as Lemma 5.2.1.) In the case under consideration, strategies of extremal aiming are determined from the conditions

$$(22)\qquad \begin{aligned} &\max_{\nu} \langle s_0(t,x),\ [\Phi(\theta,t)]_m\tilde{g}(t,\mu_0,\nu)\rangle \\ &\qquad = \min_{\mu}\max_{\nu} \langle s_0(t,x),\ [\Phi(\theta,t)]_m\tilde{g}(t,\mu,\nu)\rangle, \\ &\min_{\mu} \langle s_0(t,x),\ [\Phi(\theta,t)]_m\tilde{g}(t,\mu,\nu_0)\rangle \\ &\qquad = \max_{\nu}\min_{\mu} \langle s_0(t,x),\ [\Phi(\theta,t)]_m\tilde{g}(t,\mu,\nu)\rangle, \end{aligned}$$

where $\mu_0 = \tilde{U}_0(t,x)$, $\nu_0 = \tilde{V}_0(t,x)$. The strategies $\tilde{U}_0$ and $\tilde{V}_0$ must satisfy these relations in the region $\mathcal{E}^+$. Outside this region the functions U_0 and V_0 may be defined arbitrarily. The theorem presented below is an analog of Theorem 5.2.4.

Theorem 9.3.6. *For any $t < \theta$ let the function $\tilde{\varphi}^{(m)}(\cdot,t)$ (20) be convex. Then the programmed maximin $\tilde{\epsilon}_0$ (19) coincides everywhere on $T \times R^n$ with the value of the differential game (9), (18) considered in the class of mixed strategies. The pair of strategies of extremal aiming U_0 and V_0 defined by relations (22) forms a saddle point for the game under consideration.*

One can reformulate the remaining assertions of Chapter 5 in a manner analogous to Theorems 9.3.4 - 9.3.6. Note that here we have confined ourselves to a consideration of the method of extremal aiming for a linear system of the form (9). In the general case of a nonlinear system of the form (9.2.5) one cannot obtain the corresponding programmed constructions as a direct analog of constructions presented in Chapter 6. In this case different constructions are required which we are not going to discuss.

9.3.3. We shall describe the construction of a priori stable bridges for differential games considered in the class of mixed positional strategies. Introduce the following notation:

$$\breve{F}_1(t,x) \overset{\text{def}}{=} \bigcap_{\mu} \breve{F}_1(t,x,\mu),$$
(23)
$$\breve{F}_2(t,x) \overset{\text{def}}{=} \bigcap_{\nu} \breve{F}_2(t,x,\nu),$$

where $\nu \in \text{rpm}(Q)$, $\mu \in \text{rpm}(P)$, the sets $\breve{F}_2(t,x,\nu)$ and $\breve{F}_1(t,x,\mu)$ are defined in accordance with (9.2.1). Observe that $F_2(t,x)$ (7.1.9) $\div$ $\breve{F}_2(t,x)$ (23) and $F_1(t,x)$ (7.1.15) $\div$ $\breve{F}_1(t,x)$ (23). The following assertion is valid (cf. Theorems 7.1.1 and 7.1.2).

Theorem 9.3.7. *Let $w(\cdot)$: $[t_0,T] \to R^n$ be an absolutely continuous function satisfying the conditions*

(24) $\quad (t,w(t)) \in N$ *for* $t_0 \leqslant t \leqslant T$, $(T,w(T)) \in M$

and the differential inclusion

(25) $\quad \dot{w}(t) \in \breve{F}_2(t,w(t)), \quad t_0 \leqslant t \leqslant T.$

Then the graph of this function, i.e. the set

(26) $\quad W_1 \overset{\text{def}}{=} \{(t,w(t)): t_0 \leqslant t \leqslant T\}$

is a μ-stable bridge in a problem of an (M, N)-approach. Let $\breve{U}^e$ be a mixed strategy extremal to the set W_1, i.e. it is chosen from the condition

$$\max_{\nu} \langle x - w(t), \breve{f}(t,x,\mu^e,\nu)\rangle$$
(27)
$$= \min_{\mu} \max_{\nu} \langle x - w(t), \breve{f}(t,x,\mu,\nu)\rangle \quad (\mu^e = \breve{U}^e(t,x)).$$

Then for any motion $x(\cdot) \in X(t_0,x_0,\breve{U}^e)$ where $x_0 = w(t_0)$, the condition of an (M, N)-approach holds. Moreover, for all $x(\cdot) \in X(t_0,x_0,\breve{U}^e)$ and all $t \in [t_0,T]$ the equality $x(t) = w(t)$ is valid.

In the statement of the theorem presented below the function $\breve{\kappa}_1$ is used. This function is defined as follows:

(28) $\quad \breve{\kappa}_1(l,t,w) \overset{\text{def}}{=} \min_{\nu} \max_{\mu} \langle l,\breve{f}(t,w,\mu,\nu)\rangle.$

Note that κ_1 (7.1.17) $\div$ $\breve{\kappa}_1$ (28).

Theorem 9.3.8. *For all $(t,w) \in T \times R^n$ let the function $\tilde{\kappa}_1(\cdot,t,w)$ be convex. Then, for any initial position $(t_0,x_0) \in T \times R^n$, the following alternatives are valid: either a μ-stable path $\mathcal{W}_1$ emanates from this initial position defined by the relations (24) - (26) (in this case the mixed strategy $\tilde{U}^e$ extremal to the set $\mathcal{W}_1$ yields a solution of an $(\mathcal{M}, \mathcal{N})$- approach problem (Problem 9.2.1)), or for some $\epsilon > 0$ this position (t_0,x_0) is contained on the ν-stable integral manifold*

$$\mathcal{W}_2 \overset{\text{def}}{=} \text{cl}\{(t,w(t)): t_0 \leqslant t \leqslant t^{\epsilon}(w(\cdot)), \ w(\cdot) \in \mathcal{X}(t_0,x_0,\tilde{\mathcal{F}}_2)\}$$

and the mixed strategy $\tilde{V}^e$ extremal to the set $\mathcal{W}_2$ yields a solution of an $(\mathcal{M}, \mathcal{N})^{\epsilon}$-evasion problem (Problem 9.2.2).

This theorem is an analog of Theorem 7.2.3. Note that in the statement presented above, $\mathcal{X}(t_0,x_0,\tilde{\mathcal{F}}_2)$ is the set of all solutions of the differential inclusion (25), and the functional t^{ϵ} is defined in accordance with (2.1.7).

Other results from Sections 7.1 - 7.3 can be restated in the same manner and the corresponding assertions for differential games considered in the class of mixed positional strategies can thus be obtained.

9.3.4. We now consider examples which illustrate the theorems presented above.

Example 1. Let an inertia-free point X move on the plane R^2. Let the motion equation for this point be

$$\dot{x}(t) = L(\beta(t))u(t) + L(\alpha(t))v(t). \tag{29}$$

Here $x = (x_1,x_2)^{\mathrm{T}}$ is a two-dimensional phase vector, $u = (u_1,u_2)^{\mathrm{T}}$ and $v = (v_1,v_2)^{\mathrm{T}}$ are controlling vectors for the first and second players respectively, $L(\varphi)$ is a rotation matrix about angle φ, i.e.

$$L(\varphi) = \begin{bmatrix} \cos\varphi & -\sin\varphi \\ \sin\varphi & \cos\varphi \end{bmatrix}. \tag{30}$$

Assume that $\alpha(t)$ is a controlling parameter for the first player and $\beta(t)$ for the second. It is also assumed that controlling actions satisfy the constraints

$$u_1^2 + u_2^2 \leqslant r_1^2, \ v_1^2 + v_2^2 \leqslant r_2^2, \ |\alpha| \leqslant \rho_1, \ |\beta| \leqslant \rho_2.$$

The players control the motion of point X as follows. The velocity of the points consists of two summands. The first is obtained as a result of the rotation of vector $u(t)$ selected by the first player about an angle $\beta(t)$ stipulated by the second. The second summand is the vector $v(t)$ rotated about an angle $\alpha(t)$. The vector

$v(t)$ is chosen by the second player while the angle $\alpha(t)$ is chosen by the first. We thus obtain that disturbances of the *luft* type are acting on the controls $u(t)$ and $v(t)$ selected by the first and second players. These disturbances are created by the second and first players respectively.

The sets of admissible (pure) controls of the first and second players are determined by the relations

$$\text{(31)} \qquad \begin{aligned} \mathcal{P} &= \{(u_1,u_2,\alpha) \in R^3\colon\ u_1^2 + u_2^2 \leqslant r_1^2,\ |\alpha| \leqslant \rho_1\}, \\ \mathcal{Q} &= \{(v_1,v_2,\beta) \in R^3\colon\ v_1^2 + v_2^2 \leqslant r_2^2,\ |\beta| \leqslant \rho_2\}. \end{aligned}$$

Thus $\mathcal{P}$ and $\mathcal{Q}$ are cylinders whose bases are circles of radius r_1 and r_2 and whose heights are equal to $2\rho_1$ and $2\rho_2$ respectively. We shall assume that

$$\text{(32)} \qquad 0 \leqslant \rho_1 \leqslant \pi/2,\ \ 0 \leqslant \rho_2 \leqslant \pi/2,\ \ r_1\cos\rho_2 - r_2\cos\rho_1 \geqslant 0$$

and that the game is defined on the interval $T = (-\infty,0]$ with $\theta = 0$ being the fixed time of the termination of the game. The payoff functional is determined by the equality

$$\text{(33)} \qquad \gamma(x(\cdot)) = x_1^2(0) + x_2^2(0)$$

i.e. here $\sigma(x) = x_1^2 + x_2^2$. According to the formalization adopted in this chapter we shall consider idealized game-theoretical problems for which an instantaneous mixing of controls is admissible. Therefore the original equation (29) must be replaced by the equation

$$\text{(34)} \qquad \dot{x}(t) = \tilde{g}(\mu_t,\nu_t)$$

where

$$\text{(35)} \qquad \tilde{g}(\mu,\nu) = \int_{\mathcal{P}}\int_{\mathcal{Q}} [L(\beta)u + L(\alpha)v]\mu(d(u,\alpha))\nu(d(v,\beta)).$$

We use Theorem 9.3.1 to determine the value and the saddle point of the game. Set

$$\begin{aligned} \omega_*(t,x) &\overset{\text{def}}{=} (x_1^2 + x_2^2)^{1/2} + t(r_1\cos\rho_2 - r_2\cos\rho_1), \\ \mathcal{D}_+ &\overset{\text{def}}{=} \{(t,x) \in (-\infty,0] \times R^2\colon \omega_*(t,x) > 0\}, \\ \mathcal{D}_0 &\overset{\text{def}}{=} \{(t,x) \in (-\infty,0] \times R^2\colon \omega_*(t,x) \leqslant 0\}. \end{aligned}$$

We shall show that the function

$$\text{(36)} \qquad \omega^0(t,x) = \begin{cases} \omega_*^2(t,x) & \text{for } (t,x) \in \mathcal{D}_+ \\ 0 & \text{for } (t,x) \in \mathcal{D}_0 \end{cases}$$

is the value function of the game. Indeed, the function ω^0 is continuously differentiable and satisfies the boundary condition $\omega^0(0,x) = \sigma(x) = x_1^2 + x_2^2$. It remains to verify that the function ω^0 satisfies equation (2), which in our example is obviously satisfied in region $\mathcal{D}_0$, while in region $\mathcal{D}_+$ is expressed by the equality

$$(37) \qquad 2\omega_*(t,x)[r_1\cos\rho_2 - r_2\cos\rho_1 + \min_{\mu}\max_{\nu}(\partial\omega_*(t,x)\partial x)\tilde{g}(\mu,\nu)] = 0.$$

Here

$$\omega_*(t,x) > 0, \quad \partial\omega_*(t,x)/\partial x = (x_1/\|x\|,\ x_2/\|x\|).$$

Let $s = (s_1,s_2)^{\mathrm{T}}$ be a unit vector. Denote by μ_s the probability measure concentrated at the two points

$$(-r_1s_1,\ -r_1s_2,\ \rho_1) \text{ and } (-r_1s_1,\ -r_1s_2,\ -\rho_1)$$

with equal weights, i.e. the probability of choosing either of these points equals 1/2. Analogously we assume that ν_s is a probability measure concentrated at the two points

$$(r_2s_1,\ r_2s_2,\ \rho_2) \text{ and } (r_2s_1,\ r_2s_2,\ -\rho_2)$$

with equal weights. Taking relations (30) - (32) and (35) into account one can show that

$$(38) \qquad \begin{aligned} &\langle s,\tilde{g}(\mu_s,\nu)\rangle \leqslant \langle s,\tilde{g}(\mu_s,\nu_s)\rangle \leqslant \langle s,\tilde{g}(\mu,\nu_s)\rangle \\ &\qquad \text{for all } \mu \in \operatorname{rpm}(\mathcal{P}) \text{ and } \nu \in \operatorname{rpm}(\mathcal{Q}), \end{aligned}$$

$$(39) \qquad \begin{aligned} &\min_{\mu}\max_{\nu}\langle s,\tilde{g}(\mu,\nu)\rangle = \max_{\nu}\min_{\mu}\langle s,\tilde{g}(\mu,\nu)\rangle \\ &\qquad = -r_1\cos\rho_2 + r_2\cos\rho_1 \quad (\|s\| = 1). \end{aligned}$$

Substituting (39) into (37) we verify that the function ω^0 (36) is the value function for the differential game under consideration.

Furthermore, in accordance with (3), (4), and (38), we obtain that a saddle point in this game is formed by a pair of mixed strategies $\tilde{U}^0$ and $\tilde{V}^0$ of the following form. Strategy $\tilde{U}^0$ assigns to position (t,x) a probability measure μ^0 concentrated with equal weights at the two points

$$(-r_1x_1/\|x\|,\ -r_1x_2/\|x\|,\ \rho_1) \text{ and } (-r_1x_1/\|x\|,\ -r_1x_2/\|x\|,\ -\rho_1),$$

while strategy $\tilde{V}^0$ assigns to position (t,x) a probability measure ν^0 concentrated with equal weights at the points

$$(r_2x_1/\|x\|,\ r_2x_1/\|x\|,\ \rho_2) \text{ and } (r_2x_1/\|x\|,\ r_2x_2/\|x\|,\ -\rho_2).$$

In this manner $\tilde{U}^0$ and $\tilde{V}^0$ are defined everywhere except at those positions where $x = 0$. If, however $x = 0$, then arbitrary probability measures $\mu \in \text{rpm}(\mathcal{P})$ and $\nu \in \text{rpm}(\mathcal{Q})$ may be used for the values of $\tilde{U}^0(t,x)$ and $\tilde{V}^0(t,x)$, respectively.

Example 2. Now let X be a mass point whose motion is described by the equation

$$(40)\qquad \ddot{x}(t) = L(\beta(t))u(t) + L(\alpha(t))v(t).$$

The vector appearing on the r.h.s. of this equation can be interpreted as a force acting on point X. As is the case in the preceding example, this vector is the sum of two summands which are obtained as a result of the rotation of the controls $u(t)$ and $v(t)$ about the angles $\beta(t)$ and $\alpha(t)$ respectively. Players' controls are selected from the sets $\mathcal{P}$ and $\mathcal{Q}$ (31), parameters r_1, r_2, ρ_1, and ρ_2 satisfy conditions (32). The payoff functional is defined by

$$\gamma(x(\cdot)) = (x_1^2(0) + x_2^2(0))^{1/2}$$

where $\theta = 0$ is the fixed time of termination of the game. The game is considered in a class of mixed positional strategies.

We replace equation (40) by a system of equations in normal form and introduce mixed strategies. We obtain that the motion of the controlled system is described by the following equations:

$$(41)\qquad \dot{x}(t) = y(t), \quad \dot{y}(t) = \tilde{g}(\mu_t,\nu_t)$$

where $x = (x_1,x_2)^T$ and $y = (y_1,y_2)^T$ are two-dimensional vectors and the vector $\tilde{g}(\mu,\nu)$ is determined by equality (35).

We utilize Theorem 9.3.6. Determine the quantity $\kappa^{(m)}$ (13). In this example we have $m = 2$, $\theta = 0$

$$[\Phi(\theta,\tau)]_m = \begin{bmatrix} 1 & 0 & -\tau & 0 \\ 0 & 1 & 0 & -\tau \end{bmatrix}.$$

Taking (39) into account we obtain

$$\kappa^{(m)}(s,\tau,\theta) = \tau(r_1\cos\rho_2 - r_2\cos\rho_1)(s_1^2 + s_2^2)^{1/2}.$$

Next, in accordance with (20) we have

$$\varphi^{(m)}(s,t) = \frac{t^2}{2}(r_1\cos\rho_2 - r_2\cos\rho_1)(s_1^2 + s_2^2)^{1/2}.$$

For any t the function $\varphi^{(m)}(\cdot,t)$ is convex (cf. (32)). The programmed maximin $\tilde{\varepsilon}_0$ (19) is determined here by the relation

$$\tilde{\varepsilon}_0(t,x,y) = \max_{\|s\| \leqslant 1} [s_1(x_1 - ty_1) + s_2(x_2 - ty_2) - (t^2/2)(r_1\cos\rho_2 - r_2\cos\rho_1)(s_1^2 + s_2^2)^{1/2}]. \tag{42}$$

In the region $\mathcal{E}^+$ (21) where $\tilde{\varepsilon}_0(t,x,y) > 0$ the maximum in (42) is attained at the unique vector

$$s^0(t,x,y) = (x - ty)\|x - ty\|^{-1}. \tag{43}$$

We obtain

$$\tilde{\varepsilon}_0(t,x,y) = \max\{0, \|x - ty\| - (t^2/2)(r_1\cos\rho_2 - r_2\cos\rho_1)\}. \tag{44}$$

Note that the inequality $\tilde{\varepsilon}_0(t,x,y) > 0$ implies that $\|x - ty\| \neq 0$.

We define strategies of extremal aiming $\tilde{U}^0$ and $\tilde{V}^0$ in accordance with (22). Utilizing (38) one can obtain that the mixed strategy $\tilde{U}^0$ assigns to position (t,x) the probability measure μ_0 concentrated with equal weights at the two points

$$(-r_1s_1^0(t,s,y), -r_1s_2^0(t,x,y),\rho_1) \text{ and } (-r_1s_1^0(t,x,y), -r_1s_2^0(t,x,y), -\rho_1).$$

Strategy $\tilde{V}^0$ assigns to position (t,x,y) the probability measure ν^0 concentrated with equal weights at the two points

$$(r_2s_1^0(t,x,y), r_2s_2^0(t,x,y), \rho_2) \text{ and } (r_2s_1^0(t,x,y), r_1s_2^0(t,x,y), -\rho_2).$$

Here the vector $s^0(t,x,y)$ is defined by equality (43) for $\|x - ty\| \neq 0$. If, however, $\|x - ty\| = 0$ we formally set $s^0(t,x,y) = (0,0)^T$.

It follows from Theorem 9.3.6 that in the given example the value function of the differential game coincides with the function of the programmed maximin $\tilde{\varepsilon}_0$ (44). The strategies of extremal aiming $\tilde{U}^0$ and $\tilde{V}^0$ defined above form a saddle point of the game under consideration.

Example 3. Consider the solution of the problem of an oscillator's relaxation whose motion is described by equations

$$\dot{x}_1 = x_2, \quad \dot{x}_2 = -x_1 + u_1 + (u_2 - v)^2 - 1. \tag{45}$$

Here (u_1,u_2) are controlling actions for the first player, v is the control for the second player. The choice of these controls is restricted by the constraints $|u_1| \leqslant 1$, $|u_2| \leqslant 1$, $|v| \leqslant 1$.

In the problem under investigation it is required to obtain a mixed positional strategy $\tilde{U}^0$ which assures that the system (45) will arrive at the position $x_1 = x_2 = 0$. This problem can be viewed as an $(\mathcal{M}, \mathcal{N})$-approach problem (Problem 9.2.1) in which $\mathcal{M} = \{(t,x_1,x_2) \in R^3: x_1 = x_2 = 0\}$, $\mathcal{N} = R^3$.

We shall apply Theorem 9.3.7. One can verify that in this example the set $\mathcal{F}_2(t,x)$ (23) is defined as follows:

$$\breve{\mathcal{F}}_2(t,x) = \{(h_1,h_2): h_1 = x_2,\ h_2 = -x_1 + \psi,\ |\psi| \leqslant 1\}.$$

According to Theorem 9.3.7 each solution of equation

$$\text{(46)} \qquad \dot{w}_1 = w_2,\ \dot{w}_2 = -w_1 + \psi(t),\ |\psi(t)| \leqslant 1$$

satisfying the conditions $w_1(\tau) = w_2(\tau) = 0$ for some $\tau \geqslant t_0$, $w(t_0) = x_0$ forms a μ-stable path. Let $\psi^0(\cdot)$: $[t_0,\tau] \to [-1,1]$ be a control which transfers the system (46) from the initial position (t_0,x_0) into the set $\mathcal{M}$ and let $w^0(\cdot)$ be the corresponding solution of equations (46). One can verify that the mixed strategy U^e extremal to the path $\{(t,w^0(t)): t_0 \leqslant t \leqslant \tau\}$ is defined in this example as follows (cf. (27)): if $x_2 - w_2^0(t) \geqslant 0$, then the probability measure $\mu_e = U_e(t,x)$ is concentrated at the single point $u^0 = (-1,0)$, if $x_2 - w_2^0(t) < 0$ then, the probability measure $\mu_e = U_e(t,x)$ is concentrated with equal weights at the two points $(1,1)$ and $(1,-1)$. As Theorem 9.3.7 asserts, this strategy will solve the problem posed.

9.4. Approximation of Mixed Strategies

In the two preceding sections, idealized statements and solutions of game-theoretical control problems in a class of mixed positional strategies were considered. The idealization consisted of constructing step-by-step motions where an instantaneous mixing of controls was admitted; ordinary controls $u \in \mathcal{P}$ and $v \in \mathcal{Q}$ were replaced by controls-measures $\mu \in \text{rpm}(\mathcal{P})$ and $\nu \in \text{rpm}(\mathcal{Q})$ and the controlled system (1.2.1) by the controlled system (9.2.5). However, in a real world system, instantaneous mixing of controls is not feasible. It is therefore necessary to construct physically realizable control procedures which approximate the formal solutions obtained in Sections 9.2 and 9.3. In this section an approximation of mixed strategies is considered under the following assumptions: in step-by-step control procedures the partitions $\Delta^{(1)}$ and $\Delta^{(2)}$ selected by the first and second players coincide and the current position of the game is being measured precisely. These restrictions are removed in the next section, where a stochastic control procedure with guidance is constructed and substantiated.

9.4.1. To give a preliminary idea of the approximation of mixed strategies, we turn our attention to an $(\mathcal{M}, \mathcal{N})$-approach problem. Let a mixed strategy U^0 yield a solution of Problem 9.2.1 (an $(\mathcal{M}, \mathcal{N})$-approach problem) for a given initial position (t_0,x_0). In this case, as will be shown below, for any $\epsilon > 0$ one can construct a stochastic control procedure such that for the generated motions, the condition of an $(\mathcal{M}, \mathcal{N})^{\epsilon}$-approach will be fulfilled with probability not smaller than the number $1 - \epsilon$. We are not going to clarify further this assertion for the time being and only note that it is of a probabilistic nature.

Thus, in what follows we shall deal with random motions. Therefore to justify the subsequent probabilistic constructions it is required to define mixed positional strategies more exactly.

Recall that in Section 9.2 mixed positional strategies for the first player were identified with *arbitrary* functions $\check{U}$: $(t,x) \to \check{U}(t,x)$: $T \times R^n \to \text{rpm}(\mathcal{P})$. We shall now require that these functions be *measurable* with respect to x in the following sense. Let $\mu_{t,x} = \check{U}(t,x)$ $((t,x) \in T \times R^n)$. It is required that for any $t \in T$, and for any continuous function $u \to \varphi(u)$: $\mathcal{P} \to R^1$, the mapping

(1) $$x \to \int_{\mathcal{P}} \varphi(u)\mu_{t,x}(du): \ R^n \to R^1$$

be Borel measurable. Analogously, we shall assume that the function $\check{V}$: $T \times R^n \to \text{rpm}(\mathcal{Q})$ will also be measurable with respect to x, i.e. for any $t \in T$ and for any continuous function $v \to \psi(v)$: $\mathcal{Q} \to R^1$ the mapping

(2) $$x \to \int_{\mathcal{Q}} \psi(v)\nu_{t,x}(dv): \ R^n \to R^1$$

will be Borel measurable (here $\nu_{t,x} = \check{V}(t,x)$).

We note the following proposition. Let the functions $\check{U}$ and $\check{V}$ be measurable in the sense indicated above. Let $g : \mathcal{P} \times \mathcal{Q} \to R^1$ be an arbitrary continuous function. Then for any $t \in T$ the mapping

(3) $$x \to \int_{\mathcal{P}} \int_{\mathcal{Q}} g(u,v)\mu_{t,x}(du)\nu_{t,x}(dv): R^n \to R^1$$

will be Borel measurable.

To verify this assertion we shall proceed as follows: According to Weierstrass' theorem we define a sequence of polynomials g_k $(k = 1,2, ...)$ which converges uniformly on $\mathcal{P} \times \mathcal{Q}$ to a function g. Note that every polynomial g_k can be represented as the sum

$$g_k(u,v) = \sum_{i=1}^{m(k)} \varphi_{k,i}(u)\psi_{k,i}(v).$$

Set

(4)
$$\alpha_{k,i}(t,x) = \int_{\mathcal{P}} \varphi_{k,i}(u)\mu_{t,x}(du),$$
$$\beta_{k,i}(t,x) = \int_{\mathcal{Q}} \psi_{k,i}(v)\nu_{t,x}(dv).$$

Functions $x \to \alpha_{k,i}(t,x)$ and $x \to \beta_{k,i}(t,x)$ are Borel measurable. Therefore the function

$$x \to \iint g_k(u,v)\mu_{t,x}(du)\nu_{t,x}(dv) = \sum_{i=1}^{m(k)} \alpha_{k,i}(t,x)\beta_{k,i}(t,x)$$

is also Borel measurable. Finally, taking into account that

$$\lim_{k\to\infty} \int_{\mathcal{P}} \int_{\mathcal{Q}} [g(u,v) - g_k(u,v)]\mu_{t,x}(du)\nu_{t,x}(dv) = 0,$$

we obtain that mapping (3) is Borel measurable.

This measurable property of functions $\breve{U}$ and $\breve{V}$ can be stated in various equivalent ways. For example, the condition of measurability with respect to x of the function $\breve{U}$: $T \times R^n \to \text{rpm}(\mathcal{P})$ can be defined as follows: for any $t \in T$ and any measurable set $G \subset \mathcal{P}$ the function $x \to \mu_{t,x}(G)$ is Borel measurable.

We present a third definition of this condition. Let frm($\mathcal{P}$) be the totality of finite regular Borel measures on $\mathcal{P}$. In the vector space frm($\mathcal{P}$) one can define the so-called weak norm (cf. Section 6.1.1). This normed space is denoted by the symbol (frm($\mathcal{P}$), $|\cdot|_w$). It is known that the subset rpm($\mathcal{P}$) $\subset$ frm($\mathcal{P}$) is compact in the normed space (frm($\mathcal{P}$), $|\cdot|_w$).

We thus obtain that the function $x \to \breve{U}(t,x)$: $R^n \to \text{rpm}(\mathcal{P})$ can be viewed as a mapping of one metric space into another. Therefore, for such a function one can apply the usual notion of measurability in the Borel sense. One can prove that this definition of measurability will be equivalent to the definition presented above. Clearly the remark presented here is valid not only for functions $\breve{U}$ but also for functions $\breve{V}$.

Below, in this chapter, we shall only consider functions $\breve{U}$: $T \times R^n \to \text{rpm}(\mathcal{P})$ and $\breve{V}$: $T \times R^n \to \text{rpm}(\mathcal{Q})$ measurable with respect to x. We thus narrow down the sets of mixed positional strategies. However, it turns out that the results obtained in Sections 9.2 and 9.3, where the measurability of the functions $\breve{U}$ and $\breve{V}$ was not required, remain valid also in the narrowed down class of mixed strategies $\breve{U}$ and $\breve{V}$. In particular, one can show that Theorem 9.2.1 (the theorem on alternatives) remains valid for measurable mixed strategies. For this it is sufficient to verify that the mixed strategy $\breve{U}^e$ (or $\breve{V}^e$) extremal to the closed set $\mathcal{W}$ can be constructed so that this function will be measurable with respect to x in the sense indicated above. To verify this fact we shall use the theorem on measurable selection (cf. Section A.3).

Let $(t,x,s) \in T \times R^n \times R^n$. Define the set

$$\mathcal{U}^0(t,x,s) \overset{\text{def}}{=} \{\mu^0 \in \text{rpm}(\mathcal{P}): \max_\nu \langle s,\breve{f}(t,x,\mu^0,\nu)\rangle$$

$$= \min_\mu \max_\nu \langle s,\breve{f}(t,x,\mu,\nu)\rangle\}.$$

(Recall that the function $\breve{f}$ is defined according to (9.2.1)). One can verify that the set $\mathcal{U}^0(t,x,s)$ is compact in the space (frm($\mathcal{P}$), $|\cdot|_w$), and the multivalued mapping $(t,x,s) \to \mathcal{U}^0(t,x,s)$ is upper semicontinuous. Utilizing the theorem on measurable selection (cf. Section A.3)) we obtain that a single-valued function $\breve{U}^0$: $T \times R^n \times R^n \to \text{rpm}(\mathcal{P})$ exists which is Borel measurable (with respect to the totality of variables) and satisfies the conditions

$$\breve{U}^0(t,x,s) \in \mathcal{U}^0(t,x,s) \quad \text{for} \quad (t,x,s) \in T \times R^n \times R^n.$$

Recall that function U^0 takes on values in a normed vector-valued space (frm(P), $|\cdot|_w$). It is thus clear how the property of Borel measurability of this function is defined.

Next consider for an arbitrary $t \in T$ the multivalued mapping $x \to W[t,x]$ defined by relation (2.4.2). It is easy to verify that this mapping is upper semicontinuous (recall that the set W is closed). Utilizing once again the theorem on measurable selection we obtain that there exists a function e: $T \times R^n \to R^n$ such that $e(t,x) \in W[t,x]$ for all $(t,x) \in T \times R^n$ and for any $t \in T$ the function $x \to e(t,x)$ is Borel measurable. Set

$$(5) \qquad s^0(t,x) = x - e(t,x), \quad \tilde{U}^e(t,x) = U^0(t,x,s^0(t,x)).$$

We obtain that the function $\tilde{U}^e$: $T \times R^n \to$ rpm(P) is measurable with respect to x as a superposition of Borel functions. Function $\tilde{U}^e$ defined by relation (5) represents a mixed strategy extremal to the set W. We have thus shown that an extremal strategy can be selected to be measurable with respect to x.

9.4.2. We shall now describe stochastic control procedures which approximate mixed strategies for the players. Let the first player select a mixed positional strategy $\tilde{U}$ and a partition $\Delta = \{t_0 = \tau_0 < \tau_1 < \cdots < \tau_{k+1} = \theta\}$. A pair (U,Δ) defines a stochastic control procedure for the first player. This procedure forms a control for the first player as follows. Let (t_0,x_0) be an initial position. At the first step, the function U assigns to the position (t_0,x_0) a probability measure μ_0. We perform a trial to choose a random vector $u \in P$ by stipulating that the probability distribution of vectors $u \in P$ is defined by the measure μ_0. Assume that on the interval $[t_0,\tau_1)$ the first player selects a constant control $u(t) = u(t_0)$ $(t_0 \leqslant t < \tau_1)$ defined as a result of this random experiment. On the time interval $[t_0,\tau_1)$ in the controlled system (1.2.1), a constant control for the first player and some control for the second player are implemented. (Below we shall indicate the manner in which the control for the second player is formed.) At time $t = \tau_1$ the controlled system arrives at the position $(\tau_1,x(\tau_1))$. The mixed strategy U corresponds a probability measure μ_1 to this position. This measure again defines a probability distribution of vectors $u \in P$. We then carry out a new trial according to this distribution to select a random vector u and thus define a constant control $u(t) = u(\tau_1)$ which is used by the first player on the interval $[\tau_1,\tau_2)$. This control, together with a control for the second player, defines the motion of the system (1.2.1) on the time interval $[\tau_1,\tau_2]$. This procedure is then repeated for all the succeeding time intervals $[\tau_i,\tau_{i+1})$.

The procedure described above forms a random control for the first player and the corresponding random motions of the controlled system. However, to define the notions of random control and random motion in a rigorous manner, one is required to state precisely how the control for the second player is formed and to utilize the condition of measurability of mixed strategies.

First we shall assume that the second player uses the procedure (V,Δ) which forms a control $v(t)$ in the same manner as the procedure (U,Δ) described above. Moreover, we shall assume that the partitions selected by the first and second players coincide, i.e. the trials to select random controls are carried out by the first and second players *simultaneously at times* $t = \tau_i$. Finally, we assume that at each instant of time selections of random controls $u(\tau_i)$ and $v(\tau_i)$ are *mutually independent*. The meaning of these assumptions will be discussed below. We also note that in Section 9.5 stochastic control procedures with guidance are considered which do not require that the partitions selected by the first and second players be the same, while the independence of random controls will be a corollary of imprecision in measuring the current position. However, it is more convenient to start by considering simpler control procedures.

Thus let U and V be measurable mixed strategies for the first and second players, $\Delta = \{t_0 = \tau_0 < \tau_1 < \cdots < \tau_{k+1} = \theta\}$ be a partition of the time interval $[t_0,\theta]$. At instants $t = \tau_i$ ($i = 0,1, ..., k$) the first and second players simultaneously select random controls. Since the controls will be constant on the intervals $[\tau_i,\tau_{i+1})$, one can correspond in a one-to-one manner the set $(u_0,v_0,u_1,v_1, ..., u_k,v_k) \in (P \times Q)^{k+1}$ to the pair of controls $u(\cdot)$, $v(\cdot)$. Evidently here u_i and v_i are the values taken on by functions $u(\cdot)$ and $v(\cdot)$ on the half-interval $[\tau_i,\tau_{i+1})$. The procedures (U,Δ) and (V,Δ) generate a probability distribution on the set $(P \times Q)^{k+1}$. We shall define this distribution by induction.

At the first step, i.e. at the initial time $t = t_0 = \tau_0$, the players select the random vectors $u_0 \in P$ and $v_0 \in Q$ simultaneously. Probability distributions of these vectors are defined by the measures $\mu_0 = U(t_0,x_0)$ and $\nu_0 = V(t_0,x_0)$. Since the players select the vectors u_0 and v_0 independently, it is natural to define the probability distribution of the pairs $(u_0,v_0) \in P \times Q$ by the measure η_0, which is equal to the product of measures μ_0 and ν_0. It is known (cf. for example [46, 52]) that the measure $\eta_0 = \mu_0 \times \nu_0$ can be uniquely defined by the relation

$$\eta_0(G_1 \times G_2) = \mu_0(G_1) \cdot \nu_0(G_2),$$

where $G_1 \in P$ and $G_2 \in Q$ are arbitrary Borel sets. For any continuous function $g: P \times Q \to R^1$ the equality

$$\int_{P\times Q} g(u,v)\eta_0(du,dv) = \int_P \int_Q g(u,v)\mu_0(du)\nu_0(dv)$$

is valid.

Thus the probability distribution of pairs $(u_0,v_0) \in P \times Q$ is defined by the measure $\eta_0 = \mu_0 \times \nu_0$. We define a measure η_1 on $(P \times Q)^2$ which gives us a probability distribution of collections $(u_0 v_0,u_1,v_1)$. Set

$$h^{(1)}(u_0,v_0) \overset{\text{def}}{=} x(\tau_1,t_0,x_0,u_0,v_0), \tag{6}$$

i.e. $h^{(1)}(u_0,v_0)$ is the phase state into which the system (1.2.1) arrives at time $t = \tau_i$ from a fixed initial position $x(t_0) = x_0$ under constant controls $u(t) = u_0$ and $v(t) = v_0$ ($t_0 \leqslant t < \tau_1$). At time $t = \tau_1$ the players select, in a mutually independent manner, the vectors $u_1 \in P$ and $v_1 \in Q$. The probability distribution of pairs (u_1,v_1) is defined by the measure $\xi_1 = \mu_1 \times \nu_1$ where $\mu_1 = U(\tau_1,x(\tau_1))$, $\nu_1 = V(\tau_1,x(\tau_1))$. Here $x(\tau_1)$ is the phase state of the system (1.2.1) realized at time $t = \tau_1$; it depends on the controls u_0, v_0 selected on the interval $[\tau_0,\tau_1)$. This dependence is expressed by the function $(u_0,v_0) \to h^{(1)}(u_0,v_0)$ (6). It thus follows that the choice of measure ξ_1 depends on (u_0,v_0), therefore we shall introduce the notation $\xi_1(\cdot,u_0,v_0)$. Since the function $(u_0,v_0) \to h^{(1)}(u_0,v_0)$ is continuous and the functions $x \to U(\tau_1,x)$ and $x \to V(\tau_1,x)$ are measurable, the mapping $(u_0,v_0) \to \xi_1(\cdot,u_0,v_0)$ which assigns to the pair $(u_0,v_0) \in P \times Q$ the probability measure $\xi_1 \in \text{rpm}(P\times Q)$ will be measurable in the same sense as the measurability of functions U and V was defined above. Viewing the mapping $(u_0,v_0) \to \xi_1(\cdot,u_0,v_0)$ as a transition probability and utilizing the corresponding theorem of probability theory (cf., for example, [46, 52]) we obtain that the probability distribution of the collection $(u_0,v_0,u_1,v_1) \in (P\times Q)^2$ should be defined by the measure η_1 which is uniquely determined by the relation

$$\eta_1(G_1 \times G_2) = \int_{G_1} \xi_1(G_2,u_0,v_0)\eta_0(d(u_0,v_0)),$$

where G_1 and G_2 are arbitrary Borel subsets of the compact set $P\times Q$.

Let the probability measure η_{i-1} be defined on $(P\times Q)^i$. We shall now define the measure η_i on $(P\times Q)^{i+1}$. Introduce the following notation:

$$y_{i-1} \overset{\text{def}}{=} (u_0,v_0,u_1,v_1, \ldots, u_{i-1},v_{i-1}) \in (P\times Q)^i \overset{\text{def}}{=} Y_{i-1}, \tag{7}$$

$$h^{(i)}(y_{i-1}) \overset{\text{def}}{=} x(\tau_i,t_0,x_0,u(\cdot),v(\cdot)), \tag{8}$$

where

$$u(t) = u_j,\quad v(t) = v_j,\quad \tau_j \leqslant t < \tau_{j+1},\quad j = 0,1, \ldots, i-1,$$

$$\mu_i(\cdot,y_{i-1}) = \hat{U}(\tau_i,h^{(i)}(y_{i-1})),\quad \nu_i(\cdot,y_{i-1}) = \hat{V}(\tau_i,h^{(i)}(y_{i-1})),$$

$$\xi_i(\cdot,y_{i-1}) = \mu_i(\cdot,y_{i-1}) \times \nu_i(\cdot,y_{i-1}).$$

Observe that the function $y_{i-1} \to h^{(i)}(y_{i-1})$ is continuous and the functions $x \to U(\tau_i,x)$ and $x \to V(\tau_i,x)$ are measurable, therefore the mapping $y_{i-1} \to \xi_i(\cdot,y_{i-1})$ will be measurable. Utilizing the last mapping as a transition probability we define the probability η_i on $Y_i = (P\times Q)^{i+1}$ by the relation

$$(10) \qquad \eta_i(G_1 \times G_2) = \int_{1} \xi_i(G_2, y_{i-1})\eta_{i-1}(d(y_{i-1})),$$

where $G_1 \subset Y_{i-1} = (P \times Q)^i$, $G_2 \subset P \times Q$ are Borel sets. Relation (10) defines the measure η_i uniquely.

These constructions of probability measures $\eta_i \in \mathrm{rpm}(Y_i)$ are performed for all $i = 0,1,2, \ldots, k$. We preset a relation which is valid for $\eta_i \in \mathrm{rpm}(Y_i)$ defined by equality (10). Let $r : Y_i \to R^1$ be a Borel measurable function. The following formula is known (cf. for example, [46, 52])

$$(11) \qquad \begin{aligned} &\int_{Y_i} r(y_i)\eta_i(dy_i) \\ &= \int_{Y_{i-1}} \Big[\int_{P \times Q} r(y_{i-1}, u_i, v_i)\xi_i(d(u_i,v_i), y_{i-1})\Big]\eta_{i-1}(dy_{i-1}). \end{aligned}$$

Here $(y_{i-1}, u_i, v_i) = y_i = (u_0, v_0, \ldots, u_i, v_i)$ and on the r.h.s. of equality (11) first the integral appearing in the square brackets is computed, followed by the outer integral.

9.4.3. In Theorems 9.4.1 and 9.4.2 presented below, approximating properties of mixed strategies extremal to stable bridges are stated.

Theorem 9.4.1. *Let $(t_0, x_0) \in W$ where W is a closed μ-stable bridge in a problem of an (M, N)-approach. Let U^e be a mixed strategy extremal to the set W. Then, for any numbers $\epsilon > 0$ and $p < 1$, one can find a $\delta > 0$ such that for any partition Δ satisfying the bound* $\mathrm{diam}\,\Delta \leqslant \delta$ *and for any mixed positional strategy V, the pair of stochastic control procedures (U^e, Δ) and (V, Δ) generates a set of random motions of the system* (1.2.1) *for which the condition of an $(M^{\bar\epsilon}, N^{\bar\epsilon})$-approach is satisfied with the probability of at least p.*

Theorem 9.4.2. *Let $(t_0, x_0) \in W$ where W is a closed ν-stable bridge in a problem of an $(M, N)^{\epsilon}$-evasion. Let V^e be a mixed strategy extremal to the set W. Then, for any numbers $\alpha \in (0, \epsilon)$ and $p < 1$, one can find a number $\delta > 0$ such that for any partition Δ, satisfying the bound* $\mathrm{diam}\Delta \leqslant \delta$ *and for any mixed strategy U, the pair of stochastic procedures (U, Δ), (V^e, Δ) generates a set of random motions of the system* (1.2.1) *for which the condition of an $(M, N)^{\alpha}$-evasion is satisfied with probability of at least p.*

The proofs of Theorems 9.4.1 and 9.4.2 are the same; therefore only the proof of Theorem 9.4.1 is presented below. However, before proceeding with the proof we shall present certain clarifications.

Recall that in view of the convention stipulated in Section 9.4.1, a mixed strategy must satisfy a measurability condition. We shall explain how the probability of fulfillment of the condition an $(\overline{M^\epsilon}, \overline{N^\epsilon})$-approach for motions generated by stochastic procedures (U^e,Δ) and (V,Δ) is determined in Theorem 9.4.1. Denote by X_ϵ the set of continuous functions $x(\cdot)$: $[t_0,\theta] \to R^n$ satisfying the condition of an $(\overline{M^\epsilon}, \overline{N^\epsilon})$-approach. Since $\overline{M^\epsilon}$ and $\overline{N^\epsilon}$ are closed ϵ-neighborhoods of the sets M and N, the set X_ϵ is closed in $C_n([t_0,\theta])$. Consider the mapping

(12) $$\psi : Y_k \to C_n([t_0,\theta])$$

which we shall define as follows. Correspond to each point $y_k = (u_0,v_0, ..., u_k,v_k) \in (P \times Q)^{k+1} = Y_k$ a pair of piecewise constant controls

(13) $$u(t) = u_i, \quad v(t) = v_i, \quad \tau_i \leqslant t < \tau_{i+1}, \quad i = 0,1, ..., k.$$

For a given initial position (t_0,x_0) this pair of controls $u(\cdot)$ and $v(\cdot)$ generates a unique motion $x(\cdot,t_0,x_0,u(\cdot),v(\cdot))$ of the system (1.2.1). Assume that the function ψ assigns to the collection $y_k = (u_0,v_0, ..., u_k,v_k)$ the above mentioned motion $x(\cdot,t_0,x_0,u(\cdot),v(\cdot)) \in C_n([t_0,\theta])$. Utilizing the well known theorems on a continuous dependence of solutions of differential equations on initial conditions and parameters, we obtain that the mapping ψ (12) is continuous. Therefore, the set $\psi^{-1}(X_\epsilon)$ will be a closed subset of the compact set Y_k. The probability η_k is defined for all Borel sets $G \subset Y_k$; hence it is also defined for the closed set $\psi^{-1}(X_\epsilon)$. In view of constructions presented above, the quantity $\eta_k(\psi^{-1}(X_\epsilon))$ is the probability that the condition of an $(\overline{M^\epsilon}, \overline{N^\epsilon})$-approach for the motions generated by stochastic procedures (U,Δ) and (V,Δ) is fulfilled. The probability of the event that the condition of an $(M, N)^\epsilon$-evasion for random motions generated by stochastic procedures (U,Δ) and (V,Δ) is fulfilled is determined analogously.

We also note the following. It follows from the proof of Theorem 9.2.1 (cf. Section 9.2 and Section 2.2) that either the point (t_0,x_0) is contained on a μ-stable bridge in an (M, N)-approach problem or this point (t_0,x_0) belongs to a ν-stable bridge in an $(M, N)^\epsilon$-evasion problem where ϵ is a positive number. Therefore Theorems 9.4.1 and 9.4.2 yield alternatives which assert that for an arbitrary initial position $(t_0,x_0) \in T \times R^n$ the conditions and assertions of either Theorem 9.4.1 or 9.4.2 are fulfilled. These alternatives are an approximation form of Theorem 9.2.1.

Proof of Theorem 9.4.1. We introduce the following notation. The symbol $y_{k,i-1}$ denotes the collection consisting of the first $2i$ vectors $(u_0,v_0, ..., u_{i-1}, v_{i-1})$. Thus, here $1 \leqslant i \leqslant k+1$, $y_{k,i-1} \in Y_{i-1} = (P \times Q)^i$, and the equality

(14) $$y_k = (y_{k,i-1}, u_i, v_i, ..., u_k, v_k)$$

is valid.

Recall that the mapping ψ (12) was defined by the equality $\psi(y_k) = x(\cdot,t_0,x_0,u(\cdot),v(\cdot))$ where $u(\cdot)$ and $v(\cdot)$ are piecewise-constant controls of the form (13). Set

(15) $$\psi_i(y_k) = x(\tau_i,t_0,x_0,u(\cdot),v(\cdot))$$

or equivalently $\psi_i(y_k) = \psi(y_k)(\tau_i)$. The equality

(16) $$\psi_i(y_k) = h^{(i)}(y_{k,i-1})$$

is also valid; here the vector $h^{(i)}$ is defined by the relation (8). Note also that $\psi_{k+1}(y_k) = \psi(y_k)(\theta)$.

Let β be a positive number. Recall that $W[\tau] = \{w \in R^n\colon (\tau,w) \in W\}$. Set

$$I^{(1)}(y_k,\beta) = \{i \in \{1, ..., k+1\}\colon \text{dist}(\psi_i(y_k), W[\tau_i]) \geq \beta\},$$

$$I^{(2)}(y_k,\beta) = \{i \in \{1, ..., k+1\}\colon \text{dist}((\tau_i,\psi_i(y_k)), M) \leq 2\beta\},$$

$$I(y_k,\beta) = I^{(1)}(y_k,\beta) \cup I^{(2)}(y_k,\beta),$$

(17) $$m(y_k,\beta) = \min\{i\colon i \in I(y_k,\beta)\},$$

(18) $$\rho_i(y_k) = \begin{cases} \text{dist}(\psi_i(y_k), W[\tau_i]) & \text{if } i < m(y_k,\beta), \\ \text{dist}(\psi_m(y_k), W[\tau_m]) & \text{if } i \geq m = m(y_k,\beta). \end{cases}$$

If $W[\tau_i] = \phi$ we set formally $\text{dist}(\psi_i(y_k), W[\tau_i]) = 1$ for any $y_k \in Y_k$. We remark that the sets and quantities introduced here will be considered for $\beta < 1$.

We clarify the meaning of the notation introduced herein. Let $x(t)$ ($t_0 \leq t \leq \theta$) be a motion of the system (1.2.1) generated by controls $u(\cdot)$ and $v(\cdot)$ of the form (13), i.e. here $x(\cdot) = \psi(y_k)$. Select one of times $t = \tau_i$ appearing in the partition Δ. Two cases are possible.

Case 1. For all $j = 1,2, ..., i$ the distance from the point $x(\tau_j)$ up to the set $W[\tau_j]$ is less than number β and the distance from the point $(\tau_j,x(\tau_j))$ up to the set M is greater than the number 2β. In this case the quantity $\rho_i(y_k)$ is the distance from the point $x(\tau_i)$ up to the set $W[\tau_i]$.

Case 2. There exists an index $j \in \{1, ..., i\}$ such that either the distance $d(x(\tau_j), W[\tau_j]) \geq \beta$ or the distance $d((\tau_j,x(\tau_j)),M) \leq 2\beta$. Let m be the smallest of the indices j for which this situation is valid. In this case set $\rho_i(y_k) = \text{dist}(x(\tau_m), W[\tau_m]) = \rho_m(y_k)$. Note that this equality is valid for all $i = m, m+1, ..., k+1$.

Recall that according to the definition of a μ-stable bridge the relations $W \subset N$ and $W(\theta) \subset M(\theta)$ are valid. Therefore $I(y_k,\beta) \neq \phi$. Moreover it follows from the inequality $\rho_{k+1}(y_k) < \beta$ that

$$(19)\qquad \begin{aligned} &(\tau_m,\psi_m(y_k)) \in M^{2\overline{\beta}},\\ &(\tau_i,\psi_i(y_k)) \in N^{\overline{\beta}},\ i = 1,2, ..., m = m(y_k,\beta). \end{aligned}$$

Let $2\beta \leqslant \epsilon/2$. Then if the diam$\Delta$ of the chosen partition is sufficiently small, it follows from (19) that the motion $x(\cdot) = \psi_{k+1}(y_k)$ satisfies the condition of an $(M^{\overline{\epsilon}}, N^{\overline{\epsilon}})$-approach.

Thus the proof of Theorem 9.4.1 is reduced to the verification of the following: for any $\beta > 0$ and $p < 1$ one can find a $\delta > 0$ such that for any partition Δ satisfying the bound diam$\Delta \leqslant \delta$ the inequality

$$(20)\qquad \eta_k(G_\beta) \geqslant p$$

will be fulfilled where $G_\beta = \{y_k \in Y_k: \rho_{k+1}(y_k) < \beta\}$.

We shall show that the functions $y_k \to \rho_i(y_k)$ are Borel measurable. In particular it will be shown that the function $y_k \to \rho_{k+1}(y_k)$ is a Borel measurable function and this will imply that G_β is a Borel set for which the probability $\eta_k(G_\beta)$ is defined. According to (18) the function ρ_i can be defined as a superposition of two mappings

$$(j,y_k) \to \text{dist}(\psi_j(y_k), W[\tau_j]),$$

$$y_k \to (\min\{i,m(y_k,\beta)\},y_k).$$

Here the first mapping is continuous and the second mapping contains a lower semicontinuous function $y_k \to m(y_k,\beta)$. Therefore the function $y_k \to \rho_i(y_k)$ is Borel measurable.

We note yet another property of the function ρ_i: $Y_k \to R^1$. It follows from definition (18) that for any two points $y_k' \in Y_k$ and $y_k'' \in Y_k$ for which $y'_{k,i-1} = y''_{k,i-1}$ (cf. (14)), the equality $\rho_i(y_k') = \rho_i(y_k'')$ is valid, i.e. the function ρ_i does not depend on the vectors u_i,v_i, ..., u_k,v_k. Therefore one can define function r_i: $Y_{i-1} \to R^1$ by setting

$$(21)\qquad r_i(y_{i-1}) = \rho_i(y_k), \text{ where } y_{i-1} = y_{k,i-1}.$$

For $i = k + 1$ we have $r_{k+1}(y_k) = \rho_{k+1}(y_k)$.

We now turn to the proof of inequality (20). In order to prove this inequality we shall utilize Chebyshev's inequality (cf. [46]), according to which we have

$$(22)\qquad \begin{aligned} \eta_k(G_\beta) &= \eta_k(\{y_k \in Y_k: \rho_{k+1}(y_k) < \beta\})\\ &= 1-\eta_k(\{y_k \in Y_k: \rho_{k+1}(y_k) \geqslant \beta\}) \geqslant 1 - \beta^{-2}E(\rho_{k+1}^2). \end{aligned}$$

Here $E(\rho_{k+1}^2)$ is the mathematical expectation

$$E(\rho_{k+1}^2) = \int_{Y_k} \rho_{k+1}^2(y_k)\eta_k(dy_k) = \int_{Y_k} r_{k+1}^2(y_k)\eta_k(dy_k).$$

Below we shall prove the recurrence estimate.

(23) $E(r_{i+1}^2) \leqslant E(r_i^2)(1 + 2\lambda(\tau_{i+1} - \tau_i)) + 2\alpha(\delta)(\tau_{i+1} - \tau_i) \quad (i = 1,2,\ldots).$

Here λ is a Lipschitz constant for function f in the region $\mathcal{D}_1 \times P \times \mathcal{Q}$ where the compact set $\mathcal{D}_1$ is defined by relations (2.3.10), $\delta = \operatorname{diam}\Delta$, $\alpha(\delta) \to 0$ as $\delta \to 0$. From the estimate (23) we obtain the inequality

(24) $E(r_{i+1}^2) \leqslant (E(r_1^2) + \alpha(\delta)/\lambda)\exp 2\lambda(\theta - \tau_1) - \alpha(\delta)/\lambda.$

We estimate the quantity $E(r_1^2)$. By the condition of Theorem 9.4.1 we have $(t_0,x_0) \in W$, i.e. $\operatorname{dist}(x_0, W[t_0]) = 0$. Let $\operatorname{dist}((t_0,x_0),M) > 2\beta$ (otherwise the assertion of Theorem 9.4.1 is self-evident). Then the condition of μ-stability of the set W implies that for $\delta = \operatorname{diam}\Delta$ sufficiently small

$$r_1(y_0) = r_1(u_0,v_0) = \operatorname{dist}(x(\tau_1,t_0,x_0,u_0,v_0), W[\tau_1]) \leqslant a\cdot\delta$$

$$\text{for all } y_0 = (u_0,v_0) \in P\times\mathcal{Q} = Y_0,$$

where a is constant. We thus obtain $E(r_1^2) \leqslant a^2\delta^2$. It follows from this estimate and the estimate (24) that for any $\beta > 0$ and $p < 1$ one can select a $\delta_0 > 0$ such that for diam $\Delta \leqslant \delta_0$ the inequality $E(r_{k+1}^2) \leqslant (1-p)\beta^2$ will be fulfilled. Substituting this inequality into (22) we arrive at the required inequality (20). Thus the proof of Theorem 9.4.1 is reduced to the proof of the estimate (23).

According to (11) we have

$$E(r_{i+1}^2) = \int r_{i+1}^*(y_{i-1})\eta_{i-1}(dy_{i-1}), \text{ where}$$

(25)

$$r_{i+1}^*(y_{i-1}) = \int_{P\times\mathcal{Q}} r_{i+1}^2(y_{i-1},u_i,v_i)\xi_i(d(u_i,v_i),y_{i-1}).$$

Recall that $\xi_i(\cdot,y_{i-1})$ is defined by relations (9). The next step of the proof consists in verification of the inequality

(26) $r_{i+1}^*(y_{i-1}) \leqslant r_i^2(y_{i-1})(1 + 2\lambda(\tau_{i+1} - \tau_i)) + 2\alpha(\delta)(\tau_{i+1} - \tau_i).$

It is easy to see that (25) and (26) imply (23).

We select and fix a collection $y_{i-1} = (u_0,v_0, \ldots, u_{i-1},v_{i-1}) \in (P \times \mathcal{Q})^{i} = Y_{i-1}$. We form the collection $y_k = (y_{i-1},u_i,v_i, \ldots, u_k,v_k)$, where $(u_i,v_i, \ldots, u_k,v_k)$ is an arbitrary element of the set $(P \times \mathcal{Q})^{k-i+1}$. We shall distinguish two cases

(27) $m(y_k,\beta) \leqslant i$

and

(28) $m(y_k,\beta) > i.$

It follows from (17) that inequalities (27) and (28) do not depend on the choice of the vectors $u_i,v_i, \ldots, u_k,v_k$.

If the inequality (27) is fulfilled then it follows from the definitions of functions r_i (21) and ρ_i (18) that

$$r_{i+1}(y_{i-1},u_i,v_i) = r_i(y_{i-1}) \text{ for all } (u_i,v_i) \in P \times Q.$$

Therefore according to (25) we have

$$r^*_{i+1}(y_{i-1}) = r^2_i(y_{i-1}).$$

Clearly in this case inequality (26) is fulfilled. Assume now that inequality (28) is valid. To simplify the subsequent calculations we shall introduce the notation

$$h^{(i)}(y_{i-1}) = x_*, \quad \tau_i = t_*, \quad \tau_{i+1} = t^*, \quad u_i = u, \quad v_i = v,$$

$$h^{(i+1)}(y_{i-1},u_i,v_i) = x^*(u,v) = x(t^*,t_*,x_*,u,v),$$

$$r_i(y_{i-1}) = r_*, \quad r_{i+1}(y_{i-1},u_i,v_i) = r(u,v).$$

Utilizing this notation we express the quantities r_* and $r(u,v)$ under the fulfillment of inequality (28) as

$$\begin{aligned} & r_* = \operatorname{dist}(x_*, W[t_*]), \\ (29) \qquad & \\ & r(u,v) = \operatorname{dist}(x^*(u,v), W[t^*]). \end{aligned}$$

Inequality (26) is now rewritten as

(30) $$\int_P \int_Q r^2(u,v)\mu_e(du)\nu_*(dv) \leqslant r^2_*(1 + 2\lambda(t^* - t_*)) + 2\alpha(\delta)(t^* - t_*).$$

Here, according to (9), $\mu_e = \breve{U}(t_*,x_*)$, $\nu_* = \breve{V}(t_*,x_*)$. From the construction of strategy $\breve{U}^e$ (cf. (9.2.10)) it follows that

(31) $$\max_\nu \langle s_*,\breve{f}(t_*,x_*,\mu_e,\nu)\rangle = \min_\mu \max_\nu \langle s_*,\breve{f}(t_*,x_*,\mu,\nu)\rangle,$$

(32) $$s_* = x_* - w_*, \quad w_* = e(t_*,x_*),$$

where the function e is defined by relations (2.4.2) and (2.4.4). In the case under consideration $W[t_*] \neq \phi$, therefore

(33) $$\|s_*\| = \operatorname{dist}(x_*, W[t_*]) = r_*.$$

We utilize the condition of μ-stability of the set $\mathcal{W}$. According to the definition of the functions r_i and ρ_i, it follows from inequality (28) that $(t_*,x_*) \notin \mathcal{M}^{2\beta}$ and $r_* = \|x_* - w_*\| \leq \beta$. Thus we have the estimate $\mathrm{dist}((t_*,w_*),\mathcal{M}) \geq \beta$ from which it follows that by selecting δ sufficiently small ($\delta = \mathrm{diam}\ \Delta$, $t^* - t_* \leq \delta$) the relation

(34) $$\{(t,w(t)) \colon t_* \leq t \leq t^*\} \cap \mathcal{M} = \phi$$

is fulfilled.

Here $w(\cdot)$ is the solution of the differential equation

(35) $$\dot{w}(t) = \tilde{f}(t,w(t),\mu_t,\nu^*), \quad w(t_*) = w_*.$$

In this equation we select $\nu^* \in \mathrm{rpm}(\mathcal{Q})$ from the condition

(36) $$\min_{\mu} \langle s_*,\tilde{f}(t_*,x_*,\mu,\nu^*)\rangle = \max_{\nu} \min_{\mu} \langle s_*,\tilde{f}(t_*,x_*,\mu,\nu)\rangle.$$

Taking relation (34) into account, it follows from the condition of μ-stability of the set $\mathcal{W}$ (cf. Section 9.2.3) that a measurable function $\mu(\cdot)\colon [t_*,t^*] \to \mathrm{rpm}(\mathcal{P})$ can be selected in such a manner that the solution of equation (35) will satisfy the condition

(37) $$w(t^*) \in \mathcal{W}[t^*].$$

We define the measure $\mu^* \in \mathrm{rpm}(\mathcal{P})$ by the equality

$$\mu^*(G) = (t^* - t_*)^{-1} \int_{t_*}^{t^*} \mu_t(G)dt,$$

where $G \subset \mathcal{P}$ is a Borel set. It can be shown that

$$\|w(t^*) - w_* - (t^* - t_*)\tilde{f}(t_*,w_*,\mu^*,\nu^*)\| \leq (t^* - t_*)\alpha_1(t^* - t_*),$$

where $\alpha_1(\delta) \to 0$ as $\delta \to 0$. It follows from (29) and (37) that

(38) $$r(u,v) \leq \|x^*(u,v) - w_* - (t^*-t_*)\tilde{f}(t_*,w_*,\mu^*,\nu^*)\| + (t^*-t_*)\alpha_1(t^*-t_*).$$

Recall that $x^*(u,v) = x(t^*,t_*,u,v)$, where $x(\cdot,t_*,x_*,u,v)$ is a solution of the differential equation

$$\dot{x}(t) = f(t,x(t),u,v), \quad x(t_*) = x_*.$$

Therefore

$$\|x^*(u,v) - x_* - (t^* - t_*)f(t_*,x_*,u,v)\| \leq (t^* - t_*)\alpha_2(t^* - t_*),$$

$$\alpha_2(\delta) \to 0 \quad \text{as} \quad \delta \to 0.$$

It follows from (32), (33), and (38) that

$$r^2(u,v) \leqslant r_*^2 + 2(t^* - t_*)\langle s_*, f(t_*,x_*,u,v)\rangle$$

$$- 2(t^* - t_*)\langle s_*, \breve{f}(t_*,w_*,\mu^*,\nu^*)\rangle + (t^* - t_*)\alpha_3(t^* - t_*),$$

$$\alpha_3(\delta) \to 0 \quad \text{as} \quad \delta \to 0.$$

If we utilize the Lipschitz condition for the function $w \to f(t_*,w,\mu^*,\nu^*)$, then the preceding inequality can be rewritten as

$$r^2(u,v) \leqslant r_*^2(1 + 2\lambda(t^* - t_*))$$

$$+ 2(t^* - t_*)\langle s_*, f(t_*,x_*,u,v) - \breve{f}(t_*,x_*,\mu^*,\nu^*)\rangle$$

$$+ (t^* - t_*)\alpha_3(t^* - t_*).$$

Utilizing notation (9.2.1) we obtain

$$\int\int r^2(u,v)\mu_e(du)\nu_*(dv) \leqslant r_*^2(1 + 2\lambda(t^* - t_*))$$

$$+ 2(t^* - t_*)\langle s_*, \breve{f}(t_*,x_*,\mu_e,\nu_*) - \breve{f}(t_*,x_*,\mu^*,\nu^*)\rangle$$

$$+ (t^* - t_*)\alpha_3(t^* - t_*).$$

It follows from (31), (36), and (9.2.11) that here

$$\langle s_*, \breve{f}(t_*,x_*,\mu_e,\nu_*) - \breve{f}(t_*,x_*,\mu^*,\nu^*)\rangle \leqslant 0.$$

We thus arrive at estimate (30). The proof of Theorem 9.4.1 is completed.

9.4.4. It was mentioned in Section 9.2 that from the alternatives formulated in Theorem 9.2.1 one can deduce the existence of saddle points in the class of mixed positional strategies for differential games with various types of payoff functionals. Assume for definiteness that the payoff functional is of the form $\gamma(x(\cdot)) = \sigma(x(\theta))$, where $\sigma : R^n \to R^1$ is a continuous function. For such payoff functionals the corresponding differential game possesses a saddle point in the class of mixed strategies $\breve{U}: T \times R^n \to \text{rpm}(\mathcal{P})$ and $\breve{V}: T \times R^n \to \text{rpm}(\mathcal{Q})$, i.e. for any initial position (t_0,x_0) there exists a pair of mixed strategies $\breve{U}_0$ and $\breve{V}_0$ such that

$$\max\{\sigma(x(\theta)): x(\cdot) \in X_1(t_0,x_0,\breve{U}_0)\}$$

$$= \min_{\breve{U}} \max\{\sigma(x(\theta)): x(\cdot) \in X_1(t_0,x_0,\breve{U})\} = \omega^0(t_0,x_0)$$

$$= \max_{\breve{V}} \min\{\sigma(x(\theta)): x(\cdot) \in X_2(t_0,x_0,\breve{V})\}$$

$$= \min\{\sigma(x(\theta)): x(\cdot) \in X_2(t_0,x_0,\breve{V}_0)\}.$$

Here $X_1(t_0,x_0,\tilde{U})$ and $X_2(t_0,x_0,\tilde{V})$ are sets of motions (deterministic ones) defined in accordance with the idealized constructions in Section 9.2.

Consider the stochastic procedure $(\tilde{U}_0,\Delta)$ and $(\tilde{V}_0,\Delta)$ which approximate the optimal mixed strategies U_0 and V_0. The following assertion is valid.

Let U_0 and V_0 be optimal mixed strategies (in the sense indicated above). Let U and V be arbitrary mixed strategies for the first and second players. It is assumed that all the strategies satisfy the condition of measurability indicated in Section 9.4.1. Then, for any numbers $\epsilon > 0$ and $p < 1$, one can find a number $\delta > 0$ such that for an arbitrary partition Δ satisfying the bound $\operatorname{diam}\Delta < \delta$ for the set of random motions generated by stochastic control procedures $(\tilde{U}_0,\Delta)$ and $(\tilde{V},\Delta)$, the inequality $\sigma(x(\theta)) \leqslant \omega^0(t_0,x_0) + \epsilon$ is satisfied with probability of at least p and for the set of random motions generated by stochastic procedures $(\tilde{U},\Delta)$ and $(\tilde{V}_0,\Delta)$ the inequality $\sigma(x(\theta)) \geqslant \omega^0(t_0,x_0) - \epsilon$ will be fulfilled with probability of at least p.

As was done in Chapter 3 for pure positional strategies, we can define here optimal mixed strategies U_0 and V_0 in the form of strategies extremal to the corresponding stable bridges. Therefore the assertion presented here follows from Theorems 9.4.1 and 9.4.2.

9.5. Stochastic Control with Guidance

In the preceding section stochastic control procedures were considered which approximate mixed positional strategies. This approximation was justified under the assumption that partitions selected by the first and the second players coincide and that the selection of random controls $u(t) = u(\tau_i)$ and $v(t) = v(\tau_i)$, $\tau_i \leqslant t < \tau_{i+1}$, are mutually independent. In this section, we shall consider stochastic procedures with guidance. It is assumed that players measure the current position of a game imprecisely and cannot determine the control selected by the opponent. Therefore we don't require here that the partitions selected by the first and second player coincide. The condition of independence of controls for the players on short time intervals will be a natural consequence of informational errors.

9.5.1. Before proceeding to the definition of stochastic control procedures with guidance we shall discuss the possibilities of stochastic procedures $(\tilde{U},\Delta^{(1)})$ and $(\tilde{V},\Delta^{(2)})$. In Section 9.4 it is assumed that functions U and V are measurable. This assumption is required for a correct construction of probability spaces of random motions. It is also assumed therein that $\Delta^{(1)} = \Delta^{(2)} = \Delta = (t_0 = \tau_0 < \tau_1 < \cdots < \tau_{k+1} = \theta)$ and on each interval $[\tau_i,\tau_{i+1})$ the first and second players select mutually independent random controls $u(t) = u_i$, $v(t) =$

v_i, $t \in [\tau_i,\tau_{i+1})$. Note that the condition of mutual independence of controls for the first and second players is a basic one. In Section 9.4.2 this condition formally expresses the fact that the probability distribution of pairs of controls (u_i,v_i) was defined by the product of probability measures μ_i and ν_i.

We shall try to relax the assumption that the subdivisions $\Delta^{(1)}$ and $\Delta^{(2)}$ coincide. Let the first and second players select partitions $\Delta^{(1)} = (t_0 = \tau_0^{(1)} < \tau_1^{(1)} < \cdots < \tau_{k+1}^{(1)} = \theta)$ and $\Delta^{(2)} = (t_0 = \tau_0^{(2)} < \tau_1^{(2)} < \cdots < \tau_{m+1}^{(2)} = \theta)$ respectively. In this case one can also correctly construct a probability space of random motions generated by the pair of stochastic procedures $(U,\Delta^{(1)})$ and $(V,\Delta^{(2)})$. The question arises: can assertions analogous to Theorems 9.4.1 and 9.4.2 be proved for stochastic control procedures $(U,\Delta^{(1)})$ and $(V,\Delta^{(2)})$ where $\Delta^{(1)} \neq \Delta^{(2)}$? Consider the following example.

Let the motion of the controlled system be described by equations

$$\dot{x}_1 = u \cdot v_1, \qquad \dot{x}_2 = u + v_2,$$

where $|u| = 1$, $|v_1| = 1$, $|v_2| \leqslant 4$. Let $t_0 = 0$, $\theta = 1$, $x_0 = (0,0)^T$. Consider the problem of an (M, N)-approach in which it is required that $x_1(\theta) = 0$, $x_2(\theta)$ be arbitrary and the phase restriction be absent. Thus

$$M = \{(t,x_1,x_2) : t = 1,\ x_1 = 0,\ x_2 \in R^1\},$$

$$N = [0,1] \times R^2.$$

It follows from the arguments presented in Section 9.1.2 that a mixed positional strategy U_0 presents a solution of the given (M, N)-approach problem. This strategy U_0 assigns to each position (t,x_1,x_2) the probability measure μ_0 concentrated with equal weights at the two points $u_* = 1$ and $u^* = -1$. However, here we have in mind an idealized solution. Let us investigate what will happen if the players select stochastic control procedures $(U,\Delta^{(1)})$ and $(V,\Delta^{(2)})$ (here one can assume, in particular, that $U = U_0$).

Let the second player select a partition $\Delta^{(2)} = \Delta(\alpha) = (0 = \tau_0(\alpha) < \tau_1(\alpha) < \cdots < \tau_{m+1}(\alpha) = 1)$ where $\tau_0(\alpha) = 0$, $\tau_{j+1}(\alpha) = \tau_j(\alpha) + \alpha^2$ for j even, and $\tau_{j+1}(\alpha) = \tau_j(\alpha) + \alpha$ for j odd. Let $|x_2| \leqslant 3\alpha$ and

$$v^{(\alpha)}(x_2) = \begin{cases} (1,\ -1 - x_2\alpha^{-1})^T, & 0 \leqslant x_2 \leqslant 3\alpha \\ (-1,\ 1 - x_2\alpha^{-1})^T, & -3\alpha \leqslant x_2 < 0. \end{cases}$$

Define probability measures $\nu_1^{(\alpha)}(x_2)$ and ν_2 on the set $Q = \{(v_1,v_2)^T : |v_1| = 1,\ |v_2| \leqslant 4\}$. Assume that the measure $\nu_1^{(\alpha)}(x_2)$ is concentrated at the point $v^{(\alpha)}(x_2)$ and the measure ν_2 is located with equal weights at two points $v_* = (+1,0)^T$ and $v^* = (-1,0)^T$. Define the strategy $(t,x_1,x_2) \to V_\alpha(t,x_1,x_2)$ for $t = \tau_j(\alpha)$, $x_1 \in R^1$ and $|x_2| \leqslant 3\alpha$ as follows:

$$\tilde{V}_\alpha(\tau_j(\alpha),x_1,x_2) = \begin{cases} \nu_1^{(\alpha)}(x_2) & \text{for } j \text{ odd} \\ \nu_2 & \text{for } j \text{ even.} \end{cases}$$

For $t \neq \tau_j(\alpha)$ and $|x_2| > 3\alpha$ the function $\tilde{V}_\alpha$ can be defined arbitrarily; for example one may assume that $\tilde{V}_\alpha(t,x_1,x_2) = \nu_2$ ($t \neq \tau_j(\alpha)$), $x_1 \in R^1$, $|x_2| > 3\alpha$).

Let the first player select a mixed strategy $\hat{U}$ and the partition $\Delta = (0 = \tau_0 < \tau_1 < \cdots < \tau_{k+1} = 1)$. Consider the random controls $u(t)$, $v(t)$ generated by the pair of stochastic control procedures $(\hat{U},\Delta)$ and $(\tilde{V}_\alpha,\Delta(\alpha))$. One can show that the partitions $\Delta(\alpha)$ and functions $\tilde{V}_\alpha$ constructed above possess the following property.

For any stochastic control procedure for the first player $(\hat{U},\Delta)$ and for any number $\beta > 0$, one can find an $\alpha > 0$ such that for random controls $u(\cdot)$ and $v(\cdot)$ generated by the pair of procedures $(\hat{U},\Delta)$ and $(\tilde{V}_\alpha,\Delta(\alpha))$ with probability 1 the equality

$$u(t) = v_1(t) \quad \text{for} \quad t \in T_\beta$$

is fulfilled (or equivalently $u(t)\cdot v_1(t) = 1$ for $t \in T_\beta$) where the measure of the set T_β is at least $1 - \beta$.

It follows from this property that the inequality $x_1(1) \geqslant 1 - \beta$ will be fulfilled with probability 1, i.e. in the given example position $(1, x(1))$ cannot approach the target set. We thus arrive at the following conclusion: if it is assumed that players can select stochastic procedures $(\hat{U},\Delta^{(1)})$ and $(\tilde{V},\Delta^{(2)})$ where partitions $\Delta^{(1)}$ and $\Delta^{(2)}$ do not coincide, then, as the example above shows, there may not exist a stochastic procedure which will approximate an ideal solution obtained in the class of positional mixed strategies. It is not difficult to understand the reason why stochastic procedures $(\tilde{V},\Delta^{(2)})$ may destroy the performance of stochastic procedures $(\hat{U},\Delta^{(1)})$. Let $[\tau_{i-1}^{(1)},\tau_i^{(1)}) \cap [\tau_{j-1}^{(2)},\tau_j^{(2)}) \neq \phi$ and let $\tau_{i-1}^{(1)} < \tau_{j-1}^{(2)}$. At time $t = \tau_{i-1}^{(1)}$ the first player selects a random control $u(t) = u_i$ $(\tau_{i-1}^{(1)} \leqslant t < \tau_i^{(1)})$. It would be advantageous for the first player if the selected control u_i were unknown to the second player. However, it may turn out that the second player at time $t = \tau_{j-1}^{(2)} > \tau_{i-1}^{(1)}$ can determine which control was selected by the first player with sufficient reliability. Therefore the second player may actually select his or her control $v(t) = v_j$ $(\tau_{j-1}^{(2)} \leqslant t < \tau_j)$ depending on the control u_i. Thus the condition of mutual independence (or a weak correlation) of controls for the players on short time intervals will be violated. This is indeed the case in our example. For the time being we shall confine ourselves to this explanation.

Thus, in order to approximate an ideal solution obtained in the class of mixed positional strategies, it is necessary that stochastic procedures $(\hat{U},\Delta^{(1)})$ and $(\tilde{V},\Delta^{(2)})$ satisfy the condition of mutual independence of controls on short time intervals. However, as was established above, this condition may be violated. How can we overcome this obstacle?

For this purpose one can propose modified stochastic control procedures $(\check{U},\Delta^{(1)},\Delta_*^{(1)})$. Here, as above, $\check{U}$: $T \times R^n \to \text{rpm}(P)$ is a measurable function and $\Delta^{(1)} = (t_0 = \tau_0^{(1)} < \tau_1^{(1)} < \cdots < \tau_{k+1}^{(1)} = \theta)$ is a partition selected by the first player. At times $t = \tau_{i-1}^{(1)}$ $(i = 1, ..., k+1)$, the first player defines the probability measure $\mu_i = \check{U}(\tau_{i-1},x(\tau_{i-1}))$. It is assumed that each time interval $[\tau_{i-1}^{(1)},\tau_i^{(1)})$ is subdivided into half-intervals $[t_{j,i-1},t_{j+1,i-1})$, $j = 0,1, ..., \ell$, $t_{0,i-1} = \tau_{i-1}$, $t_{\ell+1,i-1} = \tau_i$. The points $t_{j,i}$, $i = 0,1, ..., k+1$, $j = 0, ..., \ell+1$ form a partition $\Delta_*^{(1)}$. At each time $t = t_{j,i-1}$ $(j = 0,1, ..., \ell)$, the first player selects a constant control $u(t) = u(t_{j,i-1})$, $t_{j,i-1} \leqslant t < t_{j+1,i-1}$; for this choice the distribution of random vectors $u(t_{j,i-1})$ is determined by the probability measure μ_i. Stochastic control procedures $(\check{V},\Delta^{(1)}, \Delta_*^{(2)})$ are determined analogously. If one requires that $\operatorname{diam}\Delta_*^{(1)}$ and $\operatorname{diam}\Delta_*^{(2)}$ be substantially smaller than $\operatorname{diam}\Delta^{(2)}$ and $\operatorname{diam}\Delta^{(1)}$ respectively, then it is possible to assert for $(\check{U},\Delta^{(1)},\Delta_*^{(1)})$ and $(\check{V},\Delta^{(2)}, \Delta_*^{(2)})$ the condition of mutual independence (more precisely – of weak correlation) of the players' controls. However, we are not going to concentrate on these constructions in detail. We shall proceed to a discussion of another method of stochastic control under which the condition of independence of controls is fulfilled.

9.5.2. We begin with the following remark. Assume that the players measure the current position of the game imprecisely. Then one cannot reliably determine during a short time interval which control was realized by the opponent. In this case, the condition of independence of players' controls will be a natural consequence of informational noise. However, when introducing informational noise one should consider the stability of the solution with respect to this noise. As was shown in Section 8.1, pure positional strategies do not possess the required stability. Therefore, using our experience obtained in Chapter 8, we shall construct a stochastic procedure with guidance. This procedure will be stable relative to informational noise, the presence of which will assure that the condition of mutual independence of the players' control is fulfilled.

We shall present a description of a *stochastic control procedure with guidance* (SCPG). As in the deterministic case considered in Section 8.2.3, we shall assume that SCPG of the first player is defined by the triple

$$(1) \qquad (\check{U}, \psi, \chi).$$

Here, as before,

$$\psi: (t_1,t_2,x,w) \to \psi(t_1,t_2,x,w): T_+^2 \times R^n \times R^n \to R^n$$

is the transition function of the guidance (recall that $T_+^2 = (t_1,t_2) \in T \times T$: $t_1 \leqslant t_2$). As in Section 8.2.3, the function χ : $T \times R^n \to R^n$ is used to determine the initial position of the guidance. Function $\tilde{U}$ assigns a probability distribution on the set $\mathcal{P}$ according to which the first player will implement the selection of random controls. Thus, we have here

$$\tilde{U}: (t,x,w) \to \tilde{U}(t,x,w): T \times R^n \times R^n \to \mathrm{rpm}(\mathcal{P}).$$

When constructing probability spaces of random motions generated by SCPG, the following condition of measurability is utilized: for any $t \in T$ and any $(t_1,t_2) \in T_+^2$ the functions

$$(x,w) \to \tilde{U}(t,x,w): R^n \times R^n \to \mathrm{rpm}(\mathcal{P}),$$

$$(x,w) \to \psi(t_1,t_2,x,w): R^n \times R^n \to R^n,$$

$$x \to \chi(t,x): R^n \to R^n$$

are Borel measurable. Note that the measurability of the first function is determined in the same manner as in Section 9.4.1 for the functions $x \to \tilde{U}(t,x)$ and $x \to \tilde{V}(t,x)$.

Assumptions of measurability of functions $\tilde{U}$, ψ, and χ are required for mathematical formalization of the problems under investigation. From physical aspects these assumptions do not narrow down the range of problems that could be incorporated into the proposed set-up since the content of these problems allows us to discretize the probabilistic constructions utilized herein and then the above stated assumptions will always be fulfilled.

We shall describe how an SCPG forms the control for the first player. Let $\Delta^{(1)} = (t_0 = \tau_0^{(1)} < \tau_1^{(1)} < \cdots < \tau_{k+1}^{(1)} = \theta)$ be a partition selected by the first player. At the initial time $t = t_0 = \tau_0^{(1)}$ the first player determines, using the function χ, the initial position of the guidance

$$w(t_0) = \chi(t_0,h(t_0)).$$

Here $h(t_0)$ is the result of an imprecise measurement, by the first player, of the initial position $x(t_0)$ of the controlled system (1.2.1). The constant control $u(t) = u_0$, $t_0 \leqslant t < \tau_1^{(1)}$ on the first step is determined by a random trial for which the probability distribution of vectors $u_0 \in P$ is defined by the measure

$$\mu_0 = \tilde{U}(t_0,h(t_0),w(t_0)).$$

The control $u(t) = u_0$ $(t_0 \leqslant t < \tau_1)$ selected in this manner coupled with a control for the second player generates the motion of the system (1.2.1) on the time interval $[t_0,\tau_1^{(1)}]$.

The position of the guidance at time $t = \tau_i^{(1)}$ $(i = 1,2, ..., k)$ is determined according to a transition function ψ by the equality

$$w(\tau_i^{(1)}) = \psi(\tau_{i-1}^{(1)},\tau_i^{(1)},h(\tau_{i-1}^{(1)}),w(\tau_{i-1}^{(1)})).$$

Here $h(\tau_{i-1}^{(1)})$ is an n-dimensional vector obtained as a result of imprecise measurement of the point $x(\tau_{i-1}^{(1)})$, $w(\tau_{i-1}^{(1)})$ is the position of the guidance at time $t = \tau_{i-1}^{(1)}$. At time $t = \tau_i^{(1)}$, the first player determines a probability measure

$$\mu_i = \widehat{U}(\tau_i^{(1)}, h(\tau_i^{(1)}), w(\tau_i^{(1)})),$$

using the function $\widehat{U}$ where $h(\tau_i^{(1)})$ is the result of the imprecise measurement carried out at time $t = \tau_i^{(1)}$. The first player selects the constant control $u(t) = u_i$, $\tau_i^{(1)} \leqslant t < \tau_{i+1}^{(1)}$, as a result of a random trial in which the probability distribution of vectors $u_i \in P$ is given by the measure μ_i. This control, together with a control $v(t)$ selected by the second player, generates the motion of the controlled system (1.2.1) on the interval $[\tau_i^{(1)}, \tau_{i+1}^{(1)}]$. This control procedure is implemented successively on the time intervals $[\tau_0^{(1)}, \tau_1^{(1)})$, $[\tau_1^{(1)}, \tau_2^{(1)})$, ..., $[\tau_k^{(1)}, \tau_{k+1}^{(1)} = \theta)$.

Observe that in an SCPG the informational noise affects only the measurement of positions of the controlled system; the locations of the guidance are calculated precisely.

We shall assume that the second player selects the partition $\Delta^{(2)} = (t_0 = \tau_0^{(2)} < \tau_1^{(2)} < \cdots < \tau_{m+1}^{(2)} = \theta)$ and forms the piecewise-constant controls $v(t) = v_j$, $\tau_j^{(2)} \leqslant t < \tau_{j+1}^{(2)}$, $j = 0,1,2, \ldots, m$. Below we shall clarify how the vectors $v_j \in Q$ are selected; for the time being, however, this assumption will be sufficient.

Consider the union $\Delta \overset{\text{def}}{=} \Delta^{(1)} \cup \Delta^{(2)}$. Enumerate the numbers belonging to the set Δ in increasing order. We thus obtain that $\Delta \overset{\text{def}}{=} (t_0 < t_1 < \cdots < t_{\ell+1} = \theta)$. For any number $t_j \in \Delta$ one of the following cases is possible: either $t_j \in \Delta^{(1)} \setminus \Delta^{(2)}$, $t_j \in \Delta^{(2)} \setminus \Delta^{(1)}$, or $t_j \in \Delta^{(1)} \cap \Delta^{(2)}$. In accordance with these three possibilities, we subdivide the set of indices $T = \{0,1, \ldots, \ell\}$ into three subsets:

$$T_1 \overset{\text{def}}{=} \{j \in T : t_j \in \Delta^{(1)} \setminus \Delta^{(2)}\},$$

$$T_2 \overset{\text{def}}{=} \{j \in T : t_j \in \Delta^{(2)} \setminus \Delta^{(1)}\},$$

$$T_3 \overset{\text{def}}{=} \{j \in T : t_j \in \Delta^{(1)} \cap \Delta^{(2)}\}.$$

Set

$$Z_j \overset{\text{def}}{=} \begin{cases} P \times R^n & \text{for } j \in T_1, \\ Q & \text{for } j \in T_2, \\ P \times R^n \times Q & \text{for } j \in T_3, \end{cases}$$

$$Y_i \overset{\text{def}}{=} \prod_{j=0}^{i} Z_j.$$

Definitions of the sets $\mathcal{Z}_j$ and $\mathcal{Y}_i$ can be explained as follows. The collections $y_i = (z_0, z_1, \ldots, z_i)$ are the *elements* of the set $\mathcal{Y}_i$. These collections contain information about the controlled process on the time interval $[t_0, t_i]$. Let $t_j \leqslant t_i$. First assume that $j \in T_1$. Then at time t_j, the first player measures the current state of the system and obtains a vector $h(t_j) \in R^n$. Moreover, the first player selects at this time the control $u(t) = u_j$ $(t_j = \tau_s^{(1)} \leqslant t < \tau_{s+1}^{(1)})$. The pair of vectors $(u_j, h(t_j))$ forms an element $z_j \in \mathcal{Z}_j$ contained in the collection $y_i =$

$(z_0, z_1, \ldots, z_i)$. If $j \in T_2$, then at time t_j the control $v(t) = v_j$ $(t_j = \tau_r^{(2)} \leqslant t < \tau_{r+1}^{(2)})$ is selected. In this case we obtain the vector $v_j = z_j$. Finally, if $j \in T_3$ then at time $t = t_j$ we arrive at the pair of vectors $(u_j, h(t_j))$ and at the same instant, the second player selects the control $v(t) = v_j$. The triple of vectors $(u_j, h(t_j), v_j)$ forms an element z_j contained in the collection y_i.

Note the following: when describing the information associated with the actions of the first player we specify a pair $(u_j, h(t_j))$ consisting of the control u_j and the result of a measurement $h(t_j)$. Information associated with actions of the second player is characterized only by his control v_j. This is because the constructions presented here are required for solutions of problems posed to the first player. Therefore the control procedure for the first player is described in more detail than the admissible actions of his opponent – the second player. Observe also that the class of admissible control procedures for the second player is sufficiently broad, and in particular contains SCPG.

Below we shall define the probability spaces

$$(\mathcal{Y}_i, \mathbf{B}_i, \eta_i).$$

Here $\mathbf{B}_i$ is a σ-algebra of Borel sets $G \subset \mathcal{Y}_i$ and η_i is a probability measure on $\mathbf{B}_i$. The construction of these probability spaces is implemented with the aid of transition probabilities. Let $G_1 \subset \mathcal{P}$, $G_2 \subset R^n$, and $G_3 \subset \mathcal{Q}$ be Borel sets. Consider the Borel measurable functions

$$(2) \qquad (h,w) \to \mu_i(G_1, h, w)\colon R^n \times R^n \to [0,1],$$

$$(3) \qquad y_{i-1} \to \zeta_i(G_2, y_{i-1})\colon \mathcal{Y}_{i-1} \to [0,1],$$

$$(4) \qquad y_{i-1} \to \nu_i(G_3, y_{i-2})\colon \mathcal{Y}_{i-1} \to [0,1].$$

We shall show how these functions are defined. Set

$$\mu_i(\cdot, h, w) = \breve{U}(t_i, h, w),$$

where $\breve{U}$ is the function appearing into the triple $(\breve{U}, \psi, \chi)$ (1). According to the description of SCPG (1), the function μ_i (2) will be measurable.

Define the function ζ_i (3). According to the control procedure described above, if $i \in T_1$ or $i \in T_3$ then at time $t = t_i$ the first player performs an imprecise measurement. We assume that for the results of these measurements – the vectors h – a probability distribution is available. It is natural to assume that this probability distribution depends in some way on the true position of the system – the vector $x(t_i)$, or on the motion of the system $x(t)$ realized on the time interval $[t_0,t_i]$. For example, consider the case when the vector h is uniformly distributed on the sphere of radius $\alpha > 0$ with the center at point $x(t_i)$, or one may consider the case when $h = x(t_i - \beta)$ where a probability distribution is defined on the lag β. Since the initial position (t_0,x_0) is fixed, by knowing the controls of players on the time interval $[t_0,t_i)$, one can determine the motion of the system on the interval $[t_0,t_i]$. Next, if the collection y_{i-1} is known so are the piecewise constant controls of the players for $t \in [t_0,t_i)$. Therefore one can assume that a function ζ_i (3) is defined which assigns to the collection y_{i-1} a probability measure characterizing the distribution of the vectors h. The condition of measurability of this function is a purely technical requirement. Actually, as a rule, this function satisfies a stronger condition of continuity.

Finally, consider the function ν_i (4). At time $t = t_i$ ($i \in T_2$ or $i \in T_3$) the second player selects random controls $v \in \mathcal{Q}$ whose probability distribution is assigned by the measure $\nu_i^* \in \text{rpm}(\mathcal{Q})$. It may be assumed that the measure ν_i^* is selected by the second player as a result of information on the controlled process, i.e. a function $g_i \to \nu_i^*(\cdot,g_i)$ is defined where the argument g_i characterizes the information available to the second player. For example, one may consider the case $g_i = (x^*(t),\ t_0 \leqslant t \leqslant t_i)$, where $x^*(\cdot)$ is the result of an imprecise measurement of the trajectory of the system (1.2.1) on the interval $[t_0,t_i]$. We shall assume that the probability distribution of elements g_i depends on the actual motion $x(t)$ ($t_0 \leqslant t \leqslant t_i$) of the system (1.2.1). Since this motion is uniquely determined by the collection y_{i-1}, one can assume that a function $y_{i-1} \to \zeta_i^*(dg_i,y_{i-1})$ exists. We thus obtain the integral

$$(5) \qquad \nu_i(G,y_{i-1}) = \int \nu_i^*(G,g_i)\zeta_i^*(dg_i,y_{i-1}),$$

where $G \subset \mathcal{Q}$, is a Borel set and the integration is carried out over the whole informational space. We are assuming that here all the conditions required for the validity of the integration are fulfilled and that the function ν_i (4) is measurable. These are again purely technical conditions which are fulfilled for a wide variety of problems.

Below, a solution of an $(\mathcal{M}, \mathcal{N})$-approach problem is described and it will be shown how the function μ_i is determined in this problem. As far as functions ζ_i and ν_i are concerned, it is essential only that they exist and satisfy certain requirements.

We now proceed to the construction of the probability spaces $(\mathcal{Y}_i, \mathbf{B}_i, \eta_i)$. For $i = 0$ we have $\mathcal{Y}_0 = \mathcal{Z}_0 = \mathcal{P} \times R^n \times \mathcal{Q}$. At the initial time we have $\zeta_0 \in \text{rpm}(R^n)$ and $\nu_0 \in \text{rpm}(\mathcal{Q})$. Moreover, the functions $\tilde{U}$: $T \times R^n \times R^n \to \text{rpm}(\mathcal{P})$ and $\chi : T \times R^n \to R^n$ define a mapping $h \to \mu_0(\cdot, h, \chi(t_0, h)) = \tilde{U}(t_0, h, \chi(t_0, h))$. Set

$$(6) \qquad \eta_0(G_1 \times G_2 \times G_3) = \left[\int_2 \mu_0(G_1, h, \chi(t_0, h))\zeta_0(dh)\right] \cdot \nu_0(G_3)$$

where $G_1 \subset \mathcal{P}$, $G_2 \subset R^n$, $G_3 \subset \mathcal{Q}$ are Borel sets. The function $h \to \mu_0(G_1, h, \chi(t_0, h))$ is Borel measurable, therefore the integration in (6) makes sense. Relation (6) defines the unique measure $\eta_0 \in \text{rpm}(\mathcal{P} \times R^n \times \mathcal{Q})$. It follows from (6) that the random controls for the second player at time $t = t_0$ are selected independently of the random controls for the first player and the results of the measurement of the initial position obtained by the first player. This is expressed by the fact that the measure $\nu_0(G_3)$ is multiplied by the integral appearing in (6) in square brackets.

Next we continue the construction by indication. Let a probability space $(\mathcal{Y}_{i-1}, \mathbf{B}_{i-1}, \eta_{i-1})$ be given. First assume that $i \in T_1$. In this case $\mathcal{Z}_i = \mathcal{P} \times R^n$. Define the transition probability (cf. [46, 52])

$$(7) \qquad \xi_i: \mathcal{Y}_{i-1} \to \text{rpm}(\mathcal{Z}_i)$$

such that for any Borel set $G \subset \mathcal{Z}_i = \mathcal{P} \times R^n$ the function $y_{i-1} \to \xi_i(G, y_{i-1})$ is Borel measurable.

Consider first the function

$$w^{(i)}: \mathcal{Y}_{i-1} \to R^n$$

which assigns to the collection $y_{i-1} = (z_0, z_1, \ldots, z_{i-1})$ the position of the guidance at time $t = t_i$. It is evident from the description of a SCPG and the properties of the transition function ψ that the collection y_{i-1} uniquely determines the position of the guidance at time $t = t_i$ and the function $y_{i-1} \to w^{(i)}(y_{i-1})$ turns out to be Borel measurable.

Now, for any $y_{i-1} \in \mathcal{Y}_{i-1}$ and any Borel sets $G_1 \subset R^n$, $G_2 \subset R^n$, one can define the probability $\xi_i^*(G_1 \times G_2, y_{i-1})$ of the event $(h, w) \in G_1 \times G_2$. Here h is the result of an imprecise measurement carried out at time $t = t_i$, w is the position of the guidance at time $t = t_i$. The event $(h, w) \in G_1 \times G_2$ is considered under the condition that in the controlled process, the collection y_{i-1} has been realized. Since a unique point $w = w^{(i)}(y_{i-1})$ corresponds to this collection y_{i-1} we thus obtain that

$$(8)\qquad \xi_i^*(G_1 \times G_2, y_{i-1}) = \begin{cases} \zeta_i(G_1, y_{i-1}) & \text{for } w^{(i)}(y_{i-1}) \in G_2, \\ 0 & \text{for } w^{(i)}(y_{i-1}) \notin G_2, \end{cases}$$

where the probability $\zeta_i(G_1, y_{i-1})$ is determined by the function ζ (3). For an arbitrary $y_{i-1} \in Y_{i-1}$, equality (8) uniquely defines a measure $\xi_i^* \in \text{rpm}(R^n \times R^n)$. For any Borel set $G \subset R^n \times R^n$, the function $y_{i-1} \to \xi_i^*(G, y_{i-1})$ is Borel measurable.

To determine the transition probability ξ_i (7) for arbitrary Borel sets $G_1 \subset P$ and $G_2 \subset R^n$, we set

$$(9)\qquad \xi_i(G_1 \times G_2, y_{i-1}) = \int_{G_2 \times R^n} \mu_i(G_1, h, w)\xi_i^*(d(h,w), y_{i-1}).$$

This relation uniquely defines the transition probability (7).

Recall that $y_i = (y_{i-1}, z_i)$, $Y_i = Y_{i-1} \times Z_i$. The required probability measure $\eta_i \in \text{rpm}(Y_i)$ is defined by the relation

$$(10)\qquad \eta_i(G_1 \times G_2) = \int_{G_1} \xi_i(G_2, y_{i-1})\eta_{i-1}(dy_{i-1}),$$

where $G_1 \subset Y_{i-1}$ and $G_2 \subset Z_i = P \times R^n$ are Borel sets.

Assume that $i \in T_2$. In this case $Z_i = Q$. Using function ν_i (4) for the transition probability, we immediately obtain that

$$(11)\qquad \eta_i(G_1 \times G_2) = \int_{G_1} \nu_i(G_2, y_{i-1})\eta_{i-1}(dy_{i-1}),$$

where $G_1 \subset Y_{i-1}$, $G_2 \subset Z_i = Q$ are Borel sets.

Finally, if $i \in T_3$, then $Z_i = P \times R^n \times Q$. In this case for any Borel sets $G_1 \subset Y_{i-1}$, $G_2 \subset P \times R^n$, and $G_3 \subset Q$ we set

$$(12)\qquad \eta_i(G_1 \times G_2 \times G_3) = \int_{G_1} \xi_i(G_2, y_{i-1})\nu_i(G_3, y_{i-1})\eta_{i-1}(dy_{i-1}),$$

where transition probabilities ξ_i and ν_i are defined by the relations (9) and (4) respectively.

Relations (10) - (12) uniquely determine the probability measure $\eta_i \in \text{rpm}(Y_i)$. The definition of the probability η_i corresponds to the description of the controlled process presented above. Note that in the case when $i \in T_3$ the probability η_i is defined by relation (12) in which the product of the probabilities $\xi_i(G_2, y_{i-1})$ and $\nu_i(G_3, y_{i-1})$ appears under the integral sign. This means that the events $(u,h) \in G_2$ and $v \in G_3$ are mutually independent. Recall that in the case when $i \in T_3$, the first player carries out a measurement at time $t = t_i$, obtains a random vector $h = h(t_i)$, and selects a random control $u \in P$. At the very same time, the second player selects his or her control v. In this case it is natural to assume that these events are mutually independent.

Thus, if a SCPG problem (1) is defined and partitions $\Delta^{(1)}$, $\Delta^{(2)}$ and transition probabilities ζ_i (3), ν_i (4) are given, then this data uniquely determines the probability spaces $(Y_i, \mathbf{B}_i, \eta_i)$ $(i = 0,1, ..., \ell)$.

9.5.3. We shall define a SCPG for the first player which would, under the conditions stipulated below, yield an approximating solution for an (M, N)-approach problem.

Let W be a closed μ-stable bridge in an (M, N)-approach problem. The set $W[t,x]$ is defined by the relation (2.4.2). Note that the mapping $x \to W[t,x]$ is upper semicontinuous for any $t \in T$. Therefore one can define a function e: $T \times R^n \to R^n$ such that $e(t,x) \in W[t,x]$ for all $(t,x) \in T \times R^n$ and such that for any $t \in T$ the function $x \to e(t,x)$ will be Borel measurable (cf. Section A.3). Set

(13) $$\chi^0(t,x) = e(t,x), \quad (t,x) \in T \times R^n.$$

To define the transition function for the guidance, the function $\breve{V}_*$: $T \times R^n \times R^n \to \text{rpm}(Q)$ will be required; this function is selected from the condition

(14) $$\min_{\mu} \langle x - w, \breve{f}(t,x,\mu,\nu_*)\rangle = \max_{\nu} \min_{\mu} \langle x - w, \breve{f}(t,x,\mu,\nu)\rangle, \quad (\nu_* = \breve{V}_*(t,x,w)).$$

Here the vector $\breve{f}(t,x,\mu,\nu)$ is defined according to (9.2.1). Let $(t_*,w_*) \in T \times R^n$; denote by $\breve{X}_2(t_*,w_*,\nu_*)$ the set of solutions of the differential inclusion (9.2.3) satisfying the initial condition $x(t_*) = w_*$. Let $\nu_* = \breve{V}_*(t_*,x_*,w_*)$ and $t^* \in (t_*,\theta]$. Assume first that $(t_*,w_*) \in W$. Select a solution $w(\cdot) \in \breve{X}_2(t_*,w_*,\nu_*)$ which satisfies either the condition $(t^*,w(t^*)) \in W$ or the condition $\{(t,w(t))\colon t_* \leq t \leq t^*\} \cap M \neq \phi$. Such a solution exists according to the definition of a μ-stable bridge (cf. Section 9.2.3). Assume that the transition function of the guidance ψ^0 assigns to the collection (t_*,t^*,x_*,w_*) a point $w^* = w(t^*)$. If, however, $(t_*,w_*) \notin W$ we then select an arbitrary solution $w(\cdot) \in \breve{X}_2(t_*,w_*,\nu_*)$ and again we assume that $\psi^0(t_*,t^*,x_*,w_*) = w(t^*)$.

Note that function $\breve{V}_*$ satisfying condition (14) can be selected to be measurable. Furthermore, utilizing the closure of the set W, the transition function of the guidance can be constructed in such a manner so that for any $(t_*,t^*) \in T^2_+$ the function $(x_*,w_*) \to \psi^0(t_*,t^*,x_*,w_*)$ is Borel measurable.

Finally, the function $\breve{U}^0$: $T \times R^n \times R^n \to \text{rpm}(P)$ is selected from the condition

(15) $$\max_{\nu} \langle x - w, \breve{f}(t,x,\mu^0,\nu)\rangle = \min_{\mu} \max_{\nu} \langle x - w, \breve{f}(t,x,\mu,\nu)\rangle, \quad (\mu^0 = \breve{U}^0(t,x,w)).$$

Note that this function $(t,x,w) \to \mu^0(\cdot,t,x,w) = \breve{U}(t,x,w)$ can be selected in such a manner that for any Borel set $G \subset P$ the function $(t,x,w) \to$

$\mu^0(G,t,x,w)$ will be Borel measurable.

Thus for a given closed μ-stable bridge a SCPG for the first player

(16) $(\tilde{U}^0,\psi^0,\chi^0)$

is defined.

We now state the conditions under which the SCPG (16) will be considered. We shall assume that informational errors satisfy the constraints

$$\int_{R^n} \|h - x_0\|^2 \zeta_0(dh) \leqslant \beta_1,$$

(17)

$$\int_{R^n} \|h - x^{(i)}(y_{i-1})\| \zeta_i(dh,y_{i-1}) \leqslant \beta_2 \quad \text{for all } y_{i-1} \in Y_{i-1},\ i = 1,2, ..., \ell.$$

Here β_1, β_2 are positive parameters. Recall that $\zeta_0 \in \text{rpm}(R^n)$ is the probability measure which characterizes the results of the measurement $h(t_0)$ carried out at the initial time $t = t_0$. The probability measure $\zeta_i(dh,y_{i-1})$ defines the distribution of results of the measurement $h(t_i)$ carried out at time $t = t_i$ under the condition that the collection y_{i-1} is realized. The function $y_{i-1} \to x^{(i)}(y_{i-1})$ assigns to the collection y_{i-1} a point $x(t_i)$ into which the system (1.2.1) arrives at time $t = t_i$. Inequality (17) estimates the mathematical expectation of the deviation of the result of an observation from the actual position of the system.

We shall state the condition which imposes certain constraints on the informational possibilities for the second player. Let the first and second players select partitions $\Delta^{(1)}$ and $\Delta^{(2)}$. We form the set $\Delta = \Delta^{(1)} \cup \Delta^{(2)} = (t_0 < t_1 < \cdots t_{\ell+1} = \theta)$. Let $1 \leqslant i \leqslant \ell + 1$, $i \in T_2 \cup T_3$, $\delta > 0$. Define the set $\mathcal{E}_{i-1}(\delta) \subset Y_{i-1} \times Y_{i-1}$. Assume that $(y^{(1)}_{i-1}, y^{(2)}_{i-1}) \in \mathcal{E}_{i-1}(\delta)$ if and only if the collections $y^{(1)}_{i-1} = (z^{(1)}_0,z^{(1)}_1, ..., z^{(1)}_{i-1})$ and $y^{(2)}_{i-1} = (z^{(2)}_0,z^{(2)}_1, ..., z^{(2)}_{i-1})$ satisfy the conditions

$$z^{(1)}_j = z^{(2)}_j \quad \text{if} \quad t_j < t_i - \delta,$$

$$v^{(1)}_j = v^{(2)}_j \quad \text{if} \quad j \in T_2 \cup T_3, \quad j \leqslant i - 1.$$

Recall that $z^{(s)}_j = (u^{(s)}_j,h^{(s)}_j)$ for $j \in T_1$, $z^{(s)}_j = v^{(s)}_j$ for $j \in T_2$, $z^{(s)}_j = (u^{(s)}_j,h^{(s)}_j,v^{(s)}_j)$ for $j \in T_3$; here $s = 1, 2$.

We shall assume that there exists a function $\delta \to \beta(\delta)$ ($\beta(\delta) \to 0$ as $\delta \to +0$) such that for any $i \in \{1,2, ..., \ell + 1\}$, $(y^{(1)}_{i-1},y^{(2)}_{i-1}) \in \mathcal{E}_{i-1}(\delta)$ and a Borel set $G \subset Q$ the estimate

(18) $|\nu_i(G,y^{(1)}_{i-1}) - \nu_i(G,y^{(2)}_{i-1}| \leqslant \beta(\delta)$

is fulfilled.

We shall explain this condition. Relation (18) means that the second player is unable to recover with sufficient speed and precision the current controls and measurements for the first player. Measure ν_i "weakly depends" on which controls $u(t_j)$ and measurements $h(t_j)$ were realized in the SCPG for the first player when $t_j \in [t_i - \delta, t_i)$ (δ is small). Therefore random controls $v(t_i)$ will be weakly correlated with random controls $u(t_j)$ and measurements $h(t_j)$, $t_j \in [t_i - \delta, t_i)$. This condition is stated in a form convenient for further utilization. However, it should be recalled that the measure $\nu_i(G, y_{i-1})$ was defined by relation (5). Therefore estimate (18) follows from an analogous estimate

(19) $$|\zeta_i^*(G^*, y_{i-1}^{(1)}) - \zeta_i^*(G^*, y_{i-1}^{(2)})| \leqslant \beta(\delta).$$

Here the probability measure ζ_i^* defines the distribution of random results of measurements carried out by the second player. Condition (19) means that the observations are weakly correlated with random events $u(t_j)$ and $h(t_j)$, $t_j \in [t_i - \delta, t_i)$ in the SCPG for the first player. Condition (19) is fulfilled for a wide class of models describing distributions of informational errors. In particular, estimate (19) is valid when the measuring procedure involves a lag.

In order not to discriminate against the second player, one can assume that informational possibilities for the first player satisfy analogous constraints and the estimate

(20) $$|\zeta_i(G, y_{i-1}^{(1)}) - \zeta_i(G, y_{i-1}^{(2)})| \leqslant \beta(\delta),$$

where $G \subset R^n$ is a Borel set (recall that ζ is a probability measure which determines the distribution of measurement results carried out by the first player). Thus when estimates (19) and (20) are fulfilled, the random controls for the players will be mutually independent (more precisely, weakly correlated). These conditions are a corollary of imprecise measurements of locations of the system.

We state the following basic result.

Let the initial position (t_0, x_0) be contained in a closed μ-stable bridge W in an (M, N)-approach problem. Assume that informational possibilities for the second player are constrained by condition (18) (or (19)). Then, for any numbers $\epsilon > 0$ and $\rho < 1$, one can find numbers β_1 and β_2 and $\delta > 0$ such that when the estimates (17) are fulfilled and when the first player chooses partition $\Delta^{(1)}$ such that diam $\Delta^{(1)} \leqslant \delta$, the condition of an $(\overline{M}^\epsilon, \overline{N}^\epsilon)$-approach will be fulfilled with probability of at least p.

A proof of this assertion can be obtained by combining the arguments presented in the proofs of Theorems 9.4.1 and 8.2.1.

Analogously, one can define a SCPG for the second player and state assertions about approximating solutions of the an $(M, N)^\epsilon$-evasion problem.

Chapter 10
LOWER AND UPPER DIFFERENTIAL GAMES

10.1. Counterstrategies

In this chapter, we continue the study of differential games in the case when the saddle point condition in a small game (condition (1.2.8)) is not satisfied. It was noted in Section 9.1 that when condition (1.2.8) is violated it is important that the nature of information available to the players is specified precisely. In Section 9.1.1, three basic types of information available to the players about their opponent's current controls (or the disturbance) were mentioned. In the preceding chapter, a situation was considered, Case II, in which both players are ignorant of their opponent's current control. In this case the use of mixed positional strategies was suggested. A meaningful (physically realizable) approximation of mixed strategies was obtained in the form of stochastic control procedures which assure a close to optimal outcome of the game with probability near 1.

We now turn to an investigation of the differential games in Cases I and III (cf. Section 9.1.1). In this section, counterstrategies will be defined. This is a basic new notion which will be required in this chapter. In the next section, it is established that differential games considered in the class of positional strategies for one player and counterstrategies for the other possess an equilibrium situation. Next, in Section 10.3 a brief survey of theorems modified from Chapters 4, 5, and 7 is presented. In Section 10.4 control procedures with guidance are defined for the types of information availability considered herein. In the conclusion of this chapter, a unification of differential games is proposed in Section 10.5; in particular, it is shown that various types of game-theoretical control problems can be reduced to a control problem considered in the class of pure positional strategies.

10.1. First determine the counterstrategies and the motions generated. Assume first that Case I is valid, i.e. at each time t the first player knows the position $(t,x(t))$ and moreover, he or she knows which control $v(t)$ is realized at that time by his or her opponent. In those cases when we are dealing with the problem of the controlling system (1.2.1) subject to a disturbance, this assumption is transformed as follows: (cf. Case I* in Section 9.1.1): at any current time t the position $(t,x(t))$ is known as well as the disturbance $v(t)$ affecting the controlled system.

As was mentioned in Section 9.1.2, one should not neglect any additional information concerning controls (or disturbance). Therefore in Cases I or I*, we use counterstrategies which we introduce as follows.

Let a function

$$(1)\qquad U^c : (t,x,v) \to U^c(t,x,v): T \times R^n \times \mathcal{Q} \to \mathcal{P}$$

be given. We shall assume that for all $(t,x) \in T \times R^n$ the function $v \to U^c(t,x,v)$: $\mathcal{Q} \to \mathcal{P}$ is Borel measurable. The upper index ^{c} is introduced here in order to distinguish counterstrategies from positional strategies $(t,x) \to U(t,x)$.

Let an initial position $(t_0,x_0) \in T \times R^n$ be given, and let the first player select a partition $\Delta = (t_0 = \tau_0 < \tau_1 < \cdots < \tau_{k+1} = \theta)$. Let $v(t)$: $[t_0,\theta] \to \mathcal{Q}$ be a measurable function. Define a step-by-step motion $x(\cdot t_0,x_0,U^c,v(\cdot),\Delta)$ as an absolutely continuous solution of the differential equation

$$(2)\qquad \dot{x}(t) = f(t,x(t),u(t),v(t)), \quad t_0 \leqslant t \leqslant \theta, \quad x(t_0) = x_0,$$

where

$$(3)\qquad u(t) = U^c(\tau_i,x(\tau_i),v(t)), \quad \tau_i \leqslant t < \tau_{i+1}, \quad i = 0,1, \dots, k.$$

It is known (cf. for example, [36]) that the function $t \to u(t)$ defined by equality (3) is measurable since on each half-interval $[\tau_i,\tau_{i+1})$ this function is a superposition of the Borel function $v \to U^c(\tau_i,x(\tau_i),v)$ and the measurable (in either Borel or Lebesgue's sense) function $t \to v(t)$. For any collection $(t_0,x_0,U^c,v(\cdot),\Delta)$ indicated above there exists a unique solution of the differential equation (2). Note also that in these constructions $v(\cdot)$ is the realization of the control for the second player; for the time being we do not specify how the function $t \to v(t)$ is formed and require only its measurability (cf. an analogous remark in Section 1.2.1).

Fix (t_0,x_0,U^c,Δ), sort (over) all the possible measurable functions $v(\cdot)$: $[t_0,\theta] \to \mathcal{Q}$, and form the corresponding set of step-by-step motions $X_1(t_0,x_0,U^c,\Delta)$

(4) $$X_1(t_0,x_0,U^c,\Delta) \overset{\text{def}}{=} \{x(\cdot,t_0,x_0,U^c,v(\cdot),\Delta):\ v(\cdot) \in \mathfrak{L}_1([t_0,\theta], \mathcal{Q})\}.$$

Recall that $\mathfrak{L}_1([t_0,\theta],\mathcal{Q})$ is the space of measurable functions $v(\cdot)$: $[t_0,\theta] \to \mathcal{Q}$.

Next, as usual, approaching the limit as diam $\Delta \to 0$, we introduce the set of constructive motions. Define

(5) $$\begin{aligned} X_1(t_0,x_0,U^c) &\overset{\text{def}}{=} \{x(\cdot):\ \exists x_\ell(\cdot) \in X_1(t_0,x_0,U^c,\Delta_\ell) \\ \ell = 1,2,\ldots,\quad & x_\ell(\cdot) \to x(\cdot),\quad \text{diam}\ \Delta_\ell \to 0\ \text{ as }\ \ell \to \infty\}. \end{aligned}$$

We thus arrive at the following definition.

Definition. *A function of the form* (1) *such that for any* $(t,x) \in T \times R^n$ *the mapping* $v \to U^c(t,x,v)$ *is Borel measurable is called a counterstrategy for the first player. The counterstrategy* U^c *generates -- in the manner indicated above -- sets of step-by-step and constructive motions.*

For the sets of step-by-step and constructive motions defined herein the following two assertions are valid.

For any initial position $(t_0,x_0) \in T \times R^n$, and any counterstrategy U^c, the set of constructive motions $X_1(t_0,x_0,U^c)$ is a nonempty compact set in the space $C_n([t_0,\theta])$. Any constructive motion $x(\cdot) \in X_1(t_0,x_0,U^e)$ is an absolutely continuous function.

For any number $\epsilon > 0$, any initial position $(t_0,x_0) \in T \times R^n$, and any counterstrategy U^c one can select a number $\delta > 0$ such that the inclusion

$$X_1(t_0,x_0,U^c,\Delta) \subset X_1^\epsilon(t_0,x_0,U^c)$$

will be fulfilled for all partitions Δ satisfying the estimate diam $\Delta \leqslant \delta$. Here X^ϵ is an ϵ-neighborhood of the set X, i.e.

$$X^\epsilon \overset{\text{def}}{=} \{y(\cdot) + z(\cdot):\ y(\cdot) \in X,\ z(\cdot) \in C_n([t_0,\theta])\},\quad |z(\cdot)| < \epsilon\ .$$

These assertions are analogous to Lemmas 1.3.1 and 1.3.2. Proofs of these assertions are actually the same as proofs of Lemmas 1.3.1 and 1.3.2. The second assertion may be used to approximate formal results obtained for constructive motions. It is done according to a well known scheme which has been used repeatedly for solving game-theoretical control problems in the class of positional strategies.

10.1.2. Any counterstrategy for the first player is compatible with an arbitrary positional strategy $(t,x) \to V(t,x)$ for the second player, i.e. for any pair (U^c,V) and for any initial position (t_0,x_0) one can determine the corresponding step-by-step and constructive motions. Indeed let

$$\Delta^{(1)} = (t_0 = \tau_0^{(1)} < \tau_1^{(1)} < \cdots < \tau_{k+1}^{(1)} = \theta)$$

and

$$\Delta^{(2)} = (t_0 = \tau_0^{(2)} < \tau_1^{(2)} < \cdots < \tau_{\ell+1}^{(2)} = \theta)$$

be partitions selected by the first and second players respectively. Consider the solution of the differential equation (2), where

$$\begin{aligned} v(t) &= V(\tau_j^{(2)},x(\tau_j^{(2)})), \quad \tau_j^{(2)} \leqslant t < \tau_{j+1}^{(2)}, \quad j = 0,1, \ldots, \ell, \\ u(t) &= U^c(\tau_i^{(1)},x(\tau_i^{(1)}),v(t)), \quad \tau_i^{(1)} \leqslant t < \tau_{i+1}^{(1)}, \quad i = 0,1, \ldots, k. \end{aligned} \tag{6}$$

This solution is denoted by $x(\cdot,t_0,x_0,U^c,V,\Delta^{(1)},\Delta^{(2)})$ and will be called a step-by-step motion generated by a counterstrategy U^c for the first player and a positional strategy V for the second player. The motion $x(\cdot,t_0,x_0,U^c,V,\Delta^{(1)},\Delta^{(2)})$ exists and is unique. Indeed, in accordance with (6), the strategy V forms a piecewise constant function $t \to v(t)$. The control $t \to u(t)$ formed by the counterstrategy U^c will also be piecewise constant in this situation. It follows from (6) that the switching of control $u(t)$ from one constant value to another may take place at times $t = \tau_i^{(1)}$ as well as at times $t = \tau_j^{(2)}$ when the control $v(t)$ is switched.

Define the set

$$\begin{aligned} X(t_0,x_0,U^c,V) \overset{\text{def}}{=} \{x(\cdot)\colon \exists x_m(\cdot) = x(\cdot,t_0,x_0,U^c,V,\Delta_m^{(1)},\Delta_m^{(2)}), \\ m = 1,2, \ldots, x_m(\cdot) \to x(\cdot), \ \operatorname{diam} \Delta_m^{(1)} \to 0, \\ \operatorname{diam} \Delta_m^{(2)} \to 0 \text{ as } m \to \infty\}. \end{aligned} \tag{7}$$

Elements of this set are called constructive motions generated by the counterstrategy U^c and the positional strategy V.

Let the set $X_1(t_0,x_0,U^c)$ be defined in accordance with (5) and the set $X_2(t_0,x_0,V)$ be the totality of constructive motions generated by the positional strategy V (cf. Section 1.5.1). It is easy to verify that

$$\begin{aligned} &X(t_0,x_0,U^c,V) \neq \phi \quad \text{and} \\ &X(t_0,x_0,U^c,V) \subset X_1(t_0,x_0,U^c) \cap X_2(t_0,x_0,V). \end{aligned} \tag{8}$$

Relations (8) show that any pair U^e,V is *compatible*.

Let a payoff functional $x(\cdot) \to \gamma(x(\cdot))$ be defined on the motions of the controlled system (2). As usual, we shall assume that the first player is solving the minimax problem. In this case we are dealing with a solution of this problem in the class of counterstrategies. Following the plan proposed in Section 1.4 we define a quality index for an arbitrary counterstrategy U^c. Set

$$(9) \qquad \Gamma_1(U^c) \overset{\text{def}}{=} \sup\{\gamma(x(\cdot)): x(\cdot) \in X_1(t_0,x_0,U^c)\}.$$

Problem 10.1.1. *For a given initial position* (t_0,x_0) *it is required to select a counterstrategy* U_0^c *such that*

$$(10) \qquad \Gamma_1(U_0^c) = \min_{U^c} \Gamma_1(U^c).$$

If the minimum in (10) is not attained we shall consider the following problem.

Problem 10.1.1$^\epsilon$. *For a given initial position* (t_0,x_0) *and a given number* $\epsilon > 0$ *it is required to find a counterstrategy* U_ϵ^c *such that*

$$(11) \qquad \Gamma_1(U_\epsilon^c) < \inf_{U^c} \Gamma_1(U^c) + \epsilon.$$

The minimax problems (10) and (11) will be considered jointly with Problems 1.5.1 and 1.5.1$^\epsilon$, respectively, i.e. with maximin problem for the second player the solution to which is required to be determined in the class of positional strategies. A differential game formed by Problem 10.1.1 (or 10.1.1$^\epsilon$) and Problem 1.5.1 (or 1.5.1$^\epsilon$) will be called a *lower differential game.* Thus a lower differential game is defined in the class of counterstrategies for the first player and positional (pure) strategies for the second player.

Recall that lower and upper multistage games were mentioned in Section 1.5.3. In a lower multistage game a partition $\Delta = \{t_0 = \tau_0 < \tau_1 < \cdots < \tau_{k+1} = \theta\}$ is given and it is assumed that at each instant of time $t = \tau_i$ the first player knows the realized position of the game $(\tau_i,x(\tau_i))$ as well as a constant control $v(t) = v(\tau_i)$ $(\tau_i \leqslant t < \tau_{i+1})$ selected by the second player. This constitutes the information advantage of the first player. In a lower differential game the first player knows at time $\tau \in [t_0,\theta]$ the control $v(\tau)$ selected by the second player at that time. However, at time τ the first player does not know which control $v(t)$ will be realized by the opponent at $t > \tau$. In a lower differential game (more precisely in Problems 10.1.1 or 10.1.1$^\epsilon$) it is assumed that the second player can realize any measurable control $t \to v(t)$.

Taking into account that any pair (U^c,V) is compatible and utilizing the arguments presented in Section 1.5.2 we obtain that

$$(12)\qquad \inf_{U^c} \Gamma_1(U^c) \geqslant \sup_{V} \Gamma_2(V).$$

Here the quantity $\Gamma_1(U^c)$ is defined by relation (9) and the inf is computed over all the counterstrategies for the first player; the quantity $\Gamma_2(V)$ is defined according to (1.5.5) and the sup is taken over the set of positional strategies for the second player.

If equality holds in inequality (12), we then say that there exists *the value of a lower differential game*

$$(13)\qquad \omega^0 = \inf_{U^c} \Gamma_1(U^c) = \sup_{V} \Gamma_2(V).$$

If, moreover, there exists a pair (U_0^c,V_0) such that

$$\Gamma_1(U_0^c) = \min_{U^c} \Gamma_1(U^c) = \max_{V} \Gamma_2(V) = \Gamma_2(V_0),$$

then this pair is called a *saddle point for a lower differential game.*

In the next section we shall prove an alternative for a lower differential game of approach-evasion which is defined in the class of counterstrategies for the first player and positional strategies for the second. The existence of the value for lower differential games with various types of payoff functionals will follow from this alternative.

10.1.3. Let the current position $(t,x(t))$ of the controlled system (1.2.1) be known, and assume that the disturbance $v(t)$ currently affecting the system is unknown. Moreover, assume that the disturbance $v(t)$ depends somehow on the control $u(t)$ selected at the current time t. Thus, let the conditions of Case III* be fulfilled. In this situation control problems will be solved in the class of positional strategies $(t,x) \to U(t,x)$.

There exists yet another situation where it makes sense to utilize positional strategies in a control problem. Assume that the current control $u(t)$ and the disturbance $v(t)$ are mutually independent. In this case one could use mixed positional strategies U. However, it is known from the preceding chapter that stochastic control procedures which approximate mixed strategies cannot assure the required quality of the controlled process for *all possible* motions. Under stochastic approximation "bad" motions may be realized. Although the probability of a "bad" motion being realized is small, the result guaranteed by a stochastic control procedure may be unsatisfactory if the substance of the problem requires us to obtain a deterministic guaranteed result. Also, in such cases one should utilize positional strategies.

We shall thus return to a minimax control problem posed in the class of positional strategies $(t,x) \to U(t,x)$: $T \times R^n \to \mathcal{P}$. The statement of this problem presented in Section 1.4 is carried over without any changes. In Chapters 1 - 8 the minimax Problem 1.4.1 was considered in conjunction with the maximin Problem 1.5.1 which was defined in the class of positional strategies V: $T \times R^n \to \mathcal{Q}$. Under fulfillment of the saddle point condition in a small game -- i.e. condition (1.2.8) -- the value of the differential games formed from Problems 1.4.1 and 1.5.1 exists in each case. If condition (1.2.8) is violated in the case presently under consideration then the dual problem for the minimax Problem 1.4.1 will not be the Problem 1.5.1 considered in the class of positional strategies V, but rather a maximin problem considered in the class of *counterstrategies*

$$(14) \qquad V^c\colon T \times R^n \times \mathcal{P} \to \mathcal{Q}.$$

It is assumed that for any $(t,x) \in T \times R^n$, the function $u \to V^c(t,x,u)$: $\mathcal{P} \to \mathcal{Q}$ is Borel measurable.

For any counterstrategy V^c and any initial position $(t_0,x_0) \in T \times R^n$, one can define a set of step-by-step motions $X_2(t_0,x_0,V^c,\Delta)$ and a set of constructive motions $X_2(t_0,x_0,V^c)$. These definitions actually do not differ from the corresponding definitions presented in Section 10.1.2. Furthermore, one can verify that any pair (U,V^c) consisting of a positional strategy for the first player U: $T \times R^n \to$ and a counterstrategy V^c (14) for the second player is compatible. Recall that a pair of counterstrategies U^c and V^c may turn out to be incompatible (cf. Section 1.5.2).

Thus in Cases III or III* we shall consider a differential game formed from a minimax Problem 1.4.1 (or 1.4.1$^\epsilon$) and maximin Problem 10.1.2 formulated as follows.

Problem 10.1.2. *For a given initial position* (t_0,x_0) *it is required to construct a counterstrategy* V_0^c *such that*

$$(15) \qquad \Gamma_2(V_0^c) = \max_{V^c} \Gamma_2(V^c)$$

where

$$(16) \qquad \Gamma_2(V^c) = \inf\{\gamma(x(\cdot))\colon\ x(\cdot) \in X_2(t_0,x_0,V^c)\}.$$

If in (15) the maximum is not attained then the following problem is considered.

Problem 10.1.2$^\epsilon$. *For a given initial position* (t_0,x_0) *and a given number* $\epsilon > 0$, *it is required to find a counterstrategy* V^c_ϵ *such that*

$$(17)\qquad \Gamma_2(V^c_\epsilon) \geqslant \sup_{V^c} \Gamma_2(V^c) - \epsilon.$$

Since the pair (U,V^c) is compatible it follows that the inequality

$$(18)\qquad \inf_U \Gamma_1(U) \geqslant \sup_{V^c} \Gamma_2(V^c)$$

holds in the differential game formed by Problems 1.4.1$^\epsilon$ and 10.1.2$^\epsilon$; here the quantities $\Gamma_1(U)$ and $\Gamma_2(V^c)$ are defined by relations (1.4.1) and (16). If equality is attained in (18), then the game under consideration possesses the value

$$(19)\qquad \omega^0 = \inf_U \Gamma_1(U) = \sup_{V^c} \Gamma_2(V^c).$$

A differential game formed by Problem 1.4.1 (or 1.4.1$^\epsilon$) and Problem 10.1.2 (or 10.1.2$^\epsilon$) will be called an *upper differential game.*

10.2. Alternatives

In this section alternatives for lower and upper differential games of approach-evasion are proved. Proofs of these alternatives are carried out according to the outline in Chapter 2 and in Section 9.2. Therefore below we shall only briefly indicate the main points of these proofs. As usual, a theorem on an alternative yields the existence of the value for various types of a payoff functional.

10.2.1. We shall formulate a *lower differential game of approach-evasion.* This game is formed by an approach problem whose solution should be determined in a class of counterstrategies for the first player and an evasion problem where the second player is allowed a choice of positional strategies. As usual, we shall assume that two closed sets $\mathcal{M}$ and $\mathcal{N}$, with $\mathcal{M} \subset \mathcal{N}$ are defined in the space $T \times R^n$.

Problem 10.2.1. *For a given initial position* (t_0,x_0) *it is required to obtain a counterstrategy* U^c_0 *such that for any motion* $x(\cdot) \in$

$\mathcal{X}_1(t_0,x_0,U_0^c)$ *the condition of an* $(\mathcal{M},\mathcal{N})$*-approach holds.*

Recall that the condition of an $(\mathcal{M},\mathcal{N})$-approach was defined by relations (2.1.3). Problem 10.2.1 differs from Problem 2.1.1 by replacing the class of positional strategies with a class of counterstrategies. A class of counterstrategies contains a class of positional strategies. Therefore, the first player acquires additional possibilities to assure that the condition of $(\mathcal{M},\mathcal{N})$-approach is fulfilled as we pass from Problem 2.1.1 to Problem 10.2.1. If the condition of a saddle point in a small game, i.e. condition (1.2.8) is violated, then it is easy to find examples for which Problem 10.2.1 is solvable while Problem 2.1.1 is not. If, however, the controlled system satisfies condition (1.2.8), then Problems 2.1.1 and 10.2.1 are equivalent, i.e. each time Problem 10.2.1 possesses a solution, so does Problem 2.1.1. This was noted above and will be discussed in Section 10.5.2.

Thus we shall be dealing with an $(\mathcal{M},\mathcal{N})$-approach problem in the class of counterstrategies for the first player. Jointly with this problem we shall consider an $(\mathcal{M},\mathcal{N})^\epsilon$-evasion problem in the class of positional strategies for the second player, i.e. Problem 2.1.2, whose formulation is carried over verbatim from Section 2.1.1. A game formed by Problems 10.2.1 and 2.1.2 will be called *a lower differential game of approach-evasion.* Below we shall prove that the following theorem on an alternative is valid for this game.

Theorem 10.2.1. *For any closed sets* $\mathcal{M} \subset \mathcal{N} \subset T \times R^n$ *and any initial position* $(t_0,x_0) \in T \times R^n$ *one and only one of the following two assertions is valid*: (a) *there exists a counterstrategy* U_0^c *which yields a solution for an* $(\mathcal{M},\mathcal{N})$*-approach problem* (*Problem* 10.2.1); (b) *there exist an* $\epsilon > 0$ *and a positional strategy* V_ϵ *which yields a solution of an* $(\mathcal{M},\mathcal{N})^\epsilon$*-evasion problem* (*Problem* 2.1.2).

10.2.2. We shall formulate the theorem on an alternative for the *upper differential game of approach-evasion* formed by Problem 2.1.1 and the following $(\mathcal{M},\mathcal{N})^\epsilon$-evasion in the class counterstrategies for the second player.

Problem 10.2.2. *Let the initial position* (t_0,x_0), *the closed sets* $\mathcal{M} \subset \mathcal{N} \subset T \times R^n$, *and a number* $\epsilon > 0$ *be given. It is required to define a counterstrategy* V_ϵ^c *such that for any motion* $x(\cdot) \in \mathcal{X}(t_0,x_0,V_\epsilon^c)$ *the condition of an* $(\mathcal{M},\mathcal{N})^\epsilon$*-evasion is valid.*

Recall that this condition was defined by relations (2.1.8). The following theorem is valid.

Theorem 10.2.2. *For any closed sets* $\mathcal{M} \subset \mathcal{N} \subset T \times R^n$, *and for any* *initial position* $(t_0,x_0) \in T \times R^n$, *one and only one of the following two assertions hold:*

(a) *there exists a positional strategy* U_0 *which yields a solution of an* $(\mathcal{M},\mathcal{N})$*-approach problem (Problem* (2.1.1));

(b) *there exist an* $\epsilon > 0$ *and a counterstrategy* V_ϵ^c *which yields a solution to an* $(\mathcal{M},\mathcal{N})^\epsilon$*-evasion problem (Problem* 10.2.2).

10.2.3. We shall describe the basic points of the proof of Theorem 10.2.1. As was mentioned above, the proof follows the outline considered in Chapter 2 and utilized in Section 9.2. We shall thus concentrate only on the differences between the proofs of Theorems 2.1.1 and 10.2.1.

Introduce the following auxiliary notions and notation.

A Borel measurable function

(1) $$u[\cdot]: \mathcal{Q} \to \mathcal{P}$$

is called a *countercontrol for the first player.* The totality of countercontrols $u[\cdot]$ will be denoted by $\mathbf{U}_{cc}$. For $(t,x) \in T \times R^n$ and $u[\cdot] \in \mathbf{U}_{cc}$ we set

(2) $$\mathcal{F}_1(t,x,u[\cdot]) = \overline{co}\{f(t,x,u[v],v): v \in \mathcal{Q}\};$$

here $\overline{co}\ \mathcal{D}$ is a convex closed hull of the set $\mathcal{D}$. The multivalued mapping $(t,x) \to \mathcal{F}_1(t,x,u[\cdot])$ satisfies all the conditions stated in Section A.4. Therefore, there exists an absolutely-continuous solution $x(\cdot): [t_0,\theta] \to R^n$ of the differential inclusion

(3) $$\dot{x}(t) \in \mathcal{F}_1(t,x(t),u[\cdot])$$

satisfying the initial condition $x(t_*) = x_*$. The totality of these solutions will be denoted by $X_1(t_*,x_*,u[\cdot])$.

Let $\mathcal{K} \subset \mathcal{L} \subset T \times R^n$.

Definition. *A set* $\mathcal{W} \subset T \times R^n$ *is called a* v_**-stable bridge in a problem of a* $(\mathcal{K},\mathcal{L})$*-approach provided this set satisfies the two conditions*

(1_{v*}) $\mathcal{W} \subset \mathcal{L}$, $\mathcal{W}(\theta) \subset \mathcal{K}$,

(2_{v*}) *for any* $(t_*,x_*) \in \mathcal{W}$, $t^* \in [t_*,\theta]$ *and* $u[\cdot] \in \mathbf{U}_{cc}$ *there exists a solution* $x(\cdot) \in X_1(t_*,x_*,u[\cdot])$ *such that either* $(t^*,x(t^*)) \in \mathcal{W}$; *or* $\{(t,x(t)): t_* \leqslant t \leqslant t^*\} \cap \mathcal{K} \neq \emptyset$.

Condition (2_{v*}) is called the *condition of* v_**-stability* of the set $\mathcal{W}$. It differs from the condition of v-stability defined in Section 2.2.3: here controls $u \in \mathcal{P}$ are replaced by countercontrols $u[\cdot] \in \mathbf{U}_{cc}$.

Let an $\epsilon > 0$ and sets $\mathcal{K}_\epsilon$ and $\mathcal{L}_\epsilon$ be defined in accordance with

(2.1.9). Recall that K_ϵ and $\mathfrak{L}_\epsilon$ are closed and $K_\epsilon \subset \mathfrak{L}_\epsilon$. Denote by $\mathbf{W}_\epsilon^{2,1}$ the totality of v_*-stable bridges in a problem of a $(K_\epsilon, \mathfrak{L}_\epsilon)$-approach. In the index 2,1 the first digit stipulates that the problem is posed before the second player while the second digit denotes the type of the differential game under consideration. Recall that a lower differential game is considered here while the players' information structure belongs to Type I (cf. the three types described in Section 9.1.1). Set

$$(4) \qquad W_\epsilon^{2,1} = \bigcup_{W \in \mathbf{W}_\epsilon^{2,1}} W .$$

Here $W_\epsilon^{2,1}$ is the *maximal v_*-stable bridge in a $(K_\epsilon, \mathfrak{L}_\epsilon)$-approach problem.* This assertion corresponds to Lemma 2.2.6.

We shall also be considering u-stable bridges in a problem of an (M, N)-approach. The definition of a u-stable bridge remains the same as that given in Section 2.1.1. The totality of u-stable bridges in an (M, N)-approach problem will be denoted by $\mathbf{W}^{1,1}$. The index 1,1 indicates that the problem is posed to the first player in a differential game of the first type. Set

$$(5) \qquad W_0^{1,1} = \bigcup_{W \in \mathbf{W}^{1,1}} W .$$

Note that this set was defined earlier in Section 2.2.2 by the relation (2.2.12) but was denoted there by the symbol W_0^1. Recall also that the set $W_0^{1,1}$ is closed and it is the *maximal u-stable bridge in an (M, N)-approach problem* (cf. Lemmas 2.2.1 and 2.2.6).

The v_*-bridges introduced herein and the u-stable bridges defined in Section 2.1.1 satisfy a duality relation expressed by the equality

$$(6) \qquad T \times R^n = W_0^{1,1} \cup \left[\bigcup_{\epsilon > 0} W_\epsilon^{2,1} \right].$$

This assertion is an analog of Theorem 2.2.1.

The proof of Theorem 2.2.1 was based on Lemmas 2.2.1 - 2.2.6. Equality (6) is deduced from assertions analogous to these lemmas. Lemmas 2.2.1 and 2.2.4 are used here in the same form as they were stated in Section 2.2.

The statement of Lemma 2.2.2 should be modified as follows:

Let $(t_1, y) \in T \times R^n$, $t_2 \in (t_1, \theta]$, $u[\cdot] \in \mathbf{U}_{cc}$, $W_* \in \mathbf{W}_\epsilon^{1,1}$. Set

$$Z \overset{\text{def}}{=} \{(t, x(t)) \colon t_1 \leqslant t \leqslant t_2,\ x(\cdot) \in X_1(t_1, y, u[\cdot])\}.$$

Assume that

$$Z \subset N^\epsilon \text{ and } Z(t_2) \subset W_*,$$

Let the set $\mathcal{W}^* \subset [t_1,\theta] \times R^n$ be defined by relation

$$\mathcal{W}^*(t) \overset{\text{def}}{=} \begin{cases} \mathcal{Z}(t) & \text{for } t_1 \leqslant t < t_2 \\ \mathcal{W}_*(t) & \text{for } t_2 \leqslant t \leqslant \theta . \end{cases}$$

Then the set $\mathcal{W}^*$ will belong to the class $\mathbf{W}_\epsilon^{1,1}$. Here the symbol $\mathbf{W}_\epsilon^{1,1}$ denotes the totality of u-stable bridges in an $(\mathcal{M}^\epsilon, \mathcal{N}^\epsilon)$-approach problem. Note that this totality was considered earlier in Sections 2.2.1 and 2.2.2 but was denoted there by the symbol $\mathbf{W}_\epsilon^1$.

This assertion is proved in the same manner as Lemma 2.2.2. Here one should use the fact that for any $(t_*,x_*) \in T \times R^n$ and $v \in \mathcal{Q}$, relation $X_2(t_*,x_*,v) \cap X_1(t_*,x_*,u[\cdot]) \neq \phi$ is valid. Here $X_2(t_*,x_*,v)$ and $X_1(t_*,x_*,u[\cdot])$ are sets of solutions of differential inclusions (2.2.3) and (3) respectively.

In the statements and proofs of Lemmas 2.2.3, 2.2.5, and 2.2.6 the condition of v-stability should be replaced by the condition of v_*-stability. Thus utilizing Lemmas 2.2.1, 2.2.4 and applying the above indicated modifications of Lemmas 2.2.2, 2.2.3, 2.2.5, and 2.2.6 one can arrive at equality (6).

We now proceed to the next stage which corresponds to the content of Sections 2.3 and 2.4. Let $\mathcal{W}$ be a closed set in $T \times R^n$. In accordance with (2.4.2) - (2.4.4) define the function

$$(t,x) \to e(t,x): T \times R^n \to R^n.$$

For any $(t,x,v) \in T \times R^n \times \mathcal{Q}$ we set

$$(7) \qquad \begin{aligned} \mathcal{U}_e(t,x,v) &\overset{\text{def}}{=} \{u_e \in \mathcal{P}: \langle x - e(t,x), f(t,x,u_e,v)\rangle \\ &= \min_{u \in \mathcal{P}} \langle x - e(t,x), f(t,x,u,v)\rangle\}. \end{aligned}$$

It is easy to verify that for any $(t,x) \in T \times R^n$ the multivalued mapping $v \to \mathcal{U}_e(t,x,v)$ is upper semicontinuous. Therefore according to Theorem A.3.2 one can select a function

$$(8) \qquad U_e^c: (t,x,v) \to U_e^c(t,x,v): T \times R^n \times \mathcal{Q} \to \mathcal{P}$$

such that

$$(9) \qquad U_e^c(t,x,v) \in \mathcal{U}_e(t,x,v) \quad \text{for all} \quad (t,x,v) \in T \times R^n \times \mathcal{Q}$$

and for all $(t,x) \in T \times R^n$ the function $U_e^c(t,x,\cdot)$ will be Borel measurable.

Definition. *A function* U_e^c (8) *defined by relations* (7), (9) *will be called a counterstrategy extremal to the set* $\mathcal{W}$.

The following assertion is valid.

Theorem 10.2.3. *Let* $\mathcal{W}$ *be a closed u-stable bridge in an* $(\mathcal{M},\mathcal{N})$-*approach problem. Let* U_e^c *be a counterstrategy extremal to the set* $\mathcal{W}$. *If* $(t_0,x_0) \in \mathcal{W}$ *then for any motion* $x(\cdot) \in X_1(t_0,x_0,U_e^c)$ *the condition of an* $(\mathcal{M},\mathcal{N})$-*approach is valid.*

Theorem 10.2.3 is an analog of Theorem 2.4.1. The proof of Theorem 2.4.1 is based on Lemma 2.3.1. For the proof of Theorem 10.2.3, Lemma 2.3.1 should be modified as follows. In place of the control u_0 selected from the condition (2.3.19) we shall define a countercontrol $u_0[\cdot]$ from the condition

$$\langle x_* - w_*,\ f(t_*,x_*,u_0[v],v)\rangle = \min_{u\in\mathcal{P}} \langle x_* - w_*,\ f(t_*,x_*,u,v)\rangle.$$

The differential inclusion (2.3.21) is replaced by the inclusion

$$(10) \qquad \dot{x}(t) \in \mathcal{F}_1(t,x(t),u_0[\cdot]) \overset{\text{def}}{=} \overline{\text{co}}\{f(t,x(t),u_0[v],v)\colon v \in \mathcal{Q}\}.$$

Finally the condition of a saddle point in a small game (condition (1.2.8)) should be replaced by the equality

$$(11) \qquad \begin{aligned} \min_{u[\cdot]} \max_{v} \langle s,f(t,x,u[v],v)\rangle &= \max_{v} \min_{u[\cdot]} \langle s,f(t,x,u[v],v)\rangle \\ &= \max_{v} \min_{u} \langle s,f(t,x,u,v)\rangle \\ ((s,t,x) \in R^n \times T \times R^n). \end{aligned}$$

Here $u[\cdot] \in \mathbf{U}_{cc}$, $v \in \mathcal{Q}$. This equality is obtained as follows. It is known that for any function $(u,v) \to \varphi(u,v)$ the relation

$$\inf_{u[\cdot]} \sup_{v} \varphi(u[v],v) = \sup_{v} \inf_{u[\cdot]} \varphi(u[v],v) = \sup_{v} \inf_{u} \varphi(u,v)$$

is valid. If a function φ is continuous then the operations inf and sup can be replaced by min and max respectively. Setting $\varphi(u,v) = \langle s,f(t,x,u,v)\rangle$ we arrive at (11).

Thus, taking into account the modifications indicated above, one obtains the required estimate for solutions of differential inclusions (2.3.22) and (10), which is analogous to the estimate (2.3.23). Next, utilizing this estimate and repeating the arguments presented in the proof of Theorem 2.4.1, one arrives at Theorem 10.2.3.

We now state the last assertion which will be required for the proof of Theorem 10.2.1.

Theorem 10.2.4. *Let* $\mathcal{W}$ *be a closed* v_**-stable bridge in a* $(K_\epsilon, \mathcal{L}_\epsilon)$*-approach problem. Let* V_e *be a positional strategy extremal to the set* $\mathcal{W}$*. If* $(t_0,x_0) \in \mathcal{W}$*, then for any motion* $x(\cdot) \in X_2(t_0,x_0,V_e)$ *the condition of a* $(K_\epsilon, \mathcal{L}_\epsilon)$*-approach is valid, or equivalently, the strategy* V_e *yields a solution of an* $(\mathcal{M}, \mathcal{N})^\epsilon$*-evasion problem (Problem 2.1.2).*

Recall that a strategy V_e extremal to the set $\mathcal{W}$ was defined in 2.4.3. The proof of this theorem, as well as of the preceding one, is based on an estimate of the form (2.3.23). In turn, this estimate is valid in view of equality (11).

Now it is easy to complete the proof of Theorem 10.2.1. According to (6), any initial position (t_0,x_0) either belongs to a closed u-stable bridge $\mathcal{W}_0^{1,1}$ in a problem of an $(\mathcal{M}, \mathcal{N})$-approach, or it belongs to the set $\mathcal{W}_\epsilon^{2,1}$ where ϵ is a positive number and $\mathcal{W}_\epsilon^{2,1}$ is a closed v_*-stable bridge in a problem of a $(K_\epsilon, \mathcal{L}_\epsilon)$-approach. In the first case it follows from Theorem 10.2.3 that Problem 10.2.1 (the problem of an $(\mathcal{M}, \mathcal{N})$-approach in the class of counterstrategies for the first player) is solvable. In the second case we obtain from Theorem 10.2.4 that Problem 2.1.2 (the problem of an $(\mathcal{M}, \mathcal{N})^\epsilon$-evasion in the class of positional strategies for the second player) is solvable. Evidently Problems 2.1.2 and 10.2.1 cannot be solved for the very same initial position (t_0,x_0). Thus Theorem 10.2.1 is proved.

Note that equality (6) can be supplemented by the relation

$$(12) \qquad \mathcal{W}_0^{1,1} \cap \mathcal{W}_\epsilon^{2,1} = \phi \text{ for all } \epsilon > 0$$

which follows from the fact that Problems 2.1.2 and 10.2.1 cannot be solved simultaneously for the same initial position.

We note the following important point. The existence of an equilibrium in the corresponding small game plays a crucial role in the proof of the existence of an equilibrium in an approach-evasion differential game. In Chapter 2, the small game was considered on the set $\mathcal{P} \times \mathcal{Q}$ and the equilibrium was expressed by equality (1.2.8). In Section 9.2, the small game was defined on the product rpm($\mathcal{P}$) × rpm($\mathcal{Q}$) and equality (9.2.11) signifies the existence of the value for such a game. In this case a small game was defined on the set $\mathbf{U}_{cc} \times \mathcal{Q}$ and the equilibrium is expressed by equality (11). It is easy to observe that the selection of a set on which a small game is considered is determined by the class of strategies in which the initial differential game of approach-evasion is posed.

10.2.4. Very briefly we shall present the basic facts for the proof of Theorem 10.2.2.

A Borel measurable function

$$(13)\qquad v[\cdot]: \mathcal{P} \to \mathcal{Q}$$

is called a *countercontrol for the second player.* The totality of countercontrols $v[\cdot]$ is denoted by $\mathbf{V}_{cc}$. For $(t_0,x_0) \in T \times R^n$, $v[\cdot] \in \mathbf{V}_{cc}$ the symbol $\mathcal{X}_2(t_*,x_*,v[\cdot])$ denotes the set of solutions of the differential inclusion

$$(14)\qquad \dot{x}(t) \in \mathcal{F}_2(t,x(t),v[\cdot])$$

satisfying the initial condition $x(t_*) = x_*$. Here

$$(15)\qquad \mathcal{F}_2(t,x,v[\cdot]) \overset{\text{def}}{=} \overline{\text{co}}\{f(t,x,u,v[u]): u \in \mathcal{P}\}.$$

Definition. *The set $\mathcal{W} \subset T \times R^n$ is called a u_*-stable bridge in a problem of an $(\mathcal{M},\mathcal{N})$-approach provided this set satisfies the following two conditions:*

(1_{u*}) $\mathcal{W} \subset \mathcal{N}$, $\mathcal{W}(\theta) \subset \mathcal{M}$,

(2_{u*}) for any $(t_*,x_*) \in \mathcal{W}$, $t^* \in (t_*,\theta]$, $v[\cdot] \in \mathbf{V}_{cc}$ *there exists a solution* $x(\cdot) \in \mathcal{X}_2(t_*,x_*,v[\cdot])$ *such that either* $(t^*,x(t^*)) \in \mathcal{W}$, *or* $\{(t,x(t)): t_* \leqslant t \leqslant t^*\} \cap \mathcal{M} \neq \phi$.

Let $\mathbf{W}^{1,3}$ be the totality of u_*-stable bridges in a problem of an $(\mathcal{M}, \mathcal{N})$-approach. Set

$$\mathcal{W}_0^{1,3} = \bigcup_{\mathcal{W}\in\mathbf{W}^{1,3}} \mathcal{W}.$$

Next, let $\mathbf{W}_\epsilon^{2,3}$ be the totality of v-stable bridges in a $(K_\epsilon,\mathfrak{L}_\epsilon)$-approach problem (this concept remains the same as given in Section 2.2.3). Set

$$\mathcal{W}_\epsilon^{2,3} = \cup\ \mathcal{W} \quad \text{for } \mathcal{W} \in \mathbf{W}_\epsilon^{2,3}.$$

This set coincides with the set introduced in Lemma 2.2.6.

The equality

$$(16)\qquad T \times R^n = \mathcal{W}_0^{1,3} \cup \Big[\bigcup_{\epsilon>0} \mathcal{W}_\epsilon^{2,3}\Big]$$

is valid.

Theorem 10.2.5. *Let $\mathcal{W}$ be a closed u_*-stable bridge in an $(\mathcal{M}, \mathcal{N})$-approach problem and let $(t_0,x_0) \in \mathcal{W}$. Then a positional strategy U_e extremal to the set $\mathcal{W}$ yields a solution for Problem* 2.1.1.

Recall that a strategy U_e extremal to the set $\mathcal{W}$ was defined in Section 2.4.1.

Theorem 10.2.6. *Let $\mathcal{W}$ be a closed v-stable bridge in a problem of a $(K_\epsilon, \mathcal{L}_\epsilon)$-approach and let $(t_0,x_0) \in \mathcal{W}$. Then the counterstrategy V_e^c extremal to the set $\mathcal{W}$ yields a solution for Problem* 10.2.2.

Here the counterstrategy V_e^c is determined from the condition

$$\langle e(t,x) - x,\ f(t,x,u,v_e)\rangle = \max_{v \in \mathcal{Q}} \langle e(t,x) - x,\ f(t,x,u,v)\rangle$$

$$(v_e = V_e^c(t,x,u))$$

and the function $(t,x) \to e(t,x)$ is defined in accordance with (2.4.2), (2.4.4).

Theorem 10.2.2 follows from Theorems 10.2.5, 10.2.6, and equality (16). Note that the proof of Theorems 10.2.5 and 10.2.6 is based on an estimate analogous to the estimate (2.3.23), which in turn follows from equality

$$\begin{aligned}\min_u \max_{v[\cdot]} \langle s, f(t,x,u,v[u])\rangle &= \max_{v[\cdot]} \min_u \langle s, f(t,x,u,v[u])\rangle \\ &= \min_u \max_v \langle s, f(t,x,u,v)\rangle\end{aligned}$$

where $u \in \mathcal{P}$, $v[\cdot] \in \mathbf{V}_{cc}$. This equality plays a role equivalent to the condition of a saddle point in a small game (condition (1.2.8) in Chapter 2).

10.2.5. Based on Theorems 10.2.1 and 10.2.2 and utilizing the constructions described in Chapter 3, one can prove the existence of the value for lower and upper differential games. For definiteness we present formulations of the corresponding results for differential games in which the payoff functional is defined by the equality

$$\gamma(x(\cdot)) = \sigma(x(\theta))$$

where $\sigma: R^n \to R^1$ is a continuous function. The following assertions are valid.

Theorem 10.2.7. *For any initial position $(t_0,x_0) \in T \times R^n$ there exists a counterstrategy U_0^c for the first player and a positional strategy V_0*

for the second player such that the equalities

$$\min_{U^c} \max\{\sigma(x(\theta)): x(\cdot) \in X_1(t_0,x_0,U^c)\}$$
$$= \max\{\sigma(x(\theta)): x(\cdot) \in X_1(t_0,x_0,U_0^c)\} = \omega_1^0(t_0,x_0)$$
(17)
$$= \min\{\sigma(x(\theta)): x(\cdot) \in X_2(t_0,x_0,V_0)\}$$
$$= \max_{V} \min\{\sigma(x(\theta)): x(\cdot) \in X_2(t_0,x_0,V)\}$$

are valid.

Theorem 10.2.8. *For any initial position $(t_0,x_0) \in T \times R^n$ there exists a positional strategy for the first player U_0 and a counterstrategy for the second player V_0^c such that*

$$\min_{U} \max\{\sigma(x(\theta)): x(\cdot) \in X_1(t_0,x_0,U)\}$$
$$= \max\{\sigma(x(\theta)): x(\cdot) \in X_1(t_0,x_0,U_0)\} = \omega_3^0(t_0,x_0)$$
(18)
$$= \min\{\sigma(x(\theta)): x(\cdot) \in X_2(t_0,x_0,V_0^c)\}$$
$$= \max_{V^c} \min\{\sigma(x(\theta)): x(\cdot) \in X_2(t_0,x_0,V^c)\}.$$

In relations (17) and (18), $\omega_1^0(t_0,x_0)$ and $\omega_3^0(t_0,x_0)$ are the values of the lower and upper differential games respectively (cf. Case I and Case III in Section 9.1.1). In accordance with the material in Chapter 9 for the given payoff functional, one can also define the value $\omega_2^0(t_0,x_0)$ of a differential game considered in the class of mixed positional strategies. Thus, to the three types of differential games correspond the three values $\omega_1^0(t_0,x_0)$, $\omega_2^0(t_0,x_0)$, and $\omega_3^0(t_0,x_0)$. In Section 10.5.2, a comparison of these quantities is carried out.

10.3. Efficient Solutions of Lower and Upper Differential Games

In Chapters 4, 5, and 7 several types of game-theoretical control problems were singled out whose solutions can be reduced to sufficiently simple computational procedures. The corresponding methods of solution were thus termed efficient. Recall that in those chapters these methods were developed under the assumption that equality (1.2.8) is fulfilled -- the condition of a saddle point in a small game. In Section 9.3, these efficient methods were modified for solving differential equations in a class of mixed positional strategies. In this section, these efficient methods are carried over to lower and upper differential games.

10.3.1. As was done in Section 9.3, we shall introduce the substitutions which should be made so that the results from Chapters 4, 5, and 7 can be applied to the class of problems investigated herein. First consider lower differential games.

Note the following: in Section 10.1.1 the counterstrategy for the first player was defined as the function

$$U^c\colon (t,x,v) \to U^c(t,x,v)\colon T \times R^n \times \mathcal{Q} \to \mathcal{P}.$$

The mapping

$$(1) \qquad (t,x) \to U^c(t,x,\cdot)\colon T \times R^n \to \mathbf{U}_{cc}$$

corresponds to this function. Recall that $\mathbf{U}_{cc}$ is the totality of countercontrols for the first player (cf. Section 10.2.3). In this section we shall interpret a counterstrategy for the first player as a mapping of the form (1). According to this interpretation, we shall introduce a function

$$f^c\colon T \times R^n \times \mathbf{U}_{cc} \times \mathcal{Q} \to R^n,$$

which we shall define by the equality

$$(2) \qquad f^c(t,x,u[\cdot],v) \overset{\text{def}}{=} f(t,x,u[v],v).$$

Using this notation, the step-by-step motion $x(\cdot,t_0,x_0,U^c,v(\cdot),\Delta)$ can be defined as a solution of the differential equation

$$(3) \qquad \dot{x}(t) = f^c(t,x(t),u_i[\cdot],v(t)), \quad x(t_0) = x_0$$

where

$$t \in [\tau_i,\tau_{i+1}), \quad u_i[\cdot] = U^c(\tau_i,x(\tau_i),\cdot), \quad i = 0,1, \ldots, k.$$

For the function f^c equality (10.2.11) can be rewritten as follows:

$$\begin{aligned} &\max_{v\in\mathcal{Q}} \min_{u[\cdot]\in\mathbf{U}_{cc}} \langle s,f^c(t,x,u[\cdot],v)\rangle \\ (4) \qquad &= \min_{u[\cdot]\in\mathbf{U}_{cc}} \max_{v\in\mathcal{Q}} \langle s,f^c(t,x,u[\cdot],v)\rangle \\ &= \max_{v\in\mathcal{Q}} \min_{u\in\mathcal{P}} \langle s,f(t,x,u,v)\rangle. \end{aligned}$$

Now we can state the substitution mentioned above. To the left of the sign $\div$ we shall write symbols, conditions, equations, etc. which were utilized in Chapters 4, 5, and 7, and to the right of this sign the corresponding symbols, conditions, equations, etc. required for the study of lower differential games will be located. We thus have

$$u \div u[\cdot], \quad \mathcal{P} \div \mathbf{U}_{cc}, \quad v \div v, \quad \mathcal{Q} \div \mathcal{Q},$$

$$f(t,x,u,v) \div f^c(t,x,u[\cdot],v),$$

equation (1.2.1) $\div$ equation (3),

equality (1.2.8) $\div$ equality (4),

$$U: T \times R^n \to \mathcal{P} \div U^c: T \times R^n \to \mathbf{U}_{cc},$$

(5)

$$V: T \times R^n \to \mathcal{Q} \div V: T \times R^n \to \mathcal{Q}.$$

Let the payoff functional be given by equality (3.1.1) and consider a lower differential game. For this game Theorem 4.1.1 can be rewritten as follows.

Theorem 10.3.1. *Let the continuous function* ω_1^0: $T \times R^n \to R^1$ *satisfy the boundary condition* $\omega_1^0(\theta,x) = \sigma(x)$ $(x \in R^n)$. *Let this function be continuously-differentiable in the region* $T^0 \times R^n$ *and satisfy the equation*

$$\frac{\partial\omega_1^0(t,x)}{\partial t} + \max_{v\in\mathcal{Q}} \min_{u\in\mathcal{P}} \left[\frac{\partial\omega_1^0(t,x)}{\partial x} f(t,x,u,v)\right] = 0 \tag{6}$$

in this region. Then this function will be the value function of the differential game under consideration. Let counterstrategy U_0^c *and positional strategy* V_0 *be selected from the conditions*

$$\frac{\partial\omega_1^0(t,x)}{\partial x} f(t,x,u_0,v) = \min_{u\in\mathcal{P}} \frac{\partial\omega_1^0(t,x)}{\partial x} f(t,x,u,v),$$

$$u_0 = U_0^c(t,x,v),$$

(7)

$$\min_{u\in\mathcal{P}} \frac{\partial\omega_1^0(t,x)}{\partial x} f(t,x,u,v_0) = \max_{v\in\mathcal{Q}} \min_{u\in\mathcal{P}} \frac{\partial\omega_1^0(t,x)}{\partial x} f(t,x,u,v),$$

$$v_0 = V_0(t,x).$$

Then for any initial position $(t_0,x_0) \in T \times R^n$ *the quantity* $\omega_1^0(t_0,x_0)$ *will be the value of the game under consideration and the pair* (U_0^c,V_0) *forms a saddle point for this game.*

Note that in equation (6) operations max and min cannot be interchanged. However, according to (4) the second summand in (6) can be written as follows:

$$\max_{v\in \mathcal{Q}} \min_{u\in \mathcal{P}} \frac{\partial \omega_1^0(t,x)}{\partial x} f(t,x,u,v)$$

$$= \max_{v\in \mathcal{Q}} \min_{u[\cdot]\in \mathbf{U}_{cc}} \frac{\partial \omega_1^0(t,x)}{\partial x} f^c(t,x,u[\cdot],v)$$

$$= \min_{u[\cdot]\in \mathbf{U}_{cc}} \max_{v\in \mathcal{Q}} \frac{\partial \omega_1^0(t,x)}{\partial x} f^c(t,x,u[\cdot],v).$$

The proof of Theorem 10.3.1 is carried out in basically the same manner as the proof of Theorem 4.1.1. Note, however, that the set $\mathcal{H}(t,x)$ which appears in the proof of Theorem 4.1.1 is replaced here by the set

$$\mathcal{H}(t,x) \overset{\text{def}}{=} \{f(t,x, \mathcal{U}^0(t,x,v),v): v \in \mathcal{Q}\},$$

where $\mathcal{U}^0(t,x,v)$ is the set of vectors u_0 satisfying condition (7).

We present statements of two theorems to which Theorems 4.4.1 and 4.4.2 correspond.

Theorem 10.3.2. *Let the payoff functional be given by equality* (3.1.1). *Assume that the function* σ: $R^n \to R^1$ *is locally Lipschitzian. Consider a lower differential game for this payoff functional. In order that the function* ω: $T \times R^n \to R^1$ *be the value function for the differential game under consideration, it is necessary and sufficient that the following set of conditions be fulfilled*:

(a) ω *is locally-Lipschitzian*,

(b) $\omega(\theta,x) = \sigma(x) \quad (x \in R^n)$,

(8) (c) $$\max_{v} \min_{h_2} D^{-}\omega(t,x) \mid (1,h_2) \leq 0,$$

(9) $$\min_{u[\cdot]} \max_{h_1} D^{+}\omega(t,x) \mid (1,h_1) \geq 0$$

for all $t < \theta$, $x \in R^n$. *In the inequalities* (8) *and* (9), $v \in \mathcal{Q}$, $u[\cdot] \in \mathbf{U}_{cc}$, $h_2 \in \mathcal{F}_2(t,x,v)$, $h_1 \in \mathcal{F}_1(t,x,u[\cdot])$, *where*

(10) $$\mathcal{F}_2(t,x,v) \overset{\text{def}}{=} \text{co}\{f(t,x,u,v): u \in \mathcal{P}\},$$

(11) $$\mathcal{F}_1(t,x,u[\cdot]) \overset{\text{def}}{=} \overline{\text{co}}\{f(t,x,u[v],v): v \in \mathcal{Q}\}.$$

Theorem 10.3.3. *Let the payoff functional be given by equality* (3.3.1). *Assume that the function* ρ: $T \times R^n \to R^1$ *is locally-Lipschitzian. In order that the function* ω: $T \times R^n \to R^1$ *be the value function for a lower differential game, it is necessary and sufficient that the* *following set of conditions be fulfilled.*

(a) ω *is locally-Lipschitzian;*
(b) $\omega(t,x) \leqslant \rho(t,x)$, $\omega(\theta,x) = \rho(\theta,x)$, *for* $t \in T$, $x \in R^n$;
(c) *in the region* $\{(t,x) \in T^0 \times R^n\colon \omega(t,x) < \rho(t,x)\}$ *the inequality* (8) *is valid;*
(d) *inequality* (9) *is valid for all* $(t,x) \in T^0 \times R$.

10.3.2. We shall show how the results of Chapter 5 are carried over to the case of a lower differential game. Let the motion of the controlled system be described by the equation

$$\dot{x}(t) = A(t)x(t) + g(t,u(t),v(t)). \tag{12}$$

Introduce the following notation:

$$\kappa(\ell,\tau,\theta) \overset{\text{def}}{=} \max_{v\in \mathcal{Q}} \min_{u\in \mathcal{P}} \langle \ell,\Phi(\theta,\tau)g(\tau,u,v)\rangle. \tag{13}$$

Unlike the case in Chapter 5, condition (5.1.36) is not assumed to be fulfilled here, therefore one cannot interchange the operations of max and min. Recall that the matrix function $\tau \to \Phi(\theta,\tau)$ satisfies equation (1.2.15). The quantity κ defined by equality (13) corresponds to the quantity κ introduced in Chapter 5 by equality (5.1.37).

Let the payoff functional be given by equality (3.1.1). Assume that the function $\sigma : R^n \to R^1$ is convex and satisfies the Lipschitz condition. Let $\sigma^*\colon R^n \to (-\infty,\infty]$ be a function conjugate to the function σ and let $\operatorname{dom} \sigma^* \overset{\text{def}}{=} \{\ell \in R^n\colon \sigma^*(\ell) < \infty\}$.

Set

$$\psi(\ell,t) \overset{\text{def}}{=} \sigma^*(\ell) - \int_t^\theta \kappa(\ell,\tau,\theta)d\tau, \tag{14}$$

$$\omega_p(t,x) \overset{\text{def}}{=} \max_{\ell\in\operatorname{dom}\sigma^*} [\langle \ell,\Phi(\theta,t)x\rangle - \psi(\ell,t)], \tag{15}$$

$$\mathfrak{L}_0(t,x) \overset{\text{def}}{=} \{\ell_0 \in \operatorname{dom} \sigma^*\colon \langle \ell_0,\Phi(\theta,t)x\rangle - \psi(\ell_0,t) = \omega_p(t,x)\}. \tag{16}$$

Utilizing the sign $\div$ one can write

$$\psi(\ell,t)\ (5.1.22) \div \psi(\ell,t)\ (14)$$

$$\omega_p(t,x)\ (5.1.23) \div \omega_p(t,x)\ (15)$$

$$\mathfrak{L}_0(t,x)\ (5.2.1) \div \mathfrak{L}_0(t,x)\ (16).$$

If, moreover, for any $t < \theta$ the function $\psi(\cdot,t)$ (14) is strictly convex, then the set $\mathfrak{L}_0(t,x)$ (16) consists of a single element $\ell_0(t,x)$. In this case, in view of the rule of extremal aiming, one should determine the counterstrategy U_0^c and the positional strategy V_0. These are selected from the conditions

(17)
$$\langle \ell_0(t,x), \Phi(\theta,t)g(t,u_0,v)\rangle = \min_{u\in P} \langle \ell_0(t,x), \Phi(\theta,t)g(t,u,v)\rangle,$$
$$u_0 = U_0^c(t,x,v),$$

(18)
$$\min_{u\in P} \langle \ell_0(t,x), \Phi(\theta,t)g(t,u,v_0)\rangle$$
$$= \max_{v\in Q} \min_{u\in P} \langle \ell_0(t,x), \Phi(\theta,t)g(t,u,v)\rangle, \quad v_0 = V_0(t,x).$$

The following theorems (analogs of Theorem 5.2.1 and 5.4.1) are valid.

Theorem 10.3.4. *Let the function* $\psi(\cdot,t)$ (14) *be strictly convex for any* $t < \theta$. *Then the function* ω_p (15) *coincides with the value function of the lower differential game* (3.1.1), (12). *The counterstrategy* U_0^c (17) *and the positional strategy* V_0 (18) *form a saddle point of the game under consideration.*

Theorem 10.3.5. *In order that the function* ω_p (15) *coincide on the whole space* $T \times R^n$ *with the value function of a lower differential game* (3.1.1), (12), *it is necessary and sufficient that the equality*

$$\max_{v\in Q} \min_{u\in P} \max_{\ell\in\mathcal{L}_0(t,x)} [\langle \ell, \Phi(\theta,t)g(t,u,v)\rangle - \kappa(\ell,t,\theta)] = 0$$

be fulfilled for all $t < \theta$, $x \in R^n$. *Here the set* $\mathcal{L}_0(t,x)$ *and the quantity* $\kappa(\ell,t,\theta)$ *are defined according to* (13) *and* (16).

10.3.3. Consider the construction of a priori stable sets for lower differential games. Introduce the following sets:

(19)
$$\mathcal{F}_2(t,x) \overset{\text{def}}{=} \bigcap_{v\in Q} \mathcal{F}_2(t,x,v),$$

(20)
$$\mathcal{F}_1(t,x) \overset{\text{def}}{=} \bigcap_{u[\cdot]\in U_{cc}} \mathcal{F}_1(t,x,u[\cdot]),$$

where the sets $\mathcal{F}_2(t,x,v)$ *and* $\mathcal{F}_1(t,x,u[\cdot])$ *are defined according to* (10) *and* (11).

Note that $\mathcal{F}_2(t,x)$ (7.1.9) $\div$ $\mathcal{F}_2(t,x)$ (19), $\mathcal{F}_1(t,x)$ (7.1.15) $\div$ $\mathcal{F}_1(t,x)$ (20). The following assertion is valid (cf. Theorems 7.1.1 and 7.1.2).

Theorem 10.3.6. *Let* $w(\cdot)$: $[t_0,\tau] \to R^n$ *be an absolutely-continuous function satisfying the conditions*

$$(t,w(t)) \in N \quad \textit{for } t_0 \leq t \leq \tau, \quad (\tau,w(\tau)) \in M$$

and the differential inclusion

$$\dot{w}(t) \in \mathcal{F}_2(t,w(t)), \quad t_0 \leqslant t \leqslant T.$$

Then the graph of this function -- the set

$$(21) \qquad W_1 \overset{\text{def}}{=} \{(t,w(t)): t_0 \leqslant t \leqslant T\}$$

will be a u-stable bridge in an $(\mathcal{M}, \mathcal{N})$-approach problem. Let U^c_e be a counterstrategy extremal to the set W_1 defined above, i.e. this counterstrategy is selected from the condition

$$(22) \qquad \begin{aligned} &\langle x - w(t), f(t,x,u_e,v)\rangle = \min_{u\in \mathcal{P}} \langle x - w(t), f(t,x,u,v)\rangle, \\ &u_e = U^c_e(t,x,v). \end{aligned}$$

Then for the initial position $(t_0,x_0) = (t_0,w(t_0))$ this counterstrategy U^c_e yields the solution of Problem 10.2.1 *(the problem of an $(\mathcal{M}, \mathcal{N})$-approach in the class of counterstrategies).*

Define the quantities

$$(23) \qquad \begin{aligned} \kappa_2(\ell,t,x) &\overset{\text{def}}{=} \min_{v\in \mathcal{Q}} \max_{u\in \mathcal{P}} \langle \ell, f(t,x,u,v)\rangle, \\ \kappa_1(\ell,t,x) &\overset{\text{def}}{=} \min_{u[\cdot]\in \mathbf{U}_{cc}} \max_{v\in \mathcal{Q}} \langle \ell, f^c(t,x,u[\cdot],v)\rangle \\ &= \max_{v\in \mathcal{Q}} \min_{u\in \mathcal{P}} \langle \ell, f(t,x,u,v)\rangle. \end{aligned}$$

Note that κ_2 (7.1.17) $\div$ κ_2 (23), κ_1 (7.1.17) $\div$ κ_1 (23). According to the lemma presented below, the sets $\mathcal{F}_2$ (19) and $\mathcal{F}_1$ (20) can be defined as intersections of half-spaces. This lemma is an analog of Lemma 7.1.2.

Lemma 10.3.1. *Let the sets $\mathcal{F}_2(t,w)$ and $\mathcal{F}_1(t,w)$ be defined in accordance with* (19) *and* (20) *respectively. The vector h belongs to the set $\mathcal{F}_i(t,w)$ $(i = 1,2)$, if and only if the inequality*

$$(24) \qquad \max_{\ell\in R^n} [\langle \ell,h\rangle - \kappa_i(\ell,t,w)] \leqslant 0$$

is fulfilled where the quantities κ_i are defined by relations (23).

Next, utilizing equality (4) and taking into account substitution (5), one can reformulate Lemma 7.2.1 as follows.

Lemma 10.3.2. *Let the sets $\mathcal{F}_2(t,x)$, $\mathcal{F}_1(t,x)$, $\mathcal{F}_2(t,x,v)$, and $\mathcal{F}_1(t,x,u[\cdot])$ be defined by relations* (19), (20), (10), *and* (11). *If the function $\kappa_2(\cdot,t,x)$: $R^n \to R^1$ is convex, then for any $u[\cdot] \in \mathbf{U}_{cc}$,*

$$\mathcal{F}_2(t,x) \cap \mathcal{F}_1(t,x,u[\cdot]) \neq \phi \ .$$

If the function $\kappa_1(\cdot,t,x)$: $R^n \to R^1$ *is convex then for any* $v \in \mathcal{Q}$,

$$\mathcal{F}_1(t,x) \cap \mathcal{F}_2(t,x,v) \neq \phi \ .$$

From this lemma in particular follows:

Theorem 10.3.7. *Let the function* $\kappa_2(\cdot,t,x)$ (23) *be convex for all* $(t,x) \in T \times R^n$. *Then, for any initial position* $(t_0,x_0) \in T \times R^n$, *the following alternative is valid: either a u-stable path* $\mathcal{W}_1$ *emanates from this initial position, defined according to* (21) *and in which case the counterstrategy* U_e^c (22) *yields a solution to Problem* 10.2.1 (*the problem of an* $(\mathcal{M}, \mathcal{N})$*-approach*); *or for some* $\epsilon > 0$, *this position is contained on a* v_**-stable bridge in a problem of an* $(\mathcal{M}, \mathcal{N})^\epsilon$*-evasion* (*on a* v_**-stable integral manifold*), *which is defined by the relation*

$$(25) \qquad \mathcal{W}_2 \overset{\text{def}}{=} cl\,\{(t,w(t)):\ t_0 \leqslant t \leqslant t^\epsilon(w(\cdot)),\ w(\cdot) \in X(t_0,x_0,\mathcal{F}_2)\},$$

and the positional strategy V_e *extremal to the set* $\mathcal{W}_2$ *yields the solution to Problem* 2.1.2.

This theorem is an analog of Theorem 7.2.3. In relation (25) the functional $t^\epsilon(w(\cdot))$ is defined according to (2.1.7), $X(t_0,x_0,\mathcal{F}_2)$ is the set of solutions of the differential inclusion $\dot{x}(t) \in \mathcal{F}_2(t,x(t))$, $x(t_0) = x_0$.

We shall confine ourselves to the above stated Theorems 10.3.1 - 10.3.7 which give a sufficiently complete idea of how the results in Chapters 4, 5, and 7 are being carried over to the case of lower differential games.

10.3.4. Consider very briefly upper differential games. We are thus assuming now that the differential games are considered in the class of positional strategies for the first player and counterstrategies for the second one. Without presenting complete formulations, we shall only indicate the alterations which are to be incorporated in the above stated theorems.

In Theorem 10.3.1 equation (6) is replaced by the equation

$$\frac{\partial\omega(t,x)}{\partial t} + \min_{u\in\mathcal{P}} \max_{v\in\mathcal{Q}} \frac{\partial\omega(t,x)}{\partial x} f(t,x,u,v) = 0.$$

Let the positional strategy U_0 and the counterstrategy V_0^c be selected from the conditions

$$\max_{v\in\mathcal{P}} \frac{\partial\omega(t,x)}{\partial x} f(t,x,u_0,v) = \min_{u\in\mathcal{P}} \max_{v\in\mathcal{Q}} \frac{\partial\omega(t,x)}{\partial x} f(t,x,u,v),$$

$$u_0 = U_0(t,x),$$

$$\frac{\partial\omega(t,x)}{\partial x} f(t,x,u,v_0) = \max_{v\in\mathcal{Q}} \frac{\partial\omega(t,x)}{\partial x} f(t,x,u,v),$$
$$v_0 = V_0^c(t,x,u).$$

Then the pair (U_0,V_0^c) forms a saddle point for an upper differential game.

Furthermore, when proceeding from lower differential games to the upper ones in the statements of Theorems 10.3.2 and 10.3.3, inequalities (8) and (9) should be replaced by the inequalities

$$\max_{v[\cdot]} \min_{h_2} D^-\omega(t,x) \mid (1,h_2) \leqslant 0, \tag{26}$$

$$\min_{u} \max_{h_1} D^+\omega(t,x) \mid (1,h_1) \geqslant 0. \tag{27}$$

Here

$$v[\cdot] \in \mathbf{V}_{cc},\quad u \in \mathcal{P},\quad h_2 \in \mathcal{F}_2(t,x,v[\cdot]),\quad h_1 \in \mathcal{F}_1(t,x,u),$$

$$\mathcal{F}_2(t,x,v[\cdot]) = \overline{\text{co}}\{f(t,x,u,v[u]): u \in \mathcal{P}\}, \tag{28}$$

$$\mathcal{F}_1(t,x,u) = \text{co}\{f(t,x,u,v): v \in \mathcal{Q}\}. \tag{29}$$

We now proceed to the results of Section 10.3.2. To obtain from Theorem 10.3.4 the corresponding assertion for an upper differential game one should replace the quantity $\kappa(\ell,\tau,\theta)$ (13) by the quantity

$$\kappa(\ell,\tau,\theta) \overset{\text{def}}{=} \min_{u\in\mathcal{P}} \max_{v\in\mathcal{Q}} \langle \ell, \Phi(\theta,\tau)g(\tau,u,v)\rangle \tag{30}$$

and define the optimal positional strategy U_0 and the counterstrategy V_0^c from the conditions

$$\begin{aligned}&\max_{v\in\mathcal{Q}} \langle \ell_0(t,x), \Phi(\theta,t)g(t,u_0,v)\rangle \\ &\quad= \min_{u\in\mathcal{P}} \max_{v\in\mathcal{Q}} \langle \ell_0(t,x), \Phi(\theta,t)g(t,u,v)\rangle, \quad u_0 = U_0(t,x),\end{aligned} \tag{31}$$

$$\begin{aligned}&\langle \ell_0(t,x), \Phi(\theta,t)g(t,u,v_0)\rangle = \max_{v\in\mathcal{Q}} \langle \ell_0(t,x), \Phi(\theta,t)g(t,u,v)\rangle, \\ &v_0 = V_0^c(t,x,u).\end{aligned} \tag{32}$$

When proceeding from a lower game considered in Theorem 10.3.5 to an upper game, one should replace the quantity κ (13) by the quantity (30) and the equality presented in Theorem 10.3.5 should be replaced by the equality

$$\min_{u\in\mathcal{P}} \max_{v\in\mathcal{Q}} \max_{\ell\in\mathcal{L}_0(t,x)} [\langle \ell,\Phi(\theta,t)g(t,u,v)\rangle - \kappa(\ell,t,\theta)] = 0.$$

We shall now state a result dealing with a differential game with the payoff of the form

$$(33)\qquad \gamma(x(\cdot)) = \mathrm{dist}([x(\theta)]_m, M).$$

Recall that $[x]_m$ is a vector consisting of the first m components of an n-dimensional vector x and M is a convex compact set in R^m. As above, we shall assume that the motion of the system is described by equation (12) and that an upper differential game is being considered. For this game we shall define the quantity $\epsilon_0(t,x)$ which is an analog of the programmed maximin introduced in Section 5.1 (cf. the equality (5.1.31)). We thus set

$$(34)\qquad \epsilon_0(t,x) \overset{\mathrm{def}}{=} \max_{\|s\| \leqslant 1} [\langle s,[\Phi(\theta,t)]_m x\rangle - \varphi^{(m)}(s,t)],$$

where

$$(35)\qquad \begin{aligned} \varphi^{(m)}(s,t) &= \max_{z\in M} \langle s,z\rangle - \int_t^\theta \kappa^{(m)}(s,\tau,\theta)d\tau,\\ \kappa^{(m)}(s,\tau,\theta) &\overset{\mathrm{def}}{=} \min_{u\in P} \max_{v\in Q} \langle s, [\Phi(\theta,\tau)]_m g(\tau,u,v)\rangle. \end{aligned}$$

Recall that $[\Phi(\theta,\tau)]_m$ is an $m \times n$ matrix consisting of the first m rows of the $n \times n$ matrix $\Phi(\theta,\tau)$ and s is an m-dimensional vector.

For any $t \leqslant \theta$ let the function $\varphi^m(\cdot,\tau)$ (35) be convex. Then in the region

$$(36)\qquad \mathcal{E}^+ = \{(t,x) \in T^0 \times R^n: \epsilon_0(t,x) > 0\},$$

the maximum in (34) is attained at the unique vector $s_0(t,x)$ (this assertion is verified in the same manner as Lemma 5.2.1). In region $\mathcal{E}_+$ the positional strategy U_0 and the counterstrategy V_0^c are determined from the conditions

$$(37)\qquad \begin{aligned} &\max_{v\in Q} \langle s_0(t,x), [\Phi(\theta,t)]_m g(t,u_0,v)\rangle\\ &= \min_{u\in P} \max_{v\in Q} \langle s_0(t,x), [\Phi(\theta,t)]_m g(t,u,v)\rangle, \quad u_0 = U_0(t,x), \end{aligned}$$

$$(38)\qquad \begin{aligned} &\langle s_0(t,x), [\Phi(\theta,t)]_m g(t,u,v_0)\rangle\\ &= \max \langle s_0(t,x), [\Phi(\theta,t)]_m g(t,u,v)\rangle, \quad v_0 = V_0^c(t,x,u). \end{aligned}$$

Outside the region $\mathcal{E}^+$, the strategy U_0 and the counterstrategy V_0^c can be defined arbitrarily.

The following theorem (an analog of Theorem 5.2.4) is valid.

Theorem 10.3.8. *Let the function $\varphi^{(m)}(\cdot,t)$ (35) be convex for all $t \leqslant \theta$. Then, for all $(t,x) \in T \times R^n$, the quantity $\epsilon_0(t,x)$ coincides with the value* *of the upper differential game* (12), (33). *The strategy* U_0 (37) *and the counterstrategy* V_c^0 (38) *form a saddle point for the game under consideration.*

We shall describe the construction of a priori stable sets for differential games considered in the class of positional strategies for the first player and counterstrategies for the second. Replace the sets $\mathcal{F}_2(t,x)$ (19) and $\mathcal{F}_1(t,x)$ (20) by the following sets:

$$\mathcal{F}_2(t,x) \overset{\text{def}}{=} \bigcap_{v[\cdot]\in \mathbf{V}_{cc}} \mathcal{F}_2(t,x,v[\cdot]), \tag{39}$$

$$\mathcal{F}_1(t,x) \overset{\text{def}}{=} \bigcap_{u\in \mathcal{P}} \mathcal{F}_1(t,x,u), \tag{40}$$

where the sets $\mathcal{F}_2(t,x,v[\cdot])$ and $\mathcal{F}_1(t,x,u)$ are defined according to (28) and (29). In place of quantities κ_1 and κ_2 (23) we shall introduce the quantities

$$\begin{aligned} \kappa_2(\ell,t,x) &\overset{\text{def}}{=} \max_{u\in\mathcal{P}} \min_{v\in\mathcal{Q}} \langle \ell, f(t,x,u,v)\rangle, \\ \kappa_1(\ell,t,x) &\overset{\text{def}}{=} \min_{u\in\mathcal{P}} \max_{v\in\mathcal{Q}} \langle \ell, f(t,x,u,v)\rangle. \end{aligned} \tag{41}$$

The following assertions are valid (cf. Lemmas 7.1.2, 7.2.1, 10.3.1, and 10.3.2).

Let the sets $\mathcal{F}_2(t,x)$ and $\mathcal{F}_1(t,x)$ be defined by relations (39) and (40) and quantities κ_1 and κ_2 by equalities (41). The vector h belongs to the set $\mathcal{F}_i(t,x)$ ($i = 1,2$) if and only if

$$\max_{\|\ell\|=1} [\langle \ell,h\rangle - \kappa_i(\ell,t,x)] \leqslant 0. \tag{42}$$

If the function $\kappa_2(\cdot,t,x)$ is convex then

$$\mathcal{F}_2(t,x) \cap \mathcal{F}_1(t,x,u) \neq \emptyset \quad \text{for any } u \in \mathcal{P}.$$

If the function $\kappa_1(\cdot,t,x)$ is convex then

$$\mathcal{F}_1(t,x) \cap \mathcal{F}_2(t,x,v[\cdot]) \neq \emptyset \quad \text{for any } v[\cdot] \in \mathbf{V}_{cc}.$$

The theorems presented below are analogs of Theorems 7.1.1, 7.1.2, 10.3.6, and 7.2.3, 10.3.7.

Theorem 10.3.9. *Let* $w(\cdot)$: $[t_0,\tau] \to R^n$ *be an absolutely-continuous solution of the differential inclusion*

$$\dot{w}(t) \in \mathcal{F}_2(t,w(t)) \tag{43}$$

satisfying the conditions

$$(t,w(t)) \in N \quad \textit{for } t_0 \leqslant t \leqslant \tau, \quad (\tau,w(\tau)) \in M, \quad \tau \leqslant \theta. \tag{44}$$

Then the positional strategy U_e *selected from the condition*

$$(45)\qquad \begin{aligned} &\max_{v\in\mathcal{Q}} \langle x - w(t), f(t,x,u_e,v)\rangle \\ &\quad = \min_{u\in\mathcal{P}} \max_{v\in\mathcal{Q}} \langle x - w(t), f(t,x,u,v)\rangle, \quad u_e = U_e(t,x) \end{aligned}$$

yields a solution to Problem 2.1.1 (*the problem of an* $(\mathcal{M}, \mathcal{N})$*-approach*).

Theorem 10.3.10. *Let the function* $\kappa_1(\cdot,t,x)$ (41) *be convex for all* $(t,x) \in T \times R^n$. *Then, for any initial position* $(t_0,x_0) \in T \times R^n$, *the following alternative is valid: either a solution of the differential inclusion* (43) *emanates from this initial position satisfying conditions* (44), *in which case the positional strategy* U_e (45) *yields a solution to Problem* 2.1.1; *or for some* $\epsilon > 0$, *this position is located on a v-stable bridge in an* $(\mathcal{M}, \mathcal{N})^{\epsilon}$*-evasion problem which is defined by the relation*

$$\mathcal{W}_2 \overset{\text{def}}{=} cl\,\{(t,w(t)):\ t_0 \leqslant t \leqslant t^{\epsilon}(w(\cdot)),\ \ w(\cdot) \in X(t_0,x_0,\mathcal{F}_2)\}$$

and the counterstrategy V_e^c *extremal to the set* $\mathcal{W}_2$ *yields a solution to Problem* 10.2.2. (*Here* $X(t_0,x_0,\mathcal{F}_2)$ *is the set of solutions of the differential inclusion* (43) *and the functional* t^{ϵ} *is defined according to* (2.1.7).)

10.3.5. Example. Consider a controlled system whose motion is described by the equation

$$(46)\qquad \ddot{x}(t) = L(\beta(t))u(t) + v(t).$$

Here x, u, and v are two-dimensional vectors and $L(\beta)$ is a rotation matrix (cf. (9.3.30)). We set

$$(47)\qquad \begin{aligned} \mathcal{P} &= \{(u_1,u_2) \in R^2:\ u_1^2 + u_2^2 \leqslant r_1^2\}, \\ \mathcal{Q} &= \{(v_1,v_2,\beta) \in R^3:\ v_1^2 + v_2^2 \leqslant r_2^2,\ |\beta| \leqslant \rho\}. \end{aligned}$$

Let the target functional

$$\gamma(x(\cdot)) = (x_1^2(0) + x_2^2(0))^{1/2}$$

be given where $t = 0$ is the termination time. It is required to determine a positional strategy U_0 which yields for the example under consideration a solution to Problem 3.1.1, i.e. it guarantees the smallest (minimax) value for the target functional. Recall that an analogous problem was considered in Section 9.3.4 (cf. Example 2) where, inter alia, the physical meaning of an equation of the form (46) was clarified.

We rewrite equation (46) in the normal form

$$\dot{x}(t) = y(t),\quad \dot{y}(t) = L(\beta(t))u(t) + v(t).$$

To solve this example we shall use Theorem 10.3.8. First determine the quantity $\varphi^{(m)}$ (35). In this example

$$M = \{(0,0)^T\}, \quad m = 2, \quad \theta = 0,$$

$$[\Phi(\theta,\tau)]_m = \begin{bmatrix} 1 & 0 & -\tau & 0 \\ 0 & 1 & 0 & -\tau \end{bmatrix}.$$

Therefore according to (47), (9.3.30) we obtain

$$\kappa^{(m)}(s,\tau,\theta) = \tau(r_1 \cos\rho - r_2)(s_1^2 + s_2^2)^{1/2}.$$

Next it follows from (35) that

$$\varphi^{(m)}(s,t) = \frac{t^2}{2}(r_1 \cos\rho - r_2)(s_1^2 + s_2^2)^{1/2}.$$

We shall assume that

$$(48) \qquad r_1 \cos\rho - r_2 \geqslant 0.$$

Under this condition the function $\varphi(\cdot,t)$ is convex. The quantity ϵ_0 (34) is defined here by the relation

$$(49) \qquad \begin{aligned} \epsilon_0(t,x,y) = \max_{\|s\| \leqslant 1} [&s_1(x_1 - ty_1) + s_2(x_2 - ty_2) \\ &- \frac{t^2}{2}(r_1 \cos\rho - r_2)(s_1^2 + s_2^2)^{1/2}]. \end{aligned}$$

In the region $\mathcal{E}^+$ where $\epsilon_0(t,x,y) > 0$, the maximum in (49) is attained at the unique vector

$$(50) \qquad s_0(t,x,y) = (x - ty)/\|x - ty\|^{-1}.$$

Note that for $(t,x,y) \in \mathcal{E}^+$ we have $\|x - ty\| > 0$. Therefore

$$(51) \qquad \epsilon_0(t,x,y) = \max\{0, \|x - ty\| - \frac{t^2}{2}(r_1 \cos\rho - r_2)\}.$$

Defining the positional strategy U_0 in accordance with (37) we obtain

$$(52) \qquad U_0(t,x,y) = -r_1 s_0(t,x,y).$$

Vector $s_0(t,x,y)$ is defined by relation (50) in the region $\mathcal{E}^+$. If, however $(t,x,y) \notin \mathcal{E}^+$ we then formally set $s_0(t,x,y) = 0$.

Under condition (48) we have, in view of Theorem 10.3.8, that the strategy U_0 yields a solution to the problem under consideration. In Fig. 10.3.1 trajectories $(x_1(t),x_2(t))$ are presented which are obtained when the strategy U_0 (52) is realized coupled with various types of

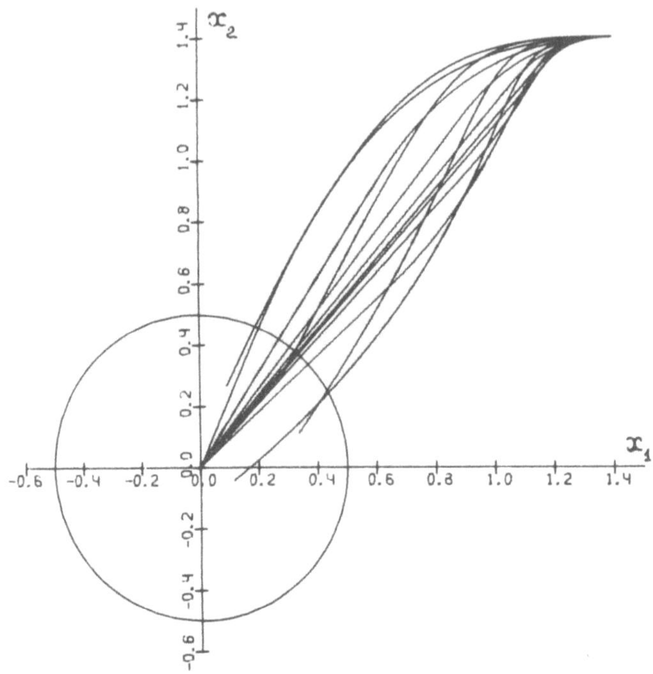

Figure 10.3.1

disturbance. Here the following parameters and initial conditions are chosen:

$$r_1 = 1.74, \quad r_2 = 1, \quad \rho = \pi/6, \quad t_0 = -2, \quad \theta = 0,$$

$$x_1(t_0) = 1.4, \quad x_2(t_0) = 1.4, \quad y_1(t_0) = -1, \quad y_2(t_0) = 0.$$

For those values we obtain that $\epsilon_0(t_0, x_0, y_0) = 0.5$.

10.4. Control Procedures with Guidance

In this section control procedures with guidance are described which are stable relative to information noise. These procedures approximate solutions of game-theoretical control problems posed in the class of positional strategies and counterstrategies.

10.4.1. Consider first a control procedure with guidance which yields a correct solution to an $(\mathcal{M}, \mathcal{N})$-approach problem in the case when the first player does not know his or her opponent's current control. One can motivate the necessity of the transition from positional strategies to the corresponding control procedures with guidance using the very same arguments as those presented in Sections 8.1 and 8.2.

Recall that in subsection 8.2.3 a general definition of control procedures with guidance was given. According to this definition a control procedure with guidance is a triple (U,ψ,χ) where U: $T \times R^n \times R^n \to \mathcal{P}$ is a function which forms controls; ψ: $T^2_+ \times R^n \times R^n \to R^n$ is the transition function of the guidance; function χ: $T \times R^n \to R^n$ assigns the initial position of the guidance. In subsection 8.2.3, step-by-step motions formed by the control procedure (U,ψ,χ) for a partition Δ selected by the first player were also defined. These general definitions carry over verbatim to the case under consideration.

We now describe the construction of a procedure (U^0,ψ^0,e) which will yield a correct solution to a problem of an $(\mathcal{M}, \mathcal{N})$-approach. We shall assume that the controlled system (1.2.1) satisfies conditions (i) - (iii) stipulated in Section 1.2.2. Unlike the case considered in Section 8.2, it is not required here that condition (iv) -- the saddle point condition in a small game -- be fulfilled.

Let $\mathcal{W}$ be a closed u_*-stable bridge in a problem of an $(\mathcal{M}, \mathcal{N})$-approach (cf. the corresponding definition in Section 10.2.4). We shall assume that the function e: $T \times R^n \to R^n$ assigns to the position (t,x) a point $e(t,x) \in \mathcal{W}[t] = \{w \in R^n : (t,w) \in \mathcal{W}\}$ which is closest to the point x. Recall that the function e is defined by relations (2.4.2), (2.4.4). Function U^0 assigns to the point (t,x,w) a point $u^0 = U^0(t,x,w) \in \mathcal{P}$ satisfying the condition

(1) $$\max_{v\in\mathcal{Q}} \langle x - w,\ f(t,x,u^0,v)\rangle = \min_{u\in\mathcal{P}} \max_{v\in\mathcal{Q}} \langle x - w,\ f(t,x,u,v)\rangle.$$

Thus the functions U^0 and e are defined here in the same manner as in the control procedure with guidance described in subsection 8.2.1.

It remains to define the guidance transition function ψ^0. Its definition will differ from the definition of the guidance transition function given in subsection 8.2.1. Let the function V^c_*: $T \times R^n \times R^n \times \mathcal{P} \to \mathcal{Q}$ assign to the collection (t,x,w,u) a point $v_* = V^c_*(t,x,w,u)$ satisfying the condition

(2) $$\langle x - w,\ f(t,x,u_*,v_*)\rangle = \max_{v\in\mathcal{Q}} \langle x - w,\ f(t,x,u,v)\rangle.$$

The function V^c_* can be selected in a manner such that the mapping $V^c_*(t,x,w,\cdot)$: $\mathcal{P} \to \mathcal{Q}$ will be Borel measurable for all $(t,x,w) \in T \times R^n \times R^n$. Let $(t_*,x_*,w_*) \in T \times R^n \times R^n$,

$$v_*[\cdot] = V^c_*(t_*,x_*,w_*,\cdot).$$

Assume that the set $\mathcal{F}_2(t,w,v_*[\cdot])$ is defined according to (10.2.15). Consider the differential inclusion

(3) $$\dot w(t) \in \mathcal{F}_2(t,w(t),v_*[\cdot]).$$

Suppose that $(t_*,w_*) \in \mathcal{W}$, $t^* \in [t_*,\theta]$. Since $\mathcal{W}$ is a u_*-stable bridge in a problem of an $(\mathcal{M}, \mathcal{N})$-approach, there exists a solution $w(\cdot)$: $[t_*,t^*] \to R^n$ of the differential inclusion (3) with the initial condition $w(t_*) =$

w_* which satisfies either the condition $(t^*,w(t^*)) \in \mathcal{W}$ or the condition $\{(t,w(t)): t_* \leqslant t \leqslant t^*\} \cap \mathcal{M} \neq \phi$. Assume that in the case under consideration, i.e. for $(t_*,w_*) \in \mathcal{W}$, the guidance transition function assigns to the collection (t_*,t^*,x_*,w_*) a point $w^* = w(t^*)$ at which the solution $w(\cdot)$ of the differential inclusion (3) arrives at time $t = t^*$. Let $(t_*,w_*) \notin \mathcal{W}$, $t^* \in [t_*,\theta]$, $x_* \in R^n$. In this case we select an arbitrary solution $w(\cdot)$: $[t_*,t^*] \to R^n$ of the differential inclusion (3) satisfying the initial condition $w(t_*) = w_*$ and set $\psi^0(t_*,t^*,x_*,w_*) = w(t^*)$.

If the number diamΔ (where Δ is the partition selected by the first player) and the information noise are sufficiently small, then the control procedure with guidance defined herein assures the motion of the controlled system in the vicinity of an u_*-stable bridge until the point $(t,x(t))$ hits a neighborhood of the target set. In this manner this control procedure yields a solution of an $(\mathcal{M}, \mathcal{N})$-approach problem which is stable relative to information noise. An exact statement of the corresponding assertion coincides verbatim with the statement of Theorem 8.2.1 and is therefore not repeated here. In the case under consideration the proof is carried out, on the whole, according to the outline described in subsection 8.2.2. The difference is that in place of condition (1.2.8) the equality (10.2.19) should be used.

10.4.2. Consider a solution of an $(\mathcal{M}, \mathcal{N})$-approach problem in the case when the first player knows both the current position and the control $v(t)$ selected by the second player. We shall describe the control procedure with guidance corresponding to this case. We start with a general definition. The triple

$$(U^c,\ \psi,\ \chi) \tag{4}$$

is called a control procedure with guidance. Here $U^c : T \times R^n \times R^n \times \mathcal{Q} \to \mathcal{P}$ is a function such that the mapping $v \to U^c(t,x,w,v)$ is Borel measurable for any $(t,x,w) \in T \times R^n \times R^n$. General definitions of the guidance transition function ψ and of the initial position function χ remain the same as in subsection 8.2.3.

We shall describe how the control procedure (U^c,ψ,χ) (4) forms step- by-step motions of the system (1.2.1). Let the first player select a partition $\Delta = (t_0 = \tau_0 < \tau_1 < \cdots < \tau_{k+1} = \theta)$. At time $t = \tau_i$ the first player obtains the vector $x^*(\tau_i)$ as a result of an inaccurate measurement of the phase position of the system $x(\tau_i)$. We shall also assume that the information about the opponent's control is subject to an error. As usual we shall denote the control selected by the second player by $v(t)$, and this control as perceived by the first player will be denoted by $v^{\#}(t)$. Thus the difference $v^{\#}(t) - v(t)$ is the measuring error of current control by the opponent. It is assumed that the function $t \to v^{\#}(t)$ is measurable (in the Borel or Lebesgue senses). At time $t = t_0$ as a result of an inaccurate measurement of the initial position $x(t_0)$ the first player obtains point $x^*(t_0)$ and determines the initial position of the guidance $w(t_0) = \chi(t_0,x^*(t_0))$. On the time interval $[\tau_0,\tau_1)$ the control for the first player is formed according to the equality

$$u(t) = U^c(t_0,x^*(t_0),w(t_0),v^\#(t)), \quad t_0 \leqslant t < \tau_1.$$

The control $u(\cdot)$: $[t_0,\tau_1) \to \mathcal{P}$ in conjunction with the control $v(\cdot)$: $[t_0,\tau_1) \to \mathcal{Q}$ for the second player, generate the motion of the system (1.2.1) $x(t)$ for $t_0 \leqslant t \leqslant \tau_1$. At time $t = \tau_1$ the system (1.2.1) reaches position $x(\tau_1)$. The position of the guidance at time $t = \tau_1$ is determined by the transition function ψ, i.e. it is assumed that

$$w(\tau_1) = \psi(\tau_0,\tau_1,x^*(\tau_0),w(\tau_0)).$$

At the next time interval the first player selects the control

$$u(t) = U^c(\tau_1,x^*(\tau_1),w(\tau_1),v^\#(t)), \quad \tau_1 \leqslant t < \tau_2,$$

which, jointly with a control for the second player, transfers the controlled system to the point $x(\tau_2)$. The position of the guidance at time $t = \tau_2$ is determined by the equality

$$w(\tau_2) = \psi(\tau_1,\tau_2,x^*(\tau_1),w(\tau_1))$$

and so on.

We shall determine a procedure (U_0^c,ψ_0,e) which yields a stable solution for an $(\mathcal{M},\mathcal{N})$-approach problem in the case under consideration. Let W be a closed u stable bridge in an $(\mathcal{M},\mathcal{N})$-approach problem (cf. the definition in subsection 2.2.1). It is assumed that the function e is defined by relations (2.4.2), (2.4.4). The function U_0^c assigns to the collection (t,x,w,v) a point $u_0 = U_0^c(t,x,w,v) \in \mathcal{P}$ satisfying the condition

$$\text{(5)} \qquad \langle x - w,\ f(t,x,u_0,v)\rangle = \min_{u\in\mathcal{P}} \langle x - w,\ f(t,x,u,v)\rangle.$$

The evolution of the guidance is determined in the same manner as in Section 8.2.1, therefore a formal definition of the transition function ψ_0 is not presented.

To state the result it is necessary to indicate how the information noise is estimated. Assume that for any pair of numbers τ_i, τ_{i+1} appearing in the partition Δ, for any $u \in \mathcal{P}$ and $s \in R^n$, $\|s\| = 1$, the bound

$$\text{(6)} \qquad \int_{\tau_i}^{\tau_{i+1}} \langle s,\ f(\tau_i,x(\tau_i),u,v(t)) - f(\tau_i,x^*(\tau_i),u,v^\#(t))\rangle \leqslant \xi_i(\tau_{i+1} - \tau_i)$$

is valid where ξ_i is a positive number -- a "weak" bound on the error. Moreover, it is also assumed that

(7) $\|x(\tau_0) - x^*(\tau_0)\| \leqslant \xi^*.$

The following assertion is valid for the control procedure with guidance (U_0^c,ψ_0,e) defined above.

Let W be a closed u-stable bridge in a problem of an (M, N)-approach and let $(t_0,x_0) \in W$. Then for any $\epsilon > 0$, a $\delta > 0$ and a $\xi > 0$ can be found such that for any choice of a partition satisfying the bound diam $\Delta \leqslant \delta$ and under the condition that bounds (6) and (7) are fulfilled with $\xi_i \leqslant \xi$ and $\xi^* \leqslant \xi$, the step-by-step motion of the system (1.2.1) generated by the control procedure (U_0^c,ψ_0,e) together with an arbitrary control for the second player will satisfy the condition of an $(\overline{M^\epsilon}, \overline{N^\epsilon})$-approach.

The proof of this assertion is carried out along the lines described in subsection 8.2.2. The main difference is that condition (1.2.8) should be replaced by equality (10.2.19).

10.5. Unification of Differential Games

In the preceding and present chapters, three types of differential games were considered. A proof of the existence of an equilibrium in these games was carried out following a unified method. The efficient methods of solution described for each one of the three types of differential game have much in common. Moreover, it will be shown below that one can develop a unified setting for the game-theoretical control problems considered in the preceding and present chapters.

10.5.1. The notion of stability serves as the central concept in our constructions. More precisely, for solving various types of problems, different notions of stability were introduced. For example, u-stable, μ-stable, and u_*-stable bridges were defined which were used in approach problems posed, respectively, in classes of counterstrategies, mixed strategies, and positional strategies. The property of stability can be defined not only for the sets $W \subset T \times R^n$ but also for the functions $\omega : T \times R^n \to R^1$. Note also that stability can be defined in various equivalent forms. Which form of stability condition is appropriate is determined by the nature of the problem under consideration. Thus the notion of stability represents a sufficiently flexible tool for investigating game-theoretical control problems.

As the first step on the route to unification of differential games, yet another definition of stability is proposed. First, we shall introduce some notation. Consider a continuous function

(1) $\xi: (t,x,s) \to \xi(t,x,s): T \times R^n \times R^n \to R^1.$

We shall assume that the function ξ is positive homogeneous in s and satisfies the Lipschitz condition with respect to variable x, i.e.

$$\xi(t,x,\alpha s) = \alpha\xi(t,x,s), \quad (t,x,s) \in T \times R^n \times R^n, \quad \alpha > 0,$$

(2) $$|\xi(t,x^{(1)},s) - \xi(t,x^{(2)},s)| \leqslant \lambda(\mathfrak{D}) \|x^{(1)} - x^{(2)}\|$$

$$\text{for all} \quad (t,x^{(i)},s) \in \mathfrak{D}, \quad i = 1,2,$$

where $\mathfrak{D}$ is an arbitrary bounded region in $T \times R^n \times R^n$.

Next let a multivalued mapping $(t,x) \to \mathcal{H}(t,x)$ be given which assigns a convex compact set $\mathcal{H}(t,x) \subset R^n$ to the point $(t,x) \in T \times R^n$. Define the sets

(3) $$\mathcal{H}_1(t,x,s,\xi) \stackrel{\text{def}}{=} \{h \in \mathcal{H}(t,x) \colon \langle s,h\rangle \leqslant \xi(t,x,s)\},$$
$$\mathcal{H}_2(t,x,s,\xi) \stackrel{\text{def}}{=} \{h \in \mathcal{H}(t,x) \colon \langle s,h\rangle \geqslant \xi(t,x,s)\}$$

where $(t,x,s) \in T \times R^n \times R^n$. We shall assume that the following conditions are fulfilled:

(a) The mapping $(t,x) \to \mathcal{H}(t,x)$ is continuous (cf. A.3),

(4) (b) $\langle x,h\rangle \leqslant \kappa(1 + \|x\|^2)$ for any $(t,x) \in T \times R^n$, $h \in \mathcal{H}(t,x)$,

κ is a positive number.

(5) (c) $\mathcal{H}_1(t,x,s^{(1)},\xi) \cap \mathcal{H}_2(t,x,s^{(2)},\xi) \neq \phi$ for all $(t,x,s^{(i)}) \in T \times R^n \times R^n$.

Below we shall consider functions ξ defined by one of the following three relations:

(6)
$$\xi_1(t,x,s) \stackrel{\text{def}}{=} \max_{v \in \mathcal{Q}} \min_{u \in \mathcal{P}} \langle s,f(t,x,u,v)\rangle$$
$$\xi_2(t,x,s) \stackrel{\text{def}}{=} \min_{\mu \in \text{rpm}(\mathcal{P})} \max_{\nu \in \text{rpm}(\mathcal{Q})} \int_{\mathcal{P}} \int_{\mathcal{Q}} \langle s,f(t,x,u,v)\rangle \mu(du)\nu(dv)$$
$$= \max_{\nu \in \text{rpm}(\mathcal{Q})} \min_{\mu \in \text{rpm}(\mathcal{P})} \int_{\mathcal{P}} \int_{\mathcal{Q}} \langle s,f(t,x,u,v)\rangle \mu(du)\nu(dv),$$
$$\xi_3(t,x,s) \stackrel{\text{def}}{=} \min_{u \in \mathcal{P}} \max_{v \in \mathcal{Q}} \langle s,f(t,x,u,v)\rangle.$$

Here f is a function on the r.h.s. of the controlled system (1.2.1). It is assumed that this function satisfies conditions (i) - (iii) stipulated in subsection 1.2.2; condition (iv), i.e. the equality (1.2.8), may not be fulfilled. It is easy to verify that each one of the functions ξ_i (i = 1,2,3) is continuous, positive homogeneous in s and satisfies the Lipschitz condition with respect to x.

The multivalued mapping $\mathcal{H}$ satisfying conditions (a) - (c) for $\xi = \xi_i$ (i = 1,2,3) may be defined by

(7) $$\mathcal{H}(t,x) = \mathcal{F}(t,x) \overset{\text{def}}{=} \text{co}\{f(t,x,u,v) \colon u \in \mathcal{P},\ v \in \mathcal{Q}\}.$$

It follows from (i) and (iii) (cf. subsection 1.2.2) that the mapping $\mathcal{H}$ satisfies conditions (a) and (b) in the case under consideration. We shall show that for $\xi = \xi_i$ (i = 1,2,3) the mapping $\mathcal{H}$ (7) also satisfies condition (c). Verification of this fact is carried out for the case i = 1.

We select a countercontrol $u_1[\cdot]\colon \mathcal{Q} \to \mathcal{P}$ and a point $v_2 \in \mathcal{Q}$ from the conditions

$$\langle s^{(1)}, f(t,x,u_1[v],v)\rangle = \min_{u \in \mathcal{P}} \langle s^{(1)}, f(t,x,u,v)\rangle,$$

$$\min_{u \in \mathcal{P}} \langle s^{(2)}, f(t,x,u,v_2)\rangle = \max_{v \in \mathcal{Q}} \min_{u \in \mathcal{P}} \langle s^{(2)}, f(t,x,u,v)\rangle.$$

According to (6) we obtain

$$\langle s^{(1)}, f(t,x,u_1[v],v)\rangle \leqslant \xi_1(t,x,s^{(1)}) \quad \text{for all} \quad v \in \mathcal{Q},$$

$$\langle s^{(2)}, f(t,x,u,v_2)\rangle \geqslant \xi_1(t,x,s^{(2)}) \quad \text{for all} \quad u \in \mathcal{P}.$$

In particular we have

(8)
$$\langle s^{(1)}, f(t,x,u_1[v_2],v_2)\rangle \leqslant \xi_1(t,x,s^{(1)}),$$
$$\langle s^{(2)}, f(t,x,u_1[v_2],v_2)\rangle \geqslant \xi_2(t,x,s^{(2)}).$$

Since $f(t,x,u_1[v_2],v_2) \in \mathcal{F}(t,x) = \mathcal{H}(t,x)$, relation (8) implies (5) where $\xi = \xi_1$. Thus condition (c) is established.

Hence the mapping $\mathcal{H}$ defined by equality (7) for $\xi = \xi_i$ (i = 1,2,3) satisfies all the conditions (a), (b), (c). Note that there is a substantial freedom in selecting the mapping $\mathcal{H}$ to be used for solving different types of problems.

Along with relation (5), we shall consider the following inequalities:

(9) $$\max\{\langle s^{(2)}, h\rangle : h \in \mathcal{H}_1(t,x,s^{(1)},\xi)\} \geqslant \xi(t,x,s^{(2)}),$$

(10) $$\min\{\langle s^{(1)}, h\rangle : h \in \mathcal{H}_2(t,x,s^{(2)},\xi)\} \leqslant \xi(t,x,s^{(1)}),$$

where $(t,x,s^{(i)}) \in T \times R^n \times R^n$ (i = 1,2).

It is easy to verify that the following lemma is valid.

Lemma 10.5.1. *Relations* (5), (9), *and* (10) *are equivalent.*

We shall now state a theorem in which the conditions of u-stability, μ-stability, and u_*-stability of the set $\mathcal{W} \subset T \times R^n$ are defined in a unified fashion. Recall that initial definitions of these conditions were given in subsections 2.2.1, 9.2.3, and 10.2.4.

Theorem 10.5.1. *Let* $\mathcal{W} \subset T \times R^n$ *be a closed set satisfying the following condition: for any* $(t_*,x_*) \in \mathcal{W}$, $t^* \in (t_*,\theta]$, *and* $s_* \in S \overset{\text{def}}{=} \{s \in R^n\colon \|s\| = 1\}$ *there exists a solution* $w(\cdot)\colon [t_*,t^*] \to R^n$ *of the differential inclusion*

$$(11) \qquad \dot{w}(t) \in \mathcal{H}_2(t,w(t),s,\xi), \quad w(t_*) = x_*,$$

such that

$$(12) \qquad \{(t,w(t))\colon t_* \leqslant t \leqslant t^*\} \cap \{\mathcal{W}(t^*) \cup \mathcal{M}\} \neq \phi .$$

For $\xi = \xi_1$ *this condition is equivalent to the condition of u-stability of the set* $\mathcal{W}$, *for* $\xi = \xi_2$ *it is equivalent to the condition of* μ*-stability of the set* $\mathcal{W}$, *while for* $\xi = \xi_3$ *it is the same as the condition of* u_**-stability of the set* $\mathcal{W}$. *In relation* (12) $\mathcal{M}$ *is a closed set appearing in the problem of an* $(\mathcal{M}, \mathcal{N})$*-approach.*

Thus, in accordance with this theorem, the closed set $\mathcal{W} \subset T \times R^n$ will be a u-stable bridge in a problem of an $(\mathcal{M}, \mathcal{N})$-approach (cf. Section 2.2.1) if and only if $\mathcal{W}(\theta) \subset \mathcal{M}$, $\mathcal{W} \subset \mathcal{N}$ and for $\xi = \xi_1$ the condition indicated in Theorem 10.5.1 is fulfilled. Analogously, utilizing Theorem 10.5.1 we obtain definitions of μ-stable and u_*-stable bridges which are equivalent to the definitions presented in subsections 9.2.3 and 10.2.4.

Proof. Consider the case $\xi = \xi_1$. Assume that the set $\mathcal{W}$ satisfies the condition of u-stability (cf. Section 2.2.1). Let $(t_*,x_*) \in \mathcal{W}$, $t^* \in (t_*,\theta]$, $s_* \in S$. It is required to prove that a solution $w(\cdot)$ of the differential inclusion (11) exists satisfying condition (12).

Set

$$(13) \qquad \delta_k = (t^* - t_*)k^{-1}; \quad \tau_{i,k} = t_* + i\delta_k; \quad i = 0,1, ..., k$$

where k is a positive integer. Define the set

$$(14) \qquad G(t,w,s,s_*) \overset{\text{def}}{=} \mathcal{H}_1(t,w,s,\xi_1) \cap \mathcal{H}_2(t,w,s_*,\xi_1).$$

Let $(x_k(\cdot),w_k(\cdot))\colon [t_*,t^*] \to R^n \times R^n$ be a function satisfying the differential inclusion

$$\dot{x}_k(t) \in \mathcal{F}_2(t,x_k(t),v_k(\tau_{i,k})),$$

$$(15) \qquad \dot{w}_k(t) \in G(t,w_k(t),s_k(\tau_{i,k}),s_*),$$

$$t \in [\tau_{i,k},\tau_{i+1,k}), \quad i = 0,1, ..., k, \quad x_k(t_*) = w_k(t_*) = x_*.$$

Here

$$s_k(t) = (w_k(t) - x_k(t))$$

and the point $v_k(\tau_{i,k}) \in \mathcal{Q}$ is selected from the condition

$$\min_{u\in\mathcal{P}} \langle s_k(\tau_{i,k}),\ f(\tau_{i,k},x_k(\tau_{i,k}),u,v_k(\tau_{i,k}))\rangle$$

$$(16)\qquad = \max_{v\in\mathcal{Q}} \min_{u\in\mathcal{P}} \langle s_k(\tau_{i,k}),\ f(\tau_{i,k},x_k(\tau_{i,k}),u,v)\rangle$$

$$= \xi_1(\tau_{i,k},x_k(\tau_{i,k}),s_k(\tau_{i,k})).$$

Recall that the set $\mathcal{F}_2(t,x,v)$ is defined in accordance with (2.2.1).

The multivalued mapping

$$(t,x,w) \to \mathcal{F}_2(t,x,v) \times \mathcal{G}(t,w,s,s_*)$$

satisfies all the conditions stipulated in Section A.4; therefore the solution $(x_k(\cdot),w_k(\cdot))$ exists. Moreover it follows from the condition of u-stability of the set $\mathcal{W}$, that this solution can be selected in such a manner that the relation

$$(17)\qquad \{(t,x_k(t)):\ t_* \leqslant t \leqslant t^*\} \cap (\mathcal{W}(t^*) \cup \mathcal{M}) \neq \phi$$

is fulfilled.

We estimate the distance between the functions $x_k(\cdot)$ and $w_k(\cdot)$. It follows from (15) and (2.2.1) that

$$\langle s_k(t),\dot{x}_k(t)\rangle \geqslant \min_{u\in\mathcal{P}} \langle s_k(t),\ f(t,x_k(t),u,v_k(\tau_{i,k}))\rangle$$

$$\text{for almost all } t \in [\tau_{i,k},\tau_{i+1,k}),\ i = 0,1,\ ...,\ k.$$

Functions f and ξ_1 are continuous and functions $s_k(\cdot)$ and $x_k(\cdot)$ are uniformly bounded and equicontinuous. According to (13) we have $\delta_k \to 0$ as $k \to \infty$. Therefore, for any $\epsilon > 0$, there exists a $k(\epsilon)$ such that for all $k \geqslant k(\epsilon)$ and for almost all $t \in [\tau_{i,k},\tau_{i+1,k})$ the estimate

$$\min_{u\in\mathcal{P}} \langle s_k(t),\ f(t,x_k(t),u,v_k(\tau_{i,k}))\rangle$$

$$\geqslant \min_{u\in\mathcal{P}} \langle s_k(\tau_{i,k}),\ f(\tau_{i,k},x_k(\tau_{i,k}),u,v_k(\tau_{i,k}))\rangle - \epsilon$$

$$= \xi_1(\tau_{i,k},x_k(\tau_{i,k}),s_k(\tau_{i,k})) - \epsilon \geqslant \xi_1(t,x_k(t),s_k(t)) - 2\epsilon$$

will be valid. Observe that equality (16) was utilized here. We thus obtain

$$(18)\qquad \langle s_k(t),\dot{x}_k(t)\rangle \geqslant \xi_1(t,x_k(t),s_k(t)) - 2\epsilon$$

for $k \geqslant k(\epsilon)$, and for almost all $t \in [t_*,t^*]$.

According to (14) and (15) we have

$$\dot{w}_k(t) \in \mathcal{H}_1(t,w_k(t),s_k(\tau_{i,k}),\xi_1)$$

for almost all $t \in [\tau_{i,k},\tau_{i+1,k})$, $i = 0,1, ..., k$. From (3) it follows that

$$\langle s_k(\tau_{i,k}),\dot{w}_k(t)\rangle \leq \xi_1(t,w_k(t),s_k(\tau_{i,k})).$$

In the same manner as in the previous estimate, we obtain

$$\langle s_k(t),\dot{w}_k(t)\rangle \leq \xi_1(t,w_k(t),s_k(t)) + \epsilon \tag{19}$$

for $k \geq k(\epsilon)$ and for almost all $t \in [t_*,t^*]$.

Select a point $v_* \in \mathcal{Q}$ from condition

$$\psi_2(t,w,s,v_*) = \max_{v\in\mathcal{Q}} \psi_2(t,w,s,v),$$

where the quantity ψ_2 is defined according to (2.3.13). From (6) and (2.3.13) we have

$$\xi_1(t,w,s) - \xi_1(t,x,s) \leq \psi_2(t,w,s,v_*) - \psi_2(t,x,s,v_*).$$

Let $s = w - x$, then we obtain from (2.3.14) that

$$\xi_1(t,w_k(t),s_k(t)) - \xi_1(t,x_k(t),s_k(t)) \leq \lambda \|s_k(t)\|^2. \tag{20}$$

Utilizing this estimate and inequalities (18), (19) we have

$$\begin{aligned}\frac{d\|w_k(t) - x_k\|^2}{dt} = \frac{d\|s_k(t)\|^2}{dt} &= 2\langle s_k(t),\dot{w}_k(t) - \dot{x}_k(t)\rangle \\ &\leq 2\lambda\|s_k(t)\|^2 + 6\epsilon\end{aligned} \tag{21}$$

for $k \geq k(\epsilon)$ and for almost all $t \in [t_*,t^*]$.

One can select from the sequence $(x_k(\cdot),w_k(\cdot))$ a subsequence which converges to a limit $(x(\cdot),w(\cdot))$. Since $\|s(t_*)\| = 0$ and the $\epsilon > 0$ can be selected in the estimate (21) to be arbitrarily small, it follows that $x(t) = w(t)$ for all $t \in [t_*,t^*]$. From (14) and (15) we obtain that

$$\dot{w}_{k_\ell}(t) \in \mathcal{H}_2(t,w_{k_\ell}(t),s_*,\xi_1)$$

for $\ell = 1,2, ...$ and for almost all $t \in [t_*,t^*]$. It follows from Theorem A.4.1 that the function $w(\cdot)$ satisfies the differential inclusion (11) with $\xi = \xi_1$. Finally, it follows from (17) where $\mathcal{W}$ and $\mathcal{M}$ are closed sets that the limit function $w(\cdot) = x(\cdot)$ satisfies condition (12). Thus if the set $\mathcal{W}$ is u-stable, then for $\xi = \xi_1$ it satisfies the condition stated in Theorem 10.5.1.

Assume now that for $\xi = \xi_1$ the set $\mathcal{W}$ satisfies the condition stipulated in Theorem 10.5.1. It is required to verify that the set $\mathcal{W}$

is u-stable, i.e. for any $(t_*,x_*) \in \mathcal{W}$, $t^* \in (t_*,\theta]$ and $v \in \mathcal{Q}$, there exists a solution $x(\cdot)$: $[t_*,t^*] \to R^n$ of the differential inclusion

$$\dot{x}(t) \in \mathcal{F}_2(t,x(t),v), \quad x(t_*) = x_* \tag{22}$$

satisfying the condition

$$\{(t,x(t)): t_* \leqslant t \leqslant t^*\} \cap (\mathcal{W}(t^*) \cup M) \neq \phi . \tag{23}$$

Define the set

$$F_0(t,x,w,v) \overset{\text{def}}{=} \{h_0 \in \mathcal{F}_2(t,x,v): \langle x - w,h_0\rangle = \min_{h\in\mathcal{F}_2(t,x,v)} \langle x - w,h\rangle\}. \tag{24}$$

Let the function $(x_k(\cdot),w_k(\cdot))$: $[t_*,t^*] \to R^n \times R^n$ satisfy the differential inclusion

$$\begin{aligned} &\dot{x}_k(t) \in \mathcal{F}_0(t,x_k(t),w_k(t),v), \\ &\dot{w}_k(t) \in \mathcal{H}_2(t,w_k(t),s_k(\tau_{i,k}),\xi_1) \\ &\tau_{i,k} \leqslant t < \tau_{i+1,k}, \quad i = 0,1, \ldots, k, \quad x_k(t_*) = w_k(t_*) = x_*. \end{aligned} \tag{25}$$

Here $s_k(t) = x_k(t) - w_k(t)$, and the times $\tau_{i,k}$ are defined according to (13). A solution of the differential inclusion (25) exists. Moreover, one can select a solution such that

$$\{(t,w_k(t)): t_* \leqslant t \leqslant t^*\} \cap (\mathcal{W}(t^*) \cup M) \neq \phi . \tag{26}$$

According to (2.2.1), (24), and (25), we have

$$\langle s_k(t),\dot{x}_k(t)\rangle \leqslant \xi_1(t,x_k(t),s_k(t))$$

for almost all $t \in [t_*,t^*]$. From (3) and (25) we obtain

$$\langle s_k(t),\dot{w}_k(t)\rangle \geqslant \xi_1(t,w_k(t),s_k(\tau_{i,k})) \geqslant \xi_1(t,w_k(t),s_k(t)) - \epsilon$$

for $k \geqslant k(\epsilon)$ and almost all $t \in [t_*,t^*]$. Here ϵ is an arbitrary small number and $k(\epsilon)$ is a sufficiently large integer. Taking the inequality $\xi_1(t,x_k(t),s_k(t)) - \xi_1(t,w_k(t),s_k(t)) \leqslant \lambda\|s_k(f)\|^2$ into account, we have

$$\frac{d\|s_k(t)\|^2}{dt} = \frac{d\|x_k(t) - w_k(t)\|^2}{dt} \leqslant 2\lambda\|s_k(t)\|^2 + 2\epsilon$$

for almost all $t \in [t_*,t^*]$.

Let $(x_{k_\ell}(\cdot),w_{k_\ell}(\cdot))$, $\ell = 1,2, \ldots$ be a subsequence convergent to the limit $(x(\cdot),w(\cdot))$. It follows from the previous estimate that $x(t) = w(t)$ for all $t \in [t_*,t^*]$. We obtain from (26) that the limit function $x(\cdot) =$

$w(\cdot)$ satisfies condition (23). Since the functions $x_{k_\ell}(\cdot)$ satisfy the differential inclusion $\dot{x}_{k_\ell}(t) \in \mathcal{F}_2(t,x_{k_\ell}(t),v)$, the limit function $x(\cdot)$ also satisfies the inclusion (22). We have thus verified the existence of a function $x(\cdot)$ satisfying relations (22), (23).

Hence it is established that in the case $\xi = \xi_1$, the condition of u-stability is equivalent to the condition stipulated in Theorem 10.5.1. Verification of the assertions of Theorem 10.5.1 for the cases $\xi = \xi_2$ and $\xi = \xi_3$ can be carried out along the lines described above.

The following assertion, which is analogous to Theorem 10.5.1, is valid for v_*-stable, ν-stable, and v-stable bridges.

Theorem 10.5.2. *Let $\mathcal{W} \subset T \times R^n$ be a closed set satisfying the following condition: for any $(t_*,x_*) \in \mathcal{W}$, $t^* \in (t_*,\theta]$ and $s_* \in S \overset{\text{def}}{=} \{s \in R^n\colon \|s\| = 1\}$ there exists a solution $w(\cdot)\colon [t_*,t^*] \to R^n$ of the differential inclusion*

$$\dot{w}(t) \in \mathcal{H}_1(t,w(t),s_*,\xi), \quad w(t_*) = x_* \tag{27}$$

satisfying the relation

$$\{(t,w(t))\colon t_* \leq t \leq t^*\} \cap (\mathcal{W}(t^*) \cup K) \neq \phi\,. \tag{28}$$

This condition is equivalent: (1) for $\xi = \xi_1$ to the condition of v_*-stability of the set $\mathcal{W}$, (2) for $\xi = \xi_2$ it is equivalent to the condition of ν- stability of the set $\mathcal{W}$, and (3) for $\xi = \xi_3$ it is equivalent to the condition of v-stability of the set $\mathcal{W}$. In relation (28) K is a closed set which appears in a $(K,\mathfrak{L})$-approach problem (cf. subsections 2.1.1 and 2.2.3).

10.5.2. Utilizing Theorems 10.5.1 and 10.5.2, it is easy to carry out a comparison of the maximal stable bridges for various types of approach and evasion problems. Let the closed sets M and N be given. As usual it is assumed that $M \subset N \subset T \times R^n$. Consider the three types of an (M,N)-approach problem -- namely, Problems 10.2.1, 9.2.1, and 2.1.1. Recall that a solution of Problem 10.2.1 must be determined in the class of counterstrategies, that of Problem 9.2.1 in the class of mixed positional strategies, and of Problem 2.1.1 in the class of pure positional strategies. Denote by $\mathcal{W}_0^{1,1}$, $\mathcal{W}_0^{1,2}$, and $\mathcal{W}_0^{1,3}$ the sets of initial positions (t_0,x_0) from which Problems 10.2.1, 9.2.1, and 2.1.1, respectively, are solvable. It was shown in the present and preceding chapters that the equalities

$$\mathcal{W}_0^{1,1} = \mathcal{W}_u, \quad \mathcal{W}_0^{1,2} = \mathcal{W}_\mu, \quad \mathcal{W}_0^{1,3} = \mathcal{W}_{u*} \tag{29}$$

are valid; here $\mathcal{W}_u$, $\mathcal{W}_\mu$, and $\mathcal{W}_{u*}$ are, respectively, the maximal u-stable, μ-stable, and u_*-stable bridges in a problem of an (M,N)-approach. Comparing informational latitudes for the first player in

Problems 10.2.1, 9.2.1, and 2.1.1 (cf. subsection 9.1.1) one can conjecture that

(30) $$\mathcal{W}_0^{1,1} \supset \mathcal{W}_0^{1,2} \supset \mathcal{W}_0^{1,3}.$$

We shall utilize Theorem 10.5.1 for a formal proof of these relations. From (6) follow the inequalities

$$\xi_1(t,x,s) \leqslant \xi_2(t,x,s) \leqslant \xi_3(t,x,s).$$

Therefore

(31) $$\mathcal{H}_2(t,x,s,\xi_1) \supset \mathcal{H}_2(t,x,s,\xi_2) \supset \mathcal{H}_2(t,x,s,\xi_3).$$

These sets are defined according to (3) where it is assumed that $\mathcal{H}(t,x) = \mathcal{F}(t,x)$ (7). Utilizing Theorem 10.5.1 we arrive at relation (30) from relations (31).

Analogously we obtain that

(32) $$\mathcal{W}_\epsilon^{2,1} \subset \mathcal{W}_\epsilon^{2,2} \subset \mathcal{W}_\epsilon^{2,3}.$$

Here $\epsilon > 0$, $\mathcal{W}_\epsilon^{2,j}$ is the set of all initial positions (t_0,x_0) from which the problem of an $(\mathcal{M}, \mathcal{N})^\epsilon$-evasion is solvable in the class of pure positional strategies (for $j = 1$), mixed positional strategies (for $j = 2$), and counterstrategies (for $j = 3$).

Let the payoff functional be defined by one of the equalities (3.1.1), (3.2.6) and (3.3.1). Denote by $\omega_1^0(t_0,x_0)$ the value of a lower differential game (recall that it is defined in the class of counterstrategies for the first player and pure positional strategies for the second player). Let $\omega_2^0(t_0,x_0)$ be the value of a differential game defined in the class of mixed positional strategies for both players. Finally, we set $\omega_3^0(t_0,x_0)$ to be the value of an upper differential game defined in the class of pure positional strategies for the first player and counterstrategies for the second player. Utilizing the constructions presented in Chapter 3 and relations (30) and (32), we obtain the inequalities

(33) $$\omega_1^0(t_0,x_0) \leqslant \omega_2^0(t_0,x_0) \leqslant \omega_3^0(t_0,x_0).$$

If the function f satisfies the condition of a saddle point in a small game, i.e. the equality (1.2.8) is fulfilled, then the quantities ξ_1, ξ_2, and ξ_3 defined by equalities (6) coincide. We thus immediately obtain from Theorem 10.5.1 that conditions of u-stability, μ-stability, and u_*-stability are equivalent. Analogously, it follows from Theorem 10.5.2 that the conditions of v-stability, ν-stability, and v_*-stability are equivalent. Next, in the case when equality (1.2.8) is fulfilled, in place of inclusions (30), (32), and inequalities (33), we arrive at the equalities

$$\mathcal{W}_0^{1,1} = \mathcal{W}_0^{1,2} = \mathcal{W}_0^{1,3},$$

$$\mathcal{W}_\epsilon^{2,1} = \mathcal{W}_\epsilon^{2,2} = \mathcal{W}_\epsilon^{2,3},$$

$$\omega_1^0(t_0,x_0) = \omega_2^0(t_0,x_0) = \omega_3^0(t_0,x_0).$$

We shall explain, for example, the meaning of the equality $\omega_1^0(t_0,x_0) = \omega_3^0(t_0,x_0)$. This equality means that the first player cannot improve on (i.e. decrease) the guaranteed result by means of a transition from pure positional strategies to a wider class of counterstrategies. When the condition (1.2.8) is fulfilled, additional information about the control selected by the opponent cannot improve the optimal result obtained in the class of positional strategies.

10.5.3. We shall continue our discussion of unification of differential games. Let the function $(t,x,s) \to \xi(t,x,s)$ be continuous, positive homogeneous in s, and satisfy the Lipschitz condition. It is not required here that the function ξ be defined by one of the equalities (6). Define the sets

$$\mathcal{H}_1(t,x,u^\#) \overset{\text{def}}{=} \mathcal{H}_1(t,x,u^\#,\xi),$$
(34)
$$\mathcal{H}_2(t,x,v^\#) \overset{\text{def}}{=} \mathcal{H}_2(t,x,v^\#,\xi).$$

Here the sets appearing on the r.h.s. are defined according to (3), where one should set $u^\# = s$ and $v^\# = s$; the function ξ is omitted on the l.h.s. to simplify the notation. Below, auxiliary game-theoretical problems will be defined in which vectors $u^\#$ and $v^\#$ play the role of controls for the first and second players. These controls are selected on the unit sphere S, i.e.

$$u^\# \in P^\# \overset{\text{def}}{=} S \overset{\text{def}}{=} \{s \in R^n\colon \|s\| = 1\},$$
(35)
$$v^\# \in Q^\# \overset{\text{def}}{=} S\,.$$

Let an initial position $(t_0,x_0) \in T \times R^n$, a partition $\Delta = \{t_0 = \tau_0 < \tau_1 < \cdots < \tau_{k+1} = \theta\}$, and an arbitrary function $U^\#\colon T \times R^n \to P^\#$ be given. Denote by $X_1(t_0,x_0,U^\#,\Delta)$ the set of step-by-step motions. Absolutely continuous functions $x(\cdot)\colon [t_0,\theta] \to R^n$ satisfying the initial condition $x(t_0) = x_0$ and the differential inclusion

$$\dot{x}(t) \in \mathcal{H}_1(t,x(t),u^\#(t)), \tag{36}$$

where $u^\#(t) = U^\#(\tau_i,x(\tau_i))$, $\tau_i \leqslant t < \tau_{i+1}$, $i = 0,1, \ldots, k$, are the elements of this set. As usual the set of constructive motions $X_1(t_0,x_0,U^\#)$ is defined by a limiting transition on the sequence of step-by-step motions, i.e.

$$X_1(t_0,x_0,U^{\#}) = \{x(\cdot): [t_0,\theta] \to R^n : \ \exists x_\ell(\cdot) \in X_1(t_0,x_0,U^{\#},\Delta_\ell),$$

$$\ell = 1,2, ..., x_\ell(\cdot) \to x(\cdot), \ \text{diam}\ \Delta_\ell \to 0 \ \text{as} \ \ell \to \infty\}.$$

The function $U^{\#}: T \times R^n \to P^{\#}$ is called a unified strategy for the first player.

A unified strategy for the second player and the corresponding sets $X_2(t_0,x_0,V^{\#},\Delta)$ and $X_2(t_0,x_0,V^{\#})$ of step-by-step and constructive motions are defined analogously. Note that absolutely continuous functions satisfying the initial condition $x(t_0) = x_0$ and the differential inclusion

$$\dot{x}(t) \in \mathcal{H}_2(t,x(t),v^{\#}(t)), \tag{37}$$

where $v^{\#}(t) = V^{\#}(\tau_i,x(\tau_i))$, $\tau_i \leq t < \tau_{i+1}$, $i = 0,1, ..., k$, will be the elements of the set $X_2(t_0,x_0,V^{\#},\Delta)$.

Recall that the sets $\mathcal{H}_1(t,x,u^{\#})$ and $\mathcal{H}_2(t,x,v^{\#})$ satisfy relation (5), therefore any pair of unified strategies $U^{\#}$ and $V^{\#}$ will be compatible in the sense that

$$X_1(t_0,x_0,U^{\#}) \cap X_2(t_0,x_0,V^{\#}) \neq \phi. \tag{38}$$

Consider the following two problems.

Problem 10.5.1. *For a given initial position (t_0,x_0) it is required to determine a unified strategy $U_0^{\#}$ such that for any motion $x(\cdot) \in X_1(t_0,x_0,U_0^{\#})$ the condition of an (M, N)-approach is valid.*

Problem 10.5.2. *For a given $\epsilon > 0$ and initial position $(t_0,x_0) \in T \times R^n$ it is required to determine a unified strategy $V_\epsilon^{\#}$ such that for any motion $x(\cdot) \in X_2(t_0,x_0,V_\epsilon^{\#})$ the condition of an $(M, N)^\epsilon$-evasion is valid.*

As above, the conditions of an (M, N)-approach and $(M, N)^\epsilon$-evasion are defined by relations (2.1.3) and (2.1.8). Below, assertions are obtained which reveal the meaning of the formal constructions presented herein.

Theorem 10.5.3. *For any closed sets $M \subset N \subset T \times R^n$ and any initial position $(t_0,x_0) \in T \times R^n$, one of the following two assertions is valid: either there exists a unified strategy $U_0^{\#}$ which yields a solution to Problem 10.5.1, or for some $\epsilon > 0$ there exists a unified strategy $V_\epsilon^{\#}$ which yields a solution to Problem 10.5.2.*

The proof of this theorem falls within the framework of arguments presented in Chapter 2. It is sufficient to replace the sets $F_1(t,x,u)$ and $F_2(t,x,v)$ (2.2.1) by the sets $\mathcal{H}_1(t,x,u^{\#})$ and $\mathcal{H}_2(t,x,v^{\#})$ (34). With this substitution one can repeat almost verbatim the proofs of the assertions presented in Sections 2.2 – 24. We shall thus confine ourselves to the following remarks.

The basic result obtained in Section 2.2 remains valid in this case as well. Here, in place of u-stable and v-stable bridges, $u^{\#}$-stable and $v^{\#}$-stable sets are considered. The only difference in the corresponding definitions (cf. subsections 2.2.1 and 2.2.3) is that the sets $\mathcal{F}_1(t,x,u)$ and $\mathcal{F}_2(t,x,v)$ are replaced by the sets $\mathcal{H}_1(t,x,u^{\#})$ and $\mathcal{H}_2(t,x,v^{\#})$. Therefore the conditions of $u^{\#}$- and $v^{\#}$- stability coincide with conditions stipulated in Theorems 10.5.1 and 10.5.2 respectively. Note that in the proof of Lemma 2.2.2 the relation $X_1(t_*,x_*,u) \cap X_2(t_*,x_*,v) \neq \phi$ is used. In our case, an analogous relation is also valid. Here the sets $X_1(t_*,x_*,u^{\#})$ and $X_2(t_*,x_*,v^{\#})$ should be defined as totalities of solutions of differential inclusions $\dot{x}(t) \in \mathcal{H}_1(t,x(t),u^{\#})$, $\dot{x}(t) \in \mathcal{H}_2(t,x(t),v^{\#})$, $x(t_*) = x_*$, respectively. Since the multivalued mapping $(t,x) \to \mathcal{H}_1(t,x,u^{\#}) \cap \mathcal{H}_2(t,x,v^{\#})$ satisfies all the conditions stipulated in Theorem A.4.1 (in particular we have in accordance with (5) that $\mathcal{H}_1(t,x,u^{\#}) \cap \mathcal{H}_2(t,x,v^{\#}) \neq \phi$), it follows that

$$X_1(t_*,x_*,u^{\#}) \cap X_2(t_*,x_*,v^{\#}) \neq \phi .$$

In Section 2.3, condition (1.2.8) was used in the course of deriving the basic estimate, which implies the equality

$$\min_{u \in P} \max\{\langle s,h \rangle:\ h \in \mathcal{F}_1(t,x,u)\} = \max_{v \in Q} \min\{\langle s,h \rangle:\ h \in \mathcal{F}_2(t,x,v)\}.$$

We shall show that in the present case an analogous equality

$$\begin{aligned} &\min_{u^{\#} \in P^{\#}} \max\{\langle s,h \rangle:\ h \in \mathcal{H}_1(t,x,u^{\#})\} \\ &\quad = \max_{v^{\#} \in Q^{\#}} \min\{\langle s,h \rangle:\ h \in \mathcal{H}_2(t,x,v^{\#})\} = \xi(t,x,s) \end{aligned} \tag{39}$$

is valid. It is sufficient to verify this equality for unit vectors s. Utilizing Lemma 10.5.1 we have

$$\max\{\langle s,h \rangle : h \in \mathcal{H}_1(t,x,u^{\#})\} \geqslant \xi(t,x,s) \geqslant \min\{\langle s,h \rangle : h \in \mathcal{H}_2(t,x,v^{\#})\}$$

for all $(t,x) \in T \times R^n$, $u^{\#} \in P^{\#}$, $v^{\#} \in Q^{\#}$. It follows from (3) and (34) that

$$\max\{\langle s,h \rangle:\ h \in \mathcal{H}_1(t,x,s)\} = \xi(t,x,s) = \min\{\langle s,h \rangle:\ h \in \mathcal{H}_2(t,x,s)\}.$$

Thus equality (39) is proved.

It remains to define an extremal strategy $U_e^{\#}$ more exactly. Let W be a closed set in $T \times R^n$. Define the function $(t,x) \to e(t,x)$ in accordance with (2.4.2), (2.4.4). Assume that the unified strategy $U_e^{\#}$ extremal to the set W is defined by the condition

$$\begin{aligned} &\max\{\langle x - e(t,x),h \rangle:\ h \in \mathcal{H}_1(t,x,u_e^{\#})\} \\ &\quad = \min_{u^{\#} \in P^{\#}} \max\{\langle x - e(t,x),h \rangle:\ h \in \mathcal{H}_1(t,x,u^{\#})\} \end{aligned} \tag{40}$$

where $u_e^{\#} = U_e^{\#}(t,x)$. This condition is an analog of condition (2.4.5). It follows from the relations which were used in the proof of equality (39) that the minimum in (40) is attained for $u_e^{\#} = (x - e(t,x))/\|x - e(t,x)\|$. Therefore the strategy $U_e^{\#}$ can be defined as follows:

$$U_e^{\#}(t,x) = \begin{cases} (x - e(t,x)) \cdot \|x - e(t,x)\|^{-1} & \text{for } x \neq e(t,x), \\ s_* & \text{for } x = e(t,x) \end{cases}$$

where s_* is an arbitrary unit vector.

Problems 10.5.1 and 10.5.2 form a unified differential game of approach-evasion. The theorems presented below clarify the aim and reason for introducing unified games.

Theorem 10.5.4. *Problem* 10.5.1 *is equivalent to Problem* 10.2.1 *for* $\xi = \xi_1$, *is equivalent to Problem* 9.2.1 *for* $\xi = \xi_2$, *and is equivalent to Problem* 2.1.1 *for* $\xi = \xi_3$. *Problem* 10.5.2 *is equivalent to Problem* 2.1.2 *for* $\xi = \xi_1$, *Problem* 9.2.2 *for* $\xi = \xi_2$, *and Problem* 10.2.2 *for* $\xi = \xi_3$.

Here ξ is the function by means of which the sets $\mathcal{H}_1(t,x,u^{\#})$ and $\mathcal{H}_2(t,x,v^{\#})$ (34) were defined; functions ξ_i $(i = 1,2,3)$ were defined by the equalities (6). The equivalence of Problems 10.5.1 and 10.2.1 means that for a given initial position (t_0,x_0), Problem 10.5.1 is solvable if and only if Problem 10.2.1 is solvable for this position. The equivalence of Problems 10.5.1 and 9.2.1, 10.5.1 and 2.1.1, and so on has an analogous meaning.

We shall show that for $\xi = \xi_1$ Problems 10.5.1 and 10.2.1 are equivalent. Let W_*($W_{\#}$) be the maximal u-stable ($u_{\#}$-stable) bridge in an $(\mathcal{M},\mathcal{N})$-approach problem. As is known (cf. Section 10.2 and the proof of Theorem 10.5.3), Problem 10.2.1 (Problem 10.5.1) is solvable if and only if $(t_0,x_0) \in W_*$ ($(t_0,x_0) \in W_{\#}$)). According to Theorem 10.5.1 the sets W_* and $W_{\#}$ coincide provided $\xi = \xi_1$. Therefore for $\xi = \xi_1$, Problems 10.5.1 and 10.2.1 are equivalent. The remaining assertions of Theorem 10.5.4 can be verified analogously.

We thus arrive at the following conclusions. A differential game of any of the three types considered in the preceding and present chapters is equivalent to a differential game formed by unified Problems 10.5.1 and 10.5.2 for an appropriate choice of function ξ. A unified game is considered in the class of positional strategies $U^{\#}$ and $V^{\#}$ independently of the class of strategies in which the initial differential game was defined. The stable bridges defined for the unified Problem 10.5.1 for $\xi = \xi_i$ $(i = 1,2,3)$ can be used for the construction of extremal counterstrategies, mixed positional strategies, and pure positional strategies which provide solutions to Problems 10.2.1, 9.2.1, and 2.1.1. Analogously, constructions developed for solving the unified Problem 10.5.2 can be applied for solving the corresponding Problems 2.1.2, 9.2.2, and 10.2.2. The above stated assertion is also valid for extremal aiming methods as well as methods of a priori stable sets. These methods can be developed in unified form for solving Problems 10.5.1 and 10.5.2 and then the corresponding results for the three types of differential games are obtained as simple corollaries. We shall confine ourselves to this remark without describing unification of the methods presented in Chapters 5 – 7.

We now state yet another result which follows from previous constructions. Let the function $(t,x,s) \rightarrow \xi(t,x,s)$ be continuous, positive homogeneous in s, and satisfy the Lipschitz condition with respect to x. Denote by $\mathbf{W}(\xi)$ the totality of closed $u^{\#}$ stable bridges in Problem 10.5.1, i.e. $\mathcal{W} \in \mathbf{W}(\xi)$ if and only if the set $\mathcal{W}$ is closed, satisfies the inclusions $\mathcal{W} \subset \mathcal{N}$, $\mathcal{W}(\theta) \subset \mathcal{M}$ and also satisfies the condition stipulated in Theorem 10.5.1 (this is just the condition of $u^{\#}$-stability). In the statement of the condition of $u^{\#}$-stability the function ξ appears, thus the totality of $u^{\#}$-stable bridges depends on the chosen function ξ. This is expressed by the notation $\mathbf{W}(\xi)$.

Theorem 10.5.5. *Let* $\xi^*(t,x,s) \geqslant \xi_*(t,x,s)$ *for all* $(t,x,s) \in T \times R^n \times S$. *Then* $\mathbf{W}(\xi^*) \subset \mathbf{W}(\xi_*)$.

Proof. It follows from (3) that $\mathcal{H}_2(t,x,\xi^*) \subset \mathcal{H}_2(t,x,\xi_*)$. Thus in accordance with (11), (12), we arrive at the assertion of Theorem 10.5.5.

Utilizing Theorems 10.5.4 and 10.5.5, one can replace the initial control system by some other (simpler) system. For example, let a μ-stable bridge for the problem of an $(\mathcal{M}, \mathcal{N})$-approach be required in control system (1.2.1). We replace the controlled system (1.2.1) by an auxiliary system

$$\dot{x} = g(t,x,u^*,v^*), \quad u^* \in \mathcal{P}^*, \quad v^* \in \mathcal{Q}^*.$$

Assume that the initial and auxiliary systems are related as follows:

$$\xi_*(t,x,s) = \xi_2(t,x,s) = \min_{\mu} \max_{\nu} \int_{\mathcal{P}} \int_{\mathcal{Q}} \langle s, f(t,x,u,v)\rangle \mu(du)\nu(dv)$$
$$\leqslant \max_{v^* \in \mathcal{Q}^*} \min_{u^* \in \mathcal{P}^*} \langle s, g(t,x,u^*,v^*)\rangle = \xi^*(t,x,s)$$

for all $(t,x,s) \in T \times R^n \times S$. Suppose one succeeds in constructing for the auxiliary system a set $\mathcal{W}^*$ which is a u-stable bridge in a problem of an $(\mathcal{M}, \mathcal{N})$-approach. Then the set $\mathcal{W}^*$ will be a μ-stable bridge for the initial system. Note that in both the initial and auxiliary problems it is possible to consider various types of stable bridges and various classes of strategies. In particular, in the case presented above, the initial problem is considered in the class of mixed positional strategies, and for this problem it is required to determine a μ-stable bridge, while in the auxiliary problem a u-stable bridge is defined, i.e. the problem is considered in the class of counterstrategies for the first player.

Chapter 11
DIFFERENTIAL-FUNCTIONAL GAMES

11.1. Strategies with Complete Memory

In previous chapters, differential games were studied in which the motion of the controlled system is described by an ordinary differential equation. For such systems the theory of positional differential games was developed. Problems of existence of equilibrium situations were considered and certain approaches to the construction of optimal strategies were suggested. It was assumed that the players possess information about the realized position of the game (sometimes it was supposed that one of the players knows the control selected by the opponent at the current time (cf. Chapter 10)). The class of payoff functionals for which an equilibrium situation exists in a positional differential game is sufficiently wide and allows us to encompass a substantial number of the typical problems of control theory. However, for certain types of payoff functionals the information about the current position of the game is insufficient both for obtaining a guaranteed result -- which cannot be improved -- and for coordination of the possibilities of competing parties. In other words, for such functionals there is no equilibrium situation in the class of positional strategies. For such games it is possible, under certain conditions, to obtain an equilibrium situation by broadening the information available to the players.

In this chapter differential games are considered in which it is assumed that the player possesses complete information about the history of the game, i.e. here the selection of controls is based not only on the information about the position of the game $(t,x(t))$ realized at the current time t but also on the information concerning the motion $x(\tau)$ realized on the time interval $[t_0,t]$. As in preceding chapters, solutions of game-theoretical problems are given by means of strategies extremal to certain stable bridges; however, unlike the situation in the case of solutions of positional differential games, here the stable bridges are defined not in a finite-dimensional space of positions (t,x) but in a functional space of motions $x(\cdot)$ (more precisely in the space of continuous functions). Note also that in this chapter controlled systems of a more general type than those in the preceding chapters are considered. These systems are described by differential-functional equations, i.e. equations in which the r.h.s. depends on the motion realized at the current time. In particular, equations involving various types of lags belong to the class of differential-functional equations.

11.1.1. We now proceed to formulate the problems to be investigated. We shall consider problems in which the controlled process is implemented on a bounded time interval $[t_0,\theta]$. For notational convenience we shall assume that $0 \leqslant t_0 \leqslant \theta = 1$, i.e. the interval $T = [0,1]$ will serve as the basic time interval.

The space of continuous functions $y: T \to R^n$ will be denoted by the symbol $\mathcal{C}$ (in preceding chapters the notation was $\mathcal{C}_n(T)$). As usual we shall assume that the norm

$$|y| \overset{\text{def}}{=} \max_{0 \leqslant t \leqslant 1} \|y(t)\|$$

is defined in $\mathcal{C}$.

Let $\tau \in T$; the notation $y|[0,\tau]$ and $y|[\tau,1]$ indicates the restriction of the function $y \in \mathcal{C}$ to the intervals $[0,\tau]$ and $[\tau,1]$ respectively. Below we shall consider functions of the form $\varphi : T \times \mathcal{C} \times \mathcal{Z} \to \Phi$ where $\mathcal{Z}$ and Φ are certain sets. We say that the function φ is *non-anticipating* if for any $(\tau,y^{(i)},z) \in T \times \mathcal{C} \times \mathcal{Z}$ $(i = 1,2)$ such that $y^{(1)}|[0,\tau] = y^{(2)}|[0,\tau]$, the equality $\varphi(\tau,y^{(1)},z) = \varphi(\tau,y^{(2)},z)$ is valid. This property is also called a causality condition.

Consider the function $f: T \times \mathcal{C} \times \mathcal{P} \times \mathcal{Q} \to R^n$ where $\mathcal{P}$ and $\mathcal{Q}$ are compact sets. We shall assume that the function f satisfies the following conditions:

(i) function f is continuous and satisfies the Lipschitz condition

(1)
$$\|f(t,y^{(1)},u,v) - f(t,y^{(2)},u,v)\| \leqslant \lambda|y^{(1)} - y^{(2)}|$$
$$((t,y^{(i)},u,v) \in T \times \mathcal{C} \times \mathcal{P} \times \mathcal{Q}, \quad i = 1,2).$$

(ii) for any $(t,y) \in T \times C$ and $s \in R^n$ the equality

(2)
$$\min_{u\in\mathcal{P}} \max_{v\in\mathcal{Q}} \langle s,f(t,y,u,v)\rangle = \max_{v\in\mathcal{Q}} \min_{u\in\mathcal{P}} \langle s,f(t,y,u,v)\rangle$$

is valid (the condition of a saddle point in a small game).

(iii) function f is non-anticipating.

Let the motion of the controlled system be described by the differential-functional equation

(3)
$$\dot{y}(t) = f(t,y,u(t),v(t)).$$

Here, as above, $u(\cdot)$: $T \to \mathcal{P}$ and $v(\cdot)$: $T \to \mathcal{Q}$ are measurable functions which are called controls for the first and second players, respectively. We shall define a solution of equation (3). Let time $t_0 \in T$ and function $y_0 \in \mathcal{C}$ be given. A solution of equation (3) satisfying the initial condition

$$(4) \qquad y|[0,t_0] = y_0|[0,t_0]$$

is by definition a function $y \in \mathcal{C}$ such that its restriction to the interval $[0,t_0]$ satisfies the equality (4) and the restriction $y|[t_0,1]$ is absolutely-continuous and for almost all $t \in [t_0,1]$ the function satisfies equation (3). It is known (cf., e.g., [35]) that for any $t_0 \in T$, $y_0 \in \mathcal{C}$ and for any choice of measurable controls $u(\cdot)$: $T \to \mathcal{P}$ and $v(\cdot)$: $T \to \mathcal{Q}$ there exists a unique solution of equation (3). We shall denote this solution by $y(\cdot,t_0,y_0,u(\cdot),v(\cdot))$.

Note that the existence and uniqueness of a solution of equation (3) can be proved under conditions which are weaker than those stipulated above. For example, in place of a global Lipschitz condition (1) one could consider a local Lipschitz condition in conjunction with a condition which would assure extendability of the solution up to time $t = 1$. However, such refinements are not essential, and in what follows we shall confine ourselves to the case of a controlled system satisfying conditions (i) - (iii).

Equation (3) can be written in the form

$$(5) \qquad \dot{y}(t) = f_*(t,y|[0,t], u(t), v(t)).$$

Here the function f_* is defined by equality $f_*(t,z_t,u,v) \overset{\text{def}}{=} f(t,y,u,v)$ where z_t: $[0,t] \to R^n$ is a continuous function which coincides with $y|[0,t]$. Thus the function f_* is defined on $\mathcal{D} \times \mathcal{P} \times \mathcal{Q}$, where $\mathcal{D}$ is the totality of the pairs $(t, y|[0,t])$, $t \in T$, $y \in \mathcal{C}$. Relations (3) and (5) are two different forms for the same equation. Possibly the second form may be more intuitive, however it is less convenient since equation (5) involves functions $y|[0,t]$ defined on different time intervals $[0,t]$.

In conclusion, we note that equations with a lagging argument as well as ordinary differential equations of the form (1.2.1) may be viewed as a particular case of the differential-functional equation (3).

11.1.2. We now proceed to a definition of the strategies and the motions generated by them.

Let the first player select the function U: $T \times \mathcal{C} \to \mathcal{P}$ which is non-anticipating. Let an initial time $t_0 \in T$ and initial function $y_0 \in \mathcal{C}$ be given. First we shall define step-by-step motions. We assume that the first player selects a partition $\Delta = \{t_0 = \tau_0 < \tau_1 < \cdots < \tau_{k+1} = 1\}$. Denote by $\mathcal{Y}_1(t_0,y_0,U,\Delta)$ the set of functions $y \in \mathcal{C}$ for which the initial condition (4) is fulfilled, the restriction $y|[t_0,1]$ is absolutely-continuous and the differential inclusion

$$(6) \qquad \dot{y}(t) \in \text{co}\{f(t,y,u(t),v): v \in \mathcal{Q}\}$$

is satisfied for almost all $t \in [t_0,1]$. Here $u(t)$ is a piecewise-constant control which is formed by the rule

(7) $$u(t) = U(\tau_i,y),\ \ \tau_i \leqslant t < \tau_{i+1},\ \ i = 0,1,\ \dots,\ k.$$

Elements of the set $Y_1(t_0,y_0,U,\Delta)$ are called *step-by-step motions.* Recall that the function U is non-anticipating. Therefore the control $u(t) = u(\tau_i)$, $t \in [\tau_i,\tau_{i+1})$, selected at time $t = \tau_i$ depends on the motion $y\|[0,\tau_i]$ which is realized on the interval $[0,\tau_i]$ (more precisely it depends on the initial function $y_0\|[0,t_0]$ and the motions $y\|[t_0,\tau_i]$). The choice of control $u(\tau_i)$ does not depend on the events which may take place for $t > \tau_i$. It is easy to verify that $Y_1(t_0,y_0,U,\Delta) \neq \phi$.

We define the set of constructive motions. Set

(8) $$\begin{aligned} Y_1(t_0,y_0,U) &\overset{\text{def}}{=} \{y \in C : \exists\ y_\ell \in Y_1(t_0,y_0,U,\Delta_\ell), \\ &\ell = 1,2,\ \dots,\ \text{diam}\ \Delta_\ell \to 0,\ \ |y - y_\ell| \to 0\ \text{ as }\ \ \ell \to \infty\}. \end{aligned}$$

The elements of the set $Y_1(t_0,y_0,U)$ are called *constructive motions.* We thus arrive at the following definition.

Definition. *A non-anticipating function* U: $T \times C \to P$ *is called a strategy for the first player. For a given initial situation* $(t_0,y_0) \in T \times C$ *this strategy generates a set of constructive motions* $Y_1(t_0,y_0,U)$ *defined by relation* (8).

To emphasize that it is possible to store the realized motion in the case under consideration, the strategies introduced herein will sometimes be called *strategies with complete memory.* This terminology is seldom used however, mainly in those cases when it is required to distinguish between positional strategies and strategies with complete memory.

Strategies for the second player and the generated motions are introduced analogously.

Definition. *A non-anticipating function* V: $T \times C \to Q$ *is called a strategy for the second player (with complete memory). For a given initial situation* $(t_0,y_0) \in T \times C$ *the set of constructive motions generated by strategy* V *is defined by the relation*

(9) $$\begin{aligned} Y_2(t_0,y_0,V) &\overset{\text{def}}{=} \{y \in C : \exists\ y_\ell \in Y_2(t_0,y_0,V,\Delta_\ell), \\ &\ell = 1,2,\ \dots,\ \text{diam}\ \Delta_\ell \to 0,\ \ |y - y_\ell| \to 0\ \text{ as }\ \ \ell \to \infty\}. \end{aligned}$$

Here $Y_2(t_0,y_0,V,\Delta)$ is the set of step-by-step motions. The elements of this set are functions $y \in C$ for which the initial condition (4) is valid, the restriction $y\|[t_0,1]$ is absolutely continuous, and the differential inclusion

$$\dot{y}(t) \in \mathrm{co}\{f(t,y,u,v(t)) \colon u \in P\}$$

is satisfied where

(10) $v(t) = V(\tau_i,y),\ \ \tau_i \leqslant t < \tau_{i+1},\ \ i = 0,1,\ldots,k.$

The following assertion is valid for the sets of constructive and step-by-step motions introduced herein.

The sets $Y_1(t_0,y_0,U)$ and $Y_2(t_0,y_0,V)$ are nonempty, compact sets in . For any $\epsilon > 0$ one can find a $\delta > 0$ such that as $\mathrm{diam}\Delta \leqslant \delta$ the inclusions

$$Y_1(t_0,y_0,U,\Delta) \subset Y_1^{\epsilon}(t_0,y_0,U),\quad Y_2(t_0,y_0,V,\Delta) \subset Y_2^{\epsilon}(t_0,y_0,V)$$

are valid. Recall that Y^{ϵ} is an ϵ-neighborhood of the set Y.

The proof of this assertion coincides on the whole with the proofs of Lemma 1.3.1 and 1.3.2.

It can be shown that

(11) $Y_1(t_0,y_0,U) \cap Y_2(t_0,y_0,V) \neq \phi$

for any pair of strategies U, V and for any initial situation $(t_0,y_0) \in T \times C$. A verification of this assertion can be carried out as follows. Consider step-by-step motions generated by a pair of strategies U, V. These motions are defined as a solution of equation (3) satisfying the initial condition (4). In equation (3) controls $u(t)$ and $v(t)$ are formed in accordance with (7) and (10). The step-by-step motion under consideration belongs to the intersection of the sets $Y_1(t_0,y_0,U,\Delta)$ and $Y_2(t_0,y_0,V,\Delta)$. Next, approaching the limit as $\mathrm{diam}\,\Delta \to 0$, we obtain that the limit element of the sequence of these step-by-step motions will be contained in the intersection (11).

11.1.3. We shall define the notions of the value and a saddle point for a differential-functional game. Let the functional

(12) $\gamma\colon y \to \gamma(y)\colon C \to R$

be defined on the space C. This functional is called a *payoff functional.* We shall assume that functional γ is continuous. As usual, it is assumed that the first player desires to minimize the value of the payoff functional while the second maximizes the payoff. We shall determine the results which are assured for the first and second players when they choose the strategies U and V respectively.

Introduce the following notation:

(13) $\Gamma_1(t_0,y_0,U) = \max \gamma(y)$ for $y \in Y_1(t_0,y_0,U)$,

(14) $\Gamma_2(t_0,y_0,V) = \min \gamma(y)$ for $y \in Y_2(t_0,y_0,V)$.

Relation (11) implies the inequality

$$\Gamma_1(t_0,y_0,U) \geqslant \Gamma_2(t_0,y_0,V).$$

This inequality is valid for all strategies U and V, thus

$$\min_U \Gamma_1(t_0,y_0,U) \geqslant \max_V \Gamma_2(t_0,y_0,V). \tag{15}$$

A pair of strategies U^0 and V^0 for which

$$\begin{aligned}\Gamma_1(t_0,y_0,U^0) &= \min_U \Gamma_1(t_0,y_0,U) = \omega^0(t_0,y_0) \\ &= \max_V \Gamma_2(t_0,y_0,V) = \Gamma_2(t_0,y_0,V^0)\end{aligned} \tag{16}$$

is called a *saddle point* of the differential-functional game (3), (12) and the quantity $\omega^0(t_0,y_0)$ is called the *value* of this game.

We are not going to discuss motivation of this notion here, nor shall we consider the approximation of optimal solutions by means of step-by-step constructions since the exposition of these problems coincides with the material presented in Section 1.5. Below, in Sections 11.2 and 11.3, the existence of saddle points and the value of the game will be proved.

We shall confine ourselves to a consideration of the differential-functional games with bounded duration and with a continuous payoff functional. However, the constructions presented below can be carried over to the case when $T = [0,\infty)$ and the payoff functional is either lower or upper semicontinuous. Moreover, these constructions can also be carried over to the case when the condition of saddle point in a small game is not fulfilled.

11.2. Stable Functionals

The existence of the value for a differential-functional game will be proved in two stages. In this section, u-stable and v-stable functionals will be defined and the existence of a functional which is simultaneously a u-stable and v-stable one is proved (below, in Section 11.3 it will be shown that such a functional is unique and coincides with the value functional.) In Section 11.3, an extremal strategy for the first (second) player is constructed which assures that a u-stable (v-stable) functional is nonincreasing (nondecreasing) along the motions of the controlled system. We shall thus obtain that the extremal strategies for the first and second player form a saddle point for a differential-functional game. Note that the contents in the section correspond to the contents of Section 2.2.

11.2.1. We shall introduce the following notation. For any $(t,x,u,v) \in T \times \mathcal{C} \times \mathcal{P} \times \mathcal{Q}$ we shall define the sets

(1)
$$\mathcal{F}_1(t,y,u) \overset{\text{def}}{=} \text{co}\{f(t,y,u,v)\colon v \in \mathcal{Q}\},$$
$$\mathcal{F}_2(t,y,v) \overset{\text{def}}{=} \text{co}\{f(t,y,u,v)\colon u \in \mathcal{P}\}.$$

Note that these sets are analogs of the sets defined by the equalities (2.2.1).

Let $(t_*,y_*) \in T \times \mathcal{C}$. Denote by $\mathcal{Y}_1(t_*,y_*,u)$ the set whose elements are functions $y \in \mathcal{C}$ such that firstly the initial condition

(2)
$$y|[0,t_*] = y_*|[0,t_*]$$

is fulfilled and next the restriction $y_*|[t_*,1]$ is absolutely continuous and satisfies for almost all $t \in [t_*,1]$ the differential inclusion

(3)
$$\dot{y}(t) \in \mathcal{F}_1(t,y,u).$$

Analogously the set $\mathcal{Y}_2(t_*,y_*,v)$ is defined whose elements are functions $y \in \mathcal{C}$ satisfying the condition (2) whose restriction to $[t_*,1]$ is absolutely-continuous and satisfies the differential inclusion

(4)
$$\dot{y}(t) \in \mathcal{F}_2(t,y,v).$$

Note that a solution of the differential equation $\dot{y}(t) = f(t,y,v,u)$ $(t \in [t_*,1])$ satisfying the initial condition (2) is contained in the sets $\mathcal{Y}_1(t_*,y_*,u)$ and $\mathcal{Y}_2(t_*,y_*,v)$. Hence

(5)
$$\mathcal{Y}_1(t_*,y_*,u) \cap \mathcal{Y}_2(t_*,y_*,v) \neq \phi .$$

It is known (cf. for example [35]) that $\mathcal{Y}_1(t_*,y_*,u)$ and $\mathcal{Y}_2(t_*,y_*,v)$ are compact sets in $\mathcal{C}$.

We introduce an additional constraint which will be used in this section. The constraint stipulates that

$$\gamma(y) \in I \overset{\text{def}}{=} [-r,r] \quad (y \in \mathcal{C})$$

i.e. the range of values of the payoff functional is contained in the interval $[-r,r]$. In subsection 11.3.4, we shall get rid of this constraint.

Denote by Ω_1 the set of functionals $\omega : T \times \mathcal{C} \to I$ which satisfy the following three conditions: the functional ω is nonanticipating, for $t = 1$ the inequality

(6)
$$\omega(1,y) \geqslant \gamma(y) \quad (y \in \mathcal{C})$$

is fulfilled, and the condition

(7) $$\sup_{(t_*,y_*,v,t)} \inf_{y}[\omega(t,y) - \omega(t_*,y_*)] \leq 0$$

is satisfied where

$$t_* \in [0,1],\ \ y_* \in \mathcal{C},\ \ v \in \mathcal{Q},\ \ t \in [t_*,1],\ \ y \in Y_2(t_*,y_*,v).$$

Define also the set Ω_2 whose elements are nonanticipating functionals $\omega : T \times \mathcal{C} \to I$ satisfying the inequality

(8) $$\omega(1,y) \leq \gamma(y)$$

and the following condition

(9) $$\inf_{(t_*,y_*,u,t)} \sup_{y}[\omega(t,y) - \omega(t_*,y_*)] \geq 0$$

where

$$(t_*,y_*,v) \in [0,1) \times \mathcal{C} \times \mathcal{Q},\ \ t \in [t_*,1],\ \ y \in Y_1(t_*,y_*,u).$$

Conditions (7) and (9) are called, respectively, the *conditions of u-stability and v-stability* of the functional ω. It is shown below (cf. Theorem 11.2.1) that the sets Ω_1 and Ω_2 are not disjoint. Moreover, as is shown in Theorem 11.3.4, the only common element in the sets Ω_1 and Ω_2 is the value functional.

11.2.2. We present several auxiliary assertions.

Let $\tau \in [0,1]$, $u \in \mathcal{P}$, $\omega_\# \in \Omega_1$. Consider the functional $\omega^\#: T \times \mathcal{C} \to I$, which is defined by the relation

(10) $$\omega^\#(t_*,y_*) = \begin{cases} \omega_\#(t_*,y_*) & \text{for } t_* \in [\tau,1],\ y_* \in \mathcal{C}, \\ \sup\{\omega_\#(\tau,y):\ y \in Y_1(t_*,y_*,u)\} & \\ & \text{for } t_* \in [0,\tau),\ y_* \in \mathcal{C}. \end{cases}$$

Lemma 11.2.1. *Functional $\omega^\#$ is contained in the set Ω_1.*

Proof. Functional $\omega_\#$ and the multivalued mapping $(t_*,y_*) \to Y_1(t_*,y_*,u)$ are nonanticipating. Therefore the functional $\omega^\#$ is also nonanticipating. It follows from (10) that $\omega^\#(1,y) = \omega_\#(1,y) \geq \gamma(y) \geq$ $(y \in \mathcal{C})$, i.e. the functional $\omega^\#$ satisfies inequality (6). It remains to verify that the functional $\omega^\#$ is u-stable.

Let $(t_*,y_*,v) \in [0,1) \times \mathcal{C} \times \mathcal{Q}$, $t \in [t_*,1]$ and $\epsilon > 0$. According to (7), it is required to verify that there exists $y \in Y_2(t_*,y_*,v)$ such that

(11) $$\omega^\#(t,y) \leq \omega^\#(t_*,y_*) + \epsilon.$$

Three cases are possible: (i) $t_* \geq \tau$, (ii) $t \leq \tau$, (iii) $t_* < \tau < t$.

Case (i). The existence of an element $y \in Y_2(t_*,y_*,v)$ for which the inequality (11) is fulfilled follows from the property of u-stability of the functional $\omega_{\#}$.

Case (ii). Let $y \in Y_1(t_*,y_*,u) \cap Y_2(t_*,y_*,v)$ (in view of (5) this intersection is nonempty). The inclusion $Y_1(t,y,u) \subset Y_1(t_*,y_*,u)$ is valid. Therefore the inequality $\omega^{\#}(t,y) \leqslant \omega^{\#}(t_*,y_*)$ follows from (10). We thus have shown that there exists $y \in Y_2(t_*,y_*,v)$ for which the inequality is valid.

Case (iii). As was shown in Cases (i) and (ii), there exists $y_1 \in Y_2(t_*,y_*,v)$ and $y_2 \in Y_2(\tau,y_1,v)$ such that

$$\omega^{\#}(\tau,y_1) \leqslant \omega^{\#}(t_*,y_*), \quad \omega^{\#}(t,y_2) \leqslant \omega^{\#}(\tau,y_1) + \epsilon.$$

Hence $\omega^{\#}(t,y_2) \leqslant \omega^{\#}(t_*,y_*) + \epsilon$. It remains to note that $y_2 \in Y_2(t_*,y_*,v)$. The lemma is thus proved.

Lemma 11.2.2. *The set Ω_1 is nonempty and contains the functional ω defined by the equality*

$$(12) \qquad \omega(t_*,y_*) = \sup\{\gamma(y): y \in Y_1(t_*,y_*,u)\},$$

where u is an arbitrary point in P.

Proof. The multivalued mapping $(t_*,y_*) \to Y_1(t_*,y_*,u)$ is nonanticipating. Therefore the functional ω (12) is also nonanticipating. For $t_* = 1$ we have $Y_1(1,y_*,u) = \{y_*\}$, hence functional ω (12) satisfies the equality $\omega(1,y) = \gamma(y)$. Finally, repeating the arguments presented in the proof of Lemma 11.2.1 (Case (ii)), we obtain that the functional ω (12) is u-stable. The lemma is thus proved.

Lemma 11.2.3. *The functional ω_0 defined by the equality*

$$(13) \qquad \omega_0(t_*,y_*) = \inf\{\omega(t_*,y_*): \omega \in \Omega_1\}$$

belongs to the set Ω_1.

Proof. In view of Lemma 11.2.2 the functional ω_0 (13) is well-defined. Let $\omega \in \Omega_1$, $y_1|[0,t_*] = y_2|[0,t_*]$, then $\omega(t_*,y_1) = \omega(t_*,y_2)$. We thus obtain $\omega_0(t_*,y_1) = \omega_0(t_*,y_2)$ i.e. functional ω_0 (13) is nonanticipating. Recall that functional ω defined by equality (12) satisfies the condition $\omega(1,y) = \gamma(y)$. Therefore it follows from (6) that $\omega_0(1,y) = \gamma(y)$, $y \in C$. It remains to verify that functional ω_0 (13) is u-stable.

Let $(t_*,y_*,v) \in [0,1] \times C \times Q$, $t \in [t_*,1]$, $\epsilon > 0$ be given. According to (13) there exists $\omega \in \Omega_1$ such that $\omega(t_*,y_*) \leqslant \omega_0(t_*,y_*) + \epsilon/2$. Since the functional ω satisfies condition (7), there exists $y \in Y_2(t_*,y_*,v)$ such that $\omega(t,y) \leqslant \omega(t_*,y_*) + \epsilon/2$. From (13) we have $\omega_0(t,y) \leqslant \omega(t,y)$.

These inequalities imply that $\omega_0(t,y) \leqslant \omega_0(t_*,y_*) + \epsilon$, where $y \in Y_2(t_*,y_*,v)$. We have thus shown that the functional ω_0 is u-stable.

Lemma 11.2.4. *The functional* ω_0 *defined by equality* (13) *belongs to the set* Ω_2.

Proof. It was shown in the preceding lemma that functional ω_0 (13) is nonanticipating and satisfies the equality $\omega_0(1,y) = \gamma(y)$. It remains to verify that the functional ω_0 satisfies the condition of v-stability.

Let $(t_*,y_*,u) \in [0,1) \times C \times P$ and $\tau \in [t_*,1]$. We define the functional $\omega^\#$ by equality (10) in which we set $\omega_\# = \omega_0$. It follows from Lemmas 11.2.1 and 11.2.3 that $\omega^\# \in \Omega_1$. We thus obtain from (13) that $\omega^\#(t_*,y_*) \geqslant \omega_0(t_*,y_*)$. Using (10) with $\omega_\# = \omega_0$ this inequality can be rewritten as follows:

$$\omega^\#(t_*,y_*) = \sup\{\omega_0(\tau,y)\colon y \in Y_1(t_*,y_*,u)\} \geqslant \omega_0(t_*,y_*).$$

In view of (9) this bound means that the functional ω_0 is v-stable. Lemma 1.2.4 is thus proved.

It follows from Lemmas 11.2.3 and 11.2.4 that

Theorem 11.2.1. *The intersection* $\Omega_1 \cap \Omega_2$ *is nonempty and contains the functional* ω_0 (13).

11.2.3. We shall verify a topological property of functionals $\omega \in \Omega_1$ and $\omega \in \Omega_2$.

Let $\omega\colon T \times C \to I$ be a functional. Define the functionals $\underline{\omega}\colon T \times C \to I$ and $\overline{\omega}\colon T \times C \to I$ by the relations

$$(14) \qquad \underline{\omega}(t_*,y_*) = \varliminf_{(t,y)\to(t_*,y_*)} \omega(t,y), \quad \overline{\omega}(t_*,y_*) = \varlimsup_{(t,y)\to(t_*,y_*)} \omega(t,y).$$

Lemma 11.2.5. *For any functional* $\omega\colon T \times C \to I$ the functionals $\underline{\omega}$ and $\overline{\omega}$ defined by equalities (14) possess the following properties: The inequalities

$$(15) \qquad \underline{\omega}(t,y) \leqslant \omega(t,y) \leqslant \overline{\omega}(t,y) \text{ for all } (t,y) \in T \times C$$

are valid. The functional $\underline{\omega}\colon T \times C \to I$ *is lower semicontinuous while the functional* $\overline{\omega}\colon T \times C \to I$ *is upper semicontinuous. If* $\omega \in \Omega_1$ *then* $\underline{\omega} \in \Omega_1$. *If* $\omega \in \Omega_2$ *then* $\overline{\omega} \in \Omega_2$.

Proof. Bounds (15) and the semicontinuity of functionals $\underline{\omega}$ and $\overline{\omega}$ follow directly from definition (14).

Let $\omega \in \Omega_1$. We shall prove that $\underline{\omega} \in \Omega_1$. First we shall show that the functional $\underline{\omega}$ is nonanticipating. Let $t_* \in T$, $y_* \in C$, $z_* \in C$, and $y_*|[0,t_*] = z_*|[0,t_*]$. It is required to verify that $\underline{\omega}(t_*,y_*) = \underline{\omega}(t_*,z_*)$.

Select a sequence $(t_k,y_k) \in T \times C$ $(k = 1,2, ...)$ such that

(16) $$(t_k,y_k) \to (t_*,y_*), \quad \omega(t_k,y_k) \to \underline{\omega}(t_*,y_*) \quad \text{as} \quad k \to \infty.$$

Consider the sequence $z_k \in C$ $(k = 1,2, ...)$ which we shall define as follows:

$$z_k|[0,t_k] = y_k|[0,t_k], \quad z_k|[t_k,1] = r_k + z_*|[t_k,1],$$

where $r_k = y_k(t_k) - z_*(t_k)$. It follows from the definition of the sequence z_k $(k = 1,2, ...)$ that $z_k \to z_*$ as $k \to \infty$. Therefore, taking into account the fact that functional ω is nonanticipating, we have in view of (14) and (16) that

$$\begin{aligned}\underline{\omega}(t_*,y_*) &= \lim_{k\to\infty} \omega(t_k,y_k) = \lim_{k\to\infty} \omega(t_k,z_k) \\ &\geqslant \varliminf_{(t,z)\to(t_*,z_*)} \omega(t,z) = \underline{\omega}(t_*,z_*).\end{aligned}$$

Interchanging the roles of functions y_* and z_* we arrive at the opposite inequality $\underline{\omega}(t_*,z_*) \geqslant \underline{\omega}(t_*,y_*)$. Thus the equality $\underline{\omega}(t_*,z_*) = \underline{\omega}(t_*,y_*)$ is verified.

We shall show that the functional $\underline{\omega}$ satisfies the inequality (6). Let $y_* \in C$ and let a sequence $(t_k,y_k) \in T \times C$ be chosen such that

$$(t_k,y_k) \to (1,y_*), \quad \omega(t_k,y_k) \to \underline{\omega}(1,y_*) \quad \text{as} \quad k \to \infty .$$

Conditions (6) and (7) imply that there exists a sequence $y^{(k)} \in Y_2(t_k,y_k,\nu)$ $(k = 1,2, ...)$ such that

$$\omega(t_k,y_k) + k^{-1} \geqslant \omega(1,y^{(k)}) \geqslant \gamma(y^{(k)}).$$

Recall that the functional γ is continuous. Therefore it follows from convergence $y^{(k)} \to y_*$ as $k \to \infty$ that $\omega(t_k,y_k) \to \underline{\omega}(1,y_*) \geqslant \gamma(y_*)$ as $k \to \infty$.

It remains to verify that the condition of u-stability of functional ω implies the u-stability of functional $\underline{\omega}$. Let $(t_*,y_*,\nu) \in [0,1) \times C \times \mathcal{Q}$, $t \in (t_*,1]$, $\epsilon > 0$. Let $(t_k,y_k) \in T \times C$ $(k = 1,2, ...)$ be a sequence convergent to (t_*,y_*) and such that according to (14) equality

(17) $$\underline{\omega}(t_*,y_*) = \lim_{k\to\infty} \omega(t_k,y_k)$$

is valid. Since condition (7) is fulfilled, there exists $y^{(k)} \in Y_2(t_k,y_k,\nu)$ such that

(18) $$\omega(t,y^{(k)}) \leqslant \omega(t_k,y_k) + \epsilon \quad (k = 1,2, ...).$$

It can be assumed without loss of generality that $t_k < t$ $(k = 1,2, ...)$. It is known that one can extract from the sequence $y^{(k)}$ $(k = 1,2, ...)$ a convergent subsequence. To simplify the notation we set that $y^{(k)} \to$

y^* as $k \to \infty$. It is also known (cf., for example [35]) that $y^* \in Y_2(t_*,y_*,v)$. Approaching the limit in inequalities (18) as $k \to \infty$ and taking relations (14) and (17) into account, we obtain

$$\underline{\omega}(t,y^*) \leqslant \underline{\omega}(t_*,y_*) + \epsilon$$

where $y^* \in Y_2(t_*,y_*,v)$. We have thus shown that $\underline{\omega} \in \Omega_1$.

The proof of the fact that $\bar{\omega} \in \Omega_2$ provided $\omega \in \Omega_2$ is analogous.

11.2.4. We now define u-stable and v-stable bridges which will be required below for construction of the players' optimal strategies.

Let an initial time $t_0 \in T$ and an initial function $y_0 \in C$ be given. Denote by $Y(t_0,y_0)$ the set of solutions of a differential inclusion

$$\dot{y}(t) \in \text{co}\{f(t,y,u,v): u \in P, v \in Q\}, \quad t_0 \leqslant t \leqslant 1 \tag{19}$$

satisfying the initial condition (11.1.4). As above, the elements of the set $Y(t_0,y_0)$ are continuous functions $y \in C$ such that the equality (11.1.4) is fulfilled, while the restriction $y|[t_0,1]$ is absolutely-continuous and for almost all $t \in [t_0,1]$ satisfies inclusion (19).

It is important to note that $Y(t_0,y_0)$ is a compact set in C. Moreover, the inclusions

$$\begin{aligned} &Y_1(t_*,y_*,u) \subset Y(t_0,y_0), \quad Y_2(t_*,y_*,v) \subset Y(t_0,y_0), \\ &Y_1(t_*,y_*,U) \subset Y(t_0,y_0), \quad Y_2(t_*,y_*,V) \subset Y(t_0,y_0) \end{aligned} \tag{20}$$

are valid for any $t_* \in [t_0,1]$, $y_* \in (t_0,y_0)$, $u \in P$, $v \in Q$ and for any strategies U and V.

Let the functional ω_0: $T \times C \to I$ be defined by relation (13). Consider the functional $\underline{\omega}_0$ defined by the equality

$$\underline{\omega}_0(t_*,y_*) = \lim_{(t,y)\to(t_*,y_*)} \omega_0(t,y).$$

According to Lemmas 11.2.3 and 11.2.5 we have

$$\underline{\omega}_0 \in \Omega_1, \quad \underline{\omega}_0(t_*,y_*) \leqslant \omega_0(t_*,y_*). \tag{21}$$

Moreover, the functional $\underline{\omega}_0$: $T \times C \to I$ is lower semicontinuous. Therefore, taking into account that $Y_2(t_*,y_*,v)$ is a compact set in C, the condition of u-stability of functional $\underline{\omega}_0$ can be rewritten as follows:

$$\sup_{(t_*,y_*,v,t^*)} \min_{y} [\underline{\omega}_0(t^*,y) - \underline{\omega}_0(t_*,y_*)] \leqslant 0 \tag{22}$$

where

$$(t_*,y_*,v) \in [0,1) \times C \times Q, \quad t^* \in [t_*,1], \quad y \in Y_2(t_*,y_*,v).$$

Define the set

$$(23) \qquad \mathcal{W}^{(1)}(t) \overset{\text{def}}{=} \{y \in \mathcal{Y}(t_0,y_0) : \ \underline{\omega}_0(t,y) \leqslant \underline{\omega}_0(t_0,y_0)\}.$$

The following assertion is valid.

Lemma 11.2.6. *For any* $t \in [t_0,1]$, *the set* $\mathcal{W}^{(1)}(t)$ *is nonempty and compact in* $\mathcal{C}$. *The multivalued mapping* $t \to \mathcal{W}^{(1)}(t)$ *from* $[t_0,1]$ *into* $\mathcal{C}$ *possesses the property of u-stability, i.e. for any* $t_* \in [t_0,1)$, $y_* \in \mathcal{W}^{(1)}(t_*)$, $v \in \mathcal{Q}$ *and* $t^* \in [t_*,1]$ *relation*

$$(24) \qquad \mathcal{W}^{(1)}(t^*) \cap \mathcal{Y}_2(t_*,y_*,v) \neq \phi$$

is valid. For $t = 1$ *the equality*

$$(25) \qquad \mathcal{W}^{(1)}(1) = \{y \in \mathcal{Y}(t_0,y_0) \colon \gamma(y) \leqslant \underline{\omega}_0(t_0,y_0)\}$$

holds.

Proof. We shall first show that the condition of u-stability is fulfilled. Let (t_*,y_*,v,t^*) be a collection appearing in the statement of this condition. Note that $\underline{\omega}_0(t_*,y_*) \leqslant \underline{\omega}_0(t_0,y_0)$. In view of (22) there exists $y^* \in \mathcal{Y}_2(t_*,y_*,v)$ such that $\underline{\omega}_0(t^*,y^*) \leqslant \underline{\omega}_0(t_*,y_*) \leqslant \underline{\omega}_0(t_0,y_0)$. Taking into account that $y_* \in \mathcal{Y}(t_0,y_0)$ (cf. relations (20)) we have $y^* \in \mathcal{Y}_2(t_*,y_*,v) \subset \mathcal{Y}(t_0,y_0)$. We thus obtain that $y^* \in \mathcal{W}(t^*) \cap \mathcal{Y}_2(t_*,y_*,v)$ and the property of u-stability is proved.

Clearly $(t_0,y_0) \in \mathcal{W}_*^{(1)}(t_0)$. Setting in (24), $(t_*,y_*) = (t_0,y_0)$ we obtain $\mathcal{W}^{(1)}(t^*) \neq \phi$ for $t_0 \leqslant t^* \leqslant 1$.

Since (t_0,y_0) is a compact set in $\mathcal{C}$ and the functional $\underline{\omega}_0(t,\cdot)\colon \mathcal{C} \to I$ is lower semicontinuous, $\mathcal{W}^{(1)}(t)$ is a closed subset of the compact set $\mathcal{Y}(t_0,y_0)$. Consequently $\mathcal{W}^{(1)}(t)$ is a compact set in .

Finally, relation (25) follows from definition (23) and inequality $\underline{\omega}_0(1,y) \geqslant \gamma(y)$ (cf. inequality (6) and Lemma 11.2.5).

A multivalued mapping $t \to \mathcal{W}^{(1)}(t)$ defined by relation (23) will be called *a u-stable bridge*.

Analogously, we set

$$(26) \qquad \mathcal{W}^{(2)}(t) \overset{\text{def}}{=} \{y \in \mathcal{Y}(t_0,y_0) \colon \overline{\omega}_0(t,y) \geqslant \overline{\omega}_0(t_0,y_0)\}.$$

Here

$$\overline{\omega}_0(t_*,y_*) \overset{\text{def}}{=} \overline{\lim_{(t,y)\to(t_*,y_*)}} \ \omega_0(t,y) \quad (t_*,y_*) \in T \times \mathcal{C},$$

where the functional ω_0 is defined by equality (13). Note that

$$(27) \qquad \overline{\omega}_0 \in \Omega_2 \ \text{ and } \ \overline{\omega}_0(t_*,y_*) \geqslant \omega_0(t_*,y_*).$$

A multivalued mapping $t \to \mathcal{W}^{(2)}(t)$ from $[t_0,1]$ into C defined by equality (26) will be called a *v-stable bridge.*

The lemma presented below is proved in the same manner as Lemma 11.2.6.

Lemma 11.2.7. *For any* $t \in [t_0,1]$ *the set* $\mathcal{W}^{(2)}(t)$ *is nonempty and compact in* C. *The system of sets* $\mathcal{W}^{(2)}(t)$ $(t_0 \leqslant t \leqslant 1)$ *possesses the property of v-stability, i.e. for any* $t_* \in [t_0,1)$, $y_* \in \mathcal{W}^{(2)}(t_*)$, $u \in P$, $t^* \in [t_*,1]$ *the relation*

$$\mathcal{W}^{(2)}(t^*) \cap \mathcal{Y}_1(t_*,y_*,u) \neq \phi \tag{28}$$

is valid.

For $t = 1$ *the equality*

$$\mathcal{W}^{(2)}(1) = \{y \in \mathcal{Y}(t_0,y_0) \colon \gamma(y) \geqslant \bar{\omega}_0(t_0,y_0)\} \tag{29}$$

holds.

11.3. The Saddle Point

In this section strategies for the first and second players extremal to u-stable and v-stable bridges are defined. It is proved that these strategies form a saddle point for a differential-functional game. The content of this section corresponds to Sections 2.3 and 2.4, i.e. the results obtained in these sections for positional strategies are carried over for the case of strategies with complete memory.

11.3.1. We shall define a strategy for the first player extremal to a u-stable bridge.

First introduce the following notation: Let $t \in T$, $y \in C$, $w \in C$. Set

$$\rho(t,y,w) \overset{\text{def}}{=} \max_{0 \leqslant \tau \leqslant t} \|y(\tau) - w(\tau)\|. \tag{1}$$

Let $\mathcal{W}$ be a compact set in C. Set

$$\rho(t,y,\mathcal{W}) \overset{\text{def}}{=} \min_{w \in \mathcal{W}} \rho(t,y,w). \tag{2}$$

Observe that the quantity $\rho(t,y,w)$ is just the distance between the restrictions $y|[0,t]$ and $w|[0,t]$ in the metric of the space $C_n([0,t])$. Evidently for $t = 1$ we obtain

$$\rho(1,y,w) = |y - w| \overset{\text{def}}{=} \max_{0 \leqslant \tau \leqslant 1} \|y(\tau) - w(\tau)\|$$

$$\rho(1,y,\mathcal{W}) = \text{dist}(y,\mathcal{W}).$$

Consider the sets $\mathcal{W}^{(1)}(t)$ $(t_0 \leqslant t \leqslant 1)$ defined by equality (11.2.23). For $(t,y) \in [t_0,1] \times C$ we set

$$(3)\quad \begin{aligned} \mathcal{W}^{(1)}[t,y] &\overset{\text{def}}{=} \{w_0 \in \mathcal{W}^{(1)}(t)\colon \rho(t,y,w_0) = \rho(t,y,\mathcal{W}^{(1)}(t))\},\\ \mathcal{S}^{(1)}[t,y] &\overset{\text{def}}{=} \{s = y(t) - w_0(t) \in R^n : w_0 \in \mathcal{W}^{(1)}[t,y]\}.\end{aligned}$$

According to Lemma 11.2.6, the set $\mathcal{W}^{(1)}(t)$ is compact in $\mathcal{C}$. Therefore it follows from (2) and (3) that $\mathcal{W}^{(1)}[t,y] \neq \phi$. Functions $w_0 \in \mathcal{W}^{(1)}(t)$ closest to function y relative to the "metric" $\rho(t,\cdot,\cdot)$ (1) are the elements of the set $\mathcal{W}^{(1)}[t,y]$. To each function $w_0 \in \mathcal{W}^{(1)}[t,y]$ there corresponds a vector $s = y(t) - w_0(t) \in R^n$. These vectors s form the set $\mathcal{S}^{(1)}[t,y]$. It is easy to verify that for any $(t,y) \in [t_0,1] \times \mathcal{C}$ the sets $\mathcal{W}^{(1)}[t,y]$ and $\mathcal{S}^{(1)}[t,y]$ are compact sets in $\mathcal{C}$ and R^n, respectively, and that the multivalued mappings $(t,y) \to \mathcal{W}^{(1)}[t,y]$ and $(t,y) \to \mathcal{S}^{(1)}[t,y]$ are nonanticipating.

Let the function $s^{(1)}\colon T \times \mathcal{C} \to R^n$ be chosen in the following manner:

$$(4)\quad \begin{aligned} &s^{(1)}(t,y) = 0 &&\text{for } t \in [0,t_0),\ y \in \mathcal{C},\\ &s^{(1)}(t,y) \in \mathcal{S}^{(1)}[t,y] &&\text{for } t \in [t_0,1],\ y \in \mathcal{C}.\end{aligned}$$

The function $s^{(1)}$ can be selected in a manner so that it will be nonanticipating.

Completing the definition of an extremal strategy U^e, we select a function $U^e\colon T \times \mathcal{C} \to \mathcal{P}$ from the condition

$$(5)\quad \max_{v\in\mathcal{Q}} \langle s^{(1)}(t,y), f(t,y,U^e(t,y),v)\rangle = \min_{u\in\mathcal{P}} \max_{v\in\mathcal{Q}} \langle s^{(1)}(t,y), f(t,y,u,v)\rangle.$$

Functions $(t,y) \to s^{(1)}(t,y)$ and $(t,y,u,v) \to f(t,y,u,v)$ are nonanticipating, therefore the function U^e can also be selected to be nonanticipating. Thus a strategy for the first player has been constructed. This strategy will be called *extremal to a u-stable bridge* $t \to \mathcal{W}^{(1)}(t)$.

The basic property of an extremal strategy is expressed by the following theorem.

Theorem 11.3.1. *Let a u-stable bridge be defined by equality* (11.2.23). *Let* U^e *be a strategy extremal to this bridge. Then*

$$(6)\quad \mathcal{Y}_1(t_0,y_0,U^e) \subset \mathcal{W}^{(1)}(1),$$

where $\mathcal{Y}_1(t_0,y_0,U^e)$ *is the set of constructive motions generated by the strategy* U^e.

Proof. Let

$$(7)\quad \mathcal{D}_1 \overset{\text{def}}{=} T \times \mathcal{Y}(t_0,y_0).$$

Recall that $\mathcal{Y}(t_0,y_0)$ is a set of solutions of the differential inclusion (7.2.19). Replace the set $\mathcal{D}_1$ (2.3.10) by the set $\mathcal{D}_1$ (7) and define the quantities $\zeta(\delta)$, Λ, $\alpha(\delta)$, and $\beta(t,\delta)$ in the manner of subsection 2.3.2. In particular, we cite the relations

$$(8)\quad \begin{gathered} \|f(t_1,y_1,u,v) - f(t_2,y_2,u,v)\| \leqslant \lambda|y_1 - y_2| + \zeta(\delta), \\ (t_i,y_i) \in \mathcal{D}_1,\ (i = 1,2),\ |t_1 - t_2| \leqslant \delta, \\ \Lambda \overset{\text{def}}{=} \max\|f(t,y,u,v)\| \text{ for } (t,y) \in \mathcal{D}_1,\ (u,v) \in \mathcal{P} \times \mathcal{Q}, \\ \beta(t,\delta) \overset{\text{def}}{=} [\exp 2\lambda(t - t_0) - 1]\alpha(\delta)\lambda^{-1}. \end{gathered}$$

We now state the estimate analogous to the estimate proved in Section 2.3.

Let $t_* \in [t_0,1]$, $y_* \in \mathcal{Y}(t_0,y_0)$, $w_* \in \mathcal{Y}(t_0,y_0)$, $s_* = y_*(t_*) - w_*(t_*)$. Choose points $u_0 \in \mathcal{P}$ and $v_0 \in \mathcal{Q}$ from the conditions

$$(9)\quad \max_{v\in\mathcal{Q}} \langle s_*, f(t_*,y_*,u_0,v)\rangle = \min_{u\in\mathcal{P}} \max_{v\in\mathcal{Q}} \langle s_*, f(t_*,y_*,u,v)\rangle,$$

$$(10)\quad \min_{u\in\mathcal{P}} \langle s_*, f(t_*,y_*,u,v_0)\rangle = \max_{v\in\mathcal{Q}} \min_{u\in\mathcal{P}} \langle s_*, f(t_*,y_*,u,v)\rangle.$$

Let $\rho^2(t_*,y_*,w_*) \leqslant \beta(t_*,\delta)$. Then for any $y \in \mathcal{Y}_1(t_*,y_*,u_0)$ and $w \in \mathcal{Y}_2(t_*,w_*,v)$ the bound

$$(11)\quad \rho^2(t^*,y,w) \leqslant \beta(t^*,\delta) \text{ for } t^* \in [t_*,t_* + \delta]$$

will be valid. Recall that $\mathcal{Y}_1(t_*,y_*,u)$ and $\mathcal{Y}_2(t_*,y_*,v)$ are sets of solutions of differential inclusions (11.2.3) and (11.2.4).

The proof of estimate (9) is almost completely the same as that of Lemma 2.3.1. We shall therefore confine ourselves to the following remark. In view of (11.2.20) it follows from the conditions $y_* \in \mathcal{Y}(t_0,y_0)$ and $w_* \in \mathcal{Y}(t_0,y_0)$ that the inclusions $y \in \mathcal{Y}_1(t_*,y_*,u_0) \subset \mathcal{Y}(t_0,y_0)$ and $w \in \mathcal{Y}_2(t_*,w_*,v_0) \subset \mathcal{Y}(t_0,y_0)$ are valid. One can therefore use the majorant Λ and the module of continuity $\delta \to \zeta(\delta)$ which appear in relations (8) to prove the estimate (9).

The next stage of the proof is to verify the relations

$$(12)\quad \rho^2(\tau_i,y,\mathcal{W}^{(1)}(\tau_i)) \leqslant \beta(\tau_i,\delta) \quad (i = 0,1, \ldots, k + 1)$$

which must be fulfilled for any step-by-step motion $y \in \mathcal{Y}_1(t_0,y_0,U^e,\Delta)$ where the partition $\Delta \overset{\text{def}}{=} \{\tau_0 = t_0 < \tau_1 < \cdots < \tau_{k+1} = 1\}$ satisfies the condition $\operatorname{diam}\Delta \leqslant \delta$.

Consider a step-by-step motion $y \in \mathcal{Y}_1(t_0,y_0,U^e,\Delta)$. We shall use an induction argument to prove estimate (12). For $i = 0$, in view of (1) and (2), we have

$$\rho(t_0,y,\mathcal{W}^{(1)}(t_0)) \leqslant \rho(t_0,y,y_0) = \rho(t_0,y_0,y_0) = 0.$$

Here we use the fact that $y_0 \in \mathcal{W}^{(1)}(t_0)$ (cf. (11.2.23)) and $y|[0,t_0] = y_0|[0,t_0]$. Clearly a strict inequality is impossible here, i.e.

$$\rho(t_0,y,\ \mathcal{W}^{(1)}(t_0)) = 0.$$

It follows from (8) that $\beta(t_0,\delta) = 0$. Therefore for $i = 0$ the estimate (12) holds.

Let the estimate (12) be valid for some $i = j$. We shall show that the estimate (12) will be fulfilled for $i = j + 1$. To utilize the estimate (11) we set $\tau_j = t_*$, $\tau_{j+1} = t^*$, $y = y_*$, $U^e(\tau_j,y_*) = u_0$. Select a function $w_* \in \mathcal{W}^{(1)}[t_*,y_*]$ such that the equality $y_*(t_*) - w_*(t_*) = s^{(1)}(t_*,y_*) = s_*$ will be fulfilled, where $s^{(1)}$ is the function appearing in the definition of the strategy U^e under consideration. Note that

$$\rho^2(t_*,y_*,w_*) = \rho^2(t_*,y_*,\ \mathcal{W}^{(1)}(t_*)) \leqslant \beta(t_*,\delta).$$

Let $v_0 \in \mathcal{Q}$ satisfy condition (10). In view of (11.2.24) there exists

$$w \in \mathcal{W}^{(1)}(t^*) \cap \ \mathcal{Y}_2(t_*,w_*,v_0).$$

It follows from the definition of strategy U^e and of the step-by-step motion $y \in \mathcal{Y}_1(t_0,y_0,U^e,\Delta)$ that the function $t \to y(t)$ satisfies the differential inclusion (11.2.3) for $t \in [t_*,t^*]$ in which one sets $u = u_0$, where point $u_0 \in \mathcal{P}$ satisfies condition (9). Now one can utilize the estimate (11) from which it follows that

$$\rho^2(t^*,y,\mathcal{W}(t^*)) \leqslant \rho^2(t^*,y,w) \leqslant \beta(t^*,\delta).$$

Returning to the original notation, we obtain that the estimate (12) is valid for $i = j + 1$. The estimate (12) is thus verified.

For $i = k + 1$ we have $\tau_{k+1} = 1$. Therefore taking (1) and (2) into account one can rewrite the estimate (12) as follows:

$$\rho(1,y,\ \mathcal{W}^{(1)}(1)) = \operatorname{dist}(y,\ \mathcal{W}^{(1)}(1)) \overset{\text{def}}{=} \min_{w \in \mathcal{W}^{(1)}(1)} |y-w| \leqslant \sqrt{\beta(1,\delta)}. \tag{13}$$

This bound is valid for any step-by-step motion $y \in \mathcal{Y}_1(t_0,y_0,U^e,\Delta)$ where $\operatorname{diam}\Delta \leqslant \delta$. Let $y^* \in \mathcal{Y}_1(t_0,y_0,U^e)$ be an arbitrary constructive motion. In view of (11.1.8) there exists a sequence of step-by-step motions $y_\ell \in \mathcal{Y}_1(t_0,y_0,U^e,\Delta_\ell)$ such that for $y_\ell \to y^*$, $\operatorname{diam}\Delta_\ell = \delta_\ell \to 0$ as $\ell \to \infty$. It follows from (8) that $\beta(1,\delta_\ell) \to 0$ as $\ell \to \infty$. Since the set $\mathcal{W}^{(1)}(1)$ is closed, we have that $y^* \in \mathcal{W}^{(1)}(1)$. Theorem 11.3.1 is thus proved.

11.3.2. We shall define a strategy for the second player extremal to a v-stable bridge and state the basic property of this strategy. Let the sets $\mathcal{W}^{(2)}(t)$ $(t_0 \leqslant t \leqslant 1)$ be defined by equality (11.2.26). Let $(t,y) \in [t_0,1] \times \mathcal{C}$. Set

$$\mathcal{W}^{(2)}[t,y] \overset{\text{def}}{=} \{w_0 \in \mathcal{W}^{(2)}(t){:}\ \rho(t,y,w_0) = \rho(t,y,\ \mathcal{W}^{(2)}(t))\},$$

$$S^{(2)}[t,y] \overset{\text{def}}{=} \{s = w_0(t) - y(t) \in R^n : w_0 \in \mathcal{W}^{(2)}[t,y]\}.$$

Here the quantities $\rho(t,y,w)$ and $\rho(t,y,\mathcal{W})$ are defined in accordance with (1) and (2). Select a function $s^{(2)}: T \times C \to R^n$ in the following manner:

$$s^{(2)}(t,y) = 0 \qquad \text{for } t \in [0,t_0), \ y \in C,$$

$$s^{(2)}(t,y) \in S^{(2)}[t,y] \ \text{ for } t \in [t_0,1], \ y \in C.$$

Finally the function $V^e: T \times C \to \mathcal{Q}$ will be defined by the condition

$$\min_{u \in \mathcal{P}} \langle s^{(2)}(t,y), f(t,y,u,V^e(t,y))\rangle = \max_{v \in \mathcal{Q}} \min_{u \in \mathcal{P}} \langle s^{(2)}(t,y), f(t,y,u,v)\rangle.$$

Assume that the function V^e is nonanticipating. The feasibility of constructing such a function is justified in the same manner as in the case of strategy U^e. The strategy V^e defined herein will be called a *strategy extremal to a v-stable bridge* $t \to \mathcal{W}^{(2)}(t)$.

Theorem 11.3.2. *Let $\mathcal{W}^{(2)}$ be a v-stable bridge defined by the equality (11.2.26). Let V^e be a strategy extremal to this bridge. Then*

(14) $$\mathcal{Y}_2(t_0,y_0,V^e) \subset \mathcal{W}^{(2)}(1)$$

where $\mathcal{Y}_2(t_0,y_0,V^e)$ is the set of constructive motions generated by the strategy V^e.

The proof of this theorem is analogous to that of Theorem 11.3.1.

11.3.3. Having Theorems 11.3.1 and 11.3.2 at our disposal, one can prove without difficulty the existence of the value and of a saddle point for the differential-functional game under consideration.

From relations (6) and (11.2.25) we have

$$\Gamma_1(t_0,y_0,U^e) \overset{\text{def}}{=} \max_{y \in \mathcal{Y}_1(t_0,y_0,U^e)} \gamma(y) \leqslant \underline{\omega}_0(t_0,y_0).$$

(Here the notation (11.1.13) is used.) Analogously using notation (11.1.14), we obtain

$$\Gamma_2(t_0,y_0,V^e) \overset{\text{def}}{=} \min_{y \in \mathcal{Y}_2(t_0,y_0,V^e)} \gamma(y) \geqslant \overline{\omega}_0(t_0,y_0).$$

In view of (11.2.21) and (11.2.27) the inequalities

$$\underline{\omega}_0(t_0,y_0) \leqslant \omega_0(t_0,y_0) \leqslant \overline{\omega}_0(t_0,y_0)$$

are valid. Hence

$$\Gamma_1(t_0,y_0,U^e) \leqslant \underline{\omega}_0(t_0,y_0) \leqslant \omega_0(t_0,y_0) \leqslant \overline{\omega}_0(t_0,y_0) \leqslant \Gamma_2(t_0,y_0,V^e).$$

On the other hand, in view of (11.1.15) we have $\Gamma_1(t_0,y_0,U^e) \geqslant \Gamma_2(t_0,y_0,V^e)$. Consequently,

$$\Gamma_1(t_0,y_0,U^e) = \underline{\omega}_0(t_0,y_0) = \omega_0(t_0,y_0) = \overline{\omega}_0(t_0,y_0) = \Gamma_2(t_0,y_0,V^e). \tag{15}$$

In accordance with the definitions of a saddle point and the value of a game, we obtain the following assertion.

Theorem 11.3.3. *For any initial time $t_0 \in T$ and any initial function $y_0 \in C$ there exists the value $\omega^0(t_0,y_0)$ of the differential-functional game under consideration; this value coincides with the quantity $\omega_0(t_0,y_0)$ defined in accordance with* (11.2.13). *The strategies U^e and V^e extremal to the u-stable and v-stable bridges* (11.2.23) *and* (11.2.26), *respectively, form the saddle point of this game.*

We return to constructions presented in Section 11.2. It was established in Theorem 11.2.1 that the intersection $\Omega_1 \cap \Omega_2$ contains the functional ω_0 defined by equality (11.2.13). Now this assertion can be refined and we can prove that

$$\Omega_1 \cap \Omega_2 = \{\omega^0\} \tag{16}$$

where ω^0 is the value functional.

Indeed, let $\omega_* \in \Omega_1 \cap \Omega_2$. Replacing functional ω_0 by functional ω_* and repeating verbatim the constructions presented in subsections 11.2.4 and 11.3.1 - 11.3.3, we obtain that $\omega_* = \omega^0$. Thus the following theorem is valid.

Theorem 11.3.4. *The value functional of the differential-functional game under consideration is the only element belonging to the intersection $\Omega_1 \cap \Omega_2$.*

11.3.4. Recall that in view of the assumption adapted in subsection 11.2.1, the payoff functional γ takes on values on the bounded interval $I \overset{\text{def}}{=} [-r,r]$. This purely technical restriction was introduced in order to simplify the constructions in Section 11.2. We shall show that the above stated Theorems 11.3.1 - 11.3.4 remain valid for any continuous payoff functional $\gamma : C \to R$.

Consider a sequence of functionals γ_k $(k = 1,2,3, \ldots)$ of the form

$$\gamma_k(y) = \begin{cases} -k & \text{for } \gamma(y) \leqslant -k, \\ \gamma(y) & \text{for } \gamma(y) \in (-k,k), \\ k & \text{for } \gamma(y) \geqslant k. \end{cases} \tag{17}$$

Each one of these functionals γ_k is continuous on C and takes on values in the bounded interval $I_k \overset{\text{def}}{=} [-k,k]$. Clearly

$$\gamma(y) = \lim_{k\to\infty} \gamma_k(y) \quad (y \in \mathcal{C}).$$

In view of the preceding results, a differential-functional game in which the payoff is defined by equality (17) possesses the value. Denote the value functional of this game by ω_k^0. Set

$$(18) \qquad \omega^0(t_0,y_0) \overset{\text{def}}{=} \lim_{k\to\infty} \omega_k^0(t_0,y_0) \qquad (t_0,y_0) \in T \times \mathcal{C}.$$

This relation determines the value functional for a differential-functional game with payoff $\gamma: \mathcal{C} \to R$.

Indeed, let a point $(t_0,y_0) \in T \times \mathcal{C}$ be chosen. Define the corresponding set $\mathcal{Y}(t_0,y_0)$ whose elements are the solutions of the differential inclusion (11.2.19). As is known, $\mathcal{Y}(t_0,k_0)$ is a compact set in $\mathcal{C}$. Therefore there exists a number k_0 such that

$$|\gamma(y)| \leqslant k_0 \quad \text{for all} \quad y \in \mathcal{Y}(t_0,y_0).$$

Clearly

$$(19) \qquad \gamma_k(y) = \gamma_{k_0}(y) \quad \text{for} \quad k > k_0 \quad \text{and} \quad y \in \mathcal{Y}(t_0,y_0).$$

Next it follows from the definition of the functional ω_k^0 that

$$\omega_k^0(t_0,y_0) = \omega_{k_0}^0(t_0,y_0) \quad \text{for} \quad k > k_0.$$

Hence the limit in (18) exists. Relations (19) and (11.2.20) imply that the quantity $\omega^0(t_0,y_0) = \omega_{k_0}^0(t,y)$ is the value of the game under consideration in which $\gamma: \mathcal{C} \to R^n$ is an arbitrary continuous payoff functional.

11.4. Properties of Value Functionals

Below, certain properties of value functionals are presented. These facts are given without detailed proofs since, to a large extent, these proofs coincide with the proofs of the corresponding assertions for positional differential games.

11.4.1. Consider the continuity property and the Lipschitzian property of value functionals.

Let $\gamma: \mathcal{C} \to R$ be a continuous payoff functional. Define the functional ω_0 by the equality (11.2.13). It was established in Theorem 11.3.3 that the functional ω_0 coincides with the value functional of the game under consideration. Recall that the equalities $\underline{\omega}_0 = \omega_0 = \overline{\omega}_0$ are valid (cf. (11.3.15)) where, according to Lemma 11.2.5, $\underline{\omega}_0$ and $\overline{\omega}_0$ are lower and upper semicontinuous functionals respectively. Therefore the payoff functional $\omega^0: T \times \mathcal{C} \to R$ is continuous. Assume now that the payoff functional satisfies the Lipschitz condition

(1) $|\gamma(y_1) - \gamma(y_2)| \leqslant \lambda_\gamma |y_1 - y_2|, \quad y_i \in C, \ (i = 1,2).$

We shall show that in this case the value functional also satisfies the Lipschitz condition with respect to the variable y, i.e.

(2) $|\omega^0(t_0,y_0^{(1)}) - \omega^0(t_0,y_0^{(2)})| \leqslant \lambda_\omega |y_0^{(1)} - y_0^{(2)}|, \ t_0 \in T, \ y_0^{(i)} \in C, \ i = 1,2.$

Thus let $t_0 \in T$, $y_0^{(i)} \in C$ $(i = 1,2)$. Denote by $\mathcal{W}_1^{(i)}(t)$ $(t_0 \leqslant t \leqslant 1)$ the stable bridge defined according to (11.2.23) where we set $y_0 = y_0^{(i)}$. Let U_1^e be a strategy extremal to the bridge $\mathcal{W}_1^{(1)}(t)$. Theorem 11.3.1 states that the inclusion

(3) $\mathcal{Y}_1(t_0,y_0^{(1)},U_1^e) \subset \mathcal{W}_1^{(1)}(1)$

is valid. We shall show that the estimate

(4) $\mathrm{dist}(y, \mathcal{W}_1^{(1)})(1)) \leqslant |y_0^{(1)} - y_0^{(2)}|\exp(\lambda_f), \quad y \in \mathcal{Y}_1(t_0,y_0^{(2)},U_1^e)$

also holds, where λ_f is a Lipschitz constant for the function f (cf. (11.1.1)).

Set

(5) $\mathcal{D}_* \overset{\mathrm{def}}{=} T \times (\mathcal{Y}(t_0,y_0^{(1)}) \cup \mathcal{Y}(t_0,y_0^{(2)})).$

Replace the set $\mathcal{D}_1$ (11.3.7) by the set $\mathcal{D}_*$ (5), and after this substitution, define the corresponding quantities $\zeta(\delta)$, Λ, and $\alpha(\delta)$ (cf. (11.3.8) and Section 2.3.2). Consider the quantity

(6) $\beta_*(t,\delta) \overset{\mathrm{def}}{=} [\exp 2\lambda_f(t - t_0) - 1]\alpha(\delta)/\lambda_f + |y_0^{(1)} - y_0^{(2)}|^2 \exp 2\lambda_f(t-t_0).$

Note that the functions $\beta(\cdot,\delta)$ (11.3.8) and $\beta_*(\cdot,\delta)$ (6) satisfy the differential equation

$$\dot{\beta} = 2\lambda_f\beta + 2\alpha(\delta)$$

and conditions $\beta(t_0,\delta) = 0$, $\beta_*(t_0,\delta) = |y_0^{(1)} - y_0^{(2)}|^2$.

In the statement of the estimate (11.3.11) conditions $y_* \in \mathcal{Y}(t_0,y_0)$ and $w_* \in \mathcal{Y}(t_0,y_0)$ are replaced by the conditions $y_* \in (t_0,y_0^{(2)})$, $w_* \in \mathcal{Y}(t_0,y_0^{(1)})$ and the quantity $\beta(t,\delta)$ (11.3.8) is replaced by the quantity $\beta_*(t,\delta)$ (6). It can be shown that after these substitutions the estimate (11.3.11) still remains valid. Taking this fact into account, we obtain that for any step-by-step motion $y(\cdot) \in \mathcal{Y}_1(t_0,y_0^{(2)},U_1^e,\Delta)$ (diam $\Delta \leqslant \delta$) the inequalities

$$\rho^2(\tau_i,y, \mathcal{W}_1^{(1)}(\tau_i)) \leqslant \beta_*(\tau_i,\delta) \quad (i = 0,1, \ldots, k + 1)$$

are valid (cf. analogous inequalities (11.3.12)). In particular we have for $\tau_{k+1} = 1$

$$\rho(1,y, \mathcal{W}_1^{(1)}(1)) \overset{\text{def}}{=} \text{dist}(y, \mathcal{W}_1^{(1)}(1)) \leqslant \sqrt{\beta_*(1,\delta)}.$$

Approaching the limit as diam $\Delta \to 0$ and taking into account that in (6) the quantity $\alpha(\delta) \to 0$ as $\delta \to 0$, we obtain

$$\text{dist}(y, \mathcal{W}_1^{(1)}(1)) \leqslant |y_0^{(1)} - y_0^{(2)}|\exp(\lambda_f(1-t_0)) \leqslant |y_0^{(1)} - y_0^{(2)}|\exp(\lambda_f).$$

This estimate is valid for any constructive motion $y(\cdot) \in \mathcal{Y}_1(t_0,y_0^{(2)},U_1^e)$. Thus the estimate (4) is verified.

It follows from (1) and (4) that for any $y \in \mathcal{Y}_1(t_0,y_0^{(2)},U_1^e)$, there exists $w \in \mathcal{W}_1^{(1)}(1)$ such that

$$\gamma(y) = \gamma(w) + \gamma(y) - \gamma(w) \leqslant \gamma(w) + \lambda_\gamma|y - w|$$

$$\leqslant \gamma(w) + \lambda_\gamma|y_0^{(1)} - y_0^{(2)}|\exp(\lambda_f).$$

In view of Theorem 11.3.3 and inequality (11.2.25), we have $\gamma(w) \leqslant \omega^0(t_0,y_0^{(1)})$. Hence

$$\gamma(y) \leqslant \omega^0(t_0,y_0^{(1)}) + \lambda_\gamma|y_0^{(1)} - y_0^{(2)}|\exp(\lambda_f)$$

$$\text{for all} \quad y \in \mathcal{Y}_1(t_0,y_0^{(2)},U_1^e).$$

Furthermore, from the definition of the value (cf. (11.1.13) and (11.1.16)), we obtain

$$\omega^0(t_0,y_0^{(2)}) \leqslant \Gamma_1(t_0,y_0^{(2)},U_1^e) = \max_{y \in \mathcal{Y}_1(t_0,y_0^{(2)},U_1^e)} \gamma(y)$$

$$\leqslant \omega^0(t_0,y_0^{(1)}) + \lambda_\gamma\ |y_0^{(1)} - y_0^{(2)}|\exp(\lambda_f)$$

In this estimate the roles of points $y_0^{(1)}$ and $y_0^{(2)}$ can be interchanged. We thus arrive at inequality (2) with $\lambda_\omega = \lambda_\gamma \exp(\lambda_f)$.

11.4.2. In view of Theorems 11.3.3 and 11.3.4, in order that a functional $\omega: T \times C \to R$ be the value functional of the game under consideration, it is necessary and sufficient that it satisfy equality $\omega(1,y) = \gamma(y)$ $(y \in C)$ and the conditions of u-stability and v-stability (cf. (11.2.7) and (11.2.9)). For continuous and Lipschitzian functionals ω the conditions of u-stability and v-stability can be defined in various equivalent ways. The assertions presented below correspond to the results obtained in Chapter 4 for positional differential games.

Introduce the following conditions:

(7) $$\sup_{(t_*,y_*,v)} \inf_{y} \underline{\lim}_{\delta\to+0} [\omega(t_* + \delta,y) - \omega(t_*,y_*)]\delta^{-1} \leqslant 0,$$

(8) $$\sup_{(t_*,y_*,v)} \inf_{y} \overline{\lim}_{\delta\to+0} [\omega(t_* + \delta,y) - \omega(t_*,y_*)]\delta^{-1} \leqslant 0,$$

(9) $$\sup_{(t_*,y_*,v)} \inf_{y} \sup_{t} [\omega(t,y) - \omega(t_*,y_*)] \leqslant 0.$$

Here $(t_*,y_*,v) \in T \times C \times Q$, $y \in Y_2(t_*,y_*,v)$, $t \in (t_*,1]$. The following proposition is valid.

Let a continuous functional $\omega: T \times C \to R$ be nonanticipating. Then conditions (11.2.7), (7), (8), and (9) are equivalent.

We shall omit the proof of this assertion -- it can be carried out in the way it was done in subsection 4.2.1 for positional differential games.

Analogously for a continuous functional ω one can introduce the following set of conditions which are equivalent to the condition of v-stability:

(10) $$\inf_{(t_*,y_*,u)} \sup_{y} \overline{\lim}_{\delta\to+0} [\omega(t_* + \delta,y) - \omega(t_*,y_*)]\delta^{-1} \geqslant 0,$$

(11) $$\inf_{(t_*,y_*,u)} \sup_{y} \underline{\lim}_{\delta\to+0} [\omega(t_* + \delta,y) - \omega(t_*,y_*)]\delta^{-1} \geqslant 0,$$

(12) $$\inf_{(t_*,y_*,u)} \sup_{y} \inf_{t} [\omega(t,y) - \omega(t_*,y_*)] \geqslant 0.$$

Here $(t_*,y_*,u) \in T \times C \times P$, $y \in Y_1(t_*,y_*,u)$, $t \in (t_*,1]$.

If the functional $\omega: T \times C \to R$ is continuous and nonanticipating, then conditions (11.2.9), (10), (11), and (12) are equivalent.

To state the next condition, the following additional notation will be needed. Let $(t_*,y_*) \in [0,1) \times C$, $h \in R^n$. Denote by $\psi(t_*,y_*,h)$ the function $y \in C$ such that $y(t) = y_*(t)$ for $0 \leqslant t \leqslant t_*$, $y(t) = y_*(t_*) + h(t-t_*)$ for $t_* \leqslant t \leqslant 1$. Let a continuous functional $\omega: T \times C \to R$ be nonanticipating and satisfy the Lipschitz condition

(13) $$|\omega(t,y^{(1)}) - \omega(t,y^{(2)})| \leqslant \lambda|y^{(1)} - y^{(2)}| \quad (t \in T,\ \ y^{(i)} \in C,\ \ (i = 1,2)).$$

Set

(14) $$D^-\omega(t_*,y_*)|(h) = \underline{\lim}_{\delta\to+0} [\omega(t_*+\delta,\ \psi(t_*,y_*,h)) - \omega(t_*,y_*)]\delta^{-1},$$
$$D^+\omega(t_*,y_*)|(h) = \overline{\lim}_{\delta\to+0} [\omega(t_*+\delta,\ \psi(t_*,y_*,h)) - \omega(t_*,y_*)]\delta^{-1}.$$

Observe that the quantities $D^-\omega(t_*,y_*)|(h)$ and $D^+\omega(t_*,y_*)|(h)$ can take on the improper values $-\infty$ and $+\infty$.

The following estimates are valid.

Let a continuous function ω: $T \times C \to R$ be nonanticipating and satisfy the Lipschitz condition (13). Then for any $(t_*,y_*,u,v) \in T \times C \times P \times Q$, the inequalities

$$\inf_{y_2} \underline{\lim}_{\delta\to+0} [\omega(t_* + \delta,y_2) - \omega(t_*,y_*)]\delta^{-1} \leqslant \min_{h_2} D^-\omega(t_*,y_*)|(h_2)$$

(15)

$$\leqslant \inf_{y_2} \overline{\lim}_{\delta\to+0} [\omega(t_* + \delta,y_2) - \omega(t_*,y_*)]\delta^{-1},$$

$$\sup_{y_1} \underline{\lim}_{\delta\to+0} [\omega(t_* + \delta,y_1) - \omega(t_*,y_*)]\delta^{-1} \leqslant \max_{h_1} D^+\omega(t_*,y_*)|(h_1)$$

(16)

$$\leqslant \sup_{y_1} \overline{\lim}_{\delta\to+0} [\omega(t_* + \delta,y_1) - \omega(t_*,y_*)]\delta^{-1}$$

are valid where

$$y_1 \in Y_1(t_*,y_*,u),\quad h_1 \in F_1(t_*,y_*,u),\quad y_2 \in Y_2(t_*,y_*,v),\quad h_2 \in F_2(t_*,y_*,v).$$

The estimates (15) and (16) are verified in the same manner as the analogous assertions for positional differential games (cf. Lemma 4.4.1). Note that in Chapter 4 functions ω were considered which satisfy the Lipschitz condition (4.3.1) jointly in variables (t,x) while here it is required that functionals ω satisfy the Lipschitz condition (13) with respect to variable y only. However, in the proof of Lemma 4.4.1 the Lipschitz condition with respect to variable x is used. This fact permits us to utilize the arguments presented in subsection 4.4.1 for the proof of the estimates (15) and (16).

We shall assume that the payoff functional satisfies the Lipschitz condition (2). Furthermore, one can obtain from estimates (15) and (16) the following assertion -- in the same manner that Theorems 4.4.1 and 4.4.2 were obtained in subsection 4.4.1.

Theorem 11.4.1. *Let a payoff functional satisfy the Lipschitz condition (1). In order that the functional ω: $T \times C \to R$ be the value functional of the game under consideration, it is necessary and sufficient that this functional be continuous, nonanticipating, satisfy the Lipschitz condition (13) and equality $\omega(1,y) = \gamma(y)$ $(y \in C)$, and satisfy the inequalities*

$$\max_{v\in Q} \min_{h\in F_2(t_*,y_*,v)} D^-\omega(t_*,y_*)|(h) \leqslant 0,$$

$$\min_{u\in P} \max_{h\in F_1(t_*,y_*,u)} D^+\omega(t_*,y_*)|(h) \geqslant 0$$

for all $(t_*,y_*) \in [0,1) \times C$.

APPENDIX

This appendix contains a brief summary of facts dealing with semi-continuous functions, convex analysis, multivalued mappings, and differential inclusions. We do not strive for completeness and confine ourselves to those definitions and assertions which were used frequently in the preceding chapters. Below these, definitions and assertions are presented, proofs are omitted. However, references are cited which contain these proofs and in which a detailed exposition of the corresponding problems is given.

A.1. Semicontinuous Functions

Let Y be a metric space, $\mathrm{dist}(y^{(1)},y^{(2)})$ be the distance between the points $y^{(1)}$ and $y^{(2)}$. Denote by R the extended real line, i.e.

$$\bar{R} \overset{\text{def}}{=} R^1 \cup \{-\infty\} \cup \{+\infty\} \overset{\text{def}}{=} [-\infty,\infty].$$

Consider a function $\varphi\colon Y \to \bar{R}$. We say that a function φ is *lower semicontinuous* at point $y_* \in Y$ if for any $\alpha < \varphi(y_*)$, there exists a $\delta > 0$ such that $\mathrm{dist}(y_*,y) < \delta$ implies that $\varphi(y) \geqslant \alpha$. A function φ is called lower semicontinuous if it is lower semicontinuous at each point $y_* \in Y$:

Along with the above definition, any one of the following conditions can be used:

(1) $\underline{\lim}_{y \to y_*} \varphi(y) \geqslant \varphi(y_*)$ for all $y_* \in Y$;

(2) for any $\alpha \in R$ the set $\{y \in Y\colon \varphi(y) \leqslant \alpha\}$ is closed;

(3) $\mathit{epi}\ \varphi \overset{\text{def}}{=} \{(y,\alpha) \in Y \times R\colon \varphi(y) \leqslant \alpha\}$ is a closed set (it is called an *epigraph* of the function φ).

Function $\varphi: \mathcal{Y} \to \bar{R}$ is called *upper semicontinuous* if the function $-\varphi$ is lower semicontinuous.

The following assertion (Weierstrass' theorem) is known: A lower (upper) semicontinuous function attains its lower (upper) bound on any compact subset $\mathcal{D} \subset \mathcal{Y}$.

Corollary. *Let a function φ be lower semicontinuous on $\mathcal{Y}$. If for some $\alpha \in R$ the set $\{y \in \mathcal{Y} \mid \varphi(y) \leqslant \alpha\}$ is nonempty and compact then the function φ attains its lower bound on $\mathcal{Y}$.*

An analogous corollary is valid for an upper semicontinuous function.

In conclusion, we note that a function $\varphi: \mathcal{Y} \to R$ is continuous if and only if it is both lower and upper semicontinuous.

Definitions and assertions presented here can be found, for example, in [19,31,74] where proofs of these assertions are also presented.

A.2. Convex Functions

Let R^m be a Euclidean space, $\bar{R}$ be the extended real line. Consider the function $\varphi: R^m \to \bar{R}$ and define for it the set

$$\text{(1)} \qquad \operatorname{dom} \varphi \overset{\text{def}}{=} \{y \in R^m : \varphi(y) < \infty\},$$

$$\text{(2)} \qquad \operatorname{epi} \varphi \overset{\text{def}}{=} \{(y,\alpha) \in R^m \times R : \varphi(y) \leqslant \alpha\}.$$

The first of these two sets is an *effective set of the function* φ and the second one is named an *epigraph of the function* φ.

A function φ is called nonsingular if $\operatorname{dom} \varphi \neq \phi$ and $\varphi(y) > -\infty$ for all $y \in R^m$.

A function φ is called *convex* (*closed*) if the set $\operatorname{epi} \varphi$ is *convex* (*closed*) in the space $R^m \times R$. Note that the closure condition of function φ is equivalent to the lower semicontinuity property of this function (cf. Section A.1).

If φ is a nonsingular function then the condition of convexity of this function is equivalent to the following inequality:

$$\varphi(\lambda y^{(1)} + (1-\lambda) y^{(2)}) \leqslant \lambda \varphi(y^{(1)}) + (1-\lambda)\varphi(y^{(2)})$$

which is satisfied for all $y^{(i)} \in R^m$ and $\lambda \in [0,1]$.

The function $\varphi^*: R^m \to R$ defined by the equality

$$\text{(3)} \qquad \varphi^*(\ell) \overset{\text{def}}{=} \sup_{x \in R^n} [\langle x, \ell \rangle - \varphi(x)], \quad \ell \in R^m,$$

is called *conjugate* to function φ.

The following assertion connects a function φ with its conjugate function φ^*.

Theorem A.2.1. *Let φ: $R^m \to \bar{R}$ be a nonsingular convex closed function. Then the conjugate function φ^* will also be nonsingular, convex, and closed. For any $y \in R^m$ the equality*

(4) $$\varphi(x) = \varphi^{**}(x) \overset{\text{def}}{=} \sup_{\ell \in R^m} [\langle x,\ell\rangle - \varphi^*(\ell)]$$

is valid.

Note that the function φ^* is convex and closed for an arbitrary function φ: $R^m \to \underline{R}$.

The following assertion is valid for a convex function φ: let $\epsilon > 0$, $\varphi(y_*) < \infty$, and $\varphi(y) \geqslant \varphi(y_*)$ for all $y \in B(y_*,\epsilon)$, then $\varphi(y) \geqslant \varphi(y_*)$ for all $y \in R^m$ (recall that $B(y_*,\epsilon)$ is a sphere of radius ϵ with the center at point y_*). Thus any point of a local minimum of a convex function is a point of its global minimum.

The above assertions and definitions are available in practically every book dealing with the fundamentals of convex analysis (cf., e.g. [19,31,63,66]).

We shall note yet another fact which is used in Chapters 4 and 5. Let a convex function φ: $R^m \to R$ satisfy the Lipschitz condition uniformly in the whole space R^m, i.e.

$$|\varphi(y^{(1)}) - \varphi(y^{(2)})| \leqslant \lambda \| y^{(1)} - y^{(2)} \|, \quad y^{(i)} \in R^m, \; i = 1,2.$$

Consider a function φ^* conjugate to φ. Define the effective set of function φ^*. One can prove (cf., e.g. [66]) that if the above-stated Lipschitz condition is fulfilled, then the set $\mathrm{dom}\varphi^*$ is bounded.

A.3. Multivalued Mappings

Let Y and Z be metric spaces. The distance between points y_* and y^* in the space Y will be denoted by $\rho_1(y_*,y^*)$ while the distance between points z_* and z^* in the space Z will be denoted by $\rho_2(z_*,z^*)$.

Denote by comp(Z) the set of all nonempty compact subsets of the space Z. Let $z_* \in \mathcal{Z}$, $\mathcal{Z} \in \mathrm{comp}(Z)$, $\alpha > 0$. Set

$$\rho_2(z_*, \mathcal{Z}) \overset{\text{def}}{=} \inf_{z \in \mathcal{Z}} \rho_2(z_*,z),$$

$$\mathcal{Z}^\alpha \overset{\text{def}}{=} \{z_* \in Z;\ \rho_2(z_*,\mathcal{Z}) < \alpha\}.$$

The quantity $\rho_2(z_*, \mathcal{Z})$ is called the distance from the point z_* to the set $\mathcal{Z}$. The set $\mathcal{Z}^\alpha$ is an α-neighborhood of the set $\mathcal{Z}$.

One can define a metric on the set comp(Z) by setting

$$\mathrm{Dist}(Z_1, Z_2) = \inf\{\alpha > 0: \ Z_1 \subset Z_2^{\alpha}, \ Z_2 \subset Z_1^{\alpha}\},$$

$$(Z_i \in \mathrm{comp}(Z) \ \ (i = 1,2)).$$

The quantity $\mathrm{Dist}(Z_1, Z_2)$ is called the Hausdorff distance between the sets Z_1 and Z_2.

Consider the mapping

$$\Phi: y \to \Phi(y): \ \ \mathbf{Y} \to \mathrm{comp}(\mathbf{Z})$$

which assigns to the point $y \in \mathbf{Y}$ a nonempty compact set $\Phi(y) \subset \mathbf{Z}$. Such a mapping Φ will be called a *multivalued (many-valued) mapping* from $\mathbf{Y}$ into $\mathbf{Z}$.

Throughout this book we shall often deal with continuous and upper semicontinuous multivalued mappings. These mappings are defined as follows.

A multivalued mapping Φ is called *upper semicontinuous* at point $y_* \in \mathbf{Y}$ if for any $\epsilon > 0$ there exists a $\delta > 0$ such that

$$\Phi(y) \subset \Phi^{\epsilon}(y_*) \quad \text{for} \quad \rho_1(y_*,y) < \delta.$$

(Here $\Phi^{\epsilon}(y_*)$ is an ϵ-neighborhood of the compact set $\Phi(y_*)$.)

A multivalued mapping Φ is *continuous at point* $y_* \in \mathbf{Y}$ if for any $\epsilon > 0$ there exists a $\delta > 0$ such that

$$\mathrm{Dist}(\Phi(y_*),\Phi(y)) < \epsilon \quad \text{for} \quad \rho_1(y_*,y) < \delta.$$

A multivalued mapping Φ is called upper semicontinuous (continuous) if it is upper semicontinuous (continuous) at each point $y_* \in \mathbf{Y}$.

Consider the set

$$\mathrm{gr}\ \Phi \overset{\mathrm{def}}{=} \{(y,z) \in \mathbf{Y} \times \mathbf{Z} : z \in \Phi(y), \ \ y \in \mathbf{Y}\},$$

which is called the graph of a multivalued mapping Φ. A multivalued mapping Φ is called *closed* if its graph is closed in $\mathbf{Y} \times \mathbf{Z}$.

The following assertion is valid.

Theorem A3.1. *Let* $\Phi: \ \mathbf{Y} \to \mathrm{comp}(\mathbf{Z})$ *be a multivalued mapping from a metric space* $\mathbf{Y}$ *into a metric space* $\mathbf{Z}$. *If the space* $\mathbf{Z}$ *is compact, then the mapping* Φ *is upper semicontinuous if and only if it is closed.*

Thus if $\mathbf{Z}$ is a compact metric space then the condition of upper semicontinuity of the mapping Φ is equivalent to the following condition: for any sequences $y_k \in \mathbf{Y}$, $z_k \in \mathbf{Z}$ $(k = 1,2,\ldots)$ such that $z_k \in \Phi(y_k)$, $y_k \to y_*$, $z_k \to z_*$ as $k \to \infty$, the inclusion $z_* \in \Phi(y_*)$ is fulfilled.

An example is considered below in which upper semicontinuous multivalued mappings are defined. Many situations in which upper semicontinuous mappings, considered herein, occur are confined within the framework of this example.

Example. Let **Y** be a metric space, **Z** be a compact metric space, q: **Y** × **Z** → R^1 be a continuous function. Let $y \to \Phi_*(y)$: **Y** → comp(**Z**) be a continuous multivalued mapping from **Y** into **Z**. Consider multivalued mappings $y \to \Phi_0(y)$ and $y \to \Phi^0(y)$ defined as follows:

$$\Phi_0(y) \overset{\text{def}}{=} \{z_0 \in \Phi_*(y)\colon \ q(y, z_0) = \min_{z\in\Phi_*(y)} q(y,z)\},$$

$$\Phi^0(y) \overset{\text{def}}{=} \{z^0 \in \Phi_*(y)\colon \ q(y,z^0) = \max_{z\in\Phi_*(y)} q(y,z)\}.$$

It is easy to verify that the multivalued mappings Φ_0 and Φ^0 are upper semicontinuous.

A (single-valued) function φ: **Y** → **Z** is a *selector of a multivalued mapping* Φ, if $\varphi(y) \in \Phi(y)$ for all $y \in$ **Y**. The following existence theorem is valid.

Theorem A.3.2. *Let* **Y** *be a metric space,* **Z** *be a compact metric space,* Φ *be an upper semicontinuous mapping from* **Y** *into* **Z**. *Then there exists a Borel measurable function* φ: **Y** → **Z** *such that* $\varphi(y) \in \Phi(y)$ *for all* $y \in$ **Y**.

We note that there are many theorems about the existence of measurable selectors. However, for our purposes the above stated theorem is sufficient. The facts concerning multivalued mappings presented herein are available in numerous publications (cf., e.g. [1,44,47,74]).

A.4. Differential Inclusions

Let $T = [t_{\#},\theta] \in R^1$. Consider a multivalued mapping $(t,x) \to \mathcal{F}(t,x)$ from $T \times R^n$ into R^n satisfying the following conditions:

(a) $\mathcal{F}(t,x)$ is a convex compact set in R^n for all $(t,x) \in T \times R^n$;

(b) the multivalued mapping $(t,x) \to \mathcal{F}(t,x)$ is upper semicontinuous;

(c) there exists a number $\xi > 0$ such that for all $(t,x) \in T \times R^n$, the bound

$$(5) \qquad \max\{\langle x,h\rangle : h \in \mathcal{F}(t,x)\} \leqslant \xi(1 + \|x\|^2)$$

is valid.

Note that the last condition assures that the solutions could be extended over the whole interval T. This condition may be stated in various ways. For example, inequality (5) can be replaced by either one of the inequalities below:

$$\|h\| \leqslant \xi(1 + \|x\|),$$

$$\|h\| \leqslant r(t).$$

(These inequalities should be fulfilled for all $h \in \mathcal{F}(t,x)$ and $(t,x) \in T \times R^n$; here ξ is a positive number, $r(\cdot)$: $T \to R^1$ is a summable function.)

Let $(t_0,x_0) \in T \times R^n$. Denote by $X(t_0,x_0)$ the set of absolutely continuous functions $x(\cdot)$: $[t_0,\theta] \to R^n$, which satisfy the initial condition $x(t_0) = x_0$ and the differential inclusion

(6) $\dot{x}(t) \in \mathcal{F}(t,x(t))$ for almost all $t \in [t_0,\theta]$.

The elements of the set $X(t_0,x_0)$ are called *solutions of the differential inclusion* (6).

Solutions $x(\cdot) \in X(t_0,x_0)$ are defined on the interval $[t_0,\theta]$. Such solutions are considered practically always when one deals with differential inclusions. Along with this definition we shall consider yet another set of solutions. We shall denote this set by $X_*(t_0,x_0)$. Elements of the set $X_*(t_0,x_0)$ are absolutely continuous functions $x(\cdot)$: $T \to R^n$ (defined on the whole interval T) which satisfy the equality $x(t_0) = x_0$ and the differential inclusion

(7) $\dot{x}(t) \in \mathcal{F}(t,x(t))$ for almost all $t \in T$.

We note that differential inclusions were studied by many authors (cf., e.g. [14,23,75,76]). Statements given below are contained in these works or are simple corollaries of the results presented therein.

The following assertion is valid.

The set $X_*(t_0,x_0)$ is nonempty and compact in $C_n(T)$. A multivalued mapping $(t_0,x_0) \to X_*(t_0,x_0)$ from $T \times R^n$ into $C_n(T)$ is upper semicontinuous.

From this assertion an analogous assertion for the set $X(t_0,x_0)$ follows immediately; for reference's convenience, we shall state it as

Theorem A.4.1. *Let a multivalued mapping $(t,x) \to \mathcal{F}(t,x)$ satisfy the conditions* (a) - (c). *Then for any initial position $(t_0,x_0) \in T \times R^n$ the set $X(t_0,x_0)$ whose elements are solutions of the differential inclusion* (6) *will be a nonempty compact subset of the space $C_n([t_0,\theta])$. Let $(t_k,x_k) \in T \times R^n$, $x_k(\cdot) \in X(t_k,x_k)$ $(k = 1,2, \ldots)$ and $(t_k,x_k) \to (t_*,x_*)$ as $k \to \infty$. Then one can extract from the sequence of functions $x_k(\cdot)$ $(k = 1,2, \ldots)$ a convergent subsequence whose limit is contained in $X(t_*,x_*)$.*

We shall clarify the following fact. Functions $x_k(\cdot)$ are in general defined on different time intervals $[t_k,\theta]$. It is therefore required to stipulate exactly the meaning of convergence of subsequences of these functions. We shall assume that functions $x_k(\cdot)$ are redefined on the whole interval T and satisfy the differential inclusion (7). It follows from the semicontinuity of the multivalued mapping $(t_0,x_0) \to \mathcal{X}_*(t_0,x_0)$ that there exists a subsequence $x_{k_i}(\cdot) \in \mathcal{X}_*(t_{k_i},x_{k_i})$ which converges uniformly on T to the limit function $x_*(\cdot)$ belonging to the set $\mathcal{X}_*(t_*,x_*)$. An element of the set $\mathcal{X}(t_*,x_*)$ is then a restriction of functions $x_*(\cdot)$ to the interval $[t_*,\theta]$.

Let $(t_0,x_0) \in T \times R^n$ and be a compact set in $T \times R^n$. Define the sets

$$\mathcal{D}(t_0,x_0) \overset{\text{def}}{=} \{(t,x(t)) \in T \times R^n:\ t_0 \leqslant t \leqslant \theta,\ \ x(\cdot) \in \mathcal{X}(t_*,x_*)\},$$

$$D(G) \overset{\text{def}}{=} \bigcup_{(t_0,x_0)\in G} \mathcal{D}(t_0,x_0).$$

From the assertions presented above one easily obtains

Corollary A.4.1. *The sets $\mathcal{D}(t_0,x_0)$ and $\mathcal{D}(G)$ are compact in $T \times R^n$.*

We state yet another proposition. Let a function f: $T \times R^n \times \mathcal{P} \times \mathcal{Q} \to R^n$ satisfy the conditions (i) - (iii) stipulated in subsection 1.2.2. Let a multivalued mapping $(t,x) \to$ (t,x) be defined which assigns to a point $(t,x) \in T \times R^n$ a nonempty compact subset of the set $\mathcal{P}$. We shall assume that this mapping is upper semicontinuous. Select a function U: $T \times R^n \to \mathcal{P}$ satisfying the inclusion

$$U(t,x) \in \mathcal{U}(t,x) \quad \text{for all} \quad (t,x) \in T \times R^n.$$

We shall define for this function a set $\mathcal{X}_1(t_0,x_0,U)$ whose elements are constructive motions generated by the strategy U and emanating from the initial position (t_0,x_0) (cf. subsection 1.3.1). Next we define the set $\mathcal{X}(t_0,x_0,\mathcal{U})$ whose elements are the solutions $x(\cdot)$: $[t_0,\theta] \to R^n$ $(x(t_0) = x_0)$ of the differential inclusion

$$\dot{x}(t) \in \mathcal{F}_{\mathcal{U}}(t,x(t))$$

where

$$\mathcal{F}_{\mathcal{U}}(t,x) \overset{\text{def}}{=} \text{co}\{f(t,x,u,v):\ u \in \mathcal{U}(t,x),\ \ v \in \mathcal{Q}\}.$$

Note that the multivalued mapping $(t,x) \to \mathcal{F}_{\mathcal{U}}(t,x)$ satisfies the conditions (a) - (c) stated at the beginning of this section.

Theorem A.4.2. *The following inclusion is valid:*

$$X_1(t_0,x_0,U) \subset X(t_0,x_0,U).$$

Recall that the set $X_1(t_0,x_0,U)$ was defined by a limit transition of step-by-step motions. Limit transitions of step-by-step motions (Euler's broken lines) of this kind are utilized in numerous works for proofs of existence of solutions of differential inclusions. Modifying somewhat these constructions (cf., e.g. [14,23]), one could prove Theorem A.4.2.

BIBLIOGRAPHY

Translator's Remark. Whenever possible, the English version of the paper or book is presented. All the references to the literature in the text are indicated by ordinal numbers.

1. Arkin, V. I. and Levin, V. L. (1972). Convexity of values of vector integrals, theorems of measurable sampling, and variational problems, *Russian Math. Surveys*, 27 (3), 21-86.

2. Barbashin, E. A. (1970). *Introduction to the Theory of Stability*, Wolters-Nooordhoff, Groningen.

3. Barbashin, E. A. and Alimov, Yu. I. (1962). On the theory of relay differential equations, *Izvest. Vys. Uch. Zav. Matematika* (Kazan State Univ. Publishing House), 1 (26), 3-13 (in Russian).

4. Bellman, R. (1967), (1971). *Introduction to the Mathematical Theory of Control Processes*, Academic Press, New York, Vols. 1, 2.

5. Berkovitz, L. D. (1964). A variational approach to differential games, *Adv. in Game Theory*, *Ann. of Math. Studies*, 52, Princeton University Press, 127-174.

6. Blaquière, A., Gérard, F., and Leitman, G. (1969). *Quantitative and Qualitative Games*, Academic Press, New York.

7. Bohnenblust, H. F. and Karlin, S. (1950). On a theorem of Ville. *Contributions to the theory of games*, I, *Ann. of Math. Studies*, Princeton, 155-160.

8. Botkin, N. D. (1982). Evaluation of numerical construction error in differential game with fixed terminal time, *Problems of Control and Information Theory*, **11**, No. 4, 283-295.

9. Botkin, N. D. and Patsko, V. S. (1983). Positional control in a linear differential game, *Engineering Cybernetics*, **21**, No. 4, 69-75.

10. Centsov, A. G. (1975). On the structure of a game problem of convergence, *Soviet Math. Dokl.* **16**, 1404-1406.

11. Cencov, A. G. (1976). On a game problem of guidance, *Soviet Mathematics*, **17**, No. 1, 73-77.

12. Cencov, A. G. (1976). On a game problem of converging at a given instant of time, *Math. of the USSR Sbornik*, **28** (3), 353-376.

13. Coddington, E. A. and Levinson, N. (1955). *Theory of Ordinary Differential Equations*, McGraw-Hill, New York.

14. Davy, J. L. (1972). Properties of the solution set of a generalized differential equation, *Bull. Austral. Math. Soc.* **6**, 379-398.

15. Dem'janov, V. F. (1974). *Minimax: Directional Differentiation*, Izdat. Leningrad Gos. Univ., Leningrad (in Russian).

16. Dem'janov, V. F. and Malozemov, V. N. (1974). *Introduction to the Minimax*, Halsted Press, New York.

17. Dieudonné, J. (1969). *Foundations of Modern Analysis*, Academic Press, New York.

18. Dunford, N. and Schwarz, J. T. (1958). *Linear Operators*, Part I: *General Theory*, Interscience, New York.

19. Ekeland, I. and Temam, R. (1976). *Convex Analysis and Variational Problems*; *Studies in Mathematics and its Applications*, Vol. 1, North-Holland, Amsterdam.

20. Elliott, R. J. and Kalton, N. J. (1972). The existence of value in differential games of pursuit and evasion, *J. Different. Equat.* **12**, No. 3, 504-523.

21. Fan, K. (1953). Minimax theorems, *Proc. Nat. Acad. Sci. USA*, **39**, No. 1, 42-47.

22. Filippov, A. F. (1962). On certain questions in the theory of optimal control, *SIAM J. Control*, **1**, 76-84.

23. Filippov, A. F. (1964). Differential equations with discontinuous right-hand side, *Amer. Math. Soc. Transl.*, Ser 2, **42**, 199-231, Providence, R. I.

24. Fleming, W. H. (1957). A note on differential games of prescribed durations, *Contributions to the Theory of Games, Ann. Math. Studies*, No. 3, 407-412.

25. Fleming, W. H. (1961). The convergence problem for differential games, *J. Math. Anal. and Appl.* **3**, 102-116.

26. Fleming, W. H. (1964). The Cauchy problem for degenerate parabolic equations, *J. of Math. and Mech.* **13**, 987-1008.

27. Friedman, A. (1971). *Differential Games*, Interscience, New York.

28. Gusjatnikov, P. B. and Nikol'skii, M. S. (1969). Optimality of pursuit time, *Soviet Math. Dokl.* **10**, 103-106.

29. Hadwiger, H. (1957). *Vorlesungen über Inhalt, Oberfläche und Isoperimetrie*, Springer-Verlag, Berlin.

30. Hájek, O. (1975). *Pursuit Games*, Mathematics in Science and Engineering, Vol. 120, Academic Press, New York.

31. Ioffe, A. D. and Tikhomirov, W. M. (1978). *Theory of Extremal Problems*, North-Holland, Amsterdam.

32. Isaacs, R. (1965). *Differential Games*, John Wiley, New York.

33. Janin, R. (1977). On sensitivity in an optimal control problem, *J. of Math. Anal. and Appl.* **60**, 631-657.

34. Kantorovich, L. V. and Akilov, G. P. (1982). *Functional Analysis*, 2nd ed. Pergamon Press, Oxford.

35. Kikuchi, N. (1971). On control problems for functional-differential equations, *Funkcialaj Ekvacioj.* **14**, 1-23.

36. Kolmogorov, A. N. and Fomin, S. V. (1970). *Introductory Real Analysis*, Prentice Hall, Englewood Cliffs, New Jersey.

37. Krasovskiĭ, A. N. (1980). A differential game for the positional functional, *Soviet Math. Dokl.* **22**, No. 1, 251-255.

38. Krasovskiĭ, N. N. (1963). On a problem of tracking, *Jour. Appl. Math. Mech.* **17**, 363-377.

39. Krasovskii, N. N. (1966). *Theory of Controlling Motions*, Nauka, Moscow (in Russian).

40. Krasovskii, N. N. (1970). *Games Problems about Contact of Motions*, Nauka, Moscow (in Russian).

41. Krasovskii, N. N. (1979). Differential games, approximation and formal models, *Math. USSR Sbornik*, **35**, No. 6, 795-822.

42. Krasovskii, N. N. and Subbotin, A. I. (1974). *Positional Differential Games*, Nauka, Moscow (in Russian).

43. Kruzhkov, S. N. (1969). Non-linear equations of the first order and differential games connected with them, *Uspekhi Mat. Nauk*, **24** (2), 227-228 (in Russian).

44. Kuratowski, K. and Ryll-Nardzewski, C. (1965). A general theorem on selectors, *Bull. Acad. Polon. Sci.*; *Ser. Sci. Math. Astr. Phys.* **13**, 397-403.

45. Kurzhanskii, A. B. (1977). *Control and Observation under Conditions of Uncertainty*, Nauka, Moscow (in Russian).

46. Loeve, M. (1960). *Probability Theory*, 2nd edition, D. van Nostrand, Princeton, New Jersey.

47. Lusin, N. (1930). Sur la problème de J. Hadamard d'uniformisation des ensembles, *Mathematica*, **4**, 54-66.

48. Malkin, I. G. (1965). *Theory of Stability of Motions*, Nauka, Moscow (in Russian).

49. Miscenko, E. F. (1971). Pursuit and evasion problems in differential games. *Engineering Cybernetics* **9**, 787-791.

50. Natanson, I. P. (1961). *Theory of Functions of a Real Variable*, F. Ungar, New York.

51. von Neumann, J. (1928). Zur Theorie der Geselshaftsspiele, *Math. Ann.* **100**, 295-320.

52. Neveu, J. (1965). *Mathematical Foundations of the Calculus of Probabilities*, Holden-Day, San Francisco, Calif.

53. Nikol'skiĭ, M. S. (1972). Application of the first direct method of Pontryagin in pursuit games, *Engineering Cybernetics*, **10**, No. 6, 984-985.

54. Osipov, Ju. S. (1971). Alternative in a differential-difference game, *Soviet Math. Dokl.* **12**, No. 2, 619-624.

55. Parthasarathy, T. and Raghavan, T. (1971). Some Topics in Two-Person Games, *Modern Analytic and Computational Methods in Science and Mathematics*, No. 22, American Elsevier, New York.

56. Petrov, N. N. (1970). On the existence of a value for pursuit games, *Soviet Math. Dokl.* **11**, 292-294.

57. Pontryagin, L. S. (1962). *Ordinary Differential Equations*, Addison-Wesley, Reading, Mass.

58. Pontryagin, L. S. (1966). On the theory of differential games, *Russian Mathematical Surveys*, **21**, No. 4, 193-246.

59. Pontryagin, L. S. (1967). Linear differential games. 1, *Soviet Math. Dokl.* **8**, 769-771.

60. Pontryagin, L. S. (1967). Linear differential games. 2, *Soviet Math. Dokl.* **8**, 910-912.

61. Pontryagin, L. S., Boltyanskii, V. G., Gamkrelidze, R. V. and Mishchenko, E. F. (1962). *The Mathematical Theory of Optimal Processes*, Interscience, New York.

62. Pshenichniĭ, B. N. (1969). The structure of differential games, *Soviet Math. Dokl.* **10**, 70-72.

63. Pshenichniĭ, B. N. (1980). *Convex Analysis and Extremal Problems*, Nauka, Moscow.

64. Pshenichniĭ, B. N. and Sagaidak, N. I. (1970). Differential games of prescribed duration, *Cybernetics*, **6** (2), 72-83.

65. Rademacher, H. (1918). Über partielle und totale Differenzierbarkeit von Funktionen mehrerer Variablem über die Transformation der Doppelintegrale, *Math. Ann.* **79**, 340-354.

66. Rockafellar, R. T. (1970). *Convex Analysis*, Princeton University Press, Princeton, New Jersey.

67. Rudin, W. (1964). *Principles of Mathematical Analysis*, 2nd edition, McGraw-Hill, New York.

68. Subbotin, A. I. (1980). A generalization of the basic equation of the theory of differential games, *Soviet Math. Dokl.* **22**, No. 2, 358-362.

69. Subbotin, A. I. and Centsov, A. G. (1981). *Optimization of a Gaurantee in Control Problems*, Nauka, Moscow (in Russian).

70. Subbotina, N. N. (1983). Universal optimal strategies in positional differential games, *Differential Equations*, **19**, No. 11, 1377-1382.

71. Tarlinskii, S. I. (1972). On a positional guidance problem, *Soviet Math. Dokl.* **13**, No. 6, 1459-1463.

72. Ushakov, V. N. (1980). On the problem of constructing stable bridges in a differential game of approach and avoidance, *Engineering Cybernetics*, **18**, No. 4, 16-23.

73. Varaiya, P. and Lin, J. (1969). Existence of saddle points in differential games, *SIAM J. Control*, **7**, No. 1, 142-157.

74. Warga, J. (1972). *Optimal Control of Differential and Functional Equations*, Academic Press, New York.

75. Wazewski, T. (1962). Sur une généralisation de la notion des solution d'une équation au contingent, *Bull. Acad. Polon. Sci.; Ser. Sci. Math. Astr. Phys.* **10**, 11-15.

76. Zaremba, S. Ch. (1936). Sur les équations au paratingent, *Bull. Sci. Math.*, Deuxieme 2 Sér. **60**, 139-160.

INDEX

E

F

G

H

I

L

MIX
Papier aus verantwortungsvollen Quellen
Paper from responsible sources
FSC® C105338

If you have any concerns about our products,
you can contact us on
ProductSafety@springernature.com

In case Publisher is established outside the EU,
the EU authorized representative is:
Springer Nature Customer Service Center GmbH
Europaplatz 3, 69115 Heidelberg, Germany

Printed by Libri Plureos GmbH
in Hamburg, Germany